METHODS IN MOLECULAR BIOLOGY

For further volumes:
http://www.springer.com/series/7651

Protein Crystallography

Methods and Protocols

Edited by

Alexander Wlodawer

Macromolecular Crystallography Laboratory, National Cancer Institute, Frederick, MD, USA

Zbigniew Dauter

Synchrotron Radiation Research Section, MCL, National Cancer Institute, Argonne National Laboratory, Argonne, IL, USA

Mariusz Jaskolski

Department of Crystallography, Faculty of Chemistry, A. Mickiewicz University, Poznan, Poland; Center for Biocrystallographic Research, Institute of Bioorganic Chemistry, Polish Academy of Sciences, Poznan, Poland

Humana Press

Editors
Alexander Wlodawer
Macromolecular Crystallography Laboratory
National Cancer Institute
Frederick, MD, USA

Zbigniew Dauter
Synchrotron Radiation Research Section
MCL, National Cancer Institute
Argonne National Laboratory
Argonne, IL, USA

Mariusz Jaskolski
Department of Crystallography
Faculty of Chemistry
A. Mickiewicz University
Poznan, Poland

Center for Biocrystallographic Research
Institute of Bioorganic Chemistry
Polish Academy of Sciences
Poznan, Poland

ISSN 1064-3745 ISSN 1940-6029 (electronic)
Methods in Molecular Biology
ISBN 978-1-4939-6998-2 ISBN 978-1-4939-7000-1 (eBook)
DOI 10.1007/978-1-4939-7000-1

Library of Congress Control Number: 2017938302

Printed on acid-free paper

This Humana Press imprint is published by Springer Nature
The registered company is Springer Science+Business Media LLC
The registered company address is: 233 Spring Street, New York, NY 10013, U.S.A.

Preface

The determination of the first protein crystal structure took Max Perutz 22 years of titanic work, and his beloved hemoglobin was in fact not the winner: Perutz was scooped by his colleague, John Kendrew (both awarded the Nobel Prize in 1962), who determined two years earlier, in 1957, the structure of the four times smaller myoglobin. Fifteen years later, when the Protein Data Bank (PDB) was created in 1971, there were only seven protein structures deposited there, whereas today the PDB holds over 130,000 experimental macromolecular structures. The overwhelming majority (90%) were determined by crystallography. Since about 6% of the PDB crystal structures contain nucleic acids, we should properly refer to this research area as macromolecular crystallography, but the historically sanctioned term, protein crystallography, is still used. After nearly three decades of slow trickle of structures, in the mid-1990s the PDB received a tremendous boost, entering an exponential growth phase. The main factors were (1) advances in computer and information technology, providing the much needed computer power for complex calculations, but also better algorithms and means of experiment automation; (2) introduction of genetic engineering for easy production of practically any protein in bacterial cell "factories"; and (3) widespread use of powerful synchrotron sources of X-rays. A strong impetus was provided by several structural proteomics projects, which, in the wake of the genomic era, set the ambitious goal of inferring the function of all proteins encoded by the sequenced genomes from their structure.

With the use of third-generation synchrotron sources and ultra-fast pixel area detectors (APD), the data collection time has been reduced to seconds, with concomitant reduction of the crystal size (to microns) and improvement of data quality. The speed of data collection and the routine use of cryogenic temperatures (100 K) have led to increased popularity of "mail-in crystallography," where cryopreserved samples are shipped to a robot-operated beamline, and the data collection is conducted remotely. Another possibility is offered by polychromatic Laue diffraction, where structural transformations within a protein crystal (e.g., during a millisecond enzymatic reaction) are mapped using a series of nanosecond snapshots of complete datasets.

This is not the last word, however, because the emerging X-ray free electron lasers (XFELs) are offering beams more than 10 orders of magnitude brighter than even the most powerful synchrotrons. With such bright pulses, crystallites as small as 100 nm are injected to the beam, and a series of still diffraction images (from objects that are destroyed femtoseconds later) are used to reconstruct the complete diffraction pattern. The next step in this direction opens the possibility of studying the structure of single macromolecules injected into the XFEL beam. Inspired by the XFEL solutions, synchrotron beamlines are also turning towards serial crystallography (SSX).

Essentially all the steps of the crystallographic process have undergone a tremendous transformation in the last 10–20 years, and the progress has changed completely the way structural biology is practiced. Although the crystallization process is still based on the familiar phase diagram with oversaturation achieved by vapor diffusion, it is nowadays handled by crystallization robots capable of reproducible setting of thousands of trials in nanoliter volumes, and allowing for remote inspection of the crystallization process. Progress in crystallization techniques is also noted in attacking the challenge of membrane protein crystallization. Nonexistent in the PDB earlier, membrane proteins started appearing in the PDB with increasing frequency from the late-1980s, led by the Nobel Prize-winning structure of the photosynthetic reaction center. The original method based on the use of detergents has been largely replaced by crystallization in mesophases, called lipidic cubic phases (LCPs).

Progress has been noted in all three basic methods for the solution of the phase problem in macromolecular crystallography. The method of multiple isomorphous replacement (MIR) is still occasionally used for novel protein structures but with new density modification algorithms it is more readily applicable in the single-wavelength SIR version, especially in combination with anomalous scattering. Additionally, the introduction of quick halide soaks has made it possible to avoid the complications and dangers of heavy metals. However, a more convenient tackling of novel structures uses the approach of multiwavelength anomalous diffraction (MAD) worked out by Wayne Hendrickson, or its single-wavelength variant (SAD). It is based on scattering of tunable synchrotron radiation by anomalous atoms such as selenium, which can be introduced into recombinant proteins in the form of Se-Met. Owing to the presence of many possible search models in the PDB, the most successful method of choice for homologous proteins is molecular replacement (MR), originally proposed by Michael Rossmann and David Blow, now available in a number of powerful algorithms, including those based on maximum likelihood (ML). The constantly improving data resolution and quality make it possible to solve protein structures using the weak anomalous signal of the natural sulfur atoms or even by direct methods.

Also the stage of structure refinement has advanced beyond recognition from the early simplistic and tedious algorithms. It is now possible to refine within minutes models with hundreds of thousands of parameters using millions of reflections. ML is usually the algorithm of choice. It uses a different probabilistic approach to model parameter optimization, asking for such a model that maximizes the probability of the concrete data set at hand. It easily incorporates prior knowledge in the form of stereochemical restraints but requires rigorous information about data statistics.

With the fast growing volume of deposits in the PDB, the problem of dubious or (very rarely) blatantly wrong models is becoming a major concern, especially with regard to complex structures, in which imagination or wishful thinking sometimes takes precedence over experimental data, particularly in ligand modeling. Such cases, however, stimulate continual development of validation tools and sensitize the community to the need of vigilance and maintenance of high standards. The absolute number of atomic resolution ($d_{min} < 1.2$ Å) structures in the PDB is quite high (>3000), but their proportion has stayed at a less impressive level of ~2% for years. The fraction of ultrahigh resolution ($d_{min} < 0.8$ Å) structures is dismally small (0.04%). These high-quality models in the PDB are, however, of paramount importance because together with data retrieved from the CSD (Cambridge Structural Database) they serve to define better standards for macromolecular structure refinement and validation. The recent developments explore the potential of machine learning and of conformation-dependent parametrization.

The most spectacular achievements of macromolecular crystallography, often crowned with Nobel Prizes, have significantly advanced our understanding of the molecular mechanisms of life as well as contributed to the development of successful medicines, therapies, or biotechnology tools. Crystallographic studies of virus structure have a long history, dating back to Stanley, Bawden, Pirie, Franklin, and Klug, are marked by two Nobel Prizes, and have amassed several hundred models in the PDB. Crystallography has played a major role in dissecting the mechanisms of a number of viral pathogens. The most outstanding example is the battle with the HIV retrovirus. The prompt determination of the crystal structures of several key HIV proteins, most notably of the protease and reverse transcriptase, provided molecular targets for unprecedented structure-guided drug development success within just a few years. Indeed, the case of HIV protease set a new paradigm for rational drug design. Currently, this approach has been extended to fragment-based drug development, where crystallography is harnessed to identify molecular-cocktail components that can be stitched together to form drug molecules against specific macromolecular targets. In the recent outbreaks of viral infections, such as SARS, MERS, Ebola, or Zika, crystallographers have been in the front line of the battle, quickly providing dependable macromolecular structures for targeted drug development.

Perhaps the most iconic achievement of macromolecular crystallography in the recent years was the determination of the structure of the ribosome, which is a huge megadalton molecular machine responsible for the synthesis of all proteins in all living cells on our planet over several billion years. The structure explained that the ribosome is a ribozyme of catalytic RNA, as well as elucidated the mechanism of a number of antibiotics targeting the ribosomes of bacterial pathogens. The Nobel Prize to Venki Ramakrishnan, Tom Steitz, and Ada Yonath (2009) for the ribosome structure, which is a gene translation machine, followed the award to Roger Kornberg (2006) for the elucidation of the molecular mechanism of gene transcription.

The field of membrane-protein crystallography, initiated with the structure of the photosynthetic reaction center, is also growing very quickly. In the recent years, Nobel Prizes were awarded to Roderick MacKinnon (2003) for the determination of the structure of membrane channels and to Brian Kobilka and Robert Lefkowitz (2012) for the structure of the membrane-bound GPCR receptors. They sense diverse signals outside the cell (such as light, odor, hormone) and activate intracellular pathways by dissociating a subunit of the so-called G-protein that is coupled to the receptor (thus the name GPCR). There are ~800 different human GPCRs and they are the targets of ~50% of all modern drugs. It should be stressed that the first GPCR structure, determined by Krzysztof Palczewski for rhodopsin, explained the complicated molecular mechanism of our vision.

Also, recently crystallography has been used to explain the molecular mechanism of the promising versatile CRISPR-Cas9 genome-editing tool, adopted from the bacterial defense system based on clustered regularly interspaced short palindromic repeats (CRISPR) and coupled with a specific Cas nuclease.

The time is therefore ripe to describe what is currently available in the palette of methods and tools of contemporary macromolecular crystallography. The chapters included in this volume have been written by acclaimed specialists in each of the topics covered. It is hoped that this volume of Methods in Molecular Biology will help to acquaint the community of practicing and potential macromolecular crystallographers with the newest advances in the field and will inform them about the currently available tools.

Frederick, MD, USA — *Alexander Wlodawer*
Argonne, IL, USA — *Zbigniew Dauter*
Poznan, Poland — *Mariusz Jaskolski*

Contents

Contributors

ENRIQUE E. ABOLA • *Department of Chemistry, Bridge Institute, University of Southern California, Los Angeles, CA, USA*

MICHAEL BECKER • *GM/CA@APS, Advanced Photon Source, Argonne National Laboratory, Argonne, IL, USA*

HELEN M. BERMAN • *Research Collaboratory for Structural Bioinformatics Protein Data Bank, Center for Integrative Proteomics Research, Institute for Quantitative Biomedicine, Rutgers, Piscataway, NJ, USA; Department of Chemistry and Chemical Biology, Rutgers, The State University of New Jersey, Piscataway, NJ, USA*

STEPHEN K. BURLEY • *Research Collaboratory for Structural Bioinformatics Protein Data Bank, Center for Integrative Proteomics Research, Institute for Quantitative Biomedicine, Rutgers, Piscataway, NJ, USA; Department of Chemistry and Chemical Biology, Rutgers, The State University of New Jersey, Piscataway, NJ, USA; Rutgers Cancer Institute of New Jersey, Robert Wood Johnson Medical School, New Brunswick, NJ, USA; Skaggs School of Pharmacy and Pharmaceutical Sciences, La Jolla, CA, USA; San Diego Supercomputer Center, University of California, San Diego, La Jolla, CA, USA*

HENRY N. CHAPMAN • *Center for Free-Electron Laser Science, DESY, Hamburg, Germany; Department of Physics, University of Hamburg, Hamburg, Germany; The Centre for Ultrafast Imaging, University of Hamburg, Hamburg, Germany*

VADIM CHEREZOV • *Department of Chemistry, Bridge Institute, University of Southern California, Los Angeles, CA, USA*

DAVID R. COOPER • *Department of Molecular Physiology and Biological Physics, University of Virginia School of Medicine, Charlottesville, VA, USA*

MIROSLAWA DAUTER • *Basic Science Program, Leidos Biomedical Research, Inc., Argonne National Laboratory, Argonne, IL, USA*

ZBIGNIEW DAUTER • *Synchrotron Radiation Research Section, MCL, National Cancer Institute, Argonne National Laboratory, Argonne, IL, USA*

MARC C. DELLER • *Stanford ChEM-H, Macromolecular Structure Knowledge Center, Stanford University, Stanford, CA, USA*

ZYGMUNT S. DEREWENDA • *Department of Molecular Physiology and Biological Physics, University of Virginia School of Medicine, Charlottesville, VA, USA*

KAY DIEDERICHS • *Department of Biology, Universität Konstanz, Konstanz, Germany*

FRANK DIMAIO • *Department of Biochemistry, Institute of Protein Design, University of Washington, Seattle, WA, USA*

ROBERT F. FISCHETTI • *GM/CA@APS, Advanced Photon Source, Argonne National Laboratory, Argonne, IL, USA*

ELSPETH F. GARMAN • *Department of Biochemistry, University of Oxford, Oxford, UK*

ADAM GODZIK • *Bioinformatics and Systems Biology Program, Sanford Burnham Prebys Medical Discovery Institute, La Jolla, CA, USA*

MAREK GRABOWSKI • *Department of Molecular Physiology and Biological Physics, University of Virginia School of Medicine, Charlottesville, VA, USA*
WAYNE A. HENDRICKSON • *Department of Biochemistry and Molecular Biophysics, Columbia University, New York, NY, USA*
ANDRII ISHCHENKO • *Department of Chemistry, Bridge Institute, University of Southern California, Los Angeles, CA, USA*
MARIUSZ JASKOLSKI • *Faculty of Chemistry, Department of Crystallography, A. Mickiewicz University, Poznan, Poland; Center for Biocrystallographic Research, Institute of Bioorganic Chemistry, Polish Academy of Sciences, Poznan, Poland*
DAVID J. KISSICK • *GM/CA@APS, Advanced Photon Source, Argonne National Laboratory, Argonne, IL, USA*
GERARD J. KLEYWEGT • *Protein Data Bank in Europe, European Molecular Biology Laboratory–European Bioinformatics Institute, Cambridge, UK*
OLEG KOVALEVSKIY • *MRC Laboratory of Molecular Biology, Cambridge, UK*
QUN LIU • *Biology Department, Brookhaven National Laboratory, Upton, NY, USA*
JOHN L. MARKLEY • *BioMagResBank, Department of Biochemistry, University of Wisconsin-Madison, Madison, WI, USA*
AIRLIE J. MCCOY • *Department of Haematology, Cambridge Institute for Medical Research, University of Cambridge, Cambridge, UK*
ALEXANDER MCPHERSON • *Department of Molecular Biology and Biochemistry, University of California, Irvine, Irvine, CA, USA*
WLADEK MINOR • *Department of Molecular Physiology and Biological Physics, University of Virginia School of Medicine, Charlottesville, VA, USA*
ABEL MORENO • *Instituto de Química, Universidad Nacional Autónoma de Mexico, Ciudad de México, Mexico*
GARIB N. MURSHUDOV • *MRC Laboratory of Molecular Biology, Cambridge, UK*
HARUKI NAKAMURA • *Protein Data Bank Japan, Institute for Protein Research, Osaka University, Suita, Osaka, Japan*
ROBERT A. NICHOLLS • *MRC Laboratory of Molecular Biology, Cambridge, UK*
CRAIG M. OGATA • *GM/CA@APS, Advanced Photon Source, Argonne National Laboratory, Argonne, IL, USA*
PRZEMYSLAW J. POREBSKI • *Department of Molecular Physiology and Biological Physics, University of Virginia School of Medicine, Charlottesville, VA, USA*
EDWIN POZHARSKI • *Department of Biochemistry and Molecular Biology, University of Maryland School of Medicine, Baltimore, MD, USA*
SREEJITH RARAN-KURUSSI • *Macromolecular Crystallography Laboratory, Center for Cancer Research, National Cancer Institute at Frederick, Frederick, MD, USA*
BERNHARD RUPP • *k.-k. Hofkristallamt, Vista, CA, USA; Department of Genetic Epidemiology, Medical University Innsbruck, Innsbruck, Austria*
RUSLAN SANISHVILI • *GM/CA@APS, Advanced Photon Source, Argonne National Laboratory, Argonne, IL, USA*
MARIUS SCHMIDT • *Kenwood Interdisciplinary Research Complex, Physics Department, University of Wisconsin-Milwaukee, Milwaukee, WI, USA*
MICHAEL C. THOMPSON • *Department of Bioengineering and Therapeutic Sciences, University of California, San Francisco, San Francisco, CA, USA*
ANDREA THORN • *Hamburg Centre for Ultrafast Imaging, Universität Hamburg, Hamburg, Germany; Diamond Light Source, Harwell Science and Innovation Campus, Oxfordshire, UK*

DUŠAN TURK • *Department of Biochemistry and Molecular and Structural Biology, Jozef Stefan Institute, Ljubljana, Slovenia; Centre of Excellence for Integrated Approaches in Chemistry and Biology of Proteins, Ljubljana, Slovenia*

SAMEER VELANKAR • *Protein Data Bank in Europe, European Molecular Biology Laboratory–European Bioinformatics Institute, Cambridge, UK*

MEITIAN WANG • *Swiss Light Source, Paul Scherrer Institute, Villigen, Switzerland*

DAVID S. WAUGH • *Macromolecular Crystallography Laboratory, Center for Cancer Research, National Cancer Institute at Frederick, Frederick, MD, USA*

MARTIN WEIK • *Institut de Biologie Structurale, University of Grenoble Alpes, CEA, CNRS, Grenoble, France*

MANFRED S. WEISS • *Helmholtz-Zentrum Berlin, Macromolecular Crystallography (HZB-MX), Berlin, Germany*

THOMAS A. WHITE • *Center for Free-Electron Laser Science, Deutsches Elektronen-Synchrotron DESY, Hamburg, Germany*

ALEXANDER WLODAWER • *Macromolecular Crystallography Laboratory, National Cancer Institute, Frederick, MD, USA*

HEPING ZHENG • *Department of Molecular Physiology and Biological Physics, University of Virginia School of Medicine, Charlottesville, VA, USA*

Chapter 1

Expression and Purification of Recombinant Proteins in *Escherichia coli* with a His_6 or Dual His_6-MBP Tag

Sreejith Raran-Kurussi and David S. Waugh

Abstract

Rapid advances in bioengineering and biotechnology over the past three decades have greatly facilitated the production of recombinant proteins in *Escherichia coli*. Affinity-based methods that employ protein or peptide based tags for protein purification have been instrumental in this progress. Yet insolubility of recombinant proteins in *E. coli* remains a persistent problem. One way around this problem is to fuse an aggregation-prone protein to a highly soluble partner. *E. coli* maltose-binding protein (MBP) is widely acknowledged as a highly effective solubilizing agent. In this chapter, we describe how to construct either a His_6- or a dual His_6-MBP tagged fusion protein by Gateway® recombinational cloning and how to evaluate their yield and solubility. We also describe a simple and rapid procedure to test the solubility of proteins after removing their N-terminal fusion tags by tobacco etch virus (TEV) protease digestion. The choice of whether to use a His_6 tag or a His_6-MBP tag can be made on the basis of this solubility test.

Key words Fusion protein, Gateway® cloning, Hexahistidine tag, His_6-MBP, His_6-tag, Inclusion body, Maltose-binding protein, MBP, Recombinational cloning, Solubility enhancer, TEV protease, Tobacco etch virus protease

1 Introduction

A major time-consuming process in nearly all structural and functional studies of proteins is their overproduction and purification. Recombinant protein production in *Escherichia coli* has become the most popular platform for researchers who require large amounts of protein. Immobilized metal affinity chromatography (IMAC) with a polyhistidine tag (usually six consecutive histidine residues) has emerged as the most common and convenient method for purifying recombinant proteins. However, many His-tagged proteins form insoluble aggregates, especially in *E. coli* [1]. Before abandoning bacterial expression in favor of more complicated and costly eukaryotic systems, we suggest employing a simple strategy that combines the solubility-enhancing benefit conferred by *E. coli* maltose-binding protein (MBP) [2, 3] with the

Alexander Wlodawer et al. (eds.), *Protein Crystallography: Methods and Protocols*, Methods in Molecular Biology, vol. 1607, DOI 10.1007/978-1-4939-7000-1_1, © Springer Science+Business Media LLC 2017

powerful advantage of IMAC [4], made possible by the use of a polyhistidine tag in a tandem configuration with MBP (His_6-MBP) [5]. In this chapter, we describe how to construct either a His_6-tagged or His_6-MBP tagged fusion protein and conduct a few simple pilot experiments that are reliable predictors of protein production success, prior to extensive resource investment. The outcome of these pilot experiments dictates which N-terminal tag (His_6 or His_6-MBP) should be used for large-scale protein production.

2 Materials

2.1 Construction of Expression Vectors by Recombinational Cloning

1. The Gateway® destination vector pDEST566 (*see* Addgene plasmid #11517).
2. The Gateway® destination vector pDEST-HisMBP (*see* Addgene plasmid #11085).
3. The Gateway® destination vector pDEST527 (*see* Addgene plasmid #11518).
4. PCR reagents, including thermostable DNA polymerase (*see* **Note 1**).
5. Synthetic oligodeoxyribonucleotide primers for PCR amplification (*see* Fig. 1).
6. TE buffer (10 mM Tris–HCl pH 8.0, 1 mM EDTA).
7. Agarose, buffer and an apparatus for submarine gel electrophoresis of DNA (*see* **Note 2**).
8. MinElute Gel Extraction Kit (Qiagen, Valencia, CA) for the extraction of DNA from agarose gels.
9. Chemically competent *ccdB* Survival™ 2 $T1^R$ cells (Life Technologies, Grand Island, NY) for propagating pDEST566, pDEST527, pDONR221, or any vector with a Gateway® cloning cassette.
10. Competent *gyrA*$^+$ cells (e.g., DH5α, MC1061, HB101) (*see* **Note 3**).
11. Gateway® PCR Cloning System (Life Technologies).
12. LB medium and LB agar plates containing ampicillin (100 μg/ml). LB medium: Add 10 g Bacto tryptone, 5 g Bacto yeast extract, and 5 g NaCl to 1 L of H_2O and sterilize by autoclaving. For LB agar, also add 12 g of Bacto agar before autoclaving. To prepare plates, allow medium to cool until flask or bottle can be held in hands without burning, then add 1 ml ampicillin stock solution (100 mg/ml in H_2O, filter-sterilized), mix by gentle swirling, and pour or pipet ca. 30 ml into each sterile petri dish (100 mm dia.).

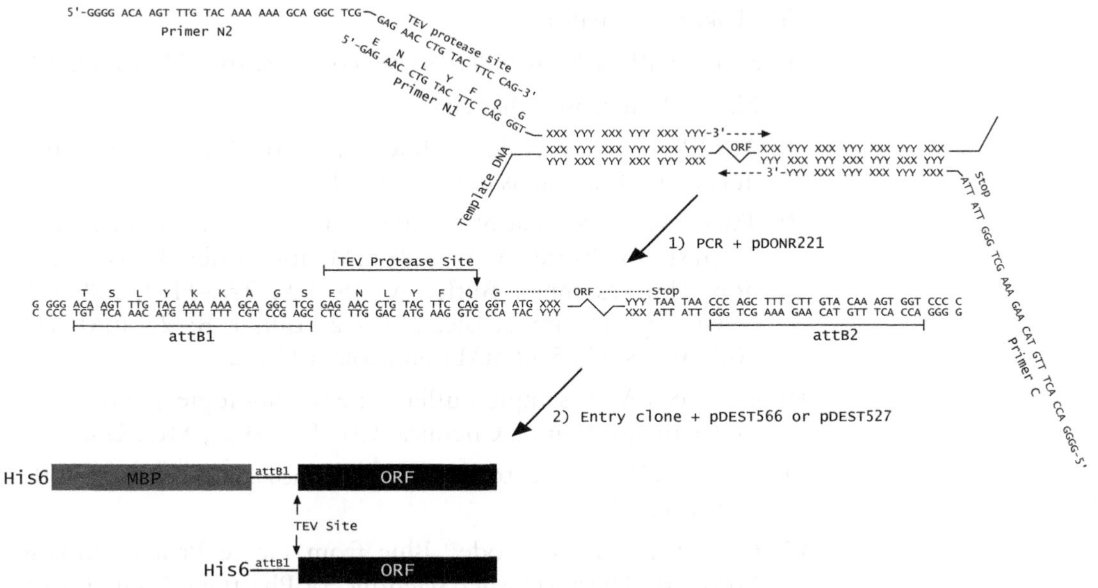

Fig. 1 Construction of a His_6- or His_6-MBP fusion vector using PCR and Gateway® cloning technology. The ORF of interest is amplified from the template DNA by PCR, using primers N1, N2, and C. Primers N1 and C are designed to base-pair to the 5′ and 3′ ends of the coding region, respectively, and contain unpaired 5′ extensions as shown. Primer N2 base pairs with the sequence that is complementary to the unpaired extension of primer N1. The final PCR product is recombined with the pDONR221 vector to generate an entry clone, via the BP reaction. This entry clone is subsequently recombined with pDEST527, pDEST566 or pDEST-HisMBP using LR Clonase to yield the final His_6- or His_6-MBP fusion vectors, respectively

13. Reagents for small-scale plasmid DNA isolation (*see* **Note 4**).
14. An incubator set at 37 °C.

2.2 Pilot Expression, Protease Cleavage, and Solubility Testing

1. Competent Rosetta™ 2(DE3) (EMD Millipore, Billerica, MA) (*see* **Notes 5** and **6**).
2. A derivative of pDEST566 and pDEST527 that produces a His_6-MBP fusion and His_6-fusion protein, respectively, with a TEV protease recognition site in the linker between the N-terminal tag(s) and the passenger protein (*see* Subheading 3.1).
3. LB agar plates and broth containing both ampicillin (100 μg/ml) and chloramphenicol (30 μg/ml). Prepare a stock solution of 30 mg/ml chloramphenicol in ethanol. Store at −20 °C for up to 6 months. (*See* Subheading 2.1, **item 11** for LB broth, LB agar, and ampicillin stock solution recipes). Dilute antibiotics 1000-fold into LB medium or molten LB agar.
4. Isopropyl-thio-β-D-galactopyranoside (IPTG), dioxane-free, ultrapure (American Bioanalytical, Natick, MA, USA). Prepare a stock solution of 200 mM concentration in H_2O and filter-sterilize. Store at −20 °C.

5. Shaker/incubator.
6. Sterile baffled-bottom flasks (Bellco Glass, Inc., Vineland, NJ).
7. Ni-NTA Agarose (Qiagen).
8. AcTEV protease (Life Technologies), or TEV protease produced and purified as described [6].
9. Two IMAC-compatible buffers that contain imidazole at 25 mM (for Buffer A) or 500 mM (for Buffer B) concentration. (For example, Buffer A: 25 mM Tris–HCl, 200 mM NaCl, 25 mM imidazole, pH 7.2; Buffer B: 25 mM Tris, 200 mM NaCl, 500 mM imidazole, pH 7.2.)
10. 4× SDS-PAGE sample buffer (Life Technologies) and 2-mercaptoethanol (Sigma Chemical Co., St. Louis, MO, USA).
11. SDS-PAGE gel, electrophoresis apparatus, and running buffer (*see* **Note** 7).
12. Gel stain (e.g., Gelcode® Blue from Pierce Protein Biology Products, Thermo Fisher Scientific, or PhastGel™ Blue R from GE Healthcare Life Sciences, Piscataway, NJ).
13. Spectrophotometer.
14. 1.5 ml microcentrifuge tubes.

3 Methods

3.1 Recombinational Cloning to Generate His_6- or His_6-MBP Fusion Vector

The Gateway® recombinational cloning system is based on the site-specific recombination reactions that mediate the integration and excision of bacteriophage λ, respectively, into and from the *E. coli* chromosome. For detailed information about this system, the reader is encouraged to consult the technical literature supplied by Thermo Fisher Scientific (Waltham, MA) (www.thermofisher.com/gateway).

3.1.1 pDEST566 and pDEST- HisMBP

To utilize the Gateway® system for the production of His_6-MBP fusion proteins, one must first construct or obtain a suitable "destination vector." Two destination vectors that can be used to produce His_6-MBP fusion proteins (pDEST566 and pDEST-HisMBP) are available from the authors or the Addgene plasmid repository (www.addgene.org, plasmids #11517 and #11085, respectively).

The Gateway® cloning cassette in pDEST566 and pDEST-HisMBP carries a gene encoding the DNA gyrase poison CcdB, which provides a negative selection against the destination vector, the donor vector, and various recombination intermediates so that only the desired recombinant is obtained when the end products of the recombinational cloning reaction are transformed into *E. coli* and grown in the presence of ampicillin. pDEST566, pDEST-HisMBP and other vectors that carry the *ccdB* gene must be propagated in a host strain with a *gyrA* mutation (e.g., *E. coli* DB3.1) that renders the cells immune to the action of CcdB or, alternatively, in a strain that produces the CcdB antidote CcdA (e.g., *ccdB* Survival™ 2 T1^R cells).

3.1.2 pDEST527

A destination vector that can be used to produce His_6-fusion proteins (pDEST527) is available from the authors or the Addgene plasmid repository (www.addgene.org, plasmid #11518). This vector is used for expression and affinity purification of proteins that are inherently soluble without the aid of solubility enhancers like MBP. It is a common practice in our laboratory to check the solubility of passenger proteins both with and without MBP and use the pDEST527-derived expression vector for large-scale expression and purification if the His_6-tagged passenger proteins do not form insoluble aggregates in *E. coli*.

3.1.3 Gateway® Cloning Protocol

To construct a His_6- or a His_6-MBP fusion expression vector, we amplify the target open reading frame (ORF) by PCR, incorporating into the primers elements that are necessary for Gateway® cloning and downstream protein production. Next we perform successive BP and LR reactions. The 3′ ends of the primers include a sufficient number of nucleotides that are complementary to the template sequence to result in a 69 °C melting temperature (by modified Breslauer's method, *see* http://www.thermoscientificbio.com/webtools/tmc/). This enables two-step PCR cycling using 72 °C as both the annealing and extension temperature. Proximal to the ORF-specific part of the forward primer, we add a sequence that encodes a TEV protease cleavage site preceded by an attB1 site to enable recombination. Because shorter primers are less expensive and because the TEV- and attB1-containing sequences are common to many of our experimental designs, we often use two overlapping forward primers, only one of which is ORF-specific (Fig. 1). An attB2 recombination site is added to the 5′ end of the ORF-specific portion of the reverse PCR primer. During early rounds of cycling, the inner, ORF-specific forward primer (N1) acts with the reverse primer (C) to create a template amplified by N2 and the same reverse primer in later rounds. To favor full-length product accumulation, the concentration of N1 is 20-fold lower than that of N2 and C (*see* **Note 8**).

1. The PCR reaction mix is prepared as follows (*see* **Note 9**): 10–25 ng template DNA, 10 μl 2× Phusion Flash PCR Master Mix (contains all necessary reaction components except primers and template), 0.025 μM primer N1, 0.5 μM primer N2, 0.5 μM primer C, H_2O (to 20 μl total volume).
2. The reaction mixture is placed in a thin-walled tube in a thermal cycler with an appropriate program, such as the following: initial denaturation for 3 min at 98 °C; 30 cycles of 98 °C for 10 s and 72 °C for 15 s, and final extension at 72 °C for 60 s (*see* **Note 10**); hold at 4 °C.

3. Purification of the PCR amplicon by agarose gel electrophoresis (*see* **Note 2**) is recommended.
4. To create the His_6-MBP fusion vector, the PCR product is recombined first into a donor vector, such as pDONR221, to yield an entry clone intermediate (BP reaction), and then into pDEST566 (LR reaction; *see* **Note 11**). Similarly, to create the His_6-fusion vector, the PCR product is recombined first into a donor vector, such as pDONR221, to yield an entry clone intermediate (BP reaction), and then into pDEST527 (LR reaction) as detailed below.
   (a) Add to a microcentrifuge tube: 100 ng of the PCR product in 1–5 μl TE or H_2O, 1.3 μl of 150 ng/μl pDONR vector DNA, and enough TE to bring the total volume to 12 μl. Mix well.
   (b) Thaw BP Clonase II enzyme mix on ice (2 min) and then vortex briefly (2 s) twice (*see* **Note 12**).
   (c) Add 3 μl of BP Clonase II enzyme mix to the components in (a) and vortex briefly; incubate the reaction at room temperature for at least 4 h (*see* **Note 13**).
   (d) Add to 10 μl of BP reaction: 2 μl of 150 ng/μl destination vector (pDEST566 or pDEST527) and 3 μl of LR Clonase II enzyme mix (*see* **Note 12**). Mix by vortexing briefly.
   (e) Incubate the reaction at room temperature for 2 h.
   (f) Add 2 μl of the proteinase K stop solution and incubate for 10 min at 37 °C.
   (g) Transform 1 μl of the reaction into 50 μl of appropriate competent *E. coli*, such as electrocompetent DH5α cells (*see* **Note 3**).
   (h) Spread the cells on an LB agar plate containing ampicillin (100 μg/ml), the selective marker for pDEST566, pDEST-HisMBP, and pDEST527. Incubate the plate at 37 °C overnight (*see* **Note 14**).
5. Plasmid DNA is isolated from saturated cultures started from individual ampicillin-resistant colonies and screened by sequencing putative clones to ensure that there are no PCR-introduced mutations.

3.2 Pilot Fusion Protein Expression, TEV Protease Cleavage, and Solubility Testing

Before investing time and resources in the large-scale expression and purification of a protein, we perform a series of pilot experiments to assess protein production, TEV protease cleavage, and target protein solubility. First, we transform the sequence-verified expression plasmid into an appropriate expression strain and induce production of the fusion protein. Following ultrasonic disruption of the cells, we confirm that the fusion protein is present in the soluble fraction of the crude cell lysate. After passing this checkpoint, we check for successful TEV cleavage and sustained solubility of the protein of interest following its liberation from His_6-MBP or His_6 tag in the crude lysate. A problem at any of these steps can be addressed before scaling-up.

3.2.1 Protein Expression

1. Transform competent Rosetta™ 2(DE3) cells (*see* **Notes 5** and **6**) with the His_6-MBP or His_6 fusion protein expression vector and spread them on an LB agar plate containing ampicillin (100 μg/ml) and chloramphenicol (30 μg/ml). Incubate the plate overnight at 37 °C.
2. Inoculate 5 ml of LB medium containing ampicillin (100 μg/ml) and chloramphenicol (30 μg/ml) in a culture tube with a single colony from the plate. Grow to saturation overnight at 37 °C with shaking.
3. The next morning, inoculate 50 ml of the same medium in a 250 ml baffled-bottom flask with 0.5 ml of the saturated overnight culture.
4. Grow the cells at 37 °C with shaking to mid-log phase (OD_{600nm} ~ 0.5).
5. Adjust the temperature to 30 °C (*see* **Note 15**) and add IPTG (1 mM final concentration).
6. After 4 h, measure the OD_{600nm} of the cultures (dilute cells 1:10 in LB to obtain an accurate reading). An OD_{600nm} of about 3–3.5 is normal, although lower or higher densities are possible.
7. Transfer 10 ml to a 15 ml conical centrifuge tube and pellet the cells by centrifugation (4000 × *g*) at 4 °C.
8. Resuspend the cell pellets in 2–4 ml of lysis buffer and then transfer the suspensions to a 1.5-ml microcentrifuge tube. Normalize the cell suspensions using absorbance values (OD_{600}) for comparisons.

Store the cell suspensions at −80 °C. Alternatively, the cells can be disrupted immediately and the procedure continued without interruption, as described below.

3.2.2 Sonication and Sample Preparation

1. Thaw the normalized cell suspensions (expressing either a His_6-tagged or His_6-MBP tagged protein) at room temperature, then place them on ice.
2. Lyse the cells by sonication (*see* **Note 16**).
3. Prepare samples of the total (T) intracellular protein from the IPTG-induced cultures for SDS-PAGE by mixing 30 μl of each sonicated cell suspension with 10 μl of 4× SDS-PAGE sample buffer containing 10% (v/v) 2-mercaptoethanol.
4. Pellet the insoluble cell debris (and proteins) by centrifuging the sonicated cell suspension from each culture at maximum speed in a microcentrifuge for 10 min at 4 °C.

5. Prepare samples of the soluble (S) intracellular protein from the IPTG-induced cultures for SDS-PAGE by mixing 30 μl of each supernatant from **step 4** with 10 μl of 4× SDS-PAGE sample buffer containing 10% (v/v) 2-mercaptoethanol.

3.2.3 TEV Protease Cleavage

To the soluble crude lysate prepared from induced cells (*see* Subheading 3.2, **step 2**) add approximately 0.05–0.10 mg/ml final concentration of pure TEV protease [6]. Mix and remove an aliquot for overnight incubation at room temperature; incubate remaining reaction at 4 °C overnight. Spin these tubes at maximum speed in a microcentrifuge for 5 min and analyze the supernatant (TEV+).

3.2.4 SDS-PAGE

We typically use precast NuPAGE gradient gels for SDS-PAGE to assess the yield and solubility of fusion proteins (*see* **Note 7**). The reader may choose any appropriate SDS-PAGE formulation appropriate for the protein size and laboratory preference.

1. Prepare new samples by mixing 30 μl of each solution with 10 μl of 4× SDS-PAGE sample buffer containing 10% (v/v) 2-mercaptoethanol.
2. Heat the T, S, and TEV+ samples at 90 °C for about 5 min and then spin them at maximum speed in a microcentrifuge for 5 min.
3. Assemble the gel in the electrophoresis apparatus, fill it with SDS-PAGE running buffer, load the samples (5–20 μl) and carry out the electrophoretic separation according to standard lab protocols. T, S, and TEV+ samples are loaded in adjacent lanes to allow easy assessment of solubility. Molecular weight standards may also be loaded on the gel, if desired.
4. Stain the proteins in the gel with GelCode® Blue reagent, PhastGel™ Blue R, or a suitable alternative.

3.2.5 Interpreting the Results

The overexpressed fusion proteins should be apparent as the predominant protein present on the gel. Examining the heaviest band relative to a molecular weight standard should confirm that the fusion is about the size of the protein of interest for His_6 tagged or plus 42 kDa (the approximate size of MBP) for His_6-MBP fusions. Placing the total and soluble fractions next to each other on the gel allows easy comparison and determination of solubility. A side-by-side analysis of His_6 tagged versus His_6-MBP tagged proteins will help to choose which tag to use for large-scale expression and purification.

Figure 2 illustrates the benefit of using MBP as a solubility enhancer. Lane 1 indicates that, upon induction, the Rosetta 2(DE3) expression strain was able to produce Chikungunya virus (ChikV) protease from a plasmid encoding the His_6-tagged protein. However, lane 2 reveals that most of the His_6-tagged protein is not found in the soluble fraction. In contrast, lanes 3 (total protein) and 4 (soluble protein) clearly demonstrate that a significant portion of

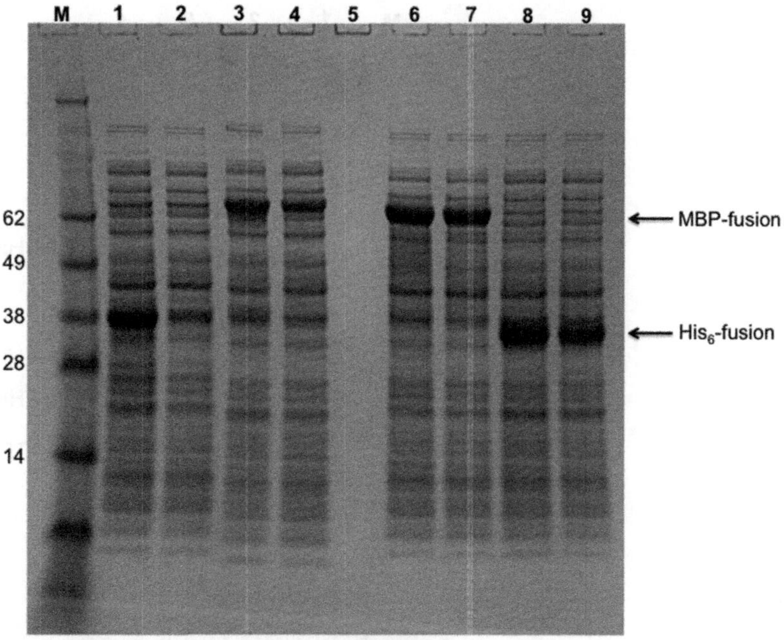

Fig. 2 Comparison of the solubility of His_6-tagged and His_6-MBP-tagged fusion proteins. *Lanes 1–4* of the SDS-PAGE gel represent protein extracted from Rosetta 2(DE3) cells expressing either His_6-tagged ChikV protease or His_6-MBP-ChikV protease from the appropriate plasmids. *Lane M*: SeeBlue Plus2 pre-stained marker standards. *Lane 1*: His_6-ChikV total protein. *Lane 2*: His_6-ChikV soluble protein. *Lane 3*: His_6-MBP-ChikV total protein. *Lane 4*: His_6-MBP-ChikV soluble protein. *Lanes 6–9* represent protein extracted from Rosetta 2(DE3) cells expressing either His_6-tagged or His_6-MBP tagged MERS-CoV $3CL^{pro}$ from the appropriate plasmids. *Lane 6*: His_6-MBP-MERS-CoV $3CL^{pro}$ total protein. *Lane 7*: His_6-MBP-MERS-CoV $3CL^{pro}$ soluble protein. *Lane 8*: His_6-MERS-CoV $3CL^{pro}$ total protein. *Lane 9*: His_6-MERS-CoV $3CL^{pro}$ soluble protein

the His_6-MBP-ChikV protease fusion protein is soluble when produced in the same strain. Lanes 6 (total protein) and 7 (soluble protein) illustrate the solubility of catalytically inactive Middle East Respiratory Syndrome coronavirus 3C-like protease (MERS-CoV $3CL^{pro}$ C148A) as a His_6-MBP tagged protein. Lanes 8 (total protein) and 9 (soluble protein) reveal that a very similar solubility is obtained even with His_6-fusions of MERS-CoV $3CL^{pro}$. Hence, MERS-CoV $3CL^{pro}$ is an example of a protein that does not require a solubility enhancer for overproduction in *E. coli*.

The soluble crude lysate fractions can be used to test the ability of TEV protease to cleave the tags in vitro. In lanes 1 and 2 of Fig. 3, which correspond to the soluble lysate and soluble products of the cleavage reaction, respectively, the band representing the fusion protein has largely disappeared, and three significant new bands have appeared: a 42-kDa band for His_6-MBP, 28-kDa band for His-TEV protease, and 37-kDa band migrating at the expected

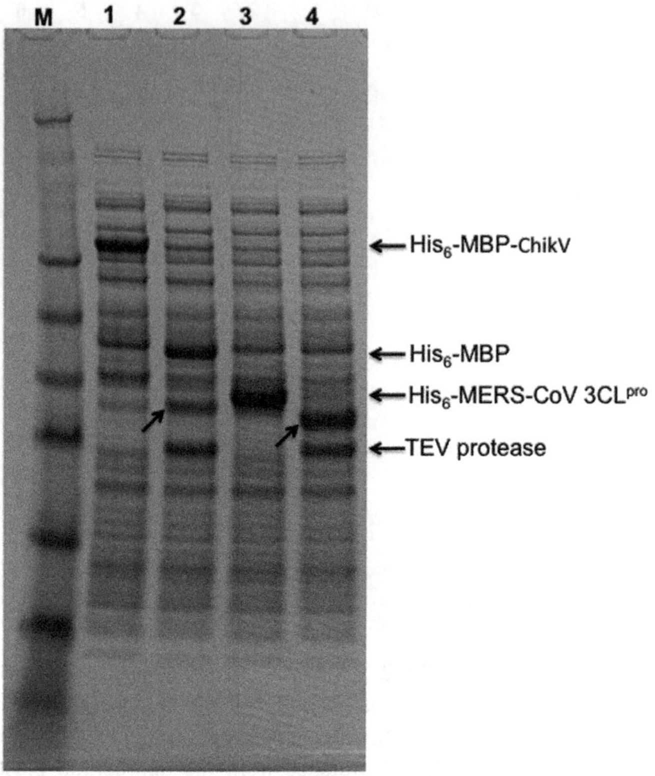

Fig. 3 Small-scale pilot expression and digestion of fusion protein with TEV protease. ChikV protease and MERS-CoV 3CL^{pro} were expressed from derivatives of pDEST-HisMBP and pDEST527, respectively, in Rosetta 2(DE3) cells as described (*see* Subheading 3.2) and analyzed by SDS-PAGE. *Lane M*: SeeBlue Plus2 prestained marker standards. *Lane 1*: His$_6$-MBP-ChikV protease soluble protein. *Lane 2*: TEV protease digest of *lane 1* sample, soluble protein. *Lane 3*: His$_6$-MERS-CoV 3CL^{pro} soluble protein. *Lane 4*: TEV protease digest of *lane 3* sample, soluble protein (*see* Subheading 3.2, **steps 4** and **5**). *Slanted arrows* indicate the positions of the liberated passenger proteins

size of ChikV protease. The prominent band corresponding to ChikV protease indicates that it remains soluble after cleavage from MBP, suggesting that it is probably properly folded. Similarly, the cleavage of His$_6$-tagged MERS-CoV 3CL^{pro} was conducted and lanes 3 and 4 indicate the soluble protein and soluble products of TEV digestion, respectively. The cleaved His$_6$-tag and the MERS-CoV 3CL^{pro} (33-kDa) are clearly visible on the gel. Had the TEV protease failed to cleave the fusions or had the target protein become insoluble after cleavage, troubleshooting would have been necessary. Otherwise, having successfully passed these diagnostic tests, the production and purification of the protein may now be scaled up as described [4].

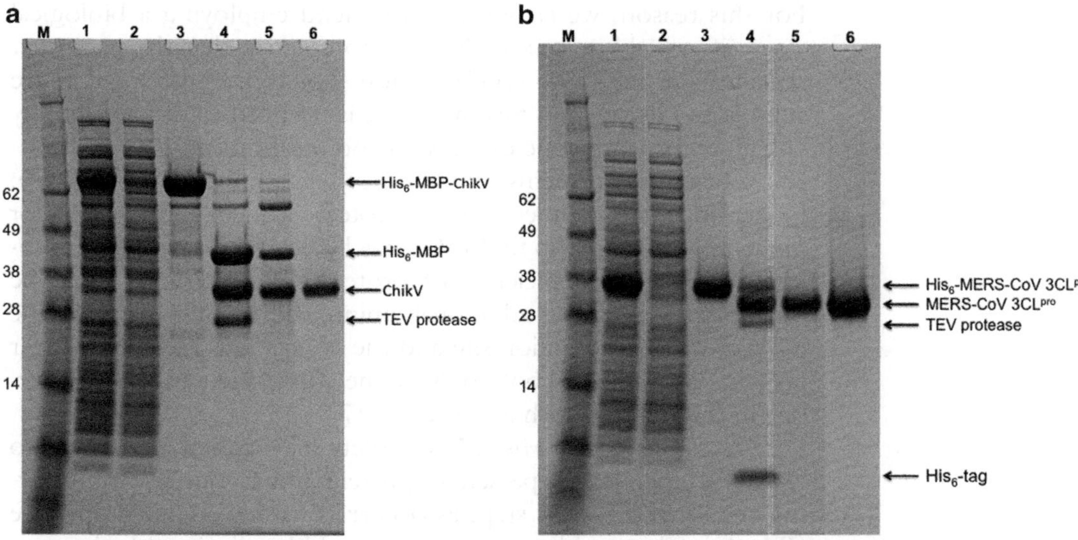

Fig. 4 Purification of fusion proteins by immobilized-metal affinity chromatography (IMAC). Purification monitored by SDS-PAGE at different stages. In panels A (ChikV protease) and B (MERS-CoV 3CL^{pro}), the following gel-loading pattern applies. *Lane M*: SeeBlue Plus2 pre-stained marker standards. *Lane 1*, soluble lysate (crude); *lane 2*, flow-through from first IMAC column (unbound); *lane 3*, eluate from first IMAC column; *lane 4*, products of TEV protease digest; *lane 5*, flow-through from second IMAC column; *lane 6*, final sample after size exclusion chromatography

A typical large-scale purification profile of a His$_6$-tagged protein (MERS-CoV 3CL^{pro}) and a His$_6$-MBP tagged protein (ChikV protease) are shown in Fig. 4. The bands representing fusion proteins and their tagless forms during the purification process are indicated in the figure.

3.2.6 Troubleshooting

Not every fusion protein (His$_6$-tagged or His$_6$-MBP tagged) will be highly soluble. However, solubility usually can be increased by reducing the temperature of the culture from 30 to 18 °C during the time that the fusion protein is accumulating in the cells (i.e., after the addition of IPTG). In some cases, the improvement can be quite dramatic. It may also be helpful to reduce the IPTG concentration to a level that will result in partial induction of the fusion protein. Under these conditions, longer induction times (18–24 h) are required to achieve a reasonable yield.

Occasionally, a passenger protein may accumulate in a soluble but biologically inactive form after intracellular processing of an MBP fusion protein. Exactly how and why this occurs is unclear, but we suspect that fusion to MBP somehow enables certain proteins to evolve into kinetically trapped folding intermediates that are no longer susceptible to aggregation. Therefore, although solubility after intracellular processing is generally a useful indicator of a passenger protein's folding state, it is not absolutely trustworthy.

For this reason, we strongly recommend employing a biological assay (if available) at an early stage to confirm that the passenger protein is in its native conformation. For those proteins that are soluble as His_6-tagged fusions, there is no need to use a solubility enhancer for large-scale expression and purification.

When fusion proteins are resistant to digestion by TEV protease, longer incubation times, higher protease concentrations, and/or higher temperature (up to 30 °C) may be helpful. In especially problematic cases, the efficiency of protease digestion can often be improved by inserting additional amino acid residues between the TEV protease recognition site and the N terminus of the passenger protein. We have used both polyglycine (Gly_3) and a FLAG-tag epitope in this position with good results [7].

Occasionally, the His_6-MBP moiety may exhibit a tendency to "stick" to the cleaved passenger protein and co-purify with it during the second IMAC step, as occurred with the ChikV protease (Fig. 4a). This problem most likely could be alleviated by increasing the salt concentration in the IMAC buffer. However, in this case the final size exclusion chromatography step separated the ChikV final product from the His_6-MBP tag.

4 Notes

1. We recommend a processive, high-fidelity polymerase such as *Phusion* (Thermo Fisher or New England Biolabs, Ipswich, MA, USA) to reduce cycling times and minimize the occurrence of mutations during PCR.
2. We typically purify fragments by horizontal electrophoresis in 1–2% Certified Molecular Biology Agarose (Bio-Rad, Hercules, CA) gels run in sodium borate solution [8] using standard submarine equipment. DNA fragments are extracted from slices of ethidium bromide-stained gel using a MinElute Gel Extraction Kit (Qiagen) in accordance with the instructions of the manufacturer.
3. Any *gyrA*+ strain of *E. coli* can be used. We prefer ElectroMAX™ DH5α-E™ Competent Cells (Life Technologies) because they are easy to use and very efficient.
4. We prefer the QIAprep™ Spin miniprep kit (Qiagen), but similar kits can be obtained from a wide variety of vendors.
5. Chemically competent cells are transformed according to the manufacturer's instructions. Electrocompetent cells can be purchased or prepared. Briefly, the cells are grown in 1 L of LB medium (with antibiotics, if appropriate) to mid-log phase (OD_{600} ~ 0.5) and then chilled on ice. The cells are pelleted at 4 °C, resuspended in 1 L of ice-cold H_2O and pelleted again. After several such washes with H_2O, the cells are resuspended

in 3–4 ml of 10% glycerol, divided into 50 μl aliquots, and immediately frozen in a dry ice–ethanol bath. Competent cells are stored at −80 °C. Electrotransformation procedures can be obtained from the electroporator manufacturers (e.g., Bio-Rad, BTX, Eppendorf). Immediately prior to electrotransformation, the cells are thawed on ice and mixed with 10–100 ng of DNA (e.g., a plasmid vector or a Gateway® reaction). The mixture is placed in an ice-cold electroporation cuvette and electroporated according to the manufacturer's recommendations (e.g., 1.5 kV pulse in a cuvette with a 1 mm gap). One milliliter of SOC medium [9] is immediately added to the cells and they are allowed to grow at 37 °C with shaking (ca. 250 rpm) for 1 h. 5–200 μl of the cells are then spread on an LB agar plate containing the appropriate antibiotic(s).

6. If the open reading frame encoding the passenger protein contains codons that are rarely used in *E. coli* (http://www.doe-mbi.ucla.edu/cgi/cam/racc.html), this can adversely affect the yield of a protein. In such cases, it is advisable that the expression strain carries an additional plasmid that codes for rare-codon-tRNA genes. The pRIL plasmid (Stratagene, La Jolla, CA) is a derivative of the p15A replicon that carries the *E. coli argU*, *ileY*, and *leuW* genes, which encode the cognate tRNAs for AGG/AGA, AUA, and CUA codons, respectively. pRIL is selected for by resistance to chloramphenicol. In addition to the tRNA genes for AGG/AGA, AUA, and CUA codons, the pRARE accessory plasmid in the Rosetta™ host strain (Novagen, Madison, WI) also includes tRNAs for the rarely used CCC and GGA codons. Like pRIL, the pRARE plasmid is a chloramphenicol-resistant derivative of the p15A replicon. Both of these tRNA accessory plasmids are compatible with derivatives of pDEST566, pDEST-HisMBP or pDEST527. Another option is to prepare the insert (cDNA of interest) synthetically, using *E. coli*-preferred codons.
7. We find it convenient to use precast SDS-PAGE gels, running buffer, molecular weight standards, and electrophoresis supplies from Life Technologies.
8. Alternatively, the PCR reaction can be performed in two separate steps, using primers N1 and C in the first step and primers N2 and C in the second step. The PCR amplicon from the first step is used as the template for the second PCR. All primers are used at the typical concentrations for PCR in the two-step protocol.
9. The PCR reaction can be modified in numerous ways to optimize results, depending on the nature of the template and primers. See Ref. 9 (Vol. 2, Chapter 8) for more information.
10. PCR cycle conditions can also be varied based on reagents and consumables chosen, template complexity and gene length. For example, when using Phusion Flash High-Fidelity PCR

Master Mix, extend the cycle for 15 s per kb of DNA. Consult the directions provided by the manufacturer of your thermostable polymerase.

11. This "one-tube" Gateway® protocol bypasses the isolation of an "entry clone" intermediate. However, the entry clone may be useful if you intend to experiment with additional Gateway® destination vectors, in which case the BP and LR reactions can be performed sequentially in separate steps; detailed instructions are included with the Gateway® PCR kit. Alternatively, entry clones can easily be regenerated from expression clones via the BP reaction, as described in the manual.
12. Clonase enzyme mixes should be thawed according to the manufacturer's directions.
13. At this point, we remove a 5 μl aliquot from the reaction and add it to 0.5 μl of proteinase K stop solution. After 10 min at 37 °C, we transform 2 μl into 50 μl of competent DH5α cells (*see* **Note 3**) and spread 100–200 μl on an LB agar plate containing kanamycin (25 μg/ml), the selective marker for pDONR221. From the number of colonies obtained, it is possible to gauge the success of the BP reaction. Additionally, entry clones can be recovered from these colonies in the event that no transformants are obtained after the subsequent LR reaction.
14. If very few or no transformants are obtained after the BP or LR reactions, the efficiency of the process can be improved by incubating the reactions overnight.
15. We have found that decreasing the induction temperature to 30 °C increases the quality and solubility of the fusion protein without significantly decreasing the yield, especially in the presence of glucose. We also test 18 °C inductions if necessary, in which case the inductions are usually longer (18–24 h).
16. We routinely disrupt cells in 1.5-ml microcentrifuge tubes on ice with two or three 30 s pulses using a VCX600 sonicator (Sonics and Materials, Inc.) with a microtip at 38% power. The cells are cooled on ice between pulses.

Acknowledgments

We thank Karina Keefe and Danielle Needle for constructing the ChikV protease and MERS-CoV 3CL^{pro}C148A expression vectors, respectively. This research was funded by the Intramural Research Program of the NIH, National Cancer Institute, Center for Cancer Research. The content of this publication does not necessarily reflect the views or policies of the Department of Health and Human Services, nor does the mention of trade names, commercial products, or organizations imply endorsement by the US Government.

References

1. Waugh DS (2005) Making the most of affinity tags. Trends Biotechnol 23:316–320
2. Kapust RB, Waugh DS (1999) *Escherichia coli* maltose-binding protein is uncommonly effective at promoting the solubility of polypeptides to which it is fused. Protein Sci 8:1668–1674
3. Fox JD, Routzahn KM, Bucher MH et al (2003) Maltodextrin-binding proteins from diverse bacteria and archaea are potent solubility enhancers. FEBS Lett 537:53–57
4. Tropea JE, Cherry S, Nallamsetty S et al (2007) A generic method for the production of recombinant proteins in *Escherichia coli* using a dual hexahistidine-maltose-binding protein affinity tag. Methods Mol Biol 363:1–19
5. Routzahn KM, Waugh DS (2002) Differential effects of supplementary affinity tags on the solubility of MBP fusion proteins. J Struct Funct Genom 2:83–92
6. Kapust RB, Tozser J, Fox JD et al (2001) Tobacco etch virus protease: mechanism of autolysis and rational design of stable mutants with wild-type catalytic efficiency. Protein Eng 14:993–1000
7. Fox JD, Waugh DS (2003) Maltose-binding protein as a solubility enhancer. Methods Mol Biol 205:99–117
8. Brody JR, Kern SE (2004) Sodium boric acid: a Tris-free, cooler conductive medium for DNA electrophoresis. BioTechniques 36:214–216
9. Sambrook J, Russell DW (2001) Molecular cloning: a laboratory manual. Cold Spring Harbor Laboratory Press, Cold Spring Harbor, NY

Chapter 2

Protein Crystallization

Alexander McPherson

Abstract

Protein crystallization was discovered by chance nearly 200 years ago and was developed in the late nineteenth century as a powerful purification tool, and a demonstration of chemical purity. The crystallization of proteins, nucleic acids, and large biological complexes, such as viruses, depends on the creation of a solution that is supersaturated in the macromolecule, but exhibits conditions that do not significantly perturb its natural state. Supersaturation is produced through the addition of mild precipitating agents such as neutral salts or polymers, and by manipulation of various parameters that include temperature, ionic strength, and pH. Also important in the crystallization process are factors that can affect the structural state of the macromolecule, such as metal ions, inhibitors, cofactors, or other conventional small molecules. A variety of approaches have been developed that combine the spectrum of factors that effect and promote crystallization, and among the most widely used are vapor diffusion, dialysis, batch, and liquid–liquid diffusion. Successes in macromolecular crystallization have multiplied rapidly in recent years due to the advent of practical, easy-to-use screening kits, and the application of laboratory robotics.

Key words Crystals, Supersaturation, Growth mechanisms, Homogeneity, X-ray diffraction, Precipitants, Crystallization methods, Vapor diffusion, Dialysis, Mother liquor

1 Introduction

Although the technologies of nuclear magnetic resonance and, more recently, cryogenic electron microscopy, have made significant inroads, presently the only technique that can yield atomic level structural images of biological macromolecules is X-ray diffraction analysis as applied to single crystals. While other methods may produce important structural and dynamic data only X-ray crystallography is adequate to precisely define atomic coordinates. The application of X-ray crystallography is absolutely dependent on crystals of the macromolecule, and not simply crystals, but crystals of sufficient size and quality to permit accurate data collection. The quality of the final structural image is directly determined by the perfection and physical properties of the crystalline specimen. The crystals, therefore, become the keystone element of the entire process, and the ultimate determinant of its success. The crystals

Alexander Wlodawer et al. (eds.), *Protein Crystallography: Methods and Protocols*, Methods in Molecular Biology, vol. 1607, DOI 10.1007/978-1-4939-7000-1_2,

themselves have no medicinal or pharmaceutical value, but provide the X-ray diffraction patterns that serve as the fundamental data, which through Fourier synthesis, allow the direct visualization of the macromolecules or their complexes composing the crystals.

When crystallizing proteins for X-ray diffraction analysis, one is usually dealing with homogenous, often exceptionally pure macromolecules, and the objective is to grow only a few large, perfect crystals. The proteins themselves may be purified from natural sources, microbes or tissues of plants and animals, or it may be produced by recombinant DNA techniques. The number of crystals needed for recording data may be few, but often the amount of protein available is severely limited. This in turn places constraints on the approaches and strategies that can be used to obtain those crystals. While new methodologies such as synchrotron radiation [1, 2] and cryocrystallography [3–5] have driven the necessary size of specimen crystals consistently downward, they have not eliminated the need for crystal perfection.

2 The Nature of Protein Crystals

Protein crystals are composed of approximately 50% solvent on average. Those seen in Fig. 1 vary from 33% solvent for monoclinic lysozyme up to 61% for concanavalin B. At the extremes one finds insulin at about 25% and tropomyosin at 90%. Protein occupies the remaining volume so that the entire crystal may be thought of as an ordered gel permeated by extensive networks of channels and interstitial spaces filled with solvent, through which small molecules can diffuse. There does not appear to be a direct correlation between the solvent volume of a protein crystal and its diffraction properties. It has, however, been noted that transitions of a crystallographic unit cell to smaller volume, with concomitant reduction of included solvent, has frequently produced an improvement in diffraction resolution [6, 7].

In proportion to molecular mass, the number of contacts (salt bridges, hydrogen bonds, hydrophobic interactions) that a conventional organic molecule forms with its neighbors in a crystal far exceeds the very few exhibited by crystalline macromolecules. Since these contacts provide the lattice interactions essential for crystal integrity, this largely explains the differences in properties between crystals of salts or small molecules and macromolecules. It may also explain why the introduction of a few additional contacts, or even one uniquely strong interaction, can profoundly affect the diffraction resolution of a protein crystal.

Living systems are based almost exclusively on aqueous chemistry within narrow ranges of temperature and pH. Macromolecules, thus, have evolved an appropriate compatibility and dependency. Serious deviations or perturbations are rarely tolerated. As a consequence, all protein crystals are grown from aqueous media, ones to

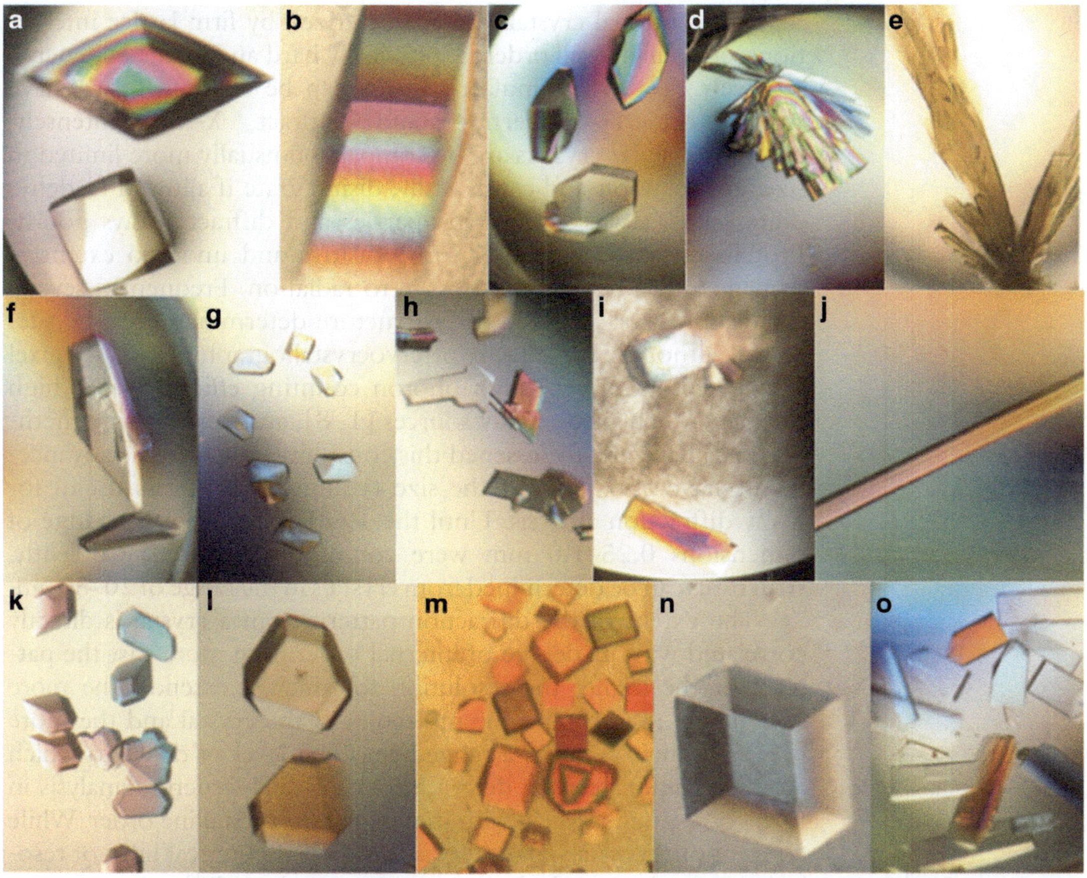

Fig. 1 An array of protein crystals showing the range of habits they may assume: in (**a**) thaumatin, (**b**) bovine trypsin, (**c**) tetragonal lysozyme, (**d**) monoclinic lysozyme, (**e**) beef liver catalase, (**f–h**) three different crystal forms of bovine RNase S, (**i**) beta-lactoglobulin, (**j**) concanavalin B, (**k**) satellite tobacco mosaic virus, (**l**) glucose isomerase, (**m**) concanavalin A, (**n**) rhombohedral canavalin, and (**o**) orthorhombic canavalin

which they are tolerant, and these solutions are called mother liquors. As described below, crystals can be made to grow from these mother liquors when the mother liquors are made supersaturated in protein.

There are important physical and chemical differences between ionic crystals, or those of most low-molecular-mass compounds, and crystals of proteins. For example, protein crystals generally have fairly simple morphologies, or habits as they are called, while conventional crystals often display very complex polyhedral or prismatic appearances. This is mainly due to the absence of centers of symmetry, mirror planes, and glide planes in protein crystals. Proteins exist in only one enantiomeric form and, therefore, cannot have such symmetry elements in their space groups. As a further consequence, protein crystals can fall into only 65 space groups rather than the 230 space groups allowed mixtures of enantiomers, and these 65 tend to have rather simple point group symmetries that are reflected in the habits.

Conventional crystals are characterized by firm lattice interactions, are usually well ordered, physically hard and brittle in general, relatively easy to manipulate, usually can be exposed to air, have strong optical properties, and diffract X-rays intensely. Macromolecular crystals are by comparison usually more limited in size, are very soft and crush easily, disintegrate if allowed to dehydrate, exhibit weak optical properties and diffract X-rays poorly. Protein crystals are temperature sensitive and undergo extensive damage after prolonged exposure to radiation. Frequently, several crystals must be analyzed for a structure determination to be successful although the advent of cryocrystallography [3–5, 8] pixel area detectors of very high photon counting efficiency [9], high intensity synchrotron X-ray sources [1, 8], and new phasing methods [10] have greatly lessened this constraint. Those same advancements have also reduced the size (volume) of crystals useful for X-ray diffraction analysis. Until the 1990s, crystals in the range of dimensions 0.25–1.0 mm were commonly required. Currently, structures can be determined from crystals in the range of 20–50 μm.

The extent of the diffraction pattern from a crystal is directly correlated with its degree of internal order. The more vast the pattern, or the higher the resolution to which it extends, the more structurally uniform are the molecules in the crystal and the more precise is their periodic arrangement. The level of detail to which atomic positions can be determined by crystal structure analysis in turn corresponds closely with that degree of crystalline order. While conventional crystals often diffract to their theoretical limit of resolution, protein crystals, by comparison, produce diffraction patterns of more limited extent. Protein crystals, all crystals in fact, are not uniform, flawless solids, but exhibit many defects and dislocations that produce a mosaic pattern of slightly misaligned sectors, or domains. Domain boundaries, often refered to as stacking faults or grain boundaries in conventional crystals, are far more numerous in protein crystals than conventional crystals, probably by several orders of magnitude [11]. These features contribute further to the limitation of diffraction quality. Some defects seen in protein and virus crystals by atomic force microscopy (AFM) are presented in Fig. 2.

The liquid channels and solvent filled cavities that permeate macromolecular crystals and the lack of order they engender are primarily responsible for the limited resolution of the diffraction patterns. Because of the relatively large solvent spaces between adjacent molecules and the consequent weak lattice forces, all molecules in the crystal may not occupy exactly equivalent orientations and positions but may vary slightly within or between unit cells. Furthermore, because of their structural complexity and their potential for conformational dynamics, protein molecules in the aqueous environment of a crystal may exhibit slight variations in the course of their polypeptide chains or the dispositions of side groups from one to another.

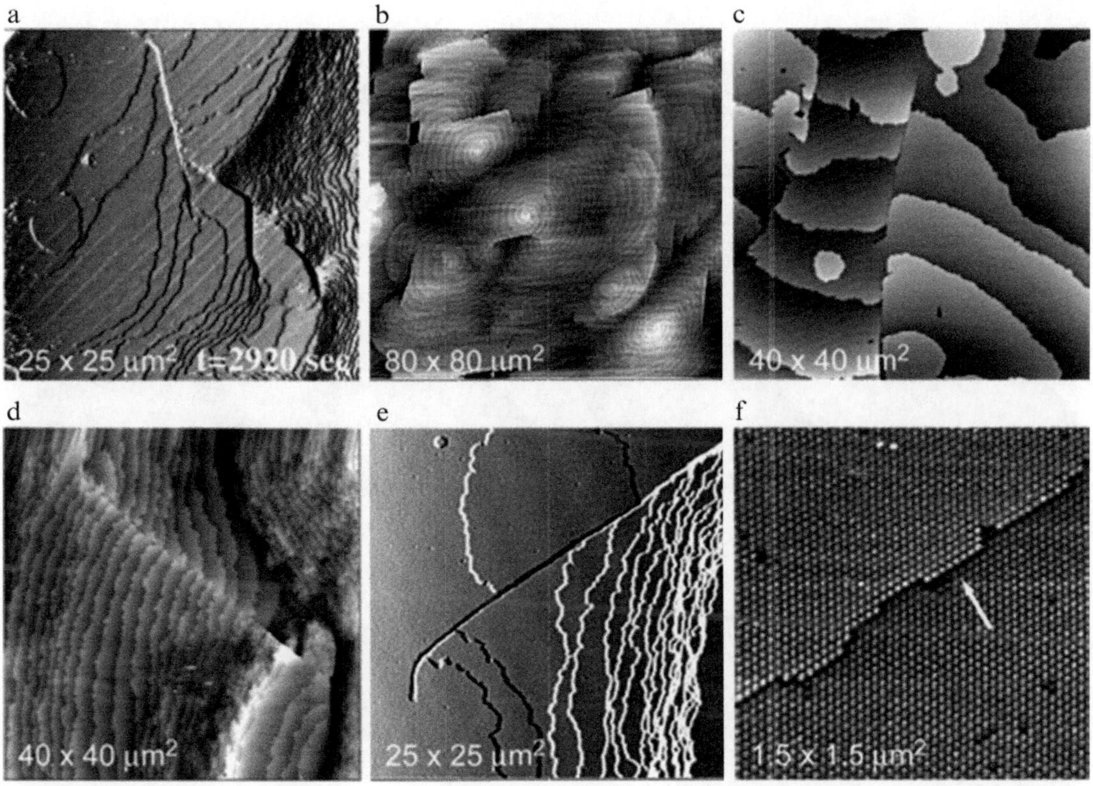

Fig. 2 Planar defects (stacking faults) in crystals of proteins and viruses. (**a**), (**c**), (**d**), and (**e**) are surfaces of satellite tobacco mosaic virus crystals, (**b**) canavalin and (**f**) a crystal of cucumber mosaic virus. The planar defects, homologous to grain boundaries in conventional crystals, divide the crystal into domains, which in turn are responsible for the mosaicity of the crystals

Although the presence of extensive solvent regions is a major contributor to the generally modest diffraction quality of protein crystals, it is also largely responsible for their value to biochemists as platforms for expermentation. Because of the high solvent content, the individual macromolecules in protein crystals are surrounded by layers of water that maintain their structure virtually unchanged from that found in solution. As a consequence, ligand binding, enzymatic activity, spectroscopic characteristics, and most other biochemical features are essentially the same as for the fully solvated molecule. Conventional chemical compounds, which may be ions, ligands, substrates, coenzymes, inhibitors, drugs, or other effector molecules, may be freely diffused into and out of the crystals. Crystalline enzymes, though immobilized, are frequently accessible for experimentation simply through alteration of the surrounding mother liquor.

Figure 3 shows, in a representative manner, how small organic molecules diffuse into a protein crystal through its network of solvent channels. The blue dye xylene cyanol, a molecule in the

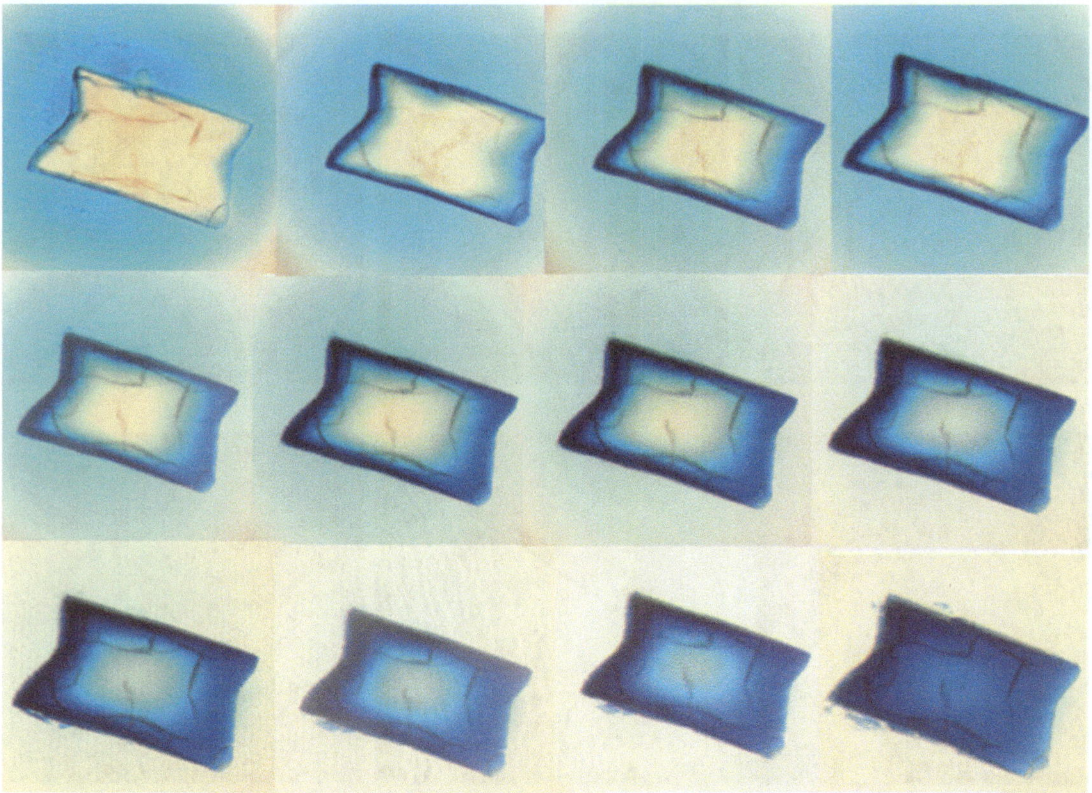

Fig. 3 A large crystal (about 1.5 mm in length) of the protein canavalin has had its mother liquor replaced with an equivalent mother liquor containing the blue dye xylene cyanol. This series of photographs taken over about 8 h shows the diffusion of the dye molecules into the protein crystal

molecular weight range of a biological coenzyme or a possible drug, was added directly to the mother liquor of a large canavalin crystal. The dye front, as it diffuses into the crystal is clearly evident, and its progress could be recorded and measured. From this it could be estimated that the dye, the small molecule, diffused through the crystal lattice at a rate of about 60 μm/h.

Figure 4 illustrates another experiment where a large thaumatin crystal was saturated with the pH sensative dye m-cresol purple at high pH (pH 8) giving it a blue color. The mother liquor was then replaced with with an equivalent one but at low pH (pH 6). As H_3O^+ ions diffused into the crystal, the dye internal to the crystal changed to a yellow color. Again, the dye transition front and its movement through the crystal was photographically recorded and measured. From this experiment it could be estimated that when a gradient of H_3O^+ of $10^{-8} > 10^{-6}$ exists between the interior of the crystal and its mother liquor, H_3O^+ ions diffuse to the center of the crystal with an average rate of about 1000 μm/h.

A diversity of crystallographic unit cells and habits that we refer to as polymorphism are common phenomena with macromolecular

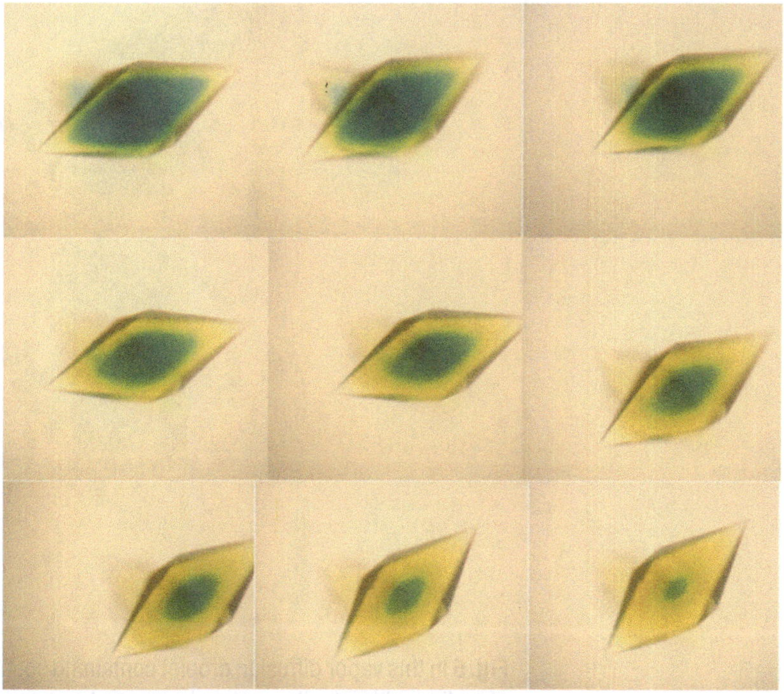

Fig. 4 A crystal of thaumatin was saturated with the pH sensitive dye m-bromocresol purple at pH 8 where the dye is *blue*. The mother liquor was then replaced with an equivalent mother liquor at pH 6. As the H_3O^+ ions diffused into the crystals, the internal m-bromocresol purple dye molecules changed color to *yellow*. The experiment, photographed over a period of about 35 min, shows the progress of the diffusion of the H_3O^+ ions into the protein crystal

crystals. In Fig. 1 alone we see crystals with three different unit cells of RNase S, two of both lysozyme and canavalin. Glucose isomerase, catalase, and trypsin can also be induced to crystallize in additional unit cells. Presumably this is a consequence of the protein's conformational dynamic range and the sensitivity of the lattice contacts involved. Thus, different unit cells and different symmetries may arise under what, by most standards, would be called identical conditions. In fact, multiple crystal forms are sometimes seen coexisting in the same sample of mother liquor as in Fig. 5.

3 Energetics, Kinetics, and Mechanisms of Protein Crystallization

There are further differences that complicate the crystallization of proteins as compared with conventional, small molecules [12–18]. First, proteins may coalesce to form several solid or dense liquid states that include amorphous precipitates, oils, or gels, as well as crystals, and most of these other forms are kinetically favored as supersaturation rises. Second, unlike most conventional crystals,

Fig. 5 In this vapor diffusion droplet containing sodium nitrate and a trace amount of sodium chloride, the protein lysozyme has crystallized in two distinctly different crystal forms. The large cluster of thin laths is the monoclinic crystal form, while the many smaller, darker crystals are of tetragonal symmetry

protein crystals nucleate, or initiate development, only at very high levels of supersaturation, often two to three orders of magnitude more than required to sustain crystal growth. This often leads to massive showers of microcrystals, or even more often, precipitate. Further, the kinetics of macromolecular crystal nucleation and growth are generally two to three orders of magnitude slower than for conventional molecules [19–23]. The latter difference arises from the considerably larger size, lowered diffusivity, and weaker association tendencies compared with small molecules or ions, as well as a lower overall probability of incorporation of an incoming protein molecule into a growth step [24].

Crystals, including protein crystals, grow by successive layer addition [18, 22, 23, 25]. The rate limiting step in crystal growth is not, however, the completion of an active layer by recruitment of molecules from solution into growth steps and kinks at the edges of expanding layers (refered to as tangential growth), as this is energetically favorable and rapid [25, 26]. The rate limiting step in crystal growth is the initiation of new, superior growth layers. This is more demanding in terms of self organization, less probable, and by far the slower process. It is refered to as growth in the normal direction.

There are four mechanisms that have been described for protein crystals to provide growth in the normal direction. These were deduced by application of AFM to actively growing crystals [27].

The various mechanisms have been treated in detail elsewhere [23, 27–30], but two of the four predominate. At higher levels of supersaturation the principal mechanism, illustrated by the examples in Fig. 6, is layer initiation by two dimensional nucleation on existing crystal surfaces. Like the formation of a three dimensional critical nucleus (*see* below), this requires molecules otherwise free in solution to self organize on a surface to form a small, crystallographically ordered array that may then expand by tangential growth. The major difference between the formation of three and two dimensional nuclei is that in the latter case the molecules are confined to a surface. This restriction of their freedom encourages their association. In addition, their self organization is guided in an epitaxial manner by the molecular lattice of the underlying, existing layer.

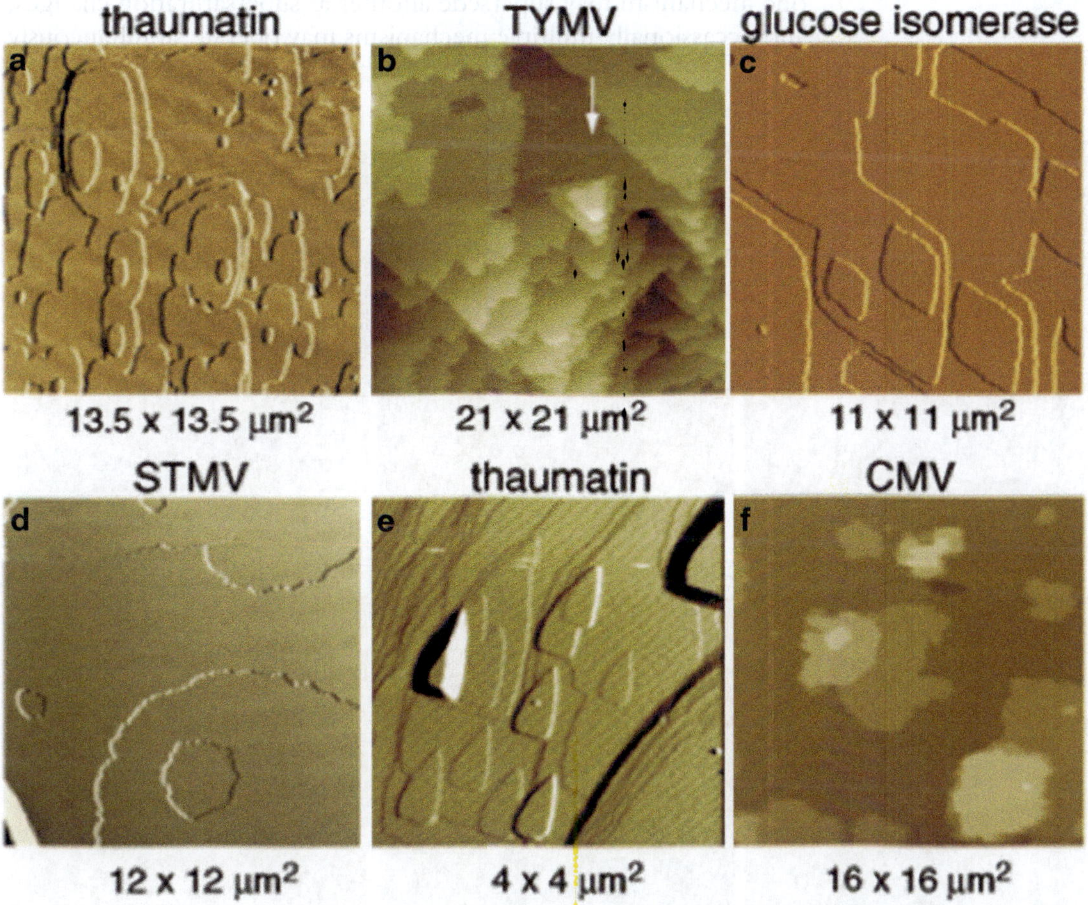

Fig. 6 A major source of growth steps and layers on the surfaces of growing macromolecular crystals, particularly at medium to high levels of supersaturation, are two dimensional nuclei that exceed critical nuclear size and subsequently develop into two dimensional islands. Shown here are two-dimensional islands on a variety of protein and virus crystals. This is the dominant mechanism for face normal growth for most macromolecular crystals. In (**b**) the *arrow* denotes a triangular nucleus that reflects the symmetry of the crystal face

The other important mechanism is nucleation of new layers in a continuous manner through the occurance and activity of screw, or spiral dislocations. Examples of such dislocations on a variety of protein crystals are shown in Fig. 7. They appear as both left and right handed spirals, as simple and compound screws, and they exhibit a variety of appearances dependent on the symmetry of the crystal and various kinetic factors. Because they do not require the improbable ordering of free molecules from solution, screw dislocations produce new layers even at low supersaturation. Together the two mechanisms of two dimensional nucleation and screw dislocation growth account for virtually all protein crystal growth.

Relevant to the practice of crystallization, the specific operable growth mechanism is principally determined by the crystallization conditions and the degree of supersaturation they produce. Often one mechanism may supersede another as supersaturation changes, and occassionally multiple mechanisms may operate simultaneously

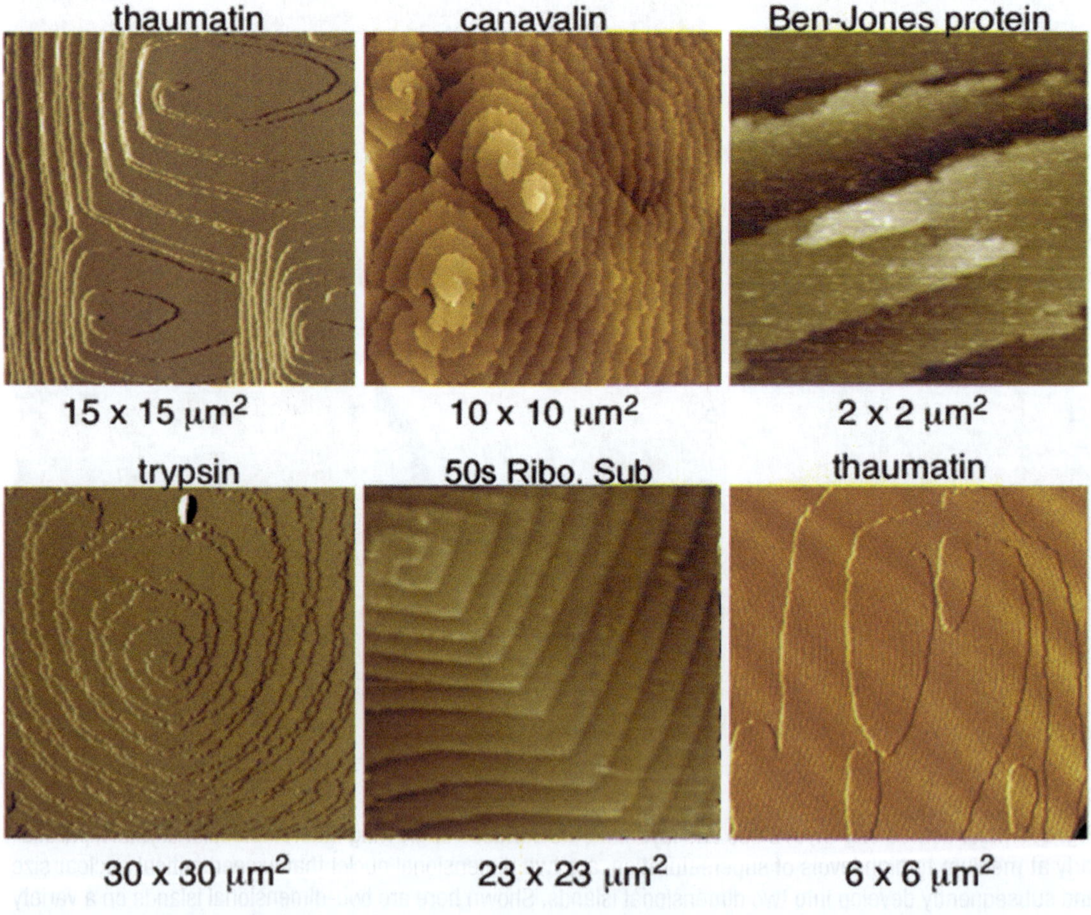

Fig. 7 A major source of growth steps on growing crystals, particularly at lower supersaturation, are screw, or spiral dislocations. Shown here are a variety of screw dislocations on the surfaces of macromolecular crystals that illustrates their diverse character

[16, 29]. The mechanisms of growth are further important because they may determine the amount and distribution of impurities incorporated in the crystal, the crystal defect structure, the ultimate size, and even certain diffraction properties such as mosaicity and resolution limit. It is important in practice to be aware that physical perturbations, such as vibrations, jarring, or temperature fluctuations can disrupt a growth mechanism or produce a shift from one mechanism to another.

4 Supersaturation, Nucleation, and Growth

Crystallization of any molecule, or any chemical species including proteins, proceeds in two distinct but inseparable steps, nucleation and growth. Nucleation is the most difficult problem to address both theoretically and experimentally because it represents a mysterious first order phase transition by which molecules pass from a wholly disordered state to an ordered one. We believe that this likely occurs through the initial formation of partially ordered, or paracrystalline intermediates, protein aggregates having only short-range order, that through internal rearrangement ultimately yield small, crystallographically ordered assemblies that we refer to as critical nuclei [16, 29, 31]. A critical nucleus is an ordered cluster of molecules that is of sufficient size (has a surface to volume ratio) such that it acquires new molecules at a rate greater than that of losing molecules.

The size, or number of molecules making up a critical nucleus is dependent on the molecular dimensions, the extent of supersaturation, and the surface free energy of molecular addition. Currently the critical nuclear size has only been described for a few proteins, and for some cases, these were only investigated in terms of two-dimensional nuclei developing on the surfaces of already existent crystals [20, 21]. Recently, a theory has emerged which attempts to explain the nucleation phenomenon in terms of statistical fluctuations in solution properties [32–34]. This idea holds that a distinctive "liquid protein phase" forms in concentrated protein solutions, and that this "phase" ultimately gives rise to critical nuclei with comprehensive order. The hypothesis is supported by observations, using both atomic force microscopy and quasi-elastic light scattering, of a third mode of crystal growth in the normal direction termed growth by three dimensional nucleation [27, 31, 35].

Growth of macromolecular crystals is a better-characterized process than nucleation, and its mechanisms are reasonably well understood. As described above, protein crystals grow principally by the classical mechanisms of dislocation growth, and growth by two-dimensional nucleation, along with two other less common mechanisms known as normal growth and three dimensional nucleation [27, 29, 31]. A common feature of nucleation and growth is that both are critically dependent on the supersaturation of the mother liquor giving rise to the crystals. Supersaturation is the variable that drives both processes and determines their occurrence, extent, and the kinetics that govern them.

Crystallization of any molecules, including proteins, absolutely requires the creation of a supersaturated state. This is best illustrated by the phase diagram for crystallization shown in Fig. 8. Supersaturation is a nonequilibrium condition in which some quantity of the macromolecule in excess of the solubility limit, under specific chemical and physical conditions, is nonetheless present in solution. Equilibrium is reestablished by formation and development of a solid state, such as crystals, as the system returns to the saturation limit. To produce a supersaturated state, the properties of an undersaturated, or saturated solution must be

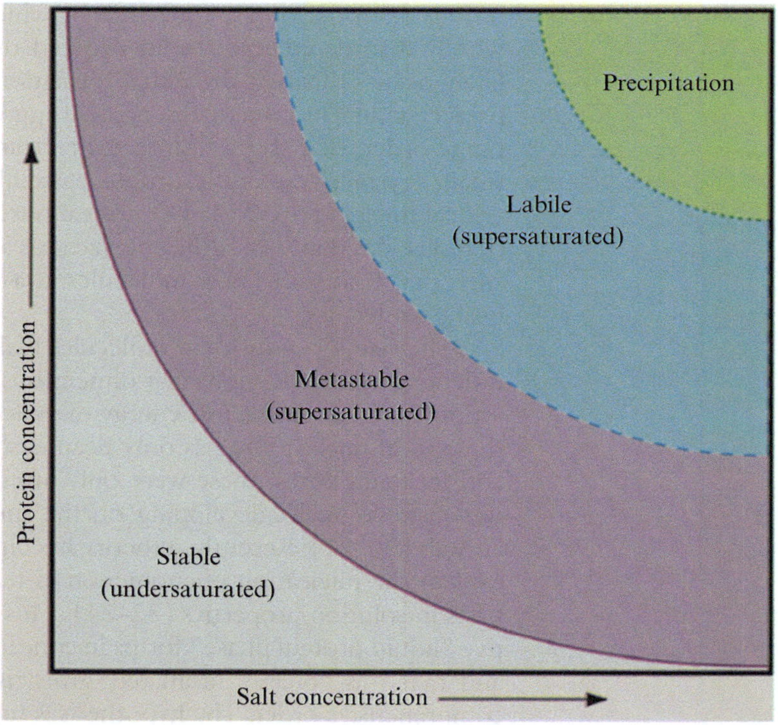

Fig. 8 The phase diagram for the crystallization of macromolecules. The solubility diagram is divided sharply into a region of undersaturation and a region of supersaturation by the line denoting maximum solubility at specific concentrations of a precipitant, which may be salt or a polymer. The *line* represents the equilibrium between existence of solid phase and free molecule phase. The region of supersaturation is further divided in a more uncertain way into the metastable and labile regions. In the metastable region nuclei will develop into crystals, but no nucleation will occur. In the labile region, both might be expected to occur. The final region, at very high supersaturation is denoted the precipitation region where that result might be most probable

modified to reduce the ability of the medium to sustain the solubility of the protein (i.e., reduce the chemical activity of the solvent). Alternatively, some property of the protein molecules must be altered to reduce protein solubility and/or increase the attraction of one protein molecule for another, thereby inducing association. In any case, relationships between solvent and solute, or between the molecules in solution, are perturbed so as to promote formation of the solid state.

If no crystals or other condensed phase is present as conditions are changed, then solute will not immediately produce a new phase, and the solution will enter and remain in the supersaturated state. The solid state, hopefully a crystal nucleus, but other incipient states as well, does not develop spontaneously as the saturation limit is exceeded. Energy, analogous to the activation energy of a chemical reaction, is required to initiate the second phase, the stable critical nucleus of a crystal, or perhaps an unfortunate precipitate. Thus, a kinetic, or energy (or probability) barrier allows conditions with time to proceed more distant from equilibrium and further into the zone of supersaturation. Once a critical nucleus does appear in a supersaturated solution, however, it will proceed to accumulate molecules from solution and grow until the system regains equilibrium at saturation. So long as nonequilibrium forces prevail and some degree of supersaturation exists to drive events, a crystal will grow or precipitate continue to form.

5 General Approach

Protein crystallization is based on a diverse set of principles, unique experiences and evolving ideas. There is no comprehensive theory, or even an organized, extensive base of fundamental data to guide an investigator, though that is an effort in progress. As a consequence, protein crystal growth is largely empirical in nature, and demands patience, perseverance and intuition.

What complicates the crystallization process, in addition to our limited understanding of the phenomena involved, is the intimidating complexity and range of the macromolecules before us. Even in the case of rather small proteins, such as cytochrome c or myoglobin for example, there are roughly a thousand atoms with hundreds of bonds and thousands of degrees of freedom. For viruses of molecular weights measured in the millions of Daltons, and for large multi-protein complexes, the possibilities for conformation, interaction, and dynamics are almost unlimited.

We are, however, beginning to develop rational approaches to protein crystallization based on an understanding of the fundamental properties of the systems. We are now increasingly using, in a systematic manner, the classical methods of physical chemistry to determine the energetic and kinetic characteristics of the mechanisms responsible for the self-organization of large biological molecules into crystal lattices. As an alternative to the precise and reasoned strategies that we commonly apply to scientific problems, we, nevertheless, still rely primarily on what is fundamentally a trial and error approach. Protein crystallization is generally a matter of searching, as systematically and intelligently as possible, the ranges of the individual parameters that influence crystal formation, finding a set, or multiple sets of factors that yield some kind of crystals, even of poor quality, and then optimizing the individual variables to obtain the best possible crystals. This is usually achieved by carrying out an extensive series, or establishing a vast matrix of crystallization trials, evaluating the results, and using what information is obtained to improve conditions in successive rounds of trials. Because the number of variables is so large, and the ranges so broad, experience and insight in designing and evaluating the individual and collective trials becomes an important consideration.

6 Screening for Initial Crystallization Conditions

As noted above, there are usually two phases in the creation of protein crystals for an X-ray diffraction investigation, and these are (a) the identification of chemical, biochemical, and physical conditions that yield a crystalline material, though it may initially be inadequate for X-ray analysis, and (b) the systematic alteration of those initial conditions by incremental amounts to obtain optimal samples for diffraction. The first of these is fraught with the greater risk, as some proteins simply refuse to form crystals, and clues as to why are elusive or absent. Optimization, however, often proves the more demanding of effort, more time consuming, and frustrating.

There are two fundamental approaches to searching for crystallization conditions. The first is a systematic variation of what are believed to be the most important variables, i.e., precipitant type (salt, polymer, organic liquid) and its concentration, pH, temperature, protein concentration, and potential ligands. The second is what we might term a shotgun approach, but a shotgun aimed with intelligence, experience, and accumulated wisdom. While far more thorough in scope and more congenial to the scientific mind, the first method usually requires more effort and a greater amount of protein. In those cases where the quantity of material is limiting, it may simply be impractical. The second technique, however, provides more opportunity for useful conditions to escape discovery. In general, though, it requires less precious material.

The second approach also has, presently at least, one other major advantage, and that is availability and convenience. There is currently on the commercial market, from numerous companies, a wide variety of crystallization screening kits. The availability and ease of use of these relatively inexpensive kits, which may be used in conjunction with a variety of crystallization methods (hanging and sitting drop vapor diffusion, dialysis, etc., *see* below) make them the most popular approach for attacking, at least initially, a new crystallization problem. With these kits, nothing more is required than combining a series of potential crystallization solutions with one's protein of interest using a micropipette, sealing the samples, and waiting for good fortune to smile. Occasionally it does, but sometimes not, and that is when the crystal grower must begin using his own intelligence to diagnose problems and devise remedies.

Once some crystals, even if only microcrystals, are observed and shown to be of protein origin, then optimization begins. Every component in the solution yielding crystals must be noted and considered (buffer, salt, ions, etc.), along with pH, temperature, and whatever other factors might have an impact on the quality of the results (*see* below). Each of these parameters or factors is then carefully incremented in additional trial matrices that encompass ranges spanning the conditions which gave the "hit." Because the problem is nonlinear, that is, one variable may be coupled to another, this process is often more complex and difficult than one might anticipate [15, 16, 36–39]. It is here that the amount of protein and the limits of the investigator's patience may prove a formidable constraint.

7 Creating Supersaturation in Practice

In practice, one begins with a solution, a potential mother liquor, which contains some concentration of the protein below its solubility limit, or alternatively at its solubility maximum (an exception being the batch method, *see* below). The objective is then to gradually alter conditions so that the solubility of the protein in the sample is significantly reduced, thereby rendering the solution supersaturated. This may be done through a variety of approaches. Principally, these depend upon (a) altering the protein itself (e.g., by change of pH, which alters the ionization state of surface amino acid residues, by binding a ligand, or by introducing mutations), (b) altering the chemical activity of the water (e.g., by addition of salt or organic solvent), (c) altering the degree of attraction of one protein molecule for another (e.g., addition of bridging ions or molecules), or (d) altering the nature of the interactions between the protein molecules and the solvent (e.g., addition of polymers such as PEG), which also reduces the chemical activity of water.

Table 1 is a compilation of approaches upon which one might develop strategies for crystallizing a protein for the first time. Indeed, there are doubtless others that hopefully emerge from the imagination and cunningness of the investigator. The details of the various approaches have been described elsewhere [15, 36, 38, 39, 41] and need receive no extensive treatment here. It is probably sufficient to

Table 1
Approaches to creating supersaturation

1. *Direct Mixing of Protein and Precipitant*: A protein and precipitant solution are thoroughly mixed so that the final solution is immediately supersaturated in protein. This relies on the energy, or probability barrier to critical nucleus formation to restrain the system and limit the number of nuclei and the time of their appearance. The most common application is in microdrops under oil
2. *Temperature Alteration*: Refers to a raising or lowering of the temperature of a protein–precipitant solution that is very near supersaturation. The temperature change is made in a direction that reduces the solubility of the protein. Most proteins in high salt solutions are more soluble at cold temperature, while protein–PEG combinations and low ionic strength solutions of protein are generally more soluble at warm temperatures
3. *Alteration of Ionic Strength*: Salt is added to a protein solution to high concentration so that competition for water lowers the solubility of the protein, referred to as "salting out." Salt ions can also be removed by dialysis to create a low ionic strength state where, because of deprivation of cations, the protein is less soluble, referred to as "salting in." *See* the phase diagram in Fig. 8 and the illustration in Fig. 10
4. *pH Alteration*: As the pH of a protein solution is changed, certain amino acid side chains on the protein molecules' surfaces alter their ionization, and therefore their charge state. As a consequence the electrostatic surface of the protein molecules change. If this produces charge complementary surfaces and additional favorable interactions, or removes unfavorable interactions, then the molecules will be encouraged to associate and the solubility of the protein will be reduced. *See* Fig. 11
5. *Ligand Binding*: The solubility of most proteins, due to both long range and local conformational changes, may be altered as a consequence of ligand binding. The ligands may be coenzymes or other prosthetic groups, inhibitors, or ions. The last of these, particularly divalent cations, can also form bridges between otherwise repulsive groups on protein molecules and transform unfavorable interactions into geometry specific, favorable interactions
6. *Alteration of the Dielectric Constant of the Medium*: This is usually effected by the direct addition, or addition by dialysis, of an organic liquid of low dielectric constant into the protein solution. This encourages electrostatic and hydrogen bonding interactions between macromolecules
7. *Direct Removal of Water*: This can be brought about by simple evaporation or by concentration of the protein that reduces the water molecules available for solvation of the protein. Any method that produces dehydration of the protein falls in this category
8. *Addition of a Polymer*: PEG is most commonly used as a polymeric precipitant and it is hypothesized to act principally through the mechanism of "volume exclusion." Because of its very large hydrodynamic radius, the disordered polymer restricts the volume of solvent that the protein can access, essentially depriving it of solvating water molecules. It effectively concentrates and dehydrates the protein and thereby reduces its solubility. There is a possibility that PEG may also act as an adhesive intermediary between protein molecules to enhance their association
9. *Removal of a Solubilizing Agent*: Some proteins can be concentrated to an enhanced degree by inclusion in the solution of some agent that increases its solubility such as a chaotrope or osmolyte [40]. Removal of the agent after concentrating then leaves the protein at reduced solubility and perhaps at supersaturation. The agent may be removed by dialysis
10. *Addition of Non-volatile Alcohols and Low Molecular Weight Polymers*: Liquid compounds such as MPD, hexanediol, PEGs 200 Da, 400 Da, and low molecular weight Jeffamines reduce the solubility of proteins, probably by competing for water and altering dielectric constants, but their true mechanism remains obscure. They may also incorporate into crystals and favorably alter the solvent structure inside the crystals

say that if a protein has a propensity to crystallize, it can probably be accomplished by variation of precipitant type, precipitant concentration, pH, to a lesser extent temperature, but with all due consideration to the biochemical properties and eccentricities of the protein under investigation. Finally, we are all advised that with real estate there are three important factors, and they are location, location, and location. With protein crystallization there are similarly three, and they are purity, purity, and homogeneity.

8 Methodology

The growth of protein crystals must be carried out in some physical apparatus that allows the investigator to reduce the solubility of the protein by altering the properties of the mother liquor, using, perhaps, one of the strategies in Table 1. Currently, these involve, almost exclusively, microtechniques. Crystallization "trials" with a matrix of 48 or 96 conditions may be carried out with volumes of only a fraction of a milliliter if done manually, a few microliters or less with some robotic or microfluidic systems. These employ plastic, multichambered trays for hanging and sitting drops, plexiglass buttons for dialysis, or microdrops under oil. Other methods are found in Table 2.

Crystallization devices and the associated methodologies have also been described in detail elsewhere [14, 36, 39, 45]. Detailed instructions and web sites are frequently provided by the manufacturers of the crystallization kits, supplies, and plasticware, along with many helpful illustrations. The hanging drop and sitting drop procedures for vapor diffusion, and the batch method using microdrops under oil are now most in favor, and are recommended for most investigations. In those cases where mother liquor components cannot be transported through the vapor phase (e.g., metal ions, detergents) then microdialysis may be the only recourse. An important point, however, is that the best method for screening conditions and obtaining an initial set of crystallization parameters may not be the best means for optimization. Thus one may start with one technique but ultimately find that another gives larger crystals of higher quality.

Vapor diffusion, in either the "sitting drop" or "hanging drop" arrangements is the most popular approach. This is illustrated for both arrangements in Fig. 9a, b. Vapor diffusion relies on the equilibration of a small droplet, 1–10 μl in volume usually, against a larger liquid reservoir of 0.5–1.0 ml. The droplet is initially a mixture, most commonly 1:1, of a stock protein solution at 10–30 mg/ml, with the reservoir that may contain, for example, a buffered, concentrated salt or PEG solution. Loss of water from the droplet to the reservoir and equilibration of the two over time, hours to days, restores the droplet (almost) to the concentration of

Table 2
Physical and chemical procedures for producing solubility minima

1. *Bulk (visual) Crystallization*: This is a term used by old biochemists to describe the direct addition of salt to a protein solution while watching intently for the appearance of an opalescent sheen (Tyndall effect) that indicates incipient crystallization. This was a skill acquired by researchers for about a century when crystallization was principally used as a powerful purification tool, and ultimately as a demonstration of protein purity
2. *Batch Method in Vials*: Once a precipitant concentration that produces crystals is identified for a particular protein, then a protein solution is simply mixed with the precipitant at 1–3% less. The protein–salt solution is then dispersed into small screw cap vials and allowed to stand. Very slow evaporation around the caps causes the precipitant concentration to gradually increase in the vials until supersaturation is achieved. This was the most widely used technique until about 1970
3. *Evaporation*: Probably the oldest method in existence, it was used to produce the first reported protein crystals, those of earthworm hemoglobin [42]. Very slow and controlled evaporation may still be used and effected through the use of narrow capillaries or wicks
4. *Dialysis and Microdialysis*: This procedure can be performed in bulk using dialysis tubing, or on a microscale using capillaries or dialysis buttons enclosed by a dialysis membrane. Dialysis has the great advantage that it allows investigation of a continuum of precipitant concentrations or a range of pH. It permits the introduction or removal of ligands, coenzymes, inhibitors or ions. It can be used with the same protein sample to carry out multiple, independent experiments simply by changing the exterior solution
5. *Concentration Dialysis*: Dialysis as described above in (4) but dissolving in the outside solution a high concentration of high molecular weight PEG (PEG 20,000 for example). Water is withdrawn as dialysis proceeds and the protein becomes increasingly concentrated as conditions are altered. Apparatus was once available for performing the concentration function by drawing a vacuum on the system simultaneous with dialysis, in which case no PEG was required
6. *Liquid Bridge*: A kind of direct, but slow mixing of a protein solution with a precipitant solution. Drops of each are placed in a sealed container maintained at 100% humidity, and a needle is used to draw out a thin connecting liquid bridge between the two drops. Protein diffuses very slowly, thus it is precipitant that gradually diffuses into the protein drop and promotes supersaturation
7. *Free Interface Diffusion*: In a tube or capillary, a lighter protein solution is gently layered upon a heavier precipitant solution (or vice versa depending on relative densities). Slow diffusion along with some convection across the interface produces a diversity of local concentration gradients that may promote crystal nucleation and support crystal growth. This technique has been particularly useful in microgravity where pure diffusive transport prevails and convection is absent
8. *Vapor Diffusion*: This refers to any arrangement, microdroplets "sitting" on a plastic support, for example, or "hanging" from a glass cover slip, equilibrating through the vapor phase with a larger volume reservoir solution containing a higher concentration of precipitant. Over time the osmolarity of the drop asymptotically approaches that of the reservoir because of water loss from the protein–precipitant drop. This approach in one manifestation or another is currently in widest and most popular use.
9. *Sequential Extraction*: This method [43] is primarily used to produce microcrystals to be later used for seeding. It depends on the sequential extraction of a salt induced, protein precipitate (centrifuge pellet) by solutions of decreasing precipitant concentration at 4 °C. The drops of extract are subsequently placed at 25 °C where the protein (in salt solution) is less soluble and supersaturation is achieved

(continued)

Table 2 (continued)

10.	*pH Induced Crystallization*: This is a powerful approach and can be accomplished through direct addition of acid or base, by dialysis, or by vapor diffusion. This takes advantage of the frequent strong dependence of protein solubility as a function of H_3O^+ concentration.
11.	*Temperature Induced Crystallization*: This method takes advantage of the difference in solubility of some proteins as a function of temperature within the range of 0 –37 °C. For most, but not all proteins, this dependence is rather weak so that the technique is used infrequently with proteins. It is extremely important in the crystallization of conventional molecules
12.	*Effector Addition*: This depends on the difference in solubility of a protein when it has a coenzyme, inhibitor, or other ligand bound to it. Dialysis or direct addition can be used to introduce or remove a ligand and thereby affect protein solubility
13.	*Crystallization in Gels*: Diffusion through gels, such as silica or agarose gels, of a mobile precipitant into a protein containing gel, essentially free interface diffusion in a gel, can be used to produce the supersaturated state. Currently in popular use for the crystallization of membrane and lipophilic proteins, the "lipidic cubic phase" [44] for crystallization takes advantage of the complex structure of the gel (mesophase) itself to induce nucleation and allow controlled growth

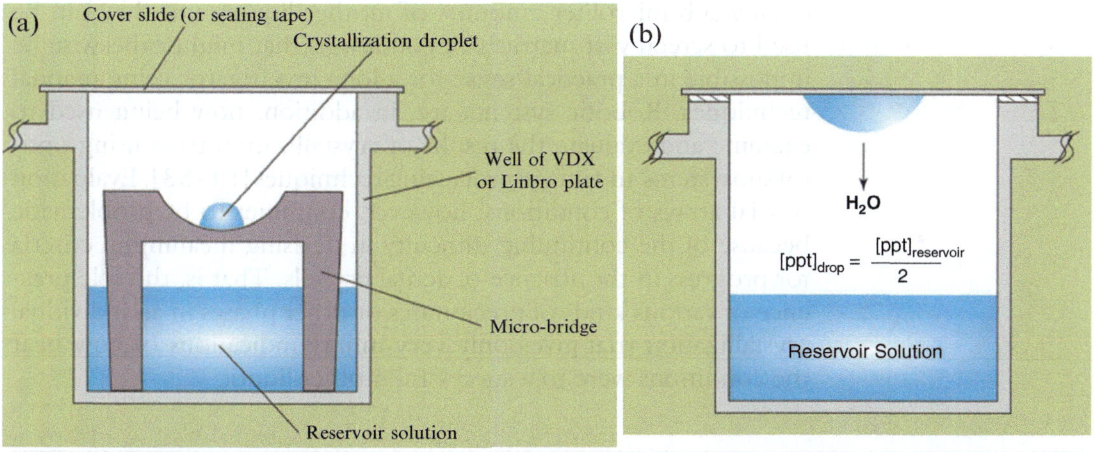

Fig. 9 The sitting drop vapor diffusion method is illustrated in this schematic diagram (**a**). The drop on the elevated platform, which is commonly 2–10 μl, consists of half stock protein solution and half the reservoir solution, which contains some concentration of a salt or polymer precipitant. About 0.5 ml of the reservoir solution is added to the bottom of the cell before sealing. By water equilibration through the vapor phase the drop ultimately approaches the reservoir in osmolarity both raising the concentration of the precipitant in the drop and increasing the protein concentration. The hanging drop vapor diffusion method is illustrated schematically in (**b**). The components of the drop and reservoir, and the physical equilibration process are the same here as for the sitting drop. The exception is that the protein drop is suspended from a cover slip over the reservoir rather than resting on a surface. Plasticware for carrying out both sitting and hanging drop vapor diffusion are widely, and commercially available in numerous formats

the protein in the stock solution, and increases the precipitant concentration to that of the reservoir. Hopefully this process produces a droplet supersaturated in protein.

Though proportions and volumes are different, dialysis has the same objective, to gradually raise the salt or PEG concentration of a protein solution to the point where it becomes saturated in protein. Drops under oil [46], or the batch method as it is called, uses no equilibration. A protein solution and a potential crystallization promoting solution (salt, PEG, buffer) are simply mixed in some reasonable proportion and droplets dispersed under oil. This method relies on the nucleation energy barrier to allow immediate establishment of supersaturated drops. Other techniques, such as free interface diffusion [47], slow diffusion and mixing of protein and precipitant solutions across a liquid–liquid interface, are also in use, but see less application than batch or vapor diffusion.

In high throughput laboratories, screening for crystallization conditions, and even optimization in some cases, has generally been consigned to robotic devices [48–50]. This is particularly true in those of large pharmaceutical companies where many proteins may be under simultaneous structural investigation. Automated systems have the advantages of exceptional record maintenance, most can deploy sub microliter amounts of mother liquor, and they can be used to screen vast matrices of conditions that might otherwise be impossible in a practical sense for a lone investigator using manual techniques. Robotic systems are, in addition, now being used to examine and evaluate the results of crystallization trials using optical subsystems and image processing techniques [51–53]. Evaluation of trial arrays of conditions, however, continues to be problematic because of the continuing difficulty in devising meaningful criteria for progress in the absence of actual crystals. That is, the sole presence of various kinds of precipitates or other phases in an individual crystallization trial gives only very murky indications of how near the conditions were to a successful mother liquor.

9 Precipitants

One of the most important components in a mother liquor intended to crystallize a protein is sometimes called the precipitant, or other times the crystallization agent. It is generally, but not always, the chemical component that reduces the solubility of the protein or reduces the chemical activity of water. Salts, such as ammonium sulfate or potassium phosphate, or polymers such as PEG are classic examples. The mechanisms by which they act (*see* below) may be different, but their essential role is to deprive the protein of solvating water and to promote protein association.

If one were to examine the reagents utilized in any of the commercial crystallization screens that are based on shotgun approaches, or examined the crystallization conditions that have been compiled into data bases [54, 55], then it would become apparent that a wide variety of precipitating (crystallizing) agents have been used. Indeed many agents have been employed, and some, such as ammonium sulfate and polyethylene glycol have produced a great number of successes. It is often necessary, however, to explore many precipitants, and it is difficult to know initially which might offer the greatest likelihood of obtaining crystals.

Individual precipitants and their properties have also been reviewed in some detail [16] and are not extensively discussed here. To summarize, however, it is possible to group the precipitants into categories based on their mechanisms for promoting crystallization. The majority of precipitants of proteins fall into four broad categories (1) salts, (2) organic solvents, (3) long chain polymers, and (4) low molecular weight polymers and non-volatile alcohols. The first two classes are typified by ammonium sulfate and ethyl alcohol respectively, and higher polymers such as polyethylene glycol 4000 are characteristic of the third. In the fourth category we might place compounds such as methylpentanediol (MPD) and polyethylene glycols of molecular weight less than about 1000.

The solubility of proteins in concentrated salt solutions is complicated, but it can be viewed naively as a competition between salt ions, principally the anions, and the protein for the binding of water molecules, which are essential for the maintenance of solubility [56–60]. At sufficiently high salt concentrations the protein molecules become so uncomfortably deprived of solvent that they seek association with one another in order to satisfy their electrostatic and amphipathic requirements. In this environment large, semi ordered aggregates that could lead to critical crystal nuclei, as well as disordered amorphous precipitate may form. Other salt ions, chiefly cations, also may be necessary to insure protein solubility. At low ionic strengths, cation availability may be insufficient to maintain protein solubility, and under those conditions too, crystals may form. The behavior of typical proteins over the entire range of salt concentrations, including both the "salting in" and "salting out" regions is illustrated by Fig. 10.

Salts exert their effect principally by dehydrating proteins through competition for water molecules, and a measure of their efficiency in this is the ionic strength, whose value is the product of the molarity of each ion in solution with the square of their valences. Thus, multivalent ions are the most efficient precipitants. Sulfates, phosphates, citrates, and more recently malonates [61] and mixtures of the salts of dicarboxylic acids have traditionally been employed.

One might anticipate little variation among different salts so long as the valences of their ions were the same. Thus there should be little expected variation between two different sulfates such as Li_2SO_4 and $(NH_4)_2SO_4$ if only ionic strength were involved. This is often observed not to be the case. In addition to salting out, which

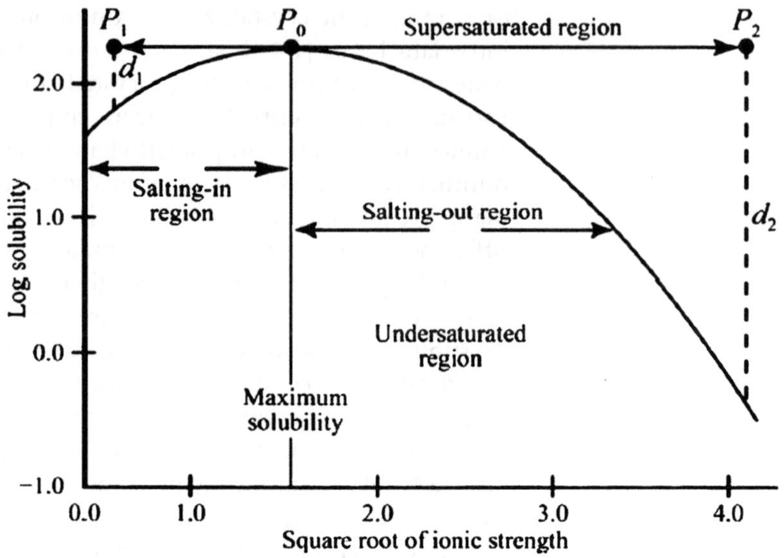

Fig. 10 The curve shown here represents a typical solubility curve for a protein and divides the region of undersaturation from that of supersaturation. It also illustrates the existence of the classical "salting in" and a "salting out" region for the protein. By taking advantage of the latter effects, supersaturation may be achieved by equilibrating a system from a point of maximum solubility (P_0) to one of reduced solubility (P_1 or P_2) by adjusting the precipitant concentration

is a general dehydration effect not really much different than evaporation or concentration (except that water is not physically removed) there are also specific protein–ion interactions that may have further consequences [58, 60]. This is perhaps not unexpected given the varied hydration properties of different ions and the unique polyvalent character of individual proteins, protein structural and dynamic complexity, and the intimate dependence of their physical properties on their surroundings. It is inadequate, therefore, when attempting to crystallize a protein to examine only one or two salts and ignore the broader range. Alternative salts can sometimes produce crystals of varied quality, morphology, and in some cases diffraction properties.

It is usually not possible to predict the degree of saturation or molarity of a precipitating agent required for the crystallization of a particular protein without some prior knowledge of its solubility behavior. In general, however, it is a concentration of the precipitant just a few percent less than that which yields an amorphous precipitate [62], and this can be determined for a macromolecule under a given set of conditions using only minute amounts of material [15]. To determine the approximate insolubility points with a particular precipitant a 10 μl droplet of a 5–15 mg/ml protein solution can be placed in the well of a depression slide and observed under a low-power light microscope as increasing amounts of saturated salt solution or organic solvent (in 1- or 2-µl increments) are added. If the well is sealed between additions with a coverslip, the increases can be made over a period of many hours.

Along with precipitant type and concentration, pH is usually the most important variable influencing the solubility of proteins. As such, it provides a powerful approach to creating supersaturated solutions, and hence effecting crystallization. Its manipulation at various ionic strengths and in the presence of diverse precipitants is a foundational concept in formulating screening matrices and discovering successful crystallization conditions. An example of how pH might be used to effect crystallization of a protein is illustrated in Fig. 11.

Organic solvents reduce the dielectric constsnt of the medium, hence the screening of the electric fields that mediate macromolecular interactions in solution. A danger, however, is that they also tend to destabilize protein structure. As the concentration of organic solvent is increased, interaction between protein molecules increases, solvent becomes less effective (the activity coefficient of water is reduced) and the solid state becomes more favored [63, 64]. Organic solvents should be used at low temperature, at or below 0 °C, and they should be added very slowly with good mixing [16]. Since they are usually volatile, vapor diffusion techniques are equally applicable. Ionic strength should, in general, be maintained

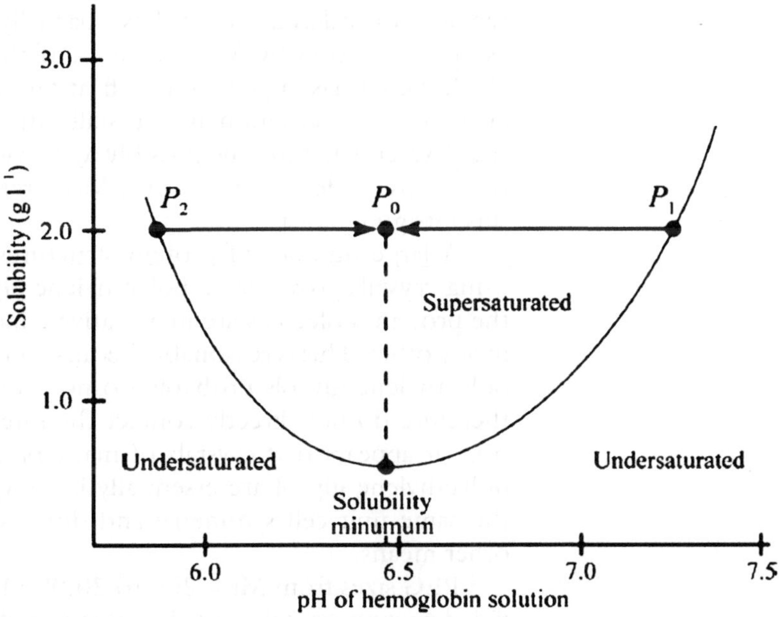

Fig. 11 As shown here, most proteins have specific solubility minima as a function of pH. One can take advantage of this property to produce supersaturation by altering a system from a pH permitting high solubility (P_1 or P_2) to a point of low solubility (P_0). This is a powerful approach to promoting crystallization of macromolecules

low and whatever means are otherwise available should be pursued to protect against denaturation.

Some polymers, among which polyethylene glycols (PEG) are most popular [65, 66], produce volume exclusion effects that induce separation of proteins from solution [65, 67]. Polymeric precipitants, unlike proteins, have no consistent, fixed conformation. They writhe and twist randomly in solution and, as a consequence, occupy far more space than their molecular weights would suggest. This effect, refered to as volume exclusion, results in less solvent available space for the protein molecules that then segregate, aggregate, and ultimately form a solid state, in favorable cases crystals. PEG is also extremely hydrophilic and binds water molecules avidly (about 2.3 water molecules per monomeric unit [68, 69]). It, like salts, competes for solvent, thereby dehydrating the protein molecules.

Evidence has emerged recently that suggests PEG, and some related polymeric precipitants, may not exert their effects on protein crystallization exclusively by the mechanism of volume exclusion and dehydration. Some observations indicate that PEG, at least fragments and lower molecular weight components (all PEG preparations exhibit a distribution of lengths about a mean) may actually co-crystallize with the protein due to positive, associative interactions [68, 69], and thus occupy interstitial spaces and channels otherwise filled by solvent alone. Inside the crystal, PEG likely remains disordered, or at best partially ordered, and thereby escapes detection by X-ray analysis. If this PEG incorporation is valid, then it has important ramifications, as PEG could well influence protein association and crystallization by both altering interstitial water structure, and possibly by providing a soft superstructure that helps guide crystal growth. More remains to be done to test this intriguing idea.

A large number of protein structures have now been solved using crystals grown from polyethylene glycol. These confirm that the protein molecules are in as native condition in this medium as in any other. This is reasonable because the larger molecular weight polyethylene glycols probably do not even enter the crystals and therefore do not directly contact the interior molecules. In addition, it appears that crystals of many proteins when grown from polyethylene glycol are essentially isomorphous with, and exhibit the same unit cell symmetry and dimensions as those grown by other means.

PEG sizes from Mr = 200 to 20,000 Da have successfully provided protein crystals, but the most useful seem to be those in the range 2000–8000 Da. A large number of reports have appeared, however, in which a protein could not easily be crystallized using this range but yielded in the presence of PEG 200 Da or 20,000 Da. The molecular weight sizes may not be completely interchangeable for a given protein even within the mid range. Some produce the best-formed and largest crystals only at, say, Mr = 4000 Da, and less perfect examples at other weights. This is a parameter that is best optimized by empirical means along with concentration. The very low molecular weight PEGs such as 200 and 400 Da are somewhat similar in character to MPD and hexanediol. There does not appear to be any correlation between the molecular weight of a protein and that of the PEG best used for its crystallization. The higher molecular weight PEGs do, however, have a proportionally greater capacity to force proteins from solution.

An advantage of PEG over most other precipitating agents is that proteins crystallize within a fairly narrow range of PEG concentration; this being from about 4% to 20% (although there are numerous examples where either higher or lower concentrations were necessary). In addition, the exact PEG concentration at which crystals form is rather insensitive. If one is within a few percent of the optimal value (in some cases even more), some success is likely to be achieved. With most crystallizations from high ionic strength solutions or from organic solvents, one must be within 1% or 2% of an optimum lying anywhere between 15% and 85% saturation. The advantage of PEG, then, is that when conducting a series of initial trials to determine what conditions will give crystals, one can use a fairly coarse selection of concentrations and over a rather narrow total range.

Since PEG solutions are not volatile, PEG must be used like salt or MPD and equilibrated with the protein by dialysis, slow mixing, free interface diffusion, or vapor equilibration. When the reservoir concentration of PEG is in the range of 5–12%, the protein solution to be equilibrated should be at an initial concentration of about half, conveniently obtained by mixing equal volumes of the reservoir and protein solution. When the final PEG concentration to be attained is much higher than 12%, it is probably advisable to initiate the mother liquor at no more than 5–10% below the desired final value.

10 Factors Affecting Crystallization

There are many factors that can affect the crystallization of proteins and these, too, have been reviewed elsewhere [16, 36, 38, 39]. They fall into categories of physical factors such as temperature, chemical influences such as pH or ionic strength, and biochemical factors that include, among many others, purity and monodispersity. Any one, or any combination may affect the liklihood of crystallization occuring at all, the nucleation probability and rate, crystal growth rate and mechanism, and the ultimate sizes and quality of the products. As noted above, pH and the concentrations of salt and other precipitants are virtually always of importance. The concentration of the protein, which may vary from as low as 2 mg/ml for viruses and large complexes [70], to as much as a hundred mg/ml for some highly soluble proteins, is an additional, significant variable.

Other parameters may be unimportant in some cases but play a crucial role in others. In particular the presence or absence of ligands, coenzymes, or inhibitors, the variety of salt or buffer, the equilibration technique used, temperature fluctuations, and the presence of detergents and chaotropes [40] are all pertinent factors. Parameters of somewhat lesser and largely obscure significance are things like gravity, electric and magnetic fields, or viscosity. It can not be predicted which of these many variables may be of importance for a particular macromolecule, and the influence of any one must, in general, be investigated through empirical trials.

An intriguing problem, or opportunity depending on one's perspective, is what additional components or compounds should be included in the mother liquor in addition to solvent, buffer, protein, and precipitating agent [16, 36, 40, 71, 72]. The most desirable effectors, it would seem, are those which maintain the protein in a single, homogeneous, and invariant state. Reducing agents such as glutathione, β-mercaptoethanol, and dithithreitol are useful to preserve sulfhydryl groups and prevent oxidation. EDTA and EGTA are effective if one wishes to protect the protein from heavy or transition metal ions. Inclusion of these components may be particularly important when crystallization requires a long period of time to reach completion. When crystallization is carried out at room temperature in polyethylene glycol or low ionic strength solutions, then attention must be given to preventing the growth of microbes. These generally secrete proteolytic enzymes that may have serious effects on the integrity of the protein under study. Inclusion of sodium azide, thymol or chlorobutanol at low levels may be necessary to suppress invasive bacteria and fungi.

Substrates, coenzymes and inhibitors often serve to maintain an enzyme in a more compact and stable form. Thus a greater degree of structural homogeneity may be imposed on a population of protein molecules and a reduced level of statistical variation achieved by complexing the protein with a natural ligand before attempting its crystallization. In some cases an apoprotein and its ligand complexes may be significantly different in their physical behavior and can, in terms of crystallization, be treated as almost entirely separate problems. Complexes may provide additional opportunities for growing crystals if the native apoprotein is refractory. It is worthwhile, therefore, when searching for crystallization conditions, to explore complexes of the macromolecule with substrates, coenzymes, and inhibitors at an early stage. Such complexes are, in addition, often inherently more interesting in a biochemical sense than the apoprotein.

Various metal ions have occasionally been observed to promote the crystallization of proteins. Bacterial glucose isomerase, for example, crystallizes readily from PEG solutions in the presence of Mg^{++}, but only with difficulty in its absence. Cadmium and some other divalent cations induce immediate crystallization of the iron storage protein ferritin [73]. In some instances ions are essential for activity. It is, therefore, reasonable to expect that they might aid in maintaining certain structural features of the molecule. There are other examples, however, where metal ions, particularly divalent metal ions of the transition series such as Ca^{++}, were found to encourage crystal growth but played no recognized role in the protein's activity or structure. They likely serve as bridging agents between molecules in the crystal lattice.

11 The Protein as a Variable

A factor of particular importance to crystallization is the homogeneity and monodispersity of the protein [74, 75] and this deserves special emphasis. Some proteins may crystallize even from very heterogeneous mixtures (egg albumin, lysozyme, canavalin, α-amylase, for example), and indeed, crystallization has long been used as a powerful purification tool. It is the reason, in fact, why it originated as a technique and has been held in such high regard. In general, however, the likelihood of success in crystal growth is greatly advanced by increased homogeneity of the protein sample. Investment in further purification is always warranted, and usually profitable. When every effort to crystallize a protein fails, the best recourse is to further purify.

Recombinant DNA technology provided an enormous impetus to crystal growth research and X-ray crystallography 35 years ago, as it provided crystallographers access to proteins found in very low abundance that nevertheless played important roles in cells. Indeed it may be on the verge of providing another advance at this very time. Arguably, the most important parameter in protein crystallization is the protein itself. Until recently we have had little or no direct control over most of the important features of that parameter. Modification at the genetic level, however, provides us that opportunity, and its possibilities are only now being realized [76–78].

Through truncations, mutations, chimeric conjugates, and many other protein engineering contrivances, the probability of crystallization has been significantly enhanced. If we can learn how to go about this in a rational and systematic manner then advances may occur in future years that match the progress of the past. Approaches to application of mutation will be addressed and elaborated by others elsewhere in this volume.

12 Optimization of Crystallization Conditions

Optimization means adjusting the parameters of crystallization conditions, initially estimated from screening matrices [16, 37, 39], with the objective of discovering improved conditions that ultimately yield the best crystals for diffraction data collection. Optimization is in a sense refinement, but it is complicated somewhat because the parameters almost certainly are not independent of one another. They may be linked or correlated. Furthermore, solubility diagrams, which would have many dimensions, do not exist for specific proteins. Every protein has a unique length and amino acid sequence, and a unique three-dimensional conformation. Every protein is an individual with its own eccentricities and peculiarities. A further complication is that there can be an "embarrassment of riches" where many "hits" are obtained initially and the question arises as to which deserve the effort required for further improvement.

Optimization, as it is often practiced is in principle relatively straightforward. The parameters that define the initial conditions are first identified (pH, precipitant type, precipitant concentration, temperature, ion concentration, etc.). Following this, solutions are made that incrementally and systematically vary the parameters about the initial values. That is, if the pH of the initial hit was 7.0, then the same mother liquor might be composed but at pH 6, 6.2, 6.4, 6.6, etc. up to pH 8.0. This does not guarantee that one will arrive at optimal conditions, parameters may be correlated, but it is the best approach that we have.

While simple in principle, optimization becomes demanding in the laboratory. First of all, the number of parameters or effecting conditions may be large [15, 16, 36, 37]. It may not be clear which parameters are actually important, or what the range for exploration should be. Thus we have as an initial goal of optimization to deduce what variables are relevant and how to prioritize each relative to another so that adjustments can be made, all the while minimizing or neglecting the least or irrelevant factors.

Optimization may require a substantial amount of protein sample, and this may be severely limited. Thus, efficiency and economy become essential, and the use of very small volume trials [48, 50, 52] will be tempting. Small volumes, however, should be treated with caution. One seldom obtains large crystals from nanoliter volumes of mother liquor, and when promising results from very small drops are scaled up to larger volumes to grow larger crystals (which larger volumes tend to yield) the increase in scale fails to materialize.

The greatest obstacle to success in optimization is most frequently an absence of sufficient commitment, or a lack of effort on the part of the investigator. Screening for new crystallization conditions can be made almost, but not quite, painless. Commercial kits can be purchased that contain precisely prepared solutions. Robotics are now employed to dispense samples into plates, further robotic devices categorize and store the plates, and automated photographic systems present images of the many drops for viewing [49, 50, 52].

Automated systems, however, cannot make optimization effortless, and that is because optimization requires composition of a vast number of solutions that must be formulated or purchased, and the use of robotics in optimization presents as many problems as it solves, at least at this point in time. Making up a myriad of solutions, adjusting their pHs to exact values, and so on is tedious.

In other words, doing a lot of basic laboratory chemistry demands a lot of hard labor. Many investigators would rather struggle with marginal, or even miserable crystals obtained from the first hit than undertake the optimization effort.

13 Membrane Proteins

Proteins that are naturally membrane associated, or that are otherwise unusually hydrophobic or lipophilic in nature present unique problems. Such proteins are, in general, only sparingly soluble in normal aqueous media, some virtually insoluble, others lose their active conformations, and this in turn makes the application of conventional protein crystallization techniques problematic. Problems are difficult but not intractable. To address these difficulties the use of detergents, particularly non-ionic detergents, has been developed [79–83]. No attempt will be made here to describe the various techniques or the combinations of detergents and accessory molecules that have been used, as that involves a number of complexities and considerations that are covered in another chapter.

An essential difficulty associated with inclusion of a solubilization agent, such as a detergent, is that it adds an additional dimension to the matrix of conditions that must otherwise be evaluated. For example, if one is content to use a standard 48 drop screen of conditions, at least initially, then the additional search for a useful detergent means that the 48 trial screen must be multiplied by the number of detergent candidates. A further problem is that there are a great number of potentially useful detergents. Hampton Research (Aliso Viejo, CA), a major source of screening reagents, offers three different detergent kits of 24 compounds each. Were one to simply apply a basic 48-well screen with each detergent, then that would require a total of 3456 individual trials. While this may actually be possible with highly automated, nanoscale systems, and where a substantial amount of material is available, it is impractical for most laboratories.

Basic crystal screens, whether they are systematic screens or shotgun screens, should not, however, be abandoned. It becomes essential though to reduce, at least initially, the number of detergents to be considered. If, for example, a set of six highly promising detergents could be identified, then less than 300 trials would be called for initially, an undertaking well within the capabilities of most labs. No one, however, has yet reduced the set to a favored few. Everyone has an opinion as to which detergents should be favored, and no consensus has yet emerged from data bases and analyses of experiments. To make matters even more challenging, it appears that some, perhaps many detergents function best when accompanied by small amphiphilic molecules such as LDAO. This would of course add yet another dimension to the screening problem.

While not as valuable as naming actual candidate detergents, the author can point to a number of useful reviews and discussions that illustrate the properties and virtues of various detergents for membrane crystallization. Reference 83 is a good review of workup until that time, and more recently, there are fine discourses by Loll [80], Caffrey [44], Garavito and Ferguson-Miller [84], Hunte, et al. [85], and Wiener [79, 86], as well as a chapter in this book.

14 Some Important Concepts

Although approaches to protein crystallization remain largely empirical, substantial progress has been made. We have now identified useful reagents, devised a host of physical–chemical techniques for studying the crystallization process, and gained a better understanding of the unique features of proteins and their complex assemblies that affect their capacity to crystallize. Some principles now stand out regarding the crystallization problem, and these are summarized in Table 3.

Table 3
Some important concepts in protein crystallization

1. *Protein Purity*—Crystallization occurs because a population of structurally and chemically homogeneous molecules are made amenable to the formation of periodic bonding arrangements. Molecular misfits create disruptions of order and inhibit critical nucleus formation and crystal growth. Efforts to make the most pure and uniform protein sample as possible are never wasted
2. *Solubility and Monodispersity*—High protein concentration generally means more reliable crystallization and a greater overall chance of success in initial screens, and this depends on solubility of the protein. Solubility, however, also implies protein monodispersity and the absence of arbitrary oligomers and aggregates in the sample that are little more than contaminants
3. *Stability*—A foundational concept in crystallization is the unchanging nature of the molecules with regard to conformation and physical–chemical properties. It is now a given that the more stable a protein, the more likely it is to crystallize. The investigator must do whatever possible to insure that the protein molecules remain in their native state
4. *Supersaturation*—This is the crucial, controlling factor in determining nucleation probability, and both the mechanisms and kinetics of crystal growth. It can be achieved in many ways, and the path by which it is reached is as important as the ultimate value. A solution supersaturated in protein is a physical necessity for crystallization
5. *Association*—Supersaturation can be reached in many cases by enhancing attractive, specific interactions between protein molecules and thereby reducing their solubility. Additives, ions, protein modifications are traditional approaches. Reducing the chemical activity of the solvent abets this process and is the mechanism by which most precipitating agents operate
6. *Nucleation*—This is essential to start the crystallization process, and it is largely dependent on probability. That in turn depends on the degree of supersaturation and the path (through the phase diagram) by which supersaturation is reached. Competition from other condensed phases, such as precipitate, is the primary adversary. Enough supersaturation is necessary; too much supersaturation is a damper.

(continued)

Table 3 (continued)

7.	*Variety*—Because of the stochastic elements involved in crystallization, chance is an important factor. The more chances one has, the more likely is success. Explore as many possibilities and opportunities as possible in terms of sample source, sample conformation, physical, chemical, and biochemical parameters
8.	*Constancy*—Physical and/or chemical perturbations can inject energy into a dynamic, crystallizing system and cause deviations of otherwise ordered growth mechanisms. Disruptions from mechanical jarring, evaporation, or from temperature fluctuations can be devastating. Maintain the crystallizing samples at an optimal state during the full course
9.	*Impurities*—The incorporation of impurities, not only molecules present in the protein sample, but in the reagents, apparatus, or from the environment can seriously contribute to unwanted nucleation and growth termination
10.	*Preservation*—Crystals vary in their long term stability once they have reached terminal size. It may be necessary to take "post crystallization" measures to insure that the crystals maintain their quality until X-ray data collection can begin. These may include lowering temperature, increasing the precipitant concentration, prevention of evaporation through plastics, addition of stabilizers, cryo-vitrification, or mounting in sealed capillaries. Shock and handling must be avoided

References

1. Helliwell JR (1992) Macromolecular crystallography with synchrotron radiation. Cambridge University Press, Cambridge, UK
2. Bingel-Erlenmeyer R, Olieric V, Grimshaw J et al (2011) SLS crystallization platform at Beamline X06DA-A fully automated pipeline enabling in situ X-ray diffraction screening. Cryst Growth Des 11:916–923
3. Garman EF, Schneider TR (1997) Macromolecular cryocrystallography. J Appl Crystallogr 30:211–237
4. Pflugrath JW (2004) Macromolecular cryo-crystallography—methods for cooling and mounting protein crystals at cryogenic temperatures. Methods 34:415–423
5. Pflugrath JW (2015) Practical macromolecular cryocrystallography. Acta Crystallogr F Struct Biol Commun 71:622–642
6. Heras B, Edeling MA, Byriel KA et al (2003) Dehydration converts DsbG crystal diffraction from low to high resolution. Structure 11:139–145
7. Kiefersauer R, Than ME (2000) A novel free-mounting system for protein crystals: transformation and improvement of diffraction power by accurately controlled humidity changes. J Appl Crystallogr 33:1223–1230
8. Pflugrath JW (1992) Developments in X-ray detectors. Curr Opin Struct Biol 2:811–815
9. Gruner SM, Eikenberry EF, Tate MW (2001) Comparison of X-ray detectors. In: Rossmann MG, Arnold E (eds) International tables for crystallography, vol F. Kluwer Academic Publishers, Dordrecht, pp 143–153
10. Rossmann MG, Arnold E (2001) Crystallography of biological macromolecules, vol F. International tables for crystallography. Dordrecht, Kluwer Academic Publishers
11. Malkin AJ, Kuznetsov YG, McPherson A (1996) Defect structure of macromolecular crystals. J Struct Biol 117:124–137
12. Feigelson RS (1988) The relevance of small molecule crystal growth theories and techniques to the growth of biological macromolecules. J Cryst Growth 90:1–13
13. Feher G (1986) Mechanisms of nucleation and growth of protein crystals. J Cryst Growth 76:545–546
14. Durbin SD, Feher G (1996) Protein crystallization. Annu Rev Phys Chem 47:171–204
15. McPherson A (1982) The preparation and analysis of protein crystals. Wiley, New York
16. McPherson A (1999) Crystallization of biological macromolecules. Cold Spring Harbor Laboratory Press, Cold Spring Harbor, NY
17. McPherson A, Malkin AJ, Kuznetsov YG (1995) The science of macromolecular crystallization. Structure 3(8):759–768

18. Rosenberger F (1986) Inorganic and protein crystal growth—similarities and differences. J Cryst Growth 76:618
19. Kuznetsov YG, Malkin AJ, Greenwood A et al (1995) Interferometric studies of growth kinetics and surface morphology in macromolecular crystal growth. Canavalin, thaumatin and turnip yellow mosaic virus. J Struct Biol 114:184–196
20. Malkin AJ, Kuznetsov YG, Glantz W et al (1996) Atomic force microscopy studies of surface morphology and growth kinetics in thaumatin crystallization. J Phys Chem 100:11736–11743
21. Malkin AJ, Kuznetsov YG, McPherson A (1997) An in situ investigation of catalase crystallization. Surf Sci 393:95–107
22. Chernov AA, Komatsu H (1995) Principles of crystal growth in protein crystallization. In: Bruinsma JPEOSL (ed) Science and technology of crystal growth. Kluwer, Dordrecht, The Netherlands, p 67
23. Vekilov PG, Chernov AA (2002) The physics of protein crystallization. Solid State Phys 57:2–147
24. Chernov AA (2003) Protein crystals and their growth. J Struct Biol 142:3–21
25. Rosenberger A (1979) Fundamentals of crystal growth. Springer-Verlag, Berlin
26. Chernov AA (1984) Modern crystallography III. Crystal growth. Springer-Verlag, Berlin.
27. Malkin AJ, Kuznetsov YG, Land TA et al (1995) Mechanisms of growth for protein and virus crystals. Nat Struct Biol 2:956–959
28. McPherson A, Malkin AJ, Kuznetsov YG et al (2001) Atomic force microscopy applications in macromolecular crystallography. Acta Crystallogr D Biol Crystallogr 57:1053–1060
29. McPherson A, Malkin AJ, Kuznetsov YG (2000) Atomic force microscopy in the study of macromolecular crystal growth. Annu Rev Biophys Biomol Struct 29:361–410
30. McPherson A, Kuznetsov YG (2014) Mechanisms, kinetics, impurities and defects: consequences in macromolecular crystallization. Acta Crystallogr F Struct Biol Commun 70:384–403
31. Kuznetsov Yu G, Malkin A, McPherson A (1998) Atomic force microscopy studies of phase separations in macromolecular systems. Phys Rev B 58:6097–6103
32. Ten Wolde R, Frenkel D (1997) Enhancement of protein crystal nucleation by critical density fluctuations. Science 277:1975–1978
33. Haas C, Drenth J (1999) Understanding protein crystallization on the basis of the phase diagram. J Cryst Growth 196:388–394
34. Piazza R (1999) Interactions in protein solutions near crystallization: a colloid physics approach. J Cryst Growth 196:415–423
35. Malkin A, McPherson A (1994) Light scattering investigations of nucleation processes and kinetics of crystallization in macromolecular systems. Acta Crystallogr D Biol Crystallogr 50:385–395
36. McPherson A, Gavira JA (2014) Introduction to protein crystallization. Acta Crystallogr F Struct Biol Commun 70:2–20
37. McPherson A, Cudney B (2014) Optimization of crystallization conditions for biological macromolecules. Acta Crystallogr F Struct Biol Commun 70:1445–1467
38. Ducruix A, Giége R (1992) Crystallization of nucleic acids and proteins, a practical approach. IRL Press, Oxford
39. Bergfors TM (1999) Protein crystallization: techniques, strategies and tips. International University Line, La Jolla, CA
40. Bolen DW (2004) Effects of naturally occurring osmolytes on protein stability and solubility: issues important in protein crystallization. Methods 34:312–322
41. McPherson A, Cudney R, Patel S (2003) The crystallization of proteins, nucleic acids, and viruses for X-ray diffraction analysis. In: Fahnestock SR, Steinbuchel A (eds) Biopolymers, vol 18(16), pp 427–468
42. Hünefeld FL (1840) Der Chemismus in der thierischen organisation. FA Brockhaus, Leipzig
43. Jacoby WB (1968) A technique for the crystallization of proteins. Anal Biochem 26:295
44. Caffrey M (2003) Membrane protein crystallization. J Struct Biol 142:108–132
45. McPherson A (1976) The growth and preliminary investigation of protein and nucleic acid crystals for X-ray diffraction analysis. Methods Biochem Anal 23:249–345
46. Chayen NE, Shaw-Stuart PD, Blow DM (1992) Microbatch crystallization under oil: a new technique allowing many small volume crystallization trials. J Cryst Growth 122:176–180
47. Salemme FR (1972) A free interface diffusion technique for the crystallization of proteins for X-ray crystallography. Arch Biochem Biophys 151:533–539
48. Bard J, Ercolani K, Svenson K, Olland A, Somers W (2004) Automated systems for protein crystallization. Methods 34:329–347
49. Hui R, Edwards A (2003) High-throughput protein crystallization. J Struct Biol 142:154–161
50. Santarsiero BD, Yegian DT, Lee CC et al (2002) An approach to rapid protein crystalli-

zation using nanodroplets. J Appl Crystallogr 35:278–281

51. Hosfield D, Palan J, Hilger M et al (2003) A fully integrated protein crystallization platform for small-molecule drug discovery. J Struct Biol 142:207–217
52. DeLucas LJ, Bray TL, Nagy L, McCombs D, Chernov N, Hamrick D, Cosenza L, Belgovskiy A, Stoops B, Chait A (2003) Efficient protein crystallization. J Struct Biol 142:188–206
53. Luft JR, Collins RJ, Fehrman NA et al (2003) A deliberate approach to screening for initial crystallization conditions of biological macromolecules. J Struct Biol 142:170–179
54. Gilliland GL, Tung M, Blakeselee DM et al (1994) Biological macromolecules crystallization database, version 3.0: new features, data and the NASA archive for protein crystal growth data. Acta Crystallogr D Biol Crystallogr 50:408–413
55. Gilliland GL (1988) A biological macromolecule crystallization database: a basis for a crystallization strategy. J Cryst Growth 90:51–60
56. Cohn EJ, Ferry JD (1943) The interactions of proteins with ions and dipolar ions. In: Cohn EJ, Edsall JT (eds) Proteins, amino acids and peptides as ions and dipolar ions. Reinhold, New York, pp 586–622
57. Cohn EJ, Edsall JT (eds) (1943) Proteins, amino acids and peptides as ions and dipolar ions. Van Nostrand-Reinhold, Princeton, NJ
58. Collins KD (2004) Ions from the Hofmeister series and osmolytes: effects on proteins in solution and in the crystallization process. Methods 34:300–311
59. Herriott RM (1942) Solubility methods in the study of proteins. Chem Rev 30:413
60. Hofmeister F (1888) Zur Lehre von der Wirkung der Saltz. Nauyn—Schniedebergs Arch Exp Pathol Pharmakol 24:247
61. McPherson A (2001) A comparison of salts for the crystallization of macromolecules. Protein Sci 10:418–422
62. Sumner JB, Somers GF (1943) The enzymes. Academic Press, New York
63. Englard S, Seifter S (1990) Precipitation techniques. Methods Enzymol 182:301–306
64. Cohn EJ, Hughes WL, Weare JH (1974) Crystallization of serum albumin from ethanol water mixtures. J Am Chem Soc 69:1753–1761
65. McPherson A (1976) Crystallization of proteins from polyethylene glycol. J Biol Chem 251:3600–6303
66. Patel S, Cudney R, McPherson A (1995) Polymeric precipitants for the crystallization of macromolecules. Biochem Biophys Res Commun 207:819–828
67. Ingham KC (1990) Precipitation of proteins with polyethylene glycol. Methods Enzymol 182:301–306
68. Israelachvili J (1997) The different faces of poly(ehylene glycol). Proc Natl Acad Sci U S A 94:8378–8379
69. Sheth SR, Leckband D (1997) Measurements of attractive forces between proteins and end-grafted poly(ethylene glycol) chains. Proc Natl Acad Sci U S A 94:8399–8404
70. McPherson A, Larson SB (2015) A guide to the crystallographic analysis of icosahedral viruses. Crystallogr Rev 21:4–55
71. Timasheff SN, Arakawa T (1988) Mechanism of protein precipitation and stabilization by co-solvents. J Cryst Growth 90:39–46
72. McPherson A, Cudney B (2006) Searching for silver bullets: an alternative strategy for crystallizing macromolecules. J Struct Biol 156:387–406
73. Granick S (1942) Ferritin: I. Physical and chemical properties of horse spleen ferritin. J Biol Chem 146:451–461
74. Giege R, Lorber B, Theobald-Dietrich A (1994) Crystallogenesis of biological macromolecules: facts and perspectives. Acta Crystallogr D Biol Crystallogr 50:339–350
75. McPherson A, Malkin A, Kuznetsov YG et al (1996) Incorporation of impurities into macromolecular crystals. J Cryst Growth 168:74–92
76. Dale GE, Oefner C, D'Arcy A (2003) The protein as a variable in protein crystallization. J Struct Biol 142:88–97
77. Derewenda ZS (2004) The use of recombinant methods and molecular engineering in protein crystallization. Methods 34:354–363
78. Derewenda ZS, Vekilov PG (2006) Entropy and surface engineering in protein crystallization. Acta Crystallogr D Biol Crystallogr 62:116–124
79. DeLucas L (ed) (2009) Membrane protein crystallization: current topics in membranes, vol 63. Elsevier, Amsterdam
80. Loll PJ (2003) Membrane protein structural biology: the high throughput challenge. J Struct Biol 142:144–153
81. Michel H (ed) (1990) General and practical aspects of membrane protein crystallization. Crystallization of membrane proteins. CRC Press, Boca Raton, FL
82. Wiener MC (2004) A pedestrian guide to membrane protein crystallization. Methods 34:364–372

83. Zulauf M (1990) Detergent phenomena in membrane protein crystallization. In: Michel H (ed) Crystallization of membrane proteins. CRC Press, Boca Raton, FL
84. Garavito RM, Ferguson-Miller S (2001) Detergents as tools in membrane biochemistry. J Biol Chem 276:32403–32406
85. Hunte C, Jagow G, Schagger H (2003) Membrane protein purification and crystallization: a practical guide. Academic Press, San Diego
86. Weiner MC (2001) Existing and emergent roles for surfactants in the three-dimensional crystallization of integral membrane proteins. Curr Opin Struct Biol 6:412–419

Chapter 3

Advanced Methods of Protein Crystallization

Abel Moreno

Abstract

This chapter provides a review of different advanced methods that help to increase the success rate of a crystallization project, by producing larger and higher quality single crystals for determination of macromolecular structures by crystallographic methods. For this purpose, the chapter is divided into three parts. The first part deals with the fundamentals for understanding the crystallization process through different strategies based on physical and chemical approaches. The second part presents new approaches involved in more sophisticated methods not only for growing protein crystals but also for controlling the size and orientation of crystals through utilization of electromagnetic fields and other advanced techniques. The last section deals with three different aspects: the importance of microgravity, the use of ligands to stabilize proteins, and the use of microfluidics to obtain protein crystals. All these advanced methods will allow the readers to obtain suitable crystalline samples for high-resolution X-ray and neutron crystallography.

Key words Electric fields, Magnetic fields, Counter-diffusion techniques, Crystal growth in gels, Protein crystallization

1 Introduction

Proteins, nucleic acids, polysaccharides, and lipids are regarded as the most important molecules of life. The function of these molecules in sustaining life depends on their three-dimensional structure and on their highly specific mutual interactions, dictated by their structure and bonding properties [1]. Getting to know the structures of macromolecules and of their complexes will enhance our understanding of biological processes of life. It will also hint at novel ways to treat a wide range of diseases, from congenital anomalies through bacterial and viral infections to autoimmunity diseases [2], or even many different types of cancers [3, 4]. X-ray crystallography is the hallmark of this search (it is the most powerful technique for structure elucidation of macromolecules), as it reaches near-atomic resolution in the most favorable cases, without a priori limitation on the size or on the complexity of the studied molecules. X-ray crystallography requires the growth of large and well-diffracting crystals (for conventional crystallography) or

Alexander Wlodawer et al. (eds.), *Protein Crystallography: Methods and Protocols*, Methods in Molecular Biology, vol. 1607, DOI 10.1007/978-1-4939-7000-1_3, © Springer Science+Business Media LLC 2017

nanocrystals (for free electron lasers, XFELs). The production of such crystals is the most intractable stage in the process of structure determination [5, 6].

There are a number of strategies, from classical techniques to advanced methods, that focus on obtaining high quality single crystals (Fig. 1) for high resolution crystallographic analyses. Despite the existence of a large variety of conventional crystallization techniques (*see* Chapters 2 and 4 by McPherson and Derewenda) and the automation of high-throughput screening systems, statistics from various structural programs indicate that only fewer than 20% of de novo overexpressed proteins yield diffracting crystals [7]. This represents a very low success rate considering the cumulative difficulties of cloning, expressing, and purifying proteins. Although we cannot fully identify why some proteins do not crystallize, this may be due to the intrinsic physico-chemical properties of the protein per se. For this reason, it will be useful to have user-friendly tools that allow the experimenter to a priori select successful protein targets for crystallization and for identifying problematic proteins. The proteins that are recalcitrant to crystallization can be highly flexible as well as completely unstructured. They will not nucleate properly for different reasons, such as propensity to aggregate in an amorphous phase or difficulty to form stable crystal contacts. Therefore, obtaining good crystals can be very tricky and often needs a combination of strategies such as protein engineering, sophisticated crystallization techniques, and a good understanding of the nucleation and crystal growth processes [8–10].

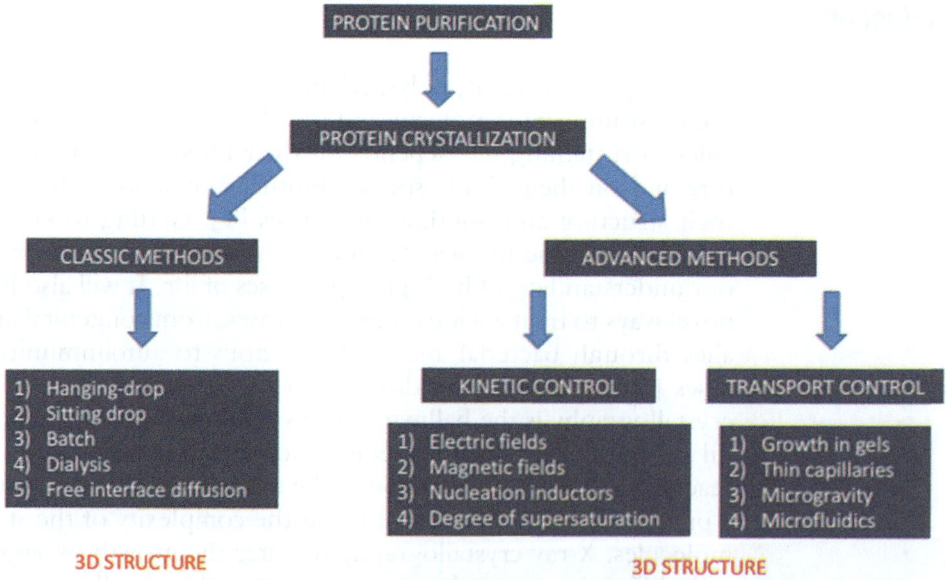

Fig. 1 A scheme representing different methods used to crystallize proteins. The classical methods are shown on the *left*, and the advanced methods, usually called nonconventional methods of protein crystallization, are shown on the *right*

In this chapter different advanced methods that help to increase the success rate of a crystallization project in order to obtain high quality single crystals for crystallographic research are discussed. The first part presents the body of knowledge regarding the crystallization process from physical and chemical perspectives. The second part introduces the reader to new approaches related to more sophisticated methods, not only for growing protein crystals but also for controlling the crystal size and orientation by electromagnetic fields, as well as through other advanced methods. Additional information including the importance of microgravity, the use of ligands to stabilize proteins, and the use of microfluidics to obtain suitable protein crystals for high-resolution X-ray crystallography, is also presented.

2 Technical Approaches

2.1 Fundamentals of Protein Crystallization Process Applied to Advanced Methods of Protein Crystal Growth

The solubility diagrams and the energetics of nucleation (Fig. 2a, b, respectively) provide vital and necessary information for the optimization of crystal growth [11–13]. In most cases, their use will lead to a reasonable strategy for obtaining protein crystals and for assuring high reproducibility. Crystal nucleation occurs in two stages: nucleation of new crystal embryos, and growth of a few nuclei into full-size diffracting crystals (Fig. 2b). It has been shown that the optimal conditions for growing high-quality crystals (large size, and minimum of imperfections) involve lower macromolecule supersaturation levels than those required for initial nucleation [14, 15]. Nucleation cannot take place at these lower supersaturations because an energy barrier of kinetic origin

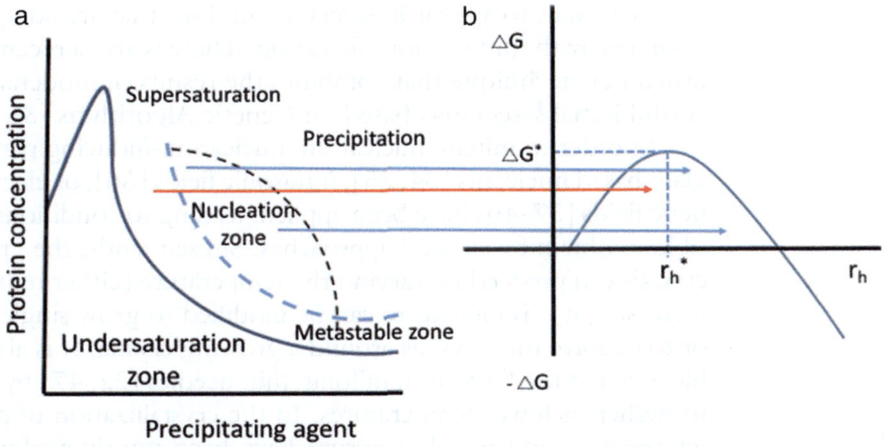

Fig. 2 (**a**) The solubility phase diagram (also known as Oswald-Miers diagram) is divided into different zones: undersaturated, supersaturated, metastable, nucleation, and precipitation. (**b**) The energetics of the system is very important to understand; it expresses the kinetics of the crystallization and allows to predict the critical size of the nucleus to be converted into crystal

(due to the energetically expensive formation of the crystal–solution interface) is involved [16, 17]. The establishment of crystallization solubility phase diagrams allows precise identification of the limits between the spontaneous nucleation and the optimal growth (often called "metastable") zones [18, 19]. That information can subsequently be used for growing crystals as close as possible to the metastable zone or for incubating the trials at nucleation conditions for a time sufficient for the formation of a few nuclei before transiting to metastable conditions for optimal growth (by changing the concentration of the precipitating agent, pH, or temperature) [18, 20–22].

There are alternative setup techniques such as microbatch under oil [23] or crystallization in capillaries [24, 25]. Often, these alternatives produce crystals under screening conditions that will be difficult to produce with other setups (e.g., standard vapor diffusion). These alternative techniques can also produce higher-quality crystals. Each technique relies on a different geometry and different way to reach supersaturation, therefore they present a kinetically different situation. These subtle differences frequently lead to different results in an unpredictable way. Tiny crystals of the same protein can start the nucleation process. There are various crystal seeding techniques, including the standard microseeding and streak-seeding into metastable conditions using microcrystals as sources of crystalline seeds [15, 26–28]. A new method called "Random Microseed Matrix Screening" and related techniques [29–32] that have been recently developed, involve crushing and preparing a seed-stock from microcrystalline material of any quality present in one or more droplets of the initial crystallization screen. This method can also dispense nano-volumes of seed stock into all the conditions of the same or other screens. This procedure allows crystals to appear in screen conditions that are adequate for crystal growth, but not for nucleation. There is also a recently published new technique that combines the results of moderately successful initial screenings based on Genetic Algorithms [33].

In order to initiate nucleation, nucleation-inducing particles or glass-based nucleants [34, 35], ultrasonic fields [36], or electromagnetic fields [37–46] have been applied, leading to conditions impossible to obtain by classical approaches. Subsequently, the growth of crystals can proceed by varying the temperature (either reducing or increasing it). Temperature can be modified to grow single crystals or to dissolve tiny crystals around a growing crystal. It is also possible to avoid the formation of long, thin needles [22, 47] by moving to higher or lower temperatures. In the crystallization of proteins, temperature and mainly pressure have been poorly explored [22, 48–50]. There are usually two temperatures available (most commonly 4 and 18 °C) for growing protein crystals. The existence of different polymorphs has been recently reported, after carefully testing a wide range of temperatures as well as other physicochemical parameters of the crystallization experiment [47, 51–54].

2.2 Protein Concentration

All concentrations should be measured in triplicate with an UV-VIS spectrophotometer, following the calibration procedures provided by the supplier. A calibration concentration plot can be obtained for each new protein, even if its extinction coefficient is not reported in the literature, for the calculation of protein concentration [55].

2.3 Gel Preparation

Agarose gel 0.6% (w/v) stock solution of low melting point agarose (T_{gel} = 297–298 K, Hampton Research HR8-092) can be prepared by dissolving 0.06 g agarose in 10 mL of water heated at 363 K up to a transparent solution with constant stirring. The solution is passed through a 0.22 μm porosity membrane filter for removing all dust particles or insoluble fibers of agarose. The gel-solution can be stored in 1.0 mL aliquots in Eppendorf tubes in the refrigerator. Prior to crystallization in agarose gels, an Eppendorf tube of 1.0 mL is heated at 363 K in order to melt the gel. Most proteins are damaged when exposed to high temperature, so it is best to mix only the precipitant agent with agarose to allow reaching the proper temperature without damaging the protein. Although in the last decade agarose has been the most popular gel for protein crystallization [56–58], there are other types of gels that have also been used for the same purpose [59–62].

3 Advanced Crystallization Methods in Practice

3.1 Experimental Setup for Constructing a Growth-Cell for Applying Electric Fields

As mentioned in Subheading 2.1, it is important to separate the nucleation and crystal growth phenomena. This can be also accomplished using electromagnetic fields. In particular, the use of electric fields has been shown to be useful for successful crystallization of proteins.

For that purpose, one can use a crystal growth cell that consists of two polished float conductive ITO (Indium Tin Oxide Electrode) glass plates, 3.0 × 2.5 cm^2, with a resistance ranging from 4 to 8 Ω (Delta Technologies, Minnesota, USA). The two electrodes are placed parallel to each other. The cell is prepared using a U-like or double well frame (for vapor diffusion set up) as shown in Fig. 3, made of elastic black rubber material, sealed with vacuum grease. Closure of the growth cell can be done by using a gun for melting silicone. The conductive ITO-coated surfaces are placed inwards, at 0.5 cm from each other, to provide appropriate connection area when applying direct (DC) or alternating current (AC) (Fig. 4a, b respectively). Each cell has a volume capacity of approximately 100 μL for precipitant (larger well) and 50 μL for protein plus precipitant (smaller well, as shown in Fig. 3 on the right), or a full volume of 200 μL when a batch configuration is used (Fig. 3, left). The sitting-drop vapor diffusion or batch crystallization conditions for each protein have to be properly established before applying the current. After closing the cell with a cover of melted silicone,

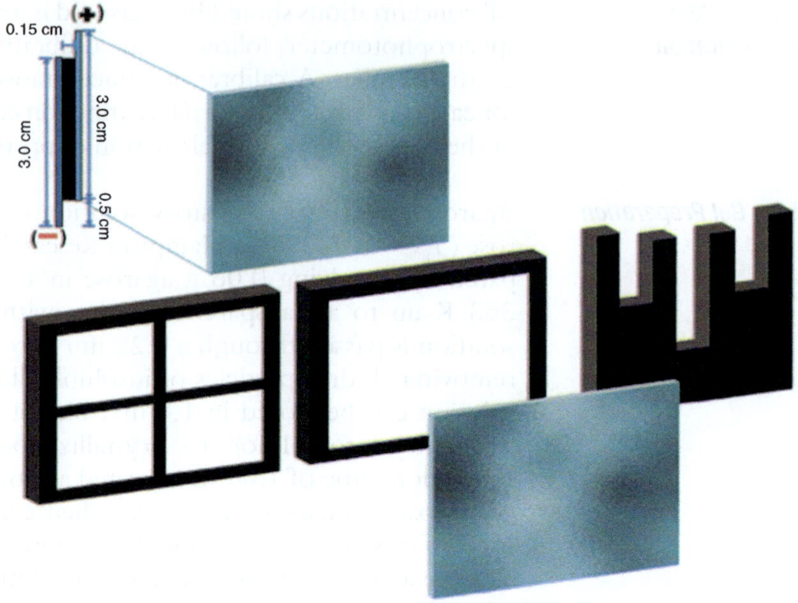

Fig. 3 Different designs of e-crystallization growth-cell for applying electric field to the crystallization process of biological macromolecules. The two frames on the *left* are useful for batch crystallization setup, and the one on the *right* is for a vapor diffusion setup

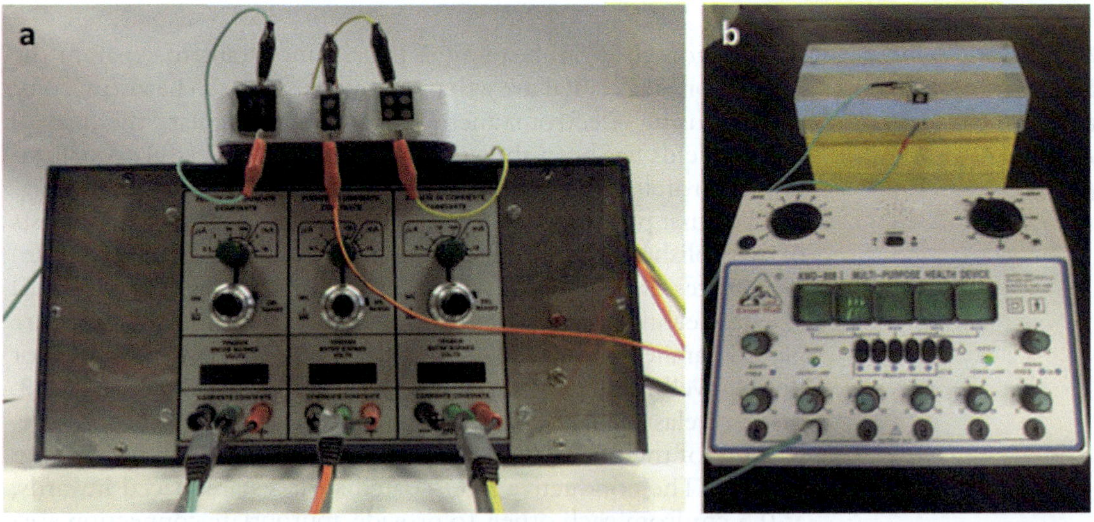

Fig. 4 Two pieces of apparatus used for e-crystallization of proteins: (**a**) for applying DC ranging from 2–6 μA, (**b**) for applying AC during the crystallization of proteins. First, nucleation is induced and then the crystal growth proceeds via vapor diffusion

the system is connected to a DC source (Fig. 4a) that supplies direct current (ranging from 2 to 6 μA) or alternating current (ranging from 2 to 8 Hz), as shown in Fig. 4b. During nucleation, the AC or DC current is turned off after 48 h, so the nuclei are

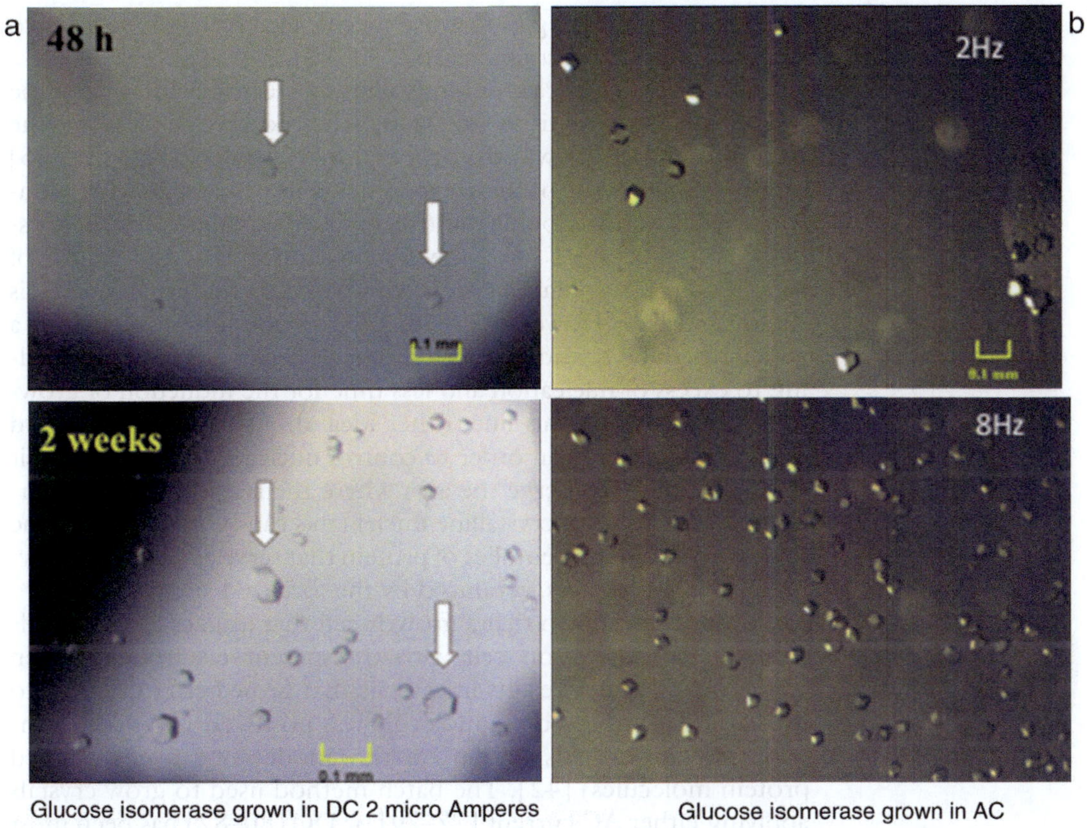

Fig. 5 Crystals of glucose isomerase grown: (**a**) when a direct current (DC) of 2 μA is applied for 48 h and subsequently the crystal growth proceeds in 2 weeks by the sitting-drop setup, and (**b**) when an alternant current (AC) of 2 and 8 Hz is applied for 48 h. The bar scale for (**b**) is the same at 2 and 8 Hz

fixed on the surface of the ITO electrodes. After that the DC growth cell is left at a constant temperature to allow crystals to grow by vapor diffusion (Fig. 5a). In the case of AC, a current of 2 Hz will produce fewer crystals and at 8 Hz will produce a higher number of crystals, although smaller in size (Fig. 5b). Thus AC of 8 Hz or higher values could be used to prepare protein nanocrystals for XFEL experiments.

3.2 The Influence of Electric Fields in the Control of Nucleation

New devices and novel methodologies to control nucleation and the size of crystals (utilizing glass beads for fragmentation of protein crystals to be analyzed in a fine mesh grid via cryo-EM) have been recently described [5, 63]. Magnetic [64, 65] or electric fields [41, 66–68] have been applied in order to obtain larger and higher quality protein single crystals either for conventional X-ray crystallography or for neutron diffraction [69]. The use of AC currents has demonstrated that there is an effect on crystal size (see above). Higher frequencies (between 10 and 50 Hz) have produced tiny crystals for seeding purposes and for crystal growth research.

There are other strategies that use specific electromagnetic fields to control transport phenomena [67, 70–73].

In the particular case of ultrasonic and electric fields, one of the pioneering contributions to study the positive effect on the nucleation processes was the proposal by Nanev and Penkova [36] in 2001. The results of these experiments in which a 25 kHz ultrasonic field (thermal double pulse technique) was applied to the crystallization process of lysozyme, demonstrated that the length of time required to obtain crystals, compared to the usual length, is reduced in half. However, the intensity of the ultrasonic field is a parameter to be considered, as the crystals broke mechanically, leading to excess of nucleation and less time for the induction of growing crystals. Along this line, other idea about using femtosecond lasers was developed in order to control nucleation [74–76]. Their use permitted to observe the area where the laser strike led to formation of only a few crystalline nuclei (this can be explained by the formation of small assemblies of protein that serve as seeds for growing nucleation centers, produced by the focalized laser radiation).

It has been shown that a growth-cell that utilizes electric fields (called e-crystallization cell with transparent electrodes), when applied to proteins, results in crystals that grow better oriented to the cathode (if the protein molecule was positively charged), compared to the crystals grown on the anode (negatively charged protein molecules) [42]. The batch method used to grow crystals applying either AC current [77–79] or DC [80–82] has been most widely used. However, in most cases, these batch crystallization conditions are not experimentally feasible to apply AC current to other proteins more than lysozyme [83]. A reengineered e-crystallization growth cell adapted to a sitting-drop setup has recently been described [42]. Another advantage has been reported for the experimental e-crystallization growth-cell, where after applying DC (to fix the nuclei on the electrode), the crystal growth process proceeds by vapor diffusion. Such a device has been used for crystallization and to search for different polymorphs of glucose isomerase [51] and lysozyme [70] at different temperatures. Along the crystal growth process, we usually obtain four different regimes: (1) induction/equilibration, (2) transient nucleation, (3) steady state nucleation and crystal growth, and (4) depletion [14]. During induction/equilibration, the sitting drop is equilibrating against the reservoir solution and becomes supersaturated when the electric field is applied; there were no nuclei visible in the light passing through the glasses of the ITO transparent electrodes. Eventually, no new crystals were formed and the existing protein crystal nuclei just continued to grow until completion of the process, reaching sizes from 100 to 300 μm, thus becoming suitable for X-ray crystallography. The crystals can be even used for diffraction experiments in situ, if the commercially available ITO electrodes made of plastic material (polyethylene) are used.

3.3 Experimental Setup for Crystallization of Proteins Under the Influence of Magnetic Fields

A majority of the advanced methods mentioned in this chapter are based on the solubility diagram, such as that shown in Fig. 2a [11, 12, 21] or that phase diagram obtained from the physical and chemistry approaches [84]. Recently, advanced methods have been developed for obtaining very high quality crystals not only by growth in gels but also in the presence of strong magnetic fields. In the particular case of magnetic fields, whether they are homogeneous or nonhomogeneous, they always act differently on samples. Nonhomogeneous magnetic fields are responsible for the reduction of gravity forces on the solution through the action of the magnetic force [46, 64, 85]. By applying a vertical magnetic field gradient, a magnetizing force is generated on the sample. If this force is opposite to the gravitational force, the result will be a reduction in the vertical acceleration (effective gravity) with subsequent decrease of natural convection [86]. Convection is practically nullified, generating a situation similar to that found under microgravity conditions [45]. Furthermore, Wakayama et al. found that, in the presence of a magnetizing force opposite to "g" (gravitational vector), fewer lysozyme crystals were obtained than in its absence [87]. The crystals that were obtained diffracted to a higher resolution, in agreement with the mathematical model [46].

For experiments of protein crystallization under the influence of magnetic forces, all proteins and precipitating agents have to be mixed according to the known batch crystallization conditions. It is important to emphasize that the preparation of the batch solution for crystallization must follow the rule that the most viscous solution must be added first, followed by the less viscous ones. Additionally, in order to guarantee highly ordered crystals, a gel can be introduced into the crystallization droplets. This must be done by mixing 1:1:1 (e.g., 5 μL + 5 μL + 5 μL) in the following order: precipitant, agar (0.60% w/v), and the protein. In the cases of standard solution, the gel might be replaced by water to preserve the same crystallization conditions as in the classic crystal growth methods. One must bear in mind that all concentrations from the stock solutions will be reduced to 1/3. Once mixed, the solution or the gelled mix is ready for the magnetic field experiments, as shown in Fig. 6. The mixture (prepared in 0.5 mL Eppendorf tubes) is drawn into a disposable 50 μL glass pipette of (Sigma-Aldrich Z-543292, 1.0-mm inner diameter), using capillarity forces. Green mounting clay from Hampton Research (HR4-326) can be used to seal both ends of the capillary tubes. Once sealed, the capillary pipettes are introduced into an NMR glass tube (8 mm in diameter) and left for at least 48 h in the presence of a magnetic field generated in a 500–700 MHz (11.7–16.5 T) NMR instrument (Fig. 6). All experiments are performed at the temperature of the control unit of the NMR probe head, usually ranging from 291 to 293 K. The sample is left in the magnetic field for at least 2 days or more. Crystals will be better and larger if the time is longer.

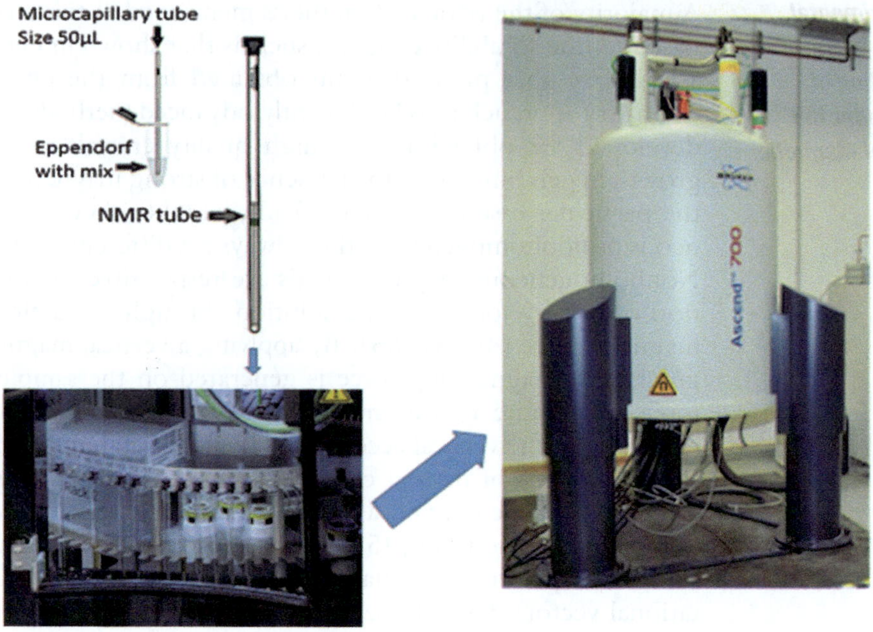

Fig. 6 A setup for experiments performed in the presence of strong magnetic field. Two types of capillaries are used: glass pipettes and NMR tubes. The magnetic field should be applied from 500 to 700 MHz (11.7–16.5 T) by using an NMR apparatus, this is that commonly used in analytical chemistry laboratories

After the end of the experiment, the NMR tube is recovered from the magnet and the capillary pipettes are extracted from it. Then, the capillary pipettes are cut at both ends in order to harvest the crystals. The cut in the capillary pipettes can be done with a glass-capillary cutting stone (Hampton Research Cod. HR4-334). Once both ends of the capillary pipettes are opened, a little air pressure (applied by using plastic latex tubing attached to a 1 mL syringe for blowing it out, or by using a pipette bulb) is sufficient to expel the solution or gel with the crystals into a few microliters of a mother liquor or cryoprotectant on either a two-well or a nine-well glass plate. When necessary, the gel can be dissected with microtools in order to release the crystals. A small incision will open the gel and liberate the crystal to permit the cryoprotectant to enter and to replace the water molecules. All crystals should be immediately mounted and flash-cooled for X-ray data collection.

3.4 The Influence of Magnetic Force to Orient and to Grow Large Protein Crystals

We could observe better quality crystals when applying strong homogeneous magnetic fields, although the field effect was different depending on the space group in which the protein crystallized. The most remarkable effect of this strong magnetic force for the growth of lysozyme crystals was when they grew in the polar space group $P2_1$ [65]. The viscosity of the solution increased when magnetic fields of 10 T were applied [88, 89]. The increase in viscosity was translated into reduced convection. In addition, an orientation effect was observed in the crystals formed under high magnetic fields. In a more recent study, decreasing the diffusion coefficient of lysozyme was assessed in a crystallization solution exposed to a homogeneous magnetic field of 10 T [44, 65, 90]. All these observations are interrelated and are due to the orienting effect of the magnetic field at a microscopic level. In a supersaturated solution, protein nuclei are in suspension in the solution and sediment when reaching an adequate size, which depends on the value of the field. These nuclei would act as blocks that hinder free diffusion of monomers, making the solution more viscous and, hence, lowering convection. Additionally, paramagnetic salts will produce multiple orientation responses to the application of strong magnetic fields [91].

Figure 7 shows the results of growing lysozyme and glucose isomerase crystals for 1 week inside a 700 MHz (16.5 Tesla) NMR magnet. To achieve this, it is necessary to know the conditions of batch crystallization of the protein under study. Once these conditions are known, the time needed to induce nucleation must be known and, finally, access to an NMR equipment of at least 500 MHz (11.7 MHz) or higher is needed to grow large protein crystals. The equipment must be available for the duration of the experiment (at least 3 consecutive days).

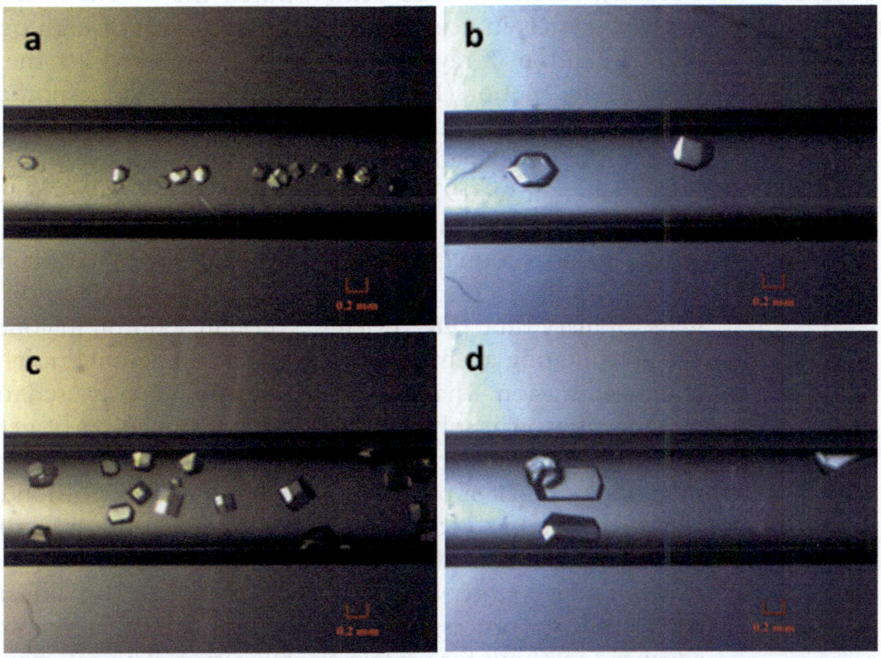

Fig. 7 Crystals of glucose isomerase: Control (**a**) and (**b**) grown in the presence of a magnetic field of 700 MHz (16.5 T). Crystals of lysozyme used as control (**c**), and (**d**) grown in the presence of a magnetic field of 700 MHz (16.5 T). The control crystals are four times smaller than those obtained inside the magnet

Studies of the influence of magnetic fields on crystal growth have been conducted during the last 15 years and they are still being continued [44, 46, 87]. There is still much to be learned about the effect of homogeneous and nonhomogeneous magnetic fields in solutions on a variety of biological macromolecules [89, 92, 93]. All these phenomena apparently favor the quality of the resulting crystals, although we still need more detailed research to understand the underlying mechanisms [64, 94]. There have been a few efforts in this respect, such as combining the positive effect of crystal growth in gels and strong magnetic fields to prove that the crystal growth kinetics is quite close to that obtained in microgravity conditions [43, 65, 86]. The effects of many physical parameters, such as electrical [66, 67, 77, 78] and magnetic fields [45, 46, 64, 65] on the control of nucleation and growth of protein crystals have been assessed. On the other hand, combining the electric and magnetic fields in order to influence crystal orientation can also benefit its homogeneous size in average of many crystals at the same time [68]. One of the main advantages of growing crystals under magnetic fields for a long time (1–3 weeks) is the ability to control their size. The large crystals obtained by applying magnetic force could be suitable not only for neutron diffraction experiments, but also for conventional X-ray crystallography, since one large crystal could yield several data sets of high quality.

3.5 Crystallization by Counter-Diffusion

Recently, several reviews demonstrated the potential of growing protein crystals in gels, which produce crystals of high quality for high-resolution X-ray crystallography compared to the crystals obtained in solution (Fig. 8a) [25, 59, 62, 95, 96]. Another way of reducing the natural convection of solutions under earth gravity is to incorporate jellified media into the solutions. Already in 1968, Zeppezauer et al. [97] described the use of micro-dialysis cells formed by capillary tubes sealed with gel caps (polyacrylamide) for reducing convection in crystallization solutions, and obtaining better crystals. In 1972, Salemme also applied crystallization inside a glass capillary tube [98], placing a protein solution in contact with the precipitating agent solution and reaching equilibrium through counter-diffusion. That technique was subsequently used to crystallize the ribosomal subunits [99]. The combination of gel-growth and the use of capillary tubes have led to the production of a considerable number of protein crystals by counter-diffusion methods [25, 95]. The historical journey of counter-diffusion methods and its fundamentals and experimental development are described below. These counter-diffusion techniques, based on diffusion-control transport processes [100–104], can also be considered as advanced methods for protein crystallization. The gel-growth technique has been used for the crystallization of inorganic salts and it was first applied for protein crystallization at the beginning of 1990s [105–107]. The counter-diffusion methods have proved efficient

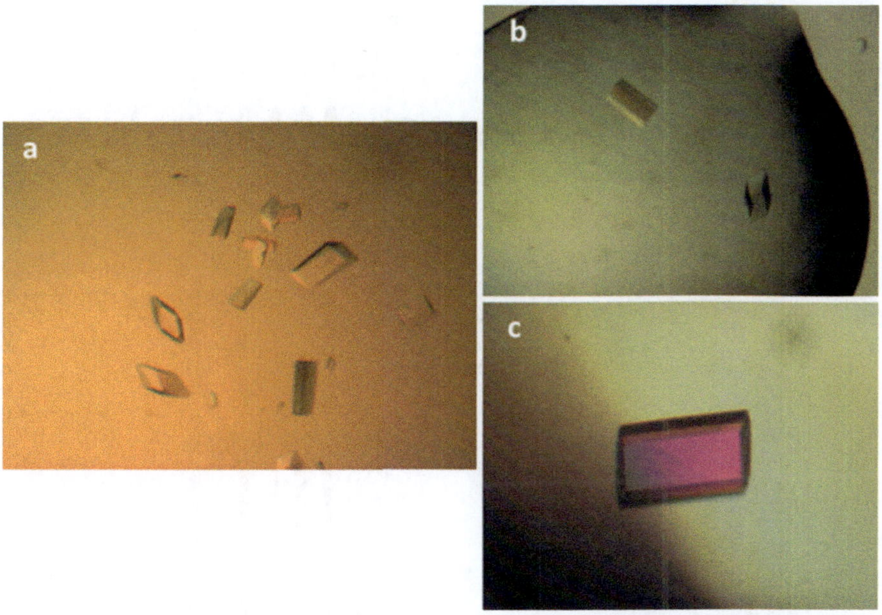

Fig. 8 Crystals of the enzyme aspartyl t-RNA synthetase grown: (**a**) in solution, (**b**) in an agarose gel (0.2% w/v), and (**c**) in a silica gel obtained by the neutralization of sodium metasilicate. As reference, the size of the larger crystal in (**a**) is 100 μm. In (**b**) both crystals are 200–250 μm, and in (**c**) ~400 μm

and effective in crystallizing a certain number of proteins, which could not be crystallized by conventional approaches [25, 95, 96].

García-Ruiz and Henisch theoretically proposed in the middle of 1980s the use jellified media to crystallize biological macromolecules by the gel-growth method (Fig. 8b, c) [102, 103]. This technique, based on the principles of reduced convection and diffusion transport, also offers an advantage of including a wide range of consecutive conditions in a single experiment [108, 109]. These advances permitted García-Ruiz and his team to develop, in 1993, the first variant of the counter-diffusion methods, called the gel acupuncture technique (Fig. 9a) [24]. This novel technique utilizes a precipitating agent that diffuses through the gel support by the capillary force inside a capillary tube filled with a protein solution, thus enabling crystallization [110]. Nowadays, this technique is better known thanks to the assessment of the different types of gels, capillaries, additives, as well as the type of precipitating agents that can be used [24, 59, 62, 111, 112]. In contrast to other techniques that use capillaries, different levels of supersaturation can exist, allowing precipitation to occur in very high supersaturation zones (nucleation occurs when supersaturation is high, and the growth of the nuclei when supersaturation diminishes), increasing the probability of finding adequate crystallization conditions [100, 108]. Another advantage includes a possibility of crystallizing proteins in capillaries smaller than 0.5 mm in diameter

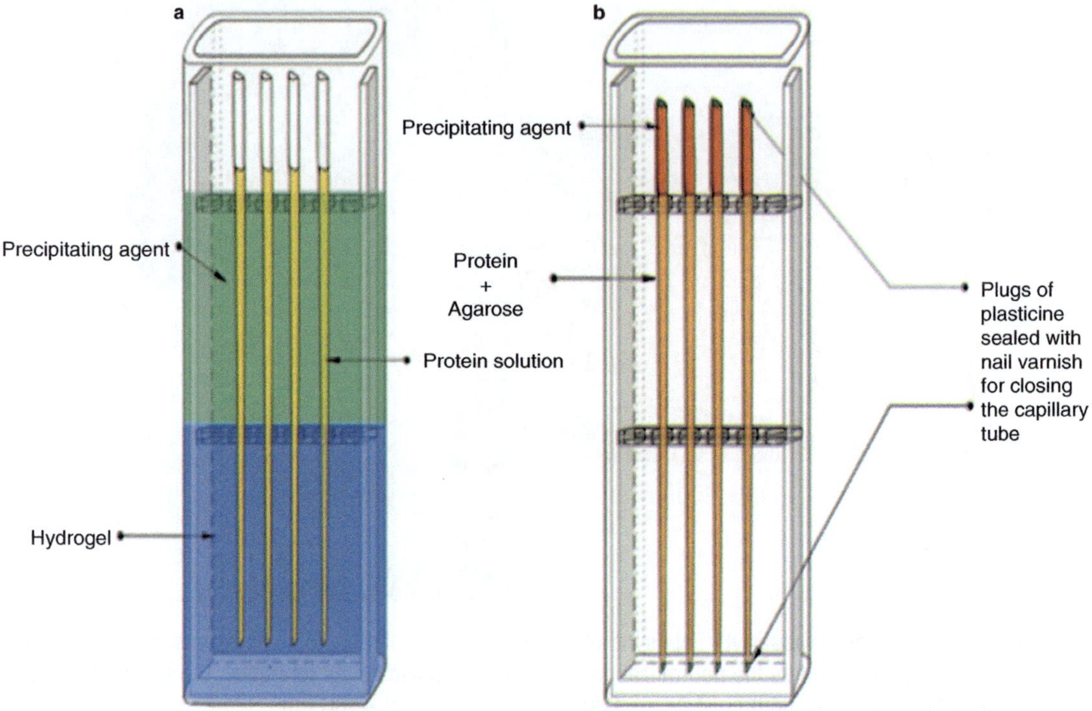

Fig. 9 Two basic experimental setups of the counter-diffusion methods. (**a**) The gel acupuncture method (known as GAME) is shown on the *left*. The capillary tubes are inserted into the gel, the protein is inside the capillaries and the precipitant on top of the gel. (**b**) The *right panel* illustrates counter diffusion in Lindemann capillary tubes, where the protein is mixed with agarose (or any other gel) and the precipitant is applied on top of the gel

(sometimes this allows to obtain cylinders of protein crystals). This helps to avoid their later physical manipulation and the risk of breakage when collecting X-ray diffraction data [56, 113]. Additionally, this method allows crystallization of macromolecules in the presence of cryoprotectant agents and/or heavy metals.

By means of crystallization strategies utilizing the counter-diffusion method (Fig. 9b) it has been possible to crystallize a variety of proteins with different molecular weights and with a wide range of isoelectric points, as well as viruses and protein-nucleic acid complexes [25, 95, 96]. In addition, as demanded by the advances in structural proteomics, there is a device that allows multiple simultaneous and independent crystallization assays suitable for an effective screening of crystallization conditions [113]. It combines the advantages of multiple conditions inside a capillary, increasing the chances of finding the optimal conditions and the possibility to obtain diffraction data directly from the crystallization device. This would turn this method into the first fully automated process, leading from the initial stages until data acquisition for structural analysis.

4 Other Practical Approaches

4.1 Crystallization in Microgravity

Years of experimenting with different crystals have confirmed that by minimizing the convective transport of mass, it is generally possible to obtain higher quality crystals, with improved mechanical and optical properties, with reduced density of defects, and larger in size.

It is natural to think that reduction or absence of gravity will lead to superior quality crystals [114–117]. Experimental observations and data support the hypothesis that convective flow introduces statistical disorder, defects, and surface dislocations into growing crystals [118–123]. Convective transport tends to be variable and erratic, generates variations in supersaturation levels around the crystal faces that are being formed, and exposes them to permanently high nutrient concentrations, equal to those inside the solution. However, in microgravity, where convection is suppressed, a reduction in nutrients concentration is produced in the crystal interphase. Transport is then purely diffusive, which is very slow for proteins. This gives rise to a "nutrients diminution zone" around the nucleus and, due to the absence of gravity, the nucleus is quasi-stable. Generally speaking, the nutrient molecules diffuse very slowly due to their size, which lengthens and extends the nucleation. Large aggregates diffuse even more slowly than the monomers that form the crystal. Hence, the vacuum zone acts as a "diffusive filter," preventing their incorporation into the growing crystal. Apparently, this is the main mechanism responsible for the improvement in the crystal quality due to microgravity. This hypothesis is not only supported by experimentation but also by a mathematical model that explains the transport process [124].

Microgravity experiments, in which the lack of convection leads to impressive results, have been evolving. It is now possible to perform a multitude of simultaneous experiments. However, there are still two criteria that can be applied for optimizing crystallization conditions: namely (1) one performs several assays that assess a wide range of conditions, consuming a large amount of material, or (2) one adjusts beforehand the preliminary conditions for future space missions. However, the consecutive missions may be delayed for months or years, which will counter the advantages of the microgravity method [125].

Utilization of microgravity has been reborn at the beginning of the twenty-first century, but only for crystallization of macromolecular complexes that have never been observed in crystalline form on Earth [126–128]. It is not surprising that most of the effort put in this new trend will offer interesting results that were difficult to get in the past due to uncontrolled experimental conditions in the rockets or during space missions (temperature variations, pressure issues, inadequate containers, etc.). We should expect specific missions for specific problems in protein crystallization in the near future (perhaps the intrinsically disordered proteins would give us some structures that are hard to obtain on Earth or never seen in a crystalline state) [129, 130].

4.2 The Use of Ligands to Stabilize and Crystallize Proteins

When we do not know the crystallization conditions for a protein, bioinformatics analysis to predict if the protein is not intrinsically disordered should be performed first. Next, one should first make simple solubility tests (precipitation with ammonium sulfate (AMS), polyethylene glycols (PEGs), to try different temperatures for crystallization experiments as well as different pH values). Nowadays, there are commercial kits available; these are tools that allow investigating many crystallization conditions based on statistical analysis of protein crystallization. They are based on sets of conditions published at the beginning of the 1990s and even more recently [131, 132]. Crystallization robots have been developed to facilitate screening of hundreds of conditions in a short time. Once an adequate crystallization condition has been found, it can be refined by screening around it. However, there is an additional limitation if the protein under investigation is intrinsically disordered. Many proteins require ligands to stabilize their fold and to allow them to crystallize more readily. The main characteristics of the strategies and limitations of how to stabilize a protein were reviewed elsewhere [133, 134]. It was shown that 100 out of 200 proteins had been crystallized thanks to the use of specific ligands, although not all of them crystallized favorably. The use of amino acids and their analogs has been widely studied and yielded promising results [134]. Details on how to use specific ligands and nucleants in order to crystallize any protein were reviewed elsewhere [135, 136]. Pharmaceutical companies have used these strategies to investigate drugs targeted for diverse diseases. The system MAESTRO (http://www.schrodinger.com) is a suite of programs based on computational chemistry, enabling the prediction of the most probable molecules and bonds that can be used to stabilize protein, RNA, DNA, or macromolecular complexes.

4.3 Application of Microfluidics to Protein Crystallization

The limited availability of many proteins is often the key impediment in crystallographic research [137], emphasizing the need for systems that require minimal amounts of protein for crystallization. This is easy when working on the scale of liters or milliliters, but the process gets complicated as we lower the scale by 5 or 6 orders of magnitude [138]. In this way, devices that use microfluidics arise as potential tools for protein crystallization due to their ability to perform many experiments in reduced volume.

Among the desired features of these "microfluidic chips," we can highlight injection of very exact solution volumes and high reproducibility of the results [139–141]. They are characterized by either a low Reynold's number, or a lack of turbulence, which allows only laminar flows, and ultra-fast diffusive mixtures [142]. Due to a density gradient, the microfluidic systems present either a low Grashof number or the absence of convection. This property demonstrates that it is possible to crystallize proteins with very effective kinetics [143, 144]. In the work of Hansen et al., [143] many parallel reactions were performed. The necessary solutions were introduced either manually or with the help of a robot, into 48 wells. The protein and the precipitating agents were placed in individual chambers that were later connected by eliminating the separating barrier. The total volume of the two chambers was 25 nL, and the relations between both species were set when designing the chip (in this case, they were 1:4, 1:1, and 4:1). Thanks to this device, 11 different macromolecules were successfully crystallized and one was used for diffraction experiments. Among the advantages of these devices [143] are the very precise measurement of the amount of solutions, the absence of the effects of viscosity that affect diffusion of molecules, the ease of harvesting the grown crystals, and the fact that liquid-liquid diffusion methods can be applied in the presence of gravity due to the absence of convection. With the use of microfluidics, equilibrium is achieved faster and the time to grow crystals is reduced. The plan for the future is to enable time-resolved serial crystallography using smaller size chips suitable for collecting X-ray data in situ [139].

However, the microfluidic chips still pose some disadvantages that must be addressed before they can be implemented for large-scale crystallization. Among the disadvantages is the permeability of the elastic connections. Another disadvantage is that it is hard to implement optimization stages, as the experiment starts with pre-mixed solutions (stocks). In the future, it would be advisable to incorporate a chip of this type that can prepare solutions and to couple it in a series [145, 146]. On the other hand, harvesting of crystals is a manual process, in which the whole device is opened, increasing the risk of losing the remaining crystals in order to extract just one.

The design of these devices has been possible thanks to advances in engineering; however, the cost is still very high compared to the traditional systems. Fabrication of chips, which are similar to integrated circuits, requires strict control of cleanliness in the process because micrometric lines are being manufactured. The equipment used for their manipulation is usually very sophisticated and costly and can be used for just one experiment. The advantages of microfluidics have been recently demonstrated for different applications using graphene and a variety of materials in the fabrication of the chips, even when applied for the crystallization of membrane proteins [142, 147–157].

4.4 Automation of Mass Crystallization (High-Throughput)

Recent advances in genomics have led to large-scale efforts in structural biology in a variety of biological samples [157], culminating in Structural Proteomics Consortia and in granting large public subsidies to scientific laboratories as well as to private enterprises, particularly pharmaceutical companies [158, 159]. Projects on metabolomics are diverse and range from studies of structure–function relationships, through mechanisms involved in protein folding and applications to biomedical research [160], to a more pragmatic focus, involving rational design of drugs based on the structure of their target molecules. The use of X-ray crystallography is critical in these studies [161].

Laboratories specializing in structural biology have, in theory, the capacity for handling large-scale projects that require maximal automation at all stages, including crystallographic research [5, 162, 163]. This is not a big issue, particularly if we consider crystallization through microfluidic techniques. In fact, there are already different types of robots on the market that perform these functions. For example, Decode Biostructures produces ROBOHTC, comprising a robot that prepares different crystallization solutions (Matrix Maker) and another robot that arranges the drops. Douglas Instruments is responsible for ORYX 6, which can be used to perform vapor diffusion assays through sitting drops, as well as microbatch assays. This company has also developed a random micro-seeding matrix screening for high throughput hints for protein crystallization conditions [30]. This robot can set up 240 cells per hour. Another commonly used robot is the TTP LabTech Mosquito. This robot contains a set of precision micropipettes mounted on a continuous band, which dispense drop volumes from 50 nL to 1.2 μL. Discarding the disposable micropipettes avoids cross-contamination and eliminates exhaustive washings time. The equipment permits sitting drop or microbatch crystallization tests, as well as hanging drop experiments. It can dispense drops into plates with 96, 384, or 1536 wells.

Up to now we have generally mentioned the advances and drawbacks of large-scale structural biology. Although most experiments have dealt with soluble proteins rather than membrane proteins, high-throughput methodologies have nonetheless been implemented also for membrane protein crystallization [149, 153, 164]. Many strategies and techniques known as in meso crystallization [165, 166] (including crystallization in lipid cubic and sponge phases) have allowed the determination of several hundred membrane protein structures [167, 168].

Finally, we can understand that the advances in the processes and automation made in the past have allowed structural biology to be developed worldwide [169]. Many laboratories are able to successfully clone, express, purify, and crystallize soluble proteins at a rate that was unthinkable some years ago. However, there is still much to be done to control and predict each of the different stages of the general process. A summary of all types of possibilities provided by the high-throughput equipment related to proteins crystallization, has been reviewed and published elsewhere [170].

5 Criteria to Analyze Crystal Quality

Beautiful crystals do not necessarily diffract X-rays to high resolution, but only a few publications have dealt with the strategies for increasing crystal quality [171]. A majority of publications were focused on proteins, though the principle can be applied to other biological macromolecules (DNA, RNA, polysaccharides, macromolecular complexes) [136, 172, 173].

The most adequate techniques to estimate the quality of a crystal are those that employ X-ray topography [174–178]. Here the diffraction equipment is placed in a very characteristic way and the crystal oriented in a preferred direction. Once this has been achieved, the diffraction of a spot is followed through the Ewald sphere (around the crystal), and its quality is characterized through rocking curves. The obtained curves are processed with specific programs that allow us to determine the crystal quality very accurately. If the curve is very fine or pointed, we can confirm that the quality of the crystal is very good; and on the contrary, when the curve is Gaussian-shaped, we can confirm that the quality of the crystal is not very good.

All the advanced methods for protein crystallization mentioned here are the result of the developments in biological crystallogenesis. The name "protein crystallogenesis" was coined by Richard Giegé in the middle of 1990s [179]. It is an outstanding science that studies all the physicochemical processes that govern the growth of crystals of biological macromolecules [180]. The methods, strategies, and devices used to obtain high quality crystals for X-ray crystallography are also part of this fascinating science.

Acknowledgments

The author acknowledges the support from the DGAPA-UNAM Project PAPIIT No. IT200215. The author also appreciates the free use of the NMR facility of LURMN-IQ-UNAM for growing crystals.

References

1. Giege R, Sauter S (2010) Biocrystallography: past, present, future. HFSP J 4:109–121
2. Baranovsky AG, Matushin VG, Vlassov AV et al (1997) DNA- and RNA-hydrolyzing antibodies from the blood of patients with various forms of viral hepatitis. Biochemistry (Moscow) 62:1358–1366
3. Viadiu H (2008) Molecular architecture of tumor suppressor p53. Curr Top Med Chem 8:1327–1334
4. Ciribilli Y, Monti P, Bisio A et al (2013) Transactivation specificity is conserved among p53 family proteins and depends on a response element sequence code. Nucl Acids Res 41:8637–8653
5. Baxter EL, Aguila R, Alonso-Mori R et al (2016) High-density grids for efficient data collection from multiple crystals. Acta Crystallogr D Biol Crystallogr 72: 2–11

6. Cohen AE, Soltis SM, Gonzalez A et al (2014) Goniometer-based femtosecond crystallography with X-ray free electron lasers. Proc Natl Acad Sci U S A 111:17122–17127
7. Giege R (2013) A historical perspective on protein crystallization from 1840 to the present day. FEBS J 280:6456–6497
8. Sanchez-Puig N, Sauter C, Lorber B et al (2012) Predicting protein crystallizability and nucleation. Protein Peptide Lett 19:725–731
9. Thygesen J, Krumbholz S, Levin I et al (1996) Ribosomal crystallography: from crystal growth to initial phasing. J Cryst Growth 168:308–323
10. Berkovitch-Yellin Z, Hansen HAS, Bennett WS et al (1991) Crystals of 70s ribosomes from thermophilic bacteria are suitable for X-ray-analysis at low resolution. J Cryst Growth 110:208–213
11. Saridakis E, Chayen NE (2003) Systematic improvement of protein crystals by determining the supersolubility curves of phase diagrams. Biophys J 84:1218–1222
12. Moretti JJ, Sandler SI, Lenhoff AM (2000) Phase equilibria in the lysozyme-ammonium sulfate-water system. Biotechnol Bioeng 70:498–506
13. Chang J, Lenhoff AM, Sandler SI (2004) Determination of fluid-solid transitions in model protein solutions using the histogram reweighting method and expanded ensemble simulations. J Chem Phys 120:3003–3014
14. Chayen NE, Saridakis E, Sear RP (2006) Experiment and theory for heterogeneous nucleation of protein crystals in a porous medium. Proc Natl Acad Sci U S A 103:597–601
15. Blow DM, Chayen NE, Lloyd LF et al (1994) Control of nucleation of protein crystals. Protein Sci 3:1638–1643
16. Vekilov PG (2010) Nucleation. Cryst Growth Des 10:5007–5019
17. Wang S, Wen X, Golen JA et al (2013) Antifreeze protein-induced selective crystallization of a new thermodynamically and kinetically less preferred molecular crystal. Chem Eur J 19:16104–16112
18. Saridakis E, Chayen NE (2000) Improving protein crystal quality by decoupling nucleation and growth in vapor diffusion. Protein Sci 9:755–757
19. Penkova A, Chayen N, Saridakis E et al (2002) Nucleation of protein crystals in a wide continuous supersaturation gradient. Acta Crystallogr D Biol Crystallogr 58:1606–1610
20. Chayen NE, Saridakis E (2008) Protein crystallization: from purified protein to diffraction-quality crystal. Nat Methods 5:147–153
21. Dumetz AC, Chockla AM, Kaler EW et al (2009) Comparative effects of salt, organic, and polymer precipitants on protein phase behavior and implications for vapor diffusion. Cryst Growth Des 9:682–691
22. Astier JP, Veesler S (2008) Using temperature to crystallize proteins: a mini-review. Cryst Growth Des 8:4215–4219
23. Chayen NE, Stewart PDS, Blow DM (1992) Microbatch crystallization under oil—a new technique allowing many small-volume crystallization trials. J Cryst Growth 122:176–180
24. Garcia-Ruiz JM, Moreno A (1994) Investigations on protein crystal-growth by the gel acupuncture method. Acta Crystallogr D Biol Crystallogr 50:484–490
25. Otalora F, Gavira JA, Ng JD et al (2009) Counterdiffusion methods applied to protein crystallization. Prog Biophys Mol Biol 101:26–37
26. Stura EA, Wilson IA (1991) Applications of the streak seeding technique in protein crystallization. J Cryst Growth 110:270–282
27. Bergfors T (2003) Seeds to crystals. J Struct Biol 142:66–76
28. Obmolova G, Malia TJ, Teplyakov A et al (2014) Protein crystallization with microseed matrix screening: application to human germline antibody Fabs. Acta Crystallogr F Struct Biol Commun 70:1107–1115
29. Malia TJ, Obmolova G, Luo J et al (2011) Crystallization of a challenging antigen-antibody complex: TLR3 ECD with three noncompeting Fabs. Acta Crystallogr F Struct Biol Commun 67:1290–1295
30. Stewart PDS, Kolek SA, Briggs RA et al (2011) Random microseeding: a theoretical and practical exploration of seed stability and seeding techniques for successful protein crystallization. Cryst Growth Des 11:3432–3441
31. Gavira JA, Hernandez-Hernandez MA, Gonzalez-Ramirez LA et al (2011) Combining counter-diffusion and microseeding to increase the success rate in protein crystallization. Cryst Growth Des 11:2122–2126
32. D'Arcy A, Villard F, Marsh M (2007) An automated microseed matrix-screening method for protein crystallization. Acta Crystallogr D Biol Crystallogr 63:550–554
33. Saridakis E (2011) Novel genetic algorithm-inspired concept for macromolecular crystal optimization. Cryst Growth Des 11:2993–2998
34. Saridakis E, Chayen NE (2009) Towards a 'universal' nucleant for protein crystallization. Trends Biotechnol 27:99–106
35. Khurshid S, Saridakis E, Govada L et al (2014) Porous nucleating agents for protein crystallization. Nature Protocols 9:1621–1633

36. Nanev CN, Penkova A (2001) Nucleation of lysozyme crystals under external electric and ultrasonic fields. J Cryst Growth 232:285–293
37. Nanev CN, Penkova A (2002) Nucleation and growth of lysozyme crystals under external electric field. Colloids Surf A 209: 139–145
38. Penkova A, Gliko O, Dimitrov IL et al (2005) Enhancement and suppression of protein crystal nucleation due to electrically driven convection. J Cryst Growth 275:e1527–e1532
39. Taleb M, Didierjean C, Jelsch C et al (1999) Crystallization of proteins under an external electric field. J Cryst Growth 200:575–582
40. Taleb M, Didierjean C, Jelsch C et al (2001) Equilibrium kinetics of lysozyme crystallization under an external electric field. J Cryst Growth 232:250–255
41. Mirkin N, Frontana-Uribe BA, Rodriguez-Romero A et al (2003) The influence of an internal electric field upon protein crystallization using the gel-acupuncture method. Acta Crystallogr D Biol Crystallogr 59:1533–1538
42. Flores-Hernandez E, Stojanoff V, Arreguin-Espinosa R et al (2013) An electrically assisted device for protein crystallization in a vapor-diffusion setup. J Appl Crystallogr 46: 832–834
43. De la Mora E, Flores-Hernandez E, Jakoncic J et al (2015) SdsA polymorph isolation and improvement of their crystal quality using nonconventional crystallization techniques. J Appl Crystallogr 48:1551–1559
44. Sazaki G, Yoshida E, Komatsu H et al (1997) Effects of a magnetic field on the nucleation and growth of protein crystals. J Cryst Growth 173:231–234
45. Wakayama NI (2003) Effects of a strong magnetic field on protein crystal growth. Cryst Growth Des 3:17–24
46. Yin DC (2015) Protein crystallization in a magnetic field. Prog Cryst Growth Charact Mater 61:1–26
47. Veesler S, Ferte N, Costes MS et al (2004) Temperature and pH effect on the polymorphism of aprotinin (BPTI) in sodium bromide solutions. Cryst Growth Des 4:1137–1141
48. Kadri A, Lorber B, Jenner G et al (2002) Effects of pressure on the crystallization and the solubility of proteins in agarose gel. J Cryst Growth 245:109–120
49. Kadri A, Lorber B, Charron C et al (2005) Crystal quality and differential crystal-growth behaviour of three proteins crystallized in gel at high hydrostatic pressure. Acta Crystallogr D Biol Crystallogr 61:784–788
50. Lorber B, Jenner G, Giege R (1996) Effect of high hydrostatic pressure on nucleation and growth of protein crystals. J Cryst Growth 158:103–117
51. Martinez-Caballero S, Cuellar-Cruz M, Demitri N et al (2016) Glucose isomerase polymorphs obtained using an ad hoc protein crystallization temperature device and a growth cell applying an electric field. Cryst Growth Des 16:1679–1686
52. Candoni N, Grossier R, Hammadi Z et al (2012) Practical physics behind growing crystals of biological macromolecules. Protein Pept Lett 19:714–724
53. Heijna MCR, van Enckevort WJP, Vlieg E (2008) Growth inhibition of protein crystals: a study of lysozyme polymorphs. Cryst Growth Des 8:270–274
54. Vera L, Antoni C, Devel L et al (2013) Screening using polymorphs for the crystallization of protein-ligand complexes. Cryst Growth Des 13:1878–1888
55. Olson BJ, Markwell J (2007) Assays for determination of protein concentration. Curr Protoc Protein Sci 3:Unit 3.4
56. Gavira JA, Garcia-Ruiz JM (2002) Agarose as crystallisation media for proteins II: trapping of gel fibres into the crystals. Acta Crystallogr D Biol Crystallogr 58:1653–1656
57. Sauter C, Balg C, Moreno A et al (2009) Agarose gel facilitates enzyme crystal soaking with a ligand analog. J Appl Crystallogr 42:279–283
58. Charron C, Robert MC, Capelle B et al (2002) X-ray diffraction properties of protein crystals prepared in agarose gel under hydrostatic pressure. J Cryst Growth 245:321–333
59. Gonzalez-Ramirez LA, Caballero AG, Garcia-Ruiz JM (2008) Investigation of the compatibility of gels with precipitating agents and detergents in protein crystallization experiments. Cryst Growth Des 8:4291–4296
60. Gavira JA, van Driessche AES, Garcia-Ruiz JM (2013) Growth of ultrastable protein-silica composite crystals. Cryst Growth Des 13:2522–2529
61. Choquesillo-Lazarte D, Garcia-Ruiz JM (2011) Poly(ethylene) oxide for small-molecule crystal growth in gelled organic solvents. J Appl Crystallogr 44:172–176
62. Pietras Z, Lin H-T, Surade S et al (2010) The use of novel organic gels and hydrogels in protein crystallization. J Appl Crystallogr 43:58–63
63. Calero G, Cohen AE, Luft JR et al (2014) Identifying, studying and making good use of macromolecular crystals. Acta Crystallogr F Struct Biol Commun 70:993–1008

64. Sazaki G (2009) Crystal quality enhancement by magnetic fields. Prog Biophys Mol Biol 101:45–55
65. Surade S, Ochi T, Nietlispach D et al (2010) Investigations into protein crystallization in the presence of a strong magnetic field. Cryst Growth Des 10:691–699
66. Hammadi Z, Veesler S (2009) New approaches on crystallization under electric fields. Prog Biophys Mol Biol 101:38–44
67. Koizumi H, Uda S, Fujiwara K et al (2015) Crystallization of high-quality protein crystals using an external electric field. J Appl Crystallogr 48:1507–1513
68. Sazaki G, Moreno A, Nakajima K (2004) Novel coupling effects of the magnetic and electric fields on protein crystallization. J Cryst Growth 262:499–502
69. Blakeley MP, Hasnain SS, Antonyuk SV (2015) Sub-atomic resolution X-ray crystallography and neutron crystallography: promise, challenges and potential. IUCrJ 2:464–474
70. Tomita Y, Koizumi H, Uda S et al (2012) Control of Gibbs free energy relationship between hen egg white lysozyme polymorphs under application of an external alternating current electric field. J Appl Crystallogr 45:207–212
71. Wakayama NI (1997) Electrochemistry under microgravity conditions. 3. Behavior of fluids under high magnetic fields. Denki Kagaku 65:179–182
72. Wang L, Zhong C, Wakayama NI (2002) Damping of natural convection in the aqueous protein solutions by the application of high magnetic fields. J Cryst Growth 237–239:312–316
73. Maki S, Oda Y, Ataka M (2004) High-quality crystallization of lysozyme by magneto-Archimedes levitation in a superconducting magnet. J Cryst Growth 261:557–565
74. Adachi H, Takano K, Yoshimura M et al (2003) Effective protein crystallization using crystal hysteresis. Jpn J Appl Phys 42: L384–L385
75. Yoshikawa HY, Murai R, Adachi H et al (2014) Laser ablation for protein crystal nucleation and seeding. Chem Soc Rev 43:2147–2158
76. Yoshikawa HY, Murai R, Sugiyama S et al (2009) Femtosecond laser-induced nucleation of protein in agarose gel. J Cryst Growth 311:956–959
77. Frontana-Uribe BA, Moreno A (2008) On electrochemically assisted protein crystallization and related methods. Cryst Growth Des 8:4194–4199
78. Uda S, Koizumi H, Nozawa J et al (2014) Crystal growth under external electric fields. AIP Conf Proc 1618:261–264
79. Pérez Y, Eid D, Acosta F et al (2008) Electrochemically assisted protein crystallization of commercial cytochrome C without previous purification. Cryst Growth Des 8:2493–2496
80. Gil-Alvaradejo G, Ruiz-Arellano RR, Owen C et al (2011) Novel protein crystal growth electrochemical cell for applications in X-ray diffraction and atomic force microscopy. Cryst Growth Des 11:3917–3922
81. Koizumi H, Uda S, Fujiwara K et al (2011) Control of effect on the nucleation rate for hen egg white lysozyme crystals under application of an external AC electric field. Langmuir 27:8333–8338
82. Nieto-Mendoza E, Frontana-Uribe BA, Sazaki G et al (2005) Investigations on electromigration phenomena for protein crystallization using crystal growth cells with multiple electrodes: effect of the potential control. J Cryst Growth 275:e1437–e1446
83. Koizumi H, Uda S, Fujiwara K et al (2013) Improvement of crystal quality for tetragonal hen egg white lysozyme crystals under application of an external alternating current electric field. J Appl Crystallogr 46:25–29
84. Rupp B (2015) Origin and use of crystallization phase diagrams. Acta Crystallogr F Struct Biol Commun 71:247–260
85. Yin DC, Wakayama NI, Harata M et al (2004) Formation of protein crystals (orthorhombic lysozyme) in quasi-microgravity environment obtained by superconducting magnet. J Cryst Growth 270:184–191
86. Wada H, Hirota S, Matsumoto S et al (2012) Application of high-field superconducting magnet to protein crystallization. Phys Procedia 36:953–957
87. Wakayama NI, Wang LB, Ataka M (2002) Effect of a strong magnetic field on protein crystal growth. Proc SPIE 4813. doi:10.1117/12.450135
88. Zhong C, Wang L, Wakayama NI (2001) Effect of a high magnetic field on protein crystal growth—magnetic field induced order in aqueous protein solutions. J Cryst Growth 233:561–566
89. Zhong CW, Wakayama NI (2001) Effect of a high magnetic field on the viscosity of an aqueous solution of protein. J Cryst Growth 226:327–332
90. Qi JW, Wakayama NI, Ataka M (2001) Magnetic suppression of convection in protein crystal growth processes. J Cryst Growth 232:132–137

91. Yin DC, Geng LQ, Lu QQ et al (2009) Multiple orientation responses of lysozyme crystals to magnetic field when paramagnetic salts are used as the crystallization agents. Cryst Growth Des 9:5083–5091
92. Moreno A, Quiroz-Garcia B, Yokaichia F et al (2007) Protein crystal growth in gels and stationary magnetic fields. Cryst Res Technol 42:231–236
93. Numoto N, Shimizu K-I, Matsumoto K et al (2013) Observation of the orientation of membrane protein crystals grown in high magnetic force fields. J Cryst Growth 367:53–56
94. Heijna MCR, Poodt PWG, Tsukamoto K et al (2007) Magnetically controlled gravity for protein crystal growth. Appl Phys Lett 90:264105
95. Garcia-Ruiz JM (2003) Counterdiffusion methods for macromolecular crystallization. Methods Enzymol 368:130–154
96. Lorber B, Sauter C, Theobald-Dietrich A et al (2009) Crystal growth of proteins, nucleic acids, and viruses in gels. Prog Biophys Mol Biol 101:13–25
97. Zeppezauer M, Eklund H, Zeppezau ES (1968) Micro diffusion cells for growth of single protein crystals by means of equilibrium dialysis. Arch Biochem Biophys 126:564–573
98. Salemme FR (1972) A free interface diffusion technique for the crystallization of proteins for X-ray crystallography. Arch Biochem Biophys 151:533–539
99. Yonath A, Mussig J, Wittmann HG (1982) Parameters for crystal growth of ribosomal subunits. J Cell Biochem 19:145–155
100. Garcia-Ruiz JM, Novella MI, Moreno R et al (2001) Agarose as crystallization media for proteins: I. Transport processes. J Cryst Growth 232:165–172
101. Garcia-Ruiz JM, Gonzalez-Ramirez LA, Gavira JA et al (2002) Granada crystallisation box: a new device for protein crystallisation by counter-diffusion techniques. Acta Crystallogr D Biol Crystallogr 58:1638–1642
102. Henisch HK, Garcia-Ruiz JM (1986) Crystal-growth in gels and liesegang ring formation. 1. Diffusion relationships. J Cryst Growth 75:195–202
103. Henisch HK, Garcia-Ruiz JM (1986) Crystal-growth in gels and liesegang ring formation. 2. Crystallization criteria and successive precipitation. J Cryst Growth 75:203–211
104. Carotenuto L, Piccolo C, Castagnolo D et al (2002) Experimental observations and numerical modelling of diffusion-driven crystallisation processes. Acta Crystallogr D Biol Crystallogr 58:1628–1632
105. Robert MC, Lefaucheux F (1988) Crystal-growth in gels—principle and applications. J Cryst Growth 90:358–367
106. Vidal O, Robert MC, Boue F (1998) Gel growth of lysozyme crystals studied by small angle neutron scattering: case of silica gel, a nucleation inhibitor. J Cryst Growth 192:271–281
107. Bonnete F, Vidal O, Robert MC et al (1996) Gel techniques and small angle X-ray scattering to follow protein crystal growth. J Cryst Growth 168:185–191
108. Garcia-Ruiz JM, Otalora F, Novella ML et al (2001) A supersaturation wave of protein crystallization. J Cryst Growth 232:149–155
109. Garcia-Ruiz JM, Otalora F, Garcia-Caballero A (2016) The role of mass transport in protein crystallization. Acta Crystallogr F Struct Biol Commun 72:96–104
110. Garcia-Ruiz JM, Moreno A, Viedma C et al (1993) Crystal quality of lysozyme single-crystals grown by the gel acupuncture method. Mater Res Bull 28:541–546
111. Bolanos-Garcia VM (2003) The use of oil in a counter-diffusive system allows to control nucleation and coarsening during protein crystallization. J Cryst Growth 253:517–523
112. Sauter C, Ng JD, Lorber B et al (1999) Additives for the crystallization of proteins and nucleic acids. J Cryst Growth 196:365–376
113. Ng JD, Gavira JA, Garcia-Ruiz JM (2003) Protein crystallization by capillary counterdiffusion for applied crystallographic structure determination. J Struct Biol 142:218–231
114. Littke W, John C (1984) Protein single crystal growth under microgravity. Science 225:203–204
115. DeLucas LJ, Smith GD, Carter DC et al (1991) Microgravity protein crystal-growth—results and hardware development. J Cryst Growth 109:12–16
116. DeLucas LJ, Moore KM, Long MM et al (2002) Protein crystal growth in space, past and future. J Cryst Growth 237:1646–1650
117. DeLucas LJ (2001) Protein crystallization—is it rocket science? Drug Discovery Today 6:734–744
118. Snyder R, Pusey M, Carter D et al (1991) Protein crystal-growth in microgravity. AIAA/IKI Microgravity Sci Symp Proc 1:202–204
119. Snell EH, Judge RA, Crawford L et al (2001) Investigating the effect of impurities on macromolecule crystal growth in microgravity. Cryst Growth Des 1:151–158

120. Kundrot CE, Judge RA, Pusey ML et al (2001) Microgravity and macromolecular crystallography. Cryst Growth Des 1:87–99
121. Carotenuto L, Cartywright J, Otalora F et al (2001) Depletion zone around sedimenting protein crystals in microgravity. ESA J 454:323–329
122. Ries-Kautt M, Broutin I, Ducruix A et al (1997) Crystallogenesis studies in microgravity with the advanced protein crystallization facility on SpaceHab-01. J Cryst Growth 181:79–96
123. Baird JK, Guo LH (1998) Free convection and surface kinetics in crystal growth from solution. J Chem Phys 109:2503–2508
124. Lin SP, Hudman M (1995) Non-equilibrium evaporation and condensation at microgravity. Microgravity Sci Technol 8:163–169
125. Judge RA, Snell EH, van der Woerd MJ (2002) Extracting trends from microgravity crystallization history. Acta Crystallogr D Biol Crystallogr 61:763–771
126. Judge RA, Snell EH, van der Woerd MJ (2005) Extracting trends from two decades of microgravity macromolecular crystallization history. Acta Crystallogr D Biol Crystallogr 61:763–771
127. Trakhanov SD, Grebenko AI, Shirokov VA et al (1991) Crystallization of protein and ribosomal particles in microgravity. J Cryst Growth 110:317–321
128. Chayen NE, Snell EH, Helliwell JR et al (1997) CCD video observation of microgravity crystallization: Apocrustacyanin C-1. J Cryst Growth 171:219–225
129. Pletser V, Bosch R, Potthast L et al (2009) The protein crystallisation diagnostics facility (PCDF) on board ESA Columbus Laboratory. Microgravity Sci Technol 21:269–277
130. Dieckmann MWM, Dierks K (2000) Characterisation of selected bio-molecules in the course of the STS-95 mission, using diagnostics developed within ESA's Technology and Research Program. Opt Dev Diagn Mater Sci 4098:11–25
131. Jancarik J, Kim SH (1991) Sparse-matrix sampling—a screening method for crystallization of proteins. J Appl Crystallogr 24:409–411
132. Kim C, Vink M, Hu M et al (2010) An automated pipeline to screen membrane protein 2D crystallization. J Struct Funct Genomics 11:155–166
133. Manjasetty BA, Turnbull AP, Panjikar S et al (2008) Automated technologies and novel techniques to accelerate protein crystallography for structural genomics. Proteomics 8:612–625
134. Vedadi M, Niesen FH, Allali-Hassani A et al (2006) Chemical screening methods to identify ligands that promote protein stability, protein crystallization, and structure determination. Proc Natl Acad Sci USA 103:15835–15840
135. Bolanos-Garcia VM, Chayen NE (2009) New directions in conventional methods of protein crystallization. Prog Biophys Mol Biol 101: 3–12
136. Ochi T, Balanos-Garcia VM, Stojanoff V et al (2009) Perspectives on protein crystallisation. Prog Biophys Mol Biol 101:56–63
137. Rupp B, Segelke BW, Krupka HI et al (2002) The TB structural genomics consortium crystallization facility: towards automation from protein to electron density. Acta Crystallogr D Biol Crystallogr 58:1514–1518
138. van der Woerd M, Ferree D, Pusey M (2003) The promise of macromolecular crystallization in microfluidic chips. J Struct Biol 142:180–187
139. Pawate AS, Srajer V, Schieferstein J et al (2015) Towards time-resolved serial crystallography in a microfluidic device. Acta Crystallogr F Struct Biol Commun 71: 823–830
140. Horstman EM, Goyal S, Pawate A et al (2015) Crystallization optimization of pharmaceutical solid forms with X-ray compatible microfluidic platforms. Cryst Growth Des 15:1201–1209
141. Liu J, Hansen C, Quake SR (2003) Solving the "world-to-chip" interface problem with a microfluidic matrix. Anal Chem 75:4718–4723
142. Perry SL, Higdon JJL, Kenis PJA (2010) Design rules for pumping and metering of highly viscous fluids in microfluidics. Lab Chip 10:3112–3124
143. Hansen C, Quake SR (2003) Microfluidics in structural biology: smaller, faster... better. Curr Opin Struct Biol 13:538–544
144. Hansen C, Leung K, Mousavil P (2007) Chipping in to microfluidics. Phys World 20:24–29
145. Sauter C, Dhouib K, Lorber B (2007) From macrofluidics to microfluidics for the crystallization of biological macromolecules. Cryst Growth Des 7:2247–2250
146. Abdallah BG, Roy-Chowdhury S, Fromme R et al (2016) Protein crystallization in an actuated microfluidic nanowell device. Cryst Growth Des 16:2074–2082
147. Mignard E, Lorber N, Sarrazin F et al (2011) Microfluidics: a new tool for research in chemistry. Actualite Chimique 353-354: 25–28

148. Gong H, Beauchamp M, Perry S et al (2015) Optical approach to resin formulation for 3D printed microfluidics. RSC Adv 5:106621–106632
149. Abdallah BG, Kupitz C, Fromme P et al (2013) Crystallization of the large membrane protein complex photosystem I in a microfluidic channel. ACS Nano 7:10534–10543
150. Abdallah BG, Zatsepin NA, Roy-Chowdhury S et al (2015) Microfluidic sorting of protein nanocrystals by size for X-ray free-electron laser diffraction. Struct Dyn 2:041719
151. Maeki M, Yamaguchi H, Tokeshi M et al (2016) Microfluidic approaches for protein crystal structure analysis. Anal Sci 32:3–9
152. Li JJ, Chen QL, Li GZ et al (2009) Research and application of microfluidics in protein crystallization. Prog Chem 21:1034–1039
153. Hunter MS, Fromme P (2011) Toward structure determination using membrane-protein nanocrystals and microcrystals. Methods 55:387–404
154. Yokoyama T, Ostermann A, Mizoguchi M et al (2014) Crystallization and preliminary neutron diffraction experiment of human farnesyl pyrophosphate synthase complexed with risedronate. Acta Crystallogr F Struct Biol Commun 70:470–472
155. Tanaka I, Kusaka K, Chatake T et al (2013) Fundamental studies for the proton polarization technique in neutron protein crystallography. J Synchrotron Radiat 20:958–961
156. Kawamura K, Yamada T, Kurihara K et al (2011) X-ray and neutron protein crystallographic analysis of the trypsin-BPTI complex. Acta Crystallogr D Biol Crystallogr 67:140–148
157. Gul S, Hadian K (2014) Protein-protein interaction modulator drug discovery: past efforts and future opportunities using a rich source of low- and high-throughput screening assays. Expert Opin Drug Discov 9:1393–1404
158. Zimmerman MD, Grabowski M, Domagalski MJ et al (2014) Data management in the modern structural biology and biomedical research environment. Methods Mol Biol 1140:1–25
159. Stewart PS, Mueller-Dieckmann J (2014) Automation in biological crystallization. Acta Crystallogr F Struct Biol Commun 70:686–696
160. de Raad M, Fischer CR, Northen TR (2016) High-throughput platforms for metabolomics. Curr Opin Chem Biol 30:7–13
161. Zheng H, Hou J, Zimmerman MD et al (2014) The future of crystallography in drug discovery. Expert Opin Drug Discovery 9:125–137
162. Russi S, Song J, McPhillips SE et al (2016) The Stanford Automated Mounter: pushing the limits of sample exchange at the SSRL macromolecular crystallography beamlines. J Appl Crystallogr 49:622–626
163. Boivin S et al (2016) An integrated pipeline for sample preparation and characterization at the EMBL@PETRA3 synchrotron facilities. Methods 95:70–77
164. Urban M, Tampe R (2016) Membranes on nanopores for multiplexed single-transporter analyses. Microchim Acta 183:965–971
165. Bogorodskiy A, Frolov F, Mishin A et al (2015) Nucleation and growth of membrane protein crystals in meso—a fluorescence microscopy study. Cryst Growth Des 15:5656–5660
166. Cherezov V, Clogston J, Papiz MZ et al (2006) Room to move: crystallizing membrane proteins in swollen lipidic mesophases. J Mol Biol 357:1605–1618
167. Liu W, Wacker D, Gati C et al (2013) Serial femtosecond crystallography of G protein-coupled receptors. Science 342:1521–1524
168. Caffrey M, Cherezov V (2009) Crystallizing membrane proteins using lipidic mesophases. Nature Protocols 4:706–731
169. Krauss IR, Merlino A, Vergara A et al (2013) An overview of biological macromolecule crystallization. Int J Mol Sci 14:11643–11691
170. Saridakis E (2012) Perspectives on high-throughput technologies applied to protein crystallization. Protein Pept Lett 19:778–783
171. Newman J (2006) A review of techniques for maximizing diffraction from a protein crystal in stilla. Acta Crystallogr D Biol Crystallogr 62:27–31
172. Helliwell JR (2008) Macromolecular crystal twinning, lattice disorders and multiple crystals. Crystallogr Rev 14:189–250
173. Boggon TJ, Helliwell JR, Judge RA et al (2000) Synchrotron X-ray reciprocal-space mapping, topography and diffraction resolution studies of macromolecular crystal quality. Acta Crystallogr D Biol Crystallogr 56: 868–880
174. Otalora F, Capelle B, Ducruix A et al (1999) Mosaic spread characterization of microgravity-grown tetragonal lysozyme single crystals. Acta Crystallogr D Biol Crystallogr 55: 644–649

175. Robert MC, Capelle B, Lorber B (2003) Growth sectors and crystal quality. Methods Enzymol 368:154–169
176. Robert MC, Capelle B, Lorber B et al (2001) Influence of impurities on protein crystal perfection. J Cryst Growth 232:489–497
177. Vidal O, Robert MC, Arnoux B et al (1999) Crystalline quality of lysozyme crystals grown in agarose and silica gels studied by X-ray diffraction techniques. J Cryst Growth 196: 559–571
178. Otalora F, Garcia-Ruiz JM, Gavira JA et al (1999) Topography and high resolution diffraction studies in tetragonal lysozyme. J Cryst Growth 196:546–558
179. Giege R, Lorber B, Theobald-Dietrich A (1994) Crystallogenesis of biological macromolecules—facts and perspectives. Acta Crystallogr D Biol Crystallogr 50:339–350
180. Gavira JA (2016) Current trends in protein crystallization. Arch Biochem Biophys 101: 3–11

Chapter 4

The "Sticky Patch" Model of Crystallization and Modification of Proteins for Enhanced Crystallizability

Zygmunt S. Derewenda and Adam Godzik

Abstract

Crystallization of macromolecules has long been perceived as a stochastic process, which cannot be predicted or controlled. This is consistent with another popular notion that the interactions of molecules within the crystal, i.e., crystal contacts, are essentially random and devoid of specific physicochemical features. In contrast, functionally relevant surfaces, such as oligomerization interfaces and specific protein–protein interaction sites, are under evolutionary pressures so their amino acid composition, structure, and topology are distinct. However, current theoretical and experimental studies are significantly changing our understanding of the nature of crystallization. The increasingly popular "sticky patch" model, derived from soft matter physics, describes crystallization as a process driven by interactions between select, specific surface patches, with properties thermodynamically favorable for cohesive interactions. Independent support for this model comes from various sources including structural studies and bioinformatics. Proteins that are recalcitrant to crystallization can be modified for enhanced crystallizability through chemical or mutational modification of their surface to effectively engineer "sticky patches" which would drive crystallization. Here, we discuss the current state of knowledge of the relationship between the microscopic properties of the target macromolecule and its crystallizability, focusing on the "sticky patch" model. We discuss state-of-the-art in silico methods that evaluate the propensity of a given target protein to form crystals based on these relationships, with the objective to design variants with modified molecular surface properties and enhanced crystallization propensity. We illustrate this discussion with specific cases where these approaches allowed to generate crystals suitable for structural analysis.

Key words Protein crystallization, Sticky patch model, Surface entropy reduction, Lysine methylation

1 Introduction

Conventional, single-crystal X-ray diffraction analysis is only feasible if the target sample (protein, protein–DNA complex, etc.) can be obtained in a crystal form capable of diffracting X-rays. Early on, pioneers of macromolecular crystallography relied on a portfolio of proteins and viruses crystallized in the 1920s and 1930s, by such biochemists as James Sumner, John Northrop, and Wendell Stanley, who received the 1946 Nobel Prize for Chemistry for this work [1]. Their efforts were never intended to generate crystals for

Alexander Wlodawer et al. (eds.), *Protein Crystallography: Methods and Protocols*, Methods in Molecular Biology, vol. 1607,
DOI 10.1007/978-1-4939-7000-1_4,

structural analyses: this was an early era of protein biochemistry, and crystallization was the final step in the purification of proteins. Given the earlier discovery of X-ray diffraction in 1912, there was obvious interest if protein crystals, known since mid-nineteenth century, also diffract X-rays. It was John Desmond Bernal, who showed in 1934 that only if protein crystals (he used pepsin) are kept within their mother liquor, they exhibit beautiful diffraction up to virtually atomic resolution [2], effectively demonstrating that these macromolecules have distinct chemical structure, in contrast to the then prevailing colloidal theory [3]. This set the stage for macromolecular crystallography, which revolutionized contemporary biology and medicine by providing tools to explore structure–function relationships in proteins, nucleic acids, intact viruses, and other complex structures, such as ribosomes [4].

From the very beginning it was obvious that proteins exhibit very different propensities to form crystals: some crystallize under a range of conditions yielding distinct crystal forms, others precipitate in an amorphous way, form gels or oils. As early as 1909, the physiologist E. T. Reichert and the mineralogist A. P. Brown (both from the University of Pennsylvania) published a remarkable volume, entitled "*The Crystallography of Haemoglobins*" in which they described hundreds of distinct morphologies of crystals of hemoglobin obtained from the blood of various vertebrates [5]. Numerous micrographs (Fig. 1) illustrate how species-dependent

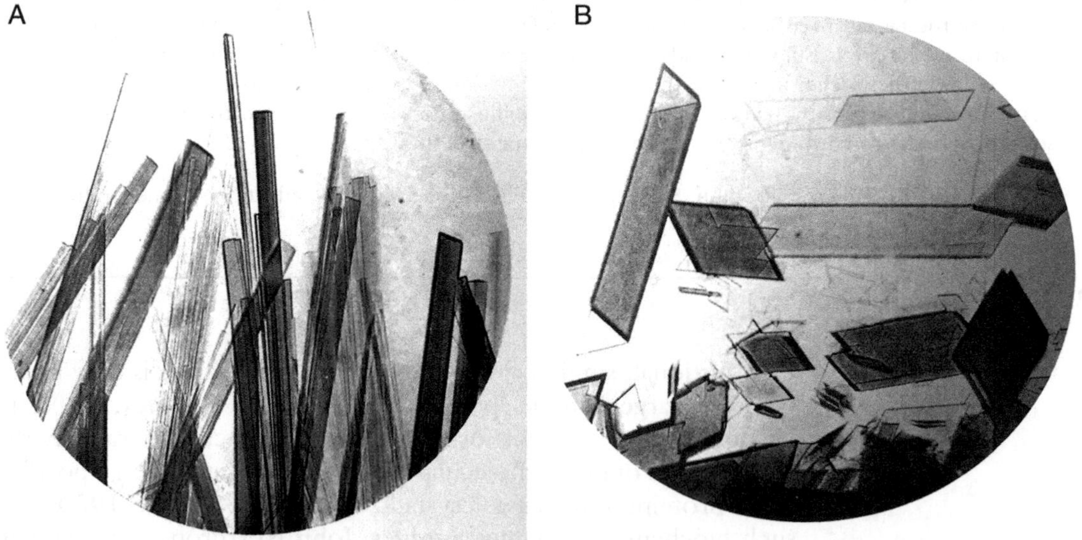

Fig. 1 Two different forms of hemoglobin crystals. **(a)** Crystals obtained from hemoglobin of the mule, and **(b)** those from the hemoglobin of Indian antelope (*Antelope cervicapra*). Both figures from Reichert, E.T. and Brown, A.P. (1909). *The differentiation and specificity of corresponding proteins and other vital substances in relation to biological classification and organic evolution: the crystallography of haemoglobin.* Carnegie Institution of Washington, Washington, DC

variations between proteins result in crystals with markedly different morphologies. At the time, the underlying chemical nature was unknown; today we understand that amino acid sequence differences are the cause. Half a century later, in 1953, John Kendrew showed that crystal forms of the same protein from different species may show very different diffraction properties; by screening myoglobin crystals from different sources he was able to select the best diffracting crystal form (sperm whale myoglobin), which eventually led to crystallographic characterization of the first protein molecule [6].

As the list of proteins known to crystallize upon purification was slowly getting exhausted, in the 1960s crystallographers faced the challenge of having to crystallize their target macromolecules first. Thus the science (or art) of protein crystallization was born. One of the first concepts to be introduced was that of screening of a spectrum of conditions, be it buffers or precipitants, in search of one where a protein would precipitate in a crystalline form as the solution passed the saturation point (*see* Chapter 2 for detailed discussion). Little attention was paid to recording data from failed experiments, and the whole process was thought to be stochastic: some proteins crystallized, others did not, and no correlation with solution conditions was apparent. The microscopic nature of the interactions underlying crystallization was virtually ignored. When crystal contacts where finally recognized as a valid target of structural analysis, early studies concluded that they are essentially random patches of protein surfaces, attesting to the stochastic process of assembly of proteins into nuclei and crystals [7–9]. This was bad news, because stochastic phenomena cannot be easily controlled and directed, and so macromolecular crystallization appeared to be destined forever to the Edisonian approach of trial and error, i.e., random screening.

The advent of molecular biology in the 1980s, and more recently of high-throughput (HT) methods that enabled the Structural Genomics initiative, brought substantial changes to the way we approach crystallization. We are no longer restricted to wild-type proteins from natural sources; in most cases the targets are recombinant proteins, and often they are custom-designed fragments of specific targets. This makes it possible, in principle, to manipulate the initial cDNA construct to enhance a protein's propensity to crystallize. Further, the HT Structural Genomics laboratories introduced highly standardized crystallization pipelines, carefully recording all outcomes, both positive and negative. This uncovered hitherto unappreciated correlations between protein properties (as encoded by the amino acid sequences) and their crystallization propensities, clearly revealing that some are much more amenable to crystallization than others [10–14]. As the number of structures deposited in the Protein Data Bank (PDB) grew at a rapid pace, new opportunities for data mining opened up.

Further, studies from fields such as soft matter physics, bioinformatics, and molecular biology began to reshape our understanding of the microscopic mechanisms underlying crystallization, coming to conclusions that are in stark contrast to the "stochastic model." In its place, a new general "sticky patch" model has emerged, emphasizing the microscopic variations of the macromolecular surface and the physicochemical phenomena behind low-affinity, yet specific intermolecular interactions, including those governing the formation of contacts in nascent crystals. In this chapter, we review the current microscopic view of crystallization based on the premise of directional, specific molecular interactions, and discuss experimental methods that exploit those concepts for the design of chemically or mutationally modified protein targets for enhanced crystallizability (NB: Macromolecular crystallography encompasses broadly the study of proteins, nucleic acids and their complexes as well as a range of chemical entities; most of this chapter focuses specifically on proteins, but the phenomena and methods described herein also apply to all kinds of protein complexes).

2 Theoretical and Experimental Evidence for the "Sticky Patch" Model

2.1 Crystallization In Silico: Lessons from Soft Matter Physics

Our current understanding of protein crystallization owes much to experimental and theoretical soft matter physics, and particularly to the study of colloids [15]. More than two decades ago, it was observed that both colloidal particles and proteins tend to crystallize when the osmotic second virial coefficient, *B2*, which depends only on the pair interaction between the particles, lies in the favorable, crystallization "slot" [16]. Studies of crystallization of isotropic spheres show that it proceeds through a slow process of nucleation, whose rate is enhanced close to the metastable liquid-vapor coexistence curve (bimodal), followed by growth [17]. Proteins may behave in this fashion, and they (like colloids) can also form amorphous aggregates that kinetically impair crystallization below the binodal, and they can be (meta)stable in the crystallization slot of the second virial coefficient without crystallization ever taking place [18–21]. The complexity of the phenomenon prompted theoretical and computational efforts to generate suitable models for phase transitions and crystallization of both these systems. Initial attempts focused on simple models with a relatively short-range interparticle attraction [22]. Subsequently, various pair potentials have been studied, allowing for variable (yet still small) range attraction and more complex potentials that included a repulsive barrier [23]. In general terms, these models required the particles in the liquid phase to be very close to each other for the attraction force to become significant. Initially this phase behavior was thought to be reasonably well suited as a starting point for simulations of globular proteins, with their roughly spherical shape, isotropic electrostatic repulsion and short-range van der Waals and effective attraction due to hydrophobicity.

One of the key assumptions underlying the early colloidal models was the isotropic nature of the interactions [22]. Most obviously, the microscopic nature of the colloid and protein solid phase is different, as illustrated by the fact that proteins do not form close packed crystals. More subtly, the overall shape of the bimodal in proteins is qualitatively incompatible with isotropic attraction. These problems suggest that additional features should be included in the minimal model. One of such features is interaction anisotropy [24–26]. In fact, a similar question has surfaced in colloid physics, given the considerable effort to design complex colloidal particles with physically patterned surfaces, or "patchy" particles [27]. It was therefore natural to extend this notion to proteins, in order to capture the orientation dependence of protein–protein interactions. Lomakin et al. [28] first developed a model taking account of the spatial variation of the protein surface, underlying varying short-range interactions. It was used, among others, by Gogelein et al. [29] to describe the phase behavior of lysozyme dispersions. This early model involves repulsive screened Coulomb interactions, with incorporated attractive surface patches that mediate interactions between molecules.

More detailed computer simulations subsequently revealed the impact of attractive surface patches on the crystal lattice, concluding that anisotropic interactions can lead to a variety of different crystal structures, depending on the geometry and strength of the patchy interactions [30]. A variant of the model, which contained competing sets of attractive patches, has been used to explain why nearly identical conditions sometimes yield different crystal forms of the same protein, specifically homodimeric and monomeric crystal forms [31]. The concept was further expanded by the introduction of a model based on spheres decorated randomly with a large number of attractive patches, to study the formation of structures with $P2_12_12_1$ symmetry, the most prevalent space group among proteins [32]. The conclusions of this study are particularly interesting. The unit cell with the lowest energy is not necessarily the one that grows fastest, because growth is favored when new particles attach through enough patches to the growth front and if particles can attach in crystallographically nonequivalent positions with the same affinity. Importantly, when nonspecific interactions that are not part of the set of crystal contacts are few and weaker than the actual crystal contacts, both nucleation and growth are successful [32]. Recently, a computational study of crystals of three proteins from the rubredoxin family characterized crystal contacts and used them to parametrize patchy particles models (Fig. 2) [33]. This first explicit bridge between soft matter physics and structural biology not only provided reasonable theoretical phase diagrams, but also microscopic-level insight into specific patterns of residues that make up crystal contacts.

Fig. 2 The "patchy model" of proteins and their interactions. The *blue* spheres are proteins on which each pair of patches corresponds to the crystal interface of the same color. From: Fusco et al. (2014) *Characterizing protein crystal contacts and their role in crystallization: rubredoxin as a case study.* Soft Matter 10 (2):290–302

To conclude, the "sticky patch" model describes crystallization as a non-stochastic process, made possible by few, attractive patches on the surface of a protein, which under specific crystallization conditions impact critically the success of nucleation and growth type as well as the crystal lattice. We now discuss how parallel advances in crystallization thermodynamics, in the chemistry and stereochemistry of crystal contacts, and in weak protein–protein interactions support and complement the "sticky patch" model.

2.2 Thermodynamics of Crystallization: A Microscopic View

The canonical, macroscopic view of crystallization thermodynamics, including phase diagrams [34] (*see* Chapter 2), has little predictive value and does not address the microscopic mechanisms of molecular interactions leading to three-dimensional order during crystal growth, or—conversely—does not explain the failure of molecules to form crystals under conditions of supersaturation, as opposed to amorphous precipitate or gel. However, recent interpretations of thermodynamic changes that accompany crystallization of macromolecules give us new insights into the microscopic aspects of the phenomenon, and taken together with the "sticky patch model" allow to answer a number of questions [35–37].

Like any equilibrium process occurring at constant pressure and temperature, crystallization is driven by the reduction in Gibbs free energy, $\mathbf{\Delta G^\circ}_{\mathrm{cryst}}$, on transfer of molecules from solution to the crystalline phase. At constant temperature $\mathbf{T}$, this is the net effect of changes in enthalpy ($\mathbf{\Delta H^\circ}_{\mathrm{cryst}}$) and entropy ($\mathbf{\Delta S^\circ}_{\mathrm{cryst}}$):

$$\mathbf{\Delta G^\circ}_{\mathrm{cryst}} = \mathbf{\Delta H^\circ}_{\mathrm{cryst}} - \mathbf{T\Delta S^\circ}_{\mathrm{cryst}}$$

Direct determination of $\mathbf{\Delta G^\circ}_{\mathrm{cryst}}$ is difficult, but available data suggest that it is modestly negative, i.e., in the range of −10 to

−100 kJ mol^{-1} [37]. This explains why crystallization is subject to "butterfly effects," because even extremely subtle phenomena (e.g., minute change of temperature) that can occur at any point during the process can shift $\mathbf{\Delta G°}_{cryst}$ into the positive or negative range, with dramatic impact on the outcome of the process.

An interesting question is if either $\mathbf{\Delta H°}_{cryst}$ (enthalpy) or $\mathbf{\Delta S°}_{cryst}$ (entropy), preferentially drive the free energy change. In the case of macromolecular crystallization, enthalpy changes cannot be large, because no strong bonds are formed. In few cases where experimental measurements of $\mathbf{\Delta H°}_{cryst}$ were made, the values were consistently small [38–40]. This suggests that entropic effects should be playing a dominant role, although the notion is counterintuitive, because the formation and growth of the three-dimensionally ordered crystal is by definition associated with significant, unfavorable decrease in entropy. Indeed, a loss of three translational and three rotational degrees of freedom per molecule is estimated to result in a value of $\mathbf{-T\Delta S°}_{cryst}$ of 30 to 100 kJ mol^{-1} [41, 42] at room temperature. However, it is when we take into account the microscopic effects associated with the formation of crystal contacts, that the situation gets much worse.

In general terms, a protein molecule can be described as having a solvent-inaccessible core with rigid secondary structure elements, and more flexible, solvent exposed loops that create the molecular surface. Much of this surface is populated by conformationally labile, long side chains, such as Lys, Arg, Glu, and Gln (NB: It has been suggested, in fact, that the presence of high-entropy side chains on protein surfaces could be the result of early evolutionary pressures [43]; given the high protein concentrations in living cells, and the associated overcrowding effects, it is reasonable to hypothesize that globular proteins have been under evolutionary pressures to avoid nonfunctional specific interactions, hence the presence of the "entropy shield" on the surface [44, 45]). When protein molecules assemble within nascent crystals, specific intermolecular contacts are formed. At these sites, many flexible side chains become sequestered and consequently ordered (Fig. 3). Although the magnitude of side chain conformational entropies of Lys, Arg, Glu, and Gln are highly dependent on the rotamer and secondary structure context, it is generally agreed that it may range at room temperature from ~2 kJ mol^{-1} in regions of defined secondary structure to ~8 kJ mol^{-1} in coil regions [46, 47]. Thus, formation of contacts that involve many such side chains is thermodynamically prohibitive. The N- and C-termini of the polypeptide chain, often disordered in solution, may also become trapped at crystal contacts, leading to further decrease in entropy. The same applies to flexible loops, sequestered upon crystallization.

To identify the specific phenomenon driving crystallization thermodynamics, we have to turn to solvent effects. Any high-resolution crystal structure of a protein reveals large numbers of ordered water molecules covering both hydrophobic and

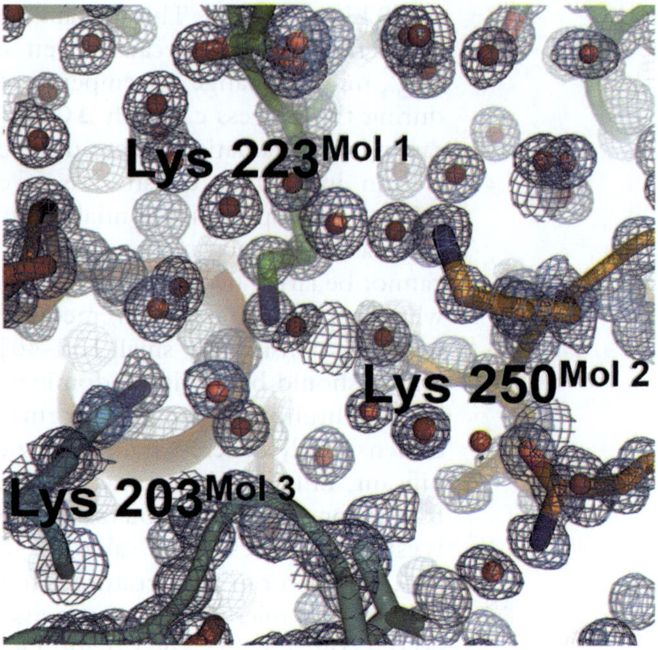

Fig. 3 Lysines sequestered at a crystal contact. Three lysine residues, each from a different molecule, are sequestered at a ternary crystal contact, and surrounded by a network of ordered water molecules, with concomitant loss of entropy. PDB code 1R6J, 0.72 Å resolution structure of the PDZ2 domain of syntenin

hydrophilic solvent-exposed surfaces [48, 49]. While lacking the dynamic aspect, these crystal structures are largely representative of the hydration shell that encapsulates macromolecules in solution and is two to three molecules deep [50, 51]. As the protein molecules become incorporated into the growing crystal, and direct contacts form between them, the structured solvent is released from the surfaces. Based on the entropy gain of transfer of one molecule of water from a clathrate, e.g., methane hydrate, or other ice-like structures into the liquid phase, it has been estimated that release of one water from a protein surface at ambient temperature into the bulk phase leads to an entropy gain of ~6 kJ mol^{-1} [52]. If a sufficient number of water molecules are released into the bulk solvent, the overall entropy gain will compensate the losses ascribed to other phenomena (*see* above) and provide the driving free energy for crystallization [36, 37, 53]. Indeed, the estimated values of $-\mathbf{T}\Delta S^{\circ}_{\text{solvent}}$ (i.e., free energy decrease due to water release) during macromolecular crystallization at ambient temperature range from ~30 kJ mol^{-1} to ~180 kJ mol^{-1}, corresponding to the release of ~5 to 30 water or solvent molecules [36, 37, 54].

It is important to note that the thermodynamic outcome of the crystallization process can only be probed experimentally on a macroscopic scale, as the combined effect of all the molecular interactions and solvent effects. In reality, a single molecule (or independent structural entity) within a crystal may form as few as three or as many as 18 interfaces with satellite molecules [55]. Such interfaces in crystal structures are identifiable solely by distance criteria (e.g., with atoms in the two molecules separated by <4.5 Å), and there is essentially no way to discriminate between cohesive and non-cohesive interactions. However, only three cohesive contacts are typically necessary for the integrity of a three-dimensional crystal, with the exception of space group $P2_12_12_1$, where only two are required [56]. The remaining contacts may have neutral, cohesive, or distinctly repulsive character and may be forced on the ensemble by the intrinsic ruggedness of the molecular shape.

In conclusion, the microscopic aspects of thermodynamics of intermolecular interaction during crystallization are consistent with a model in which the assembly of protein molecules in the nuclei and nascent crystals is orientation-dependent in order to minimize the unfavorable entropy gains stemming from loss of flexibility to fragments of protein structure (exposed side chains, loops, and flexible termini), while maximizing favorable solvent effects. Only select surface “sticky patches,” with a tendency to form cohesive interactions, serve that purpose, enforcing specific orientations.

2.3 Structural Support for the “Sticky Patch” Model

The crystal structures deposited in the PDB offer a wealth of structural data for the analysis of macromolecular packing and the nature of the protein–protein interactions (PPIs). As remarked earlier, the main effort has been to identify biologically “functional” interfaces against the background of what was believed to be random interactions. A number of methods were developed for automated in silico analysis of the interfaces and identification of functional interfaces, including those taking advantage of the evolutionary conservation as defined by Shannon entropy [57–60]. Currently, the most popular method for the analysis of protein–protein interfaces in crystals is the PISA algorithm available as a server (https://www.ebi.ac.uk/msd-srv/prot_int/) [61].

Unfortunately, a strictly binary classification of protein–protein interactions, i.e., functional vs. serendipitous crystal contacts, is overly simplistic. A more recent study utilizing a nonredundant PDB database of strictly monomeric proteins, and a regression analysis methodology, demonstrated that crystal contacts are not random, but are in fact enriched in small and hydrophobic amino acids (e.g., Ala, Val), and depleted in large and polar residues, such as Lys, Glu, and Gln (but notably not Arg) in a manner similar to functional PPIs (Fig. 4) [62]. This is an important observation, even though the dataset of interfaces that were subject to analysis by necessity had to include all contacts identified by distance criteria, regardless of whether they are thermodynamically cohesive or

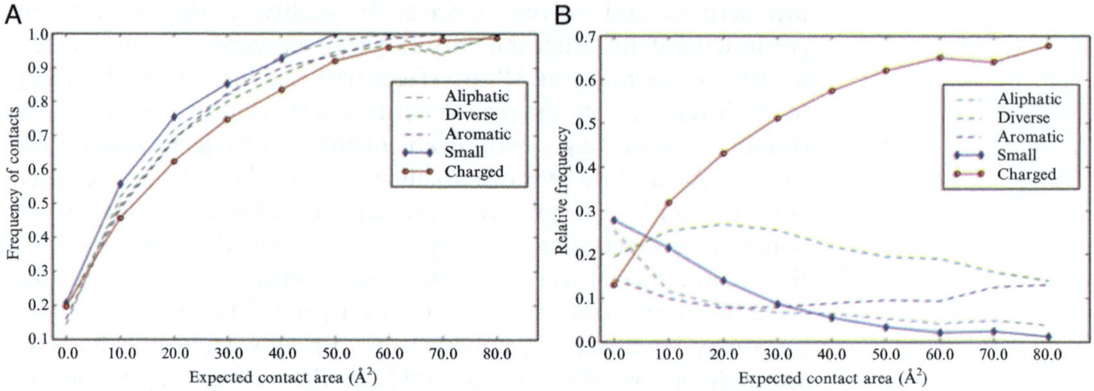

Fig. 4 Nonrandom composition of crystal contact surfaces. (**a**) Relative frequencies of five categories of amino acids, i.e., aliphatic (Val, Leu, Ile), aromatic (Trp, Phe, Tyr, His), small (Ala, Gly, Ser, Thr, Cys), charged (Lys, Arg, Glu, Asp) and other (Asn, Gln, Met, Pro), binned as a function of rECA. The relative frequency in each bin is the ratio of the number of residues of a given type to the total number of residues. (**b**) The fraction of residues involved in crystal contacts as a function of rECA plotted for the five categories as defined above. rECA is the residue expected contact area. For details *see* Cieslik and Derewenda (2009) The role of entropy and polarity in intermolecular contacts in protein crystals. Acta Cryst D 65

not. It is almost certain that if it were possible to computationally identify attractive interactions only, their amino acid content would be even more distinct. These observations are in full agreement with the simulations mentioned earlier that assessed the impact of weak, nonspecific interactions on nucleation and crystal growth [32].

Another important observation comes from the analysis and comparisons of interfaces across different crystal forms of the same, or homologous, proteins. As already noted in Chapter 2, macromolecules show considerable polymorphism and often the same protein crystallizes in various forms, sometimes from the same solution conditions. It has been recently reported that the portion of the PDB entries with at least two crystal forms is 64% [63]. Although reproducibility of crystal contacts in different crystal forms of multimeric proteins is normally taken as evidence of physiological homo-oligomerization, such functionality is often not known in vivo. Further, even taking conservatively annotated monomeric proteins into consideration, a third shows shared interfaces in different crystal forms. A striking example is the homodimeric association of the globular domain of RhoGDI (discussed in more detail in Subheading 4.2) which is reproduced across multiple crystal forms obtained using dramatically different crystallization conditions (Fig. 5). All these observations strongly support the "sticky patch model," and show that protein surfaces are decorated with distinct patches mediating specific interactions which under favorable conditions allow the formation of crystal contacts. The interfaces that mediate such contacts are not explicitly distinct

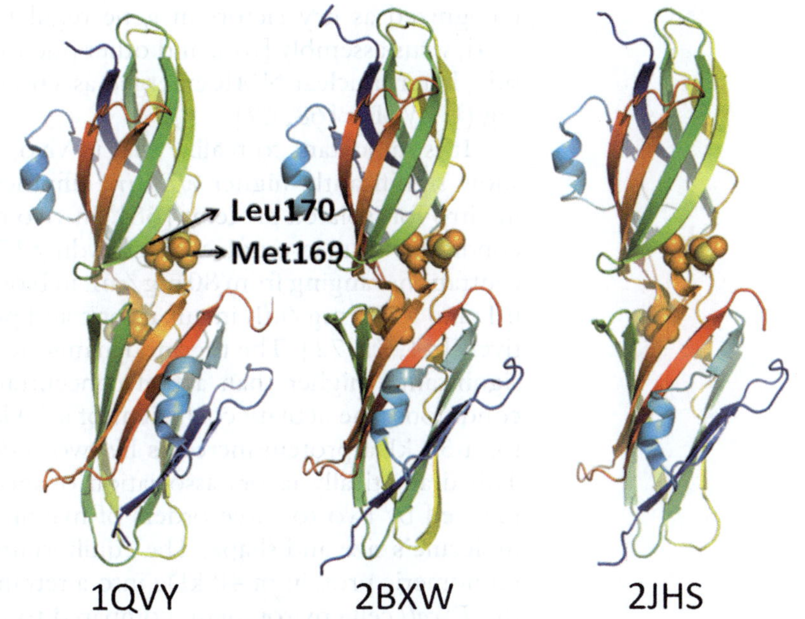

Fig. 5 Reproducibility of a homodimeric crystal contact in RhoGDI, independent of crystal form and conditions. 1QVY is a mutant containing a non-crystallographic dimer, crystallized from sodium formate and $(NH_4)_2SO_4$; 2BXW shows a crystallographic dimer, obtained from sodium citrate, with propanol and PEG; 2JHS shows a non-crystallographic dimer in crystals obtained from $(NH_4)_2SO_4$ and sodium citrate

from the so-called "functional interfaces" but the two are simply examples of opposite ends of a continuum of interactions, all of which have potential functional significance, even though in most cases we have not yet linked a particular interaction to a physiological phenomenon.

2.4 The "Sticky Patch" Model and Ultra-Weak Protein–Protein Interactions (UWPPIs)

The diversity of protein–protein interactions (PPIs) is well illustrated by the differences in their amino acid composition and size, ranging from surfaces burying in excess of 2,000 Å^2, with distinctive hydrophobic core, to small patches limited to a few amino acids of diverse nature. As a consequence, the PPIs span a huge range of affinities, from the tightest interactions with KD in the <pM range, to weak and ultra-weak interactions (WPPIs and UWPPIs) with KD > 1 μM and even >100 μM, respectively. Historically, tight and obligate interactions have always been under intense scrutiny, but WPPIs and UWPPIS were only recently appreciated as biologically important [64]. This is in part because transient and weak complexes are often very difficult to identify, isolate and evaluate by methods such as tandem affinity purification (TAP), surface plasmon resonance (SPR), or isothermal titration calorimetry (ITC). Nevertheless, (U)WPPIs are increasingly recognized as key factors in gene regulation [65], cell adhesion [66], virus assembly [67], and other phenomena. On the technical side, heteronuclear NMR emerged as a powerful method for probing (U)WPPIs [68, 69].

It is important to realize that in vivo, (U)WPPIs are likely to show significantly higher *effective* affinities than those measured in vitro for isolated proteins [68]. In vivo proteins function under conditions of macromolecular crowding [70, 71], with total concentrations ranging from 80 mg/mL in blood plasma, to ~200 mg/mL and ~400 mg/mL in eukaryotic and prokaryotic cells, respectively [44, 71, 72]. The thermodynamic activities are consequently significantly higher than actual concentrations. Under crowding conditions, the activity coefficient of a 30 kDa protein triples, and for a 50 kDa protein increases by two orders of magnitude [72]. This dramatically favors association of molecules, with KD often lowered by two to three orders of magnitude, depending on the molecule's size and shape. The equilibrium of the association of a monomeric protein of 40 kDa into a tetramer shifts by 10^3–10^5 in the *E. coli* cell environment, compared to isolated protein [72]. It is now well established that macromolecular crowding modulates (U)WPPIs in a biologically relevant manner. For example, when ubiquitin is monitored by NMR in the *E. coli* cell, it tumbles so slowly that no detectable HSQC spectra can be recorded [68]. This is due to transient, targeted (U)WPPIs interactions with large proteins or complexes, which under crowding conditions have much higher affinities than those determined in vitro. A number of computational studies aimed at characterization of (U)WPPIs revealed several differences between weak, transient interactions and tight, obligate associations. Importantly, it has been shown that total accessible surface area (ASA) and polarity of the relevant surface patches constitute critical parameters [73, 74].

Crystal contacts constitute an unexplored wealth of information regarding protein surfaces that may engage in (U)WPPIs of functional importance. For example, the 1.8 Å resolution structure of human erythrocyte ubiquitin [75] showed a crystal contact involving Leu8, Ile44, and Val70 (Fig. 6). This contact buries a modest 386 Å^2 of surface, and is only the second largest. Its biological function was recognized only significantly later [76]. Similarly, the original crystal structure of the protein tyrosine phosphatase [77] revealed a crystal contact mediated by Tyr130 and Tyr131, to which no functional significance was initially attributed. Much later, NMR titration experiments showed it to be an ultraweak (KD ~ 1.5 mM) interaction, and functional studies revealed its significance [78].

To conclude, many of the "sticky patches" mediating contacts in protein crystals may have hitherto unknown functional significance. In a more general sense, all cohesive crystal contacts represent sites where the protein may potentially interact with other

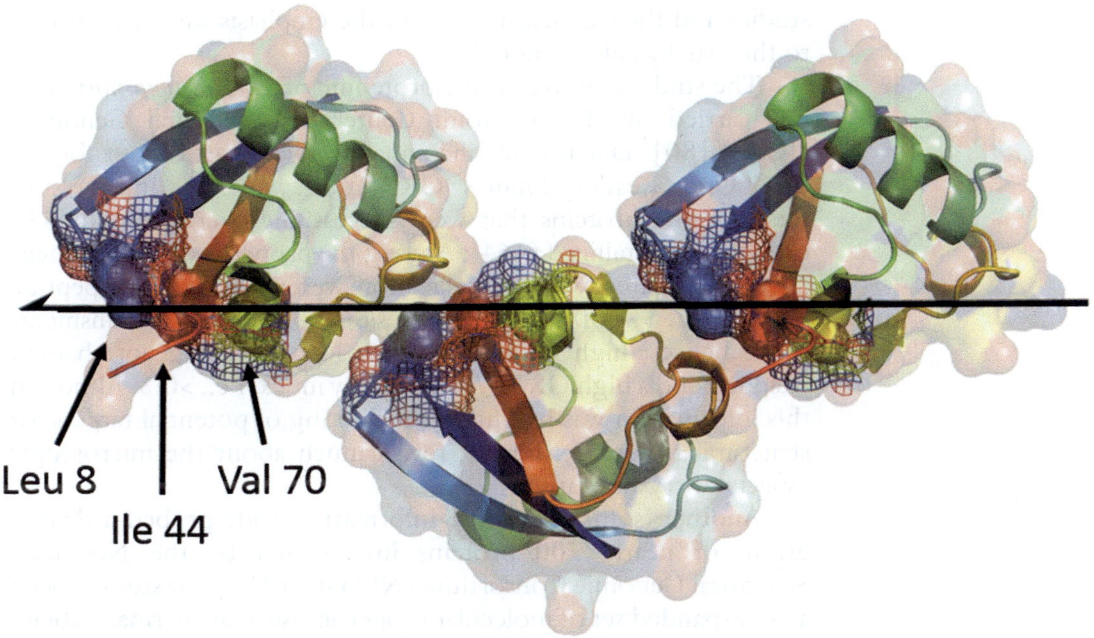

Fig. 6 A minor crystal contact in ubiquitin, mediated by a now recognized biologically active surface. The three functional amino acids are Leu8, Ile44, and Val70. The contact making surface is highlighted as a mesh. PDB code 1UBQ

proteins, especially under molecular crowding conditions. It seems that rather than attempting a binary classification of PPIs, it is safer to see these interactions as forming a continuum, from ultra-weak to high-affinity, all playing some role in protein's physiology.

2.5 Sequence-Derived Properties and Crystallization

The physicochemical properties of protein surfaces are defined singularly by the solvent-exposed amino acids, and therefore by the amino acid composition and sequence. It is therefore quite reasonable to assume that sequence based properties of proteins should be correlated with the presence and type of attractive patches, and therefore with the crystallization outcome. In other words, if "sticky patches" constitute an integral feature of an easily crystallizable protein, one should be able to detect their fingerprint using sequence analysis. Indeed, extensive in silico data mining studies have been recently made possible by the databases accumulated by HT Structural Genomics Centers. Unlike the worldwide Biomolecular Crystallization Database [79], and to some extent the PDB, these new databases contain information about both positive and negative outcomes of millions of crystallization experiments, making it possible to probe the issue of what sequence-dependent biophysical properties correlate with the binary outcome of crystallization using regression analysis and other mathematical methods. Here we briefly discuss three specific studies and their conclusions, with the emphasis on the relevance to the "sticky patch" model.

The study focusing on the proteome of *Thermatoga maritima* was carried out by the Joint Center for Structural Genomics (JCSG) [80]. Detailed in silico analyses were carried out for the 1877 Open Reading Frames (ORFs) of *T. maritima* and the subsets of those proteins that were successfully overexpressed (539 cases) and crystallized (464). Among the properties that were negatively correlated with crystallization were: excessive polypeptide length, i.e., >~600 amino acid residues; presence of transmembrane helices; high isoelectric point; low percentage of charged residues; and a high GRAVY hydropathy index, i.e., >0.3. Although this information was helpful in the filtering of potential targets for structural studies, it did not reveal much about the microscopic aspects of crystallization.

Another, somewhat more informative study probed a diverse group of nearly 700 proteins investigated by the Northeast Structural Genomic Consortium (NESGC) [81]. This study looked at an expanded set of molecular properties, such as thermal stability and oligomerization, but also at the frequency of each amino acid, mean hydrophobicity, mean side chain entropy, total and net electrostatic charge, pI, the fraction of residues predicted to be disordered, and chain length. The sequence-derived parameters were analyzed using logistic regression to evaluate the impact of a continuous variable on the binary outcome of crystallization screens. Hydropathy (GRAVY index) and side-chain entropy exhibited strong negative correlation with crystallization success rates. Interestingly, it was also discovered that high frequencies of Lys and Glu amino acids, correlated negatively with crystallization outcome (Fig. 7). These conclusions are in agreement with the thermodynamic argument that preponderance of high-conformational entropy side chains on the surface reduces the chances for suitable "sticky patches" that can mediate crystal contacts.

A third study looked in detail at the behavior of 182 proteins (also from the NESGC) which were each subject to extensive crystallization screen using 1536 conditions developed by the High-Throughput Crystallization Screening Laboratory of the Hauptman Woodward Medical Research Institute [82]. Statistical models were trained on this sample capturing trends driving crystallization. Once again, low level of side chain entropy was found to be correlated with positive crystallization outcomes. In addition, a new correlation was found between crystallization and complementary electrostatic interactions. The study concluded that crystal contacts have "specific physicochemical signature even if they are not biologically functional" [82].

Taken together, these analyses of sequence-derived properties are consistent with the "sticky patch" model, identifying side chain entropy, hydrophobicity, and electrostatics as surface properties

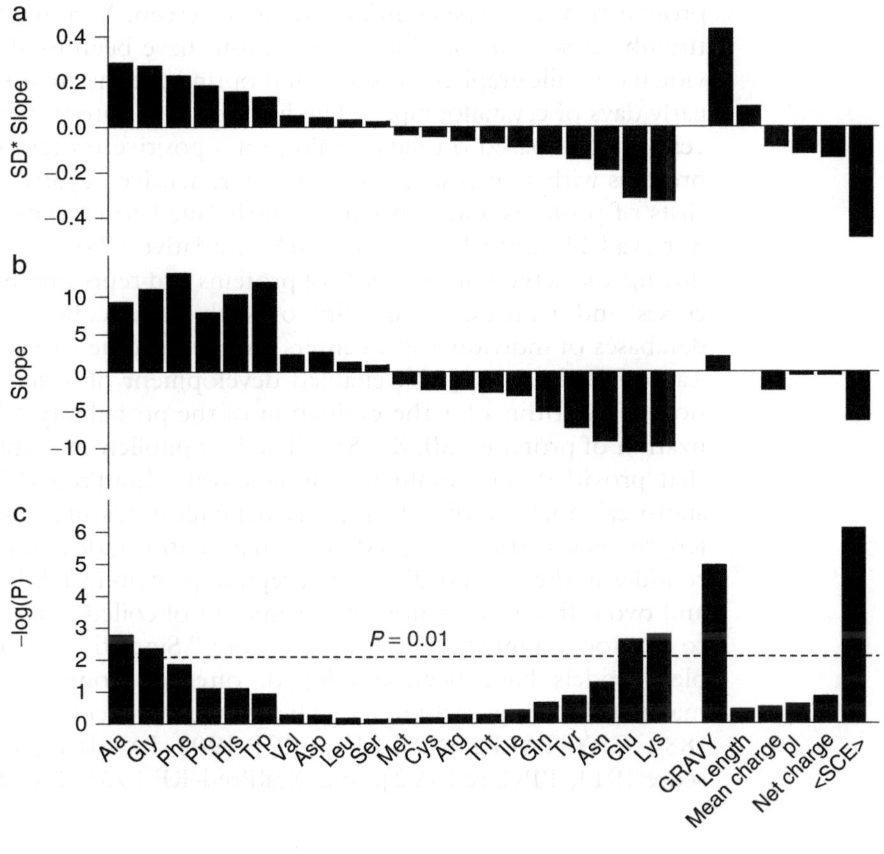

Fig. 7 Logistic regressions based on success in crystal structure determination (that is, PDB deposition) performed on a dataset of 679 proteins from the NESG protein expression and crystallization pipeline. Variables evaluated included the fractional content of each amino acid, mean residue hydrophobicity (GRAVY), chain length, mean charge (fraction including Arg, Lys, Asn, and Glu), pI, mean net charge and mean side-chain entropy (⟨SCE⟩). (**a**) Predictive value of each parameter, which is defined as the product of its logistic regression slope and the s.d. of its distribution in the dataset. (**b**) Logistic regression slope. (**c**) Negative log of logistic regression *P*-value. (From Price et al. (2009) Understanding the physical properties that control protein crystallization by analysis of large-scale experimental data. Nat Biotech. 27:51–57)

which render specific surface patches as particularly conducive to mediation of specific intermolecular interactions.

3 Prediction of Protein Propensity for Crystallization

Given the existence of detectable correlations between sequence-derived protein physicochemical properties and the propensity to crystallize, it should be possible to design predictive algorithms that evaluate in more rigorous ways the probability of a given protein to yield crystals in an extensive screen. Various "rules of thumb" based on anecdotal observations have been used by individual crystallographers to select and optimize constructs since the early days of crystallography. The first effort to automate this process [8] was based on data mining on a positive protein set (i.e., proteins with solved structures). Comprehensive negative datasets (lists of proteins and constructs which failed to crystallize) were not available until Protein Structure Initiative (PSI) started producing and screening large sets of proteins and reporting both successes and failures. Availability of such data, collected in the databases of individual PSI centers and, with some caveats, in the TargetDB database [83], enabled development of a first generation of algorithms for the evaluation of the probability of crystallization of proteins [80, 84–86]. The first publicly available server that provided such evaluation interactively, XtalPred [87], used statistical analysis of seven physicochemical features: sequence length, isoelectric point, GRAVY hydropathy index, number of residues in the longest disordered region, protein instability index, and two different measures of the amount of coiled-coil structure, to develop a single "crystallizability score." Since then, more complex models have been developed, often in conjunction with machine learning techniques. These models, including ParCrys [88], CRYSTALP2 [89], MetaPPCP [90], PXS [81], MCSG-Z score [91], PPCpred [92], and XtalPred-RF [93] (Fig. 8), have

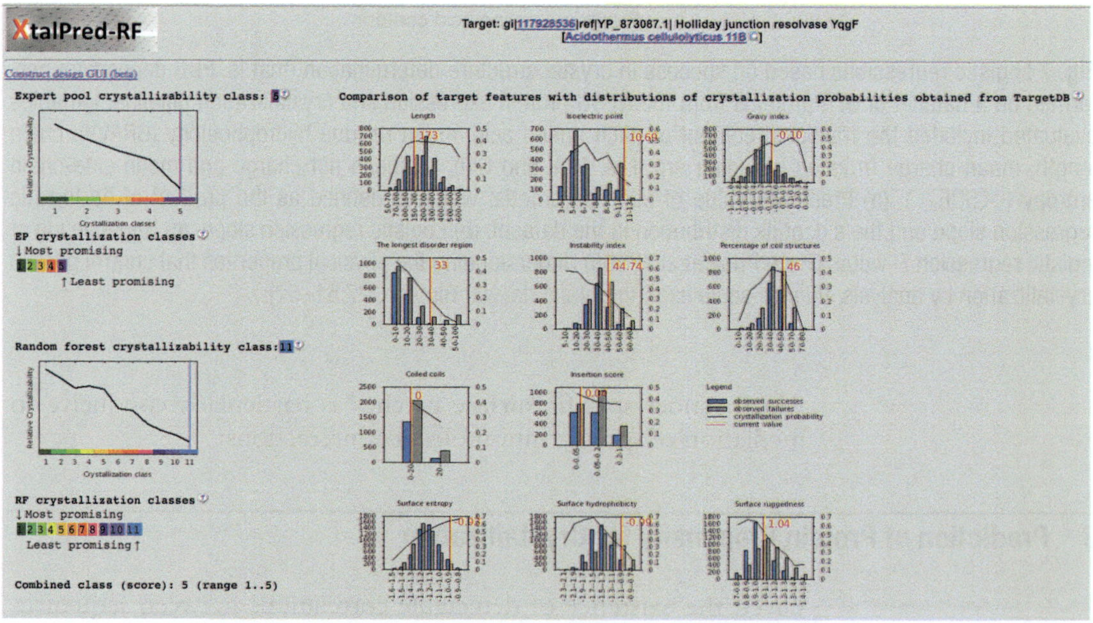

Fig. 8 Typical output from XTALPRED. The program analyzes various biophysical parameters and displays these values against statistical data for other proteins and correlation with crystallizability. Random forest scoring puts the protein into 14 categories from the easiest to most difficult for crystallization

allowed users to assess the probability of successful structure determination prior to any experimental work and to adjust their target selection strategies [94]. Such algorithms are most useful in the context of high-throughput structure determination, such as done in structural genomics centers [95–98], where typically protein families, but not specific proteins were targeted for structure determination. In such applications, several (typically 5–10) most promising candidates from a protein family, were selected for structure determination and successful structure determination of any of them was considered a success. Even modest enrichment in the number of crystallizable proteins in the target pool as compared to random selection improved overall production in the structural genomics centers and allowed them to solve thousands of protein structures, including hundreds of first representatives of novel protein fold families.

Individual structural biology groups, which often target specific, high value targets, still used such approaches [99–101] but often found them lacking the resolution needed to distinguish changes to crystallization propensities made by small changes in construct boundaries or individual mutations. New generation of algorithms, now in development, is aiming at the first task [8], while the second clearly remains out of reach of statistics-based methods.

Failure of methods based on average physicochemical features of the protein to provide more decisive improvements in selecting or designing optimal constructs for structure determination is easy to explain in the context of the "sticky patch model" of crystallization. While average values of hydrophobicity or instability can effectively predict protein solubility and recognize some features detrimental to crystallization (such as long disorder segments), they do not see individual crystal interfaces, thus methods targeting individual "sticky patches" are needed to improve the statistical models.

4 Target Protein Modification for Enhanced Crystallizability

The "sticky patch" model of crystallization opens a new, exciting possibility for direct enhancement of success rates in crystallization screens by modifying the surfaces of the target protein or complex. Briefly, if crystallization is facilitated by the existence of specific "sticky patches" on the surface of the target molecule, then it should be possible to engineer such patches by chemical modification or site-directed mutagenesis. The key question is what modifications can be effectively used, and what should they target.

Recombinant methods and protein chemistry offer a plethora of avenues for protein modification, and comprehensive discussion of all these methods is beyond the scope of this chapter (NB: some of these methods may be designed to overcome other potential bottlenecks, such as intermolecular disulfide bridges, low solubility, etc. [102, 103]). Here we are primarily interested in methods that modify very specific patches to achieve the potential for cohesive interactions, driving the formation of crystal contacts. In the most general terms there are six such strategies: (1) optimization of the recombinant construct to remove high-entropy N- and C-termini and loops; (2) the use of proteases to remove unstructured regions; (3) mutational modification of protein surface to reduce excess conformational entropy (the Surface Entropy Reduction protocol); (4) chemical modification of the protein surface by targeting specific amino acids; (5) the use of small molecule additives that specifically bind in surface crevices and modify local surface properties; (6) the use of protein chaperones which may stabilize the target protein or complex and provide additional surfaces with "sticky patches" assisting in the crystallization of the target. We briefly address each of these strategies, and refer readers to a number of extensive reviews.

4.1 Construct Optimization and Proteolytic Digestion

As is evident from our earlier discussion, the presence of disordered regions in crystallization targets, i.e., N- and C-termini and large flexible loops, is very unfavorable. This is a very important point because most target proteins under study are fragments, e.g., signaling or catalytic domains, and the correct choice of N- and C-terminal boundaries is of paramount importance. Historically, isolated stable domains were obtained using limited proteolysis and subsequent purification of the smallest functional domain. The contemporary approach is in situ proteolysis, i.e., addition of small amounts of select proteases to the crystallization mixture, so that the enzyme acts on the protein under crystallization conditions, and the proteolyzed fragment is allowed to form crystals in the same drop [104–106]. Another strategy is in silico analysis, using tools such a XtalPred [93] or DisMeta [107], to identify the boundaries of the folded stable fragment, and to clone the target fragment accordingly [108–111]. The functional core units can also be identified experimentally, following limited proteolysis, by mass spectrometry [112]. Finally, experimental methods can be used to identify the disordered regions directly, such as deuterium–hydrogen exchange coupled to mass spectrometry (DXMS) [113–115], or NMR [116]. Unfortunately, many target variants may have to be screened to identify one amenable to crystallization, because in some cases short disordered fragments may even help [117]. For example, in the case of the MAPKAP kinase 2, 16 truncation variants were assayed, all containing the catalytic domain, and shown to have dramatically different solubility and propensity for crystallization [108].

A complementary approach is to remove disordered loops, if such can be identified by other means and are believed to interfere with crystallization. For example, the variant used in the successful crystallization of the HIV gp120 envelope glycoprotein had two flexible loops which were replaced with Gly-Ala-Gly linkages to obtain a crystallizable variant [118, 119]. In the case of 8R–lipoxygenase, the replacement of a flexible Ca^{2+}-dependent membrane insertion loop consisting of five amino acids, with a Gly-Ser dipeptide resulted in crystals that diffracted to a resolution 1 Å higher than the wild-type protein [120]. An interesting variation of this approach was introduced for the preparation of crystals of the β-subunit of the signal recognition particle receptor. A 26 residue-long flexible loop was removed, but instead of replacing it with a shorter sequence, the authors connected the native N- and C-termini of the protein using a heptapeptide GGGSGGG, thus creating a circular permutation of the polypeptide chain [121].

In summary, removal of flexible fragments in the crystallization targets reduces the possibility of a prohibitive loss of entropy during crystal formation as the unstructured regions become sequestered in the crystal contacts, and exposes the cohesive patches which can mediate thermodynamically favorable crystal contacts.

4.2 Surface Entropy Reduction—Engineering "Sticky Patches"

The Surface Entropy Reduction (SER) strategy uses site-directed mutagenesis to generate protein variants with surface cohesive patches (sticky patches) designed to increase the propensity for crystallization. The concept was initially based on a broadly formulated hypothesis, consistent with the microscopic interpretation of entropy contribution to crystallization, that solvent exposed amino acids with long, flexible side chains (e.g., Lys or Glu) impede crystallization because high conformational entropy would be lost as the amino acid is sequestered in a crystal contact [122]. It was therefore suggested that surface patches enriched in these amino acids are very unlikely to mediate protein interactions, and consequently crystal contacts. Conversely, variants in which Lys and/or Glu within such patches were mutated to small residues such as Ala, should have increased probability of being involved in interactions that could consequently mediate crystal contacts. This is essentially a direct way of engineering "sticky patches" to enhance a protein's crystallizability. The hypothesis was first tested using a model system of the globular domain of the human signaling protein RhoGDI, which is unusually rich in Lys and Glu, and is recalcitrant to crystallization in its wild-type form [122–124]. The mutated variants containing single or multiple Lys → Ala or Glu → Ala mutations within a single patch (identified by close sequence proximity) have indeed shown much higher success rate in routine crystallization screens [122–124]. Importantly, the crystal structures of these variant revealed that the mutated patches directly mediate select crystal contacts, corroborating the underlying hypothesis and the "sticky patch" model. One of the interesting observations was that multiple mutations within a single patch were noticeably more effective than single mutants.

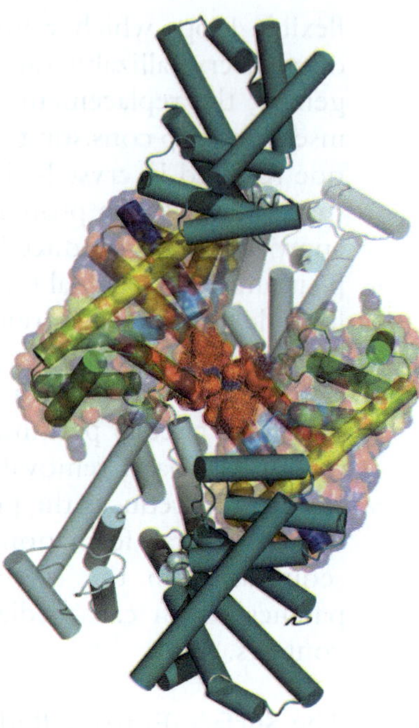

Fig. 9 A homotypic contact in the crystals of the RGS-like domain mediated by an engineered patch created by mutations K463A, E465A and E466A (*spheres*). The contact making surface is highlighted as a *mesh*. One dimer is shown in color, with full surface. The dimers below and above, arrange along a sixfold screw axis, are shown in *cyan* and *green*. PDB code 1HTJ

The SER strategy was successfully applied to a number of new proteins that were recalcitrant to crystallization in their wild-type form. The first new structure to be solved was the RGSL domain from the signaling protein PDZRhoGEF [125, 126]: a triple mutant in which two Lys and one Glu were mutated to alanine yielded good quality crystals (Fig. 9). Other successful applications quickly followed, and a number of high-profile structures were solved. Among them were: EscJ, a component of the type III secretion system, the structure of which helps to understand key aspects of virulence in gram-negative pathogens [127]; ALIX/AIP, programmed cell death 6-interacting protein, key to the understanding of mechanisms involved in retrovirus budding and endosomal protein sorting [128]; an HIV-capsid component, which helps understand the maturation of HIV [129]; a complex of the K^+ gated channel, KChIP1 with the Kv channel interacting protein (Kv4.3 T1) [130]; and the crystal structure of the BetP Na^+/betaine symporter [131]. In virtually all cases, the strategy was to target clusters of 2–3 Lys or Glu (or both) residues that were consecutive in sequence, and change them to Ala.

The basic premise of the SER strategy is strongly corroborated by the aforementioned data-mining studies showing that preponderance of high-entropy amino acids in protein sequence is negatively correlated with crystallization success [81, 82]. In an effort to better understand what mutational strategy is optimal, a more comprehensive study was carried out using the same model system of RhoGDI, expanding the target amino acids to Lys, Glu, and Gln (which has the same conformational entropy as Glu, but no charge) and replacing them with Ser, Thr and His [132]. An additional approach was tested in which Tyr was used as another alternative amino acid to replace the high-entropy residues. The rationale there was that tyrosines occur with relatively high frequency at protein–protein interfaces [133] and are known to play a crucial role at antibody–antigen recognition sites [134]. Tyrosine side chain has only two degrees of conformational freedom, compared to four in Lys and three in Glu, so that the entropy loss upon sequestration at an interface is lower, but also has a bulky hydrophobic moiety as well as a hydroxyl group capable of forming directional H-bonds. Interestingly, tyrosine was most successful target for substitution, in terms of the success rate of crystallization in a standard screen [132]. As was the case with other SER variants, those containing Tyr also crystallized with the engineered patch mediating crystal contact. However, it has also been observed that Tyr variants, especially those with two or three of these residues adjacent in sequence, display significantly lower expression yields, limiting the applicability of the method. Tyrosine remains an uncommon choice for the replacement of Lys, Gln, and Glu in the SER approach.

Currently, the choice of which residues in the target protein should be mutated is made easier by a dedicated server (SERp,

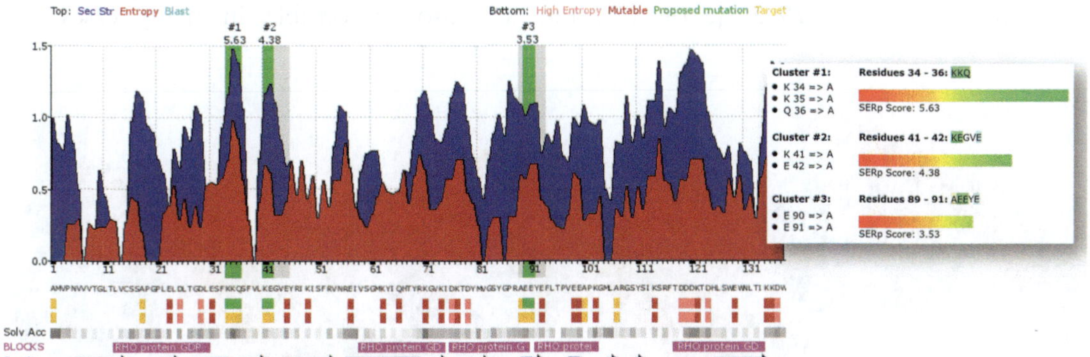

Fig. 10 Typical output from the SERp server. The program calculates a range of parameters, most important of which is the sliding window of side chain entropy and secondary structure prediction. The program identifies loops likely to be solvent exposed and suggests mutations in those loops where excess conformational entropy is likely to prevent specific contacts that may facilitate crystallization. Several variants, each with 2–3 mutations are suggested and scored. The sequence shown is that of RhoGDI; the inset shows a Table appearing elsewhere in the output scoring the suggested variants. All of the variants suggested here by the server are known to crystallize; the third variant yields crystals diffracting to 1.25 Å resolution

http://services.mbi.ucla.edu/SER/) with a predictive algorithm able to identify suitable surface sites for mutagenesis, based on amino acid sequence information only [135, 136] (Fig. 10). The server seeks to identify solvent-exposed loops with patches populated by high-entropy residues (Lys, Glu, and Gln). The investigator then makes the choice about what type of amino acid is to replace the wild-type residues (typically this is Ala).

The SERp server has been used in a multitude of studies to design crystallizable variants of many proteins and macromolecular complexes. There are currently over 150 nonredundant entries, with a total of over 450 depositions in the PDB based on SER strategy. This database allows for an interesting insight about how the engineered SER patches affect molecular packing and consequently crystallization at the microscopic level. A preliminary unpublished survey reveals that in over 90% of cases mutated patches are directly involved in crystal contacts. This is an irrefutable validation of the original notion that mutations of high-entropy residues create "sticky patches" with enhanced propensity to mediate protein–protein interactions and crystal contacts. It is also interesting to note that the SER "sticky patches" generate crystal contacts with unique topological patters. Most of them mediate homotypic contacts, i.e., interactions between two identical engineered patches in neighboring molecules (Fig. 9). This specific interaction gives rise to twofold symmetry, either crystallographic or non-crystallographic, in which case a dimer occupies the asymmetric unit. A minority of SER patches forms heterotypic contacts, in which the mutated patch interacts with a completely different, wild-type patch on the surface of a neighboring molecule. These contacts are associated with crystallographic screw axes, typically in such space groups as $P2_1$, $P2_12_12_1$, or $C2$, but are also responsible in many cases for

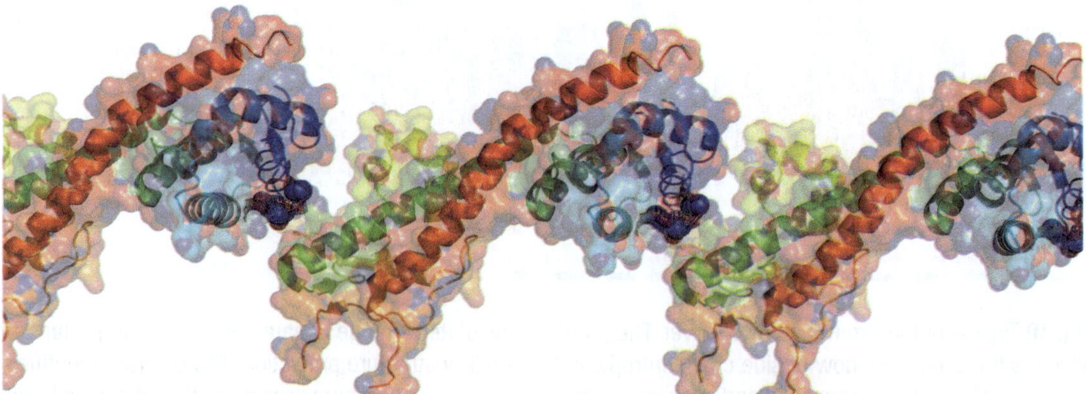

Fig. 11 A heterotypic contact in the crystals of the *Yersinia pestis* V-antigen mediated on one molecule by a patch containing the mutations K40A, D41A, K42A (*deep blue*; shown as *spheres*). The contact making surface is highlighted as a *mesh*. PDB code 1R6F

translational contacts, especially in the rather rare *P*1 space group (Fig. 11).

Although precise calculation of the **ΔG** free energy change based on structure is notoriously difficult, it is interesting to note that crystal contacts generated by SER appear to be generally thermodynamically cohesive, based on calculations by PISA [61]. This suggest that the SER contacts constitute in fact one of the minimum three cohesive interactions actually responsible for the integrity of the lattice. It is also possible that SER-mediated interactions, particularly the homotypic ones that lead to dimerization, exist in solution prior to nucleation and crystallization.

The vast majority of the successful applications of the SER strategy was limited to engineering a single patch. In a typical case, the SER crystal contact either generates a crystallographic (or non-crystallographic) oligomer (i.e., primary contact), or mediates interactions between oligomers (secondary contacts). We note that often primary contacts have significantly larger buried interfaces than secondary contacts.

The SER strategy offers also another advantage; it can be used to generate novel crystal forms with superior diffraction qualities, in those cases where wild-type protein yields poorly diffracting crystal forms. In a majority of crystallographic studies, one is typically satisfied if screens yield crystals that allow structure determination to ~2.1–2.5 Å resolution. Effort is typically invested in the optimization of crystals obtained in a screen, rather than in searching for other crystal forms. However, as has been often demonstrated, the quality of diffraction is dependent on a particular crystal form, rather than being correlated to a specific protein. Thus, if a wild-type protein crystallizes, other variants generated by the SER approach are very likely to yield novel crystal forms, often with better diffraction quality and higher resolution. This has been demonstrated early on during the studies of RhoGDI, which in its wild-type form never yields crystals diffracting to better than ~2.8 Å resolution. In contrast, the E154A, E155A double mutant resulted in crystals which allowed for collection of data and refinement of the structure to 1.3 Å resolution [123]. One of the additional advantages of having multiple crystal forms is that the packing of molecules may be quite different, with certain specific surfaces, such as active sites, open to solvent in some forms, while occluded in other forms. The availability of different forms allows choosing one that is best suited for the particular functional experiments.

The combination of the potential advantages associated with using SER as a method to produce alternative crystal forms of target proteins has made this technique very useful and popular in drug discovery. It has been incorporated into the arsenal of tools used in the pharmaceutical industry. One of the first published successful examples of this approach was the improvement in the quality of crystal of the intracellular kinase domain of the insulin-like growth factor, a potential drug target [137]; the resolution was increased from 2.7 Å to 1.5 Å. The HIV-1 reverse transcriptase was successfully engineered to yield crystals diffracting to 1.8 Å; there are 39 PDB entries for this protein as of May 2016 [138–142]. A SER variant of the β-site amyloid precursor protein cleaving enzyme (BACE-1), a target in Alzheimer's disease, was extensively used in drug-discovery research [143–152], with 29 PDB depositions. A drug discovery effort targeting *Trypanosoma brucei*, the causal agent of sleeping sickness, was made possible by SER variants of the methionyl-tRNA synthetase [153, 154], generating 29 coordinate sets of complexes. An effort to design inhibitors of the nitric oxide synthase oxygenase, serving as antibiotics against gram-positive pathogens, utilized a variant of the enzyme predicted by SERp [155–158], leading to 46 PDB entries. Finally, the epidermal growth factor receptor kinase domain, a target for non-small cell lung cancer, has been successfully engineered to yield 21 PDB depositions of complexes with drug leads [159–161].

4.3 Reductive Methylation

Although there is a whole range of protocols for chemical modification of proteins [162], only reductive methylation has become routine to enhance crystallization success rates. It is indeed the only approach that is technically facile, quick, and produces homogeneously modified samples. Also, reductive methylation selectively modifies lysines, which—as discussed above—disfavor specific interactions and formation of crystal contacts.

The method was initially introduced in the study of myosin subfragment-1, which proved to be rather challenging [163]. Detailed protocols were published by Rayment [164] and Tan et al. [165]. In this approach only the free amino groups (ε-amino groups of lysines, and the N-terminal amino group) are modified. Formaldehyde is the methylating agent with dimethylamine-trifluoroborane acting as the reducing agent. The most common outcome is dimethylation (N,N-dimethyllysine; dmLys), as the monomethylated derivative is more susceptible to modification than the non-methylated amine. As is invariably the case in crystallization screens, assessment of success rates is far from trivial. Nevertheless, HT Structural Genomics centers reported 10–30% success rates with selections of proteins recalcitrant to crystallization in unmodified form [165–167]. Why does this strategy enhance crystallization? Although methylation results in a slight increase in conformational entropy of the Lys side chain, it also increases the size of the hydration shell of the side chain, ordering a number of water molecules [168]. Upon packing at a crystal contact, the site containing dmLys is therefore likely to release more solvent molecules with favorable change in entropy. In addition, the methyl groups bound to Nε are polarized, and therefore capable of participating as donors in C–H...O hydrogen bonds [168]. A careful recent study evaluated 40 protein structures solved using crystals made from methylated samples and compared them

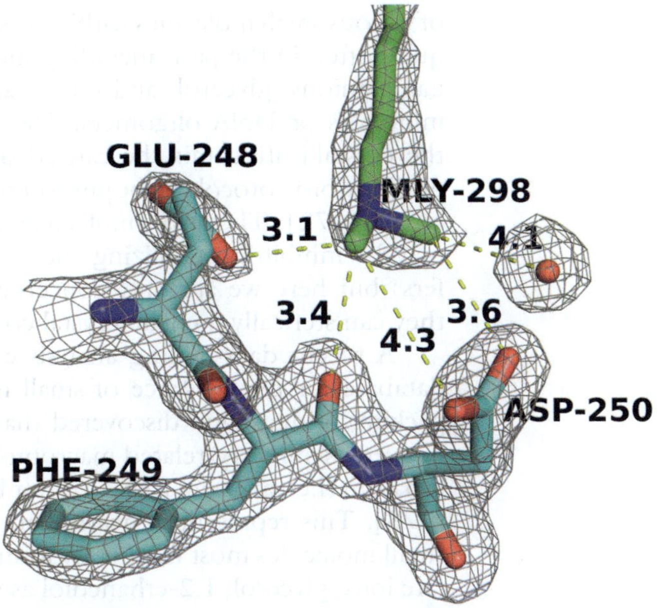

Fig. 12 A sticky patch mediated by a di-methylated lysine. A crystal contact in the *E. coli* RNA polymerase C-terminal domain (3k4g) mediated by a modified lysine. Distances are shown in Å units

with a nonredundant database of 18,972 non-methylated protein structures For 10 proteins both wild-type and methylated structures were scrutinized [169]. The results revealed that dmLys is more likely to form interactions with Glu across crystal contacts than unmodified Lys, and that this is correlated with the C–H…O directional bonds mediated by the methyl groups. In the specific case of a ParB-like nuclease, it has been shown that the methylation protocol resulted in a crystal form stabilized by intermolecular contacts that involve 44 C–H…O interactions mediated by nine dmLys residues [170]. Another effect associated with lysine methylation that may impact crystallization is a slight reduction in the isoelectric point (pI) [171] (Fig. 12).

It should be noted that methylation constitutes just one variation of reductive alkylation, which may involve introducing larger groups, such as ethyl and isopropyl [165]. However, examples of the latter are rare and have not been reported to be very successful.

To conclude, reductive methylation targets lysines on a protein's surface, and modifies them in a manner that increases the probability of these residues to form cohesive intermolecular interactions as part of "sticky patches."

4.4 The Use of Non-covalently Bound Small Molecules—"Sticky Bridges"

An alternative to the use of covalent modification is the use of small molecular weight organic or inorganic compounds which bind specifically—although typically with low affinity—in crevices of the target protein's surface, and provide modified surface patches mediating crystal contacts, i.e., "packing bridges." Sequestration of various small molecules within crystal contacts has been observed quite often in the past, including metal cations, organic and inorganic anions, glycerol, and much larger moieties such as organic inhibitors or DNA oligomers. These molecules could be part of the crystallization mix, be carried over accidentally from protein purification protocols, or be purposefully added to the crystallization screen [172]. The small molecules may be multifunctional (e.g., may be inhibitors stabilizing the enzyme or components of buffers) but here we are only concerned with the manner in which they can sterically mediate crystal contacts.

A recent data mining analysis explored a subset of the PDB database for the presence of small molecules and ions serving as packing bridges, and discovered that about 11.5% of interactions between symmetry-related macromolecules are mediated by a heteroatom (i.e., an atom that does not belong to the macromolecule) [173]. This represented nearly half (45%) of the structures. The small molecules most frequently found within the bridges were sulfate ions, glycerol, 1,2-ethanediol as well as acetate, phosphate and chloride ions and calcium ions [173]. A systematic study of the impact of diverse small molecules (other than the usual buffers and additives) on crystallization, screened 200 compounds with respect to their potential to serve as crystallization "catalysts" for 81 diverse proteins, using only two fundamental crystallization conditions [172]. Nearly 85% of the proteins crystallized, often in new crystal forms, although they were not subjected to systematic structural investigations that might reveal specific interactions mediated by the additive. However, subsequent applications of this strategy revealed explicit examples of additives promoting crystallization by acting as bridges across crystal contacts. For example, cobalamine added to the crystallization mix was found to mediate contacts between oligomers of Δ1-pyrroline-5-carboxylate dehydrogenease [174], while tellurium (VI)-centered polyoxotungstate was found to mediate contacts in crystals of hen egg-white lysozyme [175].

The key problem in this field is the unpredictability of what compounds might be helpful for crystallization, or how they might form packing bridges. The other important question is whether these interactions are indeed cohesive and contribute to the integrity of the crystal, or represent the serendipitous "trapping" of small molecules between macromolecules, which contributes little to the overall thermodynamics balance. It is very likely that many of the examples uncovered in the data mining study of the PDB [8] are indeed cases of fortuitously bound ions in contact with two molecules. Control experiments (i.e., crystallization without these small molecules) were never conducted nor reported. However, in cases where crystallization appears to be contingent on the presence of a small molecule, and if the latter is found to form a packing bridge, it is almost certain that the bridge is thermodynamically cohesive (i.e., "sticky bridge"). Similarly, if the bridge constitutes one of the primary contacts, it has to be cohesive.

There are also interesting examples showing that specific residues or motifs may be "coupled" to certain ions or compounds, and may consequently be introduced into proteins by mutagenesis. For example, crystals of the *E. coli* apo acyl carrier protein that is rich in carboxylic acids were obtained in the presence of Zn^{2+} ions, which provided bridging interactions [176]. It has been suggested that mutational introduction of aspartates on the surface of proteins with high intrinsic pI could provide a useful strategy for crystallization with metal ions. In a related example, His–Cys pairs were introduced on the surface (using T4 lysozyme as a template), allowing for coordination of Zn^{2+} ions that effectively induced dimerization, and engineering a key crystal contact [177]. In yet another case, Ca^{2+} ions were shown to form a "sticky bridge" between two molecules of the YkoF, engineered by the SER strategy [178]. Here, the removal of high entropy side chains exposed main chain carbonyls, creating a metal binding site. Interestingly, a recent theoretical study presented a general model of multivalent cation bridges as a method to activate attractive positive patches on the protein surface, bringing small molecules and ions directly into the realm of the "sticky patch" model [179].

An example that shows potential for more general applications is that of combining the use of sulfates as precipitants with surface engineering. A mutant of RhoGDI with two Arg replacing adjacent Lys residues was crystallized in the presence of amonium sulfate, and the surface ions were found to bridge the Arg-rich surface patches [124]. In this particular case, bridging sulfates may neutralize potential electrostatic repulsion, allowing this secondary contact to form, although it may not *per se* serve as a cohesive interaction.

To summarize, engineering of crystal contact bridges using small molecules or ions, either into wild-type or mutated protein, offers the possibility of creating a "sticky bridge," thermodynamically cohesive contact, or allows for creation of an interaction eliminating potential electrostatic repulsion. It is conceivable that more general recipes can be designed to couple this approach with surface mutagenesis.

4.5 Crystallization Chaperones—Using Surrogate Surfaces

Perhaps the most challenging and complex strategy of altering the surface of the target protein is using a partner protein (chaperone) that lends its surface to mediate crystal contacts, thus enabling the crystallization of the complex. (NB: chaperone proteins also serve other purposes, e.g., they may stabilize a particular conformation or enhance solubility of the target; here we focus exclusively on crystallization.) There are two options: either the chaperone is expressed in fusion with the target protein, or the chaperone is generated separately, and the complex is purified and crystallized. Below we briefly discuss the first approach, and expand more on the second, which is more popular and much more successful.

Given that many proteins are overexpressed for purposes of crystallization as fusion proteins with globular affinity tags (e.g.,

GST, MBP, thioredoxin, T4 lysozyme), the use of these fusion proteins is an obvious and straightforward option. A number of such fusion proteins have been crystallized: e.g., the DNA-binding domain of DNA replication-related element-binding factor, DREF, in fusion with GST [180] or the U2AF homology motif domain of splicing factor Puf60 in fusion with thioredoxin [181]. The drawback is that the intrinsic flexibility of a two-domain architecture may impede crystallization. A remedy is to shorten the linker between the two proteins, to achieve rigidity owing to steric restraint [182–186].

An alternative to N- or C-terminal fusions is an insertion fusion, in which the chaperone is inserted into a loop of the target. This approach has been used exclusively in membrane protein crystallization, and was initially pioneered for the *E. coli* lactose permease, in which cytochrome b562, flavodoxin, and T4 lysozyme were tested as chaperones [187, 188]. A similar insertion fusion with T4 lysozyme, replacing the third intracellular loop of the β2-adrenergic receptor was the key to successful crystallization and ultimate structure determination at 2.4 Å resolution [189, 190]. This strategy has since been used in a number of crystallographic studies of the G-protein coupled receptors (GPCRs) and other membrane proteins [191].

A more universal alternative to fusion proteins are non-covalent crystallization chaperones, i.e., binding proteins engineered to produce a high-affinity complex with the target macromolecule. The most commonly used engineered chaperones are Fab fragments of antibodies [192–198]. In its canonical version, animals are immunized with the target antigen, followed by purification of hybridoma-derived antibodies and their proteolytic digestion to obtain antigen binding fragments [192, 199]. Alternatively, the Fab fragment s directly sequenced and expressed in heterologous cells for subsequent use [200]. This strategy is costly, inefficient and prone to challenges. A far more powerful and efficient approach is in vitro selection of Fab fragments using phage display [201, 202] or, less often, ribosome display [203, 204]. Multiple templates have been used, but the most common is the one based on the herceptin scaffold. Although initially such synthetic antibodies were weaker binders than the wild-type ones [205, 206], the problem was overcome by using "reduced genetic code," which uses only select types of amino acids that produce high-affinity binders [201, 207]. Synthetic Fab fragments can be generated against a broad variety of targets, unique conformations of proteins, complexes, or weak antigens such as RNA. Automated platforms are available for high-throughput production [202]. Many targets have been successfully crystallized using synthetic Fabs based on the herceptin scaffold as chaperones. Recent examples include the Nsp1-Nup49-Nup57 channel nucleoporin heterotrimer bound to Nic96 nuclear pore complex attachment site [208]; human paxillin LD2 and LD4 motifs [209]; structure of the Get3 targeting factor with its membrane protein cargo [210]; and the prolactin receptor [211].

The in vitro display methods also allow for engineering of non-Fab scaffolds [206]. Examples include nanobodies, i.e., single chain fragments derived from camelid antibodies [212–214]; fibronectin type III domain (FN3) scaffold [134, 215]; or DARPins, i.e., designed ankyrin repeat proteins [216, 217], used in the crystallization of several proteins, including the polo-like kinase-1 [218], the integral membrane multidrug transporter AcrB [219], and the receptor-binding protein (RBP, the BppL trimer) of the baseplate complex of the lactococcal phage TP901-1 [220].

The success of the chaperones in crystallization is, of course, dependent on their ability to mediate crystal contacts in a more effective way than the target protein alone. The various chaperones described above are well studied and all show high propensity for crystallization in isolation. However, they may still be suboptimal, and could be subject to surface engineering or other modifications. The wild-type T4 lysozyme, for example, is not ideally suited because of intrinsic flexibility and recently it has been engineered for the use as an internal fusion in GPCRs by adding stabilizing disulfides, or by reducing the size of the N-lobe (miniT4). These modified molecules proved to be superior as crystallization chaperones when fused into the third loop of the M3 muscarinic receptor [221]. In another study, T4 lysozyme was also modified including the mutation of the three C-terminal residues to Ala to reduce conformational entropy [222]. However, perhaps the most relevant to this discussion is the example of the variants of MBP specifically engineered by SER mutations for enhanced crystallizability [223]. Several such variants were used as N-terminal fusion chaperones to crystallize the signal transduction regulator RACK1 from *Arabidopsis thaliana* [223]. In the crystals of the fusion protein that was crystallized, the SER patches on MBP served, as was intended, as the "sticky patches" mediating crystal contacts.

5 Conclusions and Perspectives

The challenge in the field is to obtain sufficiently detailed insight into the mechanism of the PPIs underlying crystallization, to enable us to rationally modify the crystallization experiment and its outcome. Although not long ago this would have seemed like fantasy, we are not far from realizing this goal, even if only in some specific cases. The progress is vividly illustrated by a recent success in computational design of a protein that not only self-assembles to yield macroscopic crystals, but does so yielding the expected *P6* space group symmetry [224] (Fig. 13).

The "sticky patch" model provides not only a unifying theory for a wide spectrum of PPIs, but also rationalizes many of the thermodynamic macroscopic observations, and paves the way for

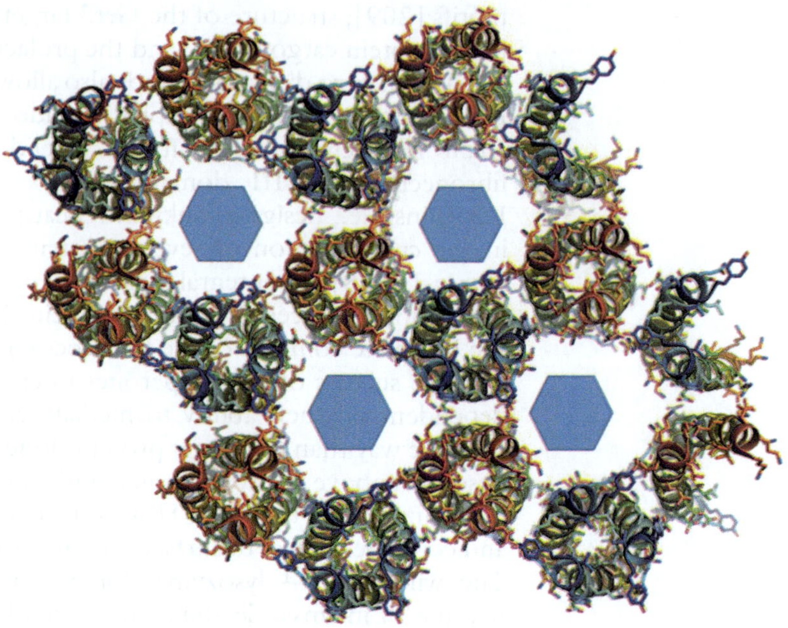

Fig. 13 A synthetic protein designed to crystallize in P_6. A single layer of trimers of a designed helical protein assembling into the P_6 lattice. The sixfold crystallographic axes are marked with *hexagons*. Threefold axes are running down the trimers (PDB code 3V86)

strategies to rationally modify the macromolecular targets to dramatically enhance their crystallizability, either by covalent or not covalent chemical modification, or by protein engineering. Nevertheless, much remains to be learned about PPIs and the ways in which we can modify them through site directed mutagenesis to control crystallization. The SER methodology for enhancing protein crystallizability has gained considerable support in recent years from various experimental, theoretical and data mining studies, which collectively generate the comprehensive "patchy model." The underlying concept—the reduction in "excess surface entropy"—is, of course, an oversimplification, because the mutations of polar, charged residues to Ala or similar smaller amino acids alter many physical properties of the protein, including electrostatic potential and solubility. Nevertheless, the distinct propensity of the SER patches to form crystal contacts, most of which are homotypic and result in transient homodimers, shows that the mutations generate the very "sticky patches" that the model invokes. A change in pI or solubility could not rationalize these effects, but it is important to take all of these properties into account as well as the role of the solvent. Although many crystal structures have been obtained using proteins modified by SER or reductive methylation, there are also numerous examples of failures of such protocols. One possibility to overcome this problem is a design of multi-patch SER strategy, which could overcome problems with particularly recalcitrant proteins with distinct paucity of attractive patches on their surfaces. Several successful examples have already been reported. The *Arabidopsis* medium/long chain prenyl pyrophosphate synthase was crystallized using a two-patch variant [225]. The structure revealed that the two SER patches assist in forming an octamer (the wild-type protein is a homodimer in solution) within the asymmetric unit, and generate secondary contacts between the octamers to allow for 3D packing. In another study, a triple-patch SER strategy was necessary to overcome extreme difficulties in the crystallization of the human vaccinia related kinase 1 (PDB code: 3OP5). Again, all patches were involved in crystal contacts.

Finally, perhaps the most intriguing questions are: When exactly the crystal contacts are formed? And how do they drive the crystallization process? The current theory of nucleation and crystal growth strongly suggests that protein nuclei form within clusters of protein-dense liquid, metastable with respect to protein solution and hundreds of nanometers in size [53, 226]. Within these clusters, overcrowding effects will contribute to significantly enhance attractive interactions between proteins molecules. Whether formation of specific oligomers, as defined by the "sticky patches" underlies nucleation, and defines the symmetry of the nascent crystal, will hopefully be elucidated by ongoing research.

Acknowledgments

We thank a number of colleagues for helpful comments: Patrick Charbonneau (Duke University), Diana Fusco (University of California, Berkeley), Peter Vekilov (University of Houston), and Urszula Derewenda (University of Virginia). Special thanks are due to the editors of this volume: Alexander Wlodawer, Mariusz Jaskolski, and Zbigniew Dauter. We also thank Heping Zheng, Jagoda Mika, and Natalya Olekhnovitch (University of Virginia) for assistance with figures. This work was supported by the NIH grant GM095847.

References

1. Manchester KL (2004) The crystallization of enzymes and virus proteins: laying to rest the colloidal concept of living systems. Endeavour 28:25–29
2. Bernal JD, Crowfoot D (1934) X-ray photographs of crystalline pepsin. Nature 133: 794–795
3. Deichmann U (2007) Collective phenomena and the neglect of molecules: a historical outlook on biology. Hist Philos Life Sci 29: 83–86
4. Jaskolski M, Dauter Z, Wlodawer A (2014) A brief history of macromolecular crystallography, illustrated by a family tree and its Nobel fruits. FEBS J 281:3985–4009
5. Reichert ET, Brown AP (1909) The differentiation and specificity of corresponding proteins and other vital substances in relation to biological classification and organic evolution: the crystallography of haemoglobin. Carnegie Institution of Washington, Washington, D.C.

6. Kendrew JC, Parrish RG, Marrack JR et al (1954) The species specificity of myoglobin. Nature 174:946–949
7. Janin J, Rodier F (1995) Protein-protein interaction at crystal contacts. Proteins 23:580–587
8. Carugo O, Argos P (1997) Protein-protein crystal-packing contacts. Protein Sci 6: 2261–2263
9. Janin J (1997) Specific versus non-specific contacts in protein crystals. Nat Struct Biol 4:973–974
10. Bonanno J (1999) Structural genomics. Curr Biol 9:R871–R872
11. Burley SK, Almo SC, Bonanno JB et al (1999) Structural genomics: beyond the human genome project. Nat Genet 23:151–157
12. Chandonia JM, Brenner SE (2006) The impact of structural genomics: expectations and outcomes. Science 311:347–351
13. Gaasterland T (1998) Structural genomics taking shape. Trends Genet 14:135
14. Skarina T, Xu X, Evdokimova E, Savchenko A (2014) High-throughput crystallization screening. Methods Mol Biol 1140:159–168
15. Fusco D, Charbonneau P (2016) Soft matter perspective on protein crystal assembly. Colloids Surf B Biointerfaces 137:22–31
16. George A, Wilson WW (1994) Predicting protein crystallization from a dilute solution property. Acta Crystallogr D Biol Crystallogr 50:361–365
17. ten Wolde PR, Frenkel D (1997) Enhancement of protein crystal nucleation by critical density fluctuations. Science 277:1975–1978
18. Wilson WW (2003) Light scattering as a diagnostic for protein crystal growth--a practical approach. J Struct Biol 142:56–65
19. Muschol M, Rosenberger F (1997) Liquid–liquid phase separation in supersaturated lysozyme solutions and associated precipitate formation/crystallization. J Chem Phys 107:1953–1958
20. Liu Y, Wang X, Ching CB (2010) Toward further understanding of lysozyme crystallization: phase diagram, protein–protein interaction, nucleation kinetics, and growth kinetics. Cryst Growth Des 10:548–558
21. Lu PJ, Zaccarelli E, Ciulla F, Schofield AB et al (2008) Gelation of particles with short-range attraction. Nature 453:499–503
22. Rosenbaum D, Zamora PC, Zukoski CF (1996) Phase behavior of small attractive colloidal particles. Phys Rev Lett 76:150–153
23. Noro MG, Frenkel D (2000) Extended corresponding-states behavior for particles with variable range attractions. J Chem Phys 113:2941–2944
24. Doye JP, Louis AA, Lin IC et al (2007) Controlling crystallization and its absence: proteins, colloids and patchy models. Phys Chem Chem Phys 9:2197–2205
25. Liu H, Kumar SK, Douglas JF (2009) Self-assembly-induced protein crystallization. Phys Rev Lett 103:018101
26. Kern N, Frenkel D (2003) Fluid–fluid coexistence in colloidal systems with short-ranged strongly directional attraction. J Chem Phys 118:9882–9893
27. Bianchi E, Blaak R, Likos CN (2011) Patchy colloids: state of the art and perspectives. Phys Chem Chem Phys 13:6397–6410
28. Lomakin A, Asherie N, Benedek GB (1999) Aeolotopic interactions of globular proteins. Proc Natl Acad Sci U S A 96:9465–9468
29. Gogelein C, Nagele G, Tuinier R et al (2008) A simple patchy colloid model for the phase behavior of lysozyme dispersions. J Chem Phys 129:085102
30. Chang J, Lenhoff AM, Sandler SI (2004) Determination of fluid-solid transitions in model protein solutions using the histogram reweighting method and expanded ensemble simulations. J Chem Phys 120:3003–3014
31. Fusco D, Charbonneau P (2014) Competition between monomeric and dimeric crystals in schematic models for globular proteins. J Phys Chem B 118:8034–8041
32. Staneva I, Frenkel D (2015) The role of non-specific interactions in a patchy model of protein crystallization. J Chem Phys 143:194511
33. Fusco D, Headd JJ, De Simone A et al (2014) Characterizing protein crystal contacts and their role in crystallization: rubredoxin as a case study. Soft Matter 10:290–302
34. Asherie N (2004) Protein crystallization and phase diagrams. Methods 34:266–272
35. Derewenda ZS, Vekilov PG (2006) Entropy and surface engineering in protein crystallization. Acta Crystallogr D Biol Crystallogr 62:116–124
36. Vekilov PG, Feeling-Taylor AR, Yau ST, Petsev D (2002) Solvent entropy contribution to the free energy of protein crystallization. Acta Crystallogr D Biol Crystallogr 58:1611–1616
37. Vekilov PG (2003) Solvent entropy effects in the formation of protein solid phases. Methods Enzymol 368:84–105
38. Yau ST, Petsev DN, Thomas BR et al (2000) Molecular-level thermodynamic and kinetic parameters for the self-assembly of apoferritin

molecules into crystals. J Mol Biol 303:667–678
39. Paunov VN, Kaler EW, Sandler SI et al (2001) A model for hydration interactions between apoferritin molecules in solution. J Colloid Interface Sci 240:640–643
40. Gliko O, Neumaier N, Pan W et al (2005) A metastable prerequisite for the growth of lumazine synthase crystals. J Am Chem Soc 127:3433–3438
41. Finkelstein AV, Janin J (1989) The price of lost freedom: entropy of bimolecular complex formation. Protein Eng 3:1–3
42. Tidor B, Karplus M (1994) The contribution of vibrational entropy to molecular association. The dimerization of insulin. J Mol Biol 238:405–414
43. Doye JPK (2004) Inhibition of protein crystallization by evolutionary negative design. Phys Biol 1:P9–P13
44. Ellis RJ (2001) Macromolecular crowding: an important but neglected aspect of the intracellular environment. Curr Opin Struct Biol 11:114–119
45. Zorrilla S, Rivas G, Acuna AU et al (2004) Protein self-association in crowded protein solutions: a time-resolved fluorescence polarization study. Protein Sci 13:2960–2969
46. Pal D, Chakrabarti P (1999) Estimates of the loss of main-chain conformational entropy of different residues on protein folding. Proteins 36:332–339
47. Chellgren BW, Creamer TP (2006) Side-chain entropy effects on protein secondary structure formation. Proteins 62:411–420
48. Lee J, Kim SH (2009) Water polygons in high-resolution protein crystal structures. Protein Sci 18:1370–1376
49. Nakasako M (2004) Water-protein interactions from high-resolution protein crystallography. Philos Trans R Soc Lond Ser B Biol Sci 359:1191–1204. discussion 1204-1196
50. Ball P (2003) Chemical physics: how to keep dry in water. Nature 423:25–26
51. Pal SK, Zewail AH (2004) Dynamics of water in biological recognition. Chem Rev 104:2099–2123
52. Dunitz JD (1994) The entropic cost of bound water in crystals and biomolecules. Science 264:670
53. Vekilov PG, Vorontsova MA (2014) Nucleation precursors in protein crystallization. Acta Crystallogr F Struct Biol Commun 70:271–282
54. Vekilov PG (2004) Dense liquid precursor for the nucleation of ordered solid phases from solution. Cryst Growth Des 4:671–685
55. Carugo O, Djinovic-Carugo K (2012) How many packing contacts are observed in protein crystals? J Struct Biol 180:96–100
56. Wukovitz SW, Yeates TO (1995) Why protein crystals favour some space-groups over others. Nat Struct Biol 2:1062–1067
57. Henrick K, Thornton JM (1998) PQS: a protein quaternary structure file server. Trends Biochem Sci 23:358–361
58. Ponstingl H, Kabir T, Thornton JM (2003) Automatic inference of protein quaternary structure from crystals. J Appl Crystallogr 36:1116–1122
59. Elcock AH, McCammon JA (2001) Calculation of weak protein-protein interactions: the pH dependence of the second virial coefficient. Biophys J 80:613–625
60. Elcock AH, McCammon JA (2001) Identification of protein oligomerization states by analysis of interface conservation. Proc Natl Acad Sci U S A 98:2990–2994
61. Krissinel E, Henrick K (2007) Inference of macromolecular assemblies from crystalline state. J Mol Biol 372:774–797
62. Cieslik M, Derewenda ZS (2009) The role of entropy and polarity in intermolecular contacts in protein crystals. Acta Crystallogr D Biol Crystallogr 65:500–509
63. Xu Q, Dunbrack RL Jr (2011) The protein common interface database (ProtCID)--a comprehensive database of interactions of homologous proteins in multiple crystal forms. Nucleic Acids Res 39:D761–D770
64. Rowe AJ (2011) Ultra-weak reversible protein-protein interactions. Methods 54:157–166
65. Huxford T, Mishler D, Phelps CB et al (2002) Solvent exposed non-contacting amino acids play a critical role in NF-κB/IκBα complex formation. J Mol Biol 324:587–597
66. van der Merwe PA, Davis SJ (2003) Molecular interactions mediating T cell antigen recognition. Annu Rev Immunol 21:659–684
67. Ceres P, Zlotnick A (2002) Weak protein-protein interactions are sufficient to drive assembly of hepatitis B virus capsids. Biochemistry 41:11525–11531
68. Wang Q, Zhuravleva A, Gierasch LM (2011) Exploring weak, transient protein--protein interactions in crowded in vivo environments by in-cell nuclear magnetic resonance spectroscopy. Biochemistry 50:9225–9236
69. Vaynberg J, Qin J (2006) Weak protein-protein interactions as probed by NMR spectroscopy. Trends Biotechnol 24:22–27
70. Zhou HX, Rivas G, Minton AP (2008) Macromolecular crowding and confinement: biochemical, biophysical, and potential physiological consequences. Annu Rev Biophys 37:375–397

71. McGuffee SR, Elcock AH (2010) Diffusion, crowding & protein stability in a dynamic molecular model of the bacterial cytoplasm. PLoS Comput Biol 6:e1000694
72. Ellis RJ (2001) Macromolecular crowding: obvious but underappreciated. Trends Biochem Sci 26:597–604
73. Nooren IM, Thornton JM (2003) Diversity of protein-protein interactions. EMBO J 22:3486–3492
74. Nooren IM, Thornton JM (2003) Structural characterisation and functional significance of transient protein-protein interactions. J Mol Biol 325:991–1018
75. Vijay-Kumar S, Bugg CE, Cook WJ (1987) Structure of ubiquitin refined at 1.8 A resolution. J Mol Biol 194:531–544
76. Dikic I, Wakatsuki S, Walters KJ (2009) Ubiquitin-binding domains - from structures to functions. Nat Rev Mol Cell Biol 10: 659–671
77. Tabernero L, Evans BN, Tishmack PA et al (1999) The structure of the bovine protein tyrosine phosphatase dimer reveals a potential self-regulation mechanism. Biochemistry 38:11651–11658
78. Akerud T, Thulin E, Van Etten RL et al (2002) Intramolecular dynamics of low molecular weight protein tyrosine phosphatase in monomer-dimer equilibrium studied by NMR: a model for changes in dynamics upon target binding. J Mol Biol 322:137–152
79. Tung M, Gallagher DT (2009) The biomolecular crystallization database version 4: expanded content and new features. Acta Crystallogr D Biol Crystallogr 65:18–23
80. Canaves JM, Page R, Wilson IA et al (2004) Protein biophysical properties that correlate with crystallization success in thermotoga maritima: maximum clustering strategy for structural genomics. J Mol Biol 344: 977–991
81. Price WN 2nd, Chen Y, Handelman SK et al (2009) Understanding the physical properties that control protein crystallization by analysis of large-scale experimental data. Nat Biotechnol 27:51–57
82. Fusco D, Barnum TJ, Bruno AE et al (2014) Statistical analysis of crystallization database links protein physico-chemical features with crystallization mechanisms. PLoS One 9:e101123
83. Chen L, Oughtred R, Berman HM et al (2004) TargetDB: a target registration database for structural genomics projects. Bioinformatics 20:2860–2862
84. Christendat D, Yee A, Dharamsi A et al (2000) Structural proteomics of an archaeon. Nat Struct Biol 7:903–909
85. Goh C-S, Lan N, Douglas SM et al (2004) Mining the structural genomics pipeline: identification of protein properties that affect high-throughput experimental analysis. J Mol Biol 336:115–130
86. Smialowski P, Schmidt T, Cox J et al (2006) Will my protein crystallize? A sequence-based predictor. Proteins 62:343–355
87. Slabinski L, Jaroszewski L, Rychlewski L, Wilson IA, Lesley SA, Godzik A (2007) XtalPred: a web server for prediction of protein crystallizability. Bioinformatics 23: 3403–3405
88. Overton IM, Padovani G, Girolami MA, Barton GJ (2008) ParCrys: a Parzen window density estimation approach to protein crystallization propensity prediction. Bioinformatics 24:901–907
89. Kurgan L, Razib AA, Aghakhani S et al (2009) CRYSTALP2: sequence-based protein crystallization propensity prediction. BMC Struct Biol 9:50
90. Mizianty MJ, Kurgan L (2009) Meta prediction of protein crystallization propensity. Biochem Biophys Res Commun 390:10–15
91. Babnigg G, Joachimiak A (2010) Predicting protein crystallization propensity from protein sequence. J Struct Funct Genom 11:71–80
92. Mizianty MJ, Kurgan L (2011) Sequence-based prediction of protein crystallization, purification and production propensity. Bioinformatics 27:I24–I33
93. Jahandideh S, Jaroszewski L, Godzik A (2014) Improving the chances of successful protein structure determination with a random forest classifier. Acta Crystallogr D Biol Crystallogr 70:627–635
94. Altan I, Charbonneau P, Snell EH (2016) Computational crystallization. Arch Biochem Biophys 602:12–20
95. Jaroszewski L, Slabinski L, Wooley J et al (2008) Genome pool strategy for structural coverage of protein families. Structure 16:1659–1667
96. Gabanyi MJ, Adams PD, Arnold K et al (2011) The structural biology knowledgebase: a portal to protein structures, sequences, functions, and methods. J Struct Funct Genom 12:45–54
97. Savitsky P, Bray J, Cooper CDO et al (2010) High-throughput production of human proteins for crystallization: the SGC experience. J Struct Biol 172:3–13

98. Xiao R, Anderson S, Aramini J et al (2010) The high-throughput protein sample production platform of the northeast structural genomics consortium. J Struct Biol 172:21–33
99. Lee CK, Cheong C, Jeon YH (2010) The N-terminal domain of human holocarboxylase synthetase facilitates biotinylation via direct interaction with the substrate protein. FEBS Lett 584:675–680
100. Oyenarte I, Lucas M, Gomez Garcia I et al (2011) Purification, crystallization and preliminary crystallographic analysis of the CBS-domain protein MJ1004 from Methanocaldococcus jannaschii. Acta Crystallogr Sect F Struct Biol Cryst Commun 67:318–324
101. Gomez-Garcia I, Stuiver M, Ereno J et al (2012) Purification, crystallization and preliminary crystallographic analysis of the CBS-domain pair of cyclin M2 (CNNM2). Acta Crystallogr Sect F Struct Biol Cryst Commun F68:1198–1203
102. Derewenda ZS (2010) Application of protein engineering to enhance crystallizability and improve crystal properties. Acta Crystallogr D Biol Crystallogr 66:604–615
103. McPherson A, Nguyen C, Cudney R et al (2011) The role of small molecule additives and chemical modification in protein crystallization. Cryst Growth Des 11:1469–1474
104. Giedroc DP, Keating KM, Williams KR et al (1986) Gene 32 protein, the single-stranded DNA binding protein from bacteriophage T4, is a zinc metalloprotein. Proc Natl Acad Sci U S A 83:8452–8456
105. Dong A, Xu X, Edwards AM et al (2007) In situ proteolysis for protein crystallization and structure determination. Nat Methods 4:1019–1021
106. Wernimont A, Edwards A (2009) In situ proteolysis to generate crystals for structure determination: an update. PLoS One 4:e5094
107. Huang YJ, Acton TB, Montelione GT (2014) DisMeta: a meta server for construct design and optimization. Methods Mol Biol 1091: 3–16
108. Malawski GA, Hillig RC, Monteclaro F et al (2006) Identifying protein construct variants with increased crystallization propensity--a case study. Protein Sci 15:2718–2728
109. Ding HT, Ren H, Chen Q et al (2002) Parallel cloning, expression, purification and crystallization of human proteins for structural genomics. Acta Crystallogr D Biol Crystallogr 58:2102–2108
110. Quevillon-Cheruel S, Leulliot N et al (2007) Production and crystallization of protein domains: how useful are disorder predictions ? Curr Protein Pept Sci 8:151–160
111. Page R (2008) Strategies for improving crystallization success rates. Methods Mol Biol 426:345–362
112. Cohen SL, Ferre-D'Amare AR, Burley SK et al (1995) Probing the solution structure of the DNA-binding protein max by a combination of proteolysis and mass-spectrometry. Protein Sci 4:1088–1099
113. Hamuro Y, Coales SJ, Southern MR et al (2003) Rapid analysis of protein structure and dynamics by hydrogen/deuterium exchange mass spectrometry. J Biomol Tech 14:171–182
114. Pantazatos D, Kim JS, Klock HE et al (2004) Rapid refinement of crystallographic protein construct definition employing enhanced hydrogen/deuterium exchange MS. Proc Natl Acad Sci U S A 101:751–756
115. Sharma S, Zheng H, Huang YPJ et al (2009) Construct optimization for protein NMR structure analysis using amide hydrogen/deuterium exchange mass spectrometry. Proteins 76:882–894
116. Gray FL, Murai MJ, Grembecka J et al (2012) Detection of disordered regions in globular proteins using (1)(3)C-detected NMR. Protein Sci 21:1954–1960
117. Carugo O (2011) Participation of protein sequence termini in crystal contacts. Protein Sci 20:2121–2124
118. Kwong PD, Wyatt R, Robinson J et al (1998) Structure of an HIV gp120 envelope glycoprotein in complex with the CD4 receptor and a neutralizing human antibody. Nature 393:648–659
119. Kwong PD, Wyatt R, Desjardins E et al (1999) Probability analysis of variational crystallization and its application to gp120, the exterior envelope glycoprotein of type 1 human immunodeficiency virus (HIV-1). J Biol Chem 274:4115–4123
120. Neau DB, Gilbert NC, Bartlett SG et al (2007) Improving protein crystal quality by selective removal of a Ca2+-dependent membrane-insertion loop. Acta Crystallogr Sect F Struct Biol Cryst Commun F63: 972–975
121. Schwartz TU, Walczak R, Blobel G (2004) Circular permutation as a tool to reduce surface entropy triggers crystallization of the signal recognition particle receptor beta subunit. Protein Sci 13:2814–2818
122. Longenecker KL, Garrard SM, Sheffield PJ et al (2001) Protein crystallization by rational mutagenesis of surface residues: Lys to Ala mutations promote crystallization of RhoGDI. Acta Crystallogr D Biol Crystallogr 57:679–688

123. Mateja A, Devedjiev Y, Krowarsch D et al (2002) The impact of Glu-->Ala and Glu-->asp mutations on the crystallization properties of RhoGDI: the structure of RhoGDI at 1.3 A resolution. Acta Crystallogr D Biol Crystallogr 58:1983–1991
124. Czepas J, Devedjiev Y, Krowarsch D et al (2004) The impact of Lys-->Arg surface mutations on the crystallization of the globular domain of RhoGDI. Acta Crystallogr D Biol Crystallogr 60:275–280
125. Garrard SM, Longenecker KL, Lewis ME et al (2001) Expression, purification, and crystallization of the RGS-like domain from the Rho nucleotide exchange factor, PDZ-RhoGEF, using the surface entropy reduction approach. Protein Expr Purif 21:412–416
126. Longenecker KL, Lewis ME, Chikumi H et al (2001) Structure of the RGS-like domain from PDZ-RhoGEF: linking heterotrimeric g protein-coupled signaling to Rho GTPases. Structure 9:559–569
127. Yip CK, Kimbrough TG, Felise HB et al (2005) Structural characterization of the molecular platform for type III secretion system assembly. Nature 435:702–707
128. Fisher RD, Chung HY, Zhai Q et al (2007) Structural and biochemical studies of ALIX/AIP1 and its role in retrovirus budding. Cell 128:841–852
129. Pornillos O, Ganser-Pornillos BK, Kelly BN et al (2009) X-ray structures of the hexameric building block of the HIV capsid. Cell 137:1282–1292
130. Pioletti M, Findeisen F, Hura GL et al (2006) Three-dimensional structure of the KChIP1-Kv4.3 T1 complex reveals a cross-shaped octamer. Nat Struct Mol Biol 13:987–995
131. Ressl S, Terwisscha van Scheltinga AC, Vonrhein C et al (2009) Molecular basis of transport and regulation in the Na(+)/betaine symporter BetP. Nature 458:47–52
132. Cooper DR, Boczek T, Grelewska K et al (2007) Protein crystallization by surface entropy reduction: optimization of the SER strategy. Acta Crystallogr D Biol Crystallogr 63:636–645
133. Conte LL, Chothia C, Janin J (1999) The atomic structure of protein-protein recognition sites. J Mol Biol 285:2177–2198
134. Koide A, Gilbreth RN, Esaki K, Tereshko V, Koide S (2007) High-affinity single-domain binding proteins with a binary-code interface. Proc Natl Acad Sci U S A 104:6632–6637
135. Goldschmidt L, Cooper DR, Derewenda ZS et al (2007) Toward rational protein crystallization: a web server for the design of crystallizable protein variants. Protein Sci 16:1569–1576
136. Goldschmidt L, Eisenberg D, Derewenda ZS (2014) Salvage or recovery of failed targets by mutagenesis to reduce surface entropy. Methods Mol Biol 1140:201–209
137. Munshi S, Hall DL, Kornienko M et al (2003) Structure of apo, unactivated insulin-like growth factor-1 receptor kinase at 1.5 A resolution. Acta Crystallogr D Biol Crystallogr 59:1725–1730
138. Bauman JD, Das K, Ho WC et al (2008) Crystal engineering of HIV-1 reverse transcriptase for structure-based drug design. Nucleic Acids Res 36:5083–5092
139. Das K, Bauman JD, Clark AD Jr et al (2008) High-resolution structures of HIV-1 reverse transcriptase/TMC278 complexes: strategic flexibility explains potency against resistance mutations. Proc Natl Acad Sci U S A 105:1466–1471
140. Frey KM, Puleo DE, Spasov KA et al (2015) Structure-based evaluation of non-nucleoside inhibitors with improved potency and solubility that target HIV reverse transcriptase variants. J Med Chem 58:2737–2745
141. Gray WT, Frey KM, Laskey SB et al (2015) Potent inhibitors active against HIV reverse transcriptase with K101P, a mutation conferring Rilpivirine resistance. ACS Med Chem Lett 6:1075–1079
142. Lee WG, Frey KM, Gallardo-Macias R et al (2015) Discovery and crystallography of bicyclic arylaminoazines as potent inhibitors of HIV-1 reverse transcriptase. Bioorg Med Chem Lett 25:4824–4827
143. Yang W, Fucini RV, Fahr BT et al (2009) Fragment-based discovery of nonpeptidic BACE-1 inhibitors using tethering. Biochemistry 48:4488–4496
144. Barrow JC, Stauffer SR, Rittle KE et al (2008) Discovery and X-ray crystallographic analysis of a spiropiperidine iminohydantoin inhibitor of beta-secretase. J Med Chem 51:6259–6262
145. McGaughey GB, Colussi D, Graham SL et al (2007) Beta-secretase (BACE-1) inhibitors: accounting for 10s loop flexibility using rigid active sites. Bioorg Med Chem Lett 17:1117–1121
146. Stauffer SR, Stanton MG, Gregro AR et al (2007) Discovery and SAR of isonicotinamide BACE-1 inhibitors that bind beta-secretase in a N-terminal 10s-loop down conformation. Bioorg Med Chem Lett 17:1788–1792
147. Lindsley SR, Moore KP, Rajapakse HA et al (2007) Design, synthesis, and SAR of macrocyclic tertiary carbinamine BACE-1 inhibitors. Bioorg Med Chem Lett 17:4057–4061

148. Moore KP, Zhu H, Rajapakse HA et al (2007) Strategies toward improving the brain penetration of macrocyclic tertiary carbinamine BACE-1 inhibitors. Bioorg Med Chem Lett 17:5831–5835
149. Stachel SJ, Coburn CA, Steele TG et al (2006) Conformationally biased P3 amide replacements of beta-secretase inhibitors. Bioorg Med Chem Lett 16:641–644
150. Rajapakse HA, Nantermet PG, Selnick HG et al (2006) Discovery of oxadiazoyl tertiary carbinamine inhibitors of beta-secretase (BACE-1). J Med Chem 49:7270–7273
151. Coburn CA, Stachel SJ, Jones KG et al (2006) BACE-1 inhibition by a series of psi[CH2NH] reduced amide isosteres. Bioorg Med Chem Lett 16:3635–3638
152. Coburn CA, Stachel SJ, Li YM et al (2004) Identification of a small molecule nonpeptide active site beta-secretase inhibitor that displays a nontraditional binding mode for aspartyl proteases. J Med Chem 47: 6117–6119
153. Pedro-Rosa L, Buckner FS, Ranade RM et al (2015) Identification of potent inhibitors of the Trypanosoma brucei methionyl-tRNA synthetase via high-throughput orthogonal screening. J Biomol Screen 20:122–130
154. Koh CY, Kim JE, Wetzel AB et al (2014) Structures of Trypanosoma brucei methionyl-tRNA synthetase with urea-based inhibitors provide guidance for drug design against sleeping sickness. PLoS Negl Trop Dis 8:e2775
155. Holden JK, Li H, Jing Q et al (2013) Structural and biological studies on bacterial nitric oxide synthase inhibitors. Proc Natl Acad Sci U S A 110:18127–18131
156. Jing Q, Li H, Fang J et al (2013) In search of potent and selective inhibitors of neuronal nitric oxide synthase with more simple structures. Bioorg Med Chem 21:5323–5331
157. Huang H, Li H, Martasek P et al (2013) Structure-guided design of selective inhibitors of neuronal nitric oxide synthase. J Med Chem 56:3024–3032
158. Yang Z, Misner B, Ji H et al (2013) Targeting nitric oxide signaling with nNOS inhibitors as a novel strategy for the therapy and prevention of human melanoma. Antioxid Redox Signal 19:433–447
159. Hanan EJ, Baumgardner M, Bryan MC et al (2016)4-Aminoindazolyl-dihydrofuro[3,4-d] pyrimidines as non-covalent inhibitors of mutant epidermal growth factor receptor tyrosine kinase. Bioorg Med Chem Lett 26:534–539
160. Heald R, Bowman KK, Bryan MC et al (2015) Noncovalent mutant selective epidermal growth factor receptor inhibitors: a lead optimization case study. J Med Chem 58:8877–8895
161. Hanan EJ, Eigenbrot C, Bryan MC et al (2014) Discovery of selective and noncovalent diaminopyrimidine-based inhibitors of epidermal growth factor receptor containing the T790M resistance mutation. J Med Chem 57:10176–10191
162. Means GE, Feeney RE (1990) Chemical modifications of proteins: history and applications. Bioconjug Chem 1:2–12
163. Rayment I, Rypniewski WR, Schmidt-Base K et al (1993) Three-dimensional structure of myosin subfragment-1: a molecular motor. Science 261:50–58
164. Rayment I (1997) Reductive alkylation of lysine residues to alter crystallization properties of proteins. Methods Enzymol 276:171–179
165. Tan K, Kim Y, Hatzos-Skintges C et al (2014) Salvage of failed protein targets by reductive alkylation. Methods Mol Biol 1140:189–200
166. Walter TS, Meier C, Assenberg R et al (2006) Lysine methylation as a routine rescue strategy for protein crystallization. Structure 14:1617–1622
167. Kim Y, Quartey P, Li H et al (2008) Large-scale evaluation of protein reductive methylation for improving protein crystallization. Nat Methods 5:853–854
168. Fan Y, Joachimiak A (2010) Enhanced crystal packing due to solvent reorganization through reductive methylation of lysine residues in oxidoreductase from Streptococcus pneumoniae. J Struct Funct Genom 11: 101–111
169. Sledz P, Zheng H, Murzyn K et al (2010) New surface contacts formed upon reductive lysine methylation: improving the probability of protein crystallization. Protein Sci 19:1395–1404
170. Shaw N, Cheng C, Tempel W et al (2007) (NZ)CH...O contacts assist crystallization of a ParB-like nuclease. BMC Struct Biol 7:46
171. Means GE (1977) Reductive alkylation of amino groups. Methods Enzymol 47:469–478
172. McPherson A, Cudney B (2006) Searching for silver bullets: an alternative strategy for crystallizing macromolecules. J Struct Biol 156:387–406
173. Carugo O, Djinovic-Carugo K (2014) Packing bridges in protein crystal structures. J Appl Crystallogr 47:458–461

174. Lagautriere T, Bashiri G, Baker EN (2015) Use of a "silver bullet" to resolve crystal lattice dislocation disorder: a cobalamin complex of Delta1-pyrroline-5-carboxylate dehydrogenase from mycobacterium tuberculosis. J Struct Biol 189:153–157
175. Bijelic A, Molitor C, Mauracher SG et al (2015) Hen egg-white lysozyme crystallisation: protein stacking and structure stability enhanced by a tellurium(VI)-centred polyoxotungstate. Chembiochem 16:233–241
176. Qiu X, Janson CA (2004) Structure of apo acyl carrier protein and a proposal to engineer protein crystallization through metal ions. Acta Crystallogr D Biol Crystallogr 60:1545–1554
177. Laganowsky A, Zhao M, Soriaga AB et al (2011) An approach to crystallizing proteins by metal-mediated synthetic symmetrization. Protein Sci 20:1876–1890
178. Devedjiev Y, Surendranath Y, Derewenda U et al (2004) The structure and ligand binding properties of the *B. subtilis* YkoF gene product, a member of a novel family of thiamin/HMP-binding proteins. J Mol Biol 343:395–406
179. Roosen-Runge F, Zhang F, Schreiber F et al (2014) Ion-activated attractive patches as a mechanism for controlled protein interactions. Sci Rep 4:7016
180. Kuge M, Fujii Y, Shimizu T et al (1997) Use of a fusion protein to obtain crystals suitable for X-ray analysis: crystallization of a GST-fused protein containing the DNA-binding domain of DNA replication-related element-binding factor, DREF. Protein Sci 6:1783–1786
181. Corsini L, Hothorn M, Scheffzek K et al (2008) Thioredoxin as a fusion tag for carrier-driven crystallization. Protein Sci 17:2070–2079
182. Smyth DR, Mrozkiewicz MK, McGrath WJ et al (2003) Crystal structures of fusion proteins with large-affinity tags. Protein Sci 12:1313–1322
183. Center RJ, Kobe B, Wilson KA et al (1998) Crystallization of a trimeric human T cell leukemia virus type 1 gp21 ectodomain fragment as a chimera with maltose-binding protein. Protein Sci 7:1612–1619
184. Monne M, Han L, Schwend T et al (2008) Crystal structure of the ZP-N domain of ZP3 reveals the core fold of animal egg coats. Nature 456:653–657
185. Wiltzius JJ, Sievers SA, Sawaya MR et al (2009) Atomic structures of IAPP (amylin) fusions suggest a mechanism for fibrillation and the role of insulin in the process. Protein Sci 18:1521–1530
186. Ke A, Wolberger C (2003) Insights into binding cooperativity of MATa1/MATalpha2 from the crystal structure of a MATa1 homeodomain-maltose binding protein chimera. Protein Sci 12:306–312
187. Prive GG, Verner GE, Weitzman C, Zen KH, Eisenberg D, Kaback HR (1994) Fusion proteins as tools for crystallization: the lactose permease from *Escherichia coli*. Acta Crystallogr D Biol Crystallogr 50:375–379
188. Engel CK, Chen L, Prive GG (2002) Insertion of carrier proteins into hydrophilic loops of the *Escherichia coli* lactose permease. Biochim Biophys Acta 1564:38–46
189. Cherezov V, Rosenbaum DM, Hanson MA et al (2007) High-resolution crystal structure of an engineered human beta(2)-adrenergic G protein-coupled receptor. Science 318: 1258–1265
190. Rosenbaum DM, Cherezov V, Hanson MA et al (2007) GPCR engineering yields high-resolution structural insights into beta(2)-adrenergic receptor function. Science 318:1266–1273
191. Chun E, Thompson AA, Liu W et al (2012) Fusion partner toolchest for the stabilization and crystallization of G protein-coupled receptors. Structure 20:967–976
192. Kovari LC, Momany C, Rossmann MG (1995) The use of antibody fragments for crystallization and structure determinations. Structure 3:1291–1293
193. Hunte C, Michel H (2002) Crystallisation of membrane proteins mediated by antibody fragments. Curr Opin Struct Biol 12:503–508
194. Prongay AJ, Smith TJ, Rossmann MG, Ehrlich LS, Carter CA, McClure J (1990) Preparation and crystallization of a human immunodeficiency virus p24-fab complex. Proc Natl Acad Sci U S A 87:9980–9984
195. Ostermeier C, Iwata S, Ludwig B et al (1995) F-V fragment mediated crystallization of the membrane-protein bacterial cytochrome-C-oxidase. Nat Struct Biol 2:842–846
196. Jiang Y, Lee A, Chen J et al (2003) X-ray structure of a voltage-dependent K+ channel. Nature 423:33–41
197. Dutzler R, Campbell EB, MacKinnon R (2003) Gating the selectivity filter in ClC chloride channels. Science 300:108–112
198. Lee SY, Lee A, Chen J, MacKinnon R (2005) Structure of the KvAP voltage-dependent K+ channel and its dependence on the lipid membrane. Proc Natl Acad Sci U S A 102:15441–15446
199. Karpusas M, Lucci J, Ferrant J et al (2001) Structure of CD40 ligand in complex with

the Fab fragment of a neutralizing humanized antibody. Structure 9:321–329

200. Nettleship JE, Ren J, Rahman N et al (2008) A pipeline for the production of antibody fragments for structural studies using transient expression in HEK 293T cells. Protein Expr Purif 62:83–89
201. Lee CV, Liang WC, Dennis MS et al (2004) High-affinity human antibodies from phage-displayed synthetic Fab libraries with a single framework scaffold. J Mol Biol 340: 1073–1093
202. Hornsby M, Paduch M, Miersch S et al (2015) A high through-put platform for recombinant antibodies to folded proteins. Mol Cell Proteomics 14:2833–2847
203. Lipovsek D, Pluckthun A (2004) In-vitro protein evolution by ribosome display and mRNA display. J Immunol Methods 290:51–67
204. Stafford RL, Matsumoto ML, Yin G et al (2014) In vitro Fab display: a cell-free system for IgG discovery. Protein Eng Des Sel 27:97–109
205. Hawkins RE, Russell SJ, Winter G (1992) Selection of phage antibodies by binding-affinity - mimicking affinity maturation. J Mol Biol 226:889–896
206. Koide S (2009) Engineering of recombinant crystallization chaperones. Curr Opin Struct Biol 19:449–457
207. Fellouse FA, Wiesmann C, Sidhu SS (2004) Synthetic antibodies from a four-amino-acid code: a dominant role for tyrosine in antigen recognition. Proc Natl Acad Sci U S A 101:12467–12472
208. Stuwe T, Bley CJ, Thierbach K et al (2015) Architecture of the fungal nuclear pore inner ring complex. Science 350:56–64
209. Nocula-Lugowska M, Lugowski M, Salgia R et al (2015) Engineering synthetic antibody inhibitors specific for LD2 or LD4 motifs of paxillin. J Mol Biol 427:2532–2547
210. Mateja A, Paduch M, Chang HY et al (2015) Protein targeting. Structure of the Get3 targeting factor in complex with its membrane protein cargo. Science 347:1152–1155
211. Rizk SS, Kouadio JL, Szymborska A et al (2015) Engineering synthetic antibody binders for allosteric inhibition of prolactin receptor signaling. Cell Commun Signal 13:1
212. Koide A, Tereshko V, Uysal S et al (2007) Exploring the capacity of minimalist protein interfaces: interface energetics and affinity maturation to picomolar K-D of a single-domain antibody with a flat paratope. J Mol Biol 373:941–953
213. Lam AY, Pardon E, Korotkov KV et al (2009) Nanobody-aided structure determination of the EpsI:EpsJ pseudopilin heterodimer from Vibrio vulnificus. J Struct Biol 166:8–15
214. Korotkov KV, Pardon E, Steyaert J et al (2009) Crystal structure of the N-terminal domain of the secretin GspD from ETEC determined with the assistance of a nanobody. Structure 17:255–265
215. Gilbreth RN, Esaki K, Koide A et al (2008) A dominant conformational role for amino acid diversity in minimalist protein-protein interfaces. J Mol Biol 381:407–418
216. Sennhauser G, Grutter MG (2008) Chaperone-assisted crystallography with DARPins. Structure 16:1443–1453
217. Batyuk A, Wu Y, Honegger A et al (2016) DARPin-based crystallization chaperones exploit molecular geometry as a screening dimension in protein crystallography. J Mol Biol 428:1574–1588
218. Bandeiras TM, Hillig RC, Matias PM et al (2008) Structure of wild-type Plk-1 kinase domain in complex with a selective DARPin. Acta Crystallogr D Biol Crystallogr 64:339–353
219. Sennhauser G, Amstutz P, Briand C et al (2007) Drug export pathway of multidrug exporter AcrB revealed by DARPin inhibitors. PLoS Biol 5:106–113
220. Veesler D, Dreier B, Blangy S et al (2009) Crystal structure and function of a DARPin neutralizing inhibitor of lactococcal phage TP901-1: comparison of DARPin and camelid VHH binding mode. J Biol Chem 284:30718–30726
221. Thorsen TS, Matt R, Weis WI et al (2014) Modified T4 lysozyme fusion proteins facilitate G protein-coupled receptor crystallogenesis. Structure 22:1657–1664
222. Notti RQ, Bhattacharya S, Lilic M et al (2015) A common assembly module in injectisome and flagellar type III secretion sorting platforms. Nat Commun 6:7125
223. Ullah H, Scappini EL, Moon AF et al (2008) Structure of a signal transduction regulator, RACK1, from *Arabidopsis thaliana*. Protein Sci 17:1771–1780
224. Lanci CJ, MacDermaid CM, Kang SG et al (2012) Computational design of a protein crystal. Proc Natl Acad Sci U S A 109: 7304–7309
225. Hsieh FL, Chang TH, Ko TP et al (2011) Structure and mechanism of an Arabidopsis medium/long-chain-length prenyl pyrophosphate synthase. Plant Physiol 155:1079–1090
226. Vorontsova MA, Maes D, Vekilov PG (2015) Recent advances in the understanding of two-step nucleation of protein crystals. Faraday Discuss 179:27–40

Chapter 5

Crystallization of Membrane Proteins: An Overview

Andrii Ishchenko, Enrique E. Abola, and Vadim Cherezov

Abstract

Membrane proteins are crucial components of cellular membranes and are responsible for a variety of physiological functions. The advent of new tools and technologies for structural biology of membrane proteins has led to a significant increase in the number of structures deposited to the Protein Data Bank during the past decade. This new knowledge has expanded our fundamental understanding of their mechanism of function and contributed to the drug-design efforts. In this chapter we discuss current approaches for membrane protein expression, solubilization, crystallization, and data collection. Additionally, we describe the protein quality-control assays that are often instrumental as a guideline for a shorter path toward the structure.

Key words Membrane protein, Expression, Crystallization, Detergent, In meso, In surfo, HiLiDe, Lipidic cubic phase, Lipidic sponge phase, Bicelle, Nanodisc, Amphipol, FRAP, Thermal shift assay

1 Introduction

Structure determination of integral membrane proteins (IMPs) has traditionally been considered challenging and has therefore mostly been carried out by specialized structural biology laboratories. However, recent breakthroughs and improvements in technologies and protocols are changing this perception and possibly expanding the number of laboratories that can include structural biology tools in their repertoire. This progress reflects a growing understanding of IMP behavior in terms of their production using recombinant expression and their stabilization outside their native membrane environment. IMPs, which comprise about 30% of the human proteome, enable interactions between the cell and its external environment and therefore play significant physiological roles. They are often implicated in human diseases and, hence, are the targets of more than 50% of currently available drugs [1]. The newly determined structures are expanding our understanding of IMPs' mechanisms of action as well as their interactions with ligands and other proteins; these structures also serve as high quality templates for structure-based drug design (SBDD).

Alexander Wlodawer et al. (eds.), *Protein Crystallography: Methods and Protocols*, Methods in Molecular Biology, vol. 1607, DOI 10.1007/978-1-4939-7000-1_5,

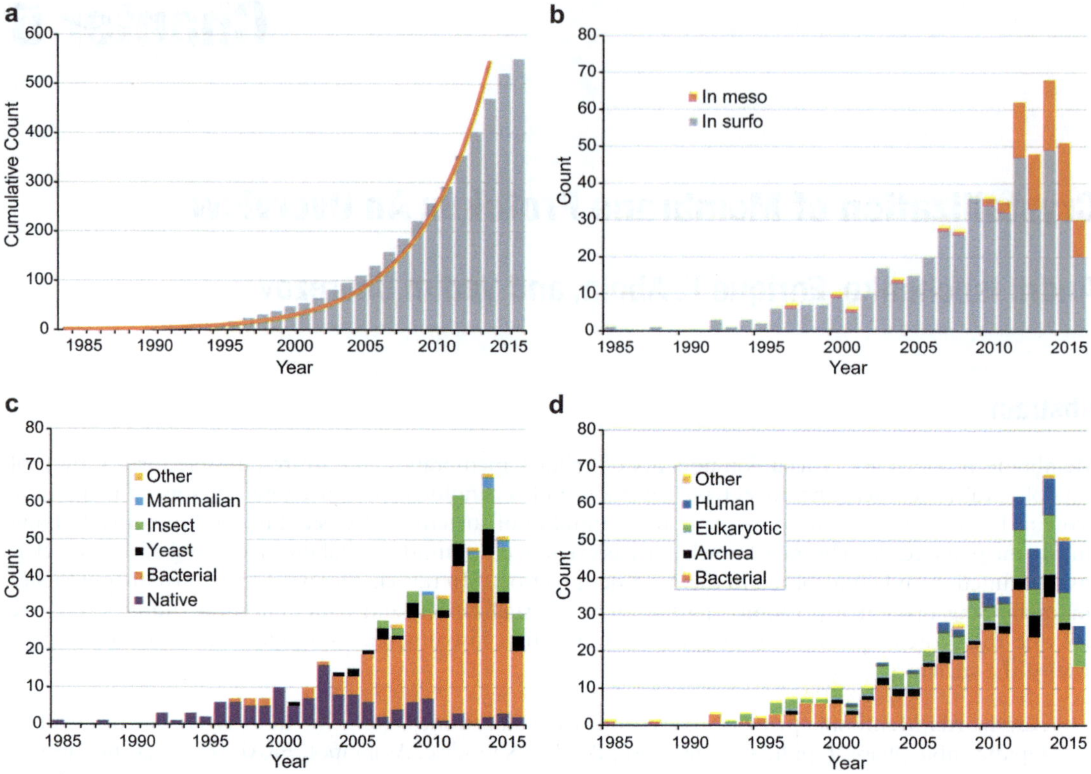

Fig. 1 IMP structure determination statistics. (**a**) Cumulative number of unique IMP structures solved by X-ray crystallography. The *orange curve* shows the best exponential fit, illustrating that the increase in the number of unique IMP structures does not follow an exponential growth. (**b**) Number of unique IMP structures solved per year using in surfo and in meso crystallization. During the last 5 years, in meso crystallization has been contributing on average about one-third of all structures. (**c**) Statistics of the use of different expression systems for unique IMP structures. "Other" data include cell-free expression and synthetic proteins. (**d**) Distribution of IMP source for unique structures. "Eukaryotic" data represent IMPs from eukaryotic organisms other than *Homo sapiens*. "Other" data include viral and man-made sequences. The data on unique IMPs were collected on May 1, 2016 from PDB and from MPSTRUCT database (http://blanco.biomol.uci.edu/mpstruc/)

There are currently over 1800 crystal structure entries in the Protein Data Bank (PDB) for ~550 unique IMPs (Fig. 1), covering all three major IMP folds: monotopic, polytopic β-barrel, and polytopic α-helical (Fig. 2a). Most structure determination studies traditionally have been carried out using vapor diffusion techniques with protein-detergent complexes (in surfo methods), however, in the last few years there is a noteworthy increase in the number of structures that have been solved using crystallization in lipidic mesophases (in meso methods) (Figs. 1b and 3).

This chapter summarizes modern approaches for crystallizing difficult IMP targets. We provide an overview of the whole process and discuss major steps, highlighting important parameters and metrics that have been used in successful studies. Detailed protocols have been published [2–5] and should be consulted for details.

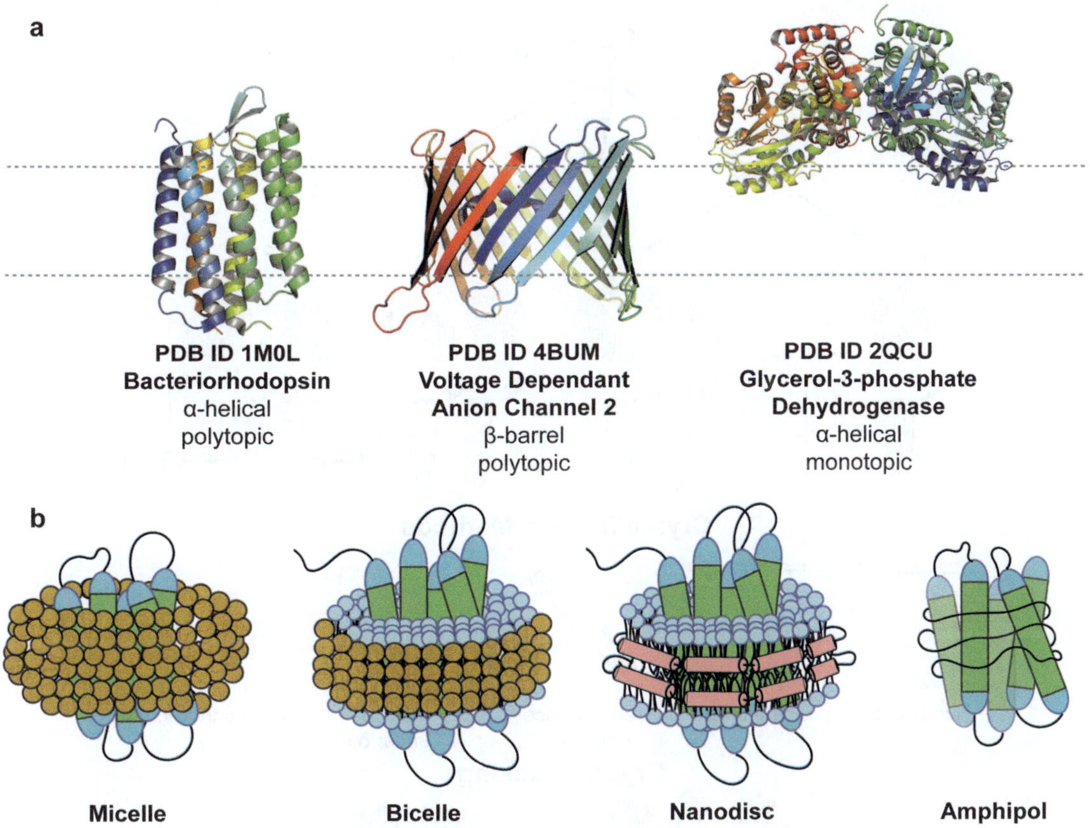

Fig. 2 (**a**) Examples of IMPs with different architecture. (**b**) Common IMP solubilization approaches

These include video-protocols [6–10], which provide an essential resource especially for those who are just starting to work in the field. This overview should therefore help assessing the possibility of including structural studies in a laboratory's attempt to understand the biological function of their protein of interest.

2 Process Overview

Until approximately 2005, most of the published IMP structures relied on the material obtained from native sources containing the target IMP in high abundance [11–13] (Fig. 1c). Recent improvements in methods for gene engineering, protein expression, solubilization, stabilization, and purification have substantially expanded the repertoire of IMPs amenable to structural studies, including human proteins (Fig. 1d). Since recombinant expression currently contributes over 95% of all published structures (Fig. 1c), approaches that involve homologous or heterologous expression will be the focus of our review. Crystallization trials of a target IMP typically require the production of several milligrams of highly

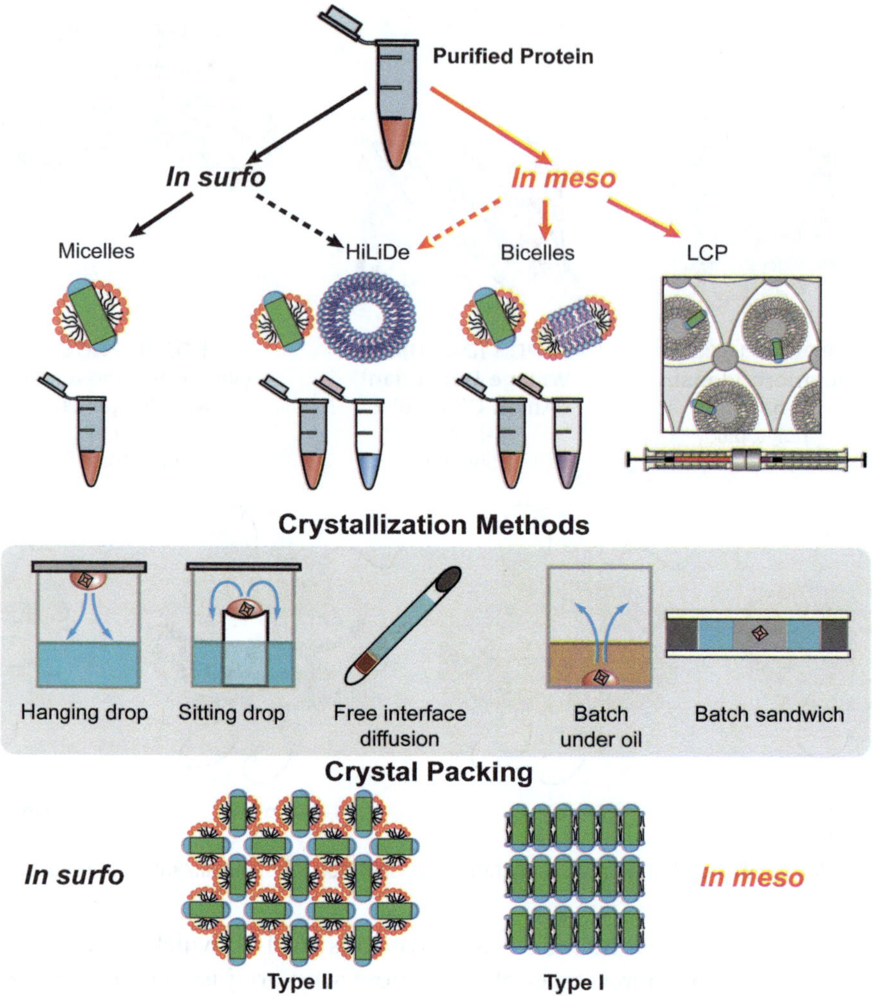

Fig. 3 A summary of IMP crystallization approaches. Depending on IMP environment, crystallization approaches can be classified as in surfo (IMP solubilized in detergent micelles) or in meso (IMP reconstituted in a lipidic mesophase). Regardless of the IMP environment, crystallization trials can be set up in any of the common formats: vapor diffusion (hanging or sitting drop), free-interface diffusion, batch. Most commonly, in surfo crystallization is performed in the vapor diffusion format, while in meso crystallization—in batch glass sandwich plates (LCP crystallization), or vapor diffusion (bicelles crystallization). In surfo crystallization typically results in type II, while in meso crystallization in type I crystal packing

purified and monodisperse protein that is stable outside its native membrane environment. Meeting these requirements often necessitates application of protein engineering. In those cases, it is essential to follow up with functional studies to ensure biological relevance of the construct that is crystallized. For example, crystallization of most G protein-coupled receptors (GPCR) required truncations of flexible termini, introduction of point mutations and/or fusion of a small soluble protein to the N terminus or grafting it into one of the intracellular loops [14]. While some of the introduced modifications may compromise signaling, most of the crystallized constructs have been shown to retain near-native ligand-binding affinities.

In general, crystallization of a new IMP target requires the execution of several major process steps starting with the design and synthesis of the gene of interest, cloning and expression using a homologous or heterologous system, purification and characterization of expressed protein, crystallization, and, finally, diffraction screening and data collection (Fig. 4). The process is iterative and several interrelated objectives have to be achieved to increase the likelihood of success. For example, identifying constructs and conditions that produce monodisperse samples often also yields improved stability and increased levels of recovery of pure protein. Important objectives include optimization of expression, extraction and purification protocols to increase protein yield and quality, stabilization of the protein to ensure conformational homogeneity, and optimization of crystallization conditions to obtain sufficiently large and well diffracting crystals for data collection. Critical metrics to follow include production of over 0.2–0.5 mg of purified protein per 1 L of biomass, high level of monodispersity as measured by analytical Size Exclusion Chromatography (aSEC), high protein melting temperature, T_m, as measured by thermal shift assays [15], and crystal size and diffraction quality. While it is impossible to set universal thresholds for the abovementioned parameters that would guarantee success in crystallization experiments, as they are strongly dependent on the nature of the target, these metrics provide important guidance during the optimization cycle and their improvement correlates with positive crystallization outcomes.

Initial expression and characterization studies are typically performed in small volumes, starting at 2–5 mL for screening constructs and increasing to 40–250 mL to measure monodispersity, T_m and other parameters. Small volume studies allow for rapid characterization of samples and, more importantly, parallel screening of multiple constructs, as well as sample conditions including the use of a variety of ligands. Once suitable constructs have been identified, large-scale expression is carried out in 1–10 L batches for crystallization trials. Small-scale pre-crystallization and crystallization assays can be initiated at any time in the sample screening process as they help develop a better understanding of the behavior of the samples. Although the process shown in Fig. 4 appears to be linear, it should be emphasized that actual work is done in a cyclic fashion, with the first cycle focusing on identifying constructs and conditions for high-level expression, the second focusing on improving protein stability and iterating all the way back to construct design if needed, and the final cycle focuses on optimizing sample conditions to generate suitable crystals for diffraction studies.

In the following sections, we provide discussion pertinent to the various major process steps and provide goals and requirements that have to be achieved and addressed.

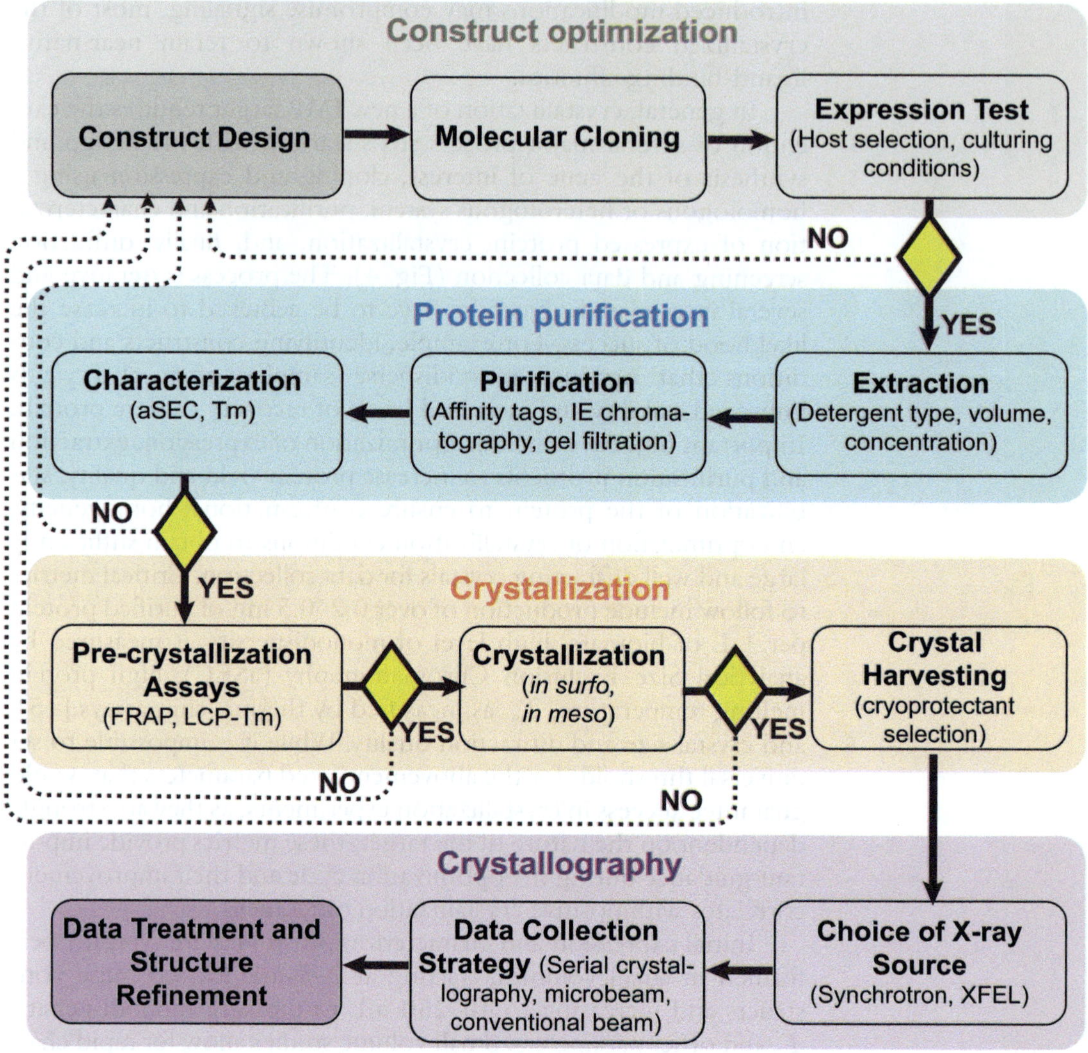

Fig. 4 Schematic diagram of a typical IMP structure determination process. *Yellow rhombic boxes* represent decision points, at which—if the sample passes certain criteria—the process proceeds to the next step, otherwise it returns back to the previous stage or to the beginning of the procedure for further optimization

3 Construct Design

The objective of this step is to produce a construct for overexpression, which requires the design of an expression vector and modifications to the sequence of the target IMP. Selection of a suitable expression vector includes a number of considerations, such as the choice of expression system, promoter, inducer, signal peptide sequence, antibiotic selection, as well as the type and placement of purification tags (His-tag, Flag-tag, etc.), which are often attached behind a cleavable sequence to either the N or C terminus, depending on functional and trafficking considerations. In the case of heterologous expression, codon optimization of the wild-type target gene for a specific expression system can be performed to account for differences between tRNA levels among species and mRNA secondary structure constraints. Additional modifications of the wild-type gene typically include truncations of flexible termini and loops, stabilizing mutations, fusions with other proteins, etc., some of which are described in more detail in Subheading 6.

4 Membrane Protein Expression

Achieving high level of functional expression of a given IMP is in general a difficult task. There is no complete understanding of all factors influencing synthesis, folding, and trafficking of IMPs, which vary substantially from one host to another. The choice of expression host, therefore, depends on the protein toxicity, desired level of expression, obligatory post-translational modifications, folding and trafficking chaperons, and, last but not least, the budget of the project.

4.1 Bacterial Expression

Several bacterial expression hosts have been used for recombinant IMP production. *Escherichia coli* (*E. coli*) is, however, by far the most popular prokaryotic expression host, often producing large quantities of IMPs with minimal effort, time, and cost [16]. The *E. coli* system has numerous advantages that make it highly successful: very fast growth kinetics (cell density doubles about every 20 min during the log phase), versatility of media and growth conditions (both minimal and rich media are possible, toxicity can be reduced by expressing at low temperatures), inexpensive consumables, and the availability of various strains and vectors to accommodate different research needs. Typically, vectors for *E. coli* expression carry a T7 promoter to drive the recombinant protein production via T7 RNA polymerase. *Lac* operon control system is used as a switch for protein expression under a T7 promoter, which is triggered by addition of isopropyl β-D-thiogalactoside (IPTG) in the middle of the log cell growth phase. Induction strength can be modulated by varying IPTG concentration, culture density, and incubation temperature after induction.

Auto-induction media are also available and were shown to be, in certain cases, more efficient than traditional IPTG-induced expression [17]. This type of media contains a mixture of lactose and glucose at a certain proportion. At the beginning of culturing, cell growth is supported by consumption of glucose as the main nutrient. As glucose depletes, bacteria switch to lactose as an energy source, which, in turn, induces protein production similarly to IPTG. In this way, the auto-induction system offers mild gradual induction at a predictable cell growth phase compared to

IPTG, reducing toxicity effects. Additionally, bacteria expressed in this media tend to have higher final cell density and higher relative yield [18].

Despite all advantages, the *E. coli* system has also its limitations. Being a prokaryotic organism, *E. coli* lacks all the machinery related to post-translational modifications, essential lipids, and molecular chaperones necessary for correct folding and trafficking of some eukaryotic IMPs.

4.2 Yeast Expression

Recombinant yeast expression was the first expression system applied to the production of eukaryotic IMPs that did not express well in *E. coli* [19]. Sharing many advantages with *E. coli*, such as rapid growth and low cost, yeast expression provides an advantage of being capable of implementing various post-translational modifications. *Saccharomyces cerevisiae* and *Pichia pastoris* are the most widely used yeast strains for IMP expression; however, there are a few others available (e.g., *Komagataella pastoris*, *Schizosaccharomyces pombe*) that can be beneficial in special cases. Yeast has proven to be a suitable expression system with about 40 unique eukaryotic IMP structures determined (Fig. 1c), including potassium channels and GPCRs [20–24]. *P. pastoris* shares the advantage of simple molecular and genetic manipulations with *S. cerevisiae,* while providing an extra advantage of 10- to 100-fold increased biomass yield out of the same culture volume [25]. A potential disadvantage of the yeast expression is the difficulty of cell disruption due to the hard cell walls.

4.3 Insect Cell Expression

Baculovirus-infected insect cells represent the most successful expression system to date for production of eukaryotic IMPs for structural studies [26]. The use of such a system has contributed to the determination of almost 70 unique structures (Fig. 1c), many of which are GPCRs. The two most popular insect cell lines used for IMP expression are *Spodoptera frugiperda* (*Sf9*) and *Trichoplusia ni* (*Hi5*). These two cell lines have varied efficiency depending on the particular target, with *Sf9* being responsible for most structures of IMPs grown in insect cells. Production of proteins via a baculovirus expression vector system (BEVS) involves two stages [26]. First, insect cells are cultured to a desired concentration (typically, $1–3 \times 10^6$ cells/mL). In the second step, the cells are infected with a baculovirus, containing the target gene. The baculovirus takes control over the gene expression machinery of the host cell that leads to the production of the target protein. In parallel to this process, the virus replicates itself using the metabolic machinery of the host. Insect cells are capable of most of the post-translational modifications that occur in mammalian cells, with a few exceptions. The N-linked glycosylation in insect cells results in glycoproteins with only simple oligo-mannose sugar chains, whereas in mammalian cells it produces glycoproteins with complex sugar groups [27].

The O-glycomes of insect cell lines can be complex and diverse; however, the O-glycosylation potential depends on the cell type. Insect cells lack sialyltransferase activity and, therefore, do not produce terminal sialic acid [28].

The most common method for generating baculovirus is based on the site-specific transposition of an expression gene of interest into a baculovirus shuttle vector (bacmid) that is proliferated in *E. coli* [29]. After culturing insect cells to a desired density, they are transfected by the purified bacmid DNA to generate a recombinant baculovirus that is used for the subsequent infection of insect cells to generate the protein of interest. The entire process takes 2–3 weeks before the first expression results are obtained. Total and cell-surface expression of the target protein can be evaluated using flow cytometry [30].

4.4 Mammalian Cell Expression

The reasons why some IMPs are well overexpressed in the above-mentioned systems while others are expressed poorly are not fully understood. Some mammalian IMPs likely require specific environment for proper folding that is not available in bacteria, yeast, and even insect cells. However, such proteins can often be produced in mammalian cells at sufficiently high levels to support structure determination [31, 32]. There are several cell lines that are relatively well studied and typically used for protein expression: Chinese hamster ovary cells (CHO), human embryonic kidney cells (HEK-293), baby hamster kidney cells (BHK-21) and monkey kidney fibroblast cells (COS-7), with HEK cells being most successful to date. Transient and stable expression modes are typically used with mammalian cells. Stable expression requires additional time for generation of stable cell lines but offers higher yield and reproducibility in long-term experiments. The disadvantages of expression in mammalian cell lines include high cost of expression media and transfection reagents, and relatively long turnover time (in the case of establishing stable expression).

5 Extraction, Solubilization, and Purification

5.1 Traditional Detergents

IMPs are most stable and functional in their native membrane environment. Extraction of IMPs from their native membranes into a soluble state, which is required for purification, is typically achieved by solubilization in detergents (surfactants) or mixed detergent/lipid systems.

Detergents are amphiphilic molecules that contain a polar head group and a hydrophobic alkyl chain tail. In aqueous solutions they spontaneously assemble into micelles with their hydrophobic tails hidden in the micelle core and polar heads exposed to the solution (Fig. 2b). This process is described by an important physicochemical parameter, known as the critical micelle concentration

(CMC), which corresponds to the minimal detergent concentration required for the formation of a micelle. At concentrations higher than the CMC, the detergents exist in a monomer–micelle equilibrium, and the concentration of free monomer detergent molecules in solution remains essentially constant [33]. As a rule-of-a-thumb, detergents are used at concentrations 10–20× CMC for IMP extraction and at least 1.5× CMC for protein purification. While CMC is mostly defined by the chemical structure of the detergent molecule, it also depends on environmental factors, such as temperature and ionic strength of the buffer. Detergent micelles comprise a few dozen to several hundred detergent molecules, and the average number of molecules in a micelle constitutes another important detergent parameter (aggregation number) [33].

The structure of the head group dictates the specific interactions of detergents with proteins, while the detergent tail length affects the CMC and the aggregation number [33]. Detergents are generally classified in three major categories, depending on the chemical structure of their headgroups (ionic, zwitterionic, nonionic). Ionic detergents bear a head group with a positive or negative net charge and a hydrocarbon tail. This type of detergents is considered relatively harsh, meaning that it is very efficient in solubilizing membrane proteins, but often fails to maintain the native IMP fold. Therefore, harsh detergents are only suitable for those applications where correct folding is not needed, such as chromatography or mass spectrometry at denaturing conditions. Examples of typical harsh ionic detergents are sodium dodecyl sulfate (SDS) or N-lauroylsarcosine (NLS). Bile salts, such as sodium cholate and CHAPS, represent a different class of milder anionic detergents with a distinguishingly different, rigid steroid core structure unlike the more standard flexible tail structure. They can efficiently extract IMPs from membranes without denaturing them. After extraction, these detergents can be exchanged with a different type of detergents that are more suitable for crystallization [34].

Non-ionic detergents can be classified in two groups: polyoxyethylene ethers and glycosidic detergents. Polyoxyethylene ethers have a short hydrophobic tail and a neutral, polar head composed of oxyethylene polymers (e.g., Brij and TWEEN) or ethyleneglycoether polymers (e.g., TRITON). Glycosidic detergents have a carbohydrate polar head, typically maltose or glucose, and a hydrophobic alkyl chain of 7–14 carbon atoms. These detergents represent the most popular class of detergents used for IMP extraction, purification, and crystallization. Notable examples include n-dodecyl-β-D-maltoside (DDM), n-decyl-β-D-maltoside (DM), and n-octyl-β-D-glucopyranoside (OG). The strength and the micelle size of detergents from this class depend on the length of their alkyl chain. For example, the short-chain nonionic detergent OG has a high extraction efficiency and forms small micelles, but it is relatively harsh and often leads to protein destabilization.

On the other hand, the long-chain DDM is a mild, stabilizing detergent, however, its larger micelles may interfere with crystallization. Mixtures of several detergents or addition of small amphiphiles, such as 1,2,3-heptanetriol, are often used to balance the micelle size and protein stability for in surfo crystallization [35].

Finally, zwitterionic detergents have a hydrophilic head with both positive and negative charges separated in space, resulting in a zero net charge. They are positioned between ionic and nonionic detergents in terms of their solubilizing efficiency. Some of them, like LDAO and CHAPSO, are relatively mild and have been successfully used for IMP crystallization using the vapor diffusion method.

Screening for the best detergent that can enable the highest yield of pure, stable and functional protein represents an important initial step of working with a new IMP target. Detergent kits are provided by several vendors for this purpose. Since a single detergent may not always be optimal for extraction, stabilization, and crystallization, one detergent, for example, could be used for extraction and later be replaced with another one for purification and/or crystallization.

5.2 Alternative Solubilizing Approaches

Since traditional detergents often impose limitations on the properties of the protein-detergent micelles, new detergents and alternatives approaches to solubilization of IMPs for crystallization and other biophysical studies have been explored. Among novel detergents, the most promising include tripod amphiphiles [36], facial amphiphiles [37], perfluorinated surfactants [38], and neopentyl glycols [39]. Neopentyl glycols feature two hydrophobic tails and two hydrophilic headgroups, typically maltose (MNG) or glucose (GNG), attached to a quaternary carbon in the center of their scaffold. Due to the presence of two hydrophobic tails, neopentyl glycols have low CMC and high stabilizing properties. For this reason, they became popular for purification of GPCRs, transporters, and other unstable IMPs for biophysical studies and in meso crystallization [40–42]. Due to the large micelle size, however, their application for in surfo crystallization is limited.

Peptide-based amphiphiles were designed to better mimic the architecture of the native membranes [43]. These lipopeptide detergents (LPDs) carry two alkyl chains attached to the ends of a peptide α-helical backbone. The length of the peptide is designed to match the typical width of the lipid bilayer (~40 Å). LPDs were shown to maintain IMPs in solution without aggregation and denaturation [44]. The disadvantage of LPDs is that they are difficult to synthesize at a large scale despite efforts to overcome this problem [45]. LPDs have not yet contributed to any IMP structure.

An alternative to the detergent solubilization approach is to use long amphiphilic polymers, known as amphipols (Apols; Fig. 2b) [46]. The polyacrylate backbone of these polymers is hybridized with various side chains, producing amphiphilic polymers with ionic, non-ionic, or zwitterionic properties. One of the most established and successful Apols is A8-35, where the letter A stands for "anionic", the first number refers to the average apparent MW (approx. 8 kDa) and the second number represents the fraction of free carboxylic groups (35%). A8-35 is unable to extract IMPs from their native membranes [47]. Therefore, extraction is performed first by a conventional detergent and then the detergent is exchanged for A8-35, typically using Bio-Beads (Bio-Rad). Apols were shown to stabilize several IMPs in solution and were successfully used for crystallization in lipidic cubic phase [48].

Lastly, the closest to the native environment for solubilized IMPs can be achieved using nanodiscs [49]. Nanodiscs are disk-like patches of lipid bilayers surrounded by an amphiphilic helical belt made of a membrane scaffold protein (MSP) (Fig. 2b). MSPs are truncated forms of human serum apolipoprotein A1, which encircle the lipid bilayer and stabilize it. There are several variants of MSPs that are available and the choice of a particular MSP construct defines the size of the nanodiscs. MSP1D1 and MSP1D1-deltaH5 are most widely used mutants that result in particles of 7–13 nm diameter comprising about 150 phospholipid molecules. MSP constructs often contain an affinity tag to facilitate purification [50, 51]. Reconstitution of IMPs into nanodiscs occurs spontaneously, when the purified protein solubilized in detergent is mixed with MSPs and phospholipids, and the excess of detergent is removed by Bio-Beads or dialysis [52]. Due to their bulk size, IMPs solubilized in nanodiscs are not well-suited for in surfo crystallization, and no high-resolution structure has yet been reported. Nanodiscs, however, can often provide an ideal native-like environment for solubilized IMPs for functional, biophysical and structural studies by NMR, surface plasmon resonance (SPR), Raman spectroscopy, single molecule fluorescence, Cryo-EM, mass spectrometry, and other methods [50, 53].

6 Approaches to IMP Stabilization

It has been well established that protein stability correlates with crystallization success [54, 55]. The choice of detergent is one of the most critical factors affecting stability of solubilized IMPs. Many IMPs are, however, unstable even in mild detergents and require additional stabilization to decrease their conformational heterogeneity and improve chances for crystallization. The most successful approaches to date include binding to a conformation-selective antibody/nanobody, fusion partners, and site-specific mutagenesis.

Monoclonal antibodies have been used for over 15 years as crystallization chaperons for obtaining high-resolution crystal structures of highly dynamic IMPs and complexes across various protein families including ion channels, membrane transporters and GPCRs [56–59]. More recently, focus has shifted toward nanobodies, which represent single-chain antibodies, produced by some animals, such as camelids and sharks, and provide certain advantages compared to conventional Fab fragments due to their high affinity, specificity and a compact size [60]. Nanobodies have been particularly successful for the stabilization of active conformations of GPCRs for LCP crystallization [61–63]. The disadvantages of this approach include the need for substantial investment of time, effort and cost on generation and selection of the right antibody/nanobody for crystallization, and the possibility that binding to an antibody/nanobody can potentially lock the protein or complex in an artificial conformation.

Fusion of the target IMP with a compact, stable soluble protein in place of flexible loops or at the N- or C-terminus has been another highly successful strategy for reducing conformational heterogeneity and increasing polar surface for crystal contacts. This approach has been instrumental for structural studies of GPCRs [57, 64, 65], and it should be applicable to other unstable IMPs as well.

Finally, it was demonstrated that strategically placed point mutations can help to stabilize IMP in a single conformational state without disrupting the overall structure. Alanine scanning mutagenesis has been developed as a systematic approach for identifying suitable mutations and extensively applied to stabilize several GPCRs in active and inactive states for biophysical and structural studies [66–68]. Several point mutations can synergistically increase protein stability, often by over 20 °C, enabling crystallization by the in surfo vapor-diffusion method. Directed evolution represents another promising approach for evolving IMPs for structural studies [69, 70]. Both of these techniques are, however, quite laborious and have their own limitations. Computer modeling combined with accumulated experimental data may eventually provide a more rational and straightforward approach for designing stabilizing mutations in IMPs.

7 Assays for Measuring IMP Stability

Since stability is a critical factor in preparation of suitable IMP samples for crystallization trials, there is a need for a simple and robust assay that can probe stability of many different IMP constructs at different conditions. Traditionally, thermal stability of proteins has been measured using biochemical assays, differential scanning calorimetry, or circular dichroism. These methods, however, are not easily amenable to a high-throughput format. Thermal Shift Assay (TSA) was initially developed for probing thermal stability of soluble proteins at a variety of conditions [71]. As the temperature is ramped up, protein unfolding is monitored by fluorescence from intrinsic tryptophan residues or from a reporter dye. Shifts in the melting temperature with respect to various factors, such as addition of ligands, changes in the protein construct, varying buffer pH and additives, etc., indicate the effect of these factors on protein stability. Measuring protein unfolding by intrinsic fluorescence is convenient, but this approach typically suffers from relatively low signal intensity. Special dyes, such as SYPRO Orange and 1,8-ANS, that increase fluorescence upon nonspecific binding to hydrophobic surfaces of an unfolded protein have been successfully employed to improve sensitivity [72, 73]. These dyes, however, are incompatible with detergent micelles, therefore a special procedure was developed for measuring stability of solubilized IMPs, making use of a thiol-reactive fluorescent probe 7-diethylamino-3-(4′-maleimidylphenyl)-4-methylcoumarin (CPM) [15]. CPM has a low fluorescence in aqueous solution (excitation/emission maxima at 384/470 nm), which increases dramatically when it reacts with cysteine residues that become exposed to solution upon protein unfolding. The only prerequisite for the assay is the availability of buried cysteines inside the IMP core. One of the limitations of this assay is its sensitivity to pH and some components of the protein buffer, such as imidazole, which could be mitigated by desalting. Another limitation is a potential overlap between excitation/emission spectra of CPM and those of certain IMP ligands. Other thiol-reactive dyes with properties similar to CPM but different excitation/emission spectra could be employed in this case [74].

Another commonly used assay for measuring the stability of solubilized IMPs is based on analytical size-exclusion chromatography (aSEC) [75]. Crystallization-quality IMP sample should ideally have a single narrow monodisperse aSEC peak. A peak at the column void volume indicates protein aggregation and should be minimized as much as possible. IMP stability can be evaluated by running the sample on an aSEC before and after short (5–10 min) incubations at elevated temperatures, and monitoring the change in the peak height corresponding to the initial monodisperse state of the protein. In another variation of this method, a green fluorescent protein (GFP) tag is fused to the IMP target, increasing the detection limit and allowing to trace the target IMP by aSEC without purification [76].

8 IMP Crystallization Methods

Since 1985, IMPs have been successfully crystallized as protein-detergent complexes (PDC) (in surfo approach) [13]. This method primarily results in so-called type II crystal packing, in which crystal contacts form only between hydrophilic parts of the protein (Fig. 3). Such packing is often associated with high solvent content and low order. An alternative method for IMP crystallization was introduced in 1996 [77], in which IMPs are incorporated in a lipidic mesophase, such as lipidic cubic phase (LCP), and crystallized directly from such native-like membrane environment. Subsequently, other types of lipid mesophases have been employed for crystallization, and, therefore, we will refer to these methods collectively as in meso crystallization. In meso crystallization produces type I crystal packing (Fig. 3), which, in general, should lead to better ordered crystals and higher-resolution diffraction. While introduced about 20 years ago, in meso methods started gaining momentum only few years ago and currently contribute about one-third of all new unique IMP structures (Fig. 1b). The lag period was required to develop new tools and technologies for manipulations with lipidic mesophase materials [78].

8.1 In Surfo Crystallization

Once solubilized in detergents, IMPs may be treated, for most purposes, as soluble proteins. Therefore, all crystallization methods developed for soluble proteins can be equally applied to IMPs. The purpose of any crystallization technique is to drive initially solubilized IMPs into a nucleation zone, where a few crystals can nucleate, and then transition into a metastable zone, where these crystals continue to grow. The difference between vapor diffusion, dialysis, batch under oil, free-interface diffusion and other methods, apart from their setup, is in the trajectory through the phase space that in each of them is achieved. Vapor diffusion has been so far the most popular method for IMP crystallization, contributing over 95% unique structures obtained by in surfo approaches.

The crystallization phase diagram for a given IMP target is affected by the choice of detergent as well as by many other factors including temperature, pH, ionic strength, type and concentration of buffer and precipitants. Since it is not possible to predict a priori which of these conditions will lead to crystal formation, crystallization trials typically involve an extensive screening. By analyzing conditions that most often led to crystallization (hot spots), special screens for IMP crystallization, such as MembFac, MemStart, MemSyS, MemGold [35], MemAdvantage [79] have been developed and made commercially available. Advancements in liquid handling and crystal imaging automation allowed using as little as 50 nL of protein solution per drop, enabling rapid and efficient screening of thousands of conditions with only 1 mg of purified IMP, thus making it possible to work with more challenging heterologously expressed human IMPs.

It is evident that detergents play an essential role during in surfo crystallization, but the requirements for suitable detergents for crystallization are different from those that are important for extraction or purification. Generally, shorter-chain detergents, although rendering IMPs less stable, have a higher likelihood of yielding crystal hits as they generate smaller PDCs allowing for tighter crystal contacts. While several dozen detergents have been used in IMP crystallization, only a few of them, such as alkyl glycosides (OG, NG) and maltosides (DDM, DM), LDAO, or polyoxyethylene detergents (e.g., C8E4, C12E8) (http://mpdb.tcd.ie; [80]), are responsible for the majority of IMP structures.

8.2 In Meso Crystallization

A common feature of all in meso crystallization methods is that IMPs are crystallized directly from the membrane environment of a lipidic mesophase. Lipids, upon mixing with aqueous solution, can self-assemble in a variety of mesophases depending on their chemical structure, temperature, hydration, and composition of the aqueous solution. Several of these mesophases, such as Lipidic Cubic Phase (LCP) [77], Lipidic Sponge Phase (LSP) [81, 82], perforated lamellar phase, obtained from bicelles [2, 83], and connected-bilayer gel [84] have been shown to be compatible with IMP crystallization. Based on accumulated experience, a general requirement for a mesophase to support crystallization is to be composed of a network of interconnected lipid bilayers and aqueous channels. IMPs, reconstituted into these bilayers, should be able to diffuse within them over long distances, eventually reaching and joining a growing crystal.

Crystallization in LCP (including LSP) is so far the most successful of all in meso methods, with overall contribution of about 15% to all unique IMP structures. Its success can be partly explained by an exceptional compatibility of LCP with a large array of precipitant conditions [85], enabling extensive screening of the crystallization space. Additionally, due to its spatial constraints, the LCP acts as a filter, prohibiting diffusion of large impurities and protein aggregates, and thus excluding them from incorporation into growing crystals [86]. The downside is that large-size target IMPs (>100 kDa) may also be excluded from diffusion in LCP. These spatial constraints can be relieved by swelling the LCP and transforming it into an LSP, which can be achieved by increasing concentrations of some common precipitants, such as PEG 400, 2-methyl-2,4-pendanediol (MPD), pentaerythritol propoxylate, and 1,4-butanediol [82]. One of the largest IMPs crystallized in a proper LCP is T4L-rhodopsin-arrestin complex with the molecular weight of ~100 kDa [87]. LSP can support crystallization of larger IMPs, such as T4L-β_2AR/Gs/nanobody complex with a total molecular weight of ~165 kDa [88].

Monoacylglycerols (MAGs) represent the most common class of host lipids used for LCP crystallization [89]. They are composed of a hydrophilic glycerol headgroup attached to a hydrophobic monounsaturated fatty acid chain via an ester bond, and are commonly referred to as N.T MAG, where "N" (neck) represents the number of hydrocarbons between the ester bond and the polar head, and "T" (tail) represents the number of such groups between the double bond and the end of the hydrophobic tail [90]. Monoolein, or 9.9 MAG, is by far the most successful host lipid for LCP crystallization, but other MAGs are available and can provide better environment for crystallization of certain IMPs [91, 92]. While MAGs are not native lipids of biological membranes, LCP generally can be doped by 5–20% of native lipids, such as phospholipids or cholesterol [93], which could specifically bind to the target IMP, increasing its conformational stability, and, thus, facilitating their crystallization.

LCP has a consistency of a viscous, transparent gel, which cannot be manipulated by a pipette and, therefore, requires special tools for handling. Many of such tools and instruments have been developed during the last 20 years. They include lipidic syringe mixer [94], repetitive nano-dispenser [95], glass-sandwich plates [96], and LCP injector for serial crystallography [97]. LCP crystallization setup and crystal imaging in LCP have been automated [98] with several instruments available on the market. IMPs can be assayed directly in LCP for their function [99], thermostability (LCP-Tm assay) [100], and mobility (LCP-FRAP assay) [101], thus allowing to quickly bypass unfavorable conditions and focus on the most promising crystallization trails. All these developments made LCP crystallization easily accessible to most structural laboratories.

Bicelle crystallization represents the second most popular method of in meso crystallization [83]. Bicelles have a disk-like shape made of a lipid bilayer membrane, typically composed of phospholipids, such as 1,2-dimyristoyl-sn-glycero-3--phosphocholine (DMPC), the rim of which is stabilized by short-chain lipids, such as 1,2-diheptanoyl-sn-glycero-3-phosphocholine (DHPC), or detergents, such as 3-((3-cholamidopropyl) dimethylammonio)-2-hydroxy-1-propanesulfonate (CHAPSO) (Fig. 2b) [102]. Bicelles provide excellent membrane-mimicking environment for solubilized IMPs, similar to nanodiscs, and have been extensively used for biophysical studies [103–105]. At right conditions, bicelles are stable in solution at low temperature (~4 °C) and transform into a gel-like perforated lamellar phase upon temperature increase [106]. It is the latter mesophase that supports IMP crystallization. The advantage of bicelles over LCP crystallization, apart from their more native-like membrane composition, is the ease of crystallization setup, in which chilled bicelles are treated like PDCs and do not require special tools for their handling. Crystallization trials are set up by pipetting or liquid-dispensing robots, typically in a vapor diffusion format. One of the serious disadvantages is that bicelle compositions have a rich and complex phase diagram, and the effect of precipitant solutions on their phase behavior is not well understood, with many common screening conditions being incompatible, leading to phase separation and false-positive crystals of lipids or detergents.

A number of IMPs have been crystallized by the bicelles method, including microbial rhodopsins, such as bacteriorhodopsin [83, 107] and xanthorhodopsin [108], GPCRs (β2 adrenergic receptor [109]), enzymes (e.g., rhomboid protease [110]), transporters (e.g., LeuT [111], maltose transporter [112]), channels (voltage-gated sodium [113] and calcium channel [114]), and β-barrels (e.g., VDAC, TamA [115, 116]). This method, therefore, should be included in the arsenal of any structural biology lab working with IMPs.

8.3 HiLiDe Method

A hybrid method that combines the advantages of in surfo and in meso approaches has recently been described, in which high concentrations of lipids and detergents are systematically screened, thus contributing to its name—HiLiDe [4]. In this method, the target IMP in the form of PDC is mixed with increasing concentrations of lipid/detergent micelles, and crystallization is set up by the vapor diffusion or batch method. Upon incubation with precipitant solutions, lipids and detergents may form a variety of lipidic mesophases similar to those encountered during in meso crystallization. In the last few years, several structures were solved using the HiLiDe method, including vitamin K epoxide reductase [117], NMDA receptor [118], glutamate-gated chloride channel GluCl [119], multihydrophobic amino acid transporter MhsT [120], P-type ATPases [121, 122], two-pore channel TPC1 [123], and a SecA-SecY protein translocation complex [124]. It appears that protein delipidation occurring during purification process can often be detrimental to protein stability, therefore addition of lipids during crystallization setup as employed in the HiLiDe method can often lead to an improved crystal formation.

9 X-Ray Diffraction Data Collection Strategies

9.1 Rotation-Based Data Collection

Crystallographic data collection is typically performed by the oscillation (or rotation) method, in which a single crystal mounted on a goniometer is oscillated (or rotated) over a small angle during each exposure. To reduce radiation damage crystals are cryocooled at 100 K, which requires selection of a suitable cryoprotectant. Considerations for the choice of the cryoprotectant in case of in surfo grown IMP crystals are similar to those applied for the crystals of soluble proteins [125, 126]. Glycerol, ethylene glycol, PEGs, some sugars and alcohols are reasonable choices, but their effects should be tested in each individual case [127]. LCP crystallization often employs PEG400 as a precipitant [128], which along with lipids protects crystals during freezing, eliminating the need for an additional cryoprotectant.

Despite tremendous progress achieved in all aspects of the IMP structure determination pipeline, generation of diffraction quality crystals still remains a major bottleneck. IMP crystals are difficult to grow and often only tiny crystals are available. Development of microcrystallography at the third generation synchrotron beamlines has enabled high-resolution data collection from IMP crystals as small as 10 μm. Advancements in software and hardware have simplified centering small crystals, namely by rastering with an attenuated minibeam [129] or by imaging techniques like SONICC [130], which is especially useful if crystals are located in opaque media, such as frozen LCP. To minimize radiation damage and extract as much resolution as possible, crystallographic data are now routinely collected on multiple crystals and merged together into a single dataset. In situ data collection is becoming popular for screening crystals for diffraction quality and even for structure determination, as it eliminates the tedious crystal harvesting step [131–133].

9.2 Serial Crystallography

The advent of new generation X-ray sources, X-ray Free Electron Lasers (XFELs), triggered the development of a new approach to crystallographic data collection, known as serial femtosecond crystallography (SFX) [134]. Extremely bright XFEL pulses of extremely short duration enabled collection of high-resolution data from tiny crystals at room temperature with negligible radiation damage. Special injectors facilitated continuous supply of microcrystals for SFX data collection [135]. A combination of LCP crystallization with the SFX method [97, 136] proved to be extremely successful, with several structures of important and challenging IMP targets solved in the last few years [87, 137, 138]. The feasibility of serial crystallography with IMP crystals grown and delivered in LCP has also been demonstrated at synchrotron sources [139].

10 Conclusion and Outlook

Significant progress has been achieved in efforts to understand the mechanism of action of IMPs using structural biology approaches. A good indication is the number of recently reported human IMP structures, tackling which, until just 5 years ago, was considered to be quite formidable. A robust process has now been established, and new technologies are continually being added to the process including the design and synthesis of new stabilizing molecules (i.e., detergents, lipids), as well as the development of new assays for guiding the crystallization process. There is, however, still the need to address two leading challenges, the relatively high cost of IMP structure determination and the high risk of failure. There is also the critical need to increase structural coverage of membrane protein complexes. Recent breakthroughs in Cryo-EM paved the way to high-resolution structures of large IMP complexes, often facilitated by combination with X-ray crystallography of individual proteins [140]. Other new promising EM-based technologies include the use of Transmission Electron Microscopy (TEM) to identify and characterize micro- and nano-sized crystals [141–143], and structure determination using electron diffraction from extremely small microcrystals (microED) [144].

One of the most exciting recent developments is focused on expanding our understanding of the role of dynamic behavior of IMPs in their biological action. Successful application of SFX at XFELs has opened new opportunities for studying conformational changes in IMPs by time-resolved crystallography. The femtosecond pulses of the XFEL beam provide an opportunity for studying fast-evolving processes in a molecular-movie fashion that could not be tracked before [145]. Taking into consideration all recent advancements described in this review, the future prospects for IMP crystallography look very bright, indeed.

References

1. Yildirim MA, Goh K-I, Cusick ME et al (2007) Drug-target network. Nat Biotechnol 25:1119–1126
2. Ujwal R, Bowie JU (2011) Crystallizing membrane proteins using lipidic bicelles. Methods 55:337–341
3. Caffrey M, Cherezov V (2009) Crystallizing membrane proteins using lipidic mesophases. Nat Protoc 4:706–731
4. Gourdon P, Andersen JL, Hein KL et al (2011) HiLiDe—systematic approach to membrane protein crystallization in lipid and detergent. Cryst Growth Des 11: 2098–2106
5. Newby ZER, O'Connell JD, Gruswitz F et al (2009) A general protocol for the crystallization of membrane proteins for X-ray structural investigation. Nat Protoc 4:619–637
6. Liu W, Cherezov V (2011) Crystallization of membrane proteins in lipidic mesophases. J Vis Exp:e2501
7. Caffrey M, Porter C (2010) Crystallizing membrane proteins for structure determination using lipidic mesophases. J Vis Exp:e1712
8. Li D, Boland C, Aragao D et al (2012) Harvesting and cryo-cooling crystals of membrane proteins grown in lipidic mesophases for structure determination by macromolecular crystallography. J Vis Exp:e4001
9. Ujwal R, Abramson J (2012) High-throughput crystallization of membrane proteins using the lipidic bicelle method. J Vis Exp:e3383
10. Li D, Boland C, Walsh K et al (2012) Use of a robot for high-throughput crystallization of membrane proteins in lipidic mesophases. J Vis Exp:e4000
11. Luecke H, Schobert B, Richter HT et al (1999) Structure of bacteriorhodopsin at 1.55 A resolution. J Mol Biol 291:899–911
12. Palczewski K, Kumasaka T, Hori T et al (2000) Crystal structure of rhodopsin: a G protein-coupled receptor. Science 289:739–745
13. Deisenhofer J, Epp O, Miki K et al (1985) Structure of the protein subunits in the photosynthetic reaction centre of Rhodopseudomonas viridis at 3 Å resolution. Nature 318:618–624
14. Ghosh E, Kumari P, Jaiman D et al (2015) Methodological advances: the unsung heroes of the GPCR structural revolution. Nat Rev Mol Cell Biol 16:69–81
15. Alexandrov AI, Mileni M, Chien EYT et al (2008) Microscale fluorescent thermal stability assay for membrane proteins. Structure 16:351–359
16. Chen R (2012) Bacterial expression systems for recombinant protein production: *E. coli* and beyond. Biotechnol Adv 30:1102–1107
17. Studier FW (2005) Protein production by auto-induction in high-density shaking cultures. Protein Expr Purif 41:207–234
18. Studier FW (2014) Stable expression clones and auto-induction for protein production in *E. coli*. Methods Mol Biol 1091:17–32
19. Newton-Vinson P, Hubalek F, Edmondson DE (2000) High-level expression of human liver monoamine oxidase B in *Pichia pastoris*. Protein Expr Purif 20:334–345
20. Jin MSM, Oldham MML, Zhang Q et al (2012) Crystal structure of the multidrug transporter P-glycoprotein from *Caenorhabditis elegans*. Nature 490:566–569

21. Tao X, Avalos JL, Chen J et al (2009) Crystal structure of the eukaryotic strong inward-rectifier K+ channel Kir2.2 at 3.1 A resolution. Science 326:1668–1674
22. Brohawn SG, del Mármol J, MacKinnon R (2012) Crystal structure of the human K2P TRAAK, a lipid- and mechano-sensitive K+ ion channel. Science 335:436–441
23. Whorton MR, MacKinnon R (2011) Crystal structure of the mammalian GIRK2 K+ channel and gating regulation by G Ppoteins, PIP2, and sodium. Cell 147:199–208
24. Shimamura T, Shiroishi M, Weyand S et al (2011) Structure of the human histamine H1 receptor complex with doxepin. Nature 475:65–70
25. He Y, Wang K, Yan N (2014) The recombinant expression systems for structure determination of eukaryotic membrane proteins. Protein Cell 5:658–672
26. Contreras-Gómez A, Sánchez-Mirón A, García-Camacho F et al (2014) Protein production using the baculovirus-insect cell expression system. Biotechnol Prog 30:1–18
27. Harrison RL, Jarvis DL (2006) Protein N-glycosylation in the baculovirus–insect cell expression system and engineering of insect cells to produce "mammalianized" recombinant glycoproteins. Adv Virus Res 68:159–191
28. Lopez M, Tetaert D, Juliant S et al (1999) O-Glycosylation potential of lepidopteran insect cell lines. Biochim Biophys Acta 1427:49–61
29. Ciccarone VC, Polayes DA, Luckow VA (1998) Generation of recombinant baculovirus DNA in *E. coli* using a baculovirus shuttle vector. Methods Mol Med 13:213–235
30. Hanson MA, Brooun A, Baker KA et al (2007) Profiling of membrane protein variants in a baculovirus system by coupling cell-surface detection with small-scale parallel expression. Protein Expr Purif 56:85–92
31. Andréll J, Tate CG (2013) Overexpression of membrane proteins in mammalian cells for structural studies. Mol Membr Biol 30:52–63
32. Tate CG (2001) Overexpression of mammalian integral membrane proteins for structural studies. FEBS Lett 504:94–98
33. Privé GG (2007) Detergents for the stabilization and crystallization of membrane proteins. Methods 41:388–397
34. Annalora AJ, Goodin DB, Hong W-X et al (2010) Crystal structure of CYP24A1, a mitochondrial cytochrome P450 involved in vitamin D metabolism. J Mol Biol 396:441–451
35. Newstead S, Iwata SO (2008) Rationalizing a-helical membrane protein crystallization. Protein Sci 17:466–472
36. Chae PS, Kruse AC, Gotfryd K et al (2013) Novel tripod amphiphiles for membrane protein analysis. Chemistry 19:15645–15651
37. Zhang Q, Ma X, Ward A et al (2007) Designing facial amphiphiles for the stabilization of integral membrane proteins. Angew Chem Int Ed 46:7023–7025
38. Ehsan M, Du Y, Scull NJ et al (2016) Highly branched pentasaccharide-bearing amphiphiles for membrane protein studies. J Am Chem Soc 138:3789–3796
39. Chae PS, Rasmussen SGF, Rana RR et al (2010) Maltose-neopentyl glycol (MNG) amphiphiles for solubilization, stabilization and crystallization of membrane proteins. Nat Methods 7:1003–1008
40. Rosenbaum DM, Zhang C, Lyons JA et al (2011) Structure and function of an irreversible agonist-β(2) adrenoceptor complex. Nature 469:236–240
41. Haga K, Kruse AC, Asada H et al (2012) Structure of the human M2 muscarinic acetylcholine receptor bound to an antagonist. Nature 482:547–551
42. Wang H, Goehring A, Wang KH et al (2013) Structural basis for action by diverse antidepressants on biogenic amine transporters. Nature 503:141–145
43. McGregor C-L, Chen L, Pomroy NC et al (2003) Lipopeptide detergents designed for the structural study of membrane proteins. Nat Biotechnol 21:171–176
44. Sadaf A, Cho KH, Byrne B, Chae PS (2015) Amphipathic agents for membrane protein study. Methods Enzymol 557:57–94
45. Zhao X, Nagai Y, Reeves PJ et al (2006) Designer short peptide surfactants stabilize G protein-coupled receptor bovine rhodopsin. Proc Natl Acad Sci U S A 103:17707–17712
46. Tribet C, Audebert R, Popot J-L (1996) Amphipols: polymers that keep membrane proteins soluble in aqueous solutions. Proc Natl Acad Sci U S A 93:15047–15050
47. Popot J-L (2010) Amphipols, nanodiscs, and fluorinated surfactants: three nonconventional approaches to studying membrane proteins in aqueous solutions. Annu Rev Biochem 79:737–775
48. Polovinkin V, Gushchin I, Sintsov M et al (2014) High-resolution structure of a membrane protein transferred from amphipol to a lipidic mesophase. J Membr Biol 247: 997–1004

49. Bayburt TH, Sligar SG (2010) Membrane protein assembly into nanodiscs. FEBS Lett 584:1721–1727
50. Hagn F, Etzkorn M, Raschle T et al (2013) Optimized phospholipid bilayer nanodiscs facilitate high-resolution structure determination of membrane proteins. J Am Chem Soc 135:1919–1925
51. Bayburt TH, Grinkova YV, Sligar SG (2002) Self-assembly of discoidal phospholipid bilayer nanoparticles with membrane scaffold proteins. Nano Lett 2:853–856
52. Ritchie TK, Grinkova YV, Bayburt TH et al (2009) Chapter 11—Reconstitution of membrane proteins in phospholipid bilayer nanodiscs. Methods Enzymol 464:211–231
53. Glück JM, Wittlich M, Feuerstein S et al (2009) Integral membrane proteins in nanodiscs can be studied by solution NMR spectroscopy. J Am Chem Soc 131:12060–12061
54. Kang HJ, Lee C, Drew D (2013) Breaking the barriers in membrane protein crystallography. Int J Biochem Cell Biol 45:636–644
55. Dupeux F, Röwer M, Seroul G et al (2011) A thermal stability assay can help to estimate the crystallization likelihood of biological samples. Acta Crystallogr D Biol Crystallogr 67:915–919
56. Hunte C, Koepke J, Lange C et al (2000) Structure at 2.3 Å resolution of the cytochrome bc1 complex from the yeast *Saccharomyces cerevisiae* co-crystallized with an antibody Fv fragment. Structure 8:669–684
57. Cherezov V, Rosenbaum DM, Hanson MA et al (2007) High-resolution crystal structure of an engineered human beta2-adrenergic G protein-coupled receptor. Science 318: 1258–1265
58. Zhou Y, Morais-Cabral JH, Kaufman A et al (2001) Chemistry of ion coordination and hydration revealed by a K+ channel-Fab complex at 2.0 A resolution. Nature 414:43–48
59. Fang Y, Jayaram H, Shane T et al (2009) Structure of a prokaryotic virtual proton pump at 3.2 A resolution. Nature 460: 1040–1043
60. De Genst E, Silence K, Decanniere K et al (2006) Molecular basis for the preferential cleft recognition by dromedary heavy-chain antibodies. Proc Natl Acad Sci U S A 103: 4586–4591
61. Rasmussen SGF, Choi H-J, Fung JJ et al (2011) Structure of a nanobody-stabilized active state of the β(2) adrenoceptor. Nature 469:175–180
62. Ring AM, Manglik A, Kruse AC et al (2013) Adrenaline-activated structure of β2-adrenoceptor stabilized by an engineered nanobody. Nature 502:575–579
63. Geertsma ER, Chang Y-N, Shaik FR et al (2015) Structure of a prokaryotic fumarate transporter reveals the architecture of the SLC26 family. Nat Struct Mol Biol 22:803–808
64. Chun E, Thompson AA, Liu W et al (2012) Fusion partner toolchest for the stabilization and crystallization of G protein-coupled receptors. Structure 20:967–976
65. Rosenbaum DM, Cherezov V, Hanson MA et al (2007) GPCR engineering yields high-resolution structural insights into beta2-adrenergic receptor function. Science 318:1266–1273
66. Serrano-Vega MJ, Magnani F, Shibata Y et al (2008) Conformational thermostabilization of the 1-adrenergic receptor in a detergent-resistant form. Proc Natl Acad Sci U S A 105:877–882
67. Warne T, Edwards PC, Leslie AGW et al (2012) Crystal structures of a stabilized β1-adrenoceptor bound to the biased agonists bucindolol and carvedilol. Structure 20:841–849
68. Magnani F, Shibata Y, Serrano-Vega MJ et al (2008) Co-evolving stability and conformational homogeneity of the human adenosine A2a receptor. Proc Natl Acad Sci U S A 105:10744–10749
69. Klenk C, Ehrenmann J, Schütz M et al (2016) A generic selection system for improved expression and thermostability of G protein-coupled receptors by directed evolution. Sci Rep 6:21294
70. Sarkar CA, Dodevski I, Kenig M et al (2008) Directed evolution of a G protein-coupled receptor for expression, stability, and binding selectivity. Proc Natl Acad Sci U S A 105:14808–14813
71. Malawski GA, Hillig RC, Monteclaro F et al (2006) Identifying protein construct variants with increased crystallization propensity—a case study. Protein Sci 15:2718–2728
72. Niesen FH, Berglund H, Vedadi M (2007) The use of differential scanning fluorimetry to detect ligand interactions that promote protein stability. Nat Protoc 2:2212–2221
73. Semisotnov GV, Rodionova NA, Razgulyaev OI et al (1991) Study of the "molten globule" intermediate state in protein folding by a hydrophobic fluorescent probe. Biopolymers 31:119–128
74. Tomasiak TM, Pedersen BP, Chaudhary S et al (2014) General qPCR and plate reader methods for rapid optimization of membrane protein purification and crystallization using thermostability assays. Curr Protoc Protein Sci 77:29.11.1–29.11.14
75. Mancusso R, Karpowich NK, Czyzewski BK et al (2011) Simple screening method for

improving membrane protein thermostability. Methods 55:324–329

76. Hattori M, Hibbs RE, Gouaux E (2012) A fluorescence-detection size-exclusion chromatography-based thermostability assay for membrane protein precrystallization screening. Structure 20:1293–1299
77. Landau EM, Rosenbusch JP (1996) Lipidic cubic phases: a novel concept for the crystallization of membrane proteins. Proc Natl Acad Sci U S A 93:14532–14535
78. Cherezov V (2011) Lipidic cubic phase technologies for membrane protein structural studies. Curr Opin Struct Biol 21:559–566
79. Parker JL, Newstead S (2012) Current trends in α-helical membrane protein crystallization: an update. Protein Sci 21:1358–1365
80. Raman P, Cherezov V, Caffrey M (2006) The membrane protein data Bank. Cell Mol Life Sci 63:36–51
81. Wadsten P, Wöhri AB, Snijder A et al (2006) Lipidic sponge phase crystallization of membrane proteins. J Mol Biol 364:44–53
82. Cherezov V, Clogston J, Papiz MZ et al (2006) Room to move: crystallizing membrane proteins in swollen lipidic mesophases. J Mol Biol 357:1605–1618
83. Faham S, Bowie JU (2002) Bicelle crystallization: a new method for crystallizing membrane proteins yields a monomeric bacteriorhodopsin structure. J Mol Biol 316:1–6
84. Rouhani S, Cartailler JP, Facciotti MT et al (2001) Crystal structure of the D85S mutant of bacteriorhodopsin: model of an O-like photocycle intermediate. J Mol Biol 313:615–628
85. Cherezov V, Fersi H, Caffrey M (2001) Crystallization screens: compatibility with the lipidic cubic phase for in meso crystallization of membrane proteins. Biophys J 81:225–242
86. Li L, Fu Q, Kors CA et al (2010) A plug-based microfluidic system for dispensing lipidic cubic phase (LCP) material validated by crystallizing membrane proteins in lipidic mesophases. Microfluid Nanofluidics 8:789–798
87. Kang Y, Zhou XE, Gao X et al (2015) Crystal structure of rhodopsin bound to arrestin by femtosecond X-ray laser. Nature 523: 561–567
88. Rasmussen SGF, DeVree BT, Zou Y et al (2011) Crystal structure of the β2 adrenergic receptor-Gs protein complex. Nature 477:549–555
89. Caffrey M (2009) Crystallizing membrane proteins for structure determination: use of lipidic mesophases. Annu Rev Biophys 38: 29–51
90. Caffrey M, Lyons J, Smyth T et al (2009) Monoacylglycerols: the workhorse lipids for crystallizing membrane proteins in mesophases. Curr Top Membr 63:83–108
91. Li D, Lee J, Caffrey M (2011) Crystallizing membrane proteins in lipidic mesophases. A host lipid screen. Cryst Growth Des 11:530–537
92. Li D, Shah STA, Caffrey M (2013) Host lipid and temperature as important screening variables for crystallizing integral membrane proteins in lipidic mesophases. Trials with diacylglycerol kinase. Cryst Growth Des 13:2846–2857
93. Cherezov V, Clogston J, Misquitta Y et al (2002) Membrane protein crystallization in meso: lipid type-tailoring of the cubic phase. Biophys J 83:3393–3407
94. Cheng A, Hummel B, Qiu H et al (1998) A simple mechanical mixer for small viscous lipid-containing samples. Chem Phys Lipids 95:11–21
95. Cherezov V, Caffrey M (2005) A simple and inexpensive nanoliter-volume dispenser for highly viscous materials used in membrane protein crystallization. J Appl Crystallogr 38:398–400
96. Cherezov V, Caffrey M (2003) Nano-volume plates with excellent optical properties for fast, inexpensive crystallization screening of membrane proteins. J Appl Crystallogr 36:1372–1377
97. Weierstall U, James D, Wang C et al (2014) Lipidic cubic phase injector facilitates membrane protein serial femtosecond crystallography. Nat Commun 5:3309
98. Cherezov V, Peddi A, Muthusubramaniam L et al (2004) A robotic system for crystallizing membrane and soluble proteins in lipidic mesophases. Acta Crystallogr D Biol Crystallogr 60:1795–1807
99. Li D, Caffrey M (2011) Lipid cubic phase as a membrane mimetic for integral membrane protein enzymes. Proc Natl Acad Sci U S A 108:8639–8644
100. Liu W, Hanson MA, Stevens RC et al (2010) LCP-Tm: an assay to measure and understand stability of membrane proteins in a membrane environment. Biophys J 98:1539–1548
101. Fenalti G, Abola EE, Wang C et al (2015) Fluorescence recovery after photobleaching in lipidic cubic phase (LCP-FRAP): a precrystallization assay for membrane proteins. Methods Enzymol 557:417–437
102. Whiles JA, Deems R, Vold RR et al (2002) Bicelles in structure-function studies of membrane-associated proteins. Bioorg Chem 30:431–442

103. Czerski L, Sanders CR (2000) Functionality of a membrane protein in bicelles. Anal Biochem 284:327–333
104. De Angelis AA, Howell SC, Nevzorov AA et al (2006) Structure determination of a membrane protein with two trans-membrane helices in aligned phospholipid bicelles by solid-state NMR spectroscopy. J Am Chem Soc 128:12256–12267
105. Sanders CR, Prosser RS (1998) Bicelles: a model membrane system for all seasons? Structure 6:1227–1234
106. Katsaras J, Harroun TA, Pencer J et al (2005) "Bicellar" lipid mixtures as used in biochemical and biophysical studies. Naturwissenschaften 92:355–366
107. Faham S, Boulting GL, Massey EA et al (2005) Crystallization of bacteriorhodopsin from bicelle formulations at room temperature. Protein Sci 14:836–840
108. Luecke H, Schobert B, Stagno J et al (2008) Crystallographic structure of xanthorhodopsin, the light-driven proton pump with a dual chromophore. Proc Natl Acad Sci U S A 105:16561–16565
109. Rasmussen SGF, Choi H-J, Rosenbaum DM et al (2007) Crystal structure of the human beta2 adrenergic G-protein-coupled receptor. Nature 450:383–387
110. Vinothkumar KR (2011) Structure of rhomboid protease in a lipid environment. J Mol Biol 407:232–247
111. Wang H, Elferich J, Gouaux E (2012) Structures of LeuT in bicelles define conformation and substrate binding in a membrane-like context. Nat Struct Mol Biol 19:212–219
112. Chen S, Oldham ML, Davidson AL et al (2013) Carbon catabolite repression of the maltose transporter revealed by X-ray crystallography. Nature 499:364–368
113. Payandeh J, Scheuer T, Zheng N et al (2011) The crystal structure of a voltage-gated sodium channel. Nature 475:353–358
114. Tang L, Gamal El-Din TM, Payandeh J et al (2014) Structural basis for Ca2+ selectivity of a voltage-gated calcium channel. Nature 505:56–61
115. Ujwal R, Cascio D, Colletier J-P et al (2008) The crystal structure of mouse VDAC1 at 2.3 A resolution reveals mechanistic insights into metabolite gating. Proc Natl Acad Sci U S A 105:17742–17747
116. Gruss F, Zähringer F, Jakob RP et al (2013) The structural basis of autotransporter translocation by TamA. Nat Struct Mol Biol 20:1318–1320
117. Liu S, Cheng W, Fowle Grider R et al (2014) Structures of an intramembrane vitamin K epoxide reductase homolog reveal control mechanisms for electron transfer. Nat Commun 5:3110
118. Lee C-H, Lü W, Michel JC et al (2014) NMDA receptor structures reveal subunit arrangement and pore architecture. Nature 511:191–197
119. Althoff T, Hibbs RE, Banerjee S et al (2014) X-ray structures of GluCl in apo states reveal a gating mechanism of Cys-loop receptors. Nature 512:333–337
120. Malinauskaite L, Quick M, Reinhard L et al (2014) A mechanism for intracellular release of Na+ by neurotransmitter/sodium symporters. Nat Struct Mol Biol 21:1006–1012
121. Wang K, Sitsel O, Meloni G et al (2014) Structure and mechanism of Zn2+-transporting P-type ATPases. Nature 514: 518–522
122. Andersson M, Mattle D, Sitsel O et al (2014) Copper-transporting P-type ATPases use a unique ion-release pathway. Nat Struct Mol Biol 21:43–48
123. Kintzer AF, Stroud RM (2016) Structure, inhibition and regulation of two-pore channel TPC1 from *Arabidopsis thaliana*. Nature 531:258–264
124. Li L, Park E, Ling J et al (2016) Crystal structure of a substrate-engaged SecY protein-translocation channel. Nature 531:395–399
125. Parkin S, Hope H (1998) Macromolecular cryocrystallography: cooling, mounting, storage and transportation of crystals. J Appl Crystallogr 31:945–953
126. Garman EF, Schneider TR (1997) Macromolecular cryocrystallography. J Appl Crystallogr 30:211–237
127. Pflugrath JW (2004) Macromolecular cryo-crystallography—methods for cooling and mounting protein crystals at cryogenic temperatures. Methods 34:415–423
128. Joseph JS, Liu W, Kunken J et al (2011) Characterization of lipid matrices for membrane protein crystallization by high-throughput small angle X-ray scattering. Methods 55:342–349
129. Cherezov V, Hanson MA, Griffith MT et al (2009) Rastering strategy for screening and centring of microcrystal samples of human membrane proteins with a sub-10 μm size X-ray synchrotron beam. J R Soc Interface 6:S587–S597
130. Kissick DJ, Dettmar CM, Becker M et al (2013) Towards protein-crystal centering using second-harmonic generation (SHG)

microscopy. Acta Crystallogr D Biol Crystallogr 69:843–851

131. Axford D, Foadi J, Hu NJ et al (2015) Structure determination of an integral membrane protein at room temperature from crystals in situ. Acta Crystallogr D Biol Crystallogr 71:1228–1237
132. Axford D, Owen RL, Aishima J et al (2012) In situ macromolecular crystallography using microbeams. Acta Crystallogr D Biol Crystallogr 68:592–600
133. Huang C-Y, Olieric V, Ma P et al (2016) In meso in situ serial X-ray crystallography of soluble and membrane proteins at cryogenic temperatures. Acta Crystallogr D Biol Crystallogr 72:93–112
134. Chapman HN, Fromme P, Barty A et al (2011) Femtosecond X-ray protein nanocrystallography. Nature 470:73–77
135. Spence JCH, Weierstall U, Chapman HN (2012) X-ray lasers for structural and dynamic biology. Rep Prog Phys 75:102601
136. Liu W, Wacker D, Gati C et al (2013) Serial femtosecond crystallography of G protein-coupled receptors. Science 342:1521–1524
137. Fenalti G, Zatsepin NA, Betti C et al (2015) Structural basis for bifunctional peptide recognition at human δ-opioid receptor. Nat Struct Mol Biol 22:265–268
138. Zhang H, Unal H, Gati C et al (2015) Structure of the angiotensin receptor revealed by serial femtosecond crystallography. Cell 161:833–844
139. Nogly P, James D, Wang D et al (2015) Lipidic cubic phase serial millisecond crystallography using synchrotron radiation. IUCrJ 2:168–176
140. Wei X, Su X, Cao P et al (2016) Structure of spinach photosystem II—LHCII supercomplex at 3.2 A resolution. Nature 534:69–74
141. Stevenson HP, Makhov AM, Calero M et al (2014) Use of transmission electron microscopy to identify nanocrystals of challenging protein targets. Proc Natl Acad Sci U S A 111:8470–8475
142. Stevenson HP, DePonte DP, Makhov AM et al (2014) Transmission electron microscopy as a tool for nanocrystal characterization pre- and post-injector. Philos Trans R Soc Lond Ser B Biol Sci 369:20130322
143. Barnes CO, Kovaleva EG, Fu X et al (2016) Assessment of microcrystal quality by transmission electron microscopy for efficient serial femtosecond crystallography. Arch Biochem Biophys 602:61–68
144. Nannenga BL, Gonen T (2014) Protein structure determination by MicroED. Curr Opin Struct Biol 27:24–31
145. Pande K, Hutchison CDM, Groenhof G et al (2016) Femtosecond structural dynamics drives the trans/cis isomerization in photoactive yellow protein. Science 352:725–729

Chapter 6

Locating and Visualizing Crystals for X-Ray Diffraction Experiments

Michael Becker, David J. Kissick, and Craig M. Ogata

Abstract

Macromolecular crystallography has advanced from using macroscopic crystals, which might be >1 mm on a side, to crystals that are essentially invisible to the naked eye, or even under a standard laboratory microscope. As crystallography requires recognizing crystals when they are produced, and then placing them in an X-ray, electron, or neutron beam, this provides challenges, particularly in the case of advanced X-ray sources, where beams have very small cross sections and crystals may be vanishingly small. Methods for visualizing crystals are reviewed here, and examples of different types of cases are presented, including: standard crystals, crystals grown in mesophase, in situ crystallography, and crystals grown for X-ray Free Electron Laser or Micro Electron Diffraction experiments. As most techniques have limitations, it is desirable to have a range of complementary techniques available to identify and locate crystals. Ideally, a given technique should not cause sample damage, but sometimes it is necessary to use techniques where damage can only be minimized. For extreme circumstances, the act of probing location may be coincident with collecting X-ray diffraction data. Future challenges and directions are also discussed.

Key words Synchrotron radiation, X-ray free electron laser (XFEL), Lipidic cubic phase (LCP), In situ crystallography, Second-order nonlinear optical imaging of chiral crystals (SONICC), Fluorescence, Micro electron diffraction (MicroED)

1 Introduction

The observation of protein crystals was first published by Friedrich Hünefeld in 1840, who serendipitously noticed crystals in blood held under glass slides, which were later determined to be crystals of the protein, hemoglobin [1]. The early study of crystals, whether mineral, chemical, or biological, was in the domain of optical crystallography, where crystals were typically categorized according to their optical properties, particularly with regard to birefringence. Following the first studies of X-ray diffraction by inorganic crystals by Max von Laue in 1912, X-ray diffraction photographs of protein crystals were first published in 1934 by John D. Bernal and Dorothy Crowfoot (later, Hodgkin), who mounted crystals of

Alexander Wlodawer et al. (eds.), *Protein Crystallography: Methods and Protocols*, Methods in Molecular Biology, vol. 1607,
DOI 10.1007/978-1-4939-7000-1_6,

pepsin in a sealed capillary, in equilibrium with a drop of mother liquor to prevent dehydration [2].

In early X-ray crystallography, it was essential to grow large crystals for work with relatively weak laboratory X-ray sources. Yet, as most crystals are too small to be seen easily with the naked eye, optical microscopy is generally used, where the resolution is theoretically limited by the lens system and wavelength of light used, according to Abbe's diffraction limit. In the laboratory, crystals were investigated with visible bright-field microscopy, and sometimes using crossed polarizers, dark-field, or eventually, phase-contrast techniques. Visualization is important both in the laboratory setting, where crystals are typically identified in crystallogenesis experiments, and for mounting at an X-ray source.

Since those seminal years, macromolecular X-ray crystallography has evolved dramatically, including crystal visualization and location techniques. Improvements in crystal locating capabilities have been challenged and driven by increases in power of very small beams, by advances in crystallization techniques, and by mounting techniques. The first synchrotron experiments with protein crystals were published from SSRL in 1976 [3], and as beam intensities from synchrotrons have increased, X-ray experiments have increasingly migrated from home sources to synchrotron sources. Microfocus beamline work, pioneered in the 1990s, notably by Christian Riekel's group at ID13 at the ESRF, has enabled work with even smaller crystals, and spurred the need for improved crystal visualization and localization techniques [4, 5]. Beam sizes have shrunk from as much as a millimeter on rotating-anode sources to current limits of ~1 μm in diameter [6–8] at third generation synchrotron sources, or X-ray free electron lasers (XFELs). Corresponding flux densities have increased from rotating anode sources, to third generation synchrotrons, to XFELs. Mini-beam usage of 5, 10, and 20-μm beam sizes is now routine, not only to locate crystals, but also to find good regions [9, 10]. To ensure centering as a crystal rotates, a goniometer must have a correspondingly small sphere of confusion.

Increases in flux density have led to increases in X-ray induced sample damage. For every X-ray photon that scatters/diffracts, about 10 are absorbed, which contribute to damage [11]. Mounting with a cryoprotectant in open polymer loops on pins enabled studies with higher X-ray doses [12–14]. However, excess "blobs" of cryosolvent give rise to lensing effects due to differences in refractive index, which is particularly evident with small crystals at high magnification. More recent arrangements for in situ studies, involving placing X-ray crystallization trays or chambers of various types for room-temperature experiments, also can lead to refractive distortions [15].

Crystallization techniques have also evolved, particularly for membrane proteins. Growing crystals in mesophase, particularly lipidic cubic phase (LCP), is increasingly common [16, 17]. When working with LCP, it is often difficult to see the crystals, which tend to be small, due to the turbid nature of the crystallization matrix. For XFEL and electron microscopy (EM) experiments, it is sometimes necessary to grow vanishingly small crystals that are invisible in a light microscope [8, 18]. Thus, methods for locating crystals have expanded to include not only optical observations, but also X-ray raster and EM techniques. This chapter reviews visualization techniques used in home and in high-throughput laboratory settings, but emphasizes applications at advanced X-ray sources, where experiments can be particularly challenging.

2 Techniques

2.1 Absorption

Bright-field microscopy has been the mainstay of crystal visualization/localization in the laboratory and at crystallographic X-ray sources. Bright-field can be challenging, however, when crystals are small and are embedded in another medium, such as cryosolvent, at high magnification. Refractive index effects can provide a distorted perception of the crystal location, shape, and size. The refractive index contains both real and imaginary components. Refraction results from different wavepacket group velocities in materials with different dielectric properties. It varies with the medium and wavelength. Bright-field is further challenged by turbidity, e.g., with LCP crystallizations [17], by crystallization plates, compartments, or capillaries used for in situ crystallography [15], and for crystals that are too small to resolve [18, 19].

In a modern beamline setting, a coaxial lens for visualizing the crystal along the X-ray beam is typically employed to remove parallax [20, 21], with a hole in the center to allow the X-ray beam to pass, and a retractable diffuse white light source that backlights the sample. There may be a low-resolution camera 90° away to assist with centering, and additional illumination may be provided from other angles. The space in crystal vicinity is congested, where typically retractable collimators and a beamstop are within a few mm of the sample. Long working distances generally require using large, expensive lenses with low numerical apertures. With regard to crystal centering in a beamline context, samples that exhibit strong refraction effects can appear to be off center (Fig. 1), whereas X-ray diffraction raster can give a definitive center. Practically, one can try to minimize refractive effects by centering loops edge on and 90° away, and also checking in symmetric small rotation increments if the loop obscures the crystal. Confocal microscopy in reflectance mode, i.e., off axis illumination and detection of scattered light, has been employed for high-resolution imaging in the visible region [22].

Whether in a laboratory or at a synchrotron beamline, bright-field illumination is usually provided by a broad-band white light

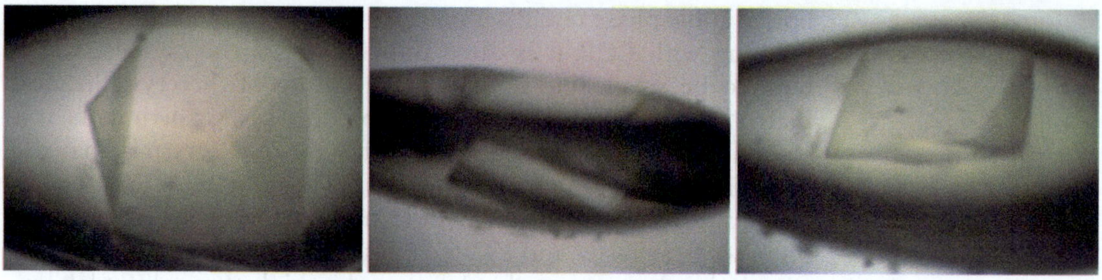

Fig. 1 Bright-field images at a synchrotron beamline of a mechanically well-centered T4 lysozyme crystal embedded in cryosolvent, viewed with 10× magnification. (*Left*) Goniometer $\omega = 0°$ – crystal seems well-centered; (*middle*) Goniometer $\omega = 90°$ – crystal seems split; (*right*) Goniometer $\omega = 135°$ – crystal seems too high. (Sample courtesy of B. Goblirsch, M. Wiener, University of Virginia)

source. To increase contrast, other wavelengths are sometimes used. Aromatic residues of proteins absorb in the ultraviolet range, and on average, ~1.0% of residues are Trp, ~3.7% Tyr, and ~4.0% Phe [23]. Ultraviolet absorption visualization has been applied in the laboratory setting to increase contrast [24, 25]. At synchrotron beamlines, UV absorption capabilities have been implemented at SSRL [26], SLS [21], and at the Photon Factory [27] using low-power LEDs. UV absorption can cause protein damage, however, and at high doses, UV illumination can even be used to generate specific chemical changes that can be used for phasing in X-ray diffraction experiments [28, 29]. Therefore, for visualization, it is desirable to expose samples to low doses of UV irradiation, which may come in short pulses.

Infrared light has also been used in the laboratory and at synchrotron beamlines. Like visible wavelengths longer than UV, infrared wavelength regions show little or no absorption, and therefore, little or no damage, unless there is a particular radiation-absorbing chromophore. Light in the mid-infrared range (3000–5000 nm) has been implemented for crystal centering in the home laboratory and at a beamline at SSRL [30]. An infrared laser providing 1064-nm light has also been implemented at a beamline at the APS, which provides IR bright-field imaging via confocal microscopy, along with other capabilities [31]. Attenuated total reflection Fourier transform infrared imaging of crystals has been tried in the laboratory setting; this technique requires crystals to be close to a surface [32], but could have potential applicability with some types of crystal supports.

Other absorption-related microscopy techniques are commonly accessed with lab microscopes, including crossed polarizers, dark-field, and phase contrast. All diminish incident light, and images can be complicated to interpret. Crossed polarizers require that a crystal be birefringent for visualization, where the refractive index along various directions within the crystal changes with the polarization of the incident light. Not all protein crystals show birefringence (for example, cubic crystals are optically isotropic). Automated birefringence studies have been implemented in a home laboratory [33]. Dark-field and phase-contrast microscopy techniques require optics upstream and downstream of the sample, and are usually implemented with short working distances. They are complementary to bright-field and provide additional visualization tools, and sometime improved localization of low-contrast specimens. These techniques are rarely used at X-ray sources, presumably because of intensity loss, and congestion is a challenge for optics-heavy microscopy techniques, where optical components are needed upstream and downstream of the sample, typically with a short working distance.

2.2 Fluorescence

Compared to absorption microscopy of colorless samples, an increase in contrast can often be achieved by selectively measuring fluorescence from aromatic amino acid residues, especially for proteins that contain tryptophan. The aromatic side chains of amino acid residues absorb significantly in the ultraviolet wavelength region and exhibit corresponding fluorescence at lower energies. Tryptophan is typically the dominant fluorophore, with fluorescence quantum yields up to ~35% [34, 35]. UV-excited UV fluorescence of crystals has been systematically studied in the laboratory context [25, 36, 37], and various complicating factors have been enumerated. These types of studies generally require relatively UV-transparent optics and materials. Self-absorption of fluorescence that is reabsorbed by the crystal can result in a gradient image, particularly for large crystals; on-axis visualization can help to mitigate gradients to some extent. Quenching of fluorescence by buffer solutions, UV fluorescence from non-crystalline proteins, and UV fluorescence from salts are among some of the other complications. Note that fluorescence imaging also depends on refractive index, but generally in a more complicated fashion than for absorbance and some other techniques, partly due to radial or complex distributions of emission [38]. At synchrotron beamlines, a UV lamp has been used for imaging [39], and pulsed UV-laser-excited UV fluorescence [40], and low-power UV LED source [27, 41] have been deployed to minimize damage from irradiation. X-rays have also been used to induce UV fluorescence [41].

Two-photon excited UV fluorescence (nonlinear excitation of aromatic residues in a thin focal plane by a green 532-nm laser) has been applied for protein crystal visualization in the laboratory [42], and is among the imaging capabilities provided by a beamline laser system at the APS [43] (Fig. 2). This technique has some advantages over conventional UV-excited fluorescence. It is a version of confocal microscopy that uses an aperture and scans the laser beam, which reduces delivered damaging out-of-plane UV dose, and also serves to suppress out-of-plane background fluorescence, thereby increasing signal-to-noise and improving effective resolution.

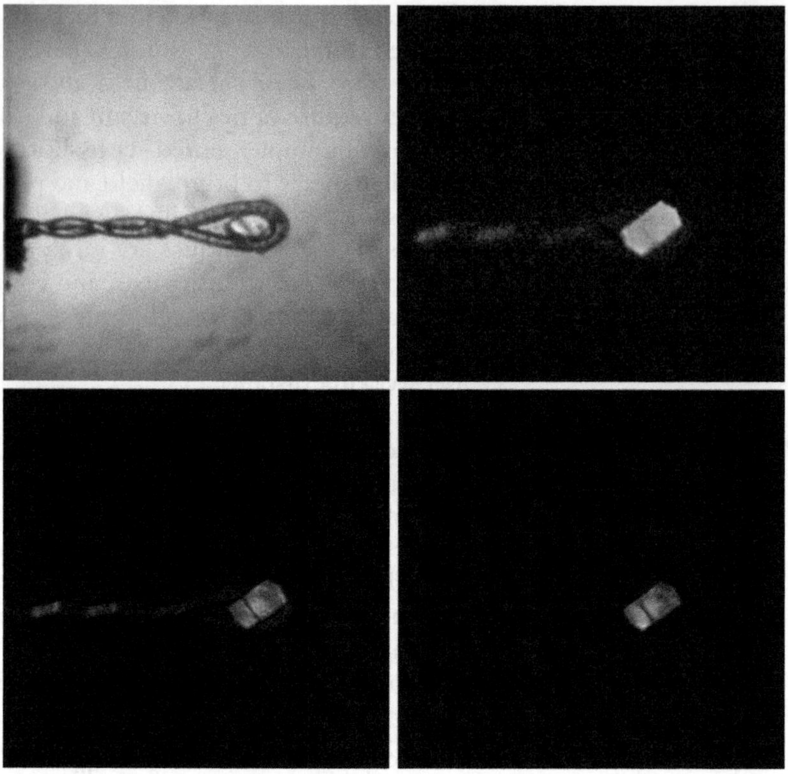

Fig. 2 Images of a lysozyme crystal. (*Upper left*) IR bright-field – incident 1064-nm light, detected transmitted IR; (*upper right, lower left, lower right*) two-photon-excited UV fluorescence – incident green, detected UV fluorescence – taken during a Z-scan, i.e., with the sample translated to different positions along the laser beam, such that the narrow, nonlinear-process focal plane intercepts the sample at different depths along the beam

The longer-wavelength excitation is relatively insensitive to scattering, which allows imaging in turbid media.

A novel fluorescence phenomenon has been observed, where UV-A (320–400 nm) excitation of protein crystals and aggregates – in a region lower in energy than for absorption by aromatic residues – generated visible blue fluorescence [44]. This has been interpreted as resulting from transitions related to peptide electrons that have delocalized through intramolecular and intermolecular hydrogen bonds. A microscopy system exploiting this phenomenon has been implemented at a beamline at SSRL [26], where most crystals show up brightly, particularly if they are dehydrated. In a recent study that may be related, visible light in the 405–480 nm region was used to excite fluorescence in the visible region for several types of protein crystals, which also showed bright fluorescence [45]. Fluorescence intensity was shown to increase dramatically as crystals were cooled to cryogenic temperatures. Hopefully more studies will provide deeper insight into the physical origin(s) of the fluorescence.

Some proteins have endogenous natural chromophores that can conveniently simplify visualization either via absorption or fluorescence. Dyes have traditionally been used to soak preformed crystals to determine if they are composed of protein or salt [46]. Dyes have also been applied for fluorescence imaging, to monitor and increase sensitivity in the crystallization process, but can in principle be applied for crystal centering in some cases, although effects on diffraction quality must be determined [47]. A non-covalent dye, ANS (8-anilinonaphthalene-1-sulfonate), has been used in fluorescence imaging via UV excitation [48, 49]. Trace labeling with a covalently attached dye has been performed for visible excitation [50, 51]. A GFP fusion protein that has been used to aid crystallization also aids crystal localization via visible excitation [52]. Confocal microscopy studies of dye-soaked crystals have been performed for high-resolution studies [22]. While super-resolution microscopy might seem attractive for some applications, these are generally near-field techniques that selectively saturate absorbers, so challenges exist for convenient application. At X-ray energies, if an anomalous scatterer is present in crystals, X-ray fluorescence can be detected while scanning or rastering samples to locate crystals [53, 54]. This can be performed with lower incident beam intensities than for X-ray diffraction raster, but does not provide information on diffraction quality.

2.3 Second Harmonic Generation

Second Harmonic Generation (SHG) microscopy, frequently referred to as Second-Order Nonlinear Optical Imaging of Chiral Crystals (SONICC) in the context of protein crystal imaging, has been shown to detect protein crystals with high sensitivity and selectivity [55, 56]. When a high-intensity, short pulsed laser is focused tightly, in this case using a beam-scanning microscope, a sample's electric field can respond anharmonically to the driving field. The net result is that the emitted light includes light of the transmitted, fundamental frequency, and light of the second-harmonic, doubled frequency. The second harmonic light is generated coherently, or "in phase", with the incident light, which restricts this process to media that are anisotropic. The key feature of natural protein crystals that makes them amenable to SHG microscopy is that their crystal structures will always be chiral. The overall response of a protein crystal can be modeled using the aggregate of the amide bonds and applying crystal symmetry [57]. While all chiral crystals will theoretically show second harmonic generation, in practice high symmetry limits the use of SHG. It has been estimated that ~84% of protein crystals may show a detectable SHG signal using current microscopes, but the signal may vary by about two orders of magnitude [58]. An attractive feature of this method is the possibility to detect sub-micron crystals and to monitor crystallization [59]. SHG is also capable of detecting crystals of membrane proteins in lipidic mesophases [60] (Fig. 3). Polarization-resolved SHG has successfully been used to identify crystal domains [61].

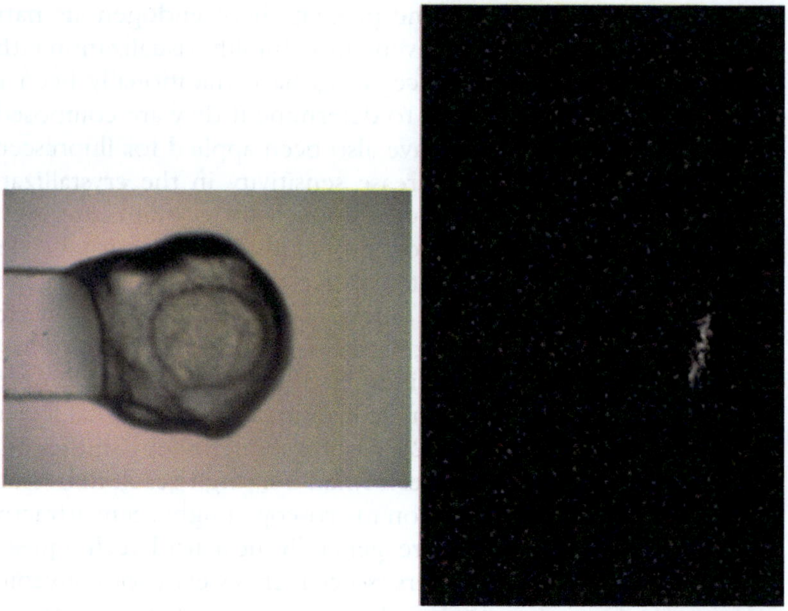

Fig. 3 Images of a GPCR crystal in LCP at a beamline. (*Left*) IR bright-field; (*right*) SHG signal. (GPCR LCP sample courtesy of V. Cherezov, formerly of The Scripps Research Institute)

Some challenges are that the system requires a laser, that crystals of high symmetry may give little or no signal, and that some salt crystals may give SHG signal as well [62]. Intercalating dyes have been studied to enhance the SHG signal [63].

A SONICC laser system has been commercialized for in-laboratory detection, where it tends to be applied to monitor growth of crystals in LCP. It was demonstrated that an SHG microscope can indeed be used for crystal centering in an X-ray beam, and potential laser-induced damage was undetectable in structures and in electron-density maps obtained from cryocooled crystals of thaumatin and myoglobin [56]. A SONICC capability, along with other laser-derived imaging modes, has been implemented at a beamline at the APS [31, 43] (Fig. 4). SONICC must be used before X-ray diffraction, i.e., samples cannot be prescreened with X-rays, as X-ray irradiation generates an artifactual SHG signal in cryocooled samples due to the polarizability of X-ray-damage-induced species [64].

2.4 X-Ray Raster

X-ray diffraction (XRD) raster is a powerful method for locating small, hard-to-find crystals that are obfuscated in cryocooled media at synchrotron beamlines, and is extensively reviewed in another chapter of this volume by Sanishvili and Fischetti. This method serves the dual purpose of crystal localization as well as a simultaneous assessment of crystal diffraction quality (Fig. 5), at the

Fig. 4 Real-time (15 Hz) images of mCherry protein crystals, which contain a chromophore with a visible absorption maximum at 587 nm. (*Left*) IR bright-field–incident 1064-nm light, detected transmitted IR, (*middle*) second harmonic generation–incident IR light, detected green SHG in narrow focal plane, (*right*) two-photon-excited UV fluorescence–incident green light, detected visible fluorescence. (Sample courtesy of C. Das group, Purdue University)

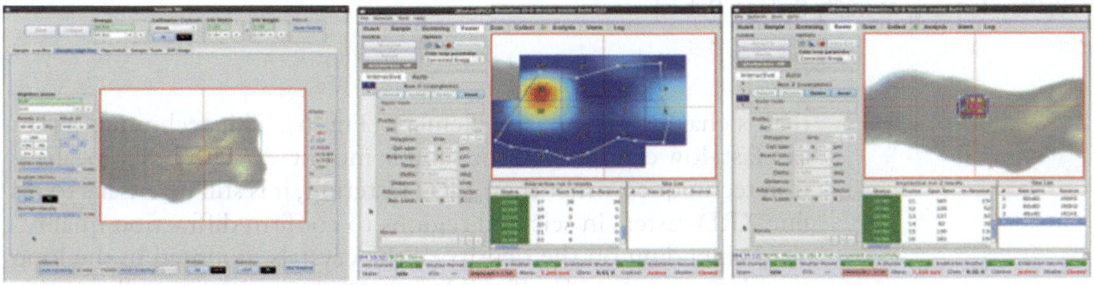

Fig. 5 A panel of three images illustrate the use of a raster search to locate a crystal grown in LCP. From left to right. (*Left*) The sample as viewed through an on-axis visualizer and displayed in the beamline graphical user interface (GUI). (*Middle*) Coarse raster to locate the crystal, color display is proportional to the number of diffraction spots found in the diffraction image corresponding to the raster pixel (*red*–highest to *blue*–lowest). (*Right*) Fine raster grid to pinpoint sample location

expense of initiating X-ray damage [65–69]. The combination of mini-beams, new detectors, and shutterless rastering with attenuation to minimize radiation damage, has elevated its status as the method of choice for LCP samples. Visualization is primarily displayed as heat map color scales proportional to the number of diffraction spots per raster pixel (Fig. 5).

X-ray fluorescence raster may also be used for crystal localization [53, 54]. A limitation of the technique is the requirement that the crystal contain an element with an absorption edge within the energy range of the beamline. Although this appears restrictive, it is applicable to the large class of selenomethionine containing crystals. In fluorescence rastering, the pixels in the raster map are proportional to the number of photons counted within a restricted energy window centered around the fluorescence emission energy

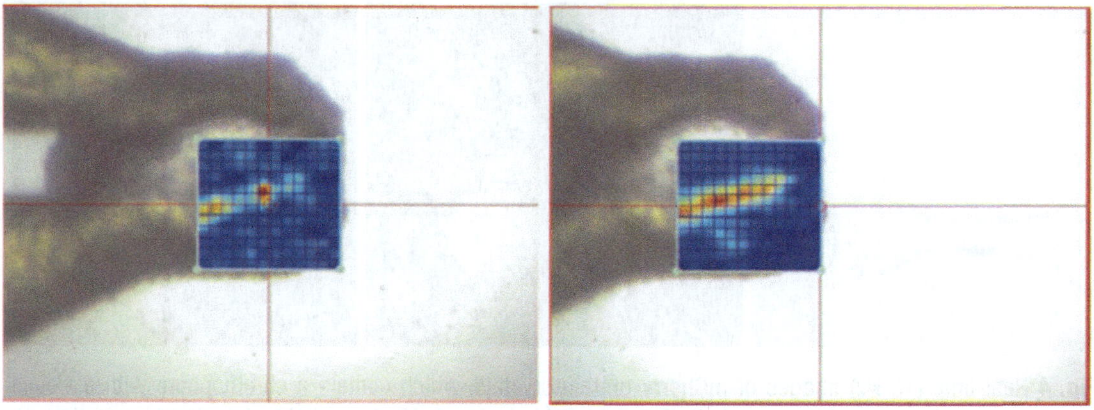

Fig. 6 Raster grids of a weakly diffracting crystal of a SeMet-containing outer membrane protein, taken with a 12.8 keV 5 μm diameter X-ray beam, and 1 s exposure per raster cell, using a CCD detector. (*Left*) XRD Raster showing a heat map scored according to Bragg candidates (*red* is highest number; *blue* is lowest) with 5-fold attenuation. (*Right*) X-ray fluorescence raster showing a heat map scored according to SeMet fluorescence counts, under the same conditions, except the attenuation was 1000-fold. (Sample courtesy of D. Aragao, D. Li, M. Caffrey, formerly of Trinity College, Dublin)

that is characteristic of the element (Fig. 6). This technique serves as a fast, low dose scan of the sample mount. As with other localization techniques, other than XRD rastering, it is still necessary to do fine XRD rasters in selected regions to confirm diffraction quality.

Depending on the density of the crystals mounted in the loop or plate, it may be feasible to apply serial crystallographic approaches, collecting single still images from multiple samples, employing rastering combined with "Fixed Target" sample delivery systems. Visualization of the sample, whether online or offline, is still required for preliminary sample characterization.

2.5 Electron Microscopy

Electron microscopy has been used to identify sub-μm crystals as "hits" to optimize crystallization conditions for growing larger crystals, and as specimens for XFEL experiments [19, 26, 70]. Staining is used to identify crystals for further optimization of crystallization conditions, but is not applied for structure determination on those specific crystals. The optimized conditions can provide unstained crystals for XFEL experiments. With the impressive recent advances in micro electron diffraction (MicroED) pioneered by the group of Tamir Gonen, unstained crystals are identified in search mode in an electron microscope [71], to be specifically targeted for electron-diffraction data collection [18, 72, 73].

2.6 Other Techniques

An assortment of additional techniques has been applied to the task of locating crystals. X-ray radiography and tomography have been applied to locating membrane-protein crystals grown in lipidic cubic phase with similar dose to XRD raster for a full tomogram, but crystal location and shape can be determined with a lower dose [74]. Ultra-high resolution optical coherence tomography has been explored, but required embedding crystals in agarose to enhance contrast [75]. Three-dimensional Raman spectroscopic imaging has been used to image crystals deposited on a nanodroplet [76]. Note that various beamlines have reported diverse spectroscopic capabilities [21, 77], where the emphasis is more on biochemical functional studies than on crystal localization. Perhaps simple adaptations of those approaches could be implemented for crystal localization.

2.7 Image Processing

Modern crystallography seeks to exploit automation, and in the context of crystal recognition and localization, this includes making advances in image processing. For robotic crystallization in laboratories, some examples of image-analysis methods for identifying crystals include using line-segment information [78], or using support vector machine-learning algorithms [79]. For complete 3D centering at X-ray crystallography beamlines, edge detection has been applied [27, 53] and procedures involving combinations of methods have been developed with the programs C3D [80] and XREC [81] to address difficult situations. These have been applied to bright-field images, and involve first identifying a sample loop, and searching for a crystal based on other algorithms. For fluorescence, sometimes simple intensities can suffice, but self-absorption effects can be significant and refractive-index effects are not necessarily absent. For techniques where excessive irradiation can potentially cause heating or damage, methods for sparse sampling can reduce sample exposure, such as those applied with SONICC [82].

3 Applications

3.1 Standard Case

Commonly, beamline users bring vitrified ("cryocooled") crystals to a synchrotron beamline mounted in various kinds of open loops made of polymer fibers, where the crystals are stabilized in place by the surrounding glassified cryosolvent. In cases where the crystal is readily visible and the volume of the surrounding cryosolvent is modest or minimal, centering based on bright-field imaging is straightforward. However, in cases where the crystal volume is small relative to that of the surrounding cryosolvent, and the magnification is high, lensing effects of the cryosolvent can lead to distorted perception of where the crystal is located (Fig. 1). Unless the cryosolvent blob is spherical, it can help to first orient the plane of the loop parallel and perpendicular to the viewing axis along the beam. In such cases, the crystal is sometimes obscured by the loop material itself, i.e., when viewing through the loop. It can help to apply small angular offsets in both directions, and choose the average of the two angles. When in doubt, use X-ray diffraction raster to be sure.

3.2 Lipidic Cubic Phase (LCP)

Growth of membrane-protein crystals in mesophase–typically, lipidic cubic phase–has become common. Unless the protein is colored, bright-field imaging of such samples commonly reveals a turbid or opaque sample, making identification of crystals challenging, at best. In the laboratory environment, crystallization drops are often viewed using bright-field imaging coupled with crossed polarizers, commercial SHG, and UV fluorescence systems. Even when the crystals are visible in the LCP media (Fig. 7, top panels), they are obscured after harvesting, mounting and freezing. At synchrotrons, X-ray diffraction rasters have been essential to the successful data collection from GPCR crystals grown in LCP. Due to the difficulty in mounting a single, small crystal from an LCP crystallization setup, the first step in data collection is the localization of multiple crystals mounted in the loop using XRD rastering (Fig. 7, bottom panels). X-ray radiography and tomography also seem promising [74]. X-rays cause damage, however, so additional

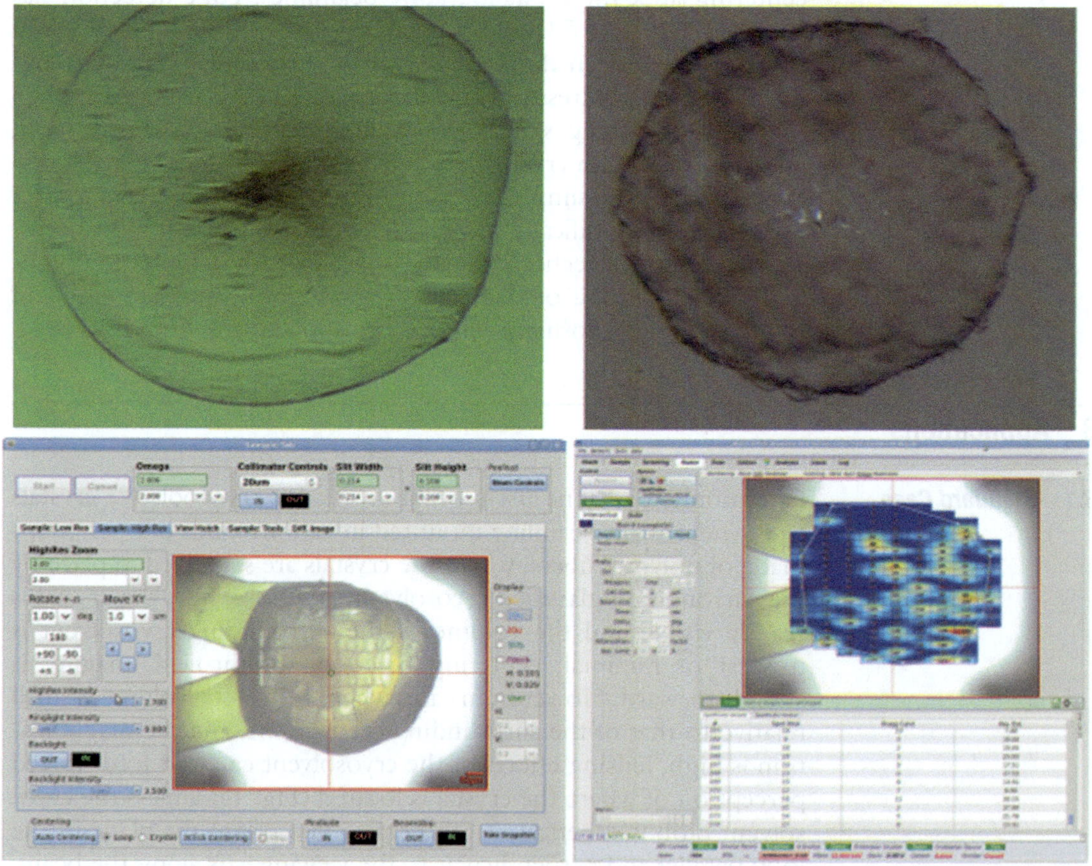

Fig. 7 Membrane protein crystals grown in LCP viewed through a microscope prior to mounting, bright-field (*upper left*), bright-field with polarizers (*upper right*). Samples viewed through an on-axis visualizer after mounting and freezing (*lower left*). Heat map after rastering (*lower right*; *red* corresponds to high, *blue* to low number of diffraction spots. (Images courtesy of C. Zhang, University of Pittsburgh School of Medicine)

techniques are desired. Fluorescence methods have applicability, as well as SONICC [31, 43], and efforts are underway to exploit these more fully.

3.3 In Situ Crystal Handling

There has been a resurgence of interest in in situ characterization and data collection. Previously, this was restricted to the use of large-format crystallization plates to screen crystal diffraction prior to mechanical perturbation by mounting in loops and chemical perturbation by the addition of cryosolvents. Off-line visualization using bright-field and UV characterization are common tools prior to in situ experiments. Commercial vendors have developed large-format plates with low X-ray absorption, decreased background scatter and good visualization (Fig. 8). Crystallization platforms at several different X-ray light sources have been used for crystal screening [15, 83, 84]. The growing interest in membrane-protein crystallization has spawned development of in situ crystallization setups for use with mesophases at room or cryogenic temperatures [85, 86]. Microfluidic crystallization and delivery formats have also emerged as small-footprint chips that provide a larger oscillation range for data collection (Fig. 9). These continue to evolve, as some lean more towards trapping crystals at fixed locations, whereas others lean more towards in situ crystal-growth chambers, or both [87–96]. These devices introduce a new layer of complexity to sample visualization. The limited range of the rotation axis of

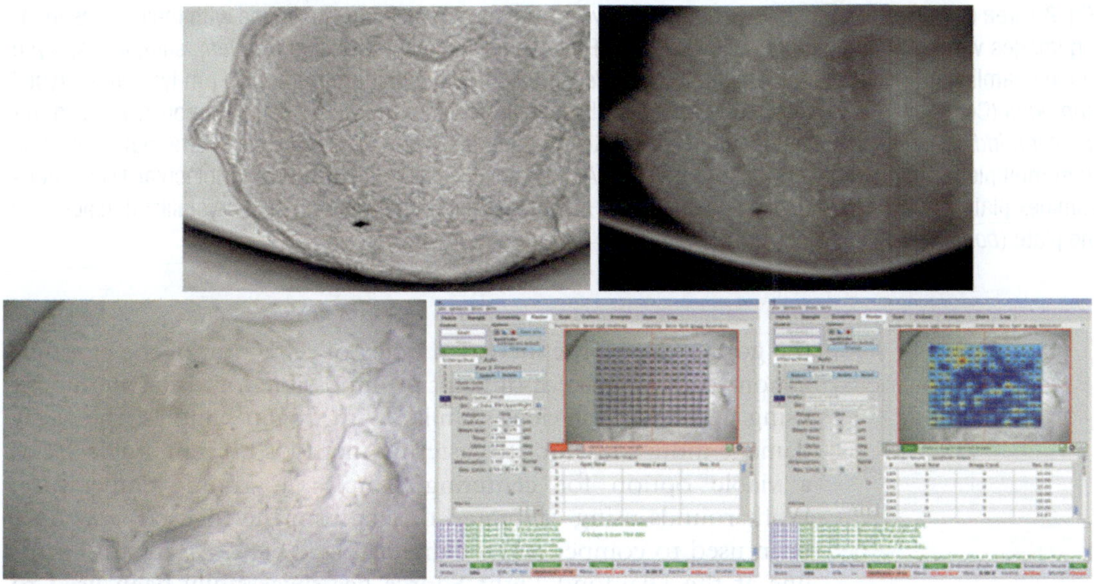

Fig. 8 Example of in situ screening for possible leads in determining crystallization conditions. Off-line optical (*upper left*) and UV (*upper right*) images of a LCP crystallization bolus. The UV image suggests the possibility for crystals. Image from the beamline on-axis visualizer (*bottom left*), setup of a raster grid (*bottom middle*), followed by positive hits (*red color* in heat map) in the raster results (*bottom right*). (Images courtesy of D. Xia and X. Bai, National Cancer Institute)

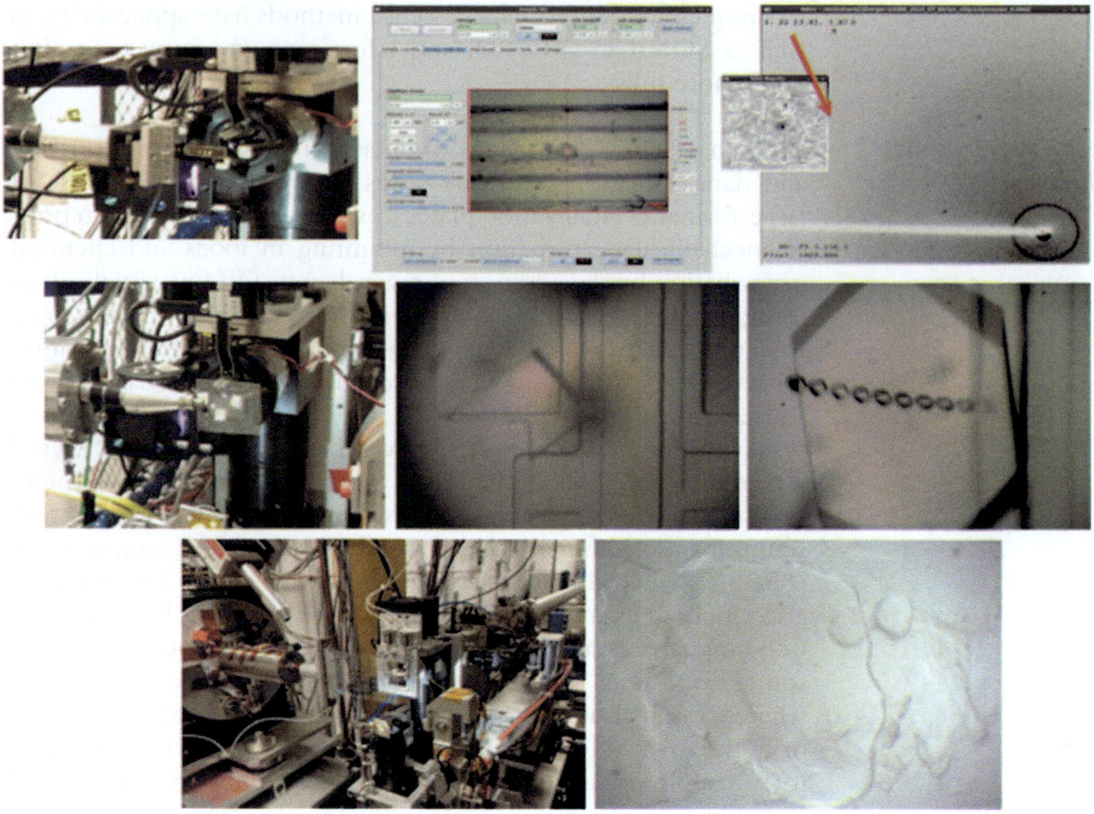

Fig. 9 Three examples of various sized sample delivery formats mounted on a beamline with their corresponding images viewed through the on-axis visualizer. Prototype sample delivery chip (*top left*), sample displayed in the beamline graphical user interface (*top middle*) and a diffraction pattern from a 20-μm lysozyme crystal (*top right*) (Courtesy of T. Murray, J. Berger). Microfluidic chip (*center left*), view through the on-axis visualizer (*center middle*), and the resultant footprints of the X-ray beam deposited on a large crystal after data collection from multiple positions (*center right*) (Courtesy of A. Pawate, J. Schieferste, P. Kenis). Molecular Dimensions Laminex plate containing LCP crystals (*bottom left*) and the on-axis view centered on a crystallization bolus on the plate (*bottom right*) (Courtesy of D. Xia, X. Bai)

the goniostat and the increased index of refraction of the crystallization window make it difficult or impossible for visual alignment using perpendicular views. Unlike with frozen samples, the X-ray damage induced by in situ rastering methods prevents their use as a viable option for centering samples at room temperature. Workarounds using calibrated offsets from the focused view have been used to compensate for the distorted view. An in situ beamline at the Diamond Light Source has successfully been used to collect enough data to solve structures [15, 97]. The limitation of the rotation range due to the size of the crystallization plate format is not a major problem due to the corresponding short lifetime of the room temperature crystal in the X-ray beam, decreasing the detrimental effects of the windows to visualization. In other words, the crystal only has to be centered for a small rotation range. Perhaps the largest boost to the possibility of in situ data collection has been the development of the serial-crystallographic-data-collection method. This approach depends on the collection of still images from a large number of randomly-oriented crystals. Sample visualization is limited to identifying the presence of crystals on the delivery media. Rastering to center a crystal is replaced by rastering to collect a data set.

3.4 X-Ray Free Electron Lasers (XFEL)

Crystals of a few microns or less on a side, even nanocrystals, have successfully been used for high-resolution structure determination at the Linac Coherent Light Source (LCLS) via serial X-ray crystallography, and some of the earliest successful results are referenced here [8, 98]. Crystals are delivered to the beam, which is circa 1 μm in diameter, via a fixed target, jet (gas dynamic virtual nozzle (GDVN)), or via a viscous injector [99–102]. Crystals for the injectors must be uniformly small, so as not to clog the injector. To help identify the presence of vanishingly small microcrystals, and to sort the crystals into size classes, SONICC, UV fluorescence, dynamic light scattering, and transmission electron microscopy of stained samples have been applied [19, 26, 70]. X-ray crystallographic experiments at the LCLS and SACLA have reinvigorated interest in sample delivery techniques and room temperature data collection. There has been a migration of initial prototype XFEL-like experiments to synchrotrons [87, 103, 104], followed by more recent feasibility studies of serial crystallographic experiments using viscous-injector technologies at third generation synchrotron sources. Slower flow rates of the viscous injector have significantly decreased the quantity of material needed to collect a complete data set. Initially, these injectors were associated with crystals grown in LCP. The efficiency of this delivery system has led to an expanded search for alternative carrier media applicable to crystals grown in hydrophilic or other conditions. Although this method of sample delivery does not require visualization of the sample during data collection (Fig. 10), preliminary examination of crystals in the growth medium and after transfer to a non-native carrier medium are essential to assess any changes in crystal appearance and to provide an estimate of crystal density prior to injector loading (Fig. 10). Fixed targets at the LCLS have rekindled interest in a variety of platforms holding multiple crystals. These targets range from crystals spread onto microfluidic chips, to patterned arrays designed to capture samples at specific locations, or combination arrays of in situ crystallization [87–96]. Once again, bright-field visualization and other forms of on-line or off-line visualization are used at the pre-screen stage prior to data collection. Data collection is usually done by automated raster methods or is programmed to take images along grid points of the array on the chip. Synchronization of sample delivery with data collection is also possible [105].

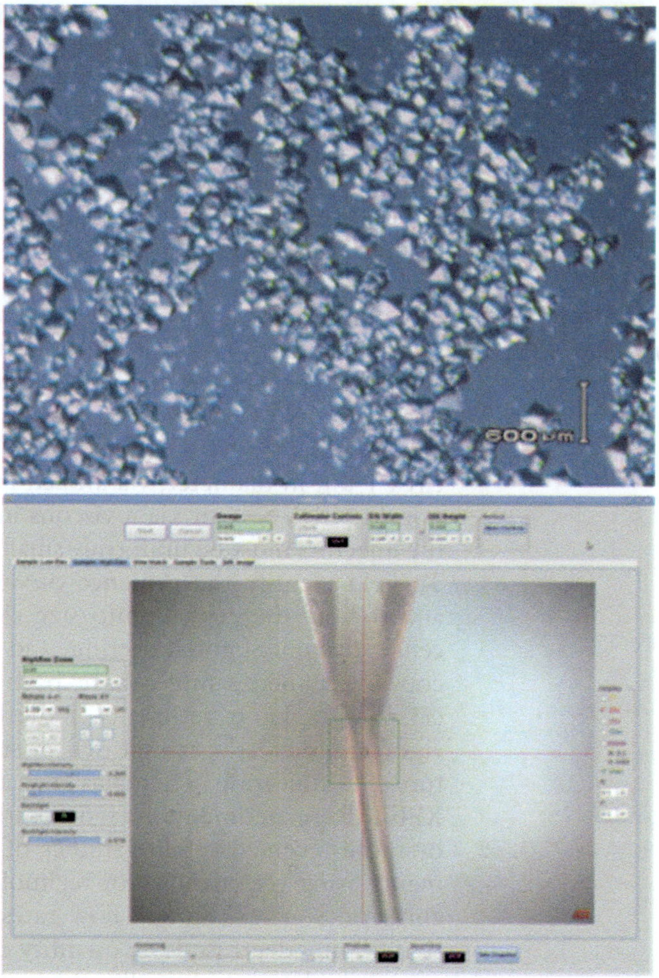

Fig. 10 Optical image of a sample of crystals transferred to viscous media prior to loading into an injector (*top*). A viscous injector in operation on a synchrotron beamline displayed on the graphical user interface (*bottom*). Visualization of the individual crystals is not needed during data collection. The on-axis visualizer is used to monitor the status and position of the sample stream relative to the beam position

3.5 Invisible Crystals

Micro Electron Diffraction (MicroED) is an exciting new development where submicron crystals are used to determine the crystal structure at high resolution via electron diffraction. Structures of small proteins, such as lysozyme, have been determined [73, 106]. Due to strong dynamical diffraction of electrons and to extinction, the crystals must be very small (maximum of ~0.1 μm in thickness). The methods applied for XFEL preparations described above can also be used to generate crystals for MicroED. A protocol for locating crystals and collecting MicroED data with an electron microscope has been described [71].

4 Summary

Numerous techniques have been applied to locate crystals for centering in the X-ray beam. The challenge in some cases is to transition from proof of concept to routine implementation. Poor contrast is sometimes an issue. Except for X-ray diffraction raster, which initiates damage, however, no other method works for all cases, so it is best to have a combination of methods for crystal localization available. Absorption and fluorescence techniques are generally applicable. Cost can be a consideration and safety and complexity issues associated with high-power lasers can influence choices. Where resources allow, SONICC and associated laser techniques can be applied. To date, X-ray raster capabilities at synchrotron beamlines have proven to be invaluable.

In the future, further advances are possible. While it is routine to express proteins with His-tags, or in fusion with extra domains to assist crystallization, routine co-expression with a convenient spectroscopic tag would be desirable. To accomplish this in a way that crystallization and diffraction are not compromised seems challenging. It is reasonable to ask whether super-resolution microscopy methods can be practically applied. Where lasers are used, the option for enhancing signal via stimulated emission from fluorophores with high fluorescence quantum yields exists. X-ray fluorescence raster with anomalous scatterers offers the opportunity to localize crystals with relatively low X-ray exposures.

For scanning relatively large sample fields for X-ray crystallography, sparse sampling and machine learning can potentially decrease damage with some techniques, and synchronization of diffraction measurement to the localization probe can increase hit-rate efficiency. With MicroED, the experimenter also has to first identify crystals for diffraction data collection in the electron microscope [71]. Sparse sampling, also known as compressive sensing, has been applied for in situ EM [107], and perhaps similar sparse-sampling techniques could be applied in some other EM circumstances as well.

Some of the methods discussed in this chapter apply to cases where crystals are essentially randomly distributed in a sample mounting system, and the task is to search for them. However, other cases exist. For example, the use of substrates or patterning mounts, where crystals are predetermined to be grown or attached in specific locations, has become feasible. Further, rastering fixed targets or streaming samples, such as at XFELs, does not strictly require seeing the crystal at all; in those cases, the probing event is also the measurement. Combining visualization methods to automatically trigger data collection events on fixed targets may increase the hit-rate efficiency at current and future synchrotron sources. Careful adaptation of dehydration methods for high-throughput experiments at room temperature might be desirable in some cases [108]. In serial crystallographic approaches, the technical need to visualize crystals during data collection has already been replaced by preliminary offline characterization and real-time statistical hit rate analysis of diffraction images.

Finally, with impressive recent gains in EM of single particles and in MicroED, and with excitement surrounding crystal injectors and single-particle diffraction with XFELs, one might ask if the need for locating crystals for X-ray crystallography might diminish. The techniques of EM and X-ray crystallography have their own merits, and they are complementary; X-ray crystallography provides an electron-density map, and EM provides a Coulomb-potential map. Comparison of the two maps often proves valuable for resolving structural and functional issues in virology and with complex macromolecular machines [109]. Such comparisons will only increase in resolution and in applicability. Further, for cases where high-quality data might be obtainable in the future by both methods from similar samples, whether 3D crystals, 2D crystals, or single particles, rich experimental data on electrostatic properties of macromolecules might be revealed [110]. Even with XFELs, relatively large 3D crystals continue to prove valuable for collecting high-quality data [111, 112], which is essential for functional studies at the chemical level.

References

1. Giegé R (2013) A historical perspective on protein crystallization from 1840 to the present day. FEBS J 280:6456–6497
2. Bernal JD, Crowfoot D (1934) X-ray photographs of crystalline pepsin. Nature 133:794–795
3. Phillips JC, Wlodawer A, Yevitz MM et al (1976) Applications of synchrotron radiation to protein crystallography: preliminary results. Proc Natl Acad Sci U S A 73:128–132
4. Riekel C (2004) Recent developments in microdiffraction on protein crystals. J Synchrotron Radiat 11:4–6
5. Nelson R, Sawaya MR, Balbirnie M et al (2005) Structure of the cross-β spine of amyloid-like fibrils. Nature 435:773–777
6. Moukhametzianov R, Burghammer M, Edwards PC et al (2008) Protein crystallography with a micrometre-sized synchrotron-radiation beam. Acta Crystallogr D Biol Crystallogr 64:158–166
7. Sanishvili R, Yoder DW, Pothineni SB et al (2011) Radiation damage in protein crystals is reduced with a micron-sized X-ray beam. Proc Natl Acad Sci U S A 108:6127–6132
8. Chapman HN, Fromme P, Barty A et al (2011) Femtosecond X-ray protein nanocrystallography. Nature 470:73–77
9. Sanishvili R, Nagarajan V, Yoder D et al (2008) A 7 microm mini-beam improves diffraction data from small or imperfect crystals of macromolecules. Acta Crystallogr D Biol Crystallogr 64:425–435
10. Fischetti RF, Xu S, Yoder DW et al (2009) Mini-beam collimator enables microcrystallography experiments on standard beamlines. J Synchrotron Radiat 16:217–225
11. Henderson R (1995) The potential and limitations of neutrons, electrons and X-rays for atomic resolution microscopy of unstained biological molecules. Q Rev Biophys 28:171–193
12. Petsko GA (1975) Protein crystallography at sub-zero temperatures: cryo-protective mother liquors for protein crystals. J Mol Biol 96:381–392
13. Teng T-Y (1990) Mounting of crystals for macromolecular crystallography in a free-standing thin film. J Appl Crystallogr 23:387–391
14. Hope H (1990) Crystallography of biological macromolecules at ultra-low temperature. Annu Rev Biophys Biophys Chem 19:107–126
15. Axford D, Owen RL, Aishima J et al (2012) In situ macromolecular crystallography using microbeams. Acta Crystallogr D Biol Crystallogr 68:592–600

16. Landau EM, Rosenbusch JP (1996) Lipidic cubic phases: a novel concept for the crystallization of membrane proteins. Proc Natl Acad Sci U S A 93:14532–14535
17. Caffrey M (2015) A comprehensive review of the lipid cubic phase or in meso method for crystallizing membrane and soluble proteins and complexes. Acta Crystallogr F Struct Biol Commun 71:3–18
18. Rodriguez JA, Ivanova MI, Sawaya MR et al (2015) Structure of the toxic core of α-synuclein from invisible crystals. Nature 525:486–490
19. Stevenson HP, Makhov AM, Calero M et al (2014) Use of transmission electron microscopy to identify nanocrystals of challenging protein targets. Proc Natl Acad Sci U S A 111:8470–8475
20. Perrakis A, Cipriani F, Castagna J-C et al (1999) Protein microcrystals and the design of a microdiffractometer: current experience and plans at EMBL and ESRF/ID13. Acta Crystallogr D Biol Crystallogr 55: 1765–1770
21. Fuchs MR, Pradervand C, Thominet V et al (2014) D3, the new diffractometer for the macromolecular crystallography beamlines of the Swiss Light Source. J Synchrotron Radiat 21:340–351
22. Khan I, Gillilan R, Kriksunov I et al (2012) Confocal microscopy on the beamline: novel three-dimensional imaging and sample positioning. J Appl Crystallogr 45:936–943
23. Gilis D, Massar S, Cerf NJ et al (2001) Optimality of the genetic code with respect to protein stability and amino-acid frequencies. Genome Biol 2:1–12
24. Lunde CS, Rouhani S, Remis JP et al (2005) UV microscopy at 280 nm is effective in screening for the growth of protein microcrystals. J Appl Crystallogr 38:1031–1034
25. Gill H (2010) Evaluating the efficacy of tryptophan fluorescence and absorbance as a selection tool for identifying protein crystals. Acta Crystallogr F Struct Biol Commun 66:364–372
26. Calero G, Cohen AE, Luft JR et al (2014) Identifying, studying and making good use of macromolecular crystals. Acta Crystallogr F Struct Biol Commun 70:993–1008
27. Chavas LMG, Yamada Y, Hiraki M et al (2011) UV LED lighting for automated crystal centring. J Synchrotron Radiat 18:11–15
28. Ravelli RBG, Leiros H-KS, Pan B et al (2003) Specific radiation damage can be used to solve macromolecular crystal structures. Structure 11:217–224
29. de Sanctis D, Zubieta C, Felisaz F et al (2016) Radiation-damage-induced phasing: a case study using UV irradiation with light-emitting diodes. Acta Crystallogr D Biol Crystallogr 72:395–402
30. Snell EH, van der Woerd MJ, Miller MD et al (2005) Finding a cold needle in a warm haystack: infrared imaging applied to locating cryocooled crystals in loops. J Appl Crystallogr 38:69–77
31. Newman JA, Zhang S, Sullivan SZ et al (2016) Guiding synchrotron X-ray diffraction by multimodal video-rate protein crystal imaging. J Synchrotron Radiat 23:959–965
32. Glassford SE, Byrne B, Kazarian SG (2013) Recent applications of ATR FTIR spectroscopy and imaging to proteins. Biochim Biophys Acta 1834:2849–2858
33. Echalier A, Glazer RL, Fulop V et al (2004) Assessing crystallization droplets using birefringence. Acta Crystallogr D Biol Crystallogr 60:696–702
34. Eftink MR (1991) Fluorescence techniques for studying protein structure. In: Suelter CH (ed) Methods of biochemical analysis: protein structure determination, vol 35. John Wiley & Sons, Inc., New York, pp 127–205
35. Callis PR, Vivian JT (2003) Understanding the variable fluorescence quantum yield of tryptophan in proteins using QM-MM simulations. Quenching by charge transfer to the peptide backbone. Chem Phys Lett 369: 409–414
36. Judge RA, Swift K, Gonzalez C (2005) An ultraviolet fluorescence-based method for identifying and distinguishing protein crystals. Acta Crystallogr D Biol Crystallogr 61:60–66
37. Desbois S, Seabrook SA, Newman J (2013) Some practical guidelines for UV imaging in the protein crystallization laboratory. Acta Crystallogr F Struct Biol Commun 69: 201–208
38. Ediger MD, Moog RS, Boxer SG et al (1982) On the refractive index correction in luminescence spectroscopy. Chem Phys Lett 88: 123–127
39. Pohl E, Ristau U, Gehrmann T et al (2004) Automation of the EMBL Hamburg protein crystallography beamline BW7B. J Synchrotron Radiat 11:372–377
40. Vernede X, Lavault B, Ohana J et al (2006) UV laser-excited fluorescence as a tool for the visualization of protein crystals mounted in loops. Acta Crystallogr D Biol Crystallogr 62:253–261
41. Gofron KJ, Duke NEC (2011) Using X-ray excited UV fluorescence for biological crystal

location. Nucl Instrum Methods A 649: 216–218

42. Madden JT, DeWalt EL, Simpson GJ (2011) Two-photon excited UV fluorescence for protein crystal detection. Acta Crystallogr D Biol Crystallogr 67:839–846
43. Madden JT, Toth SJ, Dettmar CM et al (2013) Integrated nonlinear optical imaging microscope for on-axis crystal detection and centering at a synchrotron beamline. J Synchrotron Radiat 20:531–540
44. Shukla A, Mukherjee S, Sharma S et al (2004) A novel UV laser-induced visible blue radiation from protein crystals and aggregates: scattering artifacts or fluorescence transitions of peptide electrons delocalized through hydrogen bonding? Arch Biochem Biophys 428:144–153
45. Lukk T, Gillilan RE, Szebenyi DME et al (2016) A visible-light-excited fluorescence method for imaging protein crystals without added dyes. J Appl Crystallogr 49:234–240
46. Sumner JB, Dounce AL (1937) Crystalline catalase. Science 85:366–367
47. Meyer A, Betzel C, Pusey M (2015) Latest methods of fluorescence-based protein crystal identification. Acta Crystallogr F Struct Biol Commun 71:121–131
48. Groves MR, Muller IB, Kreplin X et al (2007) A method for the general identification of protein crystals in crystallization experiments using a noncovalent fluorescent dye. Acta Crystallogr D Biol Crystallogr 63:526–535
49. Watts D, Muller-Dieckmann J, Tsakanova G et al (2010) Quantitive evaluation of macromolecular crystallization experiments using 1,8-ANS fluorescence. Acta Crystallogr D Biol Crystallogr 66:901–908
50. Forsythe E, Achari A, Pusey ML (2006) Trace fluorescent labeling for high-throughput crystallography. Acta Crystallogr D Biol Crystallogr 62:339–346
51. Pusey M, Barcena J, Morris M et al (2015) Trace fluorescent labeling for protein crystallization. Acta Crystallogr F Struct Biol Commun 71:806–814
52. Suzuki N, Hiraki M, Yamada Y et al (2010) Crystallization of small proteins assisted by green fluorescent protein. Acta Crystallogr D Biol Crystallogr 66:1059–1066
53. Karain WI, Bourenkov GP, Blume H et al (2002) Automated mounting, centering and screening of crystals for high-throughput protein crystallography. Acta Crystallogr D Biol Crystallogr 58:1519–1522
54. Stepanov S, Hilgart M, Yoder D et al (2011) Fast fluorescence techniques for crystallography beamlines. J Appl Crystallogr 44: 772–778
55. Wampler RD, Kissick DJ, Dehen CJ et al (2008) Selective detection of protein crystals by second harmonic microscopy. J Am Chem Soc 130:14076–14077
56. Kissick DJ, Dettmar CM, Becker M et al (2013) Towards protein-crystal centering using second-harmonic generation (SHG) microscopy. Acta Crystallogr D Biol Crystallogr 69:843–851
57. Moad AJ, Moad CW, Perry JM et al (2007) NLOPredict: visualization and data analysis software for nonlinear optics. J Comput Chem 28:1996–2002
58. Haupert LM, DeWalt EL, Simpson GJ (2012) Modeling the SHG activities of diverse protein crystals. Acta Crystallogr D Biol Crystallogr 68:1513–1521
59. Haupert L, Simpson G (2011) Screening of protein crystallization trials by second order nonlinear optical imaging of chiral crystals (SONICC). Methods 55:379–386
60. Kissick DJ, Gualtieri EJ, Simpson GJ et al (2010) Nonlinear optical imaging of integral membrane protein crystals in lipidic mesophases. Anal Chem 82:491–497
61. DeWalt EL, Begue VJ, Ronau JA et al (2013) Polarization-resolved second-harmonic generation microscopy as a method to visualize protein-crystal domains. Acta Crystallogr D Biol Crystallogr 69:74–81
62. Closser RG, Gualtieri EJ, Newman JA et al (2013) Characterization of salt interferences in second-harmonic generation detection of protein crystals. J Appl Crystallogr 46:1903–1906
63. Newman JA, Scarborough NM, Pogranichniy NR et al (2015) Intercalating dyes for enhanced contrast in second-harmonic generation imaging of protein crystals. Acta Crystallogr D Biol Crystallogr 71:1471–1477
64. Dettmar CM, Newman JA, Toth SJ et al (2015) Imaging local electric fields produced upon synchrotron X-ray exposure. Proc Natl Acad Sci U S A 112:696–701
65. Song J, Mathew D, Jacob SA et al (2007) Diffraction-based automated crystal centering. J Synchrotron Radiat 14:191–195
66. Cherezov V, Hanson MA, Griffith MT et al (2009) Rastering strategy for screening and centring of microcrystal samples of human membrane proteins with a sub-10 micron size X-ray synchrotron beam. J R Soc Interface 6(Suppl 5):S587–S597
67. Bowler MW, Guijarro M, Petitdemange S et al (2010) Diffraction cartography: applying microbeams to macromolecular crystallogra-

phy sample evaluation and data collection. Acta Crystallogr D Biol Crystallogr 66:855–864

68. Aishima J, Owen RL, Axford D et al (2010) High-speed crystal detection and characterization using a fast-readout detector. Acta Crystallogr D Biol Crystallogr 66:1032–1035
69. Hilgart MC, Sanishvili R, Ogata CM et al (2011) Automated sample-scanning methods for radiation damage mitigation and diffraction-based centering of macromolecular crystals. J Synchrotron Radiat 18:717–722
70. Stevenson HP, Lin G, Barnes CO et al (2016) Transmission electron microscopy for the evaluation and optimization of crystal growth. Acta Crystallogr D Biol Crystallogr 72: 603–615
71. Shi D, Nannenga BL, de la Cruz MJ et al (2016) The collection of MicroED data for macromolecular crystallography. Nat Protoc 11:895–904
72. Shi D, Nannenga BL, Iadanza MG et al (2013) Three-dimensional electron crystallography of protein microcrystals. elife 2:e01345
73. Rodriguez JA, Gonen T (2016) High-resolution macromolecular structure determination by MicroED, a cryo-EM method. Methods Enzymol 579:369–392
74. Warren AJ, Armour W, Axford D et al (2013) Visualization of membrane protein crystals in lipid cubic phase using X-ray imaging. Acta Crystallogr D Biol Crystallogr 69:1252–1259
75. Nishizawa N, Ishida S, Hirose M et al (2012) Three-dimensional, non-invasive, cross-sectional imaging of protein crystals using ultrahigh resolution optical coherence tomography. Biomed Opt Express 3:735–740
76. Nitahara S, Maeki M, Yamaguchi H et al (2012) Three-dimensional Raman spectroscopic imaging of protein crystals deposited on a nanodroplet. Analyst 137: 5730–5735
77. Owen RL, Juanhuix J, Fuchs M (2016) Current advances in synchrotron radiation instrumentation for MX experiments. Arch Biochem Biophys 602:21–31
78. Kawabata K, Takahashi M, Saitoh K et al (2006) Evaluation of crystalline objects in crystallizing protein droplets based on line-segment information in greyscale images. Acta Crystallogr D Biol Crystallogr 62:239–245
79. Pan S, Shavit G, Penas-Centeno M et al (2006) Automated classification of protein crystallization images using support vector machines with scale-invariant texture and Gabor features. Acta Crystallogr D Biol Crystallogr 62:271–279
80. Lavault B, Ravelli RBG, Cipriani F (2006) C3D: a program for the automated centring of cryocooled crystals. Acta Crystallogr D Biol Crystallogr 62:1348–1357
81. Pothineni SB, Strutz T, Lamzin VS (2006) Automated detection and centring of cryocooled protein crystals. Acta Crystallogr D Biol Crystallogr 62:1358–1368
82. Sullivan SZ, Muir RD, Newman JA et al (2014) High frame-rate multichannel beam-scanning microscopy based on Lissajous trajectories. Opt Express 22:24224–24234
83. Bingel-Erlenmeyer R, Olieric V, Grimshaw JPA et al (2011) SLS crystallization platform at beamline X06DA—a fully automated pipeline enabling in situ X-ray diffraction screening. Cryst Growth Des 11:916–923
84. Yamada Y, Hiraki M, Matsugaki N et al (2016) In-situ data collection at the photon factory macromolecular crystallography beamlines. AIP Conf Proc 1741:050023
85. Huang C-Y, Olieric V, Ma P et al (2015) In meso in situ serial X-ray crystallography of soluble and membrane proteins. Acta Crystallogr D Biol Crystallogr 71: 1238–1256
86. Huang C-Y, Olieric V, Ma P et al (2016) In meso in situ serial X-ray crystallography of soluble and membrane proteins at cryogenic temperatures. Acta Crystallogr D Biol Crystallogr 72:93–112
87. Murray TD, Lyubimov AY, Ogata CM et al (2015) A high-transparency, micro-patternable chip for X-ray diffraction analysis of microcrystals under native growth conditions. Acta Crystallogr D Biol Crystallogr 71:1987–1997
88. Lyubimov AY, Murray TD, Koehl A et al (2015) Capture and X-ray diffraction studies of protein microcrystals in a microfluidic trap array. Acta Crystallogr D Biol Crystallogr 71:928–940
89. Roedig P, Vartiainen I, Duman R et al (2015) A micro-patterned silicon chip as sample holder for macromolecular crystallography experiments with minimal background scattering. Sci Rep 5:10451
90. Kisselman G, Qiu W, Romanov V et al (2011) X-CHIP: an integrated platform for high-throughput protein crystallization and on-the-chip X-ray diffraction data collection. Acta Crystallogr D Biol Crystallogr 67: 533–539
91. Yadav MK, Gerdts CJ, Sanishvili R et al (2005) In situ data collection and structure refinement from microcapillary protein crystallization. J Appl Crystallogr 38:900–905

92. Gerdts CJ, Elliott M, Lovell S et al (2008) The plug-based nanovolume Microcapillary Protein Crystallization System (MPCS). Acta Crystallogr D Biol Crystallogr 64:1116–1122
93. Baxter EL, Aguila L, Alonso-Mori R et al (2016) High-density grids for efficient data collection from multiple crystals. Acta Crystallogr D Biol Crystallogr 72:2–11
94. Maeki M, Pawate AS, Yamashita K et al (2015) A method of cryoprotection for protein crystallography by using a microfluidic chip and its application for in situ X-ray diffraction measurements. Anal Chem 87:4194–4200
95. Pawate AS, Srajer V, Schieferstein J et al (2015) Towards time-resolved serial crystallography in a microfluidic device. Acta Crystallogr F Struct Biol Commun 71: 823–830
96. Sui S, Wang Y, Kolewe KW et al (2016) Graphene-based microfluidics for serial crystallography. Lab Chip 16:3082–3096
97. Axford D, Foadi J, Hu N-J et al (2015) Structure determination of an integral membrane protein at room temperature from crystals in situ. Acta Crystallogr D Biol Crystallogr 71:1228–1237
98. Boutet S, Lomb L, Williams GJ et al (2012) High-resolution protein structure determination by serial femtosecond crystallography. Science 337:362–364
99. DePonte DP, Weierstall U, Schmidt K et al (2008) Gas dynamic virtual nozzle for generation of microscopic droplet streams. J Phys D 41:195505
100. Johansson LC, Arnlund D, White TA et al (2012) Lipidic phase membrane protein serial femtosecond crystallography. Nat Methods 9:263–265
101. Sierra RG, Laksmono H, Kern J et al (2012) Nanoflow electrospinning serial femtosecond crystallography. Acta Crystallogr D Biol Crystallogr 68:1584–1587
102. Liu W, Wacker D, Gati C et al (2013) Serial femtosecond crystallography of G protein-coupled receptors. Science 342: 1521–1524
103. Stellato F, Oberthur D, Liang M et al (2014) Room-temperature macromolecular serial crystallography using synchrotron radiation. IUCrJ 1:204–212
104. Gati C, Bourenkov G, Klinge M et al (2014) Serial crystallography on in vivo grown microcrystals using synchrotron radiation. IUCrJ 1:87–94
105. Roessler CG, Agarwal R, Allaire M et al (2016) Acoustic injectors for drop-on-demand serial femtosecond crystallography. Structure 24:631–640
106. Nannenga BL, Shi D, Leslie AGW et al (2014) High-resolution structure determination by continuous-rotation data collection in MicroED. Nat Methods 11:927–930
107. Stevens A, Kovarik L, Abellan P et al (2015) Applying compressive sensing to TEM video: a substantial frame rate increase on any camera. Adv Struct Chem Imaging 1:1–20
108. Kiefersauer R, Grandl B, Krapp S et al (2014) IR laser-induced protein crystal transformation. Acta Crystallogr D Biol Crystallogr 70:1224–1232
109. Cheng Y (2015) Single-particle cryo-EM at crystallographic resolution. Cell 161: 450–457
110. Becker M, Weckert E (2012) On the possibility of determining structures of membrane proteins in two-dimensional crystals using X-ray free electron lasers. In: Cheng RH, Hammar L (eds) Conformational proteomics of macromolecular architecture. World Scientific, Singapore, pp 133–147
111. Hirata K, Shinzawa-Itoh K, Yano N et al (2014) Determination of damage-free crystal structure of an X-ray-sensitive protein using an XFEL. Nat Methods 11:734–736
112. Cohen AE, Soltis SM, González A et al (2014) Goniometer-based femtosecond crystallography with X-ray free electron lasers. Proc Natl Acad Sci U S A 111:17122–17127

Chapter 7

Collection of X-Ray Diffraction Data from Macromolecular Crystals

Zbigniew Dauter

Abstract

Diffraction data acquisition is the final experimental stage of the crystal structure analysis. All subsequent steps involve mainly computer calculations. Optimally measured and accurate data make the structure solution and refinement easier and lead to more faithful interpretation of the final models. Here, the important factors in data collection from macromolecular crystals are discussed and strategies appropriate for various applications, such as molecular replacement, anomalous phasing, and atomic-resolution refinement are presented. Criteria useful for judging the diffraction data quality are also discussed.

Key words Diffraction data collection, Diffraction data quality, Data collection strategy

1 Introduction

Obtaining diffraction-quality crystals is obviously a necessary precondition for solving any macromolecular structure by X-ray diffraction methods. This may be a difficult endeavor, but once appropriate crystals are obtained, it is necessary to submit them to the diffraction data collection process. This is in fact the last truly experimental stage of the crystal structure analysis, because all succeeding steps involve mainly computer calculations, and may be modified and repeated with different programs or parameters. However, the availability of high quality of diffraction data makes the subsequent steps smoother and leads to more accurate and reliable results of the structure analysis.

In the first decades of protein crystallography the data collection process was long, tedious, and required a high level of competence and attention from the experimenters. The enormous progress achieved in the last decades in the hardware and software involved in the macromolecular data collection has changed this situation. Currently diffraction data may often be successfully measured and processed by researchers who lack deep knowledge of the underlying principles, by conducting the synchrotron experiments

Alexander Wlodawer et al. (eds.), *Protein Crystallography: Methods and Protocols*, Methods in Molecular Biology, vol. 1607,
DOI 10.1007/978-1-4939-7000-1_7,

remotely from their own laboratories, using their own laptops. Nevertheless, in spite of the availability of very powerful radiation sources, highly automatic hardware controls, very efficient detectors, and intelligent processing programs, data collection is a scientific process, not a mere technicality. The suboptimal data quality, lower than the level that the crystal is capable of providing, will rebound painfully in all further steps of structure analysis.

However, in practice it is seldom possible to obtain an "ideal" set of diffraction data, characterized by very high resolution, accuracy, and completeness. Unfortunately, it is difficult to satisfy all these requirements at the same time. Measuring very weak, high-resolution reflections involves long exposure to X-rays, which introduces significant radiation damage resulting in diminished accuracy or incomplete data. Collecting and merging data from a series of crystals may alleviate this problem, if all crystals are perfectly isomorphous, otherwise the data accuracy may suffer again. In practice, the data collection process involves various compromises between several requirements, but these compromises should be chosen according to certain principles, depending on the particular intended application of diffraction data. The theory underlining the diffraction data acquisition on two-dimensional detectors can be found in several publications [1–3] and practical guidance during the experiment can be obtained from the strategy programs, such as BEST [4].

Different planned applications put different priorities on various characteristics of measured data sets. Diffraction data intended for the final atomic model refinement should extend to as high a resolution as the crystal can provide. Certain level of radiation damage can be tolerated, or data may be merged from multiple crystals. If exposures necessary to adequately measure the weak, high-resolution reflections lead to many saturated detector pixels, multiple passes of data collection are advisable with different effective exposures.

The data intended for structure solution by molecular replacement do not need to extend to high resolution, since only relatively low resolution data are used in this approach anyway. Since this method is based on the comparison of Patterson functions, the strongest reflections are especially prominent, and the completeness of the low-resolution data is therefore very important. Similarly, what is important for the identification of potential small-molecule ligands is rapid measurement of a large number of data sets, but their resolution is not so crucial. After identification of the complexes, high-resolution data may be collected afterwards.

Data to be used for phasing based on anomalous signal must be of as high accuracy as possible, since the anomalous differences are very small, on the order of a few percent of the total reflection intensities, or even smaller in case of sulfur utilized as an anomalous scatterer. Radiation damage should be avoided by limiting the exposures, or by measuring data from multiple crystals. The data collected from heavy-atom derivatives should have similar, perhaps somewhat less stringent characteristics.

Often only one set of data is collected and used for structure solution and refinement. It should then partially satisfy various, somewhat contradicting requirements. Intelligent decisions need to be made in order to achieve in such cases the optimal compromise. The following sections will discuss the most important factors influencing the quality of diffraction data collected by the single-crystal rotation mode.

2 Data Completeness

2.1 Asymmetric Unit in Reciprocal Space

A complete data set should contain all reflections within the asymmetric unit of the reciprocal space for the particular symmetry of the crystal. The concept of the asymmetric unit in the reciprocal space is different from the asymmetric unit of the cell in the crystal direct space. In direct space the asymmetric unit is a fraction of the cell that by the action of all symmetry operations of the crystal space group, completely covers the whole unit cell. The volume of such an asymmetric unit is V_{cell}/n, where n is the number of independent symmetry operators of the space group. The proposed definitions of direct cell asymmetric units are presented for each space group in the International Tables, Vol. A [5] and usually have the shape of a parallelepiped, except in cubic symmetry where the shapes are more complicated.

An asymmetric unit in the reciprocal space has the shape of a wedge with its apex at the origin, extending away to the limit of data resolution and bounded by the symmetry elements of the crystal point group (or, rather, its Laue symmetry). In the following text the term "asymmetric unit" will always refer to the reciprocal space. The definition of the reciprocal space asymmetric unit depends on the point group, not the space group, and is for example the same for crystals of $P422$, $P4_32_12$, or $I4_122$ symmetries. An example of the asymmetric unit in the reciprocal space for the crystal of 622 symmetry is shown in Fig. 1. The definitions of reciprocal space asymmetric units for all "macromolecular" (i.e., not containing centers of symmetry or mirror planes) crystal classes standardized according to CCP4 are presented in Table 1. However, these definitions apply to the native data, where the anomalous scattering is not taken into account. As a consequence of the anomalous diffraction effects, reflections related by the center of symmetry or mirrors have different intensities, hence it is necessary to record all reflections in the "anomalous asymmetric unit," comprising two native asymmetric units related by the center of symmetry or mirror planes existing in the Laue symmetry corresponding to the crystal symmetry class.

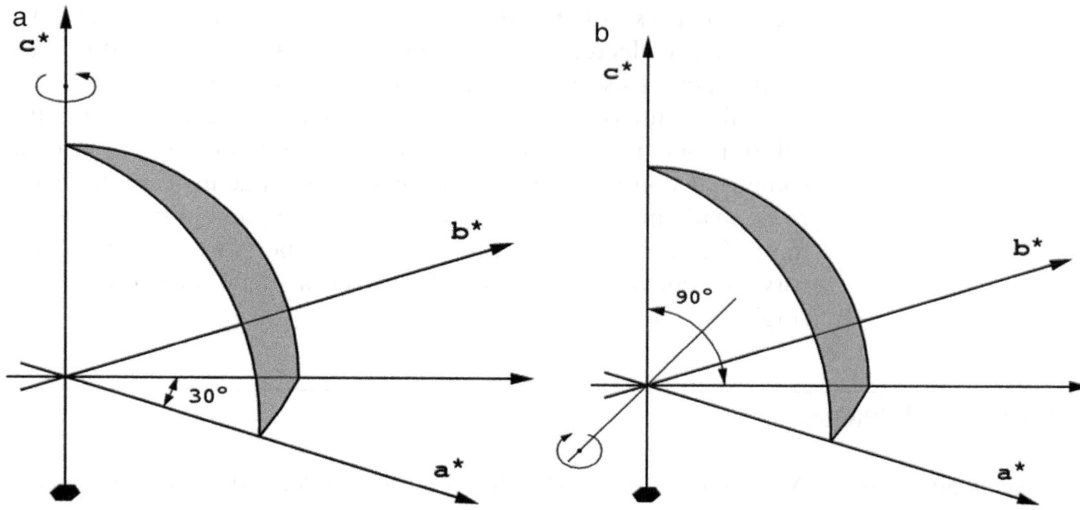

Fig. 1 The asymnmetric unit for the point group 622. (**a**) If the crystal is rotated around the sixfold axis, the complete data set may be achieved after 30° of total rotation; (**b**) if the crystal is rotated around the axis lying in the *a,b*-plane, 90° of rotation is necessary

Table 1
Definition of the reciprocal space asymmetric units according to the CCP4 standard

Crystal system	Point group	Reflection class		Conditions for indices	
Triclinic	1	*hkl*	$\pm h$	$\pm k$	$l \geq 0$
		*hk*0	$h \geq 0$	$\pm k$	$l = 0$
		0*k*0	$h = 0$	$k > 0$	$l = 0$
Monoclinic	2	*hkl*	$\pm h$	$k \geq 0$	$l \geq 0$
		*hk*0	$h \geq 0$	$k \geq 0$	$l = 0$
		0*k*0	$h = 0$	$k > 0$	$l = 0$
Orthorhombic	222	*hkl*	$h \geq 0$	$k \geq 0$	$l \geq 0$
Tetragonal	4	*hkl*	$h > 0$	$k > 0$	$l \geq 0$
		0*kl*	$h = 0$	$k \geq 0$	$l \geq 0$
	422	*hkl*	$h \geq 0$	$h \geq k \geq 0$	$l \geq 0$
Trigonal	3	*hkl*	$h \geq 0$	$k > 0$	$l \geq 0$
		00*l*	$h = 0$	$k = 0$	$l > 0$
	312	*hkl*	$h \geq 0$	$h \geq k \geq 0$	$\pm l$
		*h*0*l*	$h \geq 0$	$k = 0$	$l \geq 0$
	321	*hkl*	$h \geq 0$	$h \geq k \geq 0$	$\pm l$
		hhl	$h \geq 0$	$k = h$	$l \geq 0$
Hexagonal	6	*hkl*	$h > 0$	$k > 0$	$l \geq 0$
		0*kl*	$h = 0$	$k \geq 0$	$l \geq 0$
	622	*hkl*	$h \geq 0$	$h \geq k \geq 0$	$l \geq 0$
Cubic	23	*hkl*	$h \geq 0$	$k \geq h$	$l \geq h$
		0*kl*	$h = 0$	$k > h$	$l \geq h$
	432	*hkl*	$h \geq 0$	$k \geq l$	$l \geq h$
		0*kl*	$h = 0$	$k > l$	$l \geq h$

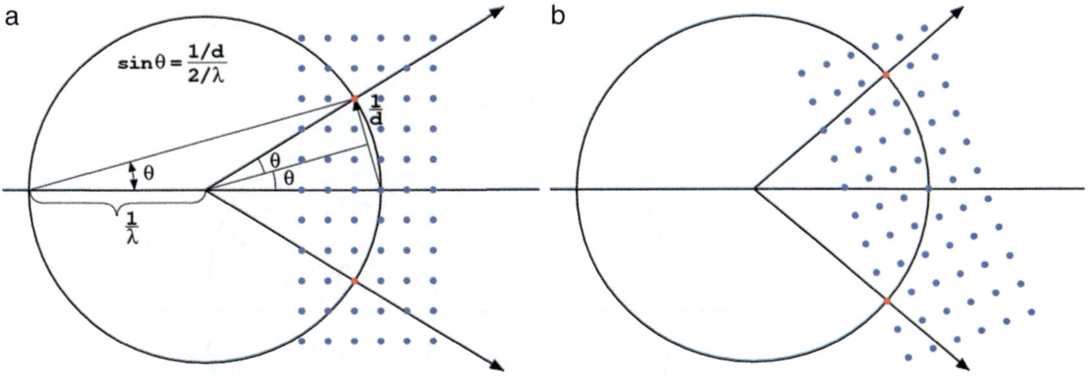

Fig. 2 The Ewald construction illustrates the Bragg's law in three dimensions. This figure shows the central section of the Ewald sphere of the radius $1/\lambda$ representing the X-radiation and reflections in reciprocal lattice representing the crystal. (**a**) If the reciprocal lattice point lies at the surface of the Ewald sphere, the trigonometric conditions corresponding to the Bragg's law are fulfilled. (**b**) To bring more reflections to the diffraction condition, the crystal has to be rotated

The diffraction condition for reflections originating from a crystal exposed to the X-ray beam is formulated by the Bragg's law, $\lambda = 2d\sin\theta$, which is conveniently illustrated by the Ewald construction, Fig. 2. If the crystal is stationary during exposure, only a few reflections are diffracting. More reflections come into diffraction condition if the crystal is rotated. This is the basis of the standard rotation method of diffraction data collection, most popular in macromolecular crystallography.

Two other approaches are also possible. Data can be acquired from a large number of exposures from many stationary crystals in random orientations, eventually acquiring highly redundant and complete set of data. This approach requires special methods for estimation of reflection intensities, since the reciprocal lattice points do not cross the surface of the Ewald sphere and the individual estimations are lower that the full reflection intensities. This method of data collection is by necessity used at the X-ray laser facilities (XFELs).

Another, Laue approach involves again a stationary crystal, but irradiated with the white, non-monochromatized X-ray radiation. In this case instead of moving the crystal and reflections to diffracting position, the particular X-ray wavelength (i.e., appropriate size of the Ewald sphere) from the continuous spectrum is adjusted to each reflection (Fig. 3). This method can be used to trace certain short-lived reaction states during chemical processes taking place in the crystal, although it has certain theoretical and practical limitations, especially influencing the completeness of low resolution reflections [6]. The Laue approach may be therefore useful for only certain special applications and will not be further addressed in this text.

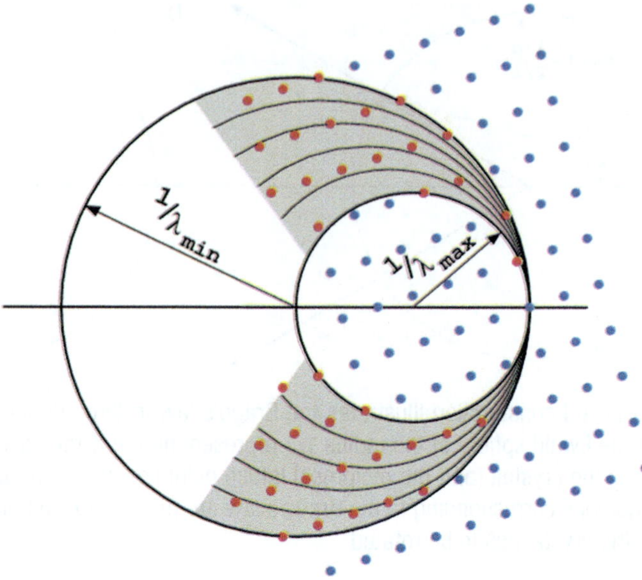

Fig. 3 If the X-rays are polychromatic ("*white*," not monochromatized), the multitude of different wavelengths invokes diffraction of many reflections, even if the crystal is stationary

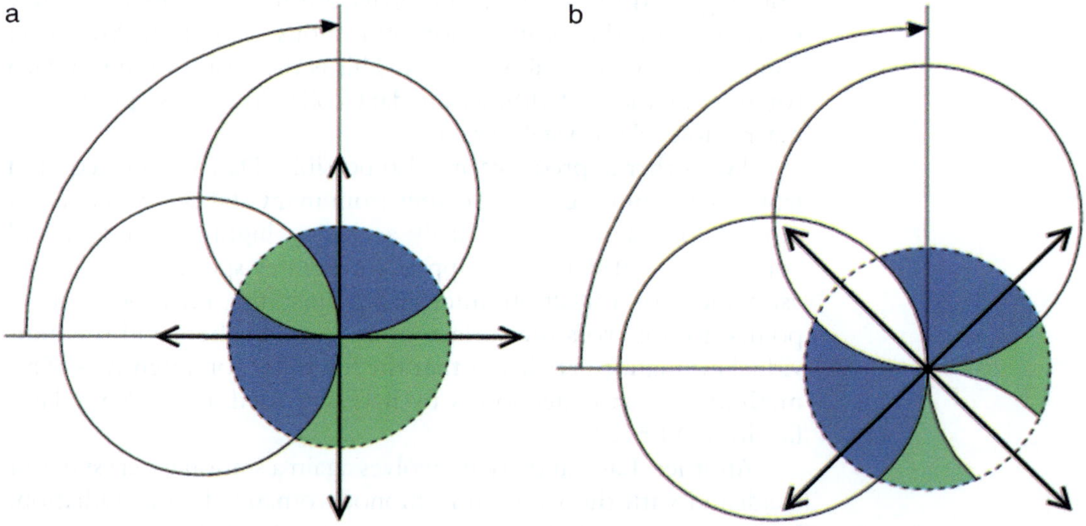

Fig. 4 If the orthorhombic crystal is rotated around one of its twofold axes, 90° of total rotation leads to the complete data set only if (**a**) it starts at the parallel orientation of its other axes with respect to the beam direction or detector plane, but in the diagonal orientation (**b**) the data will not be complete, since the region marked in *white* will not be covered

2.2 Total Rotation Range

To achieve full data completeness, all reflections within the asymmetric unit, or their symmetry equivalents, have to be measured at least once. The minimal amount of crystal rotation necessary to fully cover the asymmetric unit depends on the crystal symmetry class (Fig. 4). Table 2 summarizes these values for all

Table 2
Minimal amount of crystal rotation (°) necessary to obtain complete data depending on the crystal symmetry class and its orientation with respect to the spindle axis; (*ab*) means any direction in the *ab*-plane

Crystal class	Native data	Anomalous data
1	180 (any)	180 + $2\theta_{max}$ (any)
2	180 (*b*), 90 (*ac*)	180 (*b*), 180 + $2\theta_{max}$ (*ac*)
222	90 (*ab*, *ac*, *bc*)	90 (*ab*, *ac*, *bc*)
4	90 (*c*, *ab*)	90 (*c*), 90 + θ_{max} (*ab*)
422	45 (*c*), 90 (*ab*)	45 (*c*), 90 (*ab*)
3	60 (*c*), 90 (*ab*)	60 + $2\theta_{max}$ (*c*), 90 + θ_{max} (*ab*)
32	30 (*c*), 90 (*ab*)	60 + $2\theta_{max}$ (*c*), 90 (*ab*)
6	60 (*c*), 90 (*ab*)	60 (*c*), 90 + $2\theta_{max}$ (*ab*)
622	30 (*c*), 90 (*ab*)	30 (*c*), 90 (*ab*)
23	~60	~70
432	~35	~45

macromolecular point groups, for crystals oriented symmetrically with respect to the goniostat spindle axis. In the arbitrary crystal orientation, it is hard to estimate the necessary rotation range, and it is then better to rely on the advice of strategy programs, such as BEST [4], run on the basis of the initial test diffraction image(s).

Of course, 360° of total crystal rotation will always provide the maximum coverage possible to obtain in a single rotation pass of the crystal. That, however, does not guarantee full completeness of data, and the effect of the "blind region" will be addressed in the following section. However, 360° of crystal rotation is not necessary in most cases that include crystals with symmetry higher than *P*1.

Here one of the compromises is evident. More crystal rotation delivers more multiple measurements of the symmetry-equivalent reflections, theoretically resulting in more accurate estimation of the average intensities but, simultaneously, longer exposures lead to more radiation damage, which may spoil these benefits. The compromise depends on circumstances, such as crystal robustness, beam intensity, and detector properties. For example, if the intrinsic detector background is negligible (as is case of the photon counting pixel detectors), it may be advisable to use wider total rotation ranges with somewhat attenuated X-ray beam intensity.

2.3 Rotation Range per a Single Exposure, Mosaicity, Wide and Fine Slicing

In the rotation method of diffraction data collection, reflection intensities are recorded on a series on consecutive images recorded when a crystal is exposed to X-rays during small rotation around the goniostat spindle axis. The number of reflections recorded on each image depends on several factors. The density of reflections in the reciprocal space is constant and is related to the crystal unit cell volume. One degree rotation of a virus crystal may produce the image with thousands of reflections. On the other hand, a crystal of a small molecule may lead to only very few visible reflection spots, and this is the useful practical check whether the crystallized material is a macromolecule, or a serendipitously precipitated salt from the solution buffer.

In contrast to the precession method, in the screenless rotation method the geometry of reciprocal space is distorted on diffraction images. The straight lines of reflections in the reciprocal space are represented as hyperbolas and reflections in the individual planes in the reciprocal space are grouped on diffraction images in lunes limited by elliptical boundaries. The successive lunes become wider (in the direction perpendicular to the spindle axis) when the amount of rotation per image, $\Delta\varphi$, increases. This results from the cross-section of the cone diffracting rays by the plane of reflections projected on the flat plane of a detector, as illustrated in Fig. 5. The density of reflection profiles in each lune depends on the crystal cell dimensions in directions parallel to the plane, whereas the gap between the successive lunes depends on the distance between two consecutive reciprocal lattice planes and, therefore, the cell dimension in the direction perpendicular to the planes or, in other words, parallel to the X-ray beam. To avoid the possibility of excessive overlap of

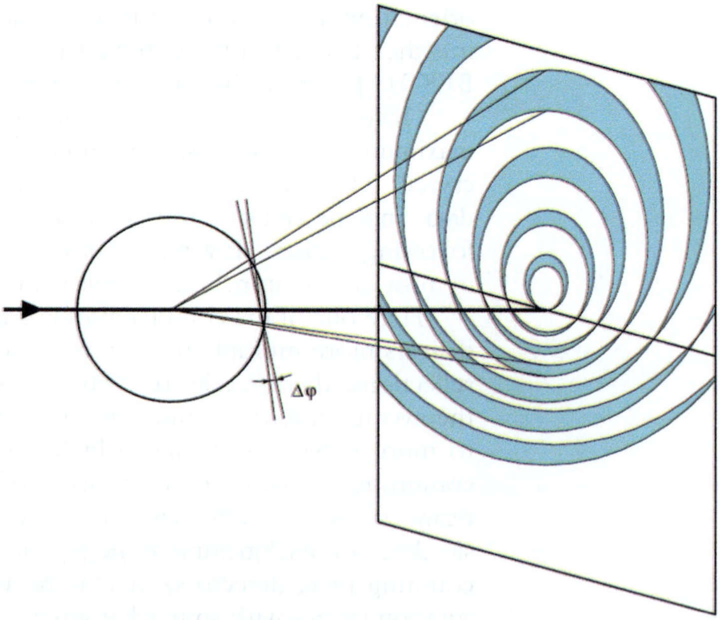

Fig. 5 Reflections in each plane of the reciprocal space will give rise to one "lune" at the detector. The width of each lune depends on the amount of crystal rotation during the exposure. Too wide rotations cause the lunes to overlap at the high diffraction angles, which may also lead to overlap of individual reflection profiles

reflection profiles, it is therefore advisable to orient the crystal at the goniostat with the longest cell dimension more or less parallel to the spindle axis, so that it never becomes parallel to the X-ray beam.

The kinematic theory of diffraction assumes that crystals are built from small mosaic blocks, slightly misoriented from each other by a small angle η. As a consequence, diffraction of each reflection from mosaic crystals is not instantaneous, but occurs during a small angular range of crystal rotation. This can be represented by the reciprocal lattice reflections having certain finite size, not being the infinitesimally small mathematical points. As a practical consequence, some reflections start diffracting on one image, but continue diffracting on the next image, while the corresponding reciprocal lattice points cross the surface of the Ewald sphere. The intensity of such partially recorded reflections (partials) are spread over spots on multiple images, in contrast to reflections fully recorded on one image. The time and angular interval spent by each reflection in crossing the Ewald sphere, and the total diffraction rocking curve, depends also on the beam divergence δ and its bandpass $\Delta\lambda/\lambda$. Although synchrotron radiation is usually highly collimated, beam divergence is not negligible and may differ in the horizontal and vertical directions, depending on the properties of the source, monochromator, and focusing mirrors of a particular beam line. These effects are illustrated in the direct and reciprocal space in Fig. 6a, b.

There are two ways of data collection, the wide slicing and fine slicing approaches, depending on the relation between the rocking width and the crystal rotation interval. In the first case some reflections are fully recorded and some are partially recorded. In the second case all reflections are multiple partials (Fig. 7). In the wide slicing approach, reflection profiles can be built from detector pixels only in two dimensions of the detector window. In the fine

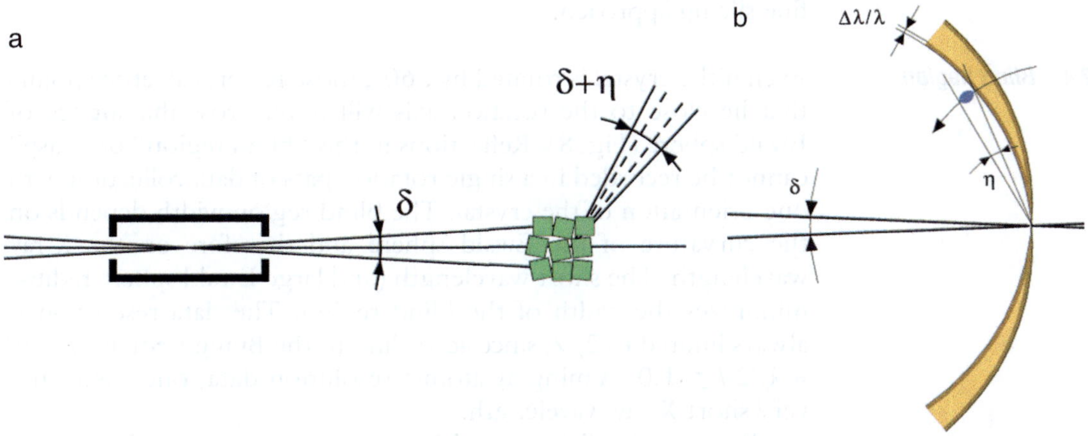

Fig. 6 A schematic representation of the beam divergence (δ), crystal mosaicity (η) and beam wavelength bandpass ($\Delta\lambda/\lambda$) in the direct (**a**) and reciprocal (**b**) space

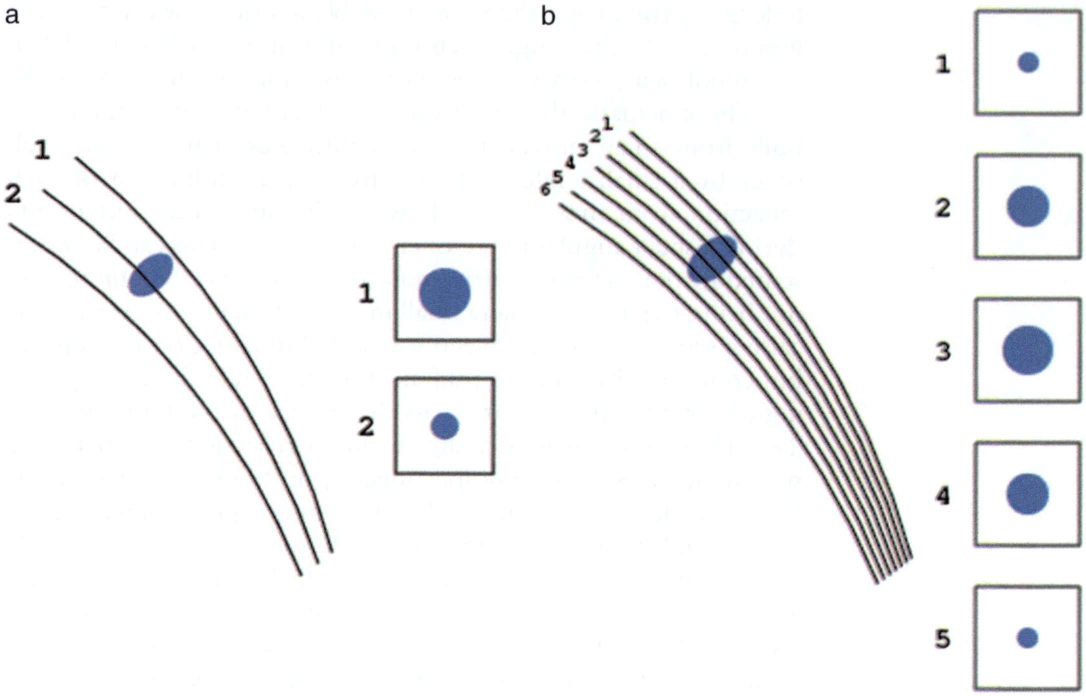

Fig. 7 The principle of the (**a**) wide slicing, when each reflection is either fully recorded on one image or split among two images, and (**b**) fine slicing, when each reflection is partially recoded on several consecutive images

slicing approach the profiles can be constructed in three dimensions in the so-called shoe-boxes, with the third direction being orthogonal to the detector plane, which may lead to more accurate estimation of the total reflection intensities. In addition, since the image width $\Delta\varphi$ is larger than the width of the rocking curve, the background accumulates during the whole exposure. Consequently, the signal-to-noise ratio in wide slicing mode is worse than in the fine slicing approach.

2.4 Blind Region

Even if the crystal is rotated by 360°, those reciprocal lattice points that lie close to the rotation axis will never cross the surface of Ewald sphere (Fig. 8). Reflections in this "blind region" or "cusp" cannot be recorded in a single rotation pass of data collection with one orientation of the crystal. The blind region width depends on the curvature of the Ewald sphere and therefore on the X-ray wavelength. The short wavelength (and large Ewald sphere radius) minimizes the width of the blind region. The data resolution is always limited to $2/\lambda$, since according to the Bragg's equation $\sin\theta = \lambda/2d \leq 1.0$. Aiming at atomic resolution data, one has to use very short X-ray wavelength.

Fortunately, if the crystal has a symmetry axis and it is misset from the direction of the spindle axis by the angle corresponding to the highest data resolution θ_{max}, all reflections within the blind

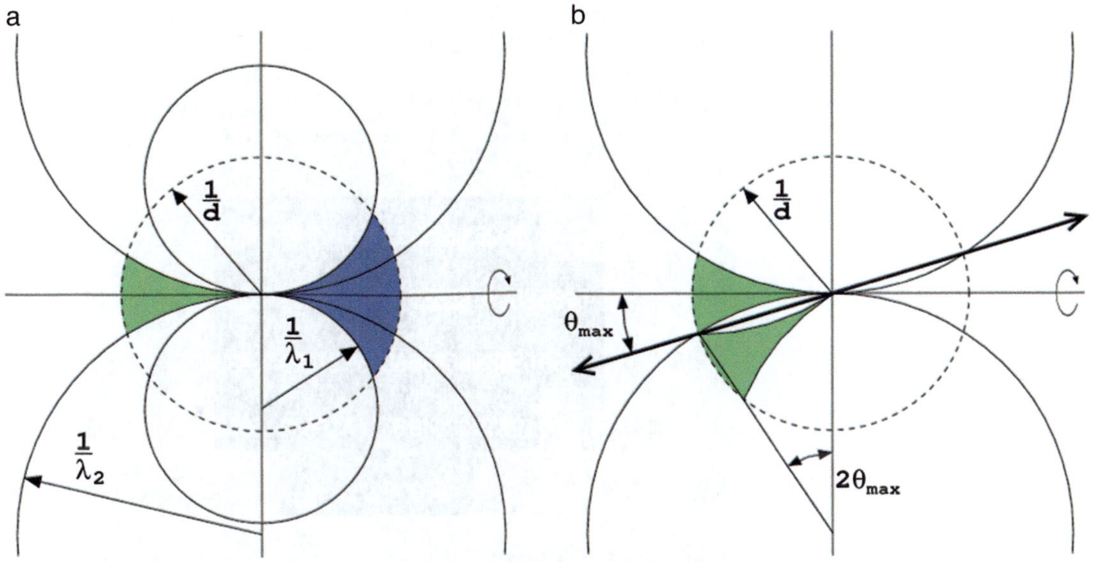

Fig. 8 Even after 360° rotation some reflections, in the blind region, close to the rotation axis, will never cross the Ewald sphere. (**a**) The blind region is wider for long wavelength than for the short wavelength, when the curvature of the Ewald sphere is lower. (**b**) If the unique symmetry axis of the crystal is misset from the spindle axis direction by more than θ_{max}, all reflections in the blind region have their symmetry equivalents measurable in other parts of reciprocal space

region have their symmetry mates in other regions of reciprocal space and full data completeness can be achieved (Fig. 8b). However, the blind region negatively affects the data completeness only if the crystal has *P*1 symmetry or it is rotated around its unique symmetry axis. The latter situation occurs in one of the approaches to collection of anomalous data, aimed at recording Bijvoet-related reflections on the same image.

2.5 Saturated Detector Pixels

Two-dimensional detectors have certain limit of intensity that can be stored in each pixel. If the electronics of the detector stores numbers as 16-bit integers, the maximum pixel values are $2^{16} - 1 = 65{,}535$, and all higher intensities are truncated, which leads to some reflections being "overloaded" (Fig. 9). Some detectors, such as PILATUS, work with 20-bit arithmetic and have therefore a much higher dynamic range.

As a result of this limitation, it is not possible to adequately record the most intense, low resolution reflections and the very weak, high resolution reflections simultaneously, on the same rotation pass with the same exposures. The strongest reflections are most important for any phasing methods and most strongly modulate all kinds of electron density maps. Missing them will negatively influence all subsequent steps of the crystal structure analysis.

A practical solution to avoid overloads is to collect data in multiple passes with different effective exposures. The "low

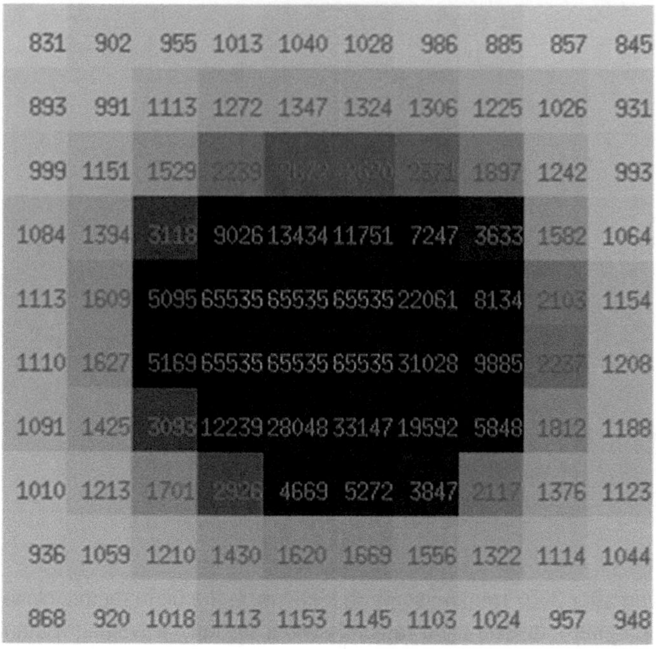

Fig. 9 Very strong reflections may have some pixels in their profiles at the detector overloaded, when very high intensity becomes truncated at a level of, for example, $2^{16} - 1 = 65{,}535$

resolution" pass should be performed first, when the crystal is not significantly radiation damaged, aiming at resolution extending only to the limit where overloads will occur in the high exposure pass. The exposure times and X-ray beam attenuation should be adjusted to avoid any overloaded pixels in the reflection profiles and the rotation amount per image may be relatively large. In the next, "high-resolution" pass, effective exposures may be increased up to ten times and the other parameters should be appropriately adjusted. All intensities from all passes are then scaled and merged together. The problem of overloads is less severe with the fine slicing mode of data collection, when intensities of strong reflections are spread over multiple images.

2.6 Beam Size

If the crystal is of high quality, it is always beneficial to expose its total volume, making use of its full diffraction potential. The beam size should preferably be adjusted to the crystal size, to avoid unnecessary excessive background on the recorded diffraction images. This is not always achievable if the crystals are shaped as plates or needles. There are, however, instances when it is advisable to use beam with a cross-section much smaller than the crystal size.

Sometimes large crystals are highly nonuniform throughout their volume, with diffraction properties (mosaicity, resolution) varying in different parts of the whole specimen. It is obviously more productive to collect data from the well-behaving part of

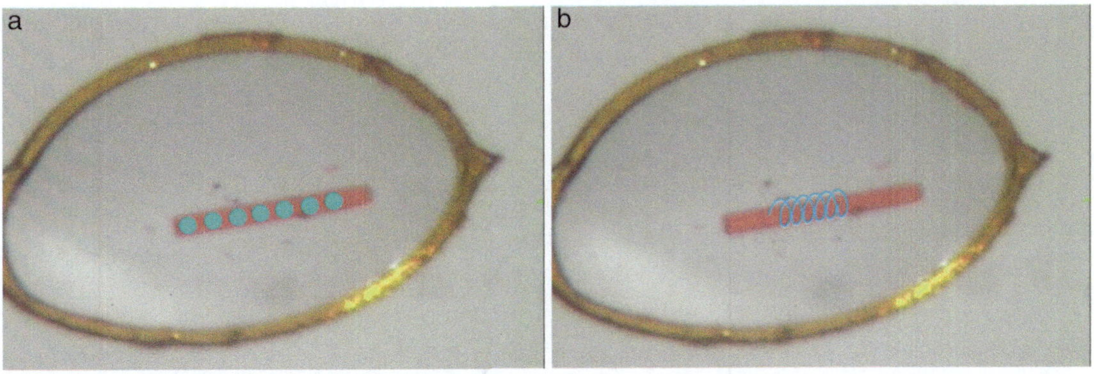

Fig. 10 Using X-ray beam collimated to small size, it is possible to collect data from several places along the long crystal (**a**), or move such a crystal, while it rotates in the so-called helical data collection protocol (**b**)

such a crystal than from the whole sample. For long, needle-like crystals it is possible to collect data with small beam size from several parts, moving it along the spindle axis after several images (Fig. 10a). Many synchrotron facilities allow for the "helical" approach, in which a crystal is moved successively while it rotates (Fig. 10b). The small beam size (and a high level of collimation) may also be beneficial if the crystal cell dimensions are very large, in order to diminish the overlap of reflection profiles at the detector window.

2.7 Radiation Damage

Radiation damage, incurred in macromolecular crystals during exposure to X-rays, has been a curse of protein crystallography from its early days. Currently, with the routine use of very intense X-ray synchrotron beam sources, radiation damage is still a very important issue, which has to be taken into account in the practice of macromolecular crystallography [7]. Even if crystals are cooled to temperatures of about 100 K, their total diffraction intensity diminishes by a factor of two after absorbing X-ray doses of 20–40 MGy. More importantly, some specific damage, in the form of decarboxylation of acidic residues, breakage of disulfide bonds, various conformational changes of amino acid side chains etc., occurs at much smaller doses, and that may lead to potential misinterpretation of various structural features and biologically important functional results.

Cryo-cooling diminishes the secondary damage effects resulting from diffusion of certain active radicals throughout the crystal. However, the primary radiation damage following absorption of X-ray quanta is inevitable. The radiation damage can only be mitigated by reduction of exposure time or attenuation of the X-ray beam intensity. A certain degree of damage may be allowed if data are to be used for final model refinement, but for anomalous phasing applications any damage must be avoided.

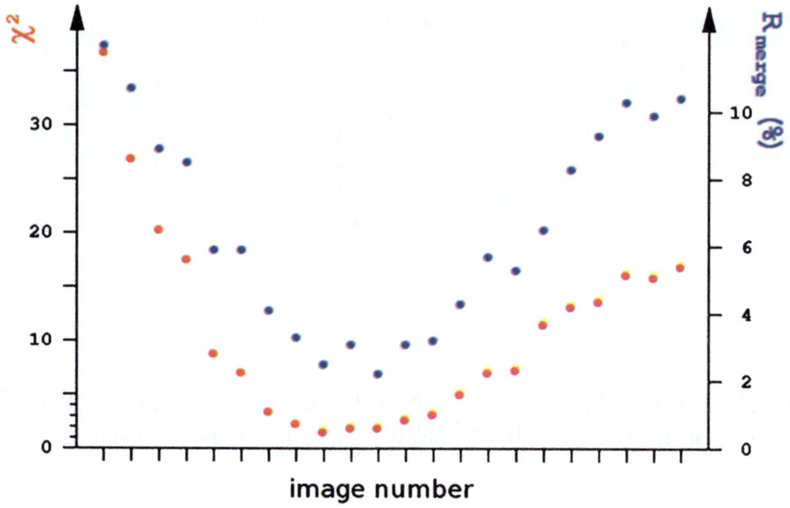

Fig. 11 Radiation damage can be visualized if the dependence R_{merge} or χ^2 values for individual images form a "smiley" curve, where the intensities differ mostly form their average at the beginning and the end of the series of images

If the data are to be used for final structure refinement, the total dose should not exceed the so-called Garman limit of 20 MGy [8]. For data used for anomalous phasing this limit should be much lower. It is advisable to evaluate radiation damage at the early stages of data collection. Some strategy programs, e.g., BEST or RADDOSE, can be used to estimate the appropriate exposures that permit to collect the complete data within the selected total absorbed dose.

The useful criteria of radiation damage are the scaling B factors and R_{merge} values. As a rule of thumb, the absorption of 1 MGy results in the increase of the scaling B-factor by about 1 A^2. Often degradation of the reflection profiles and loss of high resolution intensities can be judged by visual inspection of diffraction images. The R_{merge} and χ^2 values as a function of the image number may show characteristic "smiley" behavior, with highest values at the beginning and end of the range and lowest values in the middle (Fig. 11), since the average merged intensities are closest to those recorded in the middle of the set and most different from those measured at the start and end of the session.

2.8 Alternative Indexing and Merohedral Twinning

If the crystal point group symmetry is lower than the symmetry of the crystal lattice, reflections can be indexed in more than one way. Such cases occur when the crystal has a polar axis, when its two directions are not equivalent (Fig. 12). This affects the following crystal classes: 4, 3, 32, 6, and 23, and all space groups with various screw axes within these classes. The problem of multiple ways of indexing may also occur if certain unit cell parameters lead to lattices having by chance higher metric symmetry than the true

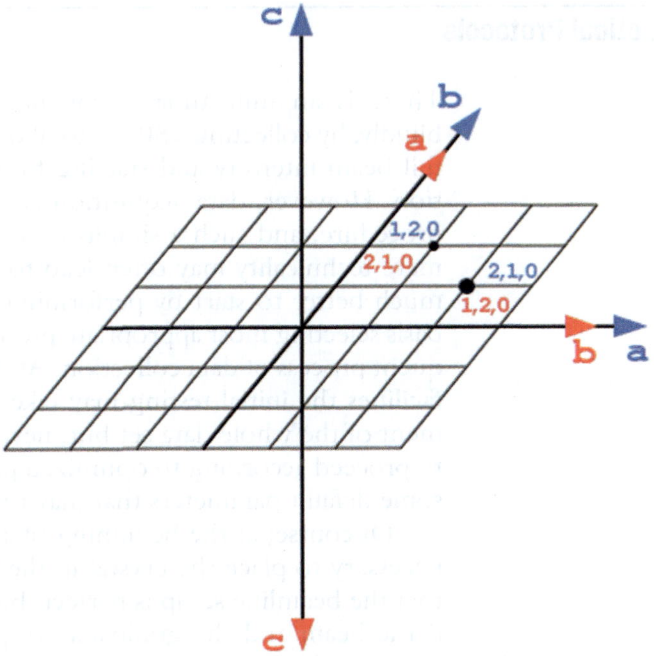

Fig. 12 In space groups possessing polar rotation axes, it is possible to index reflections in two (or more), nonequivalent ways. Both ways are correct, but if data are measured from more than one crystal, the indexing scheme has to be the same for all contributing parts

symmetry of the crystal structure. For example, a monoclinic crystal with $a = c$ will "pretend" to be orthorhombic C-centered.

The same crystal classes are also vulnerable to merohedral twinning when small, individual domains within a single crystalline specimen are mutually related by the symmetry operation belonging to the symmetry of the lattice, but not existing in the set of symmetry operations of the true point group of the crystal structure. Reflection intensities measured from perfectly merohedrally twinned crystal can be successfully merged in a higher than actual crystal symmetry. The presence of twinning can only be identified from various tests based on the statistics of reflection intensities (*see* [9] and Chapter 8 by Thompson).

If the crystal symmetry is not known, it is advisable to assume than it is lower than the full symmetry of the lattice, for example 4 instead of 422, and to adjust the strategy appropriately. For example, for a tetragonal crystal rotated around its fourfold axis, it is safer to collect the total of 90° of data with half exposure, than the minimum 45° required for the complete set in 422 symmetry, in case that the crystal will turn out to be twinned. The data can be tested for twinning early, before achieving full completeness. Such programs as xtriage [10] or POINTLESS [11] can be run even with a partial data set, when the crystal still resides at the goniostat, so that the strategy can be modified appropriately.

3 Practical Protocols

There is an unfortunate tendency to measure diffraction data blindly, by collecting 180° of total data with 0.1° wide images with full beam intensity and starting from an arbitrary crystal orientation. However, data acquisition is in fact a complicated scientific procedure, and such a simplistic treatment of data collection as a mere technicality may often lead to less than optimal results. It is much better to start by performing some initial tests and on this basis selecting most appropriate protocol and parameters for subsequent process of data collection. At the contemporary synchrotron facilities the initial testing may take more time than the measurement of the whole data set but, nevertheless, it is always beneficial to proceed according to optimized protocols rather than to rely on some default parameters that may turn out to be inappropriate.

Of course, at the beginning of any diffraction experiment it is necessary to place the crystal in the X-ray beam. One can assume that the beamline setup is perfect, but it may be advisable to check if the beam and the goniostat are properly aligned. This can be done first by centering a small object (a sharp needle, a small crystal, or an empty loop), so that it rotates around its own center while the spindle axis revolves and marking this place within the camera window. Next, a fluorescent object (a blob of fluorescent salt or a thin YAG plate) can be put at the goniostat to check if the beam is centered exactly at the rotation axis of the spindle. When the crystal is mounted for an experiment, it is important that it should rotate around its own center, to ensure the uniformity of intensities during scaling and merging procedure. If it is intended to expose a small part of a large crystal with a small beam, the crystal position must be adjusted accordingly, with the selected crystal fragment located exactly at the spindle axis. This is not always coincident with the cross-hair of the viewing camera.

3.1 Test Exposures

It is good to start the data collection session by exposing a couple of test exposures at two orthogonal orientations, such as 0° and 90°, since sometimes one of the test images may look acceptable, but the orthogonal one may disclose unacceptable characteristics. Many features can be immediately judged by eye, if the crystal is single or split, if the reflection profiles are highly diffused or overlapping, etc. One of the exposures can be recorded with relatively intense beam in order to estimate the resolution limit of diffraction. A test image should be indexed (or even integrated) and the apparent crystal symmetry established. This allows one to select the optimal data collection parameters, such as the total and per image rotation ranges, spindle axis start position, crystal-to-detector distance, X-ray beam attenuation, and exposure time. Preferably, this may be done with the use of one of the strategy programs.

3.2 Selection of Wavelength

For collecting data from native crystals it is not necessary to select any particular X-ray wavelength. At the home sources there is usually no choice, since most facilities are equipped with copper anodes delivering X-rays of 1.54 Å, or more rarely with molybdenum anodes with 0.71 Å (appropriate for high resolution data) or chromium anodes with 2.29 Å (appropriate for measuring anomalous data from lighter elements, such as S, P, Ca). At synchrotron facilities native data are usually collected with wavelengths close to 1 Å, optimal from the point of view of beam line optics and beam flux. Only aiming at very high resolution it may be necessary to use the short X-ray wavelength and as short as possible crystal-to-detector distance. Sometimes the parallel or angular detector offset from the central position may be used to increase the diffraction angles of highest resolution reflections.

If the aim is to measure anomalous data, the wavelength should be selected appropriately. For MAD work, it is necessary to record the fluorescence spectrum around the absorption edge of the selected anomalous scatterer, which can be then interpreted by the program CHOOCH [12]. One can select to measure data at three wavelengths, the peak, edge and high-energy remote (50–100 eV beyond the edge) values, or only at two wavelengths, the edge and high-energy remote, omitting the fluorescence peak value, where the absorption is the highest, especially with anomalous scatterers such as lanthanides or tantalum. For MAD data one should use modest effective exposures, to avoid incurrence of radiation damage.

For SAD work, the wavelength can be selected either at the peak value, or at the high-energy remote. The latter does not require recording the fluorescence spectra. Similarly, aiming at recording the anomalous signal from relatively light elements, such as sulfur, phosphorus or calcium that have no absorption edges in the accessible range of wavelengths on most of synchrotron beam lines, the wavelength should be set to longer values, in the vicinity of 2 Å [13].

3.3 Choice of Symmetry

Unless the crystal symmetry is known in advance, it has to be established during data collection and reflection merging. The particular space group is not important at this stage, only the point group is relevant for data collection strategy. Initial indexing may suggest the Bravais lattice of highest symmetry, but the metric of the lattice may have higher symmetry than the true symmetry of the structure. The obvious cases are the hemihedral crystal classes, such as 4 in the 422 (in fact 4/mmm) lattice, 3, 32, 6 in the 622 (6/mmm) lattice or 23 in 432 (m3m) lattice, but serendipitous agreement of the unit cell parameters with higher symmetry crystal systems sometimes may occur. The true point group may be only established at the stage of data merging, and even then perfect (pseudo)merohedral twinning may not be easily identified.

It is therefore advisable to adopt a strategy appropriate for the lower potential crystal symmetries. The integrated data can be merged even before a complete set is achieved or such partial data sets may be submitted to POINTLESS [11], to select the true symmetry operations of the crystal point group. The strategy can be then modified, for example by covering the extended total crystal rotation range.

The photon counting pixel detectors, characterized by very low intrinsic noise, offer a version of the data collection strategy where data are collected over a wide total rotation range with diminished beam intensity. Because of low noise, data quality does not suffer, but higher data multiplicity and assurance that data are complete even if the crystal symmetry is lower than apparent from the initial indexing are beneficial.

However, it is always advisable to start data collection at the optimally selected crystal orientation (spindle axis position), which ensures the earliest achievement of high completeness, even if the crystal dies because of radiation damage during the process.

3.4 Data Quality

Several criteria can be used to judge the data quality. Some are more popular than others and various criteria have different statistical validity. Some factors are global, other relate to narrow resolution bins. The traditional criteria are the data resolution limit and the R_{merge} value, calculated as $R_{\text{merge}} = (\Sigma_{hkl}\Sigma_i|I_i - \langle I \rangle|)/(\Sigma_{hkl}\Sigma_{iIi})$. In addition, in the presentation of refined structures required are the data completeness, average multiplicity of measurements of equivalent reflections, and the average ratio of intensities to their uncertainties, $I/\sigma(I)$. These values are given for all data and for the highest resolution bin.

However, none of these criteria is fully objective and statistically perfect. The R_{merge} value increases (becomes worse) with increased multiplicity, while the data quality certainly improves. It is therefore better to use more statistically valid versions, $R_{\text{meas}} = (\Sigma_{hkl}[n/(n-1)]\Sigma_i|I_i - \langle I \rangle|)/(\Sigma_{hkl}\Sigma_i I_i)$ [14] or $R_{\text{pim}} = (\Sigma_{hkl}[1/(n-1)]\Sigma_i|I_i - \langle I \rangle|)/(\Sigma_{hkl}\Sigma_i I_i)$ [15]. The average signal-to-noise ratio $I/\sigma(I)$ is a good indicator, under the condition that the uncertainties $\sigma(I)$ are estimated correctly. This is not always easy, since their evaluation depends on proper detector calibration, reflection profile and background estimation and other factors, and the proper counting statistics of the recorded X-ray quanta may not apply directly. There are ways to check and correct the level of uncertainties by comparing them with the expected statistics using, for example, the normal probability plots. It is worth paying attention to this issue since the correct estimation of uncertainties is important for all phasing and refinement methods based on statistical maximum likelihood principles. Usually required in all presentations are the overall and highest resolution data completeness, which should be high, preferably above 95% and 75%, respectively. However, highly informative is also completeness of data in the lowest resolution shell, where it may be affected by the overloaded reflections. As mentioned earlier, these strongest reflections are very important and missing them not only negatively affects the process of structure solution and refinement but also biases the other statistical indicators such as R_{merge} or $I/\sigma(I)$.

Traditionally, the accepted data resolution limit used to be point where the $I/\sigma(I)$ ratio drops below 2.0. However, detailed statistical analysis of the relationship between the accuracy and R factors of measured data (R_{meas}) and those of refined structural models (R and R_{free}) [16] shows that even weaker reflections contain useful information. It has been suggested that the most informative and statistically sound criterion to objectively judge the resolution limit of diffraction data is $CC_{1/2}$, the correlation coefficient between two, randomly split and merged groups of reflections [16]. Data resolution may be extended to a limit where $CC_{1/2}$ is still about 0.3–0.5. The $I/\sigma(I)$ ratio may then drop to values even lower than 0.5 and R_{meas} may rise above 1.0. Several practical tests confirmed that the presence of very weak reflections does not harm the quality of the refined structural models, but it is not clear if their inclusion is highly [17] or marginally [18, 19] beneficial. In fact, selection of the data resolution limit remains a rather subjective and not highly objective decision.

Anomalous signal in the data can be judged by the average Bijvet ratio $\Delta F_{anom}/F$ (as a function of resolution) and by the CC_{anom}, the correlation coefficient between signed anomalous differences in two randomly split halves of the data. Useful for phasing anomalous signal exists in resolution ranges where CC_{anom} is higher than 0.3 [20].

Diffraction data collection at contemporary synchrotron beam lines is highly automated due to the presence of very sophisticated but user-friendly control systems of hardware and software. However, it is still a scientific process, not a mere technicality. To ensure the optimal quality of data several important decisions have to be made, satisfying several, often contradictory requirements. It is beneficial, if the experimenter is aware of all the involved issues and is able to make decisions that lead to as good data quality as his/her crystals can deliver.

References

1. Arndt UW, Wonacott AJ (1977) The rotation method in crystallography. North Holland, Amsterdam
2. Dauter Z (1999) Data collection strategies. Acta Crystallogr D Biol Crystallogr 55: 1703–1717
3. Dauter Z, Wilson KS (2001) Principles of monochromatic data collection. In: Rossmann MG, Arnold E (eds) International tables for crystallography, vol. F. Kluwer Academic, Dordrecht, pp 177–195
4. Popov AN, Bourenkov GP (2003) Choice of data-collection parameters based on statistic modelling. Acta Crystallogr D Biol Crystallogr 59:1145–1153
5. Hahn T (ed) (2005) International tables for crystallography, vol A. Springer, Dordrecht

6. Cruickshank DWJ, Helliwell JR, Moffat K (1987) Multiplicity distribution of reflections in Laue diffraction. Acta Crystallogr A 43:656–674
7. Garman EF (2010) Radiation damage in macromolecular crystallography: what is it and why should we care? Acta Crystallogr D Biol Crystallogr 66:339–351
8. Owen RL, Rudiño-Pinera R, Garman EF (2006) Experimental determination of the radiation dose limit for cryocooled protein crystals. Proc Natl Acad Sci U S A 103: 4912–4917
9. Yeates TO (1997) Detecting and overcoming crystal twinning. Methods Enzymol 276: 344–358
10. Zwart PH, Grosse-Kunstleve RW, Adams PD (2005) Xtriage and Fest: automatic assessment of data quality and substructure structure factor estimation. CCP4 Newsletter 43
11. Evans PR (2006) Scaling and assessment of data quality. Acta Crystallogr D Biol Crystallogr 62:72–82
12. Evans G, Pettifer R (2001) CHOOCH: a program for deriving anomalous-scattering factors from X-ray fluorescence spectra. J Appl Crystallogr 34:82–86
13. Mueller-Dieckmann C, Panjikar S, Tucker PA, Weiss MS (2005) On the routine use of soft X-rays in macromolecular crystallography. Part III. The optimal data collection wavelength. Acta Crystallogr D Biol Crystallogr 61:1263–1272
14. Diederichs K, Karplus PA (1997) Improved R-factor for diffraction data analysis in macromolecular crystallography. Nat Struct Biol 4:269–275
15. Weiss MS, Hilgenfeld R (1997) On the use of merging R factor as a quality indicator for X-ray data. J Appl Crystallogr 30:203–205
16. Karplus PA, Diederichs K (2012) Linking crystallographic model and data quality. Science 336:1030–1033
17. Wang J (2015) Estimation of the quality of refined protein crystal structures. Protein Sci 24:661–669
18. Evans PR, Murshudov GN (2013) How good are my data and what is the resolution? Acta Crystallogr D Biol Crystallogr 69:1204–1214
19. Luo Z, Rajashankar K, Dauter Z (2014) Weak data do not make a free lunch, only a cheap meal. Acta Crystallogr D Biol Crystallogr 70:253–260
20. Schneider TR, Sheldrick GM (2002) Substructure solution with SHELXD. Acta Crystallogr D Biol Crystallogr 58:1772–1779

Chapter 8

Identifying and Overcoming Crystal Pathologies: Disorder and Twinning

Michael C. Thompson

Abstract

Macromolecular crystals are prone to a number of pathologies that result from aberrant molecular packing. Two common pathologies encountered in macromolecular crystals are rigid-body disorder and twinning. When a crystal displays one of these pathologies, its diffraction pattern is altered in a way that generally complicates structure determination. The severity of the underlying abnormalities varies from case to case, and sometimes the resulting alterations to the diffraction pattern are immediately obvious, while at other times they may go entirely unnoticed. Structure determination from a crystal that suffers from disorder or twinning may or may not be possible, depending on the specific nature of the pathology, and on how the data are handled. This chapter provides an introduction to these pathologies, with an emphasis on providing guidelines for identifying and overcoming them when they pose a threat to successful structure determination.

Key words Macromolecular crystallography, X-ray diffraction, Twinning, Disorder, Mosaicity, Crystal pathology, Pseudosymmetry, Intensity statistics

1 Introduction

As crystallographers, we strive to grow high-quality, well-diffracting crystals, in which all the molecules are perfectly ordered according to their space group symmetry. In practice, however, crystals of biological macromolecules rarely achieve such perfection. Macromolecular crystals are held together by weak and spurious intermolecular interactions [1]. The weak nature of these crystal packing interactions sometimes permits the existence of multiple, nearly isoenergetic, packing arrangements that are inconsistent with the symmetry of a given crystal's space group. In these cases, various different types of growth abnormalities can occur that introduce disorder. Additionally, when crystals do grow perfectly, the weak forces that hold them together are sometimes disrupted, leading to the introduction of disorder after growth. The presence of these pathologies has the potential to hinder successful structure determination. Consequently, it is important for practicing

Alexander Wlodawer et al. (eds.), *Protein Crystallography: Methods and Protocols*, Methods in Molecular Biology, vol. 1607, DOI 10.1007/978-1-4939-7000-1_8, © Springer Science+Business Media LLC 2017

crystallographers to develop an understanding of the various pathologies that can exist in their crystals, so that they can identify potential problems and make informed decisions about whether or not, and how, to proceed with structure determination when these problems are encountered.

Many types of abnormalities can occur in protein crystals. For most crystallographers, a deep theoretical understanding of these phenomena is unnecessary, and such details are beyond the scope of this chapter. Instead, the goal is to provide the reader with a practical understanding of various pathologies that are common in protein crystals. A brief first section describes disorder phenomena, and provides information about how to recognize the general symptoms of a disordered crystal, decide whether a crystal that presents these symptoms might still be useful for structure determination, and proceed sensibly when working with data collected from such a crystal. A longer, second section discusses a special type of crystal pathology known as "twinning," again with an emphasis on identification and proper handling of the condition. Twinning is a topic that deserves additional attention because many macromolecular crystal systems are susceptible to the pathology, and twinning occurs relatively often in macromolecular crystals [2–4]. The frequency with which twinning is possible, combined with the fact that twinning can easily go unnoticed, means that macromolecular crystallographers must always be mindful of this pathology. Fortunately, twinning can be very manageable if care is taken during data collection and reduction.

2 Disorder in Protein Crystals

The word "disorder" is somewhat ambiguous, and in crystallography it can be used in several different contexts. For example, the word "disorder" is often associated with high atomic B-factors or alternative conformations, referring to the fact that the crystallized molecules are not all in the same conformation at the same time. This type of disorder, while interesting, is not the subject of this section. Rather, the disorder discussed here refers to cases in which some molecules, or groups of molecules, undergo rigid-body displacements (translations or rotations) that violate the symmetry of the crystal to which they belong.

2.1 How Do Crystals Become Disordered?

Disorder can be introduced into protein crystals in several ways, which are not mutually exclusive. Crystal growth irregularities lead to disorder when molecules join a crystal in a way that is not consistent with the crystal's space group symmetry or with the translational relationship between unit cells. This is a possibility when molecules can add to a crystal lattice in multiple energetically favorable orientations. Irregular growth, however, is not the only source of disorder; it is common for crystals to become disordered as a result of mechanical stress, dehydration, and/or cryocooling, which emphasizes the importance of proper crystal handling.

2.2 Some Schematic Examples of Disorder—Translation and Rotation

Most cases of disorder in protein crystals present their own idiosyncrasies; however, two main phenomena, translational and rotational disorders, underlie many of these problems.

2.2.1 Translational Disorder

Rigid-body translational disorder in macromolecular crystals can occur in several distinct ways, which produce different effects on the observed X-ray diffraction.

If random translational displacements of molecules cause deviations from perfect crystallinity, then the intensity of the resulting Bragg diffraction will fall off more rapidly at higher scattering angles, and the magnitude of the average displacement can be estimated from the Wilson plot. In such cases, the individual reflections may remain relatively sharp; however, the higher resolution Bragg peaks will be weakened, resulting in an overall loss of resolution. Interestingly, if the displacements of the molecules are purely translational (and not rotational), the scattering from each individual molecule can sum incoherently, and it may be possible to measure the entire molecular transform, opening new frontiers in diffractive imaging [5].

When translational displacements between molecules are correlated in at least one dimension, the resulting pathology is commonly referred to as "lattice translocation" disorder. In specimens that suffer from translocation disorders, successive layers of the crystal are translationally displaced from one another in at least one direction, as shown in Fig. 1a. It is not uncommon for macromolecular crystals that grow as thin plates (Fig. 1b, c) to suffer from translocation disorders. Plate-like crystals often consist of stacked, two-dimensional molecular layers, such as those in space groups *P*3, *P*4, and *P*6, which are prone to displacements that disrupt the translational symmetry between unit cells. Translocation disorders can be introduced during growth, but they are also commonly the result of damage that can occur during crystal soaking, cryoprotecting, or cryocooling.

If we consider translocation disorders, the most severe type would be one in which the translational displacements between subsequent layers in the crystal are completely random. As we move toward a more ordered scenario, there is a version of the pathology that is called "order–disorder," which results from the existence of multiple packing arrangements that preserve the geometric equivalence of local crystal contacts, but break some longer-range symmetry of the crystal [6, 7]. Order–disorder pathologies were first described for protein crystals in 1954, when Bragg,

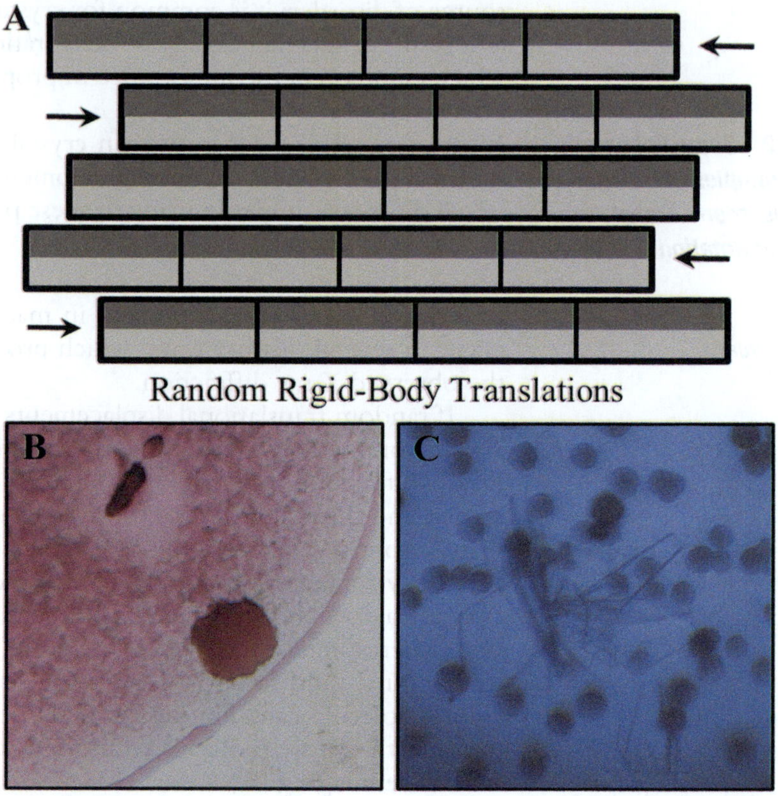

Fig. 1 Lattice translocation disorders occur when individual layers of crystallized molecules become translationally displaced relative to their neighbors, breaking the translational crystallographic symmetry (**a**). Crystals that suffer from translocation disorders often have a plate-like morphology (**b**, **c**)

Howells, and Cochran identified and characterized translocation disorder in crystals of imidazole methemoglobin [8, 9]. Interestingly, translocation disorders were studied in protein crystals several years before the earliest protein structures were determined, and these pathologies have had a somewhat rich history in macromolecular crystallography (some examples are described throughout the rest of Subheading 2).

2.2.2 Rotational Disorder

In addition to disorder involving translational displacements, there are also disorder phenomena that involve rotations of molecules or groups of molecules. As described above for random translational disorder, random rotational displacement of crystallized molecules causes a uniform fall-off in the intensity of Bragg diffraction as a function of the scattering angle, and there is a similar relationship between the magnitude of the average displacement and the slope of the Wilson plot.

If molecules are well-ordered on the short-range length scale, but there is minor long-range rotational disorder in the crystal, then a phenomenon known as mosaicity occurs [2, 10]. Mosaicity results from the fact that a single crystal actually contains multiple microscopic domains that are in very slightly different orientations (Fig. 2a), and it occurs (to some extent) in virtually all macromolecular crystals. Mosaicity can result from the incorporation of impurities during growth, as well as from cryocooling, dehydration, or mechanical stress. In a mosaic crystal, each domain can be thought of as contributing independently to the diffraction pattern, and the degree to which their reflections overlap describes their mosaicity (Fig. 2b). Highly mosaic crystals produce diffraction spots that appear as arcs (Fig. 2c, d) and extend over many successive diffraction images, which causes difficulties for data collection and processing.

In addition to random disorder and mosaicity, which occur to some degree in all crystals, other much more rare types of

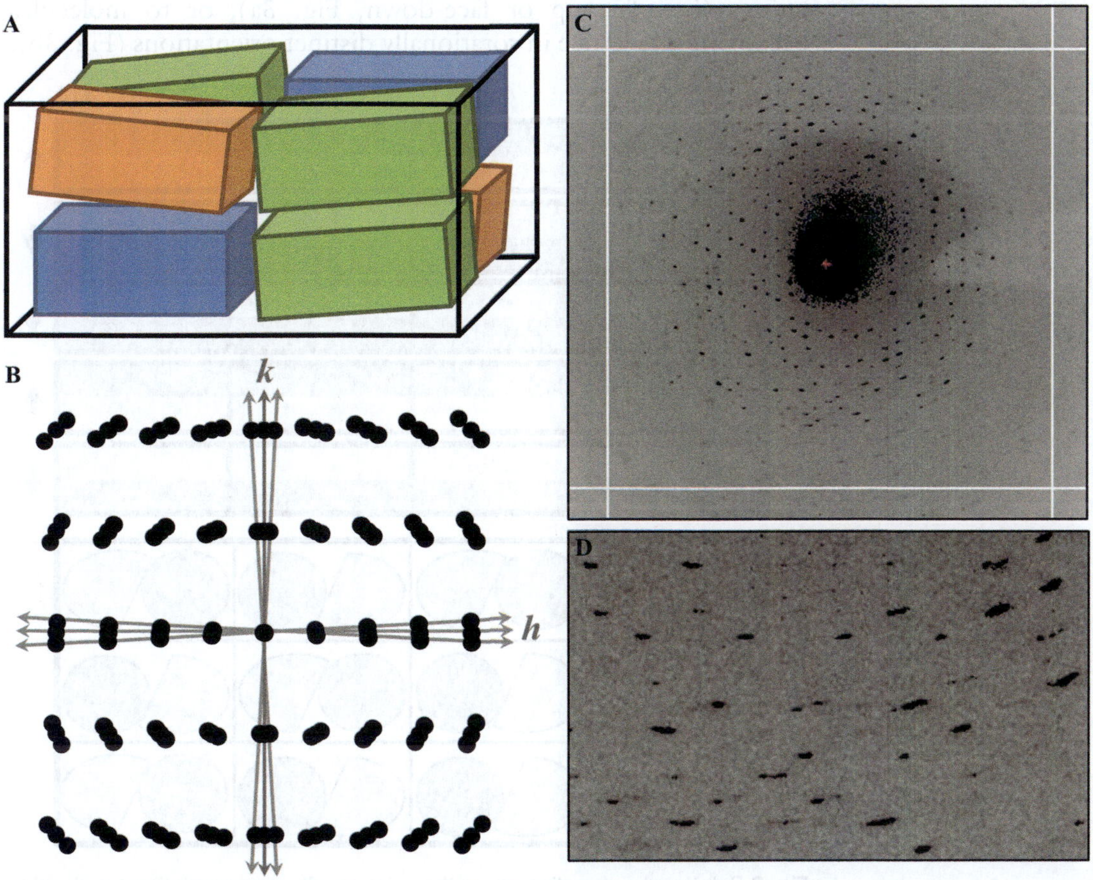

Fig. 2 In a mosaic crystal, individual domains have a high degree of short-range order but suffer from slight long-range disorder. In panel **a**, the *black outline* represents the boundary of a macroscopic crystal, with individual mosaic domains represented in color. The slight rotational offset between individual domains forms the physical basis for mosaicity. The resulting effect of crystal mosaicity on X-ray diffraction images is illustrated schematically in panel **b**, and real diffraction images from a highly mosaic crystal are shown in panels **c** and **d**. Note that the reflections appear as arcs

rotational disorder are also possible. These more exotic forms of rotational disorder are most likely introduced during crystal growth, and occur more often in crystals of molecular assemblies that have high symmetry, such as viruses [11, 12]. Reports of rotational disorder include situations in which molecules, or groups of molecules, were able to occupy their crystal lattice in additional specific orientations that were not allowed by the rotational symmetry of the space group [13–15]. When the rotational displacements are not entirely random, the situation is called "rotational order–disorder," because it is the rotational equivalent of the translational order–disorder described above. If these alternative packing arrangements are described by symmetry operations of the lattice, but not the space group, then the diffraction pattern will have statistically higher symmetry. For example, this might correspond to stacked layers in a crystal having random rotational orientations (i.e., face-up or face-down, Fig. 3a), or to molecules occupying the lattice in rotationally distinct orientations (Fig. 3b).

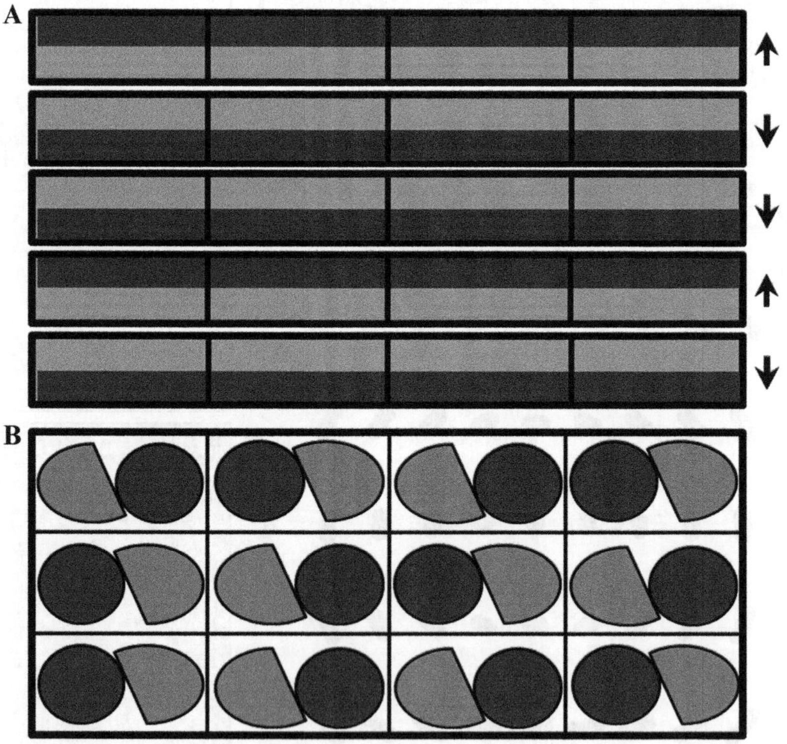

Fig. 3 Rotational order–disorder pathologies result when crystallized molecules can occupy the lattice in multiple, rotationally distinct orientations. This can happen in several different ways, including rotations of entire groups of molecules (such as layers, as pictured in panel **a**) or rotations of individual molecules (**b**). In both cases that are pictured, the existence of multiple different packing arrangements within the crystal would result in diffraction patterns with higher Laue symmetry than would be expected based on the actual space group symmetry

These "statistical crystals," whose diffraction patterns have statistically higher symmetry than their actual space group symmetry, are similar to twinned crystals, which are the subject of Subheading 3. The fact that rotational order–disorder can make a crystal appear to have higher symmetry has an interesting converse; an incorrect space group assignment can make a crystal appear as though it suffers from rotational disorder [16].

2.3 Recognizing Disorder Pathologies from Diffraction Patterns

The nature and severity of a disorder pathology ultimately determines the potential usefulness of an imperfect crystal. In some cases, disorder pathologies produce no visible effect on the X-ray diffraction from a crystal, whereas in other cases disorder can produce obvious and dramatic visual symptoms. Fortunately, the disorder pathologies that can completely hamper structure determination are often immediately apparent from a crystal's diffraction pattern, and crystals that suffer from the most serious disorder tend to produce more visually irregular diffraction images. When an experimenter prepares to collect X-ray diffraction data from a new crystal specimen, it is recommended that he or she first collect two images at 90° rotation of the crystal. The purpose of this procedure is to evaluate the quality of the crystal and determine if it is suitable to produce a complete data set. Generally, if two images separated by 90° are used for this initial analysis, the most severe types of disorder should become apparent, because disorder phenomena are sometimes only obvious at certain crystal orientations. When evaluating initial diffraction images, it is important to be able to recognize the general symptoms of crystal disorder, and to assess whether a crystal that presents these symptoms might still be useful for structure determination.

As described above, when disorder is completely random in magnitude and direction, a general loss of high-resolution Bragg diffraction is observed, although the remaining low-resolution reflections may remain sharp. The extent of the disorder dictates how rapidly the Bragg intensity decreases as a function of the scattering angle. In macromolecular crystals, which often suffer from imperfect molecular packing, it is this random translational and rotational displacement of molecules that limits the resolution of X-ray diffraction.

In the case of order–disorder pathologies, in which differently oriented molecules are related by specific operations (translational or rotational) that break the crystallographic symmetry, the diffraction pattern may or may not contain obvious symptoms of the pathology. Native Patterson maps, self-rotation functions, and analysis of unit cell contents using the Matthews' number are all critical for identifying and characterizing cases of order–disorder. In some cases, order–disorder will only become apparent when the space group symmetry or native Patterson maps indicate that an impossibly large number of molecules are packed in the unit cell, implying that different molecular orientations exist in different cells. Sometimes the presence of disorder may not be recognized until atomic refinement fails to converge to a reasonable solution. On the other hand, when the displacements between molecules or groups of molecules in a crystal occur in a nonrandom, correlated fashion, the resulting diffraction patterns can contain structured, non-Bragg scattering, appearing as "streaks" around and in between the Bragg positions. While severe streaking can hamper data collection and structure determination, mild streaking in the presence of sharp Bragg peaks can be a useful visual indication of order–disorder, and it can provide information about the physical nature of the pathology. Very detailed theory on the relationship between certain types of disorder and the resulting diffraction patterns has been described by Welberry [17].

Two examples of diffraction from disordered crystals are provided in Fig. 4. In the first example (Fig. 4a, b), the streaks in the c^* direction of reciprocal space are due to a translocation disorder, which is apparent because reflections remain sharp in the a^* and b^* directions Specifically, the individual layers of this hexagonal crystal are displaced parallel to the ab plane of the unit cell, which disrupts

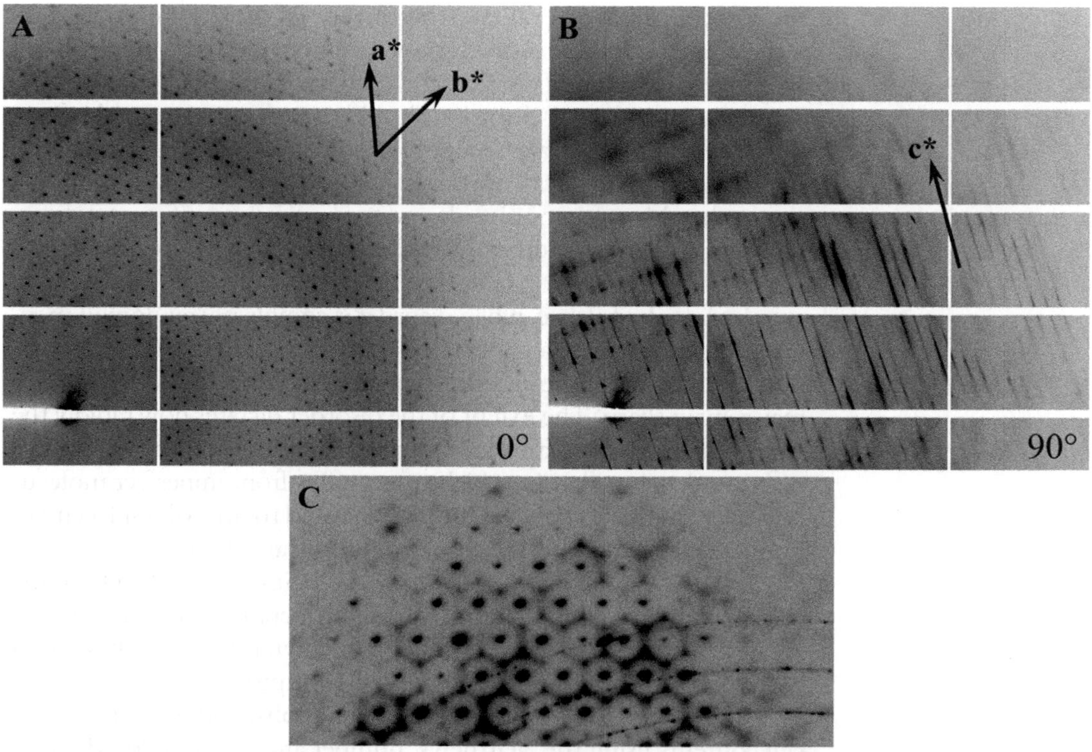

Fig. 4 Diffraction patterns with streaks around and between the Bragg positions provide indication of order–disorder phenomena. For example, crystals that suffer from lattice translocation disorder may appear to diffract well in some orientations (**a**), but poorly (and with streaks) in other directions (**b**). In a second example, correlated rotational order–disorder leads to a hexagonal pattern of streaks surrounding the Bragg positions in reciprocal space (**c**)

the translational crystallographic symmetry along the *c* axis. In this case, the disorder made the crystals useless, because the streaking was so severe in some orientations that it compromised the ability to measure all the required reflections. In the second example (Fig. 4c), a pattern of streaks is evident between the Bragg positions. In this case, a complete set of reflection intensities could still be accurately measured, because the Bragg peaks remained sharp despite the presence of the streaks.

The examples of diffraction from disordered crystals described above are simple and brief. Other examples of diffraction patterns from disordered crystals have been given by Helliwell [18], and for detailed reports of structure determination from disordered crystals, the interested reader is referred to the excellent publications cited throughout Subheading 2.

2.4 Working with Disordered Crystals

If it is possible to collect a useful X-ray data set from a disordered crystal, because the disorder has little or no effect on the measurement of Bragg intensities, then several steps can be taken in an attempt to determine a structure despite the pathology. Structure determination from disordered crystals is typically only possible for special cases of order–disorder, where there are a limited number of specific, alternative packing arrangements available to the crystallized molecules. When faced with data from such a crystal, it is first necessary to characterize the nature of the disorder. In many cases, analysis of native Patterson maps and self-rotation functions can be used to understand the physical basis for a particular disorder phenomenon (i.e., is it rotational or translational in nature, and what are the transformations that relate differently oriented molecules?). Once the nature of the disorder has been revealed, two options exist for handling the pathology and determining a structure. The first method involves identifying a set of operations that describe the disorder phenomenon, and then using those operations to correct the measured intensities [19] by removing the contribution of molecules whose positions are described by those operations. Typically, these correction methods are applied to data sets collected from crystals that exhibit lattice translocation disorders, and examples of using data corrected for translational order–disorder as input for both experimental phasing [20] and molecular replacement [21–25] have been reported. The same corrected data can be used for atomic refinement, alternating with reapplication of the correction formula [26]. A second strategy, which has been applied to crystals that suffer from rotational order–disorder [13–15], is to identify multiple overlapping orientations of the molecule by molecular replacement and the self-rotation function, or by analysis of the electron density, and then refine those overlapping molecules simultaneously with partial occupancies that must sum to unity.

If the disorder is so severe that sharp Bragg peaks can no longer be observed for all crystal orientations, then structure determination is impossible. For such crystals, small modifications of the crystallization or handling protocol may be able to eliminate the disorder pathology. For example, if disorder is introduced during crystal growth, then small changes to the crystallization conditions, such as adjusting the pH, salt concentration, or temperature of the crystallization experiment may be enough to favor one particular packing arrangement. If disorder is the result of crystal dehydration or mechanical stress, then a modification of crystal handling and cryoprotection protocols may be helpful. Crystal annealing can also be a useful method to eliminate mild disorder (although it can also make it worse) [27], and if disorder is introduced by cryocooling, then room-temperature data collection may be a viable solution. Finally, it is worth noting that in at least one case of rotational order–disorder, in crystals of cowpea mosaic virus, it was reported that subjecting crystals to high pressure (3.5 kbar) eliminated the pathology [28].

2.5 Disorder Pathologies—Summary

Rigid body translational or rotational disorder in macromolecular crystals is not uncommon. Often, this disorder is random, and it degrades the quality of the resulting diffraction patterns in a way that makes structure determination entirely impossible. In some cases though, the disorder is not entirely random, or it does not disrupt the lattice symmetry of the crystal, and diffraction data from such crystals can sometimes still be used to determine a structure if the effect of the disorder can be corrected or accounted for. It is important to be able to recognize the symptoms of disorder and to make appropriate decisions about whether or not a particular crystal or data set is useful. There is a surprisingly large body of literature describing various disorder phenomena in macromolecular crystals, which is an excellent resource when a challenging structure determination project demands a deeper understanding of these pathologies.

3 Twinning—A Special Type of Crystal Growth Abnormality

As a macromolecular crystal grows, its surface presents an ordered array of molecules, which can sometimes act as a good nucleation point for the growth of additional, differently oriented crystalline domains. When two crystals become physically conjoined, it is possible that their relative orientations obey one of several special relationships that are collectively known as "twinning." Because of the special orientational relationship between "twin domains," the reciprocal lattices corresponding to each of the two domains also become overlapped in special ways, which complicates the estimation of true structure factors from the observed intensities.

Twinning is only possible for certain types of lattices, but these lattices are fairly common in macromolecular crystallography. Notably, two independent analyses of structure factor data deposited in the Protein Data Bank (PDB) [29] revealed that approximately 30% of reported unit cells could support the existence of twinning [2, 3]. The possibility of twinning for this relatively high percentage of unit cells highlights the need for macromolecular crystallographers to be knowledgeable about this pathology. The same analyses suggested that about 2% of those crystals that could support twinning actually produced diffraction patterns with signatures of the pathology, although it is likely that analyzing the PDB underestimates the prevalence of twinning because the PDB contains no record of instances where twinning hampered structure determination altogether [2, 3]. Twinning is a potentially dangerous crystal pathology, because it can easily be overlooked, causing a structure determination effort to fail. On the other hand, if twinning is identified by applying routine tests, it is generally a manageable situation and structure determination is usually possible.

3.1 Different Types of Twinning

There are multiple types of crystal twinning, which are defined according to the specific way in which the twin domains are oriented relative to one another. When a twinned crystal is illuminated with X-rays, the twin domains diffract independently, producing overlapping diffraction patterns. As a result, each type of twinning has a unique consequence in reciprocal space. The various types of twinning are discussed below.

3.1.1 Macroscopic Twinning—A Misnomer

"Macroscopic twinning" is a phrase that is often used to describe cases in which multiple crystals grow as overlapping clusters or stacks, with individual crystals taking random orientations (Fig. 5a). In these cases, there is no special relationship between the individual crystals, thus, this so-called "macroscopic twinning" is not actually twinning at all according to the formal definition [30]. Nonetheless, since this situation is sometimes incorrectly referred to as "twinning," and because it is a type of crystal growth pathology that is common in macromolecular crystallography, it merits a brief discussion here.

Situations in which crystals overlap and/or adhere to one another can be handled in several ways. Such cases can usually be identified visually, because the individual crystals are often large enough to appear distinct under the polarizing microscope (Fig. 5c). When overlapping, clustered, or split crystals are observed (as in Fig. 5c, d), one can attempt to harvest a single specimen by gently separating it from the rest of the cluster using a cryoloop or a small needle. When differently oriented and overlapping crystals are simultaneously illuminated during X-ray data collection, this is typically quite recognizable, because the observed diffraction pattern will contain multiple, differently oriented lattices (Fig. 5b, e).

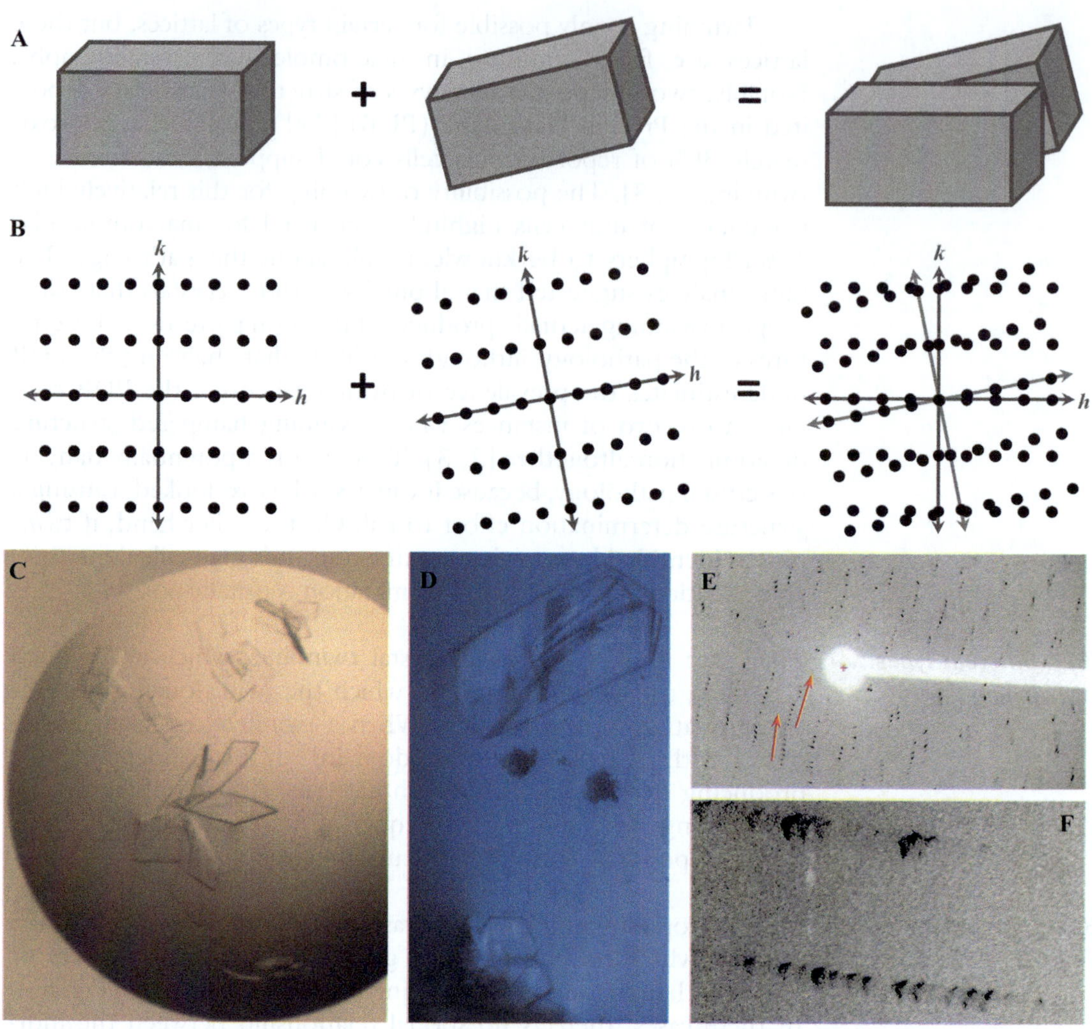

Fig. 5 Nonspecifically overlapping crystals (**a**) are sometimes referred to as "twinned" crystals, however, such situations where the relationship between the individual domains is totally random are not twins according to the formal definition. When overlapping crystals are simultaneously illuminated by the X-ray beam, they both diffract independently, resulting in diffraction images that contain multiple, distinct reciprocal lattices (**b, e, f**) Inadvertently collecting data from multiple overlapped crystals can be problematic, especially when crystals grow in clusters (**c**), or when they begin to split, fray, or crack (**d**)

Split crystals have diffraction patters with reflections so close together that an unreasonably large unit cell axis is suggested (Fig. 5f). In favorable cases, where there are few overlaps between the two sets of reciprocal lattice points, the diffraction images can be carefully examined, a single lattice can be manually selected, and a consistent set of reflections can be extracted.

3.1.2 Non-merohedral Twinning

Non-merohedral twinning describes cases in which twin domains are epitaxially related, meaning they are oriented in a way that produces a match of their lattice spacings in two-dimensions. This situation produces diffraction patterns that consist of interpenetrating reciprocal lattices, with a subset of perfectly overlapped points (Fig. 6).

Non-merohedral twinning can typically be identified in one of several ways. It can often be recognized visually from detector images, based on the absence of reflections that are not predicted to be missing in the presence of screw axes (this concept is illustrated in Fig. 6). Non-merohedral twinning generally also causes indexing to fail or produce many outliers, since no single unit cell will simultaneously predict all the observed reflections. Crystals suffering from non-merohedral twinning can sometimes still be useful; however, they should be approached with caution. Most often, it is best to try and find a crystal whose diffraction pattern does not display this pathology upon initial inspection, because non-merohedral twinning makes structure determination considerably more difficult. If untwinned crystals are not available, then in favorable cases where one lattice dominates and the overlaps are minor, non-merohedral twinning can sometimes be overcome and reasonable measurements of reflection intensities may be obtained

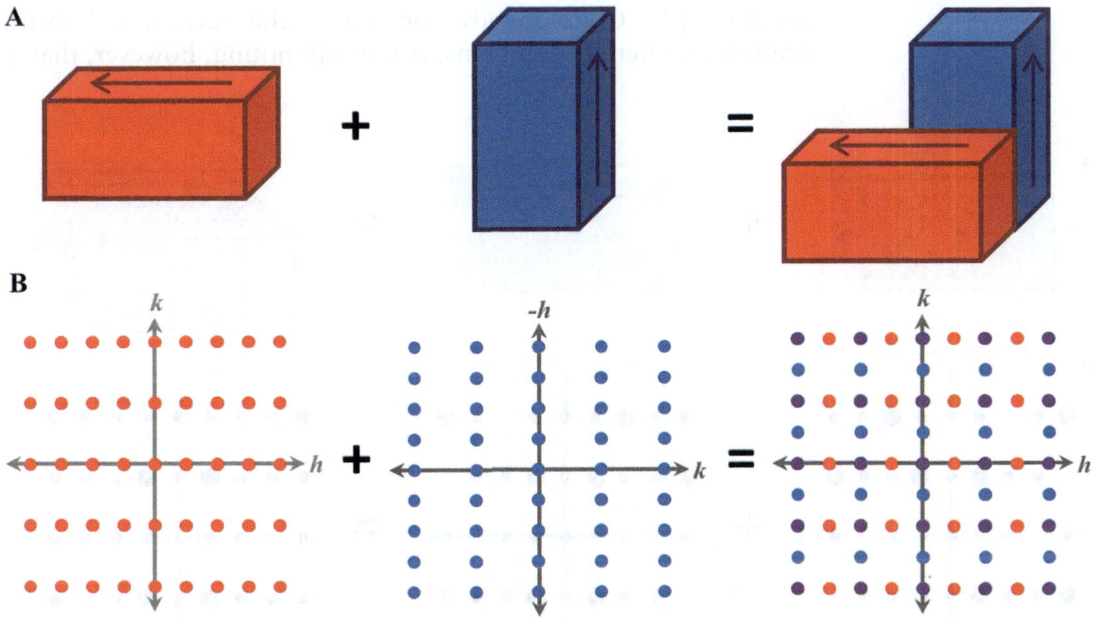

Fig. 6 When twinned crystals have epitaxially related unit cell dimensions, their reciprocal lattices will interpenetrate in a manner that causes some reflections to be perfectly superimposed—a situation known as non-merohedral twinning. In this example, the crystals have a unit cell axis which is exactly twice as long as a second axis (**a**). As a result, when the corresponding reciprocal lattices are rotated and superimposed, only some of the reflections are overlapped (**b**, *purple points*). The twinned diffraction pattern gives the impression of a larger unit cell, with two axes equal to one another; however, if this were the case, then a subset of reflections (those for which $h = 2n + 1$ and $k = 2n + 1$) would be systematically missing. A large number of systematic absences can be an indication of non-merohedral twinning

from a single lattice. Reports of non-merohedral twinning in the literature are rare, however there are a few examples [31–33].

3.1.3 Merohedral Twinning

Merohedral twinning describes the special case where twin domains are oriented in such a way that the reciprocal lattices associated with each of the individual domains are perfectly superimposable (Fig. 7). This is possible only when two conditions are simultaneously satisfied: the rotational symmetry of the crystal lattice must be higher than that of the space group, and the twin domains must be related by a transformation that is a symmetry operation of the lattice but not of the space group itself. This is possible for certain space groups in tetragonal, trigonal/hexagonal, and cubic lattices. Additionally, "pseudo-merohedral" twinning can occur for lattices with lower symmetry if their dimensions obey specific constraints. For example, an orthorhombic crystal can become pseudo-merohedrally twinned if $\mathbf{a} \approx \mathbf{b}$, making the lattice approximately tetragonal, and a monoclinic crystal can become pseudo-merohedrally twinned if $\beta \approx 90°$, making the lattice approximately orthorhombic.

In macromolecular crystallography, the large majority of reported cases of twinning are hemihedral twins, meaning that there are only two twin domains (most often related by a 180° rotation) [2]. Consequently, the rest of this section will focus primarily on hemihedral twins. It is worth noting, however, that a

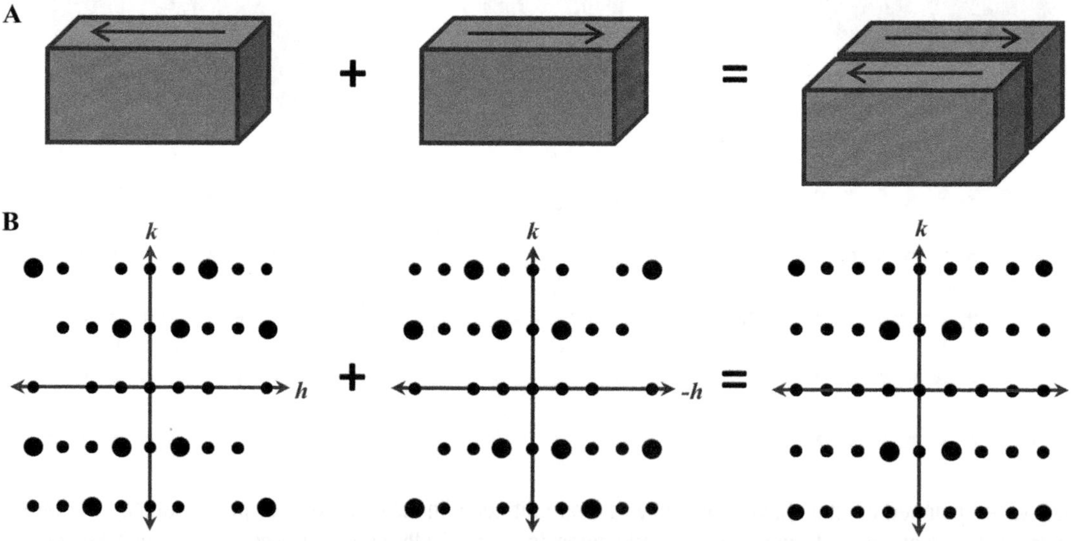

Fig. 7 In the special case of merohedral twinning, the twin domains are related by rotations that are symmetry operations of the crystal lattice, but not of its space group. In reciprocal space, merohedral twinning leads to averaging of twin-related reflections, which causes the diffraction pattern to approach higher symmetry. In this two-dimensional example, the intensities resulting from each individual twin domain have *p*2 plane group symmetry. Rotating the reciprocal lattice by 180° about the *k* axis and averaging the intensities produces a twinned diffraction pattern with *pmm* plane group symmetry

few exceptions involving more complicated forms of twinning have been reported as well. For example, several cases of tetartohedry (merohedral twin with four twin domains) have been reported [34–39], along with reports of complicated, multidomain pseudo-merohedral twinning [40, 41], and an interesting case involving a phase transition between two different unit cells [42]. Some of these more complex and exotic forms of twinning are discussed by Parsons [43] in an excellent review of the subject.

Merohedral twinning can be a particularly dangerous crystal pathology because it is easily overlooked, which can hamper structure determination efforts from a seemingly ordinary data set. Due to the perfect superposition of twin-related lattices, diffraction patterns from merohedrally twinned crystals do not appear visually abnormal, even upon careful inspection, and indexing of reflections from such crystals generally proceeds without any obvious warning signs of the pathology. In the majority of cases, merohedral twinning can only be detected by applying one of several statistical tests to the measured reflection intensities. Fortunately, these tests are routine now, and once the pathology has been identified, structure determination from merohedrally twinned crystals can often proceed without much difficulty.

3.2 Identification and Handling of Merohedral Twinning

Merohedral twinning is the most commonly reported crystal defect in macromolecular crystallography, but it is a well-characterized and manageable pathology. A great deal of work has been done to develop and implement statistical tests to identify and characterize merohedrally twinned intensity data, and in favorable cases, structure determination from twinned data proceeds easily using modern structure determination software. The following subsections explain both the theoretical and practical aspects of working with data obtained from merohedrally twinned crystals.

3.2.1 Physical Relationships between Twin Domains and their Consequence in Reciprocal Space

Two mathematical concepts, known as the twin operator and the twin fraction, are important for understanding both the relationship between twin domains in real space and the way in which merohedral twinning alters diffraction intensities in reciprocal space.

The "twin operator" (also referred to as the "twin law") is the symmetry operation that relates the two twin domains. Recall that for merohedral twins, the twin operator must be a symmetry operation of the lattice, but not of the space group. Furthermore, biological macromolecules are chiral, and so the only possible twin operators correspond to rotations. The same symmetry operator that relates twin domains in real space also describes how the two twin-related reciprocal lattices are overlapped, and as a result, twin operations are described as reciprocal (h,k,l) space operations that exchange the indices of twin-related reflections. For example, a tetragonal crystal belonging to space group $P4$ might have a twin operator $(k,h,-l)$. The reciprocal space operation $(k,h,-l)$ rotates the reciprocal lattice by 180° about an *a*, *b*-diagonal axis perpendicular to the fourfold symmetry axis. In real space, this corresponds to the lattice rotation (*y*,*x*,−*z*), which is not a symmetry operation of space group *P*4 (but is a symmetry operation of *P*422). This operation leaves the lattice unchanged, but the orientation of the molecules is different because the polar fourfold symmetry axes of the two twin domains have opposite orientation.

The "twin fraction" quantifies the fractional volume of the crystal occupied by the smaller of the two twin domains. The twin fraction (α) takes on a value between 0 and 0.5 for hemihedral twinning, and the larger of the two twin domains occupies the complementary ($1-\alpha$) volume. For any twinned crystal, the sum of the twin fractions for all twin domains must be unity. Because we are generally only interested in the portion of our crystal from which our X-ray data are collected, the volumes described by the twin fraction refer only to the part of the crystal that is illuminated by the X-ray beam. In addition to describing volumes in real space, the twin fraction has an important manifestation in reciprocal space. Because each domain in a merohedral twin diffracts X-rays independently and proportionally to its volume, each observed diffraction intensity from a twinned crystal is actually a linear combination of twin-related intensities contributed by the overlapping reciprocal lattices. The contribution to the observed diffraction intensity from each reciprocal lattice is weighted by its twin fraction, leading to the following expressions for observed intensities (*I1* and *I2*) for two reflections which are related by a hemihedral twin operator:

$$I_1 = \alpha J_1 + (1-\alpha) J_{2.}$$

$$I_2 = (1-\alpha) J_1 + \alpha J_2$$

In the above equations, α is the twin fraction, and J_1 and J_2 are the underlying "true" intensities for the same two reflections from an untwinned crystal.

Taken together, the concepts of twin operators and twin fractions explain one of the important properties of diffraction from merohedrally twinned crystals. These diffraction patterns approach higher point group symmetry than the true Laue symmetry of the crystal, because the overlap between reciprocal lattices effectively averages unrelated observations. The fact that averaged reflections are related by a defined symmetry operation means that as the twin fraction increases, the observed data will merge better in higher symmetry point groups. As a result, it is easy to interpret the space group incorrectly when unknowingly faced with twinned data. When the twin fraction is equal to 0.5, the data will merge perfectly in a point group with erroneously high symmetry, described by application of the twin operator to the true Laue symmetry of the crystal. In the example of space group *P*4 provided above, a perfect twin ($\alpha = 0.5$) would appear to belong to point group 422—the Laue symmetry would be *P*4/*mmm*, rather than *P*4/*m* expected for untwinned crystals belonging to space group *P*4.

It is important to note the relationship between merohedral twinning and the rotational order–disorder discussed in Subheading 2 and illustrated in Fig. 2. The two phenomena are very similar, in the sense that they both have the potential to produce diffraction patterns that suggest higher symmetry than the actual space group of the crystal. The primary difference between the two cases is the length scale separating differently oriented unit cells. In a twinned crystal, the molecules are perfectly ordered throughout the individual twin domains, which are larger than the coherence length of the X-ray beam. Therefore, twin domains diffract X-rays independently, and crystallographic observations must be treated as linear combinations of intensities. In the case of order–disorder, which produces a so-called "statistical crystal," the disorder occurs from one unit cell to the next, over length scales that are smaller than the coherence length. For these crystals, the observed diffraction must be treated as a linear combination of structure factors, rather than intensities.

3.2.2 When to Suspect Potential Merohedral Twinning?

It is important to understand when twinning poses a potential threat, so that an appropriate data collection strategy can be used for a given crystal. Since twinning is generally not detectable until a complete data set has been collected and integrated, and twinned crystals are often indexed in point groups with overestimated symmetry, it is not uncommon for twinned data sets to be incomplete because data were collected as if the crystal actually belonged to a higher symmetry point group. This problem can be avoided if care is taken to collect data according to the lowest possible point group symmetry given the potential existence of twinning. A list of lattices and point group symmetries that are either subject to or result from merohedral twinning is provided in Table 1, along with relevant twin operators.

Merohedral twinning is typically not evident until after a complete data set has been collected and analyzed. The pathology cannot generally be identified visually, either from the crystals themselves or from the diffraction pattern; however, certain clues may indicate the potential for twinning. Sometimes, twinned crystals will grow with concave surfaces or jagged edges (Fig. 8); however, not all crystals displaying these visual abnormalities are twinned, and not all twinned crystals provide these visual indications [2]. Therefore, the presence or absence of particular morphological features is not a good diagnostic property. Sometimes, initial indexing of twinned diffraction data identifies erroneously high symmetry, and the pathology is revealed when the space group symmetry predicts an unrealistically large number of molecules in the unit cell. This simplistic means of identifying twinning,

Table 1
Lattices and point groups that are susceptible to, or may result from, merohedral twinning

Lattice type	Observed point group	Type of twinning	Possible twin operators
Monoclinic	*2*	Partial pseudomerohedral if $\beta \approx 90°$	$(-h,-k,l)$
Orthorhombic	*222*	Partial pseudomerohedral if $a \approx b$	$(k,-h,l)$
		Perfect pseudomerohedral twin in PG2 (if $\beta \approx 90°$) appears *222*	$(-h,-k,l)$
Tetragonal	*4*	Partial merohedral twin	$(k,h,-l)$
	422	Perfect twin in PG*4* appears *422*	$(k,h,-l)$
		Perfect pseudomerohedral twin in PG *222* (if $a \approx b$) appears *422*	$(k,-h,l)$
Trigonal	*3*	Partial merohedral twin	$(-h,-k,l)$, $(k,h,-l)$, or $(-k,-h,-l)$
	312	Partial merohedral twin	$(-h,-k,l)$
		Perfect twin in PG*3* appears *312* or *32*	$(k,h,-l)$
	321	Partial merohedral twin	$(-h,-k,l)$
		Perfect twin in PG*3* appears *321*	$(-k,-h,-l)$
Hexagonal	*6*	Partial merohedral twin	$(k,h,-l)$
		Perfect twin in PG*3* appears *6*	$(-h,-k,l)$
	622	Perfect twin in PG*6* appears *622*	$(k,h,-l)$
Cubic	*23*	Partial merohedral twin	$(k,h,-l)$
	432	Perfect twin in PG*23* appears *432*	$(k,h,-l)$

When these lattices and point groups are observed in a diffraction experiment, the crystallographer should proceed with caution to ensure adequate data are collected. Additionally, twin operators corresponding to the specified type of twinning are provided

based on prediction of an overcrowded unit cell and an unreasonably small Matthews' number, is usually only relevant if the asymmetric unit contains just one molecule, and also can be indicative of short-range rotational disorder rather than twinning, as described in Subheading 2.2.2.

After a complete data set has been collected, additional evidence for merohedral twinning can be obtained by analyzing the rotational symmetry of the data set. For example, one indication of twinning is if a data set can be merged well in a low symmetry point group, and merging in a higher symmetry point group, related by a potential twin operator, yields merging statistics that are only marginally worse. Additionally, weak peaks in the self-rotation function corresponding to twin operators can also be an indication of twinning. If a structure can be determined in the presence of undiagnosed twinning, atomic refinement is likely to stall with *Rwork* and

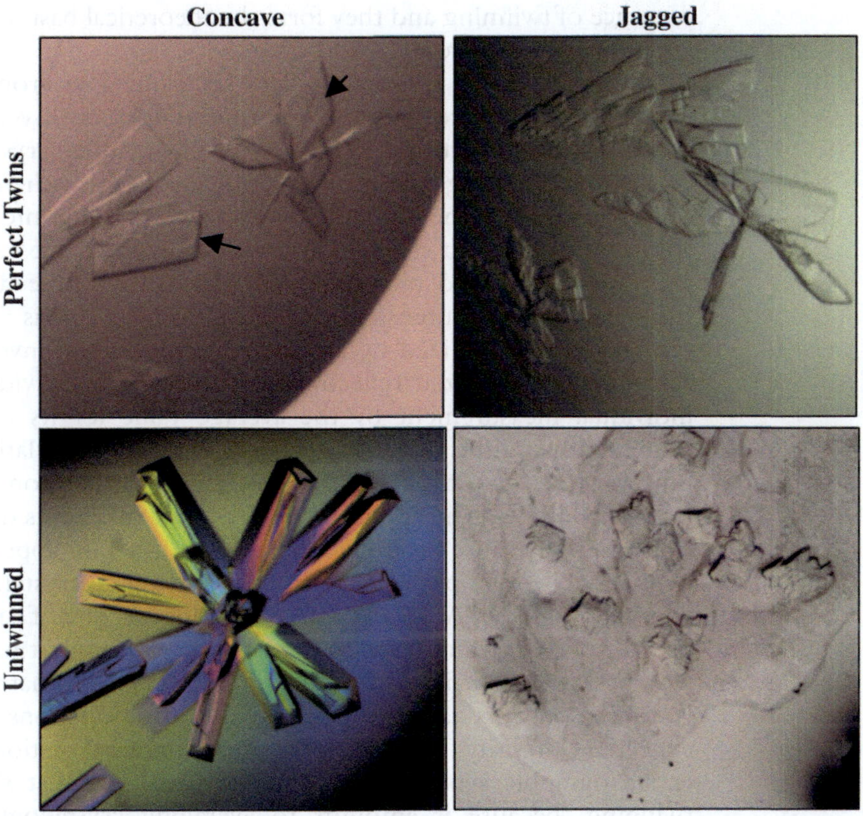

Fig. 8 Crystals with concave surfaces or jagged edges are sometimes, but not always, a sign of twinning. In the pictured examples, the crystals that have only a very mild concave features on their surfaces (*arrows*) were found to be perfectly twinned, while the crystals with deep and narrow concave surfaces, could not be twinned since their true space group is $P4_32_12$. Likewise, two cases of jagged crystals are perfectly twinned and untwinned, respectively, despite having a somewhat similar appearance. The lower left image is reproduced from Thompson et al. [87], under the original publisher's copyright agreement

Rfree at unacceptably high values (0.35–0.4), revealing the potential presence of the pathology. Identifying merohedral twinning in these mainly qualitative ways is unreliable, however, because it can be unclear whether the observed symmetry is crystallographic, due to non-crystallographic symmetry (NCS), or due to merohedral twinning. Additionally, a failed refinement could result from any number of problems, not exclusively from unidentified twinning. In order to accurately assess the presence or absence of merohedral twinning, additional statistical tests are required.

3.2.3 The Statistical Properties of Reflection Intensities Reveal Merohedral Twinning

A diagnostic consequence of merohedral (or pseudomerohedral) twinning is that it causes the observed reflection intensities to deviate from the typical Wilson distribution [44], and also changes the statistical properties of other quantities calculated from intensity data. These statistical irregularities can be used to identify the presence of twinning and they form the theoretical basis for several common twinning tests.

A typical X-ray diffraction data set from a macromolecular crystal is expected to have a certain number of both very strong and very weak reflections. In a merohedrally twinned crystal, some of these very strong reflections become averaged with very weak ones as a result of the twinning, and so the resulting intensity distribution will have a sharper peak around the average value, and the variance (second moment) will be smaller. These deviations from the expected intensity distribution form the basis for a twin test that was formalized by Rees [45, 46]. This test involves first calculating normalized reflection intensities (z) by dividing each individual measurement by the average value for its resolution shell. Following this intensity normalization, the cumulative distribution of z, $N(z)$, takes a different form depending on the twin fraction (Fig. 9). The appearance of the plot of $N(z)$ is diagnostic of twinning because for untwinned crystals, the plot appears exponential, while for a perfectly (or partially) twinned crystal, the plot appears sigmoidal. Analysis of the $N(z)$ plot was one of the earliest useful tests for twinning in crystallography, although it has several weaknesses. The main problem with the $N(z)$ plot is that twinning is not the only crystal abnormality that causes deviations from the expected intensity distribution. For example, rotational non-crystallographic symmetry (NCS) can have an effect similar to twinning, because it amounts to averaging crystallographically unrelated reflections. Also, problems such as translational NCS or

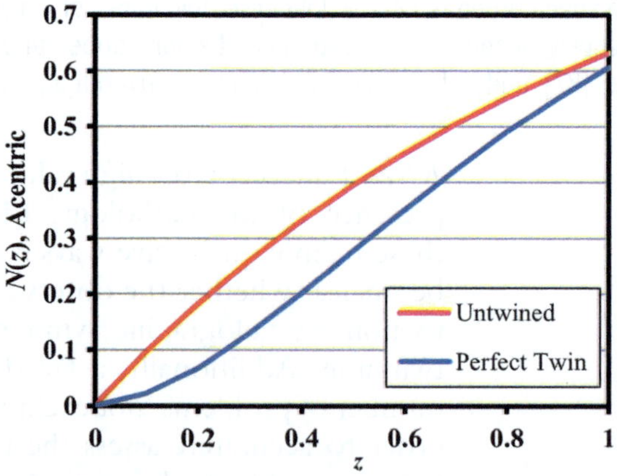

Fig. 9 Normalized cumulative intensity distributions, $N(z)$, for acentric reflections from untwinned and perfectly twinned ($\alpha = 0.5$) crystals. The plot of the cumulative distribution appears exponential for the untwinned intensities, owing to the existence of a small number of very weak and very strong reflections. In contrast, twinning combines some of the weak intensities with strong ones, which sharpens the intensity distribution and makes the plot of the cumulative distribution appear sigmoidal for twinned intensities

Table 2
Yeates' *S(H)* test for twinning

	$\langle H \rangle$	$\langle H^2 \rangle$
Acentric:	$0.5-\alpha$	$(1-2\alpha)^2/3$
Untwinned	0.5	0.333
Perfect twin	0.0	0.0
Centric:	$2(1-2\alpha)/\pi$	$(1-2\alpha)^2/2$
Untwinned	0.637	0.5
Perfect twin	0.0	0.0

Expressions are provided for H and H^2 for both acentric and centric reflections. Solutions for untwinned ($\alpha = 0$) and perfectly twinned ($\alpha = 0.5$) crystals are also given

anisotropic diffraction can have an effect on the intensity distribution that is opposite to that of twinning, due to the presence of systematic strong and weak reflections in each resolution bin. In addition to concerns arising from pseudosymmetry, the $N(z)$ plot does not provide accurate estimations of the twin fraction.

Another useful statistical test for twinning involves first calculating the ratio (H) of the difference to the sum of twin-related intensities (I_1 and I_2):

$$H = \frac{|I_1 - I_2|}{(I_1 + I_2)}$$

Yeates demonstrated [47] that the cumulative distribution of H, $S(H)$, is linear for acentric reflections, and its slope, equal to $(1\text{-}2\alpha)^{-1}$, is dependent on the twin fraction. Additionally, the averages $\langle H \rangle$ and $\langle H^2 \rangle$have characteristic values that depend on the twin fraction (Table 2).

This method of identifying merohedral twinning by analyzing the statistical properties of H is useful, because in addition to confirming the presence or absence of twinning with a simple visual test (Fig. 10), it also provides a robust estimation of the twin fraction. On the other hand, this method also has several drawbacks. First, it fails for perfectly twinned crystals, because the expression for the cumulative distribution of H becomes singular as the twin fraction approaches 0.5. Second, because H is calculated using twin-related reflections, this method requires that the correct twin operator be known, and that a significant number of twin-related reflections have been measured. Finally, the presence of rotational NCS nearly coincident with a potential twin operator has the potential to increase the estimated twin fraction.

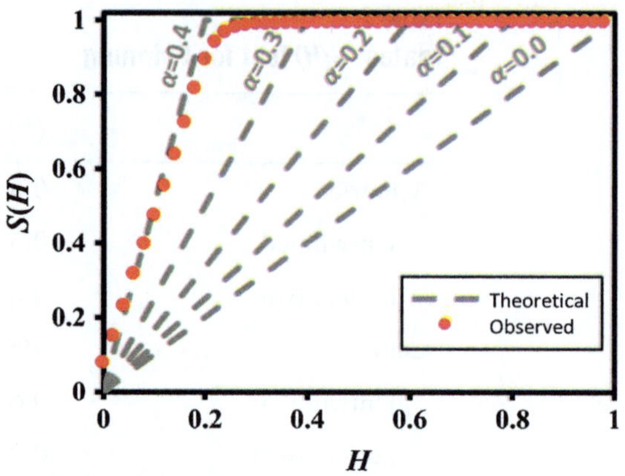

Fig. 10 The cumulative distribution of H, $S(H)$, for acentric reflections takes on a simple form that is dependent on the twin fraction. This relationship, given by $S(H) = H/(1 - 2\alpha)$, forms the basis for a twin test developed by Yeates. The expected cumulative distributions for several values of the twin fraction (α) are plotted in dashed *gray lines* (the cumulative distribution of H is given by $S(H) = H/(1 - 2\alpha)$. The *red dots* show $S(H)$ for diffraction data deposited in the PDB under accession code 4LIW. The twin operator used to calculate H was $(k,h,-l)$, and a fit of the observed data to the expression for $S(H)$ suggests a twin fraction of 0.38

The most robust statistical test for twinning in macromolecular crystallography is commonly referred to as the "L-test," or less frequently as the "Padilla–Yeates test." Like the aforementioned twin test, the L-test is based on analyzing the cumulative distribution of a ratio, $|L|$, which is calculated by selecting two intensities and dividing their difference by their sum [48]. The ratio L is defined in a manner similar to the ratio H, described above:

$$L = \frac{I_1 - I_2}{I_1 + I_2}$$

The main difference between $|L|$ and H is that the intensities used to calculate H are taken from twin-related reflections, while the intensities (I_1 and I_2) used to calculate $|L|$ correspond to intensity measurements from reflections that are proximal (locally related) in reciprocal space. The fact that the reflection intensities used to calculate $|L|$ represent locally related reflections makes the L-test robust and unaffected by the phenomena that complicate the identification of twinning. Specifically, it is insensitive to diffraction anisotropy, as well as rotational and translational pseudosymmetry (NCS), and it is capable of differentiating between twinning and rotational order–disorder. Most importantly, since L is based on locally related reflections, the L-test can be performed successfully without knowing the twin operator, and it is also insensitive to data reduction in the wrong space group. The only situation that is known to obscure the results of the L-test is when data are collected with a significant number of overlapping reflections [49]. As with H, the cumulative distribution of $|L|$, $N(|L|)$, as well as the averages $\langle |L| \rangle$ and $\langle L^2 \rangle$,are dependent on the twin fraction. The analytical expressions for $N(|L|)$, $\langle |L| \rangle$, and $\langle L^2 \rangle$ are long; however, for cases of untwinned ($\alpha = 0$) or perfectly twinned ($\alpha = 0.5$) specimens, they take on simple forms (Table 3).

Because of the dependence of $N(|L|)$ on the twin fraction, plots of $|L|$ vs $N(|L|)$ are visually diagnostic of twinning (Fig. 11). An important feature of the cumulative distribution of $|L|$ is that (unlike the cumulative distribution of H) it is defined for $\alpha = 0.5$, which means that the L-test can be applied in the presence of perfect twinning. Because of its superior properties, the L-test has supplanted most other tests for twinning in macromolecular crystallography.

The statistical tests for twinning described above are implemented in several commonly used computer programs for data reduction and analysis in macromolecular crystallography. Users of the *CCP4* suite [50] can perform these tests with the program *CTRUNCATE*, while users of the *PHENIX* suite [51] can perform twinning tests with *phenix.xtriage* [52]. In addition to performing twinning tests, if they identify possible twinning, these programs will also attempt to determine the twin operator and estimate the twin fraction. Along with the programs mentioned above, a webserver for detection of merohedral twinning is maintained by UCLA (https://services.mbi.ucla.edu/Twinning/). Even with modest computers available to most crystallographers, these tests can be performed in seconds to minutes, and therefore it is always advisable to perform them on any data set, especially when the lattice symmetry permits twinning. Once twinning has been identified, it is often not difficult to overcome.

3.3 Structure Determination from Twinned Data

Once merohedral twinning is identified in a data set, there are several potential pathways forward to successful structure determination. The best way to proceed depends primarily on the method that has been selected for phase calculation, and once an initial structure is obtained, refinement of the model can proceed with inclusion of the twin fraction as an additional parameter.

Table 3
The *L*-test

	$\langle \lvert L \rvert \rangle$	$\langle L^2 \rangle$	$N(\lvert L \rvert)$
Acentric, untwinned	1/2	1/3	$\lvert L \rvert$
Centric, untwinned	$2/\pi$	1/2	$(2/\pi)\sin^{-1}(\lvert L \rvert)$
Acentric, perfect twin	3/8	1/5	$\lvert L \rvert(3 - L^2)/2$

The table provides values of $|L|$ and L^2, as well as expressions for $N(|L|)$, for untwinned (acentric, centric) and perfectly twinned (acentric) crystals

3.3.1 Phase Calculation for Twinned Data Sets

Because the true diffraction intensities from a twinned crystal are unknown, phasing can pose a significant challenge. When working with twinned data, phase calculation by molecular replacement is generally the easiest option; however, if an acceptable search model is unavailable, experimental phasing is also possible.

Molecular replacement typically works well with twinned data, with the caveats that the contrast of the rotation function decreases with increasing twin fraction [53], and that the rotation function will typically find multiple acceptable solutions that are related by the twin operator. Generally, it is not difficult to identify just one of these solutions, and the molecular replacement module of the *PHASER* software even performs an internal twin test and applies subsequent corrections to the rotation function [54]. Sometimes, the combination of twinning and NCS can cause difficulties, but most often, twinning alone does not pose a significant problem for phasing by molecular replacement, and even perfectly twinned data can be used for molecular replacement [55]. The molecular replacement solution can subsequently be refined against the twinned data (*see* below).

In contrast to MR phasing, experimental phasing of twinned data can be quite difficult. In some uncommon cases, experimental phasing can be performed without any initial treatment of the data [56, 57]. Typically, however, the presence of merohedral twinning complicates the interpretation of isomorphous (and anomalous) difference Patterson maps, because it can lead to the disappearance of important cross-peaks and prevent the identification of heavy-atom sites. Consequently, experimental phasing of twinned data

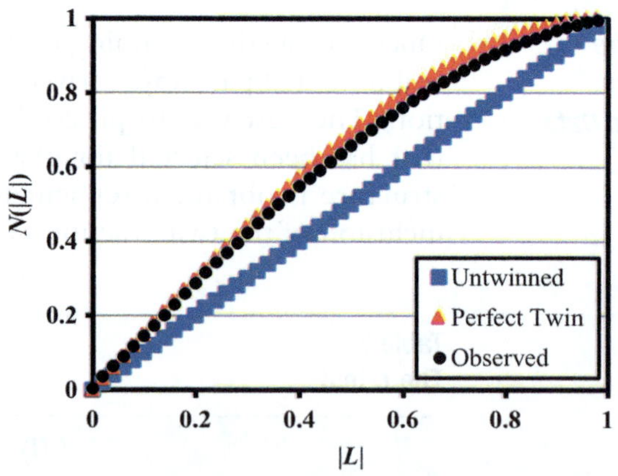

Fig. 11 The cumulative distribution of the ratio |*L*| takes on distinct and well-defined forms for untwinned and perfectly twinned data alike, which is one of the reasons it can be used as a robust twin test. The *blue* and *red curves* show the theoretical cumulative distributions of |*L*| for untwinned and perfectly twinned data respectively, for acentric reflections. The *black curve* shows the observed cumulative distribution of |*L*| for acentric diffraction data deposited in the PDB under accession code 4LIW, which has a twin fraction of 0.38

most commonly begins with a detwinning procedure. A common method of detwinning involves rearranging the two equations given in Subheading 3.2.1 to give the following relationships:

$$J_1 = \left[(1-\alpha)I_2 - \alpha I_1\right]/(1-2\alpha)$$

$$J_2 = \left[(1-\alpha)I_1 - \alpha I_2\right]/(1-2\alpha)$$

Where J_1 and J_2 represent the "detwinned" reflection intensities, and the twin fraction, α, is obtained by some estimation, such as from the cumulative distributions of H or $|L|$ (*see* Subheading 3.2.3), or from another analysis known as the "Britton plot," which uses the above equations to estimate the expected number of negative intensities as a function of the twin fraction [58, 59]. Detwinning of observed intensities can be performed easily using the *CCP4* program *DETWIN* [50]. There are two problems, however, in working with detwinned intensities. First, detwinning of perfectly twinned intensities is impossible, because the detwinning expressions are undefined for $\alpha = 0.5$. Additionally, as determined by Fisher and Sweet [59], the error for detwinned intensities is dependent on the twin fraction, as given by:

$$\sigma(J) = \sigma(I)\left[\frac{\left(1-2\alpha+2\alpha^2\right)^{1\backslash 2}}{(1-2\alpha)}\right]$$

A consequence of this relationship is that detwinned intensities can be very inaccurate for high twin fractions, which presents a challenge for experimental phasing. Despite potential difficulty, experimental phasing with detwinned intensities can be successful, and many examples exist in the literature [57, 60–62]. An initial model can be built using an experimental map calculated with detwinned intensities. Atomic refinement can be carried out using detwinned intensities, or alternatively, once an initial molecular model is obtained, it may be beneficial to continue atomic refinement of the model using the original twinned data as described below.

3.3.2 Atomic Refinement Against Twinned Data

When determining a structure from twinned data, it is essential to continue to account for the effects of twinning throughout refinement of the atomic model. Put simply, the goal of reciprocal-space refinement is to minimize the difference between observed and calculated structure factor amplitudes (F_o–F_c), and failure to account for twinning will always result in poor agreement between these two sets. Additionally, if twinned structure factor amplitudes are used for generation of electron density maps, then those maps can have worse model-phase bias because the twinned amplitudes are sometimes poor estimations of the true, untwinned amplitudes (especially for higher twin fractions) [3]. It is, of course, possible to use detwinned structure factor amplitudes as the X-ray target for atomic refinement; however, as discussed above, detwinning has the potential to introduce substantial additional error [59]. Rather than detwinning and refining a model against the resulting, inaccurate estimations of structure factors, it is better to use a "twin refinement" protocol, with the original, twinned data as the target. Twin refinement simply introduces the twin fraction as a parameter in the model used to derive calculated structure factor amplitudes, and refines it iteratively, along with the atomic coordinates, B-factors, etc. The benefits of twin refinement are twofold. First, it circumvents the need to use inaccurate, detwinned data as the refinement target. Second, the iterative refinement of the twin fraction and the atomic model provides the most accurate method of estimating the true twin fraction.

There are several practical aspects of twin refinement that should be considered before the task is undertaken. First, an acceptable program must be selected that allows the specification of a twin operator, which can then be applied to the calculated structure factor amplitudes and used to refine the twin fraction. Refinement programs that support twin refinement include *phenix.refine* [51, 63], *REFMAC5* [64], and *SHELXL* [65]. In addition to selection of a program with an appropriate twin refinement protocol, care should be taken to ensure that twin-related reflections are given the same assignment with respect to the working and free sets of reflections, in order to maintain the power of cross-validation. Finally, when performing model rebuilding into electron density maps between twin refinement cycles, one should bear in mind that map coefficients produced by twin refinement can suffer from worse model bias than maps calculated from untwinned data.

3.4 Complicated Twinning Scenarios

For partial hemihedral twins with molecules packed in relatively simple arrangements, it is generally straightforward to manage twinning if the data are treated carefully, as outlined above. In contrast, certain situations, such as perfect merohedral twinning, or twinning combined with pseudosymmetry, can still be quite difficult to handle, even when the twinning has been identified. These complications are described in more detail below. Additionally, an interesting form of artificial twinning is introduced, which has emerged from a new and exciting type of crystallographic experiment.

3.4.1 Perfect Twins

For low to moderate twin fractions, twinning is often relatively straightforward to handle. On the other hand, when the twin fraction (α) reaches 0.5, perfect twinning exists, and this situation can pose considerable difficulty [55]. Equations used to estimate the twin fraction and detwin intensities are undefined for α = 0.5, so detwinning is impossible for perfect twins. Even for partial twins with high twin fractions, detwinning can be problematic because it is increasingly inaccurate as the twin fraction approaches 0.5 [59].

Equations relating the observed, twinned intensities to the true, detwinned intensities are degenerate for the case of perfect twins, which causes multiple difficulties for experimental phasing and atomic refinement. Experimental phasing of perfect twins by MIR requires four derivatives (rather than two for untwinned data) and a complicated geometric construction, as demonstrated by Yeates and Rees [66]. Model building and refinement are also more difficult for perfect twins, because the refinement against perfectly twinned data effectively suffers a twofold reduction in the expected data–parameter ratio, and the resulting maps can be model-biased [3, 67].

3.4.2 Twinning Combined with Other Types of Pseudosymmetry

When it occurs in combination with other types of rotational or translational non-crystallographic symmetry (pseudosymmetry), identifying and characterizing twinning can become increasingly difficult. One problem, as discussed above, is that various types of pseudosymmetry can affect intensity distributions in ways that complicate twinning analyses and lead to errors in estimating the twin fraction [48]. Rotational pseudosymmetry has the potential to masquerade as twinning if the NCS operation is (nearly) coincident with a potential twin operator. When this happens, the contrast of certain twinning tests is reduced, because the pseudosymmetry reduces the difference between twinned and untwinned crystals [3, 53]. On the other hand, translational pseudosymmetry can affect intensity distributions in ways that oppose the effect of twinning, because pseudocentering operations introduce multiple classes of strong and weak reflections [48]. Additionally, the presence of pseudosymmetry is often accompanied by a large number of molecules in the asymmetric unit (ASU), also referred to as "high copy number." Even in the absence of associated pseudosymmetry, high copy number alone further complicates the identification of twinning because it undermines any arguments for or against twinning based on unit cell contents (i.e., the unit cell will be large enough to accommodate lower symmetry with more molecules in the ASU, or higher symmetry with fewer molecules in the ASU). Despite the obstacles that arise from combinations of pseudosymmetry with twinning, there are numerous inspiring reports of crystallographers overcoming these challenges and determining twinned structures with complicated packing arrangements [37, 38, 68–71].

While the co-ocurrence of non-crystallographic symmetry and twinning typically makes working with a data set more difficult, virus crystallographers have identified scenarios in which high-order rotational NCS could be used as an advantage in the detwinning process [72]. In these reports, cubic crystals of icosahedral virus particles in space group *I*23 are perfectly twinned, yielding diffraction patterns with perfect 432 point group symmetry. For these crystals, a hemihedral twin operation (90° rotation) superimposes the two- and threefold rotational symmetry axes of the icosahedral particle in the unit cell; however, the fivefold symmetry axes of the particle (rotational NCS) are not superimposed with one another after applying the twin operation. The fact that the orientation of the high-order (fivefold) NCS axis is not invariant under the twin operation allows detwinning of perfectly twinned data (which is usually impossible), using an algorithm that iteratively performs real space NCS averaging and reciprocal space scaling of twin-related intensities [12, 72]. This process has been used to determine the structures of several variants of the foot-and-mouth disease virus [73, 74], as well as a structure of *Aichi virus 1* [75]. A detailed description of the detwinning algorithm used in these studies has been published by Ginn and Stuart [12].

Interestingly, pseudosymmetry (particularly of the rotational type) and twinning are often observed together. This connection is evident from the literature—many reports of structure determination from twinned crystals also describe the existence of pseudosymmetry (see references cited above), and theory describing the relationship has been established [3, 76]. The presence of rotational pseudosymmetry has the potential to present multiple similar, but crystallographically nonequivalent, interaction interfaces for crystal growth, consistent with the observation by Zwart et al. [53] that twin domain interfaces often have packing arrangements that are similar to the crystallographic packing interfaces. The frequent co-occurrence of twinning and pseudosymmetry in macromolecular crystallography emphasizes the importance of the *L*-test [48, 76] in these difficult cases. The *L*-test is mostly immune to the existence of confounding pseudosymmetry if the reflections used to calculate *L* are chosen so that they have the same parity.

3.4.3 Intractable Cases of Merohedral Twinning

Most often, if merohedral twinning is handled with care, structure determination can still proceed successfully. There are, however, occasional cases where twinning cannot be overcome at the data analysis stage, and such situations require additional optimization of crystal growth and preparation protocols to reduce or eliminate the pathology. For example, several reports have demonstrated that slowing crystal nucleation or growth can reduce the extent of twinning [77–79]. Additionally, twinning can sometimes be defeated by slightly adjusting the crystallization conditions [18, 80]. This has been done by varying the ionic strength of the solution [81], and by addition or removal of additives such as dioxane [18] or anionic surfactants [82], either during or after crystallization. If obtaining untwinned crystals still proves impossible after optimization of crystallization conditions, it may be possible to use a microfocus X-ray beam to collect several unique data sets from spatially independent regions of a twinned crystal, in hope of identifying a small region that is less affected by the pathology [18]. Finally, as a last resort, it may be necessary to identify an entirely new crystal form that does not suffer from twinning.

3.4.4 "Computational Twinning" in Serial Crystallography

The emergence of macromolecular serial crystallography [83–85] has created a new frontier for the analysis and handling of twinning. In serial crystallography, many microcrystals (<50 μm) are each shot once with an X-ray beam, so that each crystal produces only a single diffraction image. Generally, these microcrystals are delivered to the X-ray beam in random orientations, and so unlike in single-crystal oscillation crystallography, the spatial relationship of reflections from one image to the next is unknown. As a result, each diffraction image must be indexed separately and, for space groups with indexing ambiguity, some images will be indexed in one orientation, and others will be indexed in the opposite orientation. An early serial crystallographic structure determination effort measured diffraction from crystals known to belong to space group $P6_3$ [83]. After merging together data from thousands of individual images, it was observed that the data belonged to point group 622. The data were treated as a perfect hemihedral twin in $P6_3$. In this case, the individual microcrystals were not twinned, but the data merging strategy required to process the serial crystallography data introduced twinning as a result of the ambiguity in indexing orientations. Diederichs and Brehm [67] subsequently provided a solution to this problem by developing algorithms for clustering diffraction snapshots that merge best with one another, allowing the disambiguation of indexing assignments for randomly oriented images.

3.5 Twinning—Summary

Twining occurs in macromolecular crystallography when an apparently single crystal actually consists of multiple, differently oriented twin domains that are related in one of several special ways. Non-merohedral twinning describes cases where twin-related lattices are partially superimposable. This situation is a less common form of twinning in macromolecular crystals, and it is usually visually apparent, so it is not typically concerning, since it can be identified at the data collection stage and crystals with the pathology can be abandoned. Merohedral twinning is a more dangerous crystal pathology. In the diffraction pattern from a merohedral twin, the arrangement of reciprocal lattice points has higher symmetry than the intensities, allowing the twin-related lattices to become perfectly superimposed. Merohedral twinning does not make the diffraction pattern appear abnormal, so it often goes unnoticed during data collection and sometimes twinning is not discovered from the data until it is too late to measure additional crystals or to collect data over a larger wedge of reciprocal space. Merohedrally twinned data can appear to have erroneously high symmetry, so they are often indexed in incorrect space groups, which hampers structure determination. In order to avoid these pitfalls, it is important to understand when twinning is possible, and how to proceed if it is encountered. Additionally, it is advisable to include twinning tests as a default routine in the data reduction and analysis protocols used in macromolecular crystallography [86].

4 Conclusion

Crystal pathologies such as disorder and twinning are not uncommon in macromolecular crystallography, and the threat of these pathologies is constantly looming over the field. Anecdotally, it seems that nearly all crystallographers have encountered at least one of these phenomena at some point. Crystal pathologies have a broad spectrum of characteristics. Some are completely detrimental to structure determination, while others do not present significant obstacles, even if undetected. Some are immediately obvious from the appearance of the crystals or the diffraction patterns, while others provide no visual clues and can only be identified through judicious analysis of intensity data. Understanding the many types of crystal pathologies, including how to identify and handle them, is an important part of the practicing crystallographer's knowledge, enabling the maximum amount of structural information to be obtained, even from non-ideal data sets.

Acknowledgments

I thank Todd Yeates for sharing his expertise on these subjects over the years. Additionally, I thank Tanja Kortemme, Yao-Ming Huang, Peter Cimmermančič, and Andrej Sali for sharing crystal and diffraction images, and Benjamin Barad for assistance with preparing data for figures.

References

1. Janin J, Rodier F (1995) Protein-protein interaction at crystal contacts. Proteins 23:580–587
2. TO Y, Fam BC (1999) Protein crystals and their evil twins. Structure 7:R25–R29
3. Lebedev AA, Vagin AA, Murshudov GN (2006) Intensity statistics in twinned crystals with examples from the PDB. Acta Crystallogr D Biol Crystallogr 62:83–95
4. Yeates TO (1997) Detecting and overcoming crystal twinning. Methods Enzymol 276:344–358
5. Ayyer K, Yefanov OM, Oberthür D et al (2016) Macromolecular diffractive imaging using imperfect crystals. Nature 530:202–206
6. Dornberger-Schiff K (1956) On order–disorder structures (OD-structures). Acta Crystallogr 9:593–601
7. Dornberger-Schiff K, Grell-Niemann H (1961) On the theory of order–disorder (OD) structures. Acta Crystallogr 14:167–177
8. Bragg WL, Howells ER (1954) X-ray diffraction by imidazole methaemoglobin. Acta Crystallogr 7:409–411
9. Cochran W, Howells ER (1954) X-ray diffraction by a layer structure containing random displacements. Acta Crystallogr 7:412–415
10. Rupp B (2009) Biomolecular crystallography: principles, practice, and application to structural biology. Garland Science, New York
11. Lerch TF, Xie Q, Ongley HM et al (2009) Twinned crystals of adeno-associated virus serotype 3b prove suitable for structural studies. Acta Crystallogr F Struct Biol Commun 65:177–183
12. Ginn HM, Stuart DI (2016) Recovery of data from perfectly twinned virus crystals revisited. Acta Crystallogr D Struct Biol 72:817–822
13. Pletnev S, Morozova KS, Verkhusha VV et al (2009) Rotational order-disorder structure of fluorescent protein FP480. Acta Crystallogr D Biol Crystallogr 65:906–912

14. Pletnev S, Subach FV, Verkhusha VV et al (2013) The rotational order–disorder structure of the reversibly photoswitchable red fluorescent protein rsTagRFP. Acta Crystallogr D Biol Crystallogr 70:31–39
15. Renko M, Taler-Verčič A, Mihelič M (2014) Partial rotational lattice order–disorder in stefin B crystals. Acta Crystallogr D Biol Crystallogr 70:1015–1025
16. Robbins AH, Domsic JF, Agbandje-McKenna M et al (2010) Emerging from pseudo-symmetry: the redetermination of human carbonic anhydrase II in monoclinic P2(1) with a doubled a axis. Acta Crystallogr D Biol Crystallogr 66:950–952
17. Welberry TR (2010) Diffuse X-ray scattering and models of disorder. Oxford University Press, Oxford
18. Helliwell JR (2008) Macromolecular crystal twinning, lattice disorders and multiple crystals. Crystallogr Rev 14:189–250
19. Wang J, Kamtekar S, Berman AJ et al (2005) Correction of X-ray intensities from single crystals containing lattice-translocation defects. Acta Crystallogr D Biol Crystallogr 61:67–74
20. Kamtekar S, Berman AJ, Wang J, Lázaro JM et al (2004) Insights into strand displacement and processivity from the crystal structure of the protein-primed DNA polymerase of bacteriophage φ29. Mol Cell 16:609–618
21. Tsai Y, Sawaya MR, Yeates TO (2009) Analysis of lattice-translocation disorder in the layered hexagonal structure of carboxysome shell protein CsoS1C. Acta Crystallogr D Biol Crystallogr 65:980–988
22. Rye CA, Isupov MN, Lebedev AA et al (2007) An order-disorder twin crystal of L-2-haloacid dehalogenase from Sulfolobus tokodaii. Acta Crystallogr D Biol Crystallogr 63:926–930
23. Hwang WC, Lin Y, Santelli E, Sui J et al (2006) Structural basis of neutralization by a human anti-severe acute respiratory syndrome spike protein antibody, 80R. J Biol Chem 281:34610–34616
24. Trame CB, McKay DB (2001) Structure of Haemophilus influenzae HslU protein in crystals with one-dimensional disorder twinning. Acta Crystallogr D Biol Crystallogr 57:1079–1090
25. Zhu X, Xu X, Wilson IA (2008) Structure determination of the 1918 H1N1 neuraminidase from a crystal with lattice-translocation defects. Acta Crystallogr D Biol Crystallogr 64:843–850
26. Tanaka S, Kerfeld CA, Sawaya MR et al (2008) Atomic-level models of the bacterial carboxysome shell. Science 319:1083–1086
27. Heras B, Martin JL (2005) Post-crystallization treatments for improving diffraction quality of protein crystals. Acta Crystallogr D Biol Crystallogr 61:1173–1180
28. Lin T, Schildkamp W, Brister K et al (2005) The mechanism of high-pressure-induced ordering in a macromolecular crystal. Acta Crystallogr D Biol Crystallogr 61:737–743
29. Bernstein FC, Koetzle TF, Williams GJB et al (1997) The protein data bank: a computer-based archival file for macromolecular structures. J Mol Biol 112:535–542
30. Friedel G (1926) Lecons de Cristallographie. Berger-Levrault, Paris
31. Dhillon AK, Stanfield RL, Gorny MK et al (2008) Structure determination of an anti-HIV-1 Fab 447-52D–peptide complex from an epitaxially twinned data set. Acta Crystallogr D Biol Crystallogr 64:792–802
32. Dauter Z, Botos I, LaRonde-LeBlanc N et al (2003) Pathological crystallography: case studies of several unusual macromolecular crystals. Acta Crystallogr D Biol Crystallogr 61:967–975
33. Dauter Z (2003) Twinned crystals and anomalous phasing. Acta Crystallogr D Biol Crystallogr 59:2004–2016
34. Barends TRM, de Jong RM, van Straaten KE et al (2005) *Escherichia coli* MltA: MAD phasing and refinement of a tetartohedrally twinned protein crystal structure. Acta Crystallogr D Biol Crystallogr 61:613–621
35. de Ruyck J, Schubert HL, Janczak MW et al (2014) Tetartohedral twinning in IDI-2 from Thermus thermophilus: crystallization under anaerobic conditions. Acta Crystallogr F Struct Biol Commun 70:347–349
36. Roversi P, Blanc E, Johnson S et al (2012) Tetartohedral twinning could happen to you too. Acta Crystallogr D Biol Crystallogr 68:418–424
37. Sliwiak J, Jaskolski M, Dauter Z et al (2014) Likelihood-based molecular-replacement solution for a highly pathological crystal with tetartohedral twinning and sevenfold translational noncrystallographic symmetry. Acta Crystallogr D Biol Crystallogr 70:471–480
38. Sliwiak J, Dauter Z, Kowiel M et al (2015) ANS complex of St John's wort PR-10 protein with 28 copies in the asymmetric unit: a fiendish combination of pseudosymmetry with tetartohedral twinning. Acta Crystallogr D Biol Crystallogr 71:829–843
39. Yu F, Song A, Xu C et al (2009) Determining the DUF55-domain structure of human thymocyte nuclear protein 1 from crystals partially twinned by tetartohedry. Acta Crystallogr D Biol Crystallogr 65:212–219

40. Gilski M, Drozdzal P, Kierzek R et al (2016) Atomic resolution structure of a chimeric DNA-RNA Z-type duplex in complex with Ba(2+) ions: a case of complicated multi-domain twinning. Acta Crystallogr D Biol Crystallogr 72:211–223
41. Sultana A, Alexeev I, Kursula I et al (2007) Structure determination by multiwavelength anomalous diffraction of aclacinomycin oxido-reductase: indications of multidomain pseudomerohedral twinning. Acta Crystallogr D Biol Crystallogr 63:149–159
42. Jenni S, Ban N (2009) Imperfect pseudo-merohedral twinning in crystals of fungal fatty acid synthase. Acta Crystallogr D Biol Crystallogr 65:101–111
43. Parsons S (2003) Introduction to twinning. Acta Crystallogr D Biol Crystallogr 59:1995–2003
44. Wilson AJC (1949) The probability distribution of X-ray intensities. Acta Crystallogr 2:318–321
45. Rees DC (1980) The influence of twinning by merohedry on intensity statistics. Acta Crystallogr A 36:578–581
46. Rees DC (1982) A general theory of X-ray intensity statistics for twins by merohedry. Acta Crystallogr A 38:201–207
47. Yeates TO (1988) Simple statistics for intensity data from twinned specimens. Acta Crystallogr A 44:142–144
48. Padilla JE, Yeates TO (2003) A statistic for local intensity differences: robustness to anisotropy and pseudo-centering and utility for detecting twinning. Acta Crystallogr D Biol Crystallogr 59:1124–1130
49. Knott GJ, Panjikar S, Thorn A et al (2016) A crystallographic study of human NONO ($p54^{nrb}$): overcoming pathological problems with purification, data collection and noncrystallographic symmetry. Acta Crystallogr D Struct Biol 72:761–769
50. Winn MD, Ballard CC, Cowtan KD et al (2011) Overview of the CCP4 suite and current developments. Acta Crystallogr D Biol Crystallogr 67:235–242
51. Adams PD, Afonine PV, Bunkóczi G et al (2010) PHENIX: a comprehensive Python-based system for macromolecular structure solution. Acta Crystallogr D Biol Crystallogr 66:213–221
52. Zwart PH, Grosse-Kunstleve RW, Adams PD (2005) Xtriage and Fest: automatic assessment of X-ray data and substructure structure factor estimation. CCP4 Newsl 43
53. Zwart PH, Grosse-Kunstleve RW, Lebedev AA et al (2008) Surprises and pitfalls arising from (pseudo)symmetry. Acta Crystallogr D Biol Crystallogr 64:99–107
54. McCoy AJ, Grosse-Kunstleve RW, Adams PD et al (2007) Phaser crystallographic software. J Appl Crystallogr 40:658–674
55. Redinbo MR, TO Y (1993) Structure determination of plastocyanin from a specimen with a hemihedral twinning fraction of one-half. Acta Crystallogr D Biol Crystallogr 49:375–380
56. Yang F, Forrer P, Dauter Z et al (2000) Novel fold and capsid-binding properties of the λ-phage display platform protein gpD. Nat Struct Mol Biol 7:230–237
57. Yang F, Dauter Z, Wlodawer A (2000) Effects of crystal twinning on the ability to solve a macromolecular structure using multiwavelength anomalous diffraction. Acta Crystallogr D Biol Crystallogr 56:959–964
58. Britton D (1972) Estimation of twinning parameter for twins with exactly superimposed reciprocal lattices. Acta Crystallogr A 28:296–297
59. Fisher RG, Sweet RM (1980) Treatment of diffraction data from crystals twinned by merohedry. Acta Crystallogr A 36:755–760
60. Terwisscha van Scheltinga AC, Valegård K, Hajdu J et al (2003) MIR phasing using merohedrally twinned crystals. Acta Crystallogr D Biol Crystallogr 59:2017–2022
61. Hillig RC, Renault L (2006) Detecting and overcoming hemihedral twinning during the MIR structure determination of Rna1p. Acta Crystallogr D Biol Crystallogr 62:750–765
62. Rudolph MG, Kelker MS, Schneider TR et al (2003) Use of multiple anomalous dispersion to phase highly merohedrally twinned crystals of interleukin-1β. Acta Crystallogr D Biol Crystallogr 59:290–298
63. Afonine PV, Grosse-Kunstleve RW, Echols N et al (2012) Towards automated crystallographic structure refinement with phenix.refine. Acta Crystallogr D Biol Crystallogr 68:352–367
64. Murshudov GN, Skubák P, Lebedev AA et al (2011) REFMAC 5 for the refinement of macromolecular crystal structures. Acta Crystallogr D Biol Crystallogr 67:355–367
65. Sheldrick GM (2015) Crystal structure refinement with SHELXL. Acta Crystallogr C Struct Chem 71:3–8
66. TO Y, Rees DC (1987) An isomorphous replacement method for phasing twinned structures. Acta Crystallogr A 43:30–36
67. Brehm W, Diederichs K (2014) Breaking the indexing ambiguity in serial crystallography. Acta Crystallogr D Biol Crystallogr D70:101–109

68. Guelker M, Stagg L, Wittung-Stafshede P et al (2009) Pseudosymmetry, high copy number and twinning complicate the structure determination of *Desulfovibrio desulfuricans* (ATCC 29577) flavodoxin. Acta Crystallogr D Biol Crystallogr D65:523–534
69. Barends TRM, Dijkstra BW (2003) Acetobacter turbidans alpha-amino acid ester hydrolase: merohedral twinning in $P2_1$ obscured by pseudo-translational NCS. Acta Crystallogr D Biol Crystallogr 59:2237–2241
70. Lee S, Sawaya MR, Eisenberg D (2003) Structure of superoxide dismutase from Pyrobaculum aerophilum presents a challenging case in molecular replacement with multiple molecules, pseudo-symmetry and twinning. Acta Crystallogr D Biol Crystallogr 59:2191–2199
71. Thompson MC, Yeates TO (2014) A challenging interpretation of a hexagonally layered protein structure. Acta Crystallogr D Biol Crystallogr 70:203–208
72. Lea S, Stuart D (1995) Deconvolution of fully overlapped reflections from crystals of foot-and-mouth disease virus O1 G67. Acta Crystallogr D Biol Crystallogr 51:160–167
73. Lea S, Abu-Ghazaleh R, Blakemore W et al (1995) Structural comparison of two strains of foot-and-mouth disease virus subtype O1 and a laboratory antigenic variant, G67. Structure 3:571–580
74. Kotecha A, Seago J, Scott K et al (2015) Structure-based energetics of protein interfaces guides foot-and-mouth disease virus vaccine design. Nat Struct Mol Biol 22:788–794
75. Sabin C, Plevka P (2016) The use of noncrystallographic symmetry averaging to solve structures from data affected by perfect hemihedral twinning. Acta Crystallogr F Struct Biol Commun 72:188–197
76. Dornberger-Schiff K (1959) Relation of symmetry to structure in twinning. Acta Crystallogr 12:246
77. Borshchevskiy V, Efremov R, Moiseeva E et al (2010) Overcoming merohedral twinning in crystals of bacteriorhodopsin grown in lipidic mesophase. Acta Crystallogr D Biol Crystallogr 66:26–32
78. Blow DM, Chayen NE, Lloyd LF et al (1994) Control of nucleation of protein crystals. Protein Sci 3:1638–1643
79. Chayen NE, Saridakis E (2008) Protein crystallization: from purified protein to diffraction-quality crystal. Nat Methods 5:147–153
80. Sauter C, Ng JD, Lorber B et al (1999) Additives for the crystallization of proteins and nucleic acids. J Cryst Growth 196:365–376
81. Efremov R, Moukhametzianov R, Büldt G et al (2004) Physical detwinning of hemihedrally twinned hexagonal crystals of bacteriorhodopsin. Biophys J 87:3608–3613
82. Velev OD, Pan YH, Kaler EW et al (2005) Molecular effects of anionic surfactants on lysozyme precipitation and crystallization. Cryst Growth Des 5:351–359
83. Chapman HN, Fromme P, Barty A et al (2011) Femtosecond X-ray protein nanocrystallography. Nature 470:73–77
84. Cohen AE, Soltis SM, González A et al (2014) Goniometer-based femtosecond crystallography with X-ray free electron lasers. Proc Natl Acad Sci U S A 111:17122–17127
85. Hunter MS, Segelke B, Messerschmidt M et al (2014) Fixed-target protein serial microcrystallography with an x-ray free electron laser. Sci Rep 4:6026
86. Sawaya MR (2007) Characterizing a crystal from an initial native dataset. Methods Mol Biol 364:95–120
87. Thompson MC, Crowley CS, Kopstein J et al (2014) Structure of a bacterial microcompartment shell protein bound to a cobalamin cofactor. Acta Crystallogr F Struct Biol Commun 70:1584–1590

Chapter 9

Applications of X-Ray Micro-Beam for Data Collection

Ruslan Sanishvili and Robert F. Fischetti

Abstract

Micro-diffraction tools for macromolecular crystallography, first developed at the end of 1990s and now an integral part of many synchrotron beamlines, enable some of the experiments which were not feasible just a decade or so ago. These include data collection from very small samples, just a few micrometers in size; from larger, but severely inhomogeneous samples; and from samples which are optically invisible. Improved micro-diffraction tools led to improved signal-to-noise ratio, to mitigation of radiation damage in some cases, and to better-designed diffraction experiments. Small, micron-scale beams can be attained in different ways and knowing the details of the implementation is important in order to design the diffraction experiment properly. Similarly, precision, reproducibility and stability of the goniometry, and caveats of detection systems need to be taken into account. Lastly, to make micro-diffraction widely applicable, the sophistication, robustness, and user-friendliness of these tools are just as important as the technical capabilities.

Key words Micro-beam, Micro-diffraction, Micro-focus, Raster, Small crystals, Inhomogeneous crystals, Signal-to-noise, Radiation damage, Multi-crystal data collection

1 Introduction

Micro-diffraction in macromolecular crystallography has been reviewed recently [1]. State of the art of structural biology and structure-based drug design relies more and more on structures of large, multi-domain proteins, multi-protein complexes, and membrane-associated proteins. These molecules and supramolecular structures are notorious for often yielding low quality, poorly diffracting crystals, owing to a combination of deficiencies including their small size, extreme inhomogeneity, high mosaicity and high solvent content. The resulting low resolution, poor quality data measured on in-house sources or on synchrotron beamlines without micro-diffraction tools complicate not only the structure determination and refinement, but also detailed analysis of protein–protein and protein–ligand interactions. These shortcomings necessitated the development of the experimental tools and protocols allowing better data to be measured from poor quality

Alexander Wlodawer et al. (eds.), *Protein Crystallography: Methods and Protocols*, Methods in Molecular Biology, vol. 1607, DOI 10.1007/978-1-4939-7000-1_9, © Springer Science+Business Media LLC 2017

crystals. After the first experiments demonstrating the feasibility of data collection from macromolecular micro-crystals [2–4], several beamlines around the world have implemented micro-diffraction capabilities as well [5–11]. The spectacular success of the experiments performed with micro-diffraction tools [12–18] led to further popularization of the technology and its implementation on a number of synchrotron beamlines worldwide (Table 1). As a result, micro-diffraction now is a mainstream tool in the structural biologists' tool chest.

Table 1
Micro-diffraction capabilities for macromolecular crystallography on synchrotron beamlines

Facility and beamline	Beam size (FWHM, μm)[a]	Energy range (keV)	Approach
Operating beamlines			
ALBA BL13	50 × 6–300 × 300	5–22	Direct focus
APS 14-ID-B	20	12.7	Secondary source
APS 17-ID-B	5, 10, 20	6–20	Aperture
APS 19-ID-D	5, 10, 20	6.5–19.5	Aperture
APS 22-ID-D	10, 20, 50	6–20	Aperture
APS 23-ID-B	5, 10, 20	3.5–20	Aperture
APS 23-ID-D	5, 10, 20	5–20	Aperture
APS 24-ID-C	10, 30, 70	6.5–23	Aperture
APS 24-ID-E	5, 20, 50	12.66	Aperture
APS 31-ID-D	20, 50, 100	9–13.8	Aperture
Australian MX2	7.5, 10, 20	8.5–15.5	Aperture
CHESS A1	20	19.6	Direct focus
CHESS F1	20	12.68	Direct focus
Diamond I02	10, 20, 200	5–20	Aperture
Diamond I03	20, 50, 100	5.2–21	Aperture
Diamond I04	10 × 5–100 × 100	6–18	Aperture
Diamond I04-1	10, 20, 30, 50, 70	13.53	Aperture
Diamond I24	5 × 5–50 × 40	6.4–20	Secondary source
ESRF ID13	1	5–17	Direct focus
ESRF ID23-1	10 × 10–45 × 30	6–20	Direct focus
ESRF ID23-2	10	14.2	Direct focus

(continued)

Table 1 (continued)

Facility and beamline	Beam size (FWHM, μm)[a]	Energy range (keV)	Approach
ESRF ID29	10 × 10–50 × 30	6–20	Aperture
ESRF ID30A-1	10–100	12.8	Aperture
ESRF ID30A-3	15	12.9	Direct focus
ESRF ID30B	20–200	6.5–20	Direct focus
PETRA III P13	5, 10; 30 × 20–150 × 70	4.5–17.5	Aperture, Direct focus
PETRA III P14	5 × 5–150 × 150	6–20	Secondary source
PETRA III P11	1 × 1–300 × 300	5.5–30	Secondary source
Photon Factory BL-1A	10	2.7–3.0	Aperture
Photon Factory BL-17A	20	5.9–13.8	Aperture
SPring-8 BL32XU	1–10	8.5–20	Divergence-limited source
Spring-8 BL41XU	2 × 2–35 × 50	6.5–17.7	Secondary source
SLS X06SA	5 × 5–10 × 60	5.7–17.5	Secondary source
SLS X10SA	10, 30	6–20	Aperture
SOLEIL PROXIMA2	20	5–15	Direct focus
SSRF BL18U1	10 × 5	5–18	Virtual secondary source
SSRL 12-2	10, 20	6.7–17	Aperture
Beamlines being commissioned			
MAX IV BioMAX	20 × 5	5–25	Direct focus
NSLS II FMX	1–20	5–30	Secondary source
NSLS II AMX	5–100	5–18	Direct focus
Beamlines under construction			
APS 23ID-D	1–20	6–35	Secondary source
Diamond VMXi	5 × 5–30 × 30	10–25	Direct focus
Diamond VMXm	0.5 × 0.5–4 × 5	7–25	Secondary source
NSLS II NYX	5–50	6–18	Direct focus
Sirius Manaca	0.2 × 0.2–100 × 100	2–24	Secondary source

The Aperture approach employs one set of focusing elements combined with small apertures. All other methods employ more than one set of focusing elements, described in detail in [1]

Data were collected from beamline web sites in August, 2016

[a]The focused beam typically has an elliptical cross section, and the horizontal dimension is listed first followed by vertical (HxV). Dimensions separated by a dash indicate the focal size range. A single beam size refers to beam with a circular cross section or one defined by a circular aperture. Beamlines with at least one beam size dimension measuring 20 μm or less are listed

2 Implementation of Micro-Beams

Both the definition of "micro-beam" and its actual nature have been evolving in the last 10–15 years or so. When "standard," or more typical beams were 200 μm or more in size, a 50 μm beam could have been, and sometimes still is [19] labeled micro-beam. Later, 10–20-μm beams became known as micro-beams. More recently, after the advent of dedicated micro-focused beamlines, micro-beams came to mean about 1 μm, while the beams of several, or few tens of micrometers, became better known as "mini-beams" [7, 20]. In this chapter we do not discriminate between micro- and mini-beams and consider both as part of micro-diffraction tools. Also, we use the term "micro-diffraction" whenever micro-beam is used, whether the sample is small or large.

It is useful to understand how the reported beam size is achieved. A traditional practice of closing down the slits to reduce the beam size has limitations [7]. This is because the slits are typically located at distances of few tens of centimeters upstream from the sample and the beam, which diverges after the slits, becomes larger at the sample position. The discrepancy between the slit opening and the beam size at the sample is demonstrated in Fig. 1.

Generally, the smallest beam size can be achieved by focusing the beam at the sample position. In principle, focusing is straightforward, even though it does require state-of-the-art X-ray optics. For example, the X-ray beam on a non-crystallographic beamline can be as small as few nanometers. However, this approach cannot be utilized in macromolecular crystallography due to prohibitively

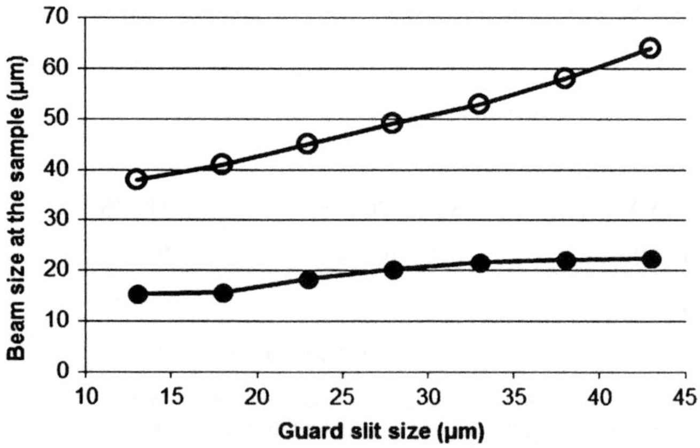

Fig. 1 Beam size at the sample position as a function of the slit size. The final horizontal (*open circles*) and vertical (*filled circles*) beam size were defined from the focal spot of 150 × 23 μm (FWHM, H × V) with guard slits located 310 mm upstream from the sample. The slit varied in the range of 13–43 μm. The beam size discrepancy at the sample position is larger in the horizontal direction than vertical because of the higher beam divergence (165 μrad vs 68 μrad)

increased divergence, which leads to undesirable spot enlargement. The spot enlargement may have detrimental effects on data quality for several reasons. For example, it can cause overlap of diffracted spots for crystals with moderate to large unit cell lengths. Using longer detector distances, if the diffraction resolution is not compromised, is desirable because it improves signal to noise ratio [21]. However, if the diffracted spot size also increases excessively due to the beam divergence, it may lead to spot overlap. Moreover, if the detector used for data collection has readout noise (for example CCD-based detectors), then enlarged spots, spreading over more pixels, would have higher readout noise by a factor of $\sqrt{(N_{long}/N_{short})}$ where N_{long} and N_{short} are the number of pixels in a Bragg spot at longer and shorter detector distances, respectively.

Another approach for achieving a small beam size, first developed by Cipriani and colleagues at beamline ID13 of the European Synchrotron Radiation Facility (ESRF) [4, 22], is to insert a beam-defining aperture of the desired size very close to the sample. Due to the short distance between the aperture and the sample, the beam diverges only slightly and the beam size at the sample position is close to the aperture size. Using the principle ideas of Cipriani's original work, several other designs for micro-beams using apertures have been implemented [7, 23] and have been since adapted on many beamlines worldwide (Table 1).

There are several benefits to using an aperture close to the sample to define the beam size. For example, the beam divergence can be much lower than in the case of direct focusing at the sample. The beam positional stability can be better than when the beam is directly focused down to a few microns at the sample. In this implementation larger beams can also be readily accessed, when needed, by simply switching to a larger aperture, or moving the aperture out of the beam path. One perceived shortcoming of this method is that the beam intensity at the sample is reduced as the aperture blocks some of the beam. In practice, both the aperture-defined and the direct-focused beams achieve comparable flux. Perhaps the reason for it is that the focusing elements of the direct-focusing method do not collect all of the incident beam, subsequently leading to some loss of flux. However, most of the modern, third generation synchrotron sources operate with low emittance (<10 nm-rad) and high brightness (10^{20} photons/s/mm^2/mrad^2/0.1% bandwidth) enabling fluxes of >10^{11} photons/s for a ~ 5 μm beam. Recently "hybrid" methods, combining both approaches, have appeared (reviewed in [1]) which can lead to even higher flux (e.g., 5 × 10^{12} photons/s on P14 beamline of PETRA III).

Another important aspect of a micro-diffraction apparatus is stable and reproducible goniometry with adequately small sphere of confusion (SOC). SOC can be defined as "minimum spherical volume enclosing the movement of a minute crystal mounted on the diffractometer when all axes are rotated through the full extent of their design limits" [24] or as a minimum sphere, which contains a centered point as it moves while each of the goniometer axis are rotated within their limits. Ideally, the beam is centered in the center of this sphere. Micro-crystallography poses stringent requirements for SOC. If, for example, a sample is about 5 μm in size and data are being collected with 5 μm beam, with SOC = 1 μm the "wobble" of the sample will be 0.5 μm around the center, moving 10% of the diffracting volume out of the beam path. The precision and reproducibility of the goniometer translations have similar effects, especially during "raster" and "vector" data collections, discussed later. Intrinsically, the fewer rotation axes and translation stages a goniometer has, the lower the SOC that can be achieved. Of practical interest in this context is how a sample is being mounted on the goniometer. Modern beamlines are typically equipped with a sample mounting robot, or automounter. If during the sample mounting the automounter exerts a force on the goniometer, the SOC can degrade over time. One solution to this problem is the mounting scheme where the automounter brings the sample close to the goniometer head and releases it without touching the goniometer, while the magnetic forces capture the sample pin base and place it on the goniometer head. On the other hand, when only a few degrees of data are collected from any given crystal, in a non-inverse beam geometry, the requirement for the small SOC can be relaxed considerably. This is because the complications from the SOC take place only when crystal rotates over rather large angles during data collection.

Another important aspect of micro-diffraction is adequate visualization of small samples, along with ability to manipulate them with high enough spatial resolution and reproducibility. A goniometer with sufficiently small SOC, precision, and sample visualization suitable for micro-diffraction was first developed by EMBL/Grenoble [4]. This diffractometer, with subsequent modifications and upgrades, has been implemented on many beamlines worldwide. New generation diffractometer, D3, has been recently developed as well at the Swiss Light Source [25]. With the aim to further reduce the SOC, a goniometer with vertical spindle axis has been implemented (http://www.embl-hamburg.de/services/mx/P14/).

3 Examples of Applications of Micro-Diffraction

While micro-beams should not be thought of as universal tools for all data collection experiments, there are a number of cases when small beams can make a difference between discarding a sample and collecting good quality data leading to successful structure determination and/or refinement.

3.1 Example 1: Improving the Signal-to-Noise Ratio by Reducing the Background

One obvious instance of using small beam is when working with equally small samples. Indeed, if the X-ray beam is substantially larger than the sample crystal, then it leads to elevated "parasitic" background—i.e., background produced by the part of the beam which misses the crystal and thus does not contribute to diffraction. Such increased background could substantially degrade data quality [20, 26]. For demonstration, let us consider I/σ_I in a simplified case when the background is the only contributor to σ. If a Bragg spot, without any background has an intensity of, say, 30 photons, then its $I/\sigma_I = 30/\sqrt{30} = 5.48$. Let us now assume that a small beam produces background of ten photons. Then $I/\sigma_I = 30/\sqrt{40} = 4.74$. If we assume that a larger beam produces background of 60 photons, then $I/\sigma_I = 30/\sqrt{90} = 3.16$. Following this trend, the detrimental effects of a large beam and associated higher background on the diffraction from small crystals becomes clearer, especially for weaker reflections at higher resolution. For example, if a Bragg peak with the intensity of two photons is superposed on a two-photon background produced by a small beam, then its I/σ_I, calculated as above, would be 1. But if the same peak is above the background of, say, 6, resulting from a larger beam, than its $I/\sigma_I = 2/\sqrt{8} \approx 0.71$ and most likely it would be rejected. In these examples, we did not take into account that because the intensity of a Bragg peak is a result of subtraction of two measurements—that of the peak and the background, their errors would add to estimate the error of the Bragg peak intensity. However, owing to the fact that background estimation is typically carried out using many pixels, its measurement error contributes little to the final σ. In practice, estimation of σ_I is not so simple and depends on many factors, but this simplified example illustrates the importance of using small beams with small crystal to avoid excess background.

The detrimental effects of background scatter should be kept in mind when manipulating samples for data collection. All possible measures should be taken to avoid excess liquid and/or protein "skin" around the sample in order to minimize the resulting background, for any size beam. For example, a sample support (loop, micro-mount etc.), much larger than the crystal should be avoided if possible, as it will carry excess liquid. Any excess liquid should be wicked away after scooping up the crystal; the protein skin could be removed from the crystallization drops prior to scooping up the crystals etc. When considering in-situ diffraction experiments, the thickness and the chemical make-up of the crystallization chambers, as well as the volume of the liquid inside, should also be considered to minimize the background.

Background reduction can be achieved in the data collection instrument as well. For example, different implementations of the same size micro-beam led to a reduction of background by a factor of 3 and thus has a significant impact on data quality (Fig. 2). If we revisit the example of the Bragg peak with two photons and background of six photons, the reduced background now would be

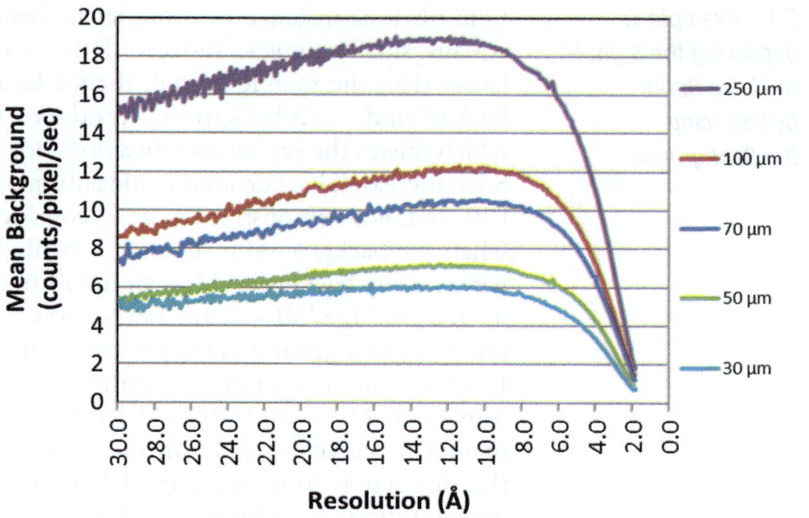

Fig. 2 The effect of exit aperture size on the scattered background intensity for a 5 μm beam-defining aperture. The 2D patterns were azimuthally integrated and divided by the number of pixels at that radius. The exit aperture size varied from 250 to 30 μm. Note the more than threefold decrease in background as the exit aperture is decreased from 250 to 30 μm. No sample was in the beam path for these measurements. (Unpublished results: S. Xu, N. Venugopalan, O. Makarov, S. Stepanov and R. F. Fischetti, GM/CA@XSD, Advanced Photon Source, Argonne National Laboratory, Argonne, IL, USA)

two photons and the $I/\sigma_I = 2/\sqrt{4} = 1$ instead of 0.71 with higher background. This example demonstrates that a simple statement of the beam size may not be sufficient and associated backgrounds are also important.

3.2 Example 2: Locating and Centering Optically Invisible Samples

Another benefit the small X-ray beams offer is locating very small samples which cannot be identified and reliably centered using only optical methods due to several reasons, such as insufficient microscope power; the crystal is veiled in a protein skin or other debris scooped up from the crystallization drop; the crystal is obstructed by the loop fiber or other supporting material. Perhaps most frequently, the sample visualization problem arises when crystals, grown in lipidic cubic phase, are cryo-cooled [27, 28]. In this case, the contrast between the lipid and the protein crystal is lost, making the latter optically invisible (Fig. 3). In such cases, scanning the sample with X-ray diffraction on a raster grid is indispensable for finding and centering the crystal [28, 29]. If the crystal diffracts well enough, a raster scan is carried out with heavily attenuated beam to avoid unnecessary exposure of the sample. However, often crystals diffract weakly necessitating the usage of a more intense beam. While radiation damage is a concern in such cases, it should be pointed out that during raster scans, a new segment of the crystal is exposed for each diffraction frame and consequently only one diffraction frame is measured per segment, per scan.

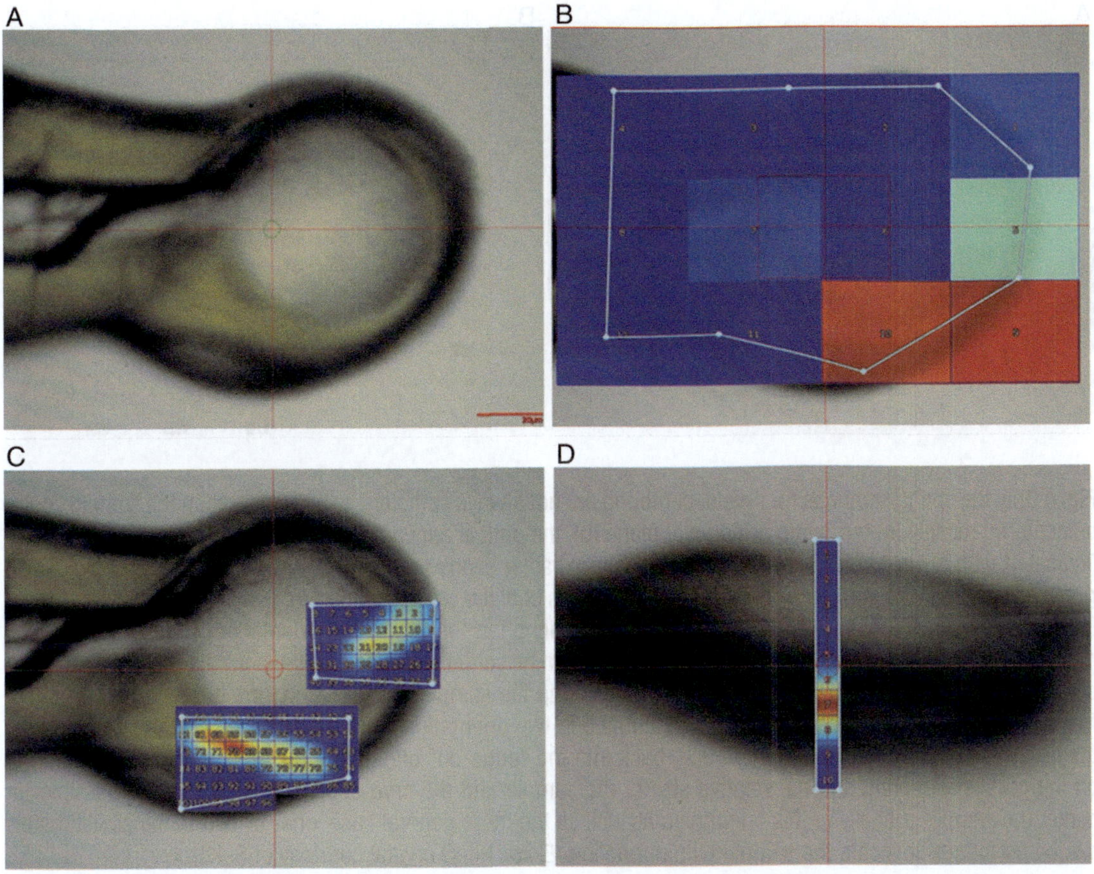

Fig. 3 Identifying and centering optically invisible crystal samples with X-ray diffraction on a raster grid. (**a**) Small crystals, cryo-cooled in lipidic cubic phase are not optically visible. (**b**) Crystals can be found and centered using X-ray diffraction raster scans. In a general case, a course raster scan with larger step size (40 × 30 μm^2—in this case) and larger beam is performed first to identify approximate location(s) of one or more crystal(s). The specific dimensions of the grid and of the beam are somewhat subjective, and are a compromise between the spatial resolution and the time spent. Color gradient in the raster scans is from *blue* with the fewest diffraction spots to *red* with the most. (**c**) Once approximate locations of the crystals are identified (*red* and *green rectangles* in **b**), a search on finer grid (5 × 5 μm^2 in this case) can be carried out. On this step, the sample can be both centered and its size measured in two directions. (**d**) Next, the sample is rotated 90° and one- or two-dimensional raster is carried out. Only one out of two crystals, found on step **c**, was centered. After this step, the sample is centered in two orthogonal planes and its size is known in three directions. With faster detectors, such as Pilatus (Dectris), Eiger (Dectris), or Rayonix HS (Rayonix), the coarse scan can be skipped and fine grid scan performed from the start

Sample centering by only optical microscopes may be problematic even if the crystal is not obstructed, for example when the liquid around the sample has a sizeable volume and it is cryo-cooled into a convex "lens" shape. In this case, if the viewing angle deviates from the optical axis of the "lens," refraction effects may lead to misplacing the sample (Fig. 4). Using a combination of optical and diffraction raster methods is a better approach to centering such samples.

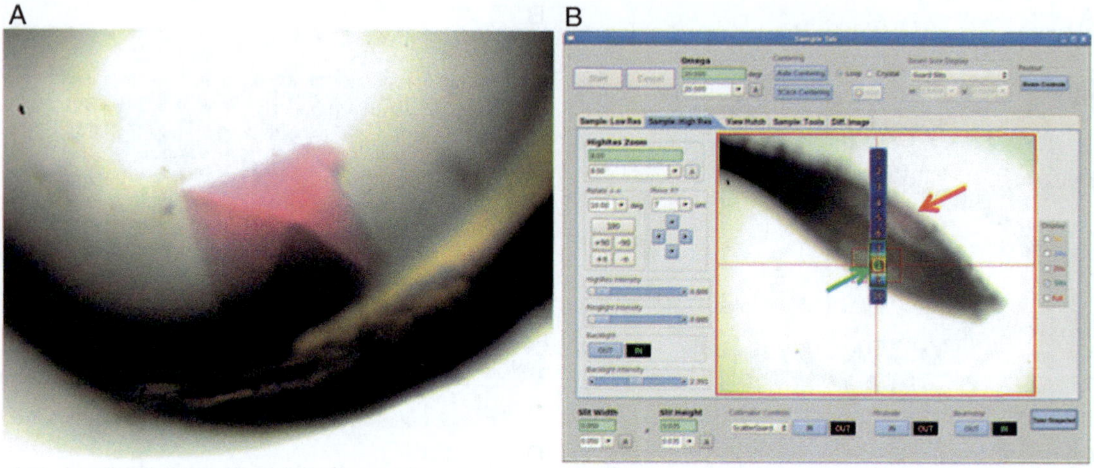

Fig. 4 Only the optical methods for crystal centering are not always sufficient. In one orientation the sample could be centered reliably (**a**) but in the orthogonal plane (**b**), the optical appearance of the sample can be misleading (indicated with a *red arrow*). In such cases, diffraction raster scans are needed (crystal location is shown with *green arrow*). First, the sample is rotated with small steps until the crystal support (e.g., nylon loop from Hampton Research, the micro-mount from MiTeGen etc.) is positioned in the "edge-on" orientation. Then the sample is rotated 90° into a "face-on" orientation (**a**) in which it can be centered optically. If the normal to the sample support plane coincides with the viewing direction, centering the sample is straightforward. If the normal deviates from the viewing angle, the sample can be identified in two orientations 180° apart and the mid-point used as the center. Then, the sample is rotated 90°, back into the "edge-on" orientation (**b**). In some cases, depending on the particular mount, and the relative size of the sample and the cryo-cooled liquid, it may be possible to center the sample optically in this orientation as well. In the more general case, one- or two-dimensional raster can be performed to locate the sample with diffraction. *E. coli* RecO crystal courtesy of S. Korolev

Raster scan-based tools were first developed on beamline ID13 of ESRF [30] and have matured over several years [28, 29, 31] paving the way for their implementation on many beamlines worldwide. Raster scans can be fully automated and included in automated data collection pipelines [32], https://epubs.stfc.ac.uk/work/63695.

In principle, raster search fields can be narrowed down by using other methods of crystal detection such as ultraviolet fluorescence microscopy, second order nonlinear imaging of chiral crystals (SONICC), etc. discussed in detail elsewhere in this volume, in the chapter by Becker et al.

3.3 Example 3: Combined Use of Small and Large X-Ray Beams for Crystal Characterization and Data Collection, Respectively

Benefits of diffraction raster scans with small beams go far beyond sample centering. Large crystals often suffer from inhomogeneity. This could be caused by growth defects, or by introducing stress due to the fact that protein crystals are not good thermal conductors, and therefore, cryo-cooling does not occur instantaneously throughout the entire crystal volume. In general, the larger the crystal volume, the greater the potential for the detrimental effects of cryo-cooling. Inhomogeneity in crystal quality can also be introduced by mechanical handling, especially when crystals adhere to the crystallization droplet support. The crystal may also dehydrate during its transfer from the crystallization droplet to the cryo-protecting solution and from the latter to the liquid nitrogen or a nitrogen cryo-stream. Whatever the root cause of the crystal inhomogeneity, if such samples are evaluated by the more traditional approach of taking one or two test diffraction images, they may be discarded as unusable when poorly ordered regions happen to be centered on the beam. Therefore, it is preferable to perform a raster grid scan with diffraction using a small beam to identify better ordered regions [20, 26, 29, 33] and measure their size. For example, a large crystal of GABA aminotransferase displayed severe inhomogeneity when tested with X-ray diffraction (Fig. 5). The raster scan revealed that the good quality region spanned about 60–100 μm. The beam size was adjusted accordingly and data collection led to successful structure solution and refinement [34].

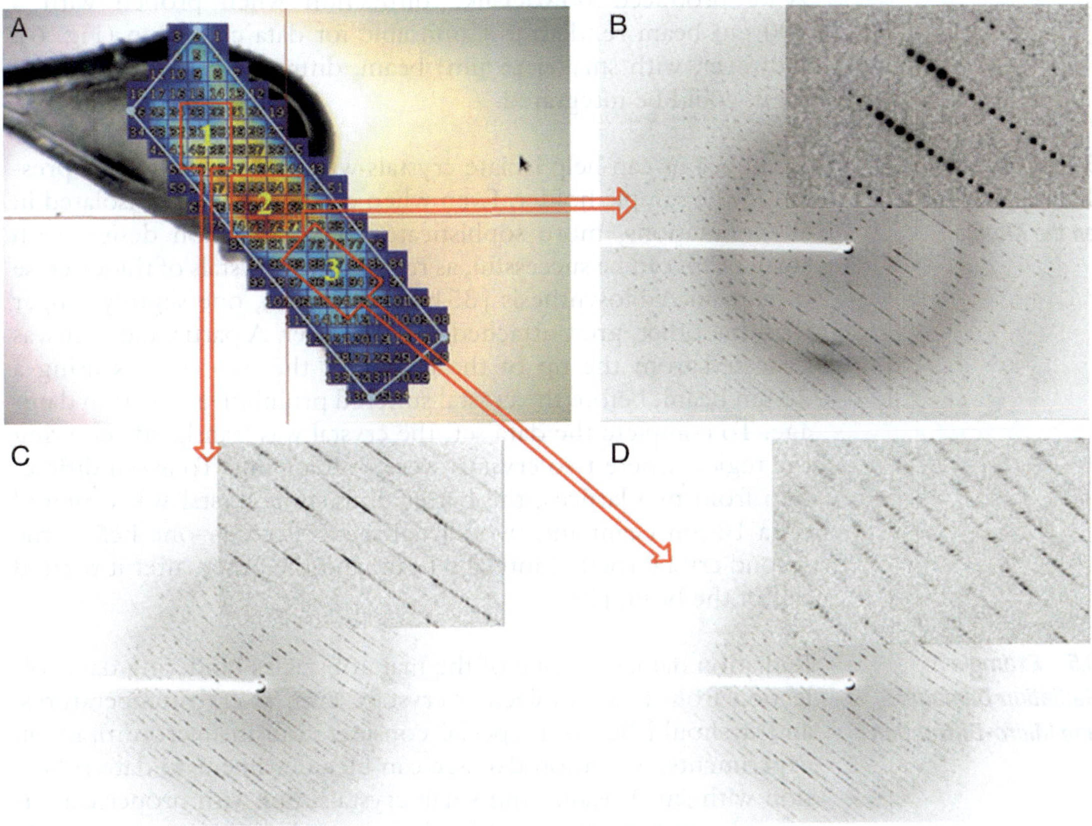

Fig. 5 X-ray diffraction raster scan of GABA aminotransferase crystal with 20 μm beam (**a**). *Red color* in the raster scan corresponds to the highest number of diffraction spots and *blue*—to the fewest. Regions 1 and 3, indicated with *yellow numbers*, displayed poor diffraction (**c**, **d**) while region 2, spanning 60–100 μm, was of very high quality (**b**)

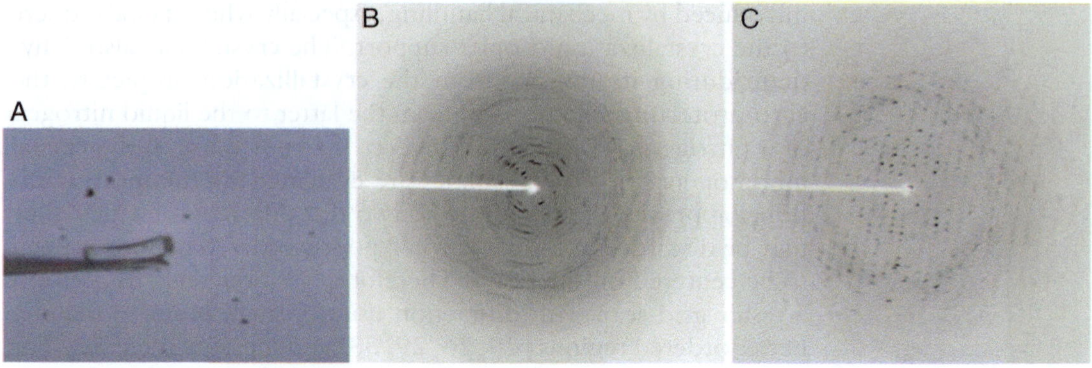

Fig. 6 Diffraction images from a bent lysozyme crystal (**a**) with 100 μm (**b**) and 5 μm X-ray beam (**c**). The diffraction quality is dramatically improved with a small X-ray beam

Even if well-diffracting region of a crystal cannot be found, using small beam can improve overall spot shapes enough that data collection becomes possible. For example, a bent crystal of lysozyme produced powder-like diffraction when probed with a 100 μm beam rendering it unusable for data collection (Fig. 6). However, with smaller (5 μm) beam, diffraction spots, while not ideal, could be integrated.

3.4 Example 4: Multiple Crystals on the Mount

Small beams can help isolate crystals when more than one is present in the sample holder. Even when crystals cannot be isolated in all orientations, more sophisticated data collection design with small beam can be successful, as reported for crystals of thioesterase of curacin A biosynthesis [35]. Two crystals, one slightly longer than the other, grew attached to each other. A partial data set was collected from the tip of the longer of the two crystals using a 20 μm beam, before the crystal suffered prohibitive radiation damage. To complete the data set, the crystal was translated centering the region where two crystals were overlapping. To avoid diffraction from two lattices, the better diffracting crystal was centered on a 10-μm beam and two partial sets collected—one before the second crystal rotated into the beam and the other, after it rotated out of the beam [35].

3.5 Example 5: Radiation Damage and Micro-Diffraction

Radiation damage is one of the main obstacles hindering data collection from macromolecular crystals, even at cryo-temperatures, and it should be given special consideration in micro-diffraction experiments. Radiation damage can be a detriment to data collection with small beams and small crystals, but with proper experimental design micro-diffraction can help mitigate it. At cryo-temperatures of typical data collection (~100 K), a majority of damage to macromolecular crystals is caused by energy deposition by the photoelectrons which are emitted after the interaction of incident X-rays with the sample [36–38].

Simulations predicted [37] and experimental data showed [39, 40] that these photoelectrons are completely reabsorbed within few microns from the primary point of emission. Consequently, radiation damage caused by energy deposition by emitted photoelectrons can extend away from the center of X-ray beam path and be reduced in the diffracting volume itself, as long as the beam cross section is smaller than the transverse distance the photoelectrons can travel, which is energy-dependent [39–41]. Thus, X-ray radiation damage to small crystals could be reduced in an energy-dependent manner. However, small beams should not be used with large crystals for the primary purpose of reducing radiation damage. Indeed, if the beam size was reduced from $100 \times 100\ \mu m^2$ to $1 \times 1\ \mu m^2$, radiation damage in the beam path would be reduced several times [39]. However, the diffraction volume would be reduced by four orders of magnitude, necessitating usage of a correspondingly more intense beam, leading to much more severe radiation damage. If a large crystal is sufficiently homogeneous, collecting data with correspondingly larger beam would result in better data [20] and less radiation damage since less intense beam would be needed in that case.

Small beams can help mitigate the radiation damage when their usage is necessitated by other reasons. For example, in the case of a very long rod-shaped crystal of the human BECN1 CCD homodimer, a raster scan identified that half of the crystal was not usable while the other half produced high quality diffraction (Fig. 7). A 20 μm beam, matching the cross-section of the crystal, was used to minimize the background and maximize the signal-to-noise ratio, as described in Subheading 3.1. Data were collected with the "vector" or "helical" protocol [8, 28], whereby the crystal was translated after measuring each frame, to expose fresh part of the sample to the beam. With this approach, the total dose was distributed evenly throughout the entire diffracting volume mapped out with the raster scan, allowing higher intensity beam to be used and yet minimize the radiation damage yielding good quality diffraction data up to 1.46 Å resolution [42].

3.6 Example 6: Small Beam Enables Data Collection from a Sample with Multiple Challenges

The benefits of micro-diffraction are not limited to small crystals or crystals with specific shapes. Often large crystals require microbeams and several advantages, offered by micro-diffraction, must be combined to enable useful data collection. In the case of large (>200–300 μm) crystals of the HOIP/E2 ~ ubiquitin complex, using a large beam always led to badly smeared diffraction spots, often from multiple lattices, rendering the crystals unusable. An added challenge for data collection was a more stringent requirement for data quality, and the need to minimize the radiation damage, since the structure solution with MR-SAD had to employ anomalous signal from Zn atoms [43]. Moreover, the crystals diffracted no further than 3.4–3.5 Å, making successful

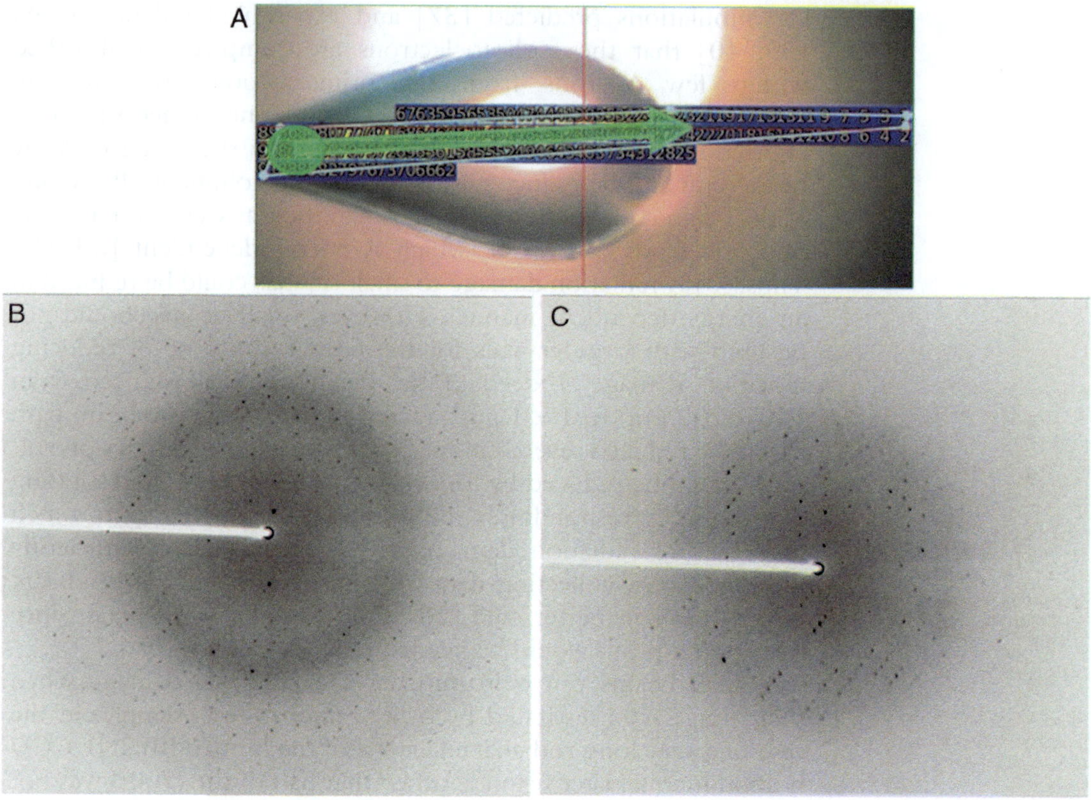

Fig. 7 Small, 20-μm beam, in combination with diffraction raster scans and vector/helical data collection helped in maximizing data quality. (**a**) Diffraction raster scan results from a long, rod-shaped crystal. *Green arrow* indicates the vector along which the crystal will be translated as it rotates during data collection. The *green circle* corresponds to 20 μm X-ray beam. (**b**) High and (**c**) low quality diffraction, corresponding to the left and right halves of the crystal

phasing and subsequent model building challenging. Therefore, maximizing the diffraction resolution was high priority despite the concern that it could lead to increased radiation damage which, in turn, could degrade the anomalous signal. Data were collected by combining several approaches described above. When probed with a 20 μm beam, most of the crystal produced poor quality diffraction (Fig. 8). Raster scans revealed that one edge was significantly better ordered than the rest of the crystal. A small beam had to be used to improve the overall spot shapes and spatial resolution. Because the smaller beam intercepted a correspondingly smaller volume of the crystal, the beam intensity had to be increased to attain the maximum possible resolution of diffraction to aid in model building. To avoid excessive radiation damage that would corrupt the anomalous signal from Zn, data collection was set up along a vector, defined with the help of raster scans. Data collection started from the orientation of the crystal producing the best quality diffraction but as the crystal rotated, the poorly diffracting parts rotated into the beam and

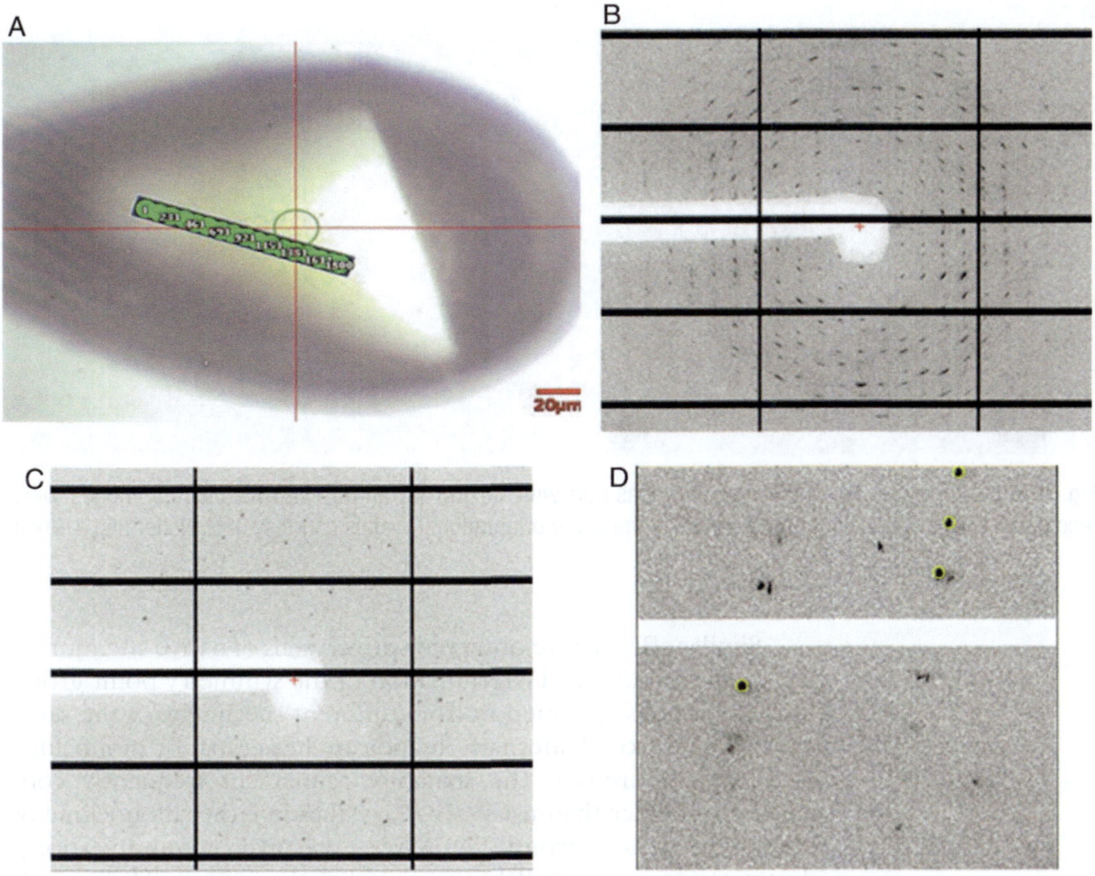

Fig. 8 Data collection from a small part of a large crystal (**a**) with very poor overall diffraction (**b**). The *green bar* on the crystal indicates where the data were collected from. (**c**) Better diffraction in the orientation at the beginning of data collection. (**d**) As crystal rotated during data collection, multiple lattices and poorly shaped spots rotated into the beam. However, they could be resolved and integrated when small beam was used

contaminated the overall data. Nevertheless, the data could be indexed using the better diffraction from the start of data collection, and all data could be integrated with the lattice parameters and orientation matrix refined from the start.

3.7 Example 7: Small Beams as Diagnostic Tools

Small beams can be useful as diagnostic tools as well. For example, it was observed that some batches of ribosome crystals diffracted well, but the structure refinement did not converge to the expected R and R_{free} values (J. Zhou, personal communication). The reason for this behavior became clear when diffraction with large and small X-ray beams were compared (Fig. 9). Diffraction with the larger, 50 μm beam (beam size typically used in this project) did not show any reason for concern. However, when the same spot of the crystal was probed with a 10 μm beam, it became apparent that there were two crystals in the beam whose diffraction overlapped almost entirely when data were collected with larger beams.

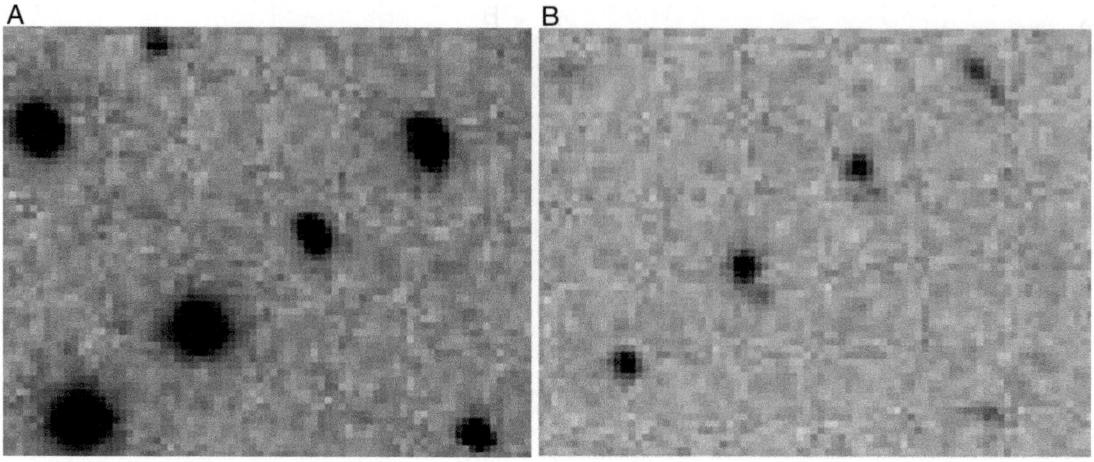

Fig. 9 Diffraction from ribosome crystals measured with 50 μm (**a**) and 10 μm (**b**) beams. Images were recorded from the same spot of the crystal in the same orientation. Crystals courtesy of H. Noller and J. Zhou

Similar effects were observed with crystals of native and mutant HIV capsid proteins. Determination of the symmetry point group was ambiguous. Diffraction from different specimens of the same crystal form would alternatively indicate hexagonal or monoclinic symmetry. Moreover, the structure refinement frequently converged to higher than usual R/R_{free} values in either monoclinic or hexagonal space group (A. Gres, personal communication). Examination with small beams revealed the reason. The crystals which grew as stack of plates produced diffraction which could be easily indexed in hexagonal point group in some orientations. However, in other orientations the diffraction spots, which were elongated, could be indexed only in monoclinic group. With small X-ray beams, some of these elongated spots separated into two, revealing that the beam was intercepting two crystals, whose diffraction overlapped in some orientations, distorting the hexagonal symmetry and corrupting some of the intensities leading to high R/R_{free} values with a seemingly correct model.

4 Recent and Near Future Developments

Several dedicated beamlines with micro-diffraction capabilities are now operating worldwide (Table 1). Perhaps an even better indicator of the popularity of micro-diffraction is its share on beamlines which offer both small and large beams with seamless, user-friendly transition between them. For example, at beamlines 23ID-D and 23ID-B at the Advanced Photon Source (APS), micro-diffraction (5, 10, and 20 μm beam) is used for approximately 65% of all data collection. This does not include the time spent on diffraction raster scans, for which micro-diffraction is used almost exclusively.

To date, crystals as small as 5 μm or less have been successfully used for data collection [32, 44].

It has been demonstrated that the sulfur SAD phasing can be successful even with poorly diffracting crystals by collecting data from a large number of samples to achieve extremely high multiplicity [45]. Micro-diffraction made this approach feasible even for small crystals [46]. To make this method more robust and user-friendly, sophisticated approaches to both data collection and processing are required. When only partial data can be measured from each sample, it is essential that the data from every new crystal is complementary to an already existing set. Otherwise the resulting overall data may be redundant in some segments of Ewald sphere but incomplete in others. Multi-crystal strategy calculations, allowing complementary data collection from each new crystal, have been successfully implemented [47, 48]. In principle, the expected radiation damage should be calculated, for example with RADDOSE [49, 50], in order to estimate how much data could be collected from each small crystal. Alternatively, radiation damage can be evaluated empirically for a representative crystal of the set. When data are collected from multiple crystals, some of the samples may not be isomorphous with the rest. These crystals need to be identified in almost real time, or soon enough to rectify the problem by collecting more data form more crystals before leaving the data collection facility. Software tools have been developed to make such monitoring easier [51–53].

There are interesting developments in non-mechanical sample handling, including optical tweezers [54] and acoustic droplet ejection [55–57]. Since these methods are primarily optimized for smaller crystals, micro-diffraction will be an indispensable tool enabling their integration in data collection facilities.

One particularly successful use of small crystals, in so-called serial crystallography, has been on X-ray Free Electron Laser (XFEL) sources. However, the supply of XFEL beamtime lags far behind the demand. Recent developments in synchrotron source technology may have a significant impact on serial crystallography of micrometer sized crystals. Currently, most synchrotron source storage rings are based on the double bend achromat lattice [58] which provide X-ray brightness up to 10^{20} photons/s/mm^2/mrad^2/0.1% bandwidth. Next generation, multi-bend achromat (MBA) storage rings [59] can provide a factor of 100 or more increase in brightness. The MBA lattice will allow the X-ray beam to be focused to a submicrometer, circular cross section with several orders of magnitude increase in intensity compared to beams available today. This can make serial crystallography on synchrotron sources an attractive complement to XFEL sources. The first of the MBA rings (MAX IV, Sweden) is operational and X-ray beamlines are being commissioned. A second ring is under construction (Sirius, Brazil). Over the next 5–10 years many storage rings will be rebuilt with an MBA lattice.

5 Conclusions

After the first proof-of-principle experiments at the end of 1990s, micro-diffraction has become highly successful in last few years. It can serve as an indispensable tool for structural biologists when crystal samples are very small; when larger samples are inhomogeneous; when diffraction spot shapes are poor; when multiple-lattice diffraction cannot be avoided; or when crystals cannot be identified or centered with optical methods alone.

Responding to the increasing demand, a number of synchrotron beamlines worldwide have implemented micro-diffraction capabilities or have been completely dedicated to it. Micro-diffraction, by enabling a more sophisticated approach to the design of diffraction experiments and by improving signal-to-noise ratio, has aided in data collection from technically challenging samples which would have been otherwise discarded.

Cutting edge structural biology has entered the realm where for more and more projects growing large and/or homogeneous crystals proved unfeasible. Micro-diffraction, used successfully in data collection from such crystals, has proved to be a valuable tool, and enabled new and important science.

Micro-diffraction is poised to become more prevalent with the advent of high brilliance MBA lattice synchrotron sources. The feasibility of serial crystallography with small beams at synchrotron sources has been demonstrated. In combination with the MBA storage rings, it has a potential to become a valuable tool for a larger number of researchers than can currently access the XFEL sources.

References

1. Smith JL, Fischetti RF, Yamamoto M (2012) Micro-crystallography comes of age. Curr Opin Struct Biol 22:602–612
2. Riekel C (2004) Recent developments in micro-diffraction on protein crystals. J Synchrotron Radiat 11:4–6
3. Cusack S, Belrhali H, Bram A et al (1998) Small is beautiful: protein micro-crystallography. Nat Struct Biol 5(Suppl): 634–637
4. Perrakis A, Cipriani F, Castagna JC et al (1999) Protein microcrystals and the design of a microdiffractometer: current experience and plans at EMBL and ESRF/ID13. Acta Crystallogr D Biol Crystallogr 55:1765–1770
5. Evans G, Alianelli L, Burt M et al (2007) Diamond beamline I24: a flexible instrument for macromolecular micro-crystallography. Synchrotron Radiat Instrum 879:836–839
6. Igarashi N, Ikuta K, Miyoshi T et al (2008) X-ray beam stabilization at BL-17A, the protein microcrystallography beamline of the photon factory. J Synchrotron Radiat 15: 292–295
7. Fischetti RF, Xu S, Yoder DW et al (2009) Mini-beam collimator enables microcrystallography experiments on standard beamlines. J Synchrotron Radiat 16:217–225
8. Flot D, Mairs T, Giraud T et al (2010) The ID23-2 structural biology microfocus beamline at the ESRF. J Synchrotron Radiat 17:107–118
9. Yamamoto M, Hirata K, Hikima T et al (2010) Protein micro-crystallography with a new micro-beam beamline. Yakugaku Zasshi 130: 641–648
10. Kawano Y, Shimizu N, Baba S et al (2009) Present status of SPring-8 macromolecular crystallography beamlines. In: Sri 2009: the 10th international conference on synchrotron radiation instrumentation, vol 1234, pp 359–362

11. de Sanctis D, Beteva A, Caserotto H et al (2012) ID29: a high-intensity highly automated ESRF beamline for macromolecular crystallography experiments exploiting anomalous scattering. J Synchrotron Radiat 19:455–461
12. Nelson R, Sawaya MR, Balbirnie M et al (2005) Structure of the cross-beta spine of amyloid-like fibrils. Nature 435:773–778
13. Rasmussen SG, Choi HJ, Rosenbaum DM et al (2007) Crystal structure of the human beta2 adrenergic G-protein-coupled receptor. Nature 450:383–387
14. Cherezov V, Rosenbaum DM, Hanson MA et al (2007) High-resolution crystal structure of an engineered human beta2-adrenergic G protein-coupled receptor. Science 318:1258–1265
15. Coulibaly F, Chiu E, Ikeda K et al (2007) The molecular organization of cypovirus polyhedra. Nature 446:97–101
16. Warne T, Serrano-Vega MJ, Baker JG et al (2008) Structure of a beta1-adrenergic G-protein-coupled receptor. Nature 454:486–491
17. Rasmussen SG, DeVree BT, Zou Y et al (2011) Crystal structure of the beta2 adrenergic receptor-Gs protein complex. Nature 477:549–555
18. Rosenbaum DM, Zhang C, Lyons JA et al (2011) Structure and function of an irreversible agonist-β(2) adrenoceptor complex. Nature 469:236–240
19. Hirano Y, Takeda K, Miki K (2016) Charge-density analysis of an iron-sulfur protein at an ultra-high resolution of 0.48 Å. Nature 534:281–284
20. Sanishvili R, Nagarajan V, Yoder D et al (2008) A 7 μm mini-beam improves diffraction data from small or imperfect crystals of macromolecules. Acta Crystallogr D Biol Crystallogr 64:425–435
21. Dauter Z (1999) Data-collection strategies. Acta Crystallogr D Biol Crystallogr 55: 1703–1717
22. Cipriani F, Felisaz F, Lavault B et al (2007) Quickly getting the best data from your macromolecular crystals with a new generation of beamline instruments. Synchrotron Radiat Instrum 879:1928–1931
23. Evans G, Axford D, Waterman D et al (2011) Macromolecular microcrystallography. Crystallogr Rev 17:105–142
24. Davis MF, Groter C, Kay HF (1968) On choosing off-line automatic X-ray diffractometers. J Appl Crystallogr 1:209–217
25. Fuchs MR, Pradervand C, Thominet V et al (2014) D3, the new diffractometer for the macromolecular crystallography beamlines of the Swiss light source. J Synchrotron Radiat 21:340–351
26. Evans G, Axford D, Owen RL (2011) The design of macromolecular crystallography diffraction experiments. Acta Crystallogr D Biol Crystallogr 67:261–270
27. Cherezov V, Hanson MA, Griffith MT et al (2009) Rastering strategy for screening and centring of microcrystal samples of human membrane proteins with a sub-10 microm size X-ray synchrotron beam. J R Soc Interface 6(Suppl 5):S587–S597
28. Hilgart MC, Sanishvili R, Ogata CM et al (2011) Automated sample-scanning methods for radiation damage mitigation and diffraction-based centering of macromolecular crystals. J Synchrotron Radiat 18:717–722
29. Aishima J, Owen RL, Axford D et al (2010) High-speed crystal detection and characterization using a fast-readout detector. Acta Crystallogr D Biol Crystallogr 66:1032–1035
30. Riekel C (2000) New avenues in X-ray microbeam experiments. Rep Prog Phys 63:233–262
31. Song J, Mathew D, Jacob SA et al (2007) Diffraction-based automated crystal centering. J Synchrotron Radiat 14:191–195
32. Zander U, Bourenkov G, Popov AN et al (2015) MeshAndCollect: an automated multi-crystal data-collection workflow for synchrotron macromolecular crystallography beamlines. Acta Crystallogr D Biol Crystallogr 71:2328–2343
33. Bowler MW, Guijarro M, Petitdemange S et al (2010) Diffraction cartography: applying microbeams to macromolecular crystallography sample evaluation and data collection. Acta Crystallogr D Biol Crystallogr 66:855–864
34. Lee H, Le HV, Wu R et al (2015) Mechanism of inactivation of GABA aminotransferase by (E)- and (Z)-(1S,3S)-3-amino-4-fluoromethylenyl-1-cyclopentanoic acid. ACS Chem Biol 10: 2087–2098
35. Gehret JJ, Gu L, Gerwick WH et al (2011) Terminal alkene formation by the thioesterase of curacin a biosynthesis: structure of a decarboxylating thioesterase. J Biol Chem 286:14445–14454
36. Teng TY, Moffat K (2000) Primary radiation damage of protein crystals by an intense synchrotron X-ray beam. J Synchrotron Radiat 7:313–317
37. Nave C, Hill MA (2005) Will reduced radiation damage occur with very small crystals? J Synchrotron Radiat 12:299–303
38. Garman EF (2010) Radiation damage in macromolecular crystallography: what is it and why should we care? Acta Crystallogr D Biol Crystallogr 66:339–351

39. Sanishvili R, Yoder DW, Pothineni SB et al (2011) Radiation damage in protein crystals is reduced with a micron-sized X-ray beam. Proc Natl Acad Sci U S A 108:6127–6132
40. Finfrock YZ, Stern EA, Yacoby Y et al (2010) Spatial dependence and mitigation of radiation damage by a line-focus mini-beam. Acta Crystallogr D Biol Crystallogr 66:1287–1294
41. Cowan JA, Nave C (2008) The optimum conditions to collect X-ray data from very small samples. J Synchrotron Radiat 15:458–462
42. Mei Y, Su M, Sanishvili R et al (2016) Identification of BECN1 and ATG14 coiled-coil interface residues important for starvation-induced autophagy. Biochemistry 55:4239–4253
43. Lechtenberg BC, Rajput A, Sanishvili R et al (2016) Structure of a HOIP/E2~ubiquitin complex reveals RBR E3 ligase mechanism and regulation. Nature 529:546–550
44. Axford D, Ji X, Stuart DI et al (2014) In cellulo structure determination of a novel cypovirus polyhedrin. Acta Crystallogr D Biol Crystallogr 70:1435–1441
45. Liu Q, Dahmane T, Zhang Z et al (2012) Structures from anomalous diffraction of native biological macromolecules. Science 336:1033–1037
46. Akey DL, Brown WC, Konwerski JR et al (2014) Use of massively multiple merged data for low-resolution S-SAD phasing and refinement of flavivirus NS1. Acta Crystallogr D Biol Crystallogr 70:2719–2729
47. Ravelli RBG, Sweet RM, Skinner JM et al (1997) STRATEGY: a program to optimize the starting spindle angle and scan range for X-ray data collection. J Appl Crystallogr 30:551–554
48. Pothineni SB, Venugopalan N, Ogata CM et al (2014) Tightly integrated single- and multi-crystal data collection strategy calculation and parallelized data processing in JBluIce beamline control system. J Appl Crystallogr 47:1992–1999
49. Paithankar KS, Garman EF (2010) Know your dose: RADDOSE. Acta Crystallogr D Biol Crystallogr 66:381–388
50. Paithankar KS, Owen RL, Garman EF (2009) Absorbed dose calculations for macromolecular crystals: improvements to RADDOSE. J Synchrotron Radiat 16:152–162
51. Foadi J, Aller P, Alguel Y et al (2013) Clustering procedures for the optimal selection of data sets from multiple crystals in macromolecular crystallography. Acta Crystallogr D Biol Crystallogr 69:1617–1632
52. Terwilliger TC, Bunkoczi G, Hung LW et al (2016) Can I solve my structure by SAD phasing? Planning an experiment, scaling data and evaluating the useful anomalous correlation and anomalous signal. Acta Crystallogr D Biol Crystallogr 72:359–374
53. Akey DL, Terwilliger TC, Smith JL (2016) Efficient merging of data from multiple samples for determination of anomalous substructure. Acta Crystallogr D Biol Crystallogr 72:296–302
54. Wagner A, Duman R, Stevens B et al (2013) Microcrystal manipulation with laser tweezers. Acta Crystallogr D Biol Crystallogr 69:1297–1302
55. Cole K, Roessler CG, Mule EA et al (2014) A linear relationship between crystal size and fragment binding time observed crystallographically: implications for fragment library screening using acoustic droplet ejection. PLoS One 9:e101036
56. Teplitsky E, Joshi K, Ericson DL et al (2015) High throughput screening using acoustic droplet ejection to combine protein crystals and chemical libraries on crystallization plates at high density. J Struct Biol 191:49–58
57. Soares AS, Engel MA, Stearns R et al (2011) Acoustically mounted microcrystals yield high-resolution X-ray structures. Biochemistry 50:4399–4401
58. Fakhri AA, Kant P, Singh G et al (2015) An analytical study of double bend achromat lattice. Rev Sci Instrum 86:033304
59. Einfeld D, Plesko M, Schaper J (2014) First multi-bend achromat lattice consideration. J Synchrotron Radiat 21:856–861

Chapter 10

Serial Synchrotron X-Ray Crystallography (SSX)

Kay Diederichs and Meitian Wang

Abstract

Prompted by methodological advances in measurements with X-ray free electron lasers, it was realized in the last two years that traditional (or conventional) methods for data collection from crystals of macromolecular specimens can be complemented by synchrotron measurements on microcrystals that would individually not suffice for a complete data set. Measuring, processing, and merging many partial data sets of this kind requires new techniques which have since been implemented at several third-generation synchrotron facilities, and are described here. Among these, we particularly focus on the possibility of in situ measurements combined with in meso crystal preparations and data analysis with the XDS package and auxiliary programs.

Key words Serial synchrotron crystallography (SSX), Microcrystal, Lipidic cubic phase (LCP), In meso in situ, Room temperature (RT), Cryogenic temperature, Data collection, Data quality, Merging, XDS, XSCALE

1 Introduction

Macromolecular crystallography (MX) has been constantly evolving since the very first X-ray structure determinations of protein molecules in the 1950s and 1960s. Nowadays X-ray crystal structures of biological macromolecules are determined at an unprecedented speed; this year, about one structure is deposited every hour in the Protein Data Bank (PDB). This is due in large part to developments in molecular biology, crystallization, data collection and processing and structure solution, as well as to advances in synchrotron radiation technology. However, the basic method of X-ray diffraction data collection remains unchanged; almost exclusively, diffraction data are collected on a single crystal entity with the rotation method [1] using a monochromatic X-ray beam. In this experiment, the most important measured quantity is the integrated intensity of a reflection, which is given by Darwin's formula [2, 3]:

$$I_{hkl} = I_0 \, r_e^2 \frac{V_{xtal}}{V_{cell}^2} \frac{\lambda^3}{\omega} LPA \left| F_{hkl} \right|^2 \qquad (1)$$

Alexander Wlodawer et al. (eds.), *Protein Crystallography: Methods and Protocols*, Methods in Molecular Biology, vol. 1607, DOI 10.1007/978-1-4939-7000-1_10, © Springer Science+Business Media LLC 2017

where I_{hkl} is the integrated intensity of reflection *hkl*, and is proportional to the square of the structure factor (F_{hkl}), I_0 is the intensity of the incident X-ray beam, r_e is the classic electron radius, λ is the X-ray wavelength, ω is the angular velocity of the crystal, *L* is the Lorentz factor, *P* is the polarization factor, and *A* is the X-ray transmission. For our purpose, we can leave various correction factors and constants out and assume that the squared structure factor is proportional to the content of the unit cell. Then Eq. 1 can be written as [4]:

$$I_{hkl} \sim I_0 \frac{V_{xtal}}{V_{cell}} \tag{2}$$

In Eq. 2, the measured intensity is proportional to both the intensity of the incident beam (I_0) and the diffraction volume illuminated by X-rays (V_{xtal}) and is inversely proportional to the unit cell volume (V_{cell}). This means, simply, that the intensities of the measured reflections decrease as the crystal size gets smaller, or as the unit cell size gets larger. Another important aspect in diffraction data collection is radiation damage, which limits the maximum obtainable data resolution for a given diffraction volume. This has a significant consequence: there is a lower limit to the size of crystals one can reliably extract diffraction data from, prior to the onset of radiation damage [4].

In the pioneering work of macromolecular crystallography in the 1950s and 1960s, experiments were carried out with well diffracting large crystals at room temperature [5, 6]. Good crystalline order and large diffraction volume allowed acquiring diffraction data with sufficient accuracy at room temperature. In the 1970s and 1980s, the structural study of viruses presented new challenges in diffraction data collection. Because of the very large unit cell (one or two magnitudes larger than for average proteins), the intensity of reflections decreases accordingly (Eq. 2). Only one or two diffraction patterns could be obtained from each crystal and the complete data set had to be assembled from many crystals [7]. This could be considered as the very first serial crystallography (SX) work although the term SX did not exist at that time. From the 1990s to 2000s, synchrotron radiation started to play an important role in modern macromolecular crystallography (http://biosync.sbkb.org). The brightness and energy tunability of synchrotron radiation enabled study of crystals smaller than before and established experimental phasing with anomalous scattering [8]. Around that time, cryogenic cooling methods were also being developed that allowed data collection with greatly reduced radiation damage [9]. Since then, collecting complete data sets from a single crystal became the method of choice in MX (referred to as "conventional crystallography (CX)" here) and it works well for crystals with diffraction volume of around 10,000 μm^3 (about

$20 \times 20 \times 20\ \mu m^3$) and larger [10]. From the 2000s, advances in crystallization methods enabled crystallization of challenging targets, such as large multi-protein complexes and membrane proteins, which often only yield micro-crystals with the largest dimension below 20 μm and the smallest dimension below 5 μm. This is about one order of magnitude smaller in diffraction volume compared with crystals used in CX. Therefore, radiation damage prevents collecting complete data sets to high resolution from those crystals. At the third generation synchrotron sources using microfocused X-rays with beam size comparable to the crystal size, partial data sets could be collected from these microcrystals, which are then merged together to form a full data set. This method is referred as microcrystallography [11, 12].

The first hard X-ray free electron laser (XFEL) opened new avenues for crystallographic data collection in 2009 [13]. The extremely high intensity and femtosecond pulses of XFELs are not suited for the rotation method, because the X-ray pulse would destroy the crystal almost immediately. However, during the femtosecond pulse, the diffraction pattern resulting from the pulse can be collected prior to crystal damage, generating a still image—the so-called "diffraction before destruction" method. Hundreds to many thousands of such still images on as many isomorphous crystal entities can be merged together to generate a full data set. This method, termed serial femtosecond crystallography (SFX), has the advantage that even the smallest microcrystals, and potentially nanometer-sized crystals, can give useful diffraction data with femtosecond XFEL pulses. Since then, sample preparation and delivery, data collection and processing methods have been actively developed to facilitate SFX experiments [14] and they have inspired recent development of SX at synchrotron sources. Although the serial data collection approaches have been used in virus crystallography and microcrystallography before, the traditional sample mounting methods are simply unfeasible when thousands of crystal samples need to be measured to get a sufficiently complete data set. The new high-throughput sample delivery methods developed for SFX have enabled screening numerous crystals and collecting their diffraction data with synchrotron radiation at an unprecedented speed. This new data collection method is named serial synchrotron crystallography (SSX), which is emerging as complementary method to CX.

The diffraction principles of both methods are the same but SSX departs from CX in many aspects, ranging from sample preparation and sample delivery to data collection and processing. Below we document the experimental and computational procedures for SSX that were devised and applied at synchrotron facilities (Fig. 1). Although the method is still under extensive development, it has already delivered data with high resolution comparable to CX data collection methods and with quality sufficient for experimental phasing.

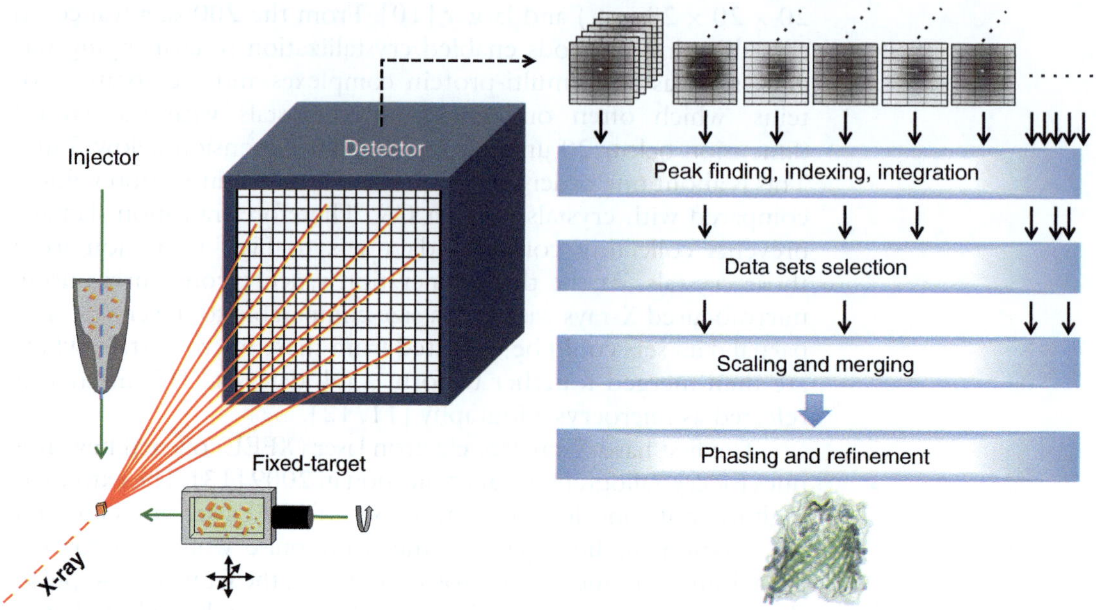

Fig. 1 Serial synchrotron crystallography (SSX): from crystals to structures

2 Sample Preparation and Delivery

In CX, single crystals are treated with cryoprotectant, harvested in loops and snap-cooled in liquid or gaseous nitrogen. Each crystal is then screened for its X-ray diffraction and the best ones are used for the final data collection. This method is effective when individual crystals have enough diffraction volume to yield a complete data set. However, it is time-consuming and cumbersome in microcrystallography and is certainly not compatible with serial crystallography, in which hundreds to thousands of crystals need to be investigated. Therefore SX calls for methods to prepare crystals in a sufficient quantity and deliver them serially into the X-ray beam in a high-throughput manner. In CX crystallization is optimized for the growth of large crystals [15]. In contrast, the demand for large numbers of small crystals in SX requires alternative sample preparation and delivery methods [16]. A variety of sample delivery systems for SX have been developed and tested at synchrotron beamlines in recent years. They can be broadly grouped into two classes—injector methods and fixed-target methods. We review them together with related crystallization developments in this section.

2.1 Injector Methods

The injector methods were originally developed for SFX application at XFELs. They come in two main variants—the gas dynamic virtual nozzle (GDVN) injector and the lipid cubic phase (LCP)

injector (and a closely related version called high viscosity extrusion (HVE) injector) [17–19]. The GDVN injector uses gas focusing to generate a liquid stream of a few micrometers in diameter and was used in the first SFX experiments with micron-sized crystals of photosystem I (0.2–2 μm) [13], lysozyme ($1 \times 1 \times 3\ \mu m^3$) [20], and cathepsin B ($0.9 \times 0.9 \times 11\ \mu m^3$) [16] at the Linear Coherent Light Source (LCLS). The GDVN injector enabled measurement of high-resolution diffraction images from nanometer to micrometer sized crystals with minimum diffraction background while still keeping the crystals fully hydrated. However, sample consumption is very high with GDVN (10–100 mg per data set) and the high flow rate of 10 m/s makes the crystal-X-ray interaction time too short to measure sufficient diffraction data with synchrotron radiation. Based on the GDVN concept, an electrospinning injector was designed with a tenfold reduction in flow rate [21]. However, the electrostatic charging from the electrospinning method may affect crystals. The LCP and HVE injectors extrude a continuous 20–50 μm diameter stream at a much slower velocity of 0.1–0.3 mm/s, which is more suitable for data collection with micro-focused X-rays at synchrotron sources. In addition to LCP and other meso phases with membrane protein crystals, the method has been extended to other high viscosity media such as grease, Vaseline, and agarose, which could be used as carrier media for soluble proteins [19, 22, 23]. The LCP injector reduces sample consumption dramatically (50- to 100-fold) compared to the GDVN injector and has enabled structure determination with SFX methods of several membrane proteins and complexes (β-adrenergic receptor [24], opioid receptor [25], angiotensin receptor [26], and rhodopsin-arrestin complex [27]). In these studies, crystals were grown in LCP and their size varied from sub-ten to a few tens of micrometers and the average protein consumption is less than 0.5 mg per data set.

Serial crystallography with LCP/HVE injectors has been explored in synchrotron sources recently. Using lysozyme as a model system, Botha et al. reported the high-resolution structure and feasibility of experimental phasing from SSX methods using an HVE injector at the PXII beamline at the Swiss Light Source (SLS) [19]. Lysozyme was crystallized with a standard protocol and crystals were introduced into LCP and vaseline by gentle mixing. The average crystal size was $10 \times 10 \times 30\ \mu m^3$ and the diameter of the LCP/Vaseline stream was 50 μm. Around the same time, Nogly et al. obtained the first SSX membrane protein structure with the LCP injector at beamline ID13 at the European Synchrotron Radiation Facility (ESRF) [28]. The test protein was bacteriorhodopsin (bR) and crystals were grown in LCP with average crystal size of $5 \times 30 \times 30\ \mu m^3$. The diameter of the LCP stream was about 50 μm. Typically 10–20 μl samples were loaded into the injector per experiment. The sample consumption is about a few hundreds micrograms of protein per data set.

The injector is essentially a container-free sample delivery method with low diffraction background and the data acquisition is simple and high-throughput. However, the method comes with its own challenges: (1) optimizing the crystallization to generate a large number of relatively homogenous crystals, whose size matches the X-ray beam and whose concentration is optimal with the flow rate of the injector stream; (2) controlling flow dynamics to minimize crystal movement when it passes through the X-ray beam; (3) processing and merging of still diffraction images.

Related to the injector method, thin-walled glass capillaries (10 μm walls and 100 μm inner-diameter) have been used as a container to accommodate a flow of crystal suspension for SSX experiments at beamline P11 at PETRAIII [29]. The combination of low flow velocity (5 mm/s) and scanning of capillary allows SSX data collection with micro-crystals in aqueous media. Lysozyme crystals of 3–6 μm in size were used as test protein. One drawback of the capillary method is the additional background scattering from capillary walls and crystallization solution inside the capillary. Another related approach is acoustic droplet ejection (ADE) [30]. Instead of delivering samples continuously, ADE offers "drop-on-demand" to eject samples only when needed. The sample consumption can be greatly reduced and the data collection can be carried out both at room and cryogenic temperatures [31, 32]. In addition to injecting droplets, acoustic force can levitate droplets in the X-ray interaction region [33]. In this case, the crystals are rotating inside the droplet and a diffraction movie is recorded with a fast frame-rate X-ray detector to trace out the orientation change.

2.2 Fixed-Target Methods (Deposition and In Situ)

Fixed-target methods were developed as an alternative approach to injector methods to improve sample hit-rate and reduce sample consumption in SFX experiments. In principle, the hit-rate could reach 100% and sample consumption could be reduced to micrograms The first demonstration was carried out with rapid encystment protein (REP24) crystals of 5 × 10 × 30 μm^3 in size deposited on silicon nitride (Si_3N_4) membranes (of 50 nm thickness) and protected by Paratone-N for room temperature measurement under vacuum at the LCLS [34]. Soon after, synchrotron goniometer-based data collection methods have been extended to SFX with crystals either positioned inside a grid or mounted with a loop or a mesh [35, 36].

At synchrotron MX beamlines, the fixed-target and related systems offer full control of the data collection and could be made compatible for measurement at both room and cryogenic temperatures. Most established data collection and processing methods for CX are readily adapted. Therefore, SSX with various fixed-target systems has been actively pursued recently at third generation synchrotron facilities as an extension to the established synchrotron micro-crystallography.

One main challenge in the fixed-target methods is to reduce X-ray background scattering from support materials and crystallization media around the crystals. If thin film materials are used to hold crystals, they should ideally be watertight, optical- and UV-transmitting, and non-birefringent. Various low X-ray background materials have been examined, such as polymer films like cyclic olefin copolymer (COC), PDMS, and PMMA in micrometer thickness, and Si_3N_4 and Si membranes in nanometer thickness. As one of the thinnest materials possible, graphene as crystal container has been successfully demonstrated recently [37–39]. Crystals can be either deposited on a chip or a grid or simply a conventional loop or mesh for cryo-crystallography or grown in situ in a crystallization plate designed with low diffraction background.

2.2.1 Deposition Methods

The deposition method is a two-step process. Crystals are grown by conventional crystallization methods first, and then transferred to the fixed-target supports for diffraction measurement. In order to expedite the serial measurement, chips with features promoting self-assembly and self-localization of crystals and grids with tailored hole sizes have been developed to allow automated data collection with predefined crystal locations. A chip with a Si mesh and polyimide film has been designed to position crystals in the prescribed locations with random orientations by exploiting the liquid-pinning potential and surface roughness [40]. With lysozyme (rod-shape, 5–50 μm) and ferritin (block-shape) as model systems, diffraction data were obtained at room temperature at beamline PXII at the SLS. To assist loading microcrystals, a micropatterned chip was developed. The chip features 150 nm thick Si_3N_4 windows to reduce X-ray background scattering and an alternating hydrophobic and hydrophilic surface pattern to assist localizing crystals in defined regions [41]. Still diffraction images were collected with lysozyme crystals of 10–50 μm in size at GM/CA beamline at the Advanced Photon Source (APS). A single-crystalline Si chip with micropores was developed to minimize background scattering [42]. The chip is made from single-crystalline Si with an active area of 1.5×1.5 mm^2 and 10 μm thickness. The size (typically a few microns), shape and pattern of micropores are controllable by micro-fabrication. In principle thousands of crystals can be loaded on a single chip. The idea is that extra crystallization solution, which contributes to background scattering, is wicked away through the micropores, while the crystals that are larger than the pores are retained. The compact format of the chip is cryo-compatible and the whole chip can be snap-cooled in liquid nitrogen. Rotation diffraction data were collected with microcrystals of CPV18 polyhedrin of size 4 μm or smaller at beamline I24 at the Diamond Light Source (DLS). In a different approach to increase hit-rate and automate serial data collection, a microfluidic chip with an array of traps was designed to collect crystals at predefined locations and semi-still diffraction images (0.02°) were collected at beamline 12-2 at the Stanford Synchrotron Radiation Lightsource (SSRL) [43]. To facilitate SX at standard MX beamlines, a compact sample-mounting grid made with polycarbonate plastic with 75 holes of 400, 200, and 125 μm in diameter was designed at the SSRL [35, 44]. The crystals can be loaded either manually or automatically with liquid-handling robots. The grid fits into the magnetic base and puck used in standard cryo-crystallography. The grids can be mounted by an automatic sample changer as for standard loop samples and the diffraction data can be collected at cryogenic temperature with cryojet cooling in the co-axial configuration.

Instead of depositing crystals in predefined locations/regions, Coquelle et al. loaded lysozyme crystals between two Si_3N_4 windows and used the continuous rastering method to collect still diffraction images from the entire Si_3N_4 assembly [45]. The crystal mounting loops and meshes commonly used in cryo-crystallography could be adapted for SSX experiments. In fact, the micro-mesh has been used to mount micro-crystals for collecting diffraction data serially for more than a decade [46–51]. Typically, a suspension containing many microcrystals is loaded on the mesh and extra solution is removed before snap-cooling. The crystals are located either visually with an on-axis microscope at the beamline or with diffraction scanning. Recently this approach has been extended to allow continuous data collection by helical scanning of a loop loaded with numerous micro-crystals. With this method, the structure of cathepsin B to 3.0 Å resolution has been determined from needle-shape crystals with diameter less than 1 μm at beamline P11 at PETRAIII [52].

2.2.2 *In Situ Methods*

The deposition process requires additional manipulation of crystals, which is time-consuming and may damage the crystal. It would be advantageous if X-ray data collection could be carried out directly with crystals in their crystallization compartment. This is the in situ method and it has many variants. Microfluidics is an attractive technology for crystallization because it allows fast screening and optimization of broad crystallization conditions with a minimum amount of protein [53]. With low X-ray scattering materials, crystallization and in situ X-ray data collection can be performed on a single microfluidic chip. Microbatch [54] and counter-diffusion [55, 56] crystallization methods have been implemented and the in situ X-ray diffraction experiments have been conducted with model proteins. The feasibility of experimental phasing with anomalous diffraction has been verified with in situ diffraction data from selenomethionine derivatized proteins, Yb derivatized crystals, and with a native protein [54, 56, 57].

Recently, microfluidic methods have also been successfully applied to in meso crystallization for membrane proteins [58]. To control crystal size and concentration, a kinetically optimized microfluidic chip was developed to crystallize proteins in emulsion droplets with one crystal per drop [59]. The microfluidic chips are suitable for SSX experiments. The room temperature SSX data from phosphonoacetate hydrolase (a soluble protein) and the photosynthetic reaction center (a membrane protein) have been collected at beamline LS-CAT at the APS [57, 58]. The SSX data from glucose isomerase has been measured at beamline F1 at the Cornell High Energy Synchrotron Source (CHESS) [59].

The 96-well crystallization plate with SBS format has been used for in situ diffraction screening at many MX beamlines [60–62]. For this purpose, 96-well crystallization plates with low background scattering were designed (CrystalQuickX and MiTeGen In-Situ-1). In addition to screening, complete data sets were obtained from both soluble and membrane proteins with in situ data collection [62–64]. The standard in situ SBS format plates still have relatively thick films. A thin polymer-film sandwich (TPFS) has been recently demonstrated with films as thin as 10 μm [65]. The whole TPFS plates can be used for in situ data collection at room temperature and each individual well can be cut out from the plate and placed under a cryojet stream for data collection at low temperatures. To bridge the in situ method and the conventional loop harvesting method, the CrystalDirect approach was developed to automate crystal harvesting directly from the crystallization plate through laser photoablation. This method allows controlled selection of crystals and removal of extra crystallization solution before harvesting [66, 67].

For membrane proteins, the lipid cubic phase (or in meso) crystallization is a very effective method [68]. The crystals grown by this method tend to be small and difficult to harvest due to the high viscosity of the mesophase. IMISX (in meso in situ serial crystallography) methods have been developed to allow efficient in situ serial data collection from microcrystals [69, 70]. The IMISX uses a double sandwich plate design where the crystallization takes place in the inner chamber made by two thin (25 μm) COC films separated by a spacer and the outer chamber consists of glass plates to avoid water loss and for easy handling and transportation. The IMISX plate can be set up either manually or robotically. Individual wells can be cut out from the plate and directly used for in situ data collection at room temperature or snap-cooled in liquid nitrogen for cryogenic data collection. The methods have been validated with several membrane proteins including enzymes, transporters, and receptors. High resolution structures have been obtained with crystals as small as 5 μm at beamline PXI at the SLS [70]. The IMISX methods are applicable for soluble proteins as well.

The recent development in IMISX and TPFS methods enabled in situ SSX for both membrane and soluble proteins at both room and cryogenic temperature. The ability to collect data at room temperature allows conformational space and dynamics of macromolecular molecules to be probed [71, 72]. At cryogenic temperatures with significantly reduced radiation damage, complete data sets can be obtained with fewer crystals, thus reducing sample consumption. It has been shown that nanogram to single-digit microgram quantities of protein can yield a high quality SSX data set. More importantly, cryogenic freezing allows us to prepare samples in advance, preserve and store them in their best state, and transport them for diffraction data collection when beamtime becomes available.

3 Data Collection

In the last decade, and primarily prompted by the advent of the PILATUS detector, crystallography has transitioned from the traditional "high dose" strategy (exposing the crystal to obtain as much signal as possible per diffraction image) to a "right dose" strategy, where the maximum attainable signal for a crystalline entity is more carefully determined to minimize the effects of radiation damage. In this section we document the boundary conditions applicable to data collection in general, and the peculiarities of partial data sets from small crystals.

3.1 Diffraction Signal and Noise

The primary goal of diffraction data collection is to obtain a complete data set with both high precision, as characterized by values of $CC_{1/2}$ and I/σ of the merged data overall, but particularly in the highest resolution shell, and high accuracy, i.e., with a minimum deviation of intensities from their true values. The precision of the data is essentially limited by a compromise between dose and the X-ray induced radiation damage, and their accuracy is often limited by the systematic errors in measurement, where—again—radiation damage is a large contributor. The strength of the diffraction signal is determined by both energy and flux of X-rays delivered on the crystal, and the intrinsic diffraction properties of the crystal, as discussed in depth by Holton and Frankel [4]. Briefly, the total diffracted signal is proportional to the diffraction volume and the average diffracted signal per reflection is inversely proportional to the unit cell volume (Eq. 2). Therefore, higher flux in the incident beam, or larger diffraction volume or both are needed for crystals with large unit cells. The background under Bragg peaks is from scattering of any material along the X-ray beam path, which could be a disordered portion of the crystal, material around the crystal (e.g., crystallization solution, lipid phase in in meso crystallization), sample support (e.g., mounting loop, films in fixed-target supports), and air. The background scattering has characteristic maxima around 3.6 and 4.5 Å for water and lipid cubic phase

(LCP), the two most common crystallization media, respectively. Reduction in X-ray background is essential for precise and accurate measurement of weak diffraction signals from micro-crystals.

The most effective approaches to improve signal-to-noise ratio in X-ray diffraction experiment with synchrotron radiation are to match X-ray beam size and crystal size and to reduce extra scattering materials in the X-ray path. Illuminating the whole diffraction volume enhances the diffraction signal with less absorbed X-ray dose, and reducing surrounding materials around the crystal minimizes the background scattering. In SSX with injector methods, one needs to find a good combination of X-ray beam size, crystal size and velocity of the stream to have most of the crystal illuminated by X-rays while it passes through. Ideally one should have a diameter of the injector stream not much bigger than the crystal size. In practice, a certain minimum stream size is needed to avoid clogging. In SSX with fixed-target methods, huge efforts have been made to develop systems with low X-ray scattering materials and less crystallization media around crystals. The goniometer or scanning stage, which holds the fixed-target samples, allows full control of crystal characterization and data collection. Crystals can be located with a grid scan, and rotation data can be collected for each selected crystals subsequently [50, 70]. Alternatively, the whole sample could be scanned either with still images or with oscillation images [45, 52].

3.2 Radiation Damage

Radiation damage limits the amount of diffraction data that can be obtained from a given diffraction volume. A protein crystal can stand doses of a few kGy at room temperature before significant loss of diffraction signal. The exact tolerable dose is case-specific and largely depends on diffusion processes of free radicals, which generate secondary damage [73]. Therefore, the damage is both dose and time (dose-rate) dependent. It has been suggested that at sufficiently high dose-rate, some "undamaged" diffraction data could be obtained before the diffusion processes start destroying crystallinity [45, 74]. At room temperature, the damage can spread well beyond the irradiated area and result in crystal deformation and cracking. In the case of cryogenically cooled crystals, the damage is only dose dependent and does not extend beyond a few μm from the irradiated area. More importantly, cryogenic cooling extends the tolerable dose to 20–30 MGy [75, 76], which is about two orders of magnitude higher compared with the dose limit at room temperature. This allows a useful amount of diffraction data to be measured from micron-sized crystals. For an average protein of several hundred amino acids, a $20 \times 20 \times 20\ \mu m^3$ crystal can yield a complete data set [10]. Therefore, it is safe to say that 3° of usable data to diffraction resolution can be obtained from one $5 \times 5 \times 5\ \mu m^3$ crystal prior to the onset of damage. If the average crystal size is known, the required number of crystals for a complete data set can be estimated. With further beamline optimization in reducing background scattering and improvement of X-ray detectors, more data can be obtained from even smaller crystals. It has been estimated that a crystal as small as 1.2 μm can produce a complete data set to 2 Å resolution under ideal experimental conditions [4].

3.3 X-Ray Beam and Detector for SSX

Recent advances in X-ray optics have made micrometer-sized X-ray beams routinely available at many MX beamlines at third generation synchrotron sources [12]. The next generation synchrotron technology promises even smaller X-ray beam with much reduced divergence and one to two orders of magnitude higher flux density. For challenging cases of weakly diffracting micro-crystals, the X-ray beam size and divergence could be tailored to increase I/σ by maximizing diffraction signal, minimizing background scattering, and reducing spot size on the detector [77]. Modern pixel array detectors are particularly suitable for the SSX experiment: the single photon sensitivity allows reliable measurement of weak diffraction signals; the small pixel size improves characterization of sharp reflections (common in room temperature and in situ measurements); large active area and high dynamic range allows accurate measurements of both strong and weak reflection spots; and high frame-rate with negligible deadtime enables continuous, shutterless data acquisition [78].

For injector methods, the "dynamics" of the interaction of X-rays and crystal can be monitored using X-ray detectors with fast frame-rate. Diffraction images with corrupted diffraction signal, due to crystal either moving out of X-ray beam or being damaged by X-rays, can be excluded in data processing. For fixed-target methods, when combined with X-ray microbeam, high flux and fast scanning stages, a fast detector allows crystal localization, diffraction characterization, and data collection in an automated workflow [50].

3.4 Data Collection Strategy

Various data collection strategies for single-crystal work with the rotation method (CX) have been used over the past decades and most of them were derived from work with image plate and CCD detectors, and influenced by the capabilities of data processing software at the time. The most common method used to be the collection of the minimum needed multiplicity in the minimum angular coverage [79, 80] with an accumulated X-ray dose below the 20 MGy limit. This method has been applied to micro-crystals in SSX as well. It is impossible to collect a complete data set from each crystal to its diffraction resolution due to its small diffraction volume. Typically intense and micro-focused X-rays are used to compensate for limited diffraction volume and to minimize background scattering, and a small wedge of data is collected until the 10–20 MGy limit is reached. The process is repeated for each crystal until the desired completeness and multiplicity are achieved

[50, 62, 63, 69, 70]. The data sets obtained in this way have sufficient quality for most molecular replacement and refinement calculations. However, the data quality may not be good enough for experimental phasing with weak anomalous scatterers, which demands data with higher accuracy and precision [81].

More recently it was realized that pixel array detectors (PADs), such as the PILATUS makes it possible to spread the tolerable X-ray dose over a larger total rotation range than the minimum required by space group symmetry and a particular crystal setting, without the penalty incurred by the readout noise of CCD detectors. This leads to high multiplicity data sets typically covering rotation ranges of 180–360°, and has the advantages of not requiring a strategy calculation, resulting in high completeness, allowing efficient scaling and outlier rejection, reducing systematic errors by averaging over their possible values, and leaving a safety margin for potentially discarding radiation-damaged frames near the end of data collection [82, 83]. Although the diffraction signal is weak in each diffraction image, it can be recorded reliably with modern detectors and extracted accurately with data processing programs. This dose distribution strategy is equally applicable for SSX. With the same amount of total dose, instead of a small wedge of data with high dose per diffraction image, a lower dose can be used to collect data with more angular coverage. This strategy will deliver complete data sets quickly with fewer crystals, which allows full characterization of the unit cell, symmetry, space group, and diffraction properties such as mosaicity, Wilson B-factor, and resolution. The intrinsic diffraction resolution of the crystals under study will be reached when more data are added, because averaging (or accumulating) will yield the same signal-to-noise ratio as that obtained with high exposure.

Another recent development in data collection is the multi-orientation and multi-crystal strategy (instead of the conventional single-crystal and single-orientation). Both multi-orientation data collection (i.e., change orientation of the crystal relative to the spindle [84, 85]) and multi-crystal merging methods [86, 87] are very powerful in reducing systematic errors and producing data with higher accuracy, which leads to better experimental phasing and potentially more accurate structures [88]. The SSX data are essentially collected in a multi-orientation and multi-crystal way with a tolerable X-ray dose per crystal. Therefore, the SSX method should be able to produce data with excellent quality as long as there is a sufficient supply of isomorphous crystals.

For data collection with still images as in injector methods and fast scanning in fixed-target methods, the highest tolerable dose could be used in a single shot aimed to use all the diffraction power that one crystal can provide. When crystal size approaches $1 \times 1 \times 1\ \mu m^3$ and smaller, the diffraction volume may be too small to yield sufficient diffraction signal above background noise for one single high-resolution image within the dose limit. The practical limit of the smallest crystal for synchrotron macromolecular crystallography has yet to be established. However, latest developments in data processing promises the possibility of extracting signals from extremely weak data (i.e., sparse data collected with much lower dose and/or from much smaller crystals.) [89].

Table 1 summarizes the protein systems and experimental conditions used in SSX to date. The smallest crystals are blocks 3–5 μm in each dimension and needles with diameter of 1 μm. The X-ray beam size as small as 150×180 nm^2 has been used with hen egg-white lysozyme crystals measuring $20 \times 20 \times 20$ μm^3. It is evident that SSX with sub-ten micrometer crystals are reachable at current synchrotron beamlines.

3.5 Data Completeness and Multiplicity in SSX

The expected completeness of the merged data depends on the symmetry, the number of data sets, their angular range and their mosaicity. The latter influence exists because reflections at the start and end of a data set's angular range are partials which are later omitted from scaling. Typically, the effective angular range of a data set is given by its nominal angular range minus two times its mosaicity.

The formula for the statistically expected distribution of multiplicities in the merged data for the case of a random orientation of crystals and a centrally positioned detector is [69]:

$$B(n*s,p,k)=\binom{n*s}{k}p^k(1-p)^{n*s-k} \quad (3)$$

The distribution is binomial, with a mean equaling the number of data sets n multiplied by two times (if Friedel's law holds) the number of non-centering symmetry operators (e.g., $s = 4$ in $C2$ and $s = 16$ in $P422$), and multiplied by the effective angular range of each data set expressed as a fraction of 180° (p).

The binomial formulation readily allows to calculate the completeness $c = 1 - B(n^*s,p,0) = 1 - (1 - p)^{n^*s}$ of the merged data. For an effective angular range of 1° ($p = 1/180$) and space group $P1$ ($s = 2$), $n = 207$ data sets are required for 90% complete merged data, and twice that number for 99% complete data. If the effective angular range is 10°, these numbers are 20 and 40, respectively. The average multiplicity corresponding to 90% and 99% completeness is about 2 * 20 * 10/180 ~ 2.2 and 2 * 40 * 10/180 ~ 4.4, respectively, independent of the space group.

The observed multiplicity distribution of acentric reflections in SSX data collected from lysozyme crystals in an IMISX plate is plotted together with the corresponding binomial distribution in Fig. 2a. The merged SSX data consist of a summed 135.6° data recorded from 113 crystals and a rotation range of 1.2°. The

Table 1
Summary of experimental setups for serial synchrotron crystallography[a]

Protein	Delivery method	Wavelength (Å)	Beam size (μm)	Flux (Ph/Sec)	Dose (MGy/crystal)[b]	Temp (K)	Crystal size (μm)	Osc (°)	No. of crystal	S.G.	Resolution (Å)	PDB ID	Beamline/reference
β_2AR-T4 L[c]	Deposition (loop)	1.033	10.6 × 11.6	2.2×10^{11}	10–15	100	4 × 8 × 25	10–15	27	*C2*	2.4	2RH1	APS GM/CA [47]
AcMNPV	Deposition (mesh)	0.9778	8 × 8	n/a	n/a	100	5–10	n/a	17	*I23*	1.84	2WUX	DLS I24 [48]
AcMNPV SeMet	Deposition (mesh)	0.9778	8 × 8	n/a	n/a	100	5–10	n/a	31	*I23*	3.0	n/a	DLS I24 [48]
Lysozyme	Deposition (chip)	1.0	n/a	n/a	n/a	293	5–50	1	~150	$P4_32_12$	2.3	n/a	SLS PXII [40]
BEV2	In situ (plate)	0.96859	20 × 20	1×10^{12}	(0.5)	297	50–60	~0.4	28	*F23*	2.1	n/a	DLS I24 [62]
FcγRIIIA	In situ (plate)	097791	20 × 20	5×10^{11}	(0.1)	297	30 × 30 × 30	~1	44	$P6_122$	2.4	n/a	DLS I24 [62]
PhnA SeMet	In situ (microfluidic)	n/a	n/a	n/a	n/a	295	100–150	10	19	$P4_32_12$	2.11	n/a	APS LS-CAT [57]
CatB in vivo	Deposition (loop)	1.2398	4 × 5	1.2×10^{12}	34	110	0.9 × 0.9 × 11	1	130	$P4_22_12$	3.0	4N4Z	PETRA P11 [52]
CPV18 in cellulo	Deposition (mesh)	0.96859	6 × 6	2×10^{11}	28	100	~5	2	20	*I23*	1.7	4OTV	DLS I24 [49]
CPV18 isolated	Deposition (mesh)	0.96859	6 × 6	2×10^{11}	21	100	~5	3	20	*I23*	1.7	4OTS	DLS I24 [49]
Lysozyme	Capillary	1.27	6 × 9	2.0×10^{12}	0.3	296	3–6	0	40,233	$P4_32_12$	2.09	4O34	PETRA P11 [29]
Glucose isomerase	In situ (microfluidic)	0.9179	100 × 100	5.53×10^{10}	(0.1)	295	30 × 40 × 50	10	72	*I222*	2.09	n/a	CHESS F1 [59]
Photosyn RC	In situ (microfluidic)	0.97950	50	n/a	n/a	298	60–100	5	23	$P4_12_12$	1.45	4TQQ	APS GM/CA [58]
Insulin	Deposition (mesh)	n/a	20 × 20	8×10^{9}	10	n/a	10	6	101	*R3*	1.7	n/a	NSLS X25 [32]
Insulin	Deposition (in situ plate)	n/a	100 × 100	n/a	n/a	294	50	6	88	*R3*	1.8	n/a	NSLS X29 [32]
Insulin	Deposition (belt)	n/a	200 × 200	n/a	n/a	100	50	6	88	*R3*	1.8	n/a	NSLS X12c [32]
Lysozyme	Injector	1.32	10 × 30	2.0×10^{12}	(0.6)	293	10 × 10 × 30	0	11,081	$P4_32_12$	1.9	4RLM	SLS PXII [19]

(continued)

Table 1
(continued)

Protein	Delivery method	Wavelength (Å)	Beam size (μm)	Flux (Ph/Sec)	Dose (MGy/crystal)[b]	Temp (K)	Crystal size (μm)	Osc (°)	No. of crystal	S.G.	Resolution (Å)	PDB ID	Beamline/reference
Lysozyme-Au (MIRAS)	Injector	1.0	10 × 30	1.0×10^{12}	(0.2)	293	10 × 10 × 30	0	11,915	$P4_32_12$	2.5	n/a	SLS PXII [19]
Lysozyme-I (MIRAS)	Injector	1.9	10 × 30	1.0×10^{12}	(0.6)	293	10 × 10 × 30	0	42,115	$P4_32_12$	2.5	n/a	SLS, PXII [19]
Lysozyme-S	Injector	2.07	10 × 30	1.0×10^{1}	(0.7)	293	15 × 15 × 60	0	106,737	$P4_32_12$	2.8	n/a	SLS, PXII [19]
Bacteriorho dopsin	Injector	0.954	2 × 3	9.1×10^{11}	0.7	294	5 × 30 × 30	0	5,691	$P6_3$	2.4	4X31	ESRF ID13 [28]
Lysozyme	Deposition (chip)	1.0	1.5 × 2.5	1×10^{11}	3.2	293	20 × 20 × 20	0	5,966	$P4_32_12$	1.95	4WL7	ESRF ID13 [45]
Lysozyme	Deposition (chip)	1.0	0.15 × 0.18	1.7×10^{10}	29.1	293	20 × 20 × 20	0	46,516	$P4_32_12$	1.85	4WL6	ESRF ID13 [45]
Lysozyme	Deposition (microfluidic)	0.98	10 × 10	2×10^{12}	2.4	294	10 × 10 × 15	0.02	232	$P4_32_12$	2.5	4WMG	SSRL BL12-2 [43]
Lysozyme	Deposition (microfluidic)	1.03318	10 × 10	n/a	n/a	293	10–15	0	324	$P4_32_12$	1.55	4Z98	APS GM/CA [41]
HiTehA	In situ (plate)	1.0	10–50 match xtal	2×10^{11}	(~0.2)	293	10–75	6–10	56	$H3$	2.3	4YCR	DLS I24 [63]
Lysozyme	In situ (IMISX)	1.0332	10 × 18	3×10^{11}	0.27	293	10 × 10 × 20	1.2	113	$P4_32_12$	1.8	4XJB	SLS PXII [69]
Lysozyme Br-SAD	In situ (IMISX)	0.9205	10 × 18	1.5×10^{10}	0.17	293	10 × 20 × 30	2	239	$P4_32_12$	1.8	4XJF	SLS PXII [69]
Lysozyme S-SAD	In situ (IMISX)	1.7	10 × 30	9×10^{9}	0.08	293	10 × 10 × 20	2	992	$P4_32_12$	2.0	4XJH	SLS PXII [69]
PepTst	In situ (IMISX)	1.0332	10 × 18	3×10^{11} 7.5×10^{10}	0.13	293	10 × 10 × 20	0.6	237 335	$C222_1$	2.8	4XNI	SLS PXII [69]
AlgE	In situ (IMISX)	1.0332	10 × 10	1.5×10^{11}	0.2	293	5 × 5 × 20	1	175	$P4_32_12$	2.0	4XNK	SLS PXII [69]
CPV18	Deposition (chip)	0.96859	7 × 7	2×10^{11}	28.5	100	<4	4	22	$I23$	1.5	4X35	DLS I24 [42]

Lysozyme	Deposition (chip)	0.96859	7 × 7	2×10^{11}	20.2	100	<4	3	73	$P4_32_12$	2.1	4X3B	DLS I24 [42]
Bacteriorhodopsin	Deposition (mesh)	0.976	10 × 10	1.5×10^{11}	6.8	100	2 × 5 × 5	10	10	$P6_3$	2.54	5A45	ESRF ID29 [50]
Maelstrom SeMet	Deposition (mesh)	0.979	10 × 10	9.5×10^{10}	4.5	100	20–50	10	45	$H32$	3.46	n/a	ESRF ID23-1 [50]
β_2AR	In situ (IMISX)	1.0332	10 × 18	3.2×10^{11}	8.4	100	5 × 10 × 30	3	104	$C2$	2.5	5D5A	SLS PXII [70]
DgkA	In situ (IMISX)	1.0	10 × 18	1.8×10^{11}	~10–20	100	20 × 20 × 50	20–40	12	$P2_12_12_1$	2.8	5D56	SLS PXII [70]
CPV1 in cellulo	Deposition (mesh)	1.0	10 × 10	$\sim 3.6 \times 10^{11}$	(~90)	100	5–15	1	9–18	$I23$	1.55	5EXY	Australian MX2 [51]

[a]Only SSX experiments with number of crystals larger than 10 are listed. The table is sorted by publication year
[b]Dose in bracket is estimated based on Eq. 5 in [90]
[c]Data on many GPCRs were collected in a similar way

observed multiplicity agrees very well with the binomial distribution, which confirms the random orientation of crystals. From the observed average multiplicity of the merged SSX data, the effective rotation range is estimated to be 0.86°. That is the average amount of data per crystal contributed to the final data set. For comparison, the multiplicity of a lysozyme data set collected from a single crystal with total rotation range equal to the summed total rotation range in SSX is also plotted (Fig. 2a). The distribution of multiplicity is broader and the average multiplicity is lower in SSX data because of the reduced effective rotation coverage.

The above considerations hold if crystal orientations are random. If, however, the crystals have preferred orientations due to their morphology, the completeness will generally be lower, and the distribution of multiplicities will differ from a binomial. The distorted multiplicity distribution features one peak and one shoulder at low and high multiplicities, respectively. This signature can be used to identify preferential orientation during SSX experiments. One such example is given in Fig. 2b. Here, plate-like crystals of the β_2-adrenoreceptor (β_2AR) were grown by the IMISX method and with a tendency to lie with their flat face parallel to the surface of the IMISX plate. The final merged data contains 104 crystals and a rotation range of 3° per crystal. With the estimated effective rotation range of 2.49°, 259° of data in total should give a completeness of 99.7% ($1 - (1 - 2.5/180)^{104*4}$) for the monoclinic space group $C2$. However, the observed overall completeness was only 95% due to the preferred orientation effect. In practice, this issue may be at least partially compensated for by increasing the number of partial data sets, by increasing the tilt

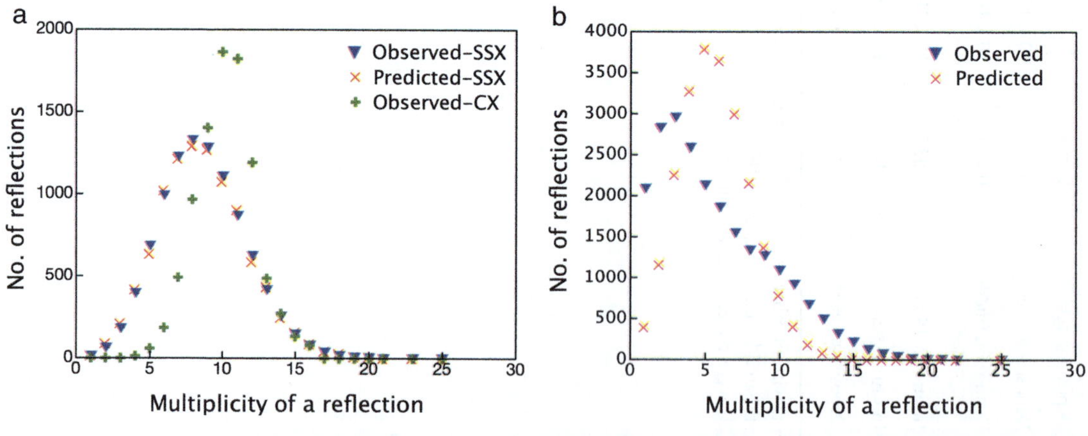

Fig. 2 Distributions of multiplicity in SSX. Figures are adapted from publications [69, 70]. (**a**) Lysozyme SSX data recorded with 113 crystals and a rotation range of 1.2° (effective rotation of 0.86°) and lysozyme CX data recorded with one crystal and a rotation range of 135°. (**b**) β_2AR SSX data recorded with 104 crystals and a rotation range of 3° (effective rotation of 2.49°)

angle of the sample support for data collection, by employing the additional degree of freedom of a kappa goniometer, or by using different mounting methods.

4 Data Processing and Merging

The next paragraphs outline the steps and considerations in data processing of partial data sets collected with the rotation method. The data processing with still images is reviewed in another chapter of this book.

4.1 Processing Individual Data Sets

Any of the commonly used data processing programs, XDS [91, 92], MOSFLM [93], or HKL [94] can in principle be used with partial data sets. However, in practice the processing of hundreds of data sets requires an automated, streamlined procedure that avoids manual intervention. XDS was chosen by us and others for this purpose since it can easily be scripted, and its operation is highly robust.

Scripts have been developed for the processing and merging of data sets collected with PILATUS and EIGER detectors at the Swiss Light Source. One script extracts header information and generates a standard XDS.INP file which is then used to process each partial data set in turn. Owing to the small number of frames in each data set, the running time of the script per data set is short, and data sets can be processed concurrently because the processing directories are uniquely assigned to each data set.

The script creates an XDS.INP file with parameters which differ from the default detector templates distributed with XDS in the following ways:

- for spot finding, the minimum number of pixels in a spot is set to 2, because most of the crystals are smaller than the detector pixels, the beam at beamline X06SA has low divergence, and the point spread function of the PILATUS and EIGER is negligible.
- approximate unit cell parameters and space group are specified if known; the symmetry information only needs to represent the correct Bravais type. Constraints on cell parameters, like equality of axes or fixed angles, increase the accuracy of spot prediction during the integration by reducing the number of degrees of freedom. In principle, space group determination may be carried out after processing all data sets in *P*1, the default if the space group is unknown. Knowledge about the correct Bravais lattice may be obtained from a single weakly exposed low-resolution data set for which the tolerable X-ray dose was spread over a wide rotation range.

- the detector distance as well as the direct beam position on the detector are given as accurately as possible, e.g., as determined with data from a good test crystal.
- the minimum fraction of indexed reflections is set below the default of 0.5 in order to index and integrate as many data sets as possible.

The indexing is usually successful for more than 90% of all partial data sets if enough reflections (50 or more; the minimum in XDS is 25) are found; if multiple adjacent or overlapping crystals contribute to a data set, the indexing usually picks up the strongest lattice.

When processing individual data sets, a reference data set (if available) can be used to resolve indexing ambiguities, which occur in some space groups, and may exist in any space group for specific relationships between cell parameters.

After the final processing, some data sets representing cracked crystals or compromised by the existence of multiple lattices, or reflections from salt, mesophase or ice, or mis-indexing can be identified and discarded. A simple way of doing this is to select those few partial data sets which have at least one cell parameter deviating by more than 3 (or 4) standard deviations from the average; this rule would falsely discard only one out of 370 (or 15,788) data sets if the cell parameters follow a Gaussian distribution. Obviously, it is prudent to start the procedure from a generous cutoff (4 standard deviations, for example), and to iterate the following steps: (a) discard the worst outliers, (b) recalculate the average cell parameters, and (c) tighten the cutoff. Histograms of cell parameter values for SSX data sets from β_2AR are shown in Fig. 3, which indeed possess an approximate Gaussian shape.

4.2 Scaling and Merging the Data Sets

After data integration, the resulting XDS_ASCII.HKL reflection files are scaled and merged in a first XSCALE run. XSCALE has undergone extensive development for serial crystallography; versions released since March 2015 support the efficient scaling and merging of thousands of partial data sets. These tasks require the existence, in each partial data set, of reflections whose unique indices also occur in other data sets. In a situation where the total rotation range of all data sets taken together is much higher than the minimum rotation range for the given space group, almost all reflections of each data set have such counterparts in other data sets, and the resulting network of scaling relationships uniquely determines the scale factor of each partial data set (except for a common arbitrary overall scale factor). However, if the number of partial data sets is so small that their total rotation range approaches the minimum rotation range, some partial data sets may have no unique reflections in common with other data sets, and therefore cannot be scaled. This situation is detected and reported by XSCALE, and those "non-overlapping" data sets have to be discarded after the first XSCALE run.

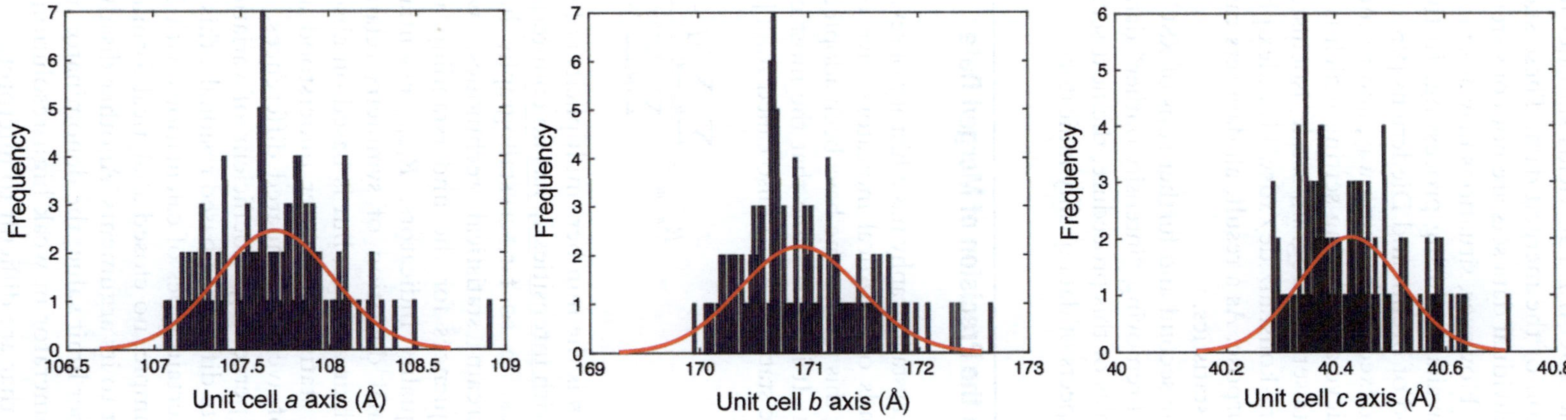

Fig. 3 Histograms of unit cell parameters in SSX data sets recorded with 111 β_2AR crystals (a 10σ peak not shown)

The XSCALE run yields statistics for the completeness and precision of the merged data. These statistics are meaningless if the individual data sets are not consistently indexed, as discussed above. However, situations may arise in which no reference data set is available during processing. In this case, one may use the method of Brehm and Diederichs [95] to identify groups of data sets indexed in the same way, and re-index all except one group to achieve an indexing setting which is consistent across all partial data sets. A program xscale_isocluster (http://strucbio.biologie.uni-konstanz.de/xdswiki/index.php/SSX) is available for this purpose. As a result, all data sets can be merged with meaningful statistics.

The second and further runs of XSCALE are devoted to finding and removing "intensity outlier" data sets. To understand the principles of this procedure, the next section first discusses important aspects of data quality indicators.

5 Assessing and Improving the Precision of Merged Data

5.1 Data Quality Indicators

X-ray crystallography has a history of several decades. Many different kinds of statistical indicators have been defined and applied during this time; some have been adopted by the community, others not. It is remarkable that the most commonly used crystallographic statistic, R_{merge} (also called R_{sym}; [96]), defined as:

$$R_{merge} = \frac{\sum_{hkl} \sum_{i}^{n} \left| I_i - \bar{I} \right|}{\sum_{hkl} \sum_{i}^{n} I_i} \tag{4}$$

where n is the number (multiplicity) of symmetry-related reflections with intensities I_i, has no counterpart in other quantitative sciences. In so far, crystallography has separated itself from mainstream statistical techniques, with some unfortunate consequences for the understanding and interpretation of its data quality indicators. R_{merge} essentially measures the mean fractional deviation of symmetry-related reflection intensities from their average, but is based on absolute differences instead of the statistically better understood and more robust square root of averaged squared differences, like those found in the PCV (percentage coefficient of variation). As with any other absolute difference based residuals, this makes it difficult to perform certain types of calculations with R_{merge}, since there exists, for example, no closed analytical formula for its derivative with respect to its arguments. Another disadvantage is that R_{merge} has no upper limit value; the denominator may become smaller than the numerator in weak high-resolution shells, and large values result, that are difficult to interpret.

Furthermore, R_{merge} calculated from a sample is a biased estimator of the population R_{merge}, in the same sense as the sample variance, when defined as $\frac{1}{n}\sum\left(I_i - \bar{I}\right)^2$, is a biased estimator of the population variance. As with sample variance, which needs to be redefined as $\frac{1}{n-1}\sum\left(I_i - \bar{I}\right)^2$ to be unbiased, R_{merge} needs to be redefined as $R_{meas} = \frac{\sum_{hkl}\sqrt{\frac{n}{n-1}}\sum_i^n\left|I_i - \bar{I}\right|}{\sum_{hkl}\sum_i^n I_i}$, an insight that was published two decades ago [97]. However, even now R_{meas} has not replaced R_{merge}, which is still being used in decision making, where its bias favors low-multiplicity over high-multiplicity data collection.

Finally, the community has not fully realized the fact that both R_{meas} and R_{merge} measure the precision of the individual measurements I_i, rather than the precision of the merged $\bar{I}$. The precision of the merged $\bar{I}$ depends on the sum of the number of photons in each of its n contributing I_i, and thus there are many different experimental strategies that result in the same precision of the merged $\bar{I}$, but yield very different values of R_{meas} and R_{merge}. This fact offers the experimenter an important degree of freedom for optimizing the experiment; favoring those experimental strategies that result in low R_{meas} or R_{merge} biases the experiment toward early radiation damage and the minimal rotation range. For a long time, this has been an unfortunate practice in CX.

There are three indicators that measure the precision of the merged data $\bar{I}$: R_{pim}, which is another variant of R_{merge} in which the factor $\sqrt{n/(n-1)}$ in the numerator of R_{meas}, is replaced by $\sqrt{1/(n-1)}$, thus accounting for the increase in precision by $\sqrt{n}$ when merging n independent observations. R_{pim} shares with R_{merge} the property that its value is unbounded and difficult to interpret.

Second, there is the average signal-to-noise ratio $\langle \bar{I}/\sigma(\bar{I})\rangle$. This indicator suffers from the fact that there are different ways and procedures to estimate $\sigma(\bar{I})$, as is reflected by the fact that different data processing programs yield quite different values for $\langle \bar{I}/\sigma(\bar{I})\rangle$ even if their estimates of the $\bar{I}$ values closely agree [98]. Furthermore, it offers no simple way to identify data sets that degrade the merged signal, because $\langle \bar{I}/\sigma(\bar{I})\rangle$ will always rise when including more observations even if the additional data are non-isomorphous.

Third, there is a correlation-coefficient based quantity called $CC_{1/2}$ which was introduced a few years ago [99], and has gained acceptance in the community because its values allow statistically well-founded decisions [100] particularly about the signal present in weak high-resolution data. Its numerical value lies in the range from −1 to 1 (but in practice only the range 0 to 1 is important) which is easily interpretable, and an analytical relationship with $\langle \bar{I} / \sigma(\bar{I}) \rangle$ exists under well-defined circumstances [101].

5.2 Identifying Outlier Data Sets

In principle, each data set in SSX has both random and systematic differences relative to all other data sets. The random component is an unavoidable consequence of the photon-counting experiment; the systematic difference is usually referred to as non-isomorphism and its size is a priori unknown. Unfortunately, there is no simple way to separate these two types of differences. This is desirable since data sets that are weak (with large random error) should not be discarded, whereas data sets that are non-isomorphous (with large systematic error) should be. It may be noted that the evaluation of unit cell parameters, as explained above, is a first filter for highly non-isomorphous data sets, but since the cell parameters of partial data sets are not very precise, the efficiency of the filter is low.

Several strategies for identifying outlier data sets have been devised and employed by us. Our first attempt [69] used the asymptotic $\langle I/\sigma(I) \rangle$ ratio (ISa), as determined by XSCALE. The ISa value is calculated from the product of the parameters a and b of the error model which XSCALE establishes for each individual data set by fitting its $\sigma(I_i)$ values to the root-mean-square difference between its intensities I_i and $\bar{I}$. One such analysis for β_2AR data is presented in Fig. 4a, and data sets with low ISa values are indicated. The problem with this approach seems to be the fact that the number of reflections in each data set is low and the spread of differences between I_i and $\bar{I}$ may be large, so that although the parameters a, b may result in reasonable $\sigma(I_i)$ estimates, their product may not be very robust.

Our second strategy is straightforward [70]. For each data set, we calculate the average of its intensity correlation coefficients ($CC_{dataset}$) against all other data sets. Finally, we discard those data sets, which display the lowest average correlation. This procedure is robust and does not depend on the $\sigma(I_i)$, but since it cannot differentiate between random and systematic error, it may discard weak data sets and may not discard non-isomorphous ones. Of course, discarding weak data sets does not compromise the merged data much; not discarding non-isomorphous data sets, however, is an issue not solved by this algorithm. Apart from its simplicity, an advantage is that this procedure, which we call "cherry-picking," can be performed before the XSCALE runs, since it does not require scaled intensities. One such correlation analysis with the β_2AR data is presented in Fig. 4b. Three data sets have $CC_{dataset}$ less than 0.9 and these are also the ones with low ISa values. The methodology of comparing CC values could also be employed in selecting the frame/dose cutoff beyond which radiation damage of a

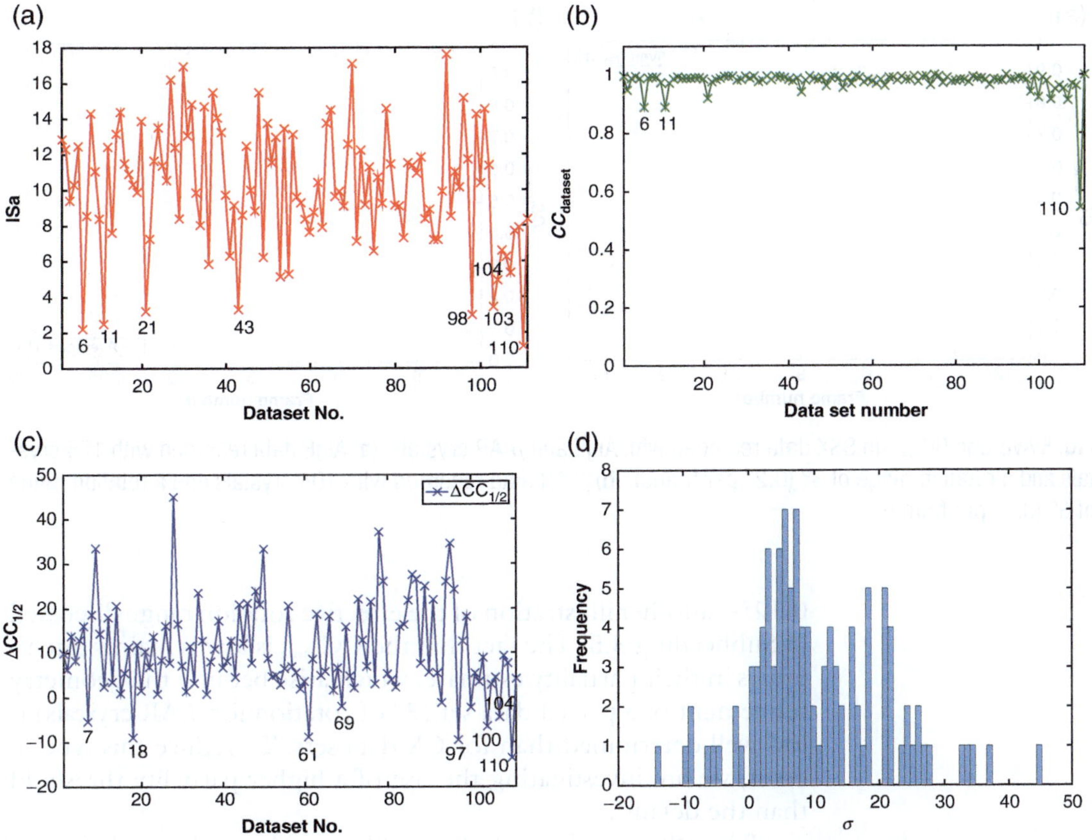

Fig. 4 Data set selection methods applied to β_2AR SSX data (111 data sets). (**a**) ISa method. Data sets with ISa value less than 5 are labeled. (**b**) $CC_{dataset}$ method. Data sets with CC_{datase} less than 0.9 are labeled. (**c**) $\Delta CC_{1/2}$ method. Data sets with negative $\Delta CC_{1/2}$ values are labeled. (**d**) Histogram of $\Delta CC_{1/2}$

partial data set is deemed problematic. As an example, we show room temperature data for the alginate transport protein, AlgE, where it is clear that after seven frames, the correlation (CC_{frame}) with the less damaged data diminishes (Fig. 5a). For these data, we chose to only merge frames 1–5 [69]. When a microbeam is used for microcrystals embedded in sample delivery media, the difficulties in crystal centering in the X-ray beam direction can result in moving part of the crystal out of X-ray beam during the rotation data collection. Such diffraction images can easily be excluded with CC_{frame}. The CC_{frame} can also be used to analyze data integration. For example, the average of all frame-based CC from 111 crystals of β_2AR displays a top-hat profile with a low CC_{frame} for reflections from the first and last frames of the rotation range (Fig. 5b). Their number is low, since most of the reflections in these frames have a partiality below the default acceptance threshold (75%), consistent with a rotation range per frame of 0.1° and average mosaicity of

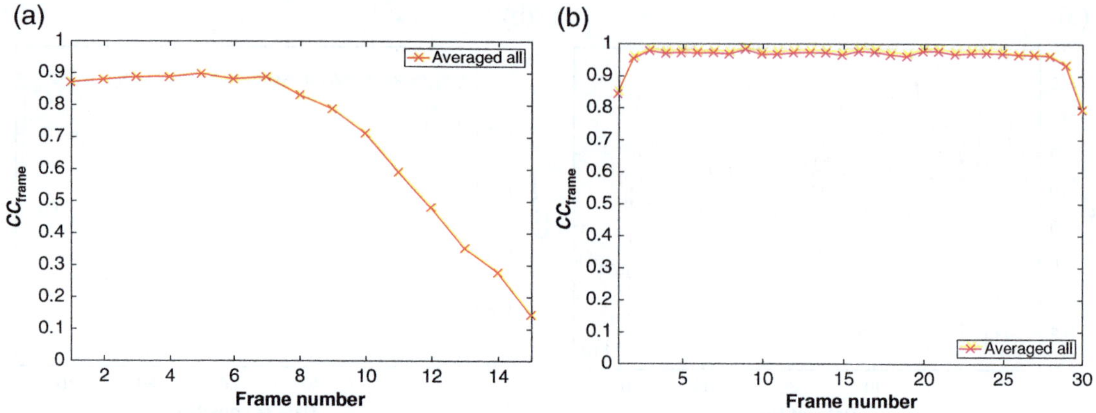

Fig. 5 Average CC_{frame} in SSX data recorded with AlgE and β_2AR crystals. (**a**) AlgE data recorded with 175 crystals and a rotation range of 3° (0.2° per frame). (**b**) β_2AR data recorded with 104 crystals and a rotation range of 3° (0.1° per frame)

0.12°—another illustration of the effective rotation range discussed in Subheading 3.5. The fact that the CC_{frame} is low is mainly due to errors in their partiality estimate, which arise because the geometry refinement of a partial data set (3° of rotation for β_2AR crystals) is less well determined than for CX data sets. To reduce this type of error, we are investigating the use of a higher partiality threshold than the default.

Recently, we showed that a "leave-one-out" calculation of $\Delta CC_{1/2,i} = CC_{1/2,\text{all}} - CC_{1/2,\text{without } i}$ can unequivocally identify non-isomorphous data sets [102]. In effect, discarding data sets with strongly negative $\Delta CC_{1/2,i}$ optimizes the target function, $CC_{1/2}$. The distinction between random and systematic error is achieved due to the fact that weak data sets (high random error) should still result in (small) positive $\Delta CC_{1/2,i}$ values, whereas non-isomorphous data sets produce negative $\Delta CC_{1/2,i}$ values. As discussed in the original work, it may be difficult to identify weak non-isomorphous data sets since their $\Delta CC_{1/2,i}$ may be indistinguishable from those of weak isomorphous data sets within the range of observed $\Delta CC_{1/2,i}$ values. In other words, a particular $\Delta CC_{1/2,i}$ value may not necessarily be statistically significant. This consideration suggests that data set outlier detection by this "$\Delta CC_{1/2}$ method" is effective only for sufficiently strong data sets; the dose and crystal size which allows this is under investigation. We applied the $\Delta CC_{1/2}$ method to the β_2AR data and the result is presented in Fig. 4c, d.

The β_2AR example illustrates the challenges in scaling and merging SSX data. Three data set selection methods are in agreement regarding the identity of the worst data set (number 110 in Fig. 4), but not beyond that. Actually, this result is not surprising since, as explained in Subheading 5.2, the three methods differ in their theoretical foundation. Rejecting unjustified outliers will increase the precision, but decrease the accuracy of the merged data. From our recent work [102], we expect that the $\Delta CC_{1/2}$ values can give a more useful ranking of non-isomorphism than the other indicators.

Automation is indispensable in SSX data processing because it is simply not practical to analyze hundreds to thousands of data sets manually. The aforementioned data processing, selection, scaling, merging, and analysis methods are robust and can be easily scripted and incorporated in data analysis pipelines at synchrotron beamlines. Fully automated pipelines from SSX data collection to the final merged data have been implemented recently at the ESRF ("MeshAndCollect" [50]), SPring-8 ("Zoo system," private communication), and the SLS.

6 Summary

In this chapter, we present a variety of methods for exposing microcrystals to the X-ray beam that are far from being voluminous enough to individually yield complete data sets, and an established and proven way for collecting, processing, and merging such data.

A new crystallographic method must be judged by its feasibility and ability to solve and refine new structures. As discussed above, the random orientation of crystals together with a modest oversampling of orientation space ensures good completeness: a coverage of the minimal rotation range (Table 1 in [79]) with about 98–99% completeness requires on average fourfold multiplicity. Since 99% completeness and an average multiplicity of 4 can likewise be considered as reasonable goals when planning single-crystal data sets, it is apparent that SSX from crystals in random orientations is an efficient means for covering reciprocal space. Additionally, SSX has the advantage over single-crystal CX, which is always limited by radiation damage, that using data from additional partial data sets will reliably and significantly improve the merged data, because the scaling is better determined, outlier intensities can be identified and rejected more efficiently, and the higher multiplicity not only results in more precise, but also more accurate data.

After following the processing steps outlined in the previous section and obtaining the merged data, the subsequent procedures for experimental phasing or molecular replacement and refinement against SSX data are, in our experience, the same as those for data collected in CX. In practice, the quality of the resulting data has enabled us to phase bromide soaks and native-SAD measurements [69, 70] with standard procedures, e.g., substructure determination with SHELXD [103], and to refine with phenix.refine [104]. Being able to phase from the anomalous signal thus attests to the high quality diffraction data and high degree of isomorphism attainable with the crystals used.

Undoubtedly, the SSX methods will continue evolving. However, current methods are mature enough for routine use. Any new crystallographic method must be compared with the existing or other developing (or alternative) methods. In this respect, SSX overcomes the limitations of single-crystal work with respect to the availability of large crystals, and the radiation damage that a single crystal tolerates. Of course there is a middle ground; in the traditional approach, several data sets from single crystals can be combined. In CX, single crystals must be harvested and mounted, one at a time, followed by X-ray diffraction screening, which makes the screening of all available crystals impractical. SSX offers attractive alternatives with innovative and automated sample delivery and serial data collection methods.

Data collection in CX provides an extreme example of the "cherry-picking" method. Screening a number of crystals and collecting a final data set from the "best crystal" is common practice, and thousands of structures have been solved this way. However, this is not necessarily best practice when it comes to SSX, because the particular diffraction geometry, chosen rotation range, and crystal peculiarity may lead to systematic measurement errors. It has been demonstrated convincingly that merging data from statistically equivalent crystals can improve both precision and accuracy of the merged data [86].

In this respect, a key assumption of the SSX method is that most crystals under investigation are statistically equivalent (isomorphous). This may not hold for systems where slight differences in molecular packing and/or composition results in crystals with significant differences in their unit cell parameters and/or reflection intensities. If these variations fall into distinct classes, clustering analysis may remedy the problem by sorting crystals into different classes and merging them separately. According to the mosaic block theory, the outcome of a CX experiment is an average structure of all mosaic blocks. The SSX experiment adds another level of averaging across all merged crystals with different levels of non-isomorphism. Based on data to date, the difference between CX and SSX structures would appear to be minor. On one hand, the individuality of each crystal can get averaged out, which can result in a lower number of observed solvent molecules in SSX structures. On the other hand, averaging can enhance common features of crystals, such as alternative side-chain conformations [24, 28, 69].

However, methods to investigate isomorphism (or rather, the lack thereof) are still in their infancy, and there are compelling scientific reasons to develop them, because the lack of dynamic information is one of the shortcomings of X-ray crystallography, which would partly be overcome by detection and analysis of groups of commonly occurring variations in macromolecular crystals.

Current methods are one-dimensional. Specifically, the ISa-based selection, the "cherry-picking" and the $\Delta CC_{1/2}$ methods all lead to a ranking of data sets relative to the average of all other data sets, rather than to a clustering of variants.

Both SSX and CX are bound by the radiation damage limit. The XFEL method has a dose advantage compared to SSX, since the femtosecond pulse can deliver a much higher dose per shot before primary radiation damage processes set in, and may result in a good signal-to-noise ratio at high resolution and ultimately "damage free" structures. For SSX, the number of photons contributing to a merged unique reflection can ultimately only be increased by exposing more crystals.

On the other hand, compared to SFX, which is limited to still images, SSX can accurately sample reflection profiles during rotation. In the case of stills, if the crystals only have a small number of mosaic blocks, the "rocking curve" of a reflection consists of the superposition (addition) of each block's individual rocking curve, which may be shifted relative to each other. If the number of mosaic blocks were large, their superposition would be (according to the Central Limit Theorem) Gaussian in shape; for small numbers however, each reflection will have a different profile, and may have several maxima and appear jagged. That means that any estimate of full intensity, which is based on a partiality estimate and the sampled portion of the jagged profile, will be in error even if the partiality estimate is correct. We believe that this effect reduces the attainable precision of XFEL data that can be compensated for only by collecting more data. Thus, while the XFEL method has the dose advantage, it suffers the disadvantage of sampling "uneven" reflection profiles, which may lower its usefulness for small crystals. Furthermore, typical protein crystals are far from ideal and their reflection profiles may exhibit non-Gaussian behavior, which also makes profile sampling with still images less efficient.

In summary, SSX has emerged as a complementary method to CX. The technologies developed for SSX and the next generation synchrotron sources make possible the acquisition of better data from smaller crystals, which was either impossible or very tedious and time consuming previously. It should be feasible to obtain high resolution structures with micrometer or even nanometer sized crystals. The serial nature of the SSX experiment makes automation indispensable, which calls for further development in workflows from crystallization, sample delivery to data collection, processing and merging. SSX is also important for screening initial hits in crystallization and for pre-characterizing samples for SFX experiments. Together with CX, SSX and SFX will broaden the horizon for X-ray based structural biology in the coming decades.

Acknowledgments

We thank Greta Assmann, Wolfgang Brehm, Martin Caffrey, Chia-Ying Huang, Vincent Olieric, Ezequiel Panepucci, Rangana Warshamanage, and all other members of the groups at the Swiss Light Source (Paul-Scherrer-Institute, Villigen, Switzerland), Trinity College (Dublin, Ireland) and University of Konstanz (Konstanz, Germany) for discussions and their contributions toward developing the methodology. We also thank Aaron Finke and Martin Caffrey for proofreading the manuscript and Rangana Warshamanage and Chia-Ying Huang for preparing the figures.

References

1. Arndt UW, Wonacott AJ (1977) The rotation method in crystallography. North-Holland Publishing Company, Amsterdam
2. Darwin CG (1914) XXXIV. The theory of X-ray reflexion. Philos Mag Ser 6 27:315–333
3. Warren BE (1969) X-ray diffraction. Addison-Wesley Pub. Co., Reading, MA
4. Holton JM, Frankel KA (2010) The minimum crystal size needed for a complete diffraction data set. Acta Crystallogr D Biol Crystallogr 66:393–408
5. Kendrew JC, Bodo G, Dintzis HM, Parrish RG, Wyckoff H, Phillips DC (1958) A three-dimensional model of the myoglobin molecule obtained by X-ray analysis. Nature 181:662–666
6. Perutz MF, Rossmann MG, Cullis AF, Muirhead H, Will G (1960) Structure of hæmoglobin: a three-dimensional fourier synthesis at 5.5-Å. resolution, obtained by X-ray analysis. Nature 185:416–422
7. Harrison SC, Olson AJ, Schutt CE, Winkler FK, Bricogne G (1978) Tomato bushy stunt virus at 2.9 A resolution. Nature 276:368–373
8. Hendrickson WA (2000) Synchrotron crystallography. Trends Biochem Sci 25: 637–643
9. Hope H (1988) Cryocrystallography of biological macromolecules: a generally applicable method. Acta Crystallogr B 44:22–26
10. Sliz P, Harrison SC, Rosenbaum G (2003) How does radiation damage in protein crystals depend on X-ray dose? Structure 11:13–19
11. Cusack S, Belrhali H, Bram A, Burghammer M, Perrakis A, Riekel C (1998) Small is beautiful: protein micro-crystallography. Nat Struct Biol 5(Suppl):634–637
12. Smith JL, Fischetti RF, Yamamoto M (2012) Micro-crystallography comes of age. Curr Opin Struct Biol 22:602–612
13. Chapman HN, Fromme P, Barty A et al (2011) Femtosecond X-ray protein nanocrystallography. Nature 470:73–77
14. Schlichting I (2015) Serial femtosecond crystallography: the first five years. IUCrJ 2:246–255
15. Gavira JA (2015) Current trends in protein crystallization. Arch Biochem Biophys 602:3–11
16. Liu W, Ishchenko A, Cherezov V (2014) Preparation of microcrystals in lipidic cubic phase for serial femtosecond crystallography. Nat Protoc 9:2123–2134
17. DePonte DP, Weierstall U, Schmidt K, Warner J, Starodub D, Spence JCH, Doak RB (2008) Gas dynamic virtual nozzle for generation of microscopic droplet streams. J Phys D Appl Phys 41:195505
18. Weierstall U, James D, Wang C et al (2014) Lipidic cubic phase injector facilitates membrane protein serial femtosecond crystallography. Nat Commun 5:3309
19. Botha S, Nass K, Barends TRM et al (2015) Room-temperature serial crystallography at synchrotron X-ray sources using slowly flowing free-standing high-viscosity microstreams. Acta Crystallogr D Biol Crystallogr 71:387–397
20. Boutet S, Lomb L, Williams GJ et al (2012) High-resolution protein structure determination by serial femtosecond crystallography. Science 337:362–364
21. Sierra RG, Laksmono H, Kern J et al (2012) Nanoflow electrospinning serial femtosecond crystallography. Acta Crystallogr D Biol Crystallogr 68:1584–1587

22. Sugahara M, Mizohata E, Nango E et al (2015) Grease matrix as a versatile carrier of proteins for serial crystallography. Nat Methods 12:61–63
23. Conrad CE, Basu S, James D et al (2015) A novel inert crystal delivery medium for serial femtosecond crystallography. IUCrJ 2:421–430
24. Liu W, Wacker D, Gati C et al (2013) Serial femtosecond crystallography of G protein-coupled receptors. Science 342:1521–1524
25. Fenalti G, Zatsepin NA, Betti C et al (2015) Structural basis for bifunctional peptide recognition at human δ-opioid receptor. Nat Struct Mol Biol 22:265–268
26. Zhang H, Unal H, Gati C et al (2015) Structure of the angiotensin receptor revealed by serial femtosecond crystallography. Cell 161:833–844
27. Kang Y, Zhou XE, Gao X et al (2015) Crystal structure of rhodopsin bound to arrestin by femtosecond X-ray laser. Nature 523:561–567
28. Nogly P, James D, Wang D, White TA, Shilova A, Nelson G, Liu H, Johansson L (2015) Lipidic cubic phase serial millisecond crystallography using synchrotron radiation. IUCrJ 2:168–176
29. Stellato F, Oberthür D, Liang M et al (2014) Room-temperature macromolecular serial crystallography using synchrotron radiation. IUCrJ 1:204–212
30. Roessler CG, Agarwal R, Allaire M et al (2016) Acoustic injectors for drop-on-demand serial femtosecond crystallography. Structure 24:631–640
31. Roessler CG, Kuczewski A, Stearns R, Ellson R, Olechno J, Orville AM, Allaire M, Soares AS, Héroux A (2013) Acoustic methods for high-throughput protein crystal mounting at next-generation macromolecular crystallographic beamlines. J Synchrotron Radiat 20:805–808
32. Soares AS, Mullen JD, Parekh RM, McCarthy GS, Roessler CG, Jackimowicz R, Skinner JM, Orville AM, Allaire M, Sweet RM (2014) Solvent minimization induces preferential orientation and crystal clustering in serial micro-crystallography on micro-meshes, in situ plates and on a movable crystal conveyor belt. J Synchrotron Radiat 21:1231–1239
33. Tsujino S, Tomizaki T (2016) Ultrasonic acoustic levitation for fast frame rate X-ray protein crystallography at room temperature. Sci Rep 6:25558
34. Hunter MS, Segelke B, Messerschmidt M et al (2014) Fixed-target protein serial microcrystallography with an X-ray free electron laser. Sci Rep 4:6026
35. Cohen AE, Soltis SM, González A et al (2014) Goniometer-based femtosecond crystallography with X-ray free electron lasers. Proc Natl Acad Sci U S A 111:17122–17127
36. Hirata K, Shinzawa-Itoh K, Yano N et al (2014) Determination of damage-free crystal structure of an X-ray-sensitive protein using an XFEL. Nat Methods 11:734–736
37. Wierman JL, Alden JS, Kim CU, McEuen PL, Gruner SM (2013) Graphene as a protein crystal mounting material to reduce background scatter. J Appl Crystallogr 46: 1501–1507
38. Warren AJ, Crawshaw AD, Trincao J, Aller P, Alcock S, Nistea I, Salgado PS, Evans G (2015) In vacuo X-ray data collection from graphene-wrapped protein crystals. Acta Crystallogr D Biol Crystallogr 71:2079–2088
39. Sui S, Wang Y, Kolewe KW, Srajer V, Henning R, Schiffman JD, Dimitrakopoulos C, Perry SL (2016) Graphene-based microfluidics for serial crystallography. Lab Chip. Advance article. doi:10.1039/C6LC00451B
40. Zarrine-Afsar A, Barends TRM, Müller C, Fuchs MR, Lomb L, Schlichting I, Miller RJD (2012) Crystallography on a chip. Acta Crystallogr D Biol Crystallogr 68:321–323
41. Murray TD, Lyubimov AY, Ogata CM, Vo H, Uervirojnangkoorn M, Brunger AT, Berger JM (2015) A high-transparency, micro-patternable chip for X-ray diffraction analysis of microcrystals under native growth conditions. Acta Crystallogr D Biol Crystallogr 71:1987–1997
42. Roedig P, Vartiainen I, Duman R et al (2015) A micro-patterned silicon chip as sample holder for macromolecular crystallography experiments with minimal background scattering. Sci Rep 5:10451
43. Lyubimov AY, Murray TD, Koehl A et al (2015) Capture and X-ray diffraction studies of protein microcrystals in a microfluidic trap array. Acta Crystallogr D Biol Crystallogr 71:928–940
44. Baxter EL, Aguila L, Alonso-Mori R et al (2016) High-density grids for efficient data collection from multiple crystals. Acta Crystallogr D Biol Crystallogr 72:2–11
45. Coquelle N, Brewster AS, Kapp U, Shilova A, Weinhausen B, Burghammer M, Colletier JP (2015) Raster-scanning serial protein crystallography using micro- and nano-focused synchrotron beams. Acta Crystallogr D Biol Crystallogr 71:1184–1196

46. Coulibaly F, Chiu E, Ikeda K, Gutmann S, Haebel PW, Schulze-Briese C, Mori H, Metcalf P (2007) The molecular organization of cypovirus polyhedra. Nature 446:97–101
47. Cherezov V, Hanson MA, Griffith MT, Hilgart MC, Sanishvili R, Nagarajan V, Stepanov S, Fischetti RF, Kuhn P, Stevens RC (2009) Rastering strategy for screening and centring of microcrystal samples of human membrane proteins with a sub-10 microm size X-ray synchrotron beam. J R Soc Interface 6(Suppl 5):S587–S597
48. Ji X, Sutton G, Evans G, Axford D, Owen R, Stuart DI (2010) How baculovirus polyhedra fit square pegs into round holes to robustly package viruses. EMBO J 29:505–514
49. Axford D, Ji X, Stuart DI, Sutton G (2014) In cellulo structure determination of a novel cypovirus polyhedrin. Acta Crystallogr D Biol Crystallogr 70:1435–1441
50. Zander U, Bourenkov G, Popov AN, de Sanctis D, Svensson O, AA MC, Round E, Gordeliy V, Mueller-Dieckmann C, Leonard GA (2015) *MeshAndCollect*: an automated multi-crystal data-collection workflow for synchrotron macromolecular crystallography beamlines. Acta Crystallogr D Biol Crystallogr 71:2328–2343
51. Boudes M, Garriga D, Fryga A, Caradoc-Davies T, Coulibaly F (2016) A pipeline for structure determination of ɪt in vivo-grown crystals using ɪt in cellulo diffraction. Acta Crystallogr D Biol Crystallogr 72:576–585
52. Gati C, Bourenkov G, Klinge M et al (2014) Serial crystallography on in vivo grown microcrystals using synchrotron radiation. IUCrJ 1:87–94
53. Li L, Ismagilov RF (2010) Protein crystallization using microfluidic technologies based on valves, droplets, and SlipChip. Annu Rev Biophys 39:139–158
54. Kisselman G, Qiu W, Romanov V, Thompson CM, Lam R, Battaile KP, Pai EF, Chirgadze NY (2011) X-CHIP: an integrated platform for high-throughput protein crystallization and on-the-chip X-ray diffraction data collection. Acta Crystallogr D Biol Crystallogr 67:533–539
55. Dhouib K, Khan Malek C, Pfleging W et al (2009) Microfluidic chips for the crystallization of biomacromolecules by counter-diffusion and on-chip crystal X-ray analysis. Lab Chip 9:1412–1421
56. Pinker F, Brun M, Morin P et al (2013) ChipX: a novel microfluidic chip for counter-diffusion crystallization of biomolecules and in situ crystal analysis at room temperature. Cryst Growth Des 13:3333–3340
57. Perry SL, Guha S, Pawate AS, Bhaskarla A, Agarwal V, Nair SK, Kenis PJA (2013) A microfluidic approach for protein structure determination at room temperature via on-chip anomalous diffraction. Lab Chip 13:3183–3187
58. Khvostichenko DS, Schieferstein JM, Pawate AS, Laible PD, Kenis PJA (2014) X-ray transparent microfluidic chip for mesophase-based crystallization of membrane proteins and on-chip structure determination. Cryst Growth Des 14:4886–4890
59. Heymann M, Opthalage A, Wierman JL, Akella S, Szebenyi DME, Gruner SM, Fraden S (2014) Room-temperature serial crystallography using a kinetically optimized microfluidic device for protein crystallization and on-chip X-ray diffraction. IUCrJ 1:349–360
60. Jacquamet L, Ohana J, Joly J et al (2004) Automated analysis of vapor diffusion crystallization drops with an X-ray beam. Structure 12:1219–1225
61. Bingel-Erlenmeyer R, Olieric V, Grimshaw JPA et al (2011) SLS crystallization platform at beamline X06DA—a fully automated pipeline enabling in situ X-ray diffraction screening. Cryst Growth Des 11:916–923
62. Axford D, Owen RL, Aishima J et al (2012) In situ macromolecular crystallography using microbeams. Acta Crystallogr D Biol Crystallogr 68:592–600
63. Axford D, Foadi J, Hu N-J, Choudhury HG, Iwata S, Beis K, Evans G, Alguel Y (2015) Structure determination of an integral membrane protein at room temperature from crystals *in situ*. Acta Crystallogr D Biol Crystallogr 71:1228–1237
64. Gelin M, Delfosse V, Allemand F, Hoh F, Sallaz-Damaz Y, Pirocchi M, Bourguet W, Ferrer JL, Labesse G, Guichou JF (2015) Combining "dry" co-crystallization and in situ diffraction to facilitate ligand screening by X-ray crystallography. Acta Crystallogr D Biol Crystallogr 71:1777–1787
65. Axford D, Aller P, Sanchez-Weatherby J, Sandy J (2016) Applications of thin-film sandwich crystallization platforms. Acta Crystallogr F Struct Biol Commun 72:313–319
66. Cipriani F, Röwer M, Landret C, Zander U, Felisaz F, Márquez JA (2012) CrystalDirect: a new method for automated crystal harvesting based on laser-induced photoablation of thin films. Acta Crystallogr D Biol Crystallogr 68:1393–1399
67. Zander U, Hoffmann G, Cornaciu I et al (2016) Automated harvesting and processing of protein crystals through laser photoabla-

tion. Acta Crystallogr D Biol Crystallogr 72:454–466

68. Caffrey M (2015) A comprehensive review of the lipid cubic phase or in meso method for crystallizing membrane and soluble proteins and complexes. Acta Crystallogr F Struct Biol Commun 71:3–18
69. Huang CY, Olieric V, Ma P, Panepucci E, Diederichs K, Wang M, Caffrey M (2015) In meso in situ serial X-ray crystallography of soluble and membrane proteins. Acta Crystallogr D Biol Crystallogr 71:1238–1256
70. Huang CY, Olieric V, Ma P et al (2016) In meso in situ serial X-ray crystallography of soluble and membrane proteins at cryogenic temperatures. Acta Crystallogr D Biol Crystallogr 72:93–112
71. Fraser JS, van den Bedem H, Samelson AJ, Lang PT, Holton JM, Echols N, Alber T (2011) Accessing protein conformational ensembles using room-temperature X-ray crystallography. Proc Natl Acad Sci U S A 108:16247–16252
72. Keedy DA, Kenner LR, Warkentin M et al (2015) Mapping the conformational landscape of a dynamic enzyme by multitemperature and XFEL crystallography. Elife 4:e07574
73. Leal RMF, Bourenkov G, Russi S, Popov AN (2013) A survey of global radiation damage to 15 different protein crystal types at room temperature: a new decay model. J Synchrotron Radiat 20:14–22
74. Owen RL, Paterson N, Axford D, Aishima J, Schulze-Briese C, Ren J, Fry EE, Stuart DI, Evans G (2014) Exploiting fast detectors to enter a new dimension in room-temperature crystallography. Acta Crystallogr D Biol Crystallogr 70:1248–1256
75. Henderson R (1990) Cryo-protection of protein crystals against radiation damage in electron and X-ray diffraction. Proc R Soc Lond B 241:6–8
76. Owen RL, Rudiño-Piñera E, Garman EF (2006) Experimental determination of the radiation dose limit for cryocooled protein crystals. Proc Natl Acad Sci U S A 103:4912–4917
77. Evans G, Axford D, Owen RL (2011) The design of macromolecular crystallography diffraction experiments. Acta Crystallogr D Biol Crystallogr 67:261–270
78. Mueller M, Wang M, Schulze-Briese C (2012) Optimal fine φ-slicing for single-photon-counting pixel detectors. Acta Crystallogr D Biol Crystallogr 68:42–56
79. Dauter Z (1999) Data-collection strategies. Acta Crystallogr D Biol Crystallogr 55:1703–1717
80. Bourenkov GP, Popov AN (2006) A quantitative approach to data-collection strategies. Acta Crystallogr D Biol Crystallogr 62:58–64
81. Borek D, Minor W, Otwinowski Z (2003) Measurement errors and their consequences in protein crystallography. Acta Crystallogr D Biol Crystallogr 59:2031–2038
82. Liu ZJ, Chen L, Wu D, Ding W, Zhang H, Zhou W, Fu ZQ, Wang BC (2011) A multidataset data-collection strategy produces better diffraction data. Acta Crystallogr A 67:544–549
83. Weinert T, Olieric V, Waltersperger S et al (2015) Fast native-SAD phasing for routine macromolecular structure determination. Nat Methods 12:131–133
84. Brockhauser S, White KI, AA MC, RBG R (2011) Translation calibration of inverse-kappa goniometers in macromolecular crystallography. Acta Crystallogr A 67:219–228
85. Waltersperger S, Olieric V, Pradervand C et al (2015) PRIGo: a new multi-axis goniometer for macromolecular crystallography. J Synchrotron Radiat 22:895–900
86. Liu Q, Dahmane T, Zhang Z, Assur Z, Brasch J, Shapiro L, Mancia F, Hendrickson WA (2012) Structures from anomalous diffraction of native biological macromolecules. Science 336:1033
87. Olieric V, Weinert T, Finke AD et al (2016) Data-collection strategy for challenging native SAD phasing. Acta Crystallogr D Biol Crystallogr 72:421–429
88. Liu Q, Hendrickson WA (2015) Crystallographic phasing from weak anomalous signals. Curr Opin Struct Biol 34:99–107
89. Ayyer K, Philipp HT, Tate MW, Wierman JL, Elser V, Gruner SM (2015) Determination of crystallographic intensities from sparse data. IUCrJ 2:29–34
90. Holton JM (2009) A beginner's guide to radiation damage. J Synchrotron Radiat 16:133–142
91. Kabsch W (2010) Xds. Acta Crystallogr D Biol Crystallogr 66:125–132
92. Kabsch W (2010) Integration, scaling, space-group assignment and post-refinement. Acta Crystallogr D Biol Crystallogr 66:133–144
93. Battye TGG, Kontogiannis L, Johnson O, Powell HR, Leslie AGW (2011) iMOSFLM: a new graphical interface for diffraction-image processing with MOSFLM. Acta Crystallogr D Biol Crystallogr 67:271–281

94. Otwinowski Z, Minor W (1997) Processing of X-ray diffraction data collected in oscillation mode. Methods Enzymol 276:307–326
95. Brehm W, Diederichs K (2013) Breaking the indexing ambiguity in serial crystallography. Acta Crystallogr D Biol Crystallogr 70: 101–109
96. Arndt UW, Crowther RA, Mallett JF (1968) A computer-linked cathode-ray tube microdensitometer for X-ray crystallography. J Sci Instrum 1:510–516
97. Diederichs K, Karplus A (1997) Improved R-factors. Nat Struct Biol 4:269–275
98. Krojer T, von Delft F (2011) Assessment of radiation damage behaviour in a large collection of empirically optimized datasets highlights the importance of unmeasured complicating effects. J Synchrotron Radiat 18:387–397
99. Karplus PA, Diederichs K (2012) Linking crystallographic model and data quality. Science 336:1030–1033
100. Diederichs K, Karplus PA (2013) Better models by discarding data? Acta Crystallogr D Biol Crystallogr 69:1215–1222
101. Karplus PA, Diederichs K (2015) Assessing and maximizing data quality in macromolecular crystallography. Curr Opin Struct Biol 34:60–68
102. Assmann G, Brehm W, Diederichs K (2016) Identification of rogue datasets in serial crystallography. J Appl Crystallogr 49:1021–1028
103. Sheldrick GM (2010) Experimental phasing with SHELXC/D/E: combining chain tracing with density modification. Acta Crystallogr D Biol Crystallogr 66:479–485
104. Adams PD, Afonine PV, Bunkóczi G et al (2010) PHENIX: a comprehensive Python-based system for macromolecular structure solution. Acta Crystallogr D Biol Crystallogr 66:213–221

Chapter 11

Time-Resolved Macromolecular Crystallography at Modern X-Ray Sources

Marius Schmidt

Abstract

Time-resolved macromolecular crystallography unifies protein structure determination with chemical kinetics. With the advent of fourth generation X-ray sources the time-resolution can be on the order of 10–40 fs, which opens the ultrafast time scale to structure determination. Fundamental motions and transitions associated with chemical reactions in proteins can now be observed. Moreover, new experimental approaches at synchrotrons allow for the straightforward investigation of all kind of reactions in biological macromolecules. Here, recent developments in the field are reviewed.

Key words Time-resolved macromolecular crystallography, Time-resolved serial femtosecond crystallography, Structure based enzymology, Chemical kinetics

1 Introduction

Macromolecular crystallography as it exists nowadays might change substantially with the advent of the brightest X-ray sources the world has ever seen, the free electron lasers for hard X-rays (X-ray FELs). The immense X-ray brilliance of these machines triggered the development of serial crystallography, where a very large number of small crystals are exposed to the X-ray beam, one by one and in random orientation. One of the goals of this chapter is to describe the advantages of this technique for time-resolved investigations on protein, especially enzyme, crystals in general at both X-ray FELs and synchrotrons. The basic concept of a time-resolved crystallographic experiment is simple: a reaction is started in the crystal and the progress of the reaction is probed at several time delays by short X-ray pulses. From the time-resolved X-ray data, macromolecular structures and their dynamics are extracted. This information is then utilized to determine the kinetic mechanism of the reaction, which in turn can be used to gain biologically, medically and pharmaceutically highly relevant insight. This chapter

Alexander Wlodawer et al. (eds.), *Protein Crystallography: Methods and Protocols*, Methods in Molecular Biology, vol. 1607,
DOI 10.1007/978-1-4939-7000-1_11,

reviews recent results and outlines a roadmap for time-resolved investigations on a broad spectrum of proteins and enzymes through a wide range of time delays at pulsed X-ray sources.

2 Time-Resolved Macromolecular Crystallography

"Kinetic crystallography" aims at studying genuine reactions in protein crystals through a wide range of methods at cryogenic and ambient temperatures [1]. Whereas at cryogenic temperatures trapping methods are employed [2, 3], at ambient, physiologic temperatures reactions must be investigated on-the-fly by time-resolved crystallography. Traditionally, time-resolved macromolecular crystallography is performed using the Laue method [4, 5] which employs polychromatic X-ray radiation consisting of a range of wavelengths. The main advantage is that the integrated reflection intensity is collected instantaneously from stills without moving the crystal. Entire data sets can be collected in a few seconds at room temperatures. If the crystal is translated quickly and data collected rapidly between each setting, a fresh, pristine crystal volume is exposed each time, especially when tightly focused, micron sized X-ray beams are employed. At least some of the radiation damage [6] can be avoided this way [7, 8]. Incidentally, the advantages of the Laue method become a disadvantage when crystals with large mosaicities are examined or when the crystals tumble, during or between exposures. In such cases, the reflections can become streaky and their intensities are difficult if not impossible to determine. Nevertheless, time-resolved crystallography using the Laue method has been successfully applied to a number of biological systems. A substantial number of original publications are covered by numerous reviews [1, 9–15].

3 Reaction Initiation

To start a reaction in a protein crystal is one of the biggest challenge to date (Table 1). Luckily there are proteins which are intrinsically sensitive to visible light in the extended wavelength range, including near ultraviolet and infrared. In these cases, reactions can be started by short optical laser pulses. There are several advantages with this approach: (1) time-resolution is limited only by the laser or X-ray pulses, whichever is longer. This means that if both laser and X-ray pulses are ultrashort, even sub-picosecond time delays can be explored. (2) In many cases photoactivated reactions end with their respective dark states, hence the reactions are cyclic. Repeated activation becomes possible to enhance the intensity of the Bragg spots in the diffraction patterns even when the X-ray

Table 1
Some methods to initiate reactions in protein crystals

Method	Time-resolution	Experimental complexity	Time to collect a data set
Laser pulses, intrinsically photosensitive	Ultrafast, <1 ps	Low	Quick
Laser pulses, caged substrates[a]	>100 ns	High	Slow
Serial crystallography, mixing and diffusion[b]	~100 μs	Low	Quick
Others such as T-jump, electric fields etc.	Moderate	Low to high	Quick

[a]With macroscopic crystals
[b]If very small crystals are used, the diffusion time might be even faster than the mixing time. The mixing time then determines the time-resolution

intensity is relatively low. (3) Time delays between X-ray and laser pulses can be controlled precisely. Even if there is jitter between these pulses, the jitter can be measured [16–19] and corrected for. Unfortunately, most reactions in proteins are not cyclic, hence they end in a different state than the dark (or initial) state and the initial state must be restored to repeat the measurement. Light-induced irreversible reactions can be found in photoswitches, such as the phytochromes [20], or other phycobiliproteins [21]. They can be switched by laser pulses of different wavelength between two different stable states ($S_{\lambda1}$ and $S_{\lambda2}$) that show distinct maxima at wavelengths λ_1 or λ_2, respectively, of their absorption spectra. Once, for example, state $S_{\lambda2}$ is formed, it must be switched back to the initial state $S_{\lambda1}$ for repeated pump-probe exposures. This can be done with a light emitting diode with an appropriate wavelength. The initial state may then be restored as demonstrated spectroscopically, for example, for the biliprotein α-phycoerythrocyanin [22]. If the protein is not intrinsically photoactive, which is true for most enzymes, there are several options. The protein is engineered to be photosensitive, for instance by genetically fusing a photoreactive domain to it [23, 24], or the so-called caged substrates are used [1, 25–29]. The caged substrates remain inactive until they are activated by intense light pulses. Both options require substantial expertise in molecular biology and chemistry, respectively. Flow cells [30] may be used to load the inactive caged substrate and wash away product after activation of the caged substrate and completed reaction (Table 1). Nevertheless, this procedure is tedious and time-consuming, and often the photoactivation yield of the caged substrate is low [27]. It requires substantial beamtime to collect an entire time-series (*see* below) with this technique.

It would be extremely beneficial for the field of enzymology if a method could be found that investigates noncyclic reactions routinely, where in addition the reaction could be initiated in the crystals easily without relying on specific chemical expertise to synthesize complex compounds. A method is outlined in the last paragraph that relies on the simple mixing of very small, micron-sized enzyme crystals with substrate. Since turnover times in enzymes are on the order of a few ms, ultrafast time-resolution is not required. Sufficient time-resolution is reached when the crystals are small and diffusion times are correspondingly fast.

4 Time-Resolved Crystallography at the Synchrotron

Third generation synchrotrons produce intense X-ray pulses containing a much larger number of photons compared to previous designs. The number of X-ray photons is large enough that the collection of a sufficiently intense Laue diffraction pattern from a small number of X-ray pulses is possible [31]. With this, the time-resolution can be as good as 100 picoseconds (ps), the duration of the X-ray pulse. The first experiment that used single X-ray pulse exposures was performed on carbonmonoxy-myoglobin (Mb-CO) crystals [32]. The time-resolution was given by the 7.5 ns pulse duration of the optical laser that initiated CO photolysis in this protein. Up to 50 single pulse X-ray exposures were necessary to produce a sufficiently intense Laue pattern. Geminate rebinding in Mb-CO as well as the photocycle of the photoactive yellow protein (PYP) are examples of cyclic reactions which can be conveniently started by a pump laser pulse and subsequently probed a time delay Δt later by a single X-ray pulse. Before the next pump-probe cycle, one must wait until the initial state is recovered. Multiple pump-probe repetitions to collect a diffraction pattern are possible and become practicable when the initial state recovery is fast. The PYP photocycle finishes after about 100 ms [33], and the geminate rebinding in myoglobin is complete after a few milliseconds [34]. With typical waiting times between exposures on the order of 1 s, it takes a few minutes to collect a single Laue pattern and about 1.5 h to collect a dataset. This time has decreased steadily. Central to it was the observation that X-ray radiation from an undulator results in better diffraction patterns because the X-ray photons are collected into a much narrower bandwidth $\Delta E/E$ of about 5% only. The reflection range of each Laue spot is excited by a much larger number of photons, whereas relatively fewer photons contribute to the background with correspondingly lower background noise [35]. In addition, beamlines became more sophisticated [36] with better focusing optics, so that all X-ray photons are focused onto the crystal and data collection is largely facilitated by ingenious software. Small beams only probe a thin layer of the laser-exposed surface of rather large crystals [36]. Excellent diffraction patterns

Table 2
Exposures and pulses

	ΔE/E bw [%]	Time-resol.	Exposures/ detector image	Pulses/ exposure	Detector images/dataset	Indexed patterns/ dataset	Mix and inject
TR-LX	2–5	100 ps	4–10	1	30–90	30–90	No
TR-SLX[a]	2–5	2–10 μs	1	20–50	10^5	100	Yes
MX[b]	0.01	NA	1	~10^5	<1000	<1000	No
(TR)-SX[c]	0.01	10–100 ms	1	10^4–10^5	10^6	5×10^4	Yes
(TR)-SFX[d]	0.1	<40 fs	1	1	10^6	5×10^4	Yes

A detector image may consist of one or multiple X-ray exposures. An X-ray exposure can employ one or multiple X-ray pulses. The X-ray time-resolution is determined by the duration of the pulse-train required for one X-ray exposure and is best when only one pulse is employed

TR-LX time-resolved Laue crystallography, *TR-SLX* time-resolved serial Laue crystallography, *MX* monochromatic, macromolecular crystallography, *(TR)-SX* (time-resolved) serial crystallography, *SFX* time-resolved serial femtosecond crystallography.

The last column indicates whether the "mix-and-inject" method (see below) is feasible

[a]Assuming a hit rate of 2%, and an indexing rate of 10%

[b]About 5×10^6 X-ray pulses per second in 24 bunch mode at Advanced Photon Source

[c]1 ms exposures, approximate hit and indexing rates from [81, 84]

[d]Assuming a hit rate of <5% and an indexing rate of 60% (of the hits)

(detector images) from single pulse exposures only are almost a reality (Table 2). A time-resolved crystallographic Laue dataset can now be collected on the order of 5 min. This speed enables the collection of multiple time delays that ideally span from the fastest delays established by the time-resolution to the end of the reaction. The consecutive alignment of time delays will result in a movie of the reaction (Fig. 1a). Since chemical kinetics is governed by exponential relaxations, the time points (frames) in such a movie are best arranged equidistantly in logarithmic time. This way fast and slow exponential processes are considered by an equal number of time points. Note: if a reaction spans many orders of magnitude in time and the delay times are arranged linearly with time across this reaction, the fastest processes are not probed at all, and almost all time delays probe only the slowest process.

5 Analysis of Time-Resolved Crystallographic Data in Terms of Structure, Kinetics and Energy Landscapes

The most comprehensive time-resolved crystallographic experiments were performed on PYP [37–39]. These investigations were facilitated by the exquisite crystal quality of PYP that allow the collection of excellent Laue diffraction patters. PYP displays a

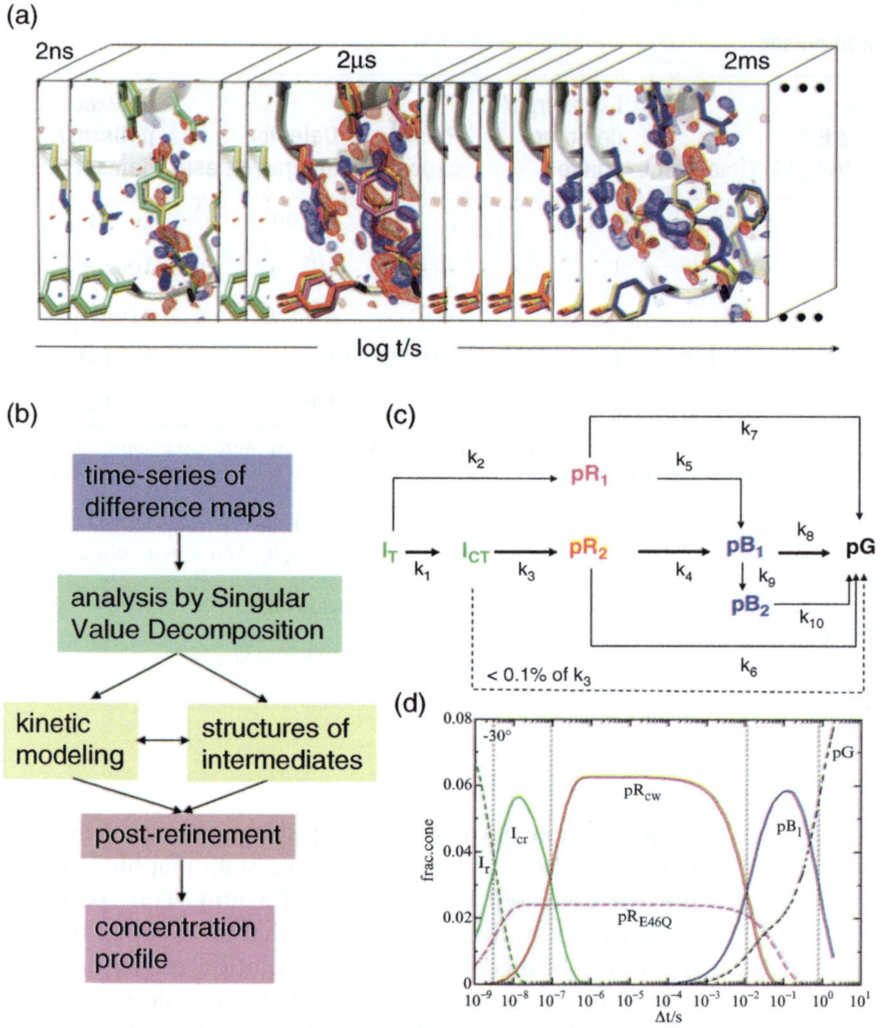

Fig. 1 Kinetic analysis of TR crystallographic data. (**a**) Schematic representation of a time-series of difference maps (structures are guides to the eye), (**b**) flow chart of the analysis driven by singular value decomposition, (**c**) chemical, kinetic mechanism of the PYP photocycle, with rate coefficients k_j, and intermediate states (in various colors); the main pathway along k_1, k_3, k_4, k_8 is shown in *bold*, (**d**) time dependent concentrations of the intermediates after post-refinement of the mechanism against the time series of DED maps. Color code in (**d**) corresponds to that of the intermediates in (**c**)

photocycle shown in Fig. 1c. Only a few pump-probe cycles (Table 2) are necessary to boost the Laue spot intensities. Accurate and complete time-resolved Laue data to around 1.5 Å can be collected using crystals with sizes on the order of 800 μm × 150 μm × 150 μm. Since small X-ray focal spot sizes are employed, the crystals can be translated multiple times to expose fresh volumes to the X-rays to prevent radiation damage [8]. The penetration depth of the optical laser light that excites the reaction is only a few μm at the absorption maximum of PYP of 446 nm.

The laser wavelength must be moved into the flanks of the absorption band. Laser pulses shifted to the blue (390 nm) or to the red (485 nm) were both successfully used to start reactions [38–40]. In these cases the penetration depth increases to about 30 μm [38] which roughly matches to vertical full-width at half-maximum (FWHM) of the X-ray beam. High laser pulse energy densities up to 4.5 mJ/mm^2 are used. It has been shown that around 30 complete Laue datasets can be collected from a single crystal without introducing too much damage by the intense laser and X-ray radiation [8], which otherwise would alter the kinetics of the reaction. By considering three time points per logarithmic decade, 30 time points would amount to a time-series spanning 10 orders of magnitude in time. Hence the entire time-range from 100 ps, the earliest time delay determined by the pulse duration at the synchrotron, to 100 ms, the end of the photocycle, can be probed with one crystal only. It is very important that during data collection the time is the fast variable. That means that all time delays, including diffraction patterns in the dark as reference, are collected at a fixed crystal setting. Only then the crystal is reoriented and translated and the process is repeated until the reciprocal space is covered. This ensures that experimental systematic variations, such as changes in the crystal thickness along the translation axis, are distributed equally through all time delays. This largely smoothens variations from time delay to time delay which might otherwise compromise the kinetic analysis. The raw data then consists of Laue diffraction patterns from about 20–30 different crystal orientations per time delay to cover reciprocal space. Hence, a time series of 30 time-delays plus the reference collected in the dark without laser excitation consists of about 700 Laue diffraction patterns.

Processing of Laue data, from indexing to scaling of intensities, while taking into account the incident X-ray spectrum, is done by specialized and sophisticated software such as "LaueView" [41] or semiautomatically by "Precognition/Epinorm" (RenzResearch Inc). As a result, time-dependent structure factor amplitudes $F_h(t)$, with h the index *hkl*, are obtained (in the following, amplitudes are denoted in normal font, structure factors with amplitude and phase are in bold). Difference structure factor amplitudes $\Delta F_h(t)$ are determined by subtracting the corresponding structure factor amplitudes $F_h(0)$ collected in the dark. The $\Delta F_h(t)$ are preferentially weighted to reduce artifacts caused by poorly measured reflections [33]; *see* Schmidt, 2008 for a detailed description. Using phases from a well refined dark state model, time-dependent difference electron density maps (DED(t)) are determined, which constitute the experimental result.

The time series of DED(t) maps (Fig. 1a) must then be interpreted in terms of structure as well as kinetics. This dual interpretation is the ultimate goal of a time-resolved crystallographic experiment. Existing software tools make use of the singular value decomposition (SVD) [42] to achieve this goal. The crystallo-kinetic software "singular value decomposition for time-resolved crystallography" (SVD4TX) can be downloaded from the author's web page (http://people.uwm.edu/smarius/). A flow chart of the SVD driven analysis is shown in Fig. 1b. The SVD separates space dependencies (difference maps) into the left singular vectors (lSV), and time dependencies into the right singular vectors (rSV). By fitting a candidate kinetic model to the significant rSVs, the time-independent DED maps (DED_i) of the $i = 1...I$ intermediate states can be determined from the corresponding lSVs by projection [10, 43]. It should be noted that it is important to distinguish the measured time-dependent DED maps in the time-series from the time-independent maps of the intermediates determined by the analysis described above. The time-dependent maps are a linear combination, or mixture, of time-independent maps of the intermediates. The linear coefficients are the time-dependent concentrations of the respective intermediate states. The structures of the intermediate states are conveniently determined using extrapolated (conventional) electron density maps calculated by extrapolated structure factors $\mathbf{F}_h^{ext}$. To calculate the $\mathbf{F}_h^{ext}$, first the DED_i map of a particular intermediate is Fourier-inverted to difference structure factors $\Delta\mathbf{F}_{h,i}$. Then, a multiple N of the $\Delta\mathbf{F}_{h,i}$ are added to dark-state structure factors which are calculated from a precisely determined reference (dark state) model [44]. From the $\mathbf{F}_h^{ext}$, extrapolated electron density ρ^{ext} is calculated. N is increased until ρ^{ext} is free of dark state electron density at positions with strong negative density features in the DED_i map [10, 45]. An initial structure is determined by real-space refining a structural model directly into the extrapolated maps with a suitable program such as Coot [46]. Refinement of the initial model proceeds in reciprocal space against the extrapolated amplitudes F_h^{ext} for example with "REFMAC" [47], and hence minimal manual intervention is necessary. Once refined structures of the intermediates are prepared, time-dependent electron density maps $DED(t)^{calc}$ are calculated using the structures of the intermediates, the dark state and the candidate kinetic mechanism [10]. The calculated $DED(t)^{calc}$ are fitted to the observed $DED(t)^{obs}$. In this way the rate coefficients in the kinetic mechanism are post-refined. The refinement also includes a scale factor that determines the extent of reaction initiation (the apparent quantum yield). From the kinetic mechanism, the corresponding post-refined rate coefficients and the extent of reaction initiation, concentration profiles for all intermediates are calculated. Figure 1 shows the result of such an analysis for the PYP photocycle measured at a reduced temperature of −30 °C: Fig. 1a schematically depicts a time-series of difference maps which are the experimental data, Fig. 1c shows a mechanism compatible with the data, and Fig. 1d shows the resulting concentration profile at this temperature. The amount of photoactivated molecules is about 10% at the beginning of the reaction. These experiments lay the foundation for "structure based kinetics," which is the simultaneous extraction of structure and kinetics from a time-series of crystallographic data. Notably, at these low temperatures, the earliest intermediate I_T is observed up to 10 ns, whereas at room temperature the determination of the I_T structure requires picosecond time resolution [39, 40].

When the temperature is increased the photocycle speeds up considerably [38]. For example, rate coefficient k_8 (Fig. 1c) changes from 0.05 s^{-1} at −40 °C to 190 s^{-1} at 40 °C, an increase by a factor of 3800. The ability to observe the PYP photocycle over a large temperature range (from −40 to 70 °C) enables the determination of barriers of activation with entropy and enthalpy differences to the transition states in protein crystals solely from time-resolved crystallographic data (Fig. 2) [38, 48]. Since in addition to space

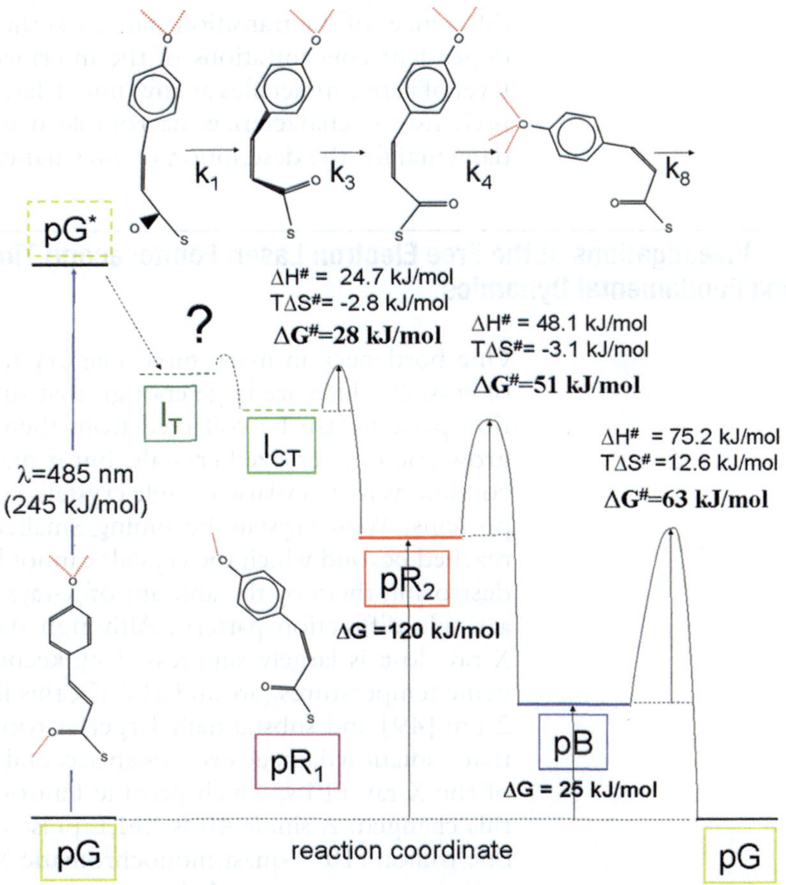

Fig. 2 Energy landscape of the main reaction pathway in the PYP photocycle [38]. The structures of the intermediates as well as entropy (at 288 K), enthalpy and free energy differences of the transition states are solely determined by five-dimensional crystallography (*see* text). Free energies of states pR_2 and pB are determined from solution and might be different in the crystal. Note that the reaction coordinate is cyclic (has periodic boundaries). Colors of the intermediates correspond to those in Fig. 1c, d

and time, now the temperature is varied, this method has been denoted five-dimensional crystallography [48]. The impact of any other parameter on the protein kinetics caused by physical factors such as pressure or exposure to X-rays [8] as well as the effect of chemical modifications caused for instance by pH [45], small molecules (drugs) and mutations may be also investigated by structure based kinetics this way.

In summary, the analysis of the time and temperature dependent DED maps will provide (1) the kinetics and temperature-dependent relaxation times, (2) the structures of the intermediates, whereas intermediates that decay faster than the time resolution at higher temperatures may still be observed at lower temperatures, (3) candidate chemical, kinetic mechanisms compatible with the data including (4) a set of (refined), temperature-dependent rate-coefficients, (5) barriers of activation with entropy and enthalpy differences of the transition states, (6) the time- and temperature-dependent concentrations of the intermediates as well as (7) the level of active molecules at any time delay. These observables comprehensively characterize macromolecular reactions, and are fundamental for the description of enzymatically catalyzed reactions.

6 Investigations at the Free Electron Laser: Femtosecond Time Scale and Fundamental Dynamics

One bottleneck in macromolecular crystallography is the growth of crystals which are large enough that sufficiently intense diffraction patterns can be collected from them. It is relatively easy to grow micrometer sized crystals, but it may take years to optimize conditions to grow larger single crystals, especially from membrane proteins. With crystals becoming smaller and smaller a limit is reached beyond which the crystals cannot be made smaller without destroying them by the amount of X-rays required to collect even a single diffraction pattern. Although damage by the deposited X-ray dose is largely suppressed by keeping the crystals at cryogenic temperatures (around 100 K), this limit seems to be around 2 μm [49] and substantially larger at room temperature. Beyond that, radiation damage exceeds an acceptable level. With the advent of the X-ray FELs, which provide femtosecond (fs) X-ray pulses, this changed. A single 40 fs XFEL pulse contains on the order of one trillion (10^{12}) quasi-monochromatic X-ray photons. For comparison, the strongest 3rd generation synchrotron beamline to date (BioCARS at the Advanced Photon Source, Argonne, IL) provides on the order 50 billion (5×10^{10}) polychromatic X-ray photons in a 100 ps pulse. At the XFEL the peak temporal photon density (number of X-ray photons per unit time per X-ray pulse) is about 50,000 times larger than at the synchrotron. This ratio is even larger for quasi-monochromatic photons, say within a 0.1% bandwidth. Within this bandwidth, the XFEL provides even 2.5 million times more quasi-monochromatic photons per unit time than the synchrotron. In addition, this large flux can be focused to an extremely small focal spot, since as the name suggests the X-ray FEL is a laser, which features a spatially coherent beam with very small divergence or crossfire. It is this small crossfire that is mainly responsible for the immense increase in brilliance, which is 9–10 orders of magnitude larger compared to the synchrotron. Focal spots as small as 100 nm without loss of X-ray photons are already routine at the Linac Coherent Light Source (LCLS) at Stanford Linear Accelerator Center (SLAC) in Menlo Park, CA. This very high flux in small spots provides the means necessary to interrogate nanocrystals. However, when a large number of X-ray photons are incident on such small crystals, the samples are irreversibly destroyed. The dose they suffer is several orders of magnitude larger than the safe-dose at which the damage can be tolerated [50]. However, a diffraction pattern is collected *before* the crystal is damaged. This is called the "diffraction-before-destruction" principle [51–53]. The crystal size limit has been overcome and there is no longer any need to grow large crystals. Crystals with only a few hundred unit cells and the edge length of a few hundred nanometers can be investigated at room temperature. Of course, since they are destroyed, the method requires a constant stream of fresh tiny crystals that are intercepted at random orientations by the XFEL beam. Since a serial stream of crystals is involved, the XFEL pulses are femtosecond long, and the crystals are nanosized, this method has been named "Serial Femtosecond Nano-Crystallography" (SFX) [53, 54].

Special injectors had to be developed to provide the stream of crystals for these experiments. There exist different types of injectors to date. Gas dynamic virtual nozzles (GDVN) [55] are the workhorses for SFX. They are used for crystals of soluble proteins as well as for membrane proteins, provided they can be maintained in liquid suspension. However, they require a relatively large number of crystals due to the high flow rate required for the formation of the jet. Another injector design, the lipidic cubic phase (LCP) injector [56], takes advantage of the slow flow rate possible with viscous media (picoL/min). It consumes little protein, but requires that the crystals be embedded in a viscous carrier medium, such as agarose [57], synthetic mineral lube [58] or, as the name suggests—lipids, such as monoolein, that form a lipidic cubic phase. A third type of injectors is the electrospun injector [59]. The jet is formed by a large electric field between the tip of the nozzle and the catcher. It also consumes very little protein. Other opportunities are provided by fixed targets, where tiny crystals are mounted on a regular grid [60–62] or deposited by other means [63] and scanned through the X-ray beam.

With all injector designs the diffraction patterns are obtained from nanometer to micrometer sized crystals in random orientation. Hence, the orientation of each and every diffraction pattern (snapshot) that contains Bragg spots must be determined (indexed) anew. Since the XFEL beam is quasi-monochromatic with a bandwidth on the order of 0.1%, only partial reflections are obtained from each snapshot. In order to reconstruct the integrated reflection intensities, a large number of these partial observations must be averaged for each reflection [64, 65]. For a good dataset, each reflection is observed more than 1000 times in as much as 60,000 indexed detector readouts (snapshots) [66]. With a high flow-rate GDVN, to obtain this number of indexeable snapshots, about two million snapshots must be collected. For comparison, the numbers from monochromatic crystallography using the rotation method are as follows: (1) the orientation has to be only determined once, all subsequent diffraction patterns are indexed from it based on known rotations. (2) Each reflection is addressed typically three to four times. (3) Subsequent diffraction patterns differ by a rotation of ~0.1°, hence the rocking curve is faithfully traced for each reflection. (4) The number of diffraction patterns collected is usually less than 1000. To make things more complicated for the XFELs, the pulse repetition rate at LCLS is currently only 120 Hz, hence there are ~8 ms gaps between the X-ray pulses. Since jet velocities with a GDVN are on the order of 10 m/s [55], 8 cm of the crystal containing jet will pass by, before it is hit by an X-ray pulse again. In addition, for typical densities of crystal suspensions, 5% of the X-ray pulses actually hit a crystal, while the remaining 95% pulses produce detector readouts without any Bragg reflections. In total, only one out of two million crystals will ever be interrogated, the rest will never see the X-ray beam and will be discarded. Thus 10^5 detector readouts containing Bragg spots (hits), of which 60% (6×10^4) can be indexed for a good dataset, require 2×10^{11} crystals. For this, 10 mL of a highly concentrated (2×10^{10} crystals/mL) crystal suspension with about 300 mg of total protein must be prepared. This amount lasts for about three quarters of a 12 h shift at the LCLS. New XFELs with higher repetition rate will come online in the near future. The European XFEL (EuXFEL) will feature a repetition rate of 27 kHz on the average, with 5 MHz bursts, and the planned new LCLS (LCLS-II) will offer repetition rates of up to 1 MHz. Rather than consuming 0.3 g of protein, less than a milligram will be required per data set in the future. It takes several hours to collect a dataset today, and it will take only a few minutes at the new machines (*see* also Chapter 12 by Chapman in this volume).

SFX can also be employed for time-resolved crystallographic experiments (TR-SFX) [19, 66–68]. TR-SFX data are collected in the same way as described above. However, a reaction in the crystals has to be initiated first, before the crystals are interrogated (and potentially destroyed) by the XFEL beam. As noted above, the fastest way to initiate a reaction is with a pulsed optical laser, which must be synchronized to the X-ray pulses. The crystals are intercepted twice in flight, the first time by the optical laser pulse that starts the reaction, and the second time after a delay Δt by the X-ray pulse. Due to the femtosecond duration of the X-ray pulses, XFELs can provide femtosecond time resolution [16, 19]. As in all other time-resolved experiments, the most challenging problem is the reaction initiation. To reach femtosecond time resolution, the reaction must be initiated by femtosecond optical laser pulses. The temporal photon density in femtosecond laser pulses is immense. First consider 4 ns blue (450 nm) laser pulses with a flux density of 0.8 mJ/mm^2 which were used in the first successful time-resolved crystallographic experiment on a protein at the LCLS with near atomic resolution [66]. The spatiotemporal photon density is about 5×10^{23} photons s^{-1} mm^{-2}. If the same energy density of 0.8 mJ/mm^2 is now produced in 100 fs by a femtosecond laser, the photon density is 2×10^{28} photons s^{-1} mm^{-2}, which is about five orders of magnitude larger than with a nanosecond laser. This enormous increase might lead to unwanted effects, such as 2-photon absorption and radical generation that may ultimately lead to the destruction of the chromophore in the protein. Careful adjustment of the laser power is necessary and compromises have to be considered. Only higher photon counts will activate enough molecules so that a difference signal can be observed, but too high laser powers will irreversibly bleach and damage the chromophore. It is advisable to investigate the reaction beforehand by ultrafast spectroscopy to explore which laser powers are acceptable.

The second most important difference between nanosecond and femtosecond excitation is that with nanosecond laser pulses each molecule in the crystals may be excited multiple times and may be pumped this way into the reaction cycle. An example makes this clear: The primary quantum yield to reach the photocycle in PYP is small (<20%) [69]. The excited state lifetime of the para-coumaric acid (pCA) chromophore in PYP is about 500 fs [19, 70, 71]. A substantial fraction of PYP does not enter the photocycle but returns back to the dark state. Within a nanosecond laser pulse, the dark state can absorb a photon again and the process can be repeated multiple times. Each time, a fraction of the molecules may reach the photocycle, boosting the number of molecules in the photocycle. The penetration depth and the extent of reaction initiation (the number of molecules in the photocycle) is then a delicate balance between absorption cross sections of the electronic ground and the excited states and occupation of these states [72]. The extent of reaction initiation achieved in a recent nanosecond experiment with PYP microcrystals was about 40% [66]. When femtosecond laser pulses are employed, however, there is only one absorption event possible. When the PYP returns to the dark state, the femtosecond laser pulse has already passed. As a consequence, the photoexcitation yield is limited to the primary yield.

The third consideration is the crystal size. Typical penetration depths are on the order of a few μm for moderately absorbing chromophores at the absorption maximum. The penetration depth should be estimated on a case-by-case basis when linear and nonlinear absorption cross sections are known [72]. As a rule of thumb, crystal sizes around 5 μm are acceptable. Then reactions are optimally started by excitation directly at the absorption maximum [22] rather than into the flanks of the absorption band, as discussed above. It is apparent that small microcrystals and nanocrystals are advantageous for femtosecond time-resolved experiments, since illumination is uniform and excitation is optimal. Injection by a GDVN is only one experimental possibility to investigate these small crystals with TR-SFX. Multiple designs might work equally well, although smaller jet velocity (mm/s) must be taken into account, so that the laser spot area does not overlap with a previously excited jet volume or a volume designated to be exposed in the dark.

Conditions for femtosecond laser excitation were established for PYP prior to the time-resolved crystallographic experiments [19]. The photocycle was investigated with TR-SFX at the CXI instrument [73] of the LCLS. A successful control experiment at a 200 ns delay showed that a sufficiently high photoactivation yield was achieved (Fig. 3a). The DED is compared to the known

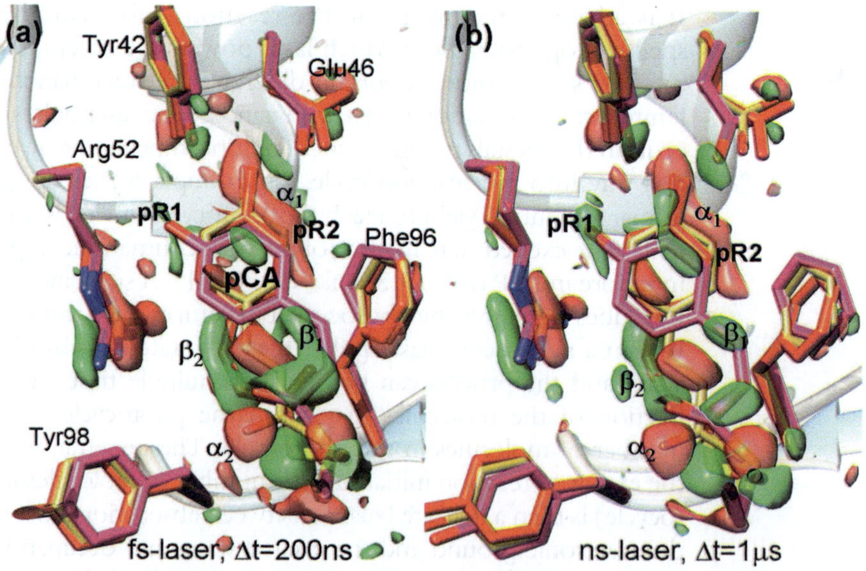

Fig. 3 Difference maps from TR-SFX after excitation with 140 fs laser pulses (**a**) and 4 ns laser pulses (**b**) at 200 ns and 1 μs delay times, respectively. Negative differences in *red* (-3σ), positive differences in *green* (3σ). Both time delays are occupied by the same mixture of pR_1 (*magenta*) and pR_2 (*red*). The dark state is shown in *yellow*. The same density features (α_1, α_2) and positive features (β_1 for pR_1) and (β_2 for pR_2) are present in both maps. The pR_1 and pR_2 structures are essentially identical in (**a**) and (**b**)

difference map (Fig. 3b) obtained by TR-SFX using nanosecond laser excitation [66]. Both difference maps show the same mixture of states pR_1 and pR_2, hence femtosecond excitation has been successfully achieved for PYP. This opened the door for experiments on the femtosecond time scale. Various femtosecond time delays were probed [19]. Figure 4a shows an example of a difference electron density map 250 fs after laser excitation using 140 fs laser pulses with a pulse energy of 0.8 mJ/mm^2 at 450 nm. The chromophore atoms are displaced by 0.7 Å on average already at this early time-delay. Spectroscopic investigations [69–71] and quantum molecular mechanics (QM/MM) calculations [74] show that the PYP is in an electronically excited state (ES). After promotion to the ES, PYP relaxes rapidly on the excited state potential energy surface (ES-PES). This result is new and exciting, as all previous TR crystallographic experiments on PYP probed only the electronic ground state (GS) dynamics. The distortion of the chromophore on the fast fs time scale was also predicted by ultrafast Raman spectroscopy [71], but the exact nature of this distortion remained obscure. After about 500 fs the *trans* to *cis* isomerization of the chromophore

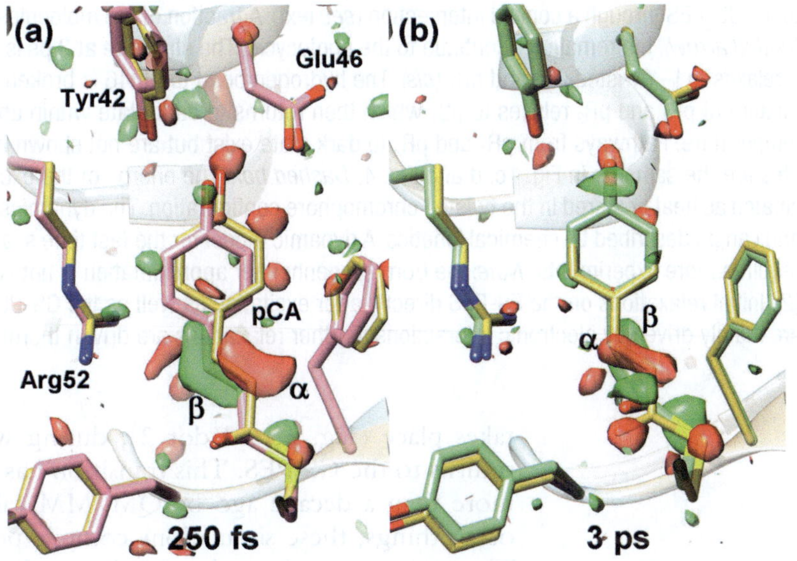

Fig. 4 Femtosecond dynamics of the PYP chromophore. Structure of the dark state in *yellow*. Some important residues and the para-coumaric acid (pCA) chromophore are marked in (**a**). Difference maps at the -3σ (*red*) and 3σ (*green*) contour level. (**a**) 250 fs delay, the chromophore (*pink*) is in the electronically excited state. Strong negative and positive features indicate the displacement of the entire chromophore. Positive feature β is kinked, the chromophore configuration is still *trans*. (**b**) 3 ps delay, the chromophore (*green*) is in the electronic ground state. Feature β is aligned along the chromophore tail axis and the tail carbonyl points out of the drawing plane. The configuration is *cis*

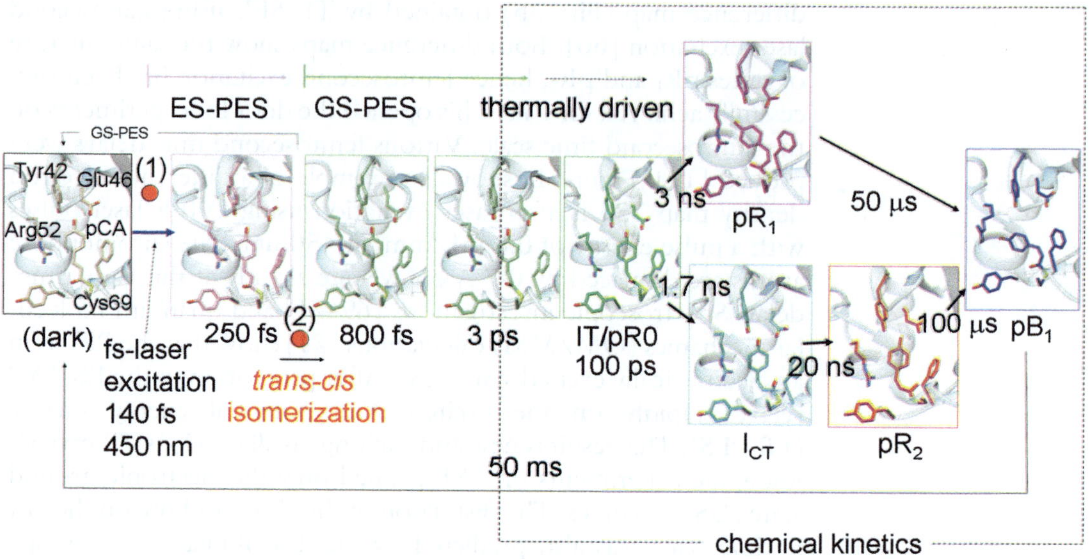

Fig. 5 The PYP photocycle comprehensively investigated from femtosecond times to the end of the reaction. Excitation by a fs laser promotes the pCA chromophore to the ES-PES. At 250 fs the pCA structure is twisted *trans*. The *trans* to *cis* isomerization happens around 550 fs. At 800 fs the chromophore is nearly *cis*. The structure relaxes on the GS-PES. *Red dot* (*1*): photoactivation, promotion to the ES-PES, *red dot* (*2*): transition to the GS-PES through a conical intersection (*see* text). A fraction of the molecules revert back to the dark state (*dotted arrow*), the remainder continue to the photocycle. The structure at 3 ps is almost identical to I_T (or pR_0). I_T relaxes to I_{CT} (twisted *cis*) and pR_1 (*cis*). The hydrogen bond to Glu46 is broken in pR_1. I_{CT} relaxes to pR_2. The mixture of pR_1 and pR_2 relaxes to pB_1 which then returns to dark state within about 50 ms depending on the temperature. Pathways from pR_1 and pR_2 to dark state exist but are not shown here. Colors of the intermediates are the same as in Fig. 1c, d and Fig. 4. *Dashed box:* The energy of the exciting blue photon is fully dissipated as heat or stored in the twisted chromophore configuration. The dynamics is driven by the thermal bath and can be described by chemical kinetics. A dynamic model for the fast time scale needs to be developed and requires more experiments. *Note*: the Born–Oppenheimer approximation is not valid at *red dot* (*1*) and *red dot* (*2*). Initial relaxations on the ES-PES directly after excitation as well as the GS-PES directly after the transition are mainly driven by electronic interactions. Further relaxations are driven thermally

takes place (Fig. 5, red dot 2), during which the chromophore returns to the GS-PES. This transition has been simulated for PYP more than a decade ago by QM/MM calculations [75]. Among other things, these simulations compile potential energy surfaces. The energy depends on the atomic coordinates of both protein and chromophore and on the state of excitation. When the ES-PES and the GS-PES meet, they form a conical intersection [74]. Although it has been contemplated by spectroscopy for some time ([76, 77], *see* [78] for a review), a direct structural observation of the transition through a conical intersection has long been sought after. Now, it has been observed for the first time [19]. Relaxation on the GS-PES is complete approximately after 3 ps where the first fast intermediate accumulates (Fig. 4b). The 3 ps structure is essentially identical to I_T which is the earliest intermediate identified by synchrotron radiation (Fig. 5). I_T and later structures (I_{CT}, pR_1, pR_2, and pB_1) are all known from time-resolved Laue crystallography [37–39]. The results of the TR-SFX experiments seamlessly integrate in the photocycle (Figs. 1c and 5). Fundamental ultrafast motions of the ES-PES that trigger chemical reactions (here the *trans* to *cis* isomerization) can now be probed in real time and understood in detail.

7 Structure-Based Enzymology

Most biological reactions cannot be initiated by light, thus other methods need to be developed as briefly mentioned above. The great advantage of serial crystallography is that cyclic and noncyclic reactions are on the same footing. Regardless of the reaction type each crystal sees the X-ray beam only once and is discarded afterwards. When crystals are small, substrate can diffuse into the crystals quite fast (Table 3) and reactions are initiated simply by diffusion. This has been named "mix-and-inject" [79]. The time-resolution is limited by either the mixing time, the diffusion time, or the time needed to transport the mixture into the X-ray interaction volume, whichever is the longest. The simplicity of the "mix-and-inject" approach (Table 1) may revolutionize time-resolved structural investigations in a sense that it provides the opportunity to routinely and seemingly effortlessly observe many biologically, pharmaceutically, and medically important enzymes in action (Fig. 6). Since turnover times in typical enzymes are in milliseconds, fast time-resolution might not be required for many enzymes, except for the fastest. This allows for somewhat larger crystals which could be interrogated also by the less brilliant and longer synchrotron X-ray pulses. Recently, X-ray focal spots on the order

Table 3
Characteristic diffusion times and tumbling times (rough guess) of different crystal sizes

Crystal size	Diffusion time[a]	Tumbling time τ[b]
0.5 μm	17 μs	230 μs/degree
2 μm	270 μs	14 ms/degree
10 μm	6.5 ms	1.8 s/degree

Laue exposure times must be much faster than the tumbling times to avoid streaks. Tumbling might be favorable for monochromatic data collection, since crystals rotate through the Ewald sphere and more full reflections are collected

[a]With $D = 5 \times 10^{-6}$ cm^2 s^{-1} for glucose in water at 25 °C

[b]With viscosity η of water 0.8×10^{-3} Pa s at 288 °C, $\tau = \frac{4}{3}\pi\eta\frac{R^3}{k_B T}$ in s/rad

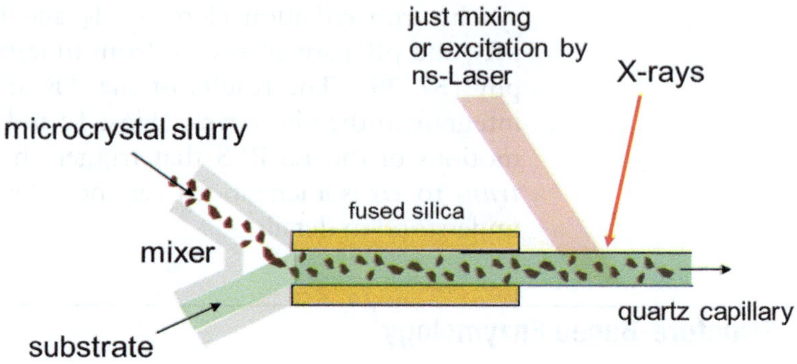

Fig. 6 Principle of "mix-and-inject." A slurry of small crystals is mixed with the substrate which is allowed to diffuse into the crystals. The enzymatically catalyzed reaction is probed by the X-ray beam. The experiment can also be performed with inactive, caged substrates which can be activated by ns-laser pulses shortly before the X-rays probe the crystals

of 20 μm were produced at BioCARS using advanced focusing optics. This would allow the measurement of 10 μm sized crystals where sufficient substrate concentrations in the crystal can be reached quite fast [9], on the order of 5 ms, depending on the substrate concentration and crystal packing (Table 3). In addition, experiments could conveniently be performed with caged substrates that are mixed with the crystals and activated shortly before they are probed (Fig. 6). On these longer time-scales XFELs are not strictly required and the experiments could be conducted at more conventional, stable synchrotron light sources with high repetition rates. An X-rays exposure consisting of multiple X-ray pulses can be used for a snapshot [80–83] as long as the characteristic tumbling times of the crystals [9] are much longer than the exposure (Table 3). At the Advanced Photon Source operated in the 24 bunch mode, X-ray pulses arrive about every 150 ns. A 10 μs exposure is much faster than the characteristic tumbling time of 10 μm crystals. X-ray photons from 70 pulses are combined in these 10 μs, which should be sufficient to obtain a good Laue diffraction pattern. Since the crystals are rapidly replaced after only one X-ray exposure, concerns about radiation damage are largely alleviated even outside the "diffraction-before-destruction" regime. The experiments can be performed at ambient temperature, which is necessary to observe the macromolecular dynamics. It should be stressed that fastest diffusion times to investigate fast processes are only reached with the smallest crystals, which may be beyond the reach of the synchrotron and necessarily require an XFEL. The decision which pulsed light source is appropriate has to be done on a case-by-case basis given the reaction to be probed. However, the "pump-probe" and the "mix-and-inject" techniques performed in conjunction with serial crystallography will provide the long-awaited tools to rotinely investigate a large number of important proteins/enzymes with time-resolved crystallography.

Acknowledgment

M.S. thanks Vukica Šrajer for reading, and commenting on, an earlier version of the manuscript. This work is supported by the BioXFEL Science and Technology Center (NSF grant 1231306).

References

1. Bourgeois D, Weik M (2009) Kinetic protein crystallography: a tool to watch proteins in action. Crystallogr Rev 15:87–118
2. Weik M, Colletier JP (2010) Temperature-dependent macromolecular X-ray crystallography. Acta Crystallogr D Biol Crystallogr 66:437–446
3. Nienhaus K, Ostermann A, Nienhaus GU et al (2005) Ligand migration and protein fluctuations in myoglobin mutant L29W. Biochemistry 44:5095–5105
4. Moffat K (1989) Time-resolved macromolecular crystallography. Annu Rev Biophys Biophys Chem 18:309–332
5. Moffat K, Szebenyi D, Bilderback D (1984) X-ray Laue diffraction from protein crystals. Science 223:1423–1425
6. Barker AI, Southworth-Davies RJ, Paithankar KS et al (2009) Room-temperature scavengers for macromolecular crystallography: increased lifetimes and modified dose dependence of the intensity decay. J Synchrotron Radiat 16:205–216
7. Youngblut M, Judd ET, Srajer V et al (2012) Laue crystal structure of Shewanella oneidensis cytochrome c nitrite reductase from a high-yield expression system. J Biol Inorg Chem 17:647–662
8. Schmidt M, Srajer V, Purwar N et al (2012) The kinetic dose limit in room-temperature time-resolved macromolecular crystallography. J Synchrotron Radiat 19:264–273
9. Schmidt M (2015) Time-resolved crystallography at X-ray free electron lasers and synchrotron light sources. Synchrotron Radiat News 28:25–30
10. Schmidt M (2008) Structure based enzyme kinetics by time-resolved X-ray crystallography. In: Zinth W, Braun M, Gilch P (eds) Ultrashort laser pulses in medicine and biology, Biological and medical physics, biomedical engineering. Springer, Berlin
11. Ren Z, Bourgeois D, Helliwell JR et al (1999) Laue crystallography: coming of age. J Synchrotron Radiat 6:891–917
12. Stoddard BL (1998) New results using Laue diffraction and time-resolved crystallography. Curr Opin Struct Biol 8:612–618
13. Srajer V (2013) Time-resolved macromolecular crystallography in practice at BioCARS, advanced photon source: from data collection to structures of intermediates. In: Howard JAK, Sparkes HA, Raithby PR, Churakov AV (eds) The future of dynamic structural science. Springer, New York, pp 237–251
14. Schmidt M, Ihee H, Pahl R et al (2005) Protein-ligand interaction probed by time-resolved crystallography. Methods Mol Biol 305:115–154
15. Bourgeois D, Royant A (2005) Advances in kinetic protein crystallography. Curr Opin Struct Biol 15:538–547
16. Barends TR, Foucar L, Ardevol A et al (2015) Direct observation of ultrafast collective motions in CO myoglobin upon ligand dissociation. Science 350:445–450
17. Bionta MR, Lemke HT, Cryan JP et al (2011) Spectral encoding of X-ray/optical relative delay. Opt Express 19:21855–21865
18. Hartmann N, Helml W, Galler A et al (2014) Sub-femtosecond precision measurement of relative X-ray arrival time for free-electron lasers. Nat Photonics 8:706–709
19. Pande K, Hutchison CDM, Groenhof G et al (2016) Femtosecond structural dynamics drives the trans/cis isomerization in photoactive yellow protein. Science 352:725–729
20. Auldridge ME, Forest KT (2011) Bacterial phytochromes: more than meets the light. Crit Rev Biochem Mol Biol 46:67–88

21. Schmidt M, Patel A, Zhao Y et al (2007) Structural basis for the photochemistry of alpha-phycoerythrocyanin. Biochemistry 46:416–423
22. Purwar N, Tenboer J, Tripathi S et al (2013) Spectroscopic studies of model photo-receptors: validation of a nanosecond time-resolved microspectrophotometer design using photoactive yellow protein and α-phycoerythrocyanin. Int J Mol Sci 14:18881–18898
23. Moglich A, Ayers RA, Moffat K (2010) Addition at the molecular level: signal integration in designed Per-ARNT-Sim receptor proteins. J Mol Biol 400:477–486
24. Moffat K (2014) Time-resolved crystallography and protein design: signalling photoreceptors and optogenetics. Phil Trans R Soc London B369:20130568
25. Schlichting I, Almo SC, Rapp G et al (1990) Time-resolved X-ray crystallographic study of the conformational change in Ha-Ras p21 protein on GTP hydrolysis. Nature 345:309–315
26. Adams SR, Tsien RY (1993) Controlling cell chemistry with caged compounds. Annu Rev Physiol 55:755–784
27. Goelder M, Givens R (eds) (2005) Dynamic studies in biology: phototriggers, photoswitches and caged biomolecules. Wiley-VCH, Weinheim
28. Ursby T, Weik M, Fioravanti E et al (2002) Cryophotolysis of caged compounds: a technique for trapping intermediate states in protein crystals. Acta Crystallogr D Biol Crystallogr 58:607–614
29. Bourgeois D, Weik M (2005) New perspectives in kinetic protein crystallography using caged compounds. In: Dynamic studies in biology: phototriggers, photoswitches and caged biomolecules. Wiley-VCH, Weinheim, pp 410–432
30. Kurisu G, Sugimoto A, Kai Y et al (1997) A flow cell suitable for time-resolved X-ray crystallography by the Laue method. J Appl Crystallogr 30:555–556
31. Moffat K, Chen Y, Ng KM et al (1992) Time-resolved crystallography—principles, problems and practice. Philos Trans R Soc A340:175–189
32. Srajer V, Teng TY, Ursby T et al (1996) Photolysis of the carbon monoxide complex of myoglobin: nanosecond time-resolved crystallography. Science 274:1726–1729
33. Ren Z, Perman B, Srajer V et al (2001) A molecular movie at 1.8 a resolution displays the photocycle of photoactive yellow protein, a eubacterial blue-light receptor, from nanoseconds to seconds. Biochemistry 40:13788–13801
34. Srajer V, Ren Z, Teng TY et al (2001) Protein conformational relaxation and ligand migration in myoglobin: a nanosecond to millisecond molecular movie from time-resolved Laue X-ray diffraction. Biochemistry 40:13802–13815
35. Srajer V, Crosson S, Schmidt M et al (2000) Extraction of accurate structure-factor amplitudes from Laue data: wavelength normalization with wiggler and undulator X-ray sources. J Synchrotron Radiat 7:236–244
36. Graber T, Anderson S, Brewer H et al (2011) BioCARS: a synchrotron resource for time-resolved X-ray science. J Synchrotron Radiat 18:658–670
37. Ihee H, Rajagopal S, Srajer V et al (2005) Visualizing reaction pathways in photoactive yellow protein from nanoseconds to seconds. Proc Natl Acad Sci U S A 102:7145–7150
38. Schmidt M, Srajer V, Henning R et al (2013) Protein energy landscapes determined by five-dimensional crystallography. Acta Crystallogr D Biol Crystallogr 69:2534–2542
39. Jung YO, Lee JH, Kim J et al (2013) Volume-conserving trans-cis isomerization pathways in photoactive yellow protein visualized by picosecond X-ray crystallography. Nat Chem 5:212–220
40. Schotte F, Cho HS, Kaila VR et al (2012) Watching a signaling protein function in real time via 100-ps time-resolved Laue crystallography. Proc Natl Acad Sci U S A 109:19256–19261
41. Ren Z, Moffat K (1995) Quantitative analysis of synchrotron Laue diffraction patterns in macromolecular crystallography. J Appl Crystallogr 28:461–481
42. Schmidt M, Rajagopal S, Ren Z et al (2003) Application of singular value decomposition to the analysis of time-resolved macromolecular X-ray data. Biophys J 84:2112–2129
43. Henry ER, Hofrichter J (1992) Singular value decomposition—application to analysis of experimental data. Meth Enzymol 210:129–192
44. Terwilliger TC, Berendzen J (1996) Bayesian difference refinement. Acta Crystallogr D Biol Crystallogr 52:1004–1011
45. Tripathi S, Srajer V, Purwar N et al (2012) pH dependence of the photoactive yellow protein photocycle investigated by time-resolved crystallography. Biophys J 102:325–332
46. Emsley P, Lohkamp B, Scott WG et al (2010) Features and development of coot. Acta Crystallogr D Biol Crystallogr 66:486–501
47. Murshudov GN, Skubak P, Lebedev AA et al (2011) REFMAC5 for the refinement of macromolecular crystal structures. Acta Crystallogr D Biol Crystallogr 67:355–367

48. Schmidt M, Graber T, Henning R et al (2010) Five-dimensional crystallography. Acta Crystallogr A 66:198–206
49. Holton JM, Frankel KA (2010) The minimum crystal size needed for a complete diffraction data set. Acta Crystallogr D Biol Crystallogr 66:393–408
50. Lomb L, Barends TR, Kassemeyer S et al (2011) Radiation damage in protein serial femtosecond crystallography using an X-ray free-electron laser. Phys Rev B84:214111
51. Chapman HN, Barty A, Bogan MJ et al (2006) Femtosecond diffractive imaging with a soft-X-ray free-electron laser. Nat Phys 2:839–843
52. Neutze R, Wouts R, van der Spoel D et al (2000) Potential for biomolecular imaging with femtosecond X-ray pulses. Nature 406:752–757
53. Chapman HN, Fromme P, Barty A et al (2011) Femtosecond X-ray protein nanocrystallography. Nature 470:73–77
54. Boutet S, Lomb L, Williams GJ et al (2012) High-resolution protein structure determination by serial femtosecond crystallography. Science 337:362–364
55. Weierstall U, Spence JC, Doak RB (2012) Injector for scattering measurements on fully solvated biospecies. Rev Sci Instrum 83:035108
56. Weierstall U, James D, Wang C et al (2014) Lipidic cubic phase injector facilitates membrane protein serial femtosecond crystallography. Nat Commun 5:3309
57. Conrad C, Basu S, James D et al (2015) A novel inert crystal delivery medium for serial femtosecond crystallography. IUCrJ 2:421–430
58. Sugahara M, Mizohata E, Nango E et al (2015) Grease matrix as a versatile carrier of proteins for serial crystallography. Nat Methods 12:61–63
59. Sierra RG, Laksmono H, Kern J et al (2012) Nanoflow electrospinning serial femtosecond crystallography. Acta Crystallogr D Biol Crystallogr 68:1584–1587
60. Mueller C, Marx A, Epp SW et al (2015) Fixed target matrix for femtosecond time-resolved and in situ serial micro-crystallography. Struct Dyn 2:054302
61. Hunter MS, Segelke B, Messerschmidt M et al (2014) Fixed-target protein serial microcrystallography with an X-ray free electron laser. Sci Rep 4:6026
62. Zarrine-Afsar A, Barends TRM, Muller C et al (2012) Crystallography on a chip. Acta Crystallogr D Biol Crystallogr 68:321–323
63. Roessler CG, Agarwal R, Allaire M et al (2016) Acoustic injectors for drop-on-demand serial femtosecond crystallography. Structure 24:631–640
64. Kirian RA, White TA, Holton JM et al (2011) Structure-factor analysis of femtosecond microdiffraction patterns from protein nanocrystals. Acta Crystallogr A 67:131–140
65. White TA, Kirian RA, Martin AV et al (2012) CrystFEL: a software suite for snapshot serial crystallography. J Appl Crystallogr 45:335–341
66. Tenboer J, Basu S, Zatsepin N et al (2014) Time-resolved serial crystallography captures high-resolution intermediates of photoactive yellow protein. Science 346:1242–1246
67. Aquila A, Hunter MS, Doak RB et al (2012) Time-resolved protein nanocrystallography using an X-ray free-electron laser. Opt Express 20:2706–2716
68. Kupitz C, Basu S, Grotjohann I et al (2014) Serial time-resolved crystallography of photosystem II using a femtosecond X-ray laser. Nature 513:5
69. Lincoln CN, Fitzpatrick AE, van Thor JJ (2012) Photoisomerisation quantum yield and non-linear cross-sections with femtosecond excitation of the photoactive yellow protein. Phys Chem Chem Phys 14:15752–15764
70. Nakamura R, Hamada N, Ichida H et al (2007) Coherent oscillations in ultrafast fluorescence of photoactive yellow protein. J Chem Phys 127:215102
71. Creelman M, Kumauchi M, Hoff WD et al (2014) Chromophore dynamics in the PYP photocycle from femtosecond stimulated Raman spectroscopy. J Phys Chem B118:659–667
72. Hutchison CDM, Tenboer J, Kupitz C et al (2016) Photocycle populations with femtosecond excitation of crystalline photoactive yellow protein. J Chem Phys Lett 654:63–71
73. Liang M, Williams GJ, Messerschmidt M et al (2015) The coherent X-ray imaging instrument at the Linac coherent light source. J Synchrotron Radiat 22:514–519
74. Groenhof G, Bouxin-Cademartory M, Hess B et al (2004) Photoactivation of the photoactive yellow protein: why photon absorption triggers a trans-to-cis isomerization of the chromophore in the protein. J Am Chem Soc 126:4228–4233
75. Groenhof G (2013) Introduction to QM/MM simulations. Methods Mol Biol 924:43–66
76. Polli D, Altoe P, Weingart O et al (2010) Conical intersection dynamics of the primary photoisomerization event in vision. Nature 467:440–443
77. Johnson PJ, Halpin A, Morizumi T et al (2015) Local vibrational coherences drive the

primary photochemistry of vision. Nat Chem 7:980–986

78. Blancafort L (2014) Photochemistry and photophysics at extended seams of conical intersection. Chemphyschem 15:3166–3181
79. Schmidt M (2013) Mix and inject, reaction initiation by diffusion for time-resolved macromolecular crystallography. Adv Condens Mat Phys 2013:1–10
80. Botha S, Nass K, Barends TR et al (2015) Room-temperature serial crystallography at synchrotron X-ray sources using slowly flowing free-standing high-viscosity microstreams. Acta Crystallogr D Biol Crystallogr 71:387–397
81. Stellato F, Oberthuer D, Mengning L et al (2014) Room-temperature macromolecular serial crystallography using synchrotron radiation. IUCrJ 1:204–212
82. Pawate AS, Srajer V, Schieferstein J et al (2015) Towards time-resolved serial crystallography in a microfluidic device. Acta Crystallogr F Struct Biol Commun 71:823–830
83. Perry SL, Guha S, Pawate AS et al (2014) Serial Laue diffraction on a microfluidic crystallization device. J Appl Crystallogr 47: 1975–1982
84. Nogly P, James D, Wang D et al (2015) Lipidic cubic phase serial millisecond crystallography using synchrotron radiation. IUCrJ 2:168–176

Chapter 12

Structure Determination Using X-Ray Free-Electron Laser Pulses

Henry N. Chapman

Abstract

The intense X-ray pulses from free-electron lasers, of only femtoseconds duration, outrun most of the processes that lead to structural degradation in X-ray exposures of macromolecules. Using these sources it is therefore possible to increase the dose to macromolecular crystals by several orders of magnitude higher than usually tolerable in conventional measurements, allowing crystal size to be decreased dramatically in diffraction measurements and without the need to cool the sample. Such pulses lead to the eventual vaporization of the sample, which has required a measurement approach, called serial crystallography, of consolidating snapshot diffraction patterns of many individual crystals. This in turn has further separated the connection between dose and obtainable diffraction information, with the only requirement from a single pattern being that to give enough information to place it, in three-dimensional reciprocal space, in relation to other patterns. Millions of extremely weak patterns can be collected and combined in this way, requiring methods to rapidly replenish the sample into the beam while generating the lowest possible background. The method is suited to time-resolved measurements over timescales below 1 ps to several seconds, and opens new opportunities for phasing. Some straightforward considerations of achievable signal levels are discussed and compared with a wide variety of recent experiments carried out at XFEL, synchrotron, and even laboratory sources, to discuss the capabilities of these new approaches and give some perspectives on their further development.

Key words XFEL, Serial crystallography, Radiation damage, Coherent diffractive imaging, Phasing, Microcrystallography

1 Introduction

X-ray free-electron lasers (XFELs) [1] offer a disruptive new technology for macromolecular structure determination. These sources produce extremely intense X-ray pulses of femtosecond duration that provide two distinct advantages for the investigation of biological molecules and their complexes. The first is that the pulses, if produced with short enough duration, outrun most of the processes of radiation damage, allowing for exposures that are many orders of magnitude greater than possible with other sources such as synchrotron radiation facilities or X-ray tubes. This in turn

Alexander Wlodawer et al. (eds.), *Protein Crystallography: Methods and Protocols*, Methods in Molecular Biology, vol. 1607, DOI 10.1007/978-1-4939-7000-1_12, © Springer Science+Business Media LLC 2017

means that samples can be orders of magnitude smaller in volume than those required in conventional experiments, making possible measurements from samples too small or too weakly scattering to be feasible with conventional sources. Many protein systems produce numerous crystals of micrometer size or smaller before optimal crystallization conditions can be found that produce larger ones. By removing the need for large protein crystals, the crystallization bottleneck in the structure determination process can be alleviated. On the timescale of femtoseconds, below the periods of atomic vibrations, the concept of temperature loses its meaning and the sample under investigation is effectively frozen in time. There is thus no need to cryogenically cool samples, which can therefore be investigated under physiological conditions, giving access to conformational states or solvation conditions that may not be otherwise apparent. Electron densities of protein crystal structures obtained using XFEL pulses usually appear much better than counterparts elucidated using synchrotron sources even for the same crystallographic resolution, including better definition of side chains and disulfide bridges [2, 3] or metal binding sites [3].

There is a rather serious consequence of this approach of outrunning radiation damage in that the illuminated sample is completely vaporized by the pulse, at least at pulse fluences beyond 10^8 photons/μm^2. This means that only a single snapshot diffraction pattern can be obtained per object, and that the sample must be rapidly replenished to collect many thousands of patterns, one by one, ideally at the repetition rate of the XFEL. For crystalline samples this requirement results in an experimental design that is quite different from usual protein crystallography experiments where diffraction is collected as a crystal is rotated on a goniometer. Instead, the approach of "serial crystallography" is to record snapshot diffraction patterns one at a time, each from a fresh crystal that is usually delivered to the beam in a random and unknown orientation. Many tens of thousands or even millions of such patterns can be accrued in a time that depends on the pulse repetition rate and detector frame rate. Several different schemes for introducing and replenishing the sample to the beam are currently utilized in such experiments, discussed below and shown in Fig. 1, including high-speed liquid jets [4, 5], extruded pastes or gels [6–8], aerosol beams [9], or rapid scanning of samples mounted on or across solid supports [10–12]. For structure determination, the still snapshots of the diffraction pattern cannot be treated in isolation but must be oriented in three-dimensional reciprocal space (usually by indexing the observed Bragg peaks) and combined to obtain a full three-dimensional set of structure factors from the ensemble and, if the scattering is very weak, to build up adequate signal. The data processing strategy must also contend with the fact that the patterns are recorded from crystals of different shapes and sizes, with randomly fluctuating pulse intensities

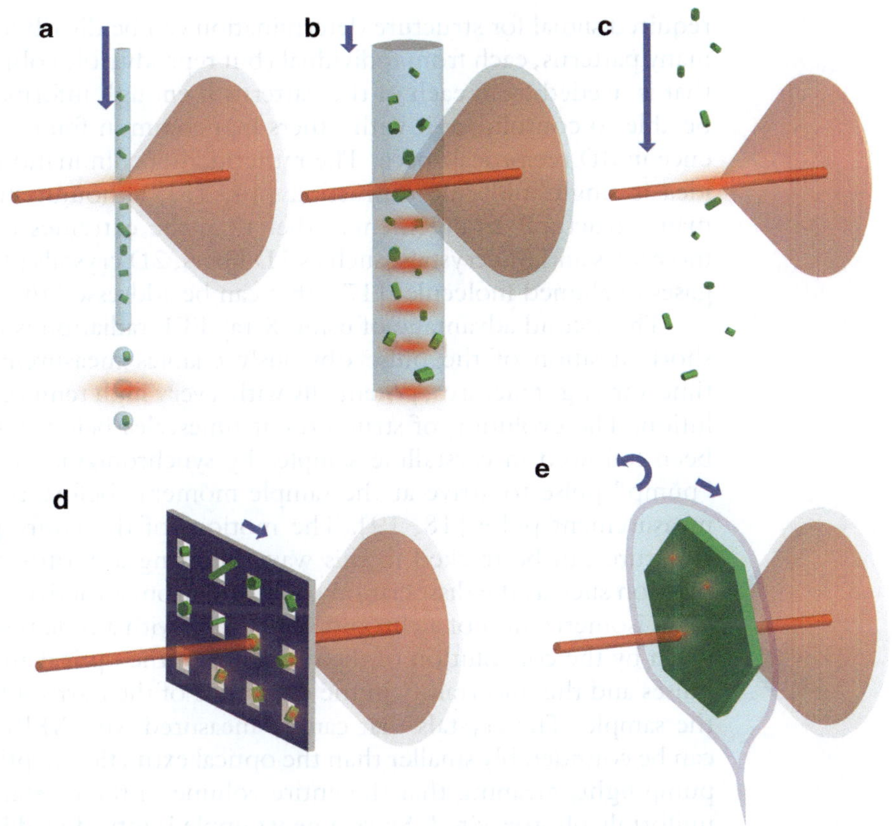

Fig. 1 Sample delivery options for serial crystallography. (**a**) Liquid micro-jet of 1–4 μm diameter gives low background and high speeds of many tens of meters per second. (**b**) Extrusion jets are slower, giving higher sample efficiency, but at the cost of higher background from about 50 μm thickness. (**c**) Aerosol injectors give the lowest background but also lowest efficiency and high speeds. (**a**–**c**) All can operate in vacuum. (**d**) Raster-scanned arrays can give 100% hit fractions for repetition rates of 120 Hz. (**e**) A large crystal mounted on a cryo-loop on a goniometer can be exposed in several places with known angular increments between pulses. Reprinted from [68]

and wavelengths (*see* Chapter 13 by White in this volume). In this sense, serial diffraction is not unlike powder diffraction (for crystalline samples) or wide-angle X-ray scattering (for single non-crystalline particles) measured one grain or particle at a time. Each Debye–Scherrer ring in a powder pattern is composed of individual reflections from different crystallites which can be integrated to average out any heterogeneities. Measuring the ensemble one crystalline grain at a time gives us the opportunity to interpret the structure factors in the three-dimensional space, merged as from an average "single" crystal, rather than collapsing data onto a less informative one-dimensional plot of intensity versus scattering angle [13], while still averaging over the ensemble. This realization leads to the possibility to decrease the specimen size even further from that attained by outrunning radiation damage. The total required signal for structure determination can be distributed over many patterns, each from individual (but reproducible) objects. All that is needed from each of the patterns is enough information to be able to consolidate it with others in a common frame of reference in 3D reciprocal space. The rather daring culmination of this idea is single-molecule diffraction [14, 15], although there are many structural arrangements other than the extremes of single molecules and 3D crystals, such as 1D fibers, 2D crystals [16], and gases of aligned molecules [17], that can be addressed this way.

The second advantage of using X-ray FEL radiation is that the short duration of the pulse obviously enables measurements of time-varying structures, potentially with a very high temporal resolution. The evolution of structures at timescales below 1 ps have been followed in crystalline samples by synchronizing an optical "pump" pulse to arrive at the sample moments before the X-ray measurement pulse [18, 19]. The motions of the entire protein structure can be tracked in this way, following a photo-activated reaction such as the dissociation of a ligand from an active site [18] or an isomerization of a chromophore [19], with a time resolution given by the convolution of the durations of the optical and X-ray pulses and the uncertainty in the difference of their arrival times at the sample. The crystals that can be measured with XFEL pulses can be considerably smaller than the optical extinction depth of the pump light, meaning that the entire volume of the crystal can be uniformly photoexcited. Since a new sample is introduced into the beam for every X-ray pulse, irreversible reactions can be studied. It would be possible to witness the initial evolution of an explosive reaction, for example—the explosion induced by the X-ray interaction would be more violent in any case. Many experiments are carried out using slurries or suspensions of small crystals that flow across the X-ray beam in the form of a liquid jet that moves at speeds of several tens of meters per second, which can be illuminated at the X-ray interaction point or further upstream of the flow, depending on the time delay. The scheme of the flowing jet also enables fast mixing experiments where a ligand is brought into contact with a protein to follow the dynamics of their binding, for example. Here again the small crystal sizes offer improved experimental conditions since the diffusion times (which set the time resolution of such a mixing measurement) in micrometer-sized crystals can be substantially less than 1 ms [20].

2 Diffraction Before Destruction

A focused X-ray pulse from a free-electron laser is so intense that it vaporizes any material, turning it into plasma. Yet it is this extreme peak intensity (defined as the number of photons per unit area and time), a billion times higher than achievable from a synchrotron radiation facility, that gives some reprieve from the effects of radiation damage that usually limit the X-ray exposure that a sample can tolerate and which otherwise require large well-diffracting crystals to overcome. This damage is unavoidable and occurs because tens of photons are absorbed in the sample for each photon that is scattered and contributes to the diffraction pattern. The photoexcited atoms emit photoelectrons which themselves carry enough energy to collisionally ionize hundreds of other atoms, leading eventually to heat generation, broken bonds, mobile radicals and solvated electrons that interact with reactive components of the molecules in the crystal, changing their structure [21]. Each photoionization imparts the energy of the photon to the sample, and the X-ray dose is measured by the total X-ray energy removed from the beam per unit mass (or number of atoms) in the sample, with SI units of Gray (1 Gy = 1 J/kg). For a given crystal size, the dose and the resulting degree of damage is thus proportional to the strength of the pattern recorded. While this unavoidable damage is a consequence of immutable atomic cross sections, it is possible to avoid many of the effects of this damage on the measured diffraction pattern by using a pulse that can "outrun" those effects [14]. In an exposure of a single XFEL pulse, any given photon interacts with atoms that could have only encountered any prior disturbance within a time less than the pulse duration, which may be 10–30 fs or less. Radicals certainly have no time to diffuse (even if created), and even if every single atom was ionized directly by a photon (which would occur in biological materials at a dose above ~50 GGy [22]), displacements of ions due to the strong Coulomb repulsion between them take some finite time to occur. The short XFEL pulse allows a dramatic increase in the strength of a diffraction pattern that can be recorded from a biological sample, albeit in a single shot. The acquisition of full three-dimensional structural information requires many serial measurements to be made on reproducible objects that are replenished on each X-ray pulse.

What is the physical limit to this concept of "diffraction before destruction"? The average scattering cross section of atoms in a protein is about 10^{-15} μm^2 for a photon energy of 8 keV [23], which means that 10^{15} photons/μm^2 would be required to scatter as many photons from a protein molecule as there are atoms in that molecule. The cross section for photoabsorption is about 30 times higher than the scattering cross section at this photon energy, yet there are not that many electrons in the atoms, so such processes will saturate. Emission of a photoelectron occurs essentially instantaneously on absorption of a photon, but there remains some time for the atom to relax after the ejection of one if its core-shell electrons. For the light elements, this primarily takes place by Auger decay, releasing yet another electron within a time of a few femtoseconds [24]. If another photoionization event takes place in an atom prior to Auger decay then the loss of both core electrons is described as a "hollow atom" whose absorption cross section is significantly reduced, frustrating further ionization. In this way, it has been predicted that incident intensities of 10^{15} photons/μm² (and hence a dose of about 1 TGy) could give rise to about 0.1 scattered photons per atom, if delivered with a 1 fs pulse [24]. However, even during this time, atoms will be ionized by collisions with photoelectrons, which can be avoided with a pulse as short as 0.1 fs. Such a short X-ray pulse is still many wavelengths in length, enough to give rise to interpretable diffraction, but the generation of X-ray pulses of this intensity is beyond current capabilities. Below fluences of 10^{14} photons/μm² (100 GGy dose) and pulse durations below 100 fs the number of scattered photons per atom is predicted to be linearly proportional to fluence [24], as assumed in Fig. 2.

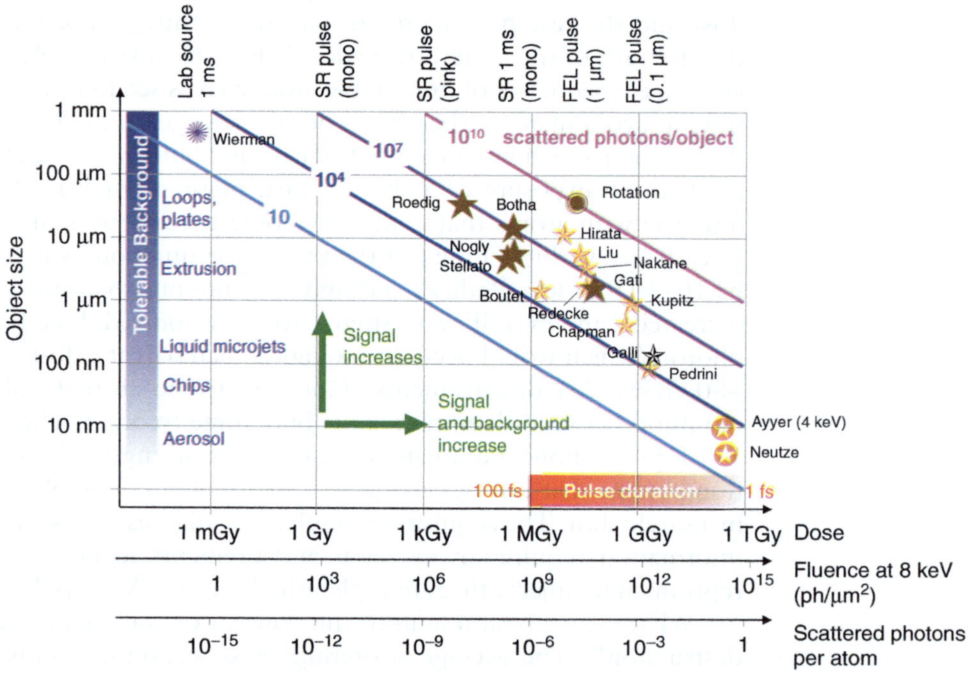

Fig. 2 A selection of serial crystallography experiments plotted on a log-log graph of crystal size versus dose. The dose is proportional to the number of X-ray interactions per atom and, for a given wavelength, the number of diffracted photons per atom, and provides a better means for comparison over different wavelengths than incident fluence. The crystal size is computed as the cube root of the illuminated volume. The total diffracted signal for a particular stoichiometry is proportional to the product of the number of diffracted photons per atom and object volume, shown as *solid lines*. *Orange* symbols correspond to XFEL experiments, *brown* to experiments at synchrotron radiation facilities, and *purple* illustrates an experiment using a laboratory source. For a given experiment, the background counts increase with X-ray fluence and hence the higher-fluence measurements from smaller crystals require delivery techniques that generate less background (described in *blue*). As fluence is increased in XFEL experiments, the pulse duration must be reduced, as indicated in *red*, requiring higher pulse powers. References are as follows: Wierman [42], Roedig [102], Stellato [103], Nogly [104], Botha [105], Boutet [106], Hirata [61], Liu [2], Nakane [55], Redecke [38], Gati [107], Chapman [73], Kupitz [108], Pedrini [109], Galli [47]. The data for the rotation series were extracted from [110]

The discussion here serves to provide a baseline to consider diffraction measurements made over a broad range of conditions, such as depicted in Fig. 2. A nanocrystal of, say, 10 × 10 × 10 unit cells should yield a similar total diffraction signal as that of a single molecule at a dose that is 10^{-3} of 100 GGy (that is, at 100 MGy), for example, giving rise to 10^{-4} diffracted photons per atom, or perhaps about 1000–10,000 total diffracted photons depending on the size of the molecule. Whether such diffraction from a single molecule or crystal could be interpreted depends on the relative contribution to the scattering pattern due to background, which is difficult to reduce with such high-fluence incident beams. Thus, most serial crystallography experiments are carried out with larger micro-crystals at lower fluence, at doses of about 10–100 MGy, with a corresponding decrease in background. The dependence of structural change during the pulse depends both on the incident beam fluence (the dose) and also the pulse duration. Longer pulses give time both for a cascade of electron collisional ionization events to take place, and for nuclear motion. Each photoelectron of 8 keV energy has the potential to create over 300 additional ionizations, over a period of about 100 fs and thus can be drastically reduced with a pulse of 10 fs or even more so with 1 fs [25]. Nuclear motion is driven by Coulomb repulsion between ions as well as the electron heating, and was observed in early experiments at LCLS to develop to about 5 Å RMS displacement after the end of a 100 fs pulse delivering 3 GGy dose [26]. Surprisingly, Bragg peaks from protein crystals could still be observed when using such long pulses. The explanation for this was that peaks were formed in the early stages of the pulse, before disorder in atom positions stifled further contribution into Bragg peaks. Thus Bragg diffraction, which is dependent only on the periodic part of the structure, is regulated by the explosion. However, given that molecular structures are not homogeneous, and in particular heavier atoms have higher photoionization cross sections, it could be expected that such disorder does not progress uniformly throughout the molecule. Some recent experiments on ferredoxin crystals at doses of up to 30 GGy show the effects of using long (80 fs) pulses in which native Fe atoms (which have larger absorption cross sections and thus undergo more photoionization events) disturb their surroundings, with correlated displacements of atoms away from the Fe atoms [27]. Modeling indicates that pulse durations below 20 fs are required. Such pulse durations would not avert nonuniform photoionization, but this effect opens up the possibility for new methods in phasing by anomalous diffraction [28, 29]. In one scheme, the difference of data collected at low and high X-ray pulse fluences could identify the positions of heavier (more easily ionized) atoms [30].

These experiments and theoretical understanding show that for a given X-ray flux, a shorter pulse is always better. Thus the highest exposures required for the strongest patterns must be delivered with pulses of high peak power (energy per unit time), focused down to submicrometer dimensions. A fluence of 10^{14} photons/μm^2 delivered in a pulse of 1 fs in a beam spot of FWHM of 0.1 μm, would require a source power of 10^{13} photons/fs, or a power of 10 TW for 8 keV photons (when accounting for the fact that beamlines cannot transport the entire XFEL output without loss). As yet, XFELs do not generate pulses of this power, but proposed schemes exist to exceed 10 TW [31]. Currently at the LCLS it is possible to deliver a pulse of about 10^{12} photons to a spot size of 0.2 μm with a pulse duration less than 20 fs, which should give rise to about 0.01 scattered photons per atom [28]. For a molecular complex like photosystem II with 72,000 atoms, this corresponds to almost 1000 photons per molecule. As seen in Subheading 3, this may be enough to provide interpretable diffraction. Defining a somewhat lower intensity regime, we can consider what dose could be tolerated for longer pulses of about 100 fs to several picoseconds. This is long enough for a fully developed electron cascade, but too short for transport of radicals [32]. Consider the case where every atom in the sample has been collisionally ionized by the end of the pulse, which implies that less than ~1% of atoms are photoionized (depending on the photon energy and if the system is large enough to trap all photoelectrons). Under this condition, most photons that interact with atoms will do so with neutral atoms; that is, with atoms that have not absorbed a photon nor been collisionally ionized. The probability of a fluorescence photon (for spectroscopy) being emitted by a perturbed atom, or an elastic scattering event (for diffraction) from a perturbed atom, will thus be small, given that the measurement is integrated over the pulse and the sample is initially neutral. The dose for this condition has been estimated at about 400 MGy for protein crystals measured with 100 fs pulses [22], compared with a tolerable dose of 30 MGy [21, 33] for cryogenically cooled samples measured with conventional sources and exposure times usually much longer than 1 ms.

Before closing this section on radiation damage, we consider some relevant points for conducting serial diffraction experiments at the much lower incident intensities of synchrotron sources since these are further discussed below. Radiation damage to protein crystals under such conditions has been extensively studied [34] although the ever-increasing brightness of these facilities, combined with beamline optics providing smaller X-ray spot sizes and improved detectors, opens up previously unexplored regimes of intensity and dose rate. The mean free path of high-energy photoelectrons in a protein crystal is on the order of 1 μm, giving crystals smaller than this size a higher dose tolerance, since many photoelectrons will deposit their energy (through the ensuing cascade of collisions) outside the crystal [35]. If the beam is bigger than the crystal then photoelectrons generated in any liquid or ice surrounding the crystal could feed into the crystal, reducing this advantage, but if the beam is substantially smaller than the crystal then the energy can be deposited into a larger volume (of crystal or surrounding) than from which the diffraction originates. Note that for a given crystal thickness and number of incident photons, the integrated Bragg intensities are independent of the spot size, meaning that higher quality data should be obtainable with an X-ray beam focus smaller than the photoelectron mean free path. This is supported by some recent experiments at the ESRF [36]. These discussions bring into relief the exact definition of dose: which mass is the energy distributed over? Given that energy can flow out of the system by a variety of means and over a large range of timescales, a suitable definition (that we use in this chapter) is the energy lost by the X-ray beam over the mass that the beam interacts with. The degree of the subsequent damage, which may occur over a longer time or greater mass than the diffracting volume, is a separate and complex issue (*see* Chapter 20 by Garman in this volume).

3 Serial Crystallography

Destructive pulses demand a strategy of replenishing the sample on every X-ray pulse, measuring single-shot diffraction patterns at the rate of those delivered pulses. This serial approach is in contrast to the best practice in conventional crystallography, which is to sweep a wedge of reciprocal space populated by many fully integrated reflections, by rotating a single crystal (or acquiring several rotation series from several crystals), to obtain accurate estimates of structure factors. Instead, the snapshots collected in serial crystallography may consist entirely of Bragg peaks that are not located in the centers of their reflecting conditions, and the patterns may be rather noisy. (For a monochromatic parallel incident beam, the 2D snapshot diffraction pattern maps to a spherical surface of 3D reciprocal space called the Ewald sphere. The Ewald sphere need not cut through the center of the reciprocal lattice nodes, which for physical crystals have finite extensions.) These deficits are made up by collecting a large number of such patterns, building up the information in a fragmented, rather than systematic, way. This approach lessens the connection between dose, crystal size, and the total collected exposure, so that it is no longer necessary to heavily expose a single crystal or to be compelled to wring the last diffracted photon from a crystal that has already suffered significant radiation damage and photo-reduction. Even with small crystals measured using synchrotron radiation, the need to cryogenically cool samples can be avoided by limiting the exposure (amounting to a dose of less than 10 kGy, for example), giving only limited diffraction information, before measuring the next crystal. Of course, if a large enough crystal is available to enable the collection of complete and accurate data at low dose, then clearly a rotation series provides the best strategy to determine the static structure.

The comparison of serial crystallography to powder diffraction further makes it clear that it is not necessary that each crystal gives strong diffraction or is exposed to accumulate its full tolerable dose. A powder pattern may consist of fewer total scattered counts than the number of crystals in the powder sample and yet have high enough signal to be measured with high accuracy. The distinction between the two techniques is that in powder diffraction there is no requirement to treat crystals separately or to determine the orientation of each crystal, since the signal is an average over all crystal orientations (at the great cost of loss of information for structure determination). In powder diffraction, dose can be reduced arbitrarily by increasing the total ensemble size, avoiding absorption effects and background. Serial crystallography usually requires a certain minimum incident fluence (and hence a certain minimum dose), however, so that orientational information of each crystal can be discerned from its pattern to enable aggregation in a common frame of reference. The achieved signal levels in several experiments are plotted in Fig. 2 as a function of dose and crystal size. The required signal is much lower than for a rotation series where Bragg intensities must be determined with high accuracy (shown by the brown circle in Fig. 2). In most implementations of serial crystallography the requirement for each pattern is that the Bragg peaks can be accurately identified as such, and that they can be indexed in order to determine the lattice orientation. Signal levels usually exceed 10^4 scattered photons from the crystal and a much higher number of photons contributing to the background. After indexing, the intensities of the indexed peaks can then be combined with those from other patterns, after estimating corrections and relative scale factors, to build up estimates of structure factors at all observable reciprocal lattice points (*see* Chapter 13 by White in this volume). This common scenario is discussed below, but it is worth to consider how much further one can go. The knowledge of the lattice orientation of an individual pattern can be used to predict where even weaker (and perhaps undetectable) peaks reside in that pattern. The undetectable peaks have signal counts, I, much less than the noise in the background, σ. Just as in the case of cryo-electron microscopy, where individual images of macromolecules can barely be identified, let alone interpreted, the process of averaging a large number of noisy observations of the same Bragg peak that all have a signal-to-noise ratio (SRN) $I/\sigma \ll 1$ should finally reveal that peak and allow the estimation of the structure factor at that point in reciprocal space. As yet, this approach has not been fully exploited at XFEL sources, primarily because even submicrometer crystals are often large enough to give detectable Bragg peaks at high resolution, and there is usually not enough beamtime available (due to limited pulse repetition rates) for experimenters to keep acquiring data that are not immediately perceivable. Nevertheless, there are plenty of systems waiting to be measured, including proteins crystallized in vivo [37–39] and natural crystals [40].

But what happens if the patterns are too weak to discern any signal at all from noise, let alone discover the orientation of the crystal? If summed together, the powder pattern would eventually emerge from enough patterns. Using ideas of "cryptotomography" developed for the case of weak single-molecule diffraction, it is indeed possible to aggregate the data in 3D reciprocal space even with signals of only a few hundred total counts per pattern (i.e., less than 0.001 photon per pixel) [41, 42]. In particular, an expectation-maximization scheme in the form of the expand–maximize–compress (EMC) algorithm [43] iteratively generates a 3D volume of diffraction intensities from noisy patterns, ideally collected with not more than a single particle or crystal contributing to a pattern—that is, with no multiple hits. During this iterative process, the current estimate of the 3D intensities is used to extract Ewald slices that would be observed at particular crystal orientations. Each noisy pattern is compared with every extracted slice to determine the probability that it is a noisy manifestation of the extracted pattern, and then the 3D volume is updated by placing the measured patterns into that volume according to the probabilities. As this converges, the merged 3D diffraction volume becomes consistent with all of the measured patterns. So far, a proof-of-principle demonstration has been made using sets of sparse diffraction patterns collected with a laboratory source [41, 42] and convergence could be reached with about 200 photons per pattern (*see* purple star in Fig. 2). At these low counts, enough patterns are required in total to eventually populate almost 10^9 voxels of reciprocal space with several photons per voxel. In the study of Wierman et al. [42], this was achieved with 8.8 million recorded diffraction patterns, which were collected from one crystal in this case, to a resolution of 1.5 Å. If it had been carried out on 8.8 million individual crystals, the dose would have been less than 1 mGy (0.001 Gy) (as graphed in Fig. 2), instead of the total accumulated dose of about 3 kGy. It is interesting to scale this to the 500 GGy doses that are tolerable using short enough XFEL pulses, whereby one could reduce the crystal volume by a factor of 10^{14}, which essentially gives a single molecule. That is, it should be feasible to carry out single molecule diffraction in a regime of about 200 scattered photons per molecule, which may suffer from about 4000 ionizations per molecule when delivered with pulses longer than atomic relaxation times. It should be noted that the EMC algorithm or related methods of manifold embedding [44], do not distinguish or index Bragg peaks, but aggregate the full diffraction volume consisting of Bragg peaks, diffuse scattering, and more. Thus, while Bragg peaks are very useful for providing the lattice orientation at high signal levels (*see* Chapter 13 by White in this volume), it should still be possible to carry out serial diffraction with non-crystalline or semi-crystalline reproducible objects.

We thus see, as summarized in Fig. 2, that serial crystallography spans a wide range of exposures and doses, covering many orders of magnitude, and ranging from the extreme case of almost as many scattered photons as atoms in the sample, to that of conventional crystallography of less than a single scattered photon per 10 or so molecules. Signal strengths range over about four orders of magnitude, depending on detector capabilities. The signals from small crystals are compensated with more intense pulses, but the background signal increases in direct proportion to incident flux, so the goal for sample delivery systems for these weakly scattering objects is to deliver them to the beam with as little extraneous material in the beam as possible (*see* Subheading 4, below). The role of background can be quite dramatic. When background dominates, halving it increases the SNR by a factor of two, requiring only $1/\sqrt{2}$ as many patterns to be collected, or having the same effect as doubling the volume of the crystal. Here we assume that the background is due to X-ray photons (obeying photon counting statistics) rather than electronic noise of the detector, or any other stray signal that would be measured when the X-ray beam is off. The overall signal-to-noise level of the merged diffraction intensities is ultimately limited by the number of patterns acquired: averaging noisy patterns is an exercise in the law of diminishing returns, depending on the square root of the number of patterns collected [45]. An example of the signal strength of diffraction of natural granulovirus particles illustrates these dependences. These virus particles consist of a crystalline shell of polyhedrin protein with a narrow size distribution and about 9000 unit cells per crystal for a crystalline volume of 0.01 μm^3 [46, 47]. Experiments carried out at the CXI instrument [48] of LCLS using a liquid micro-jet of about 3 μm diameter delivered a water suspension of granulovirus particles across the X-ray beam of 1 μm focus with 10^{12} photons per pulse and 7.9 keV photon energy [47], imparting a dose of up to 1.3 GGy (depicted in Fig. 2 as a black star). Diffraction patterns were recorded on a CS-PAD detector [49], and consisted of the diffuse background scatter from the liquid jet, as well as Bragg peaks from the polyhedrin crystal shell whenever a particle was in the focus at the arrival time of the pulse. At a resolution of 2 Å, the liquid background was about 10 photons per pixel, far in excess of the total counts in all Bragg peaks. Although Bragg peaks at this resolution could be observed occasionally, a total of 120,000 indexed patterns were needed to reach a SNR of 1, on average, in this resolution shell. As discussed above, the SNR increases with the square root of the number of patterns, and linearly with the crystal size. Since both the signal and background increase with fluence (or dose) the SNR increases with the square root of fluence (or dose), giving the empirical relationship of achievable SNR with liquid-jet background of

$$\mathrm{SNR}_{2Å} = \frac{1}{B + 0.1/\sqrt{2}} \sqrt{\frac{N_{\mathrm{patt}}}{120{,}000}} \frac{V_C}{0.01\,\mu\mathrm{m}^3} \sqrt{\frac{\mathrm{Dose}}{1.3\,\mathrm{GGy}}} \qquad (1)$$

where N_{patt} is the number of patterns and V_C is the volume of the crystal (assuming a similar unit cell volume as granulovirus, which is $(10 \text{ nm})^3$). The factor B gives the background counts per pixel relative to that generated by a 3 μm diameter liquid jet, and the factor of $0.1/\sqrt{2}$ approximates the effect of reducing the background to zero from 10 photons per pixel, although it should be noted that background counts depend on pixel size and binning. The number of patterns required to reach a given SNR at 2 Å resolution is therefore given by

$$N_{patt} = 1.2 \times 10^5 \left(B + 0.1/\sqrt{2}\right)^2 \text{SNR}_{2A}^2 \left(\frac{0.01\mu\text{m}^3}{V_C}\right)^2 \frac{1.3\text{GGy}}{\text{Dose}} \quad (2)$$

Reducing the background by a factor of 10, equivalent to increasing the crystal volume by a factor of 10, would reduce the required number of patterns by a factor of about 100. Many crystals measured at XFELs have a volume of about 1 μm³ or more, equivalent to 100 times more unit cells than granulovirus, requiring only 12 patterns to reach SNR = 1 (or 1200 patterns to reach a more desirable SNR = 10). That is, Bragg peaks of such crystals (if not disordered) can readily be observed at the LCLS, even with background from a 3 μm diameter jet. However, consider reducing the dose to just 1.3 kGy, a million times lower than in this example. For crystals of about 1 μm³, that would require about 12 million patterns to be collected just to discern peaks above noise, or 120,000 patterns if crystals were delivered to the beam with a reduced background of a single count per pixel. At 8 keV photon energy, for an average protein, a dose of 1.3 kGy would be delivered with 10^6 photons/μm², which could easily be achieved using an undulator at a synchrotron source in a single bunch and without a monochromator (pink beam). Such bunches are typically 100 ps long, allowing time-resolved serial crystallography measurements at this resolution. Certainly at 1 kGy dose, radiation damage would be low, and some further advantage over radiation damage may be gained by outrunning radiolysis processes that take place on the nanosecond timescale [32]. Novel laboratory-based sources that are under development may provide similar numbers of photons in pulses of 0.1 fs duration [50].

Equation (2) shows that the total time for a serial crystallography measurement (of a static structure or for a particular condition or time-point in a series of measurements) depends on the average brightness of the source, which is to say the time required to conduct the experiment will be shorter if more patterns are collected per second. Interestingly, the dose, proportional to the peak X-ray fluence, can be offset by collecting more patterns, so that the total scattered counts in the experiment remains constant (proportional to the dose times the number of crystals or patterns). This holds at least to the point that there are enough scattered photons per pulse to merge data in three dimensions, which might only be possible with the strongest possible pulses from XFELs, as seen in Fig. 2. The dead-time of the detector must be taken into account, and high repetition-rate sources can only be fully utilized if a detector is available that matches the repetition rate. Thus, while the highest-brightness synchrotron sources may exceed the average brightness of an XFEL operating at 120 Hz, experiments will take longer at the synchrotron without a detector operating at MHz frame rates. Here, the detectors must be integrating, not counting, devices. Even with the extremely sparse patterns that can be analysed with the EMC algorithm, signals may exceed a single count per pixel [42], and only integrating detectors could collect such signals. One of the highest frame-rate detectors currently under development is the AGIPD [51], capable of reading 3520 frames per second, in bursts separated by only 220 ns corresponding to the pulse pattern of the European XFEL, and as such this combination would provide the highest experiment brightness for serial crystallography. The future upgrade of the LCLS will likewise increase the repetition rates. With such source and detector combinations, measurements that take 10 h today at 120 Hz frame rate (such as low SNR measurement of 0.01 μm^3 crystals) will be completed in 20 min. Full datasets using crystals larger than 1 μm^3 could be acquired in tens of seconds.

Presently, most room-temperature serial crystallography experiments are carried out with crystals large enough to give detectable peaks at near the highest resolution of the final merged dataset. In these cases, the requirement on the number of patterns is to completely populate 3D reciprocal space with measurements and to average over fluctuations of the beam fluence and variations in crystal shape, size, and quality. The volume of reciprocal space that needs to be measured depends on the symmetry of the crystal. Symmetry operations of the diffraction intensities (or Patterson symmetry) are applied to each pattern, reducing the required number of measurements by the number of unique operations. (Some space groups cannot be unambiguously indexed based on the locations of the reciprocal lattice peaks alone—in this case the intensities must be compared to avoid creating a twinned dataset ([52] also *see* Chapter 13 by White in this volume).) In some cases fewer than 6000 indexed patterns could be used to obtain good estimates of structure factors [53]. 60,000 patterns were enough to produce high enough accuracy for phasing by single-wavelength anomalous diffraction [54] at LCLS using crystals of lysozyme in complex with a gadolinium (Gd) containing compound. The crystal volumes were smaller than 2 μm^3 and the dose was less than 30 MGy. Nakane et al. [55] required 150,000 indexed patterns from <1000 μm^3 crystals (with an illuminated crystal volume of about 20 μm^3, delivered in a grease matrix) and a dose of about 50 MGy to carry out native sulfur SAD phasing at 1.77 Å wavelength (7 keV photon energy), an impressive feat of reaching the necessary low convergence errors. Some valuable lessons on how to obtain higher accuracies are given by Nass et al. [56]. As described in detail in Chapter 13 by White in this volume, metrics such as R-split can be used to monitor the precision of intensities determined from an ensemble of crystals measured serially. The R-split metric estimates the precision of the full dataset by comparing intensities derived from two random halves of the dataset (Nakane et al. achieved R-split =3.1% over the resolution range of 40–2.1 Å).

In general, crystal diffraction patterns are recorded with no more than a single crystal per shot, so that all Bragg peaks belong to a single reciprocal lattice. Methods for indexing multiple lattices have been developed [57–59], however, which allow a better experiment efficiency with the possibility to index more crystals than patterns. As the number of lattices per shot increases, so does the prevalence of overlapping or near-overlapping peaks, and the gain in efficiency is only obtained for a few crystals per shot. But when diffraction intensities other than Bragg peaks are to be used for analysis, such as the continuous diffraction from a disordered crystal (*see* below), then there must not be more than a single crystal per pattern. At XFELs the intrinsic bandwidth for SASE radiation is about 0.1%, and patterns are essentially treated as monochromatic. In fact, a broader bandwidth of up to about 4% is thought to provide better peak integration requiring fewer patterns (the volume of reciprocal space spanned by Ewald spheres of the wavelength range exceeds Bragg widths, especially at higher resolutions) [60]. Somewhat paradoxically, reducing the wavelength jitter by "seeding" the FEL generation processes does not improve convergence. At synchrotron radiation facilities with a suitable undulator, it would be possible to increase beam fluences more than 100-fold by eliminating the monochromator or by using a multilayer monochromator of a few percent bandwidth, enabling exposures in microseconds or even with single bunches (~100 ps exposures). However, broader-bandwidth Laue diffraction patterns are more difficult to index in an automated fashion than monochromatic patterns.

A substantial reduction in the required number of patterns can be achieved if it is possible to acquire multiple patterns from the same crystal in more than one orientation. This could be the case with a crystal large enough, so that multiple (destructive) exposures are acquired with a spacing larger than the distance the X-ray damage is able to travel within the crystal [61, 62], or with several extremely low-dose pulses measured at a synchrotron, for example. Since the damage propagation distance is much larger at room temperature than at cryogenic temperatures (and may extend over the entire crystal [3]), such experiments are best carried out with cryo-cooled samples [3, 61]. In both cases, the crystals must be mounted in such a way that they can be rotated by a known amount between shots, increasing the complexity of the experiment. Such methods are a step towards the rotation series and the additional dependent measurements allow better estimation of parameters such as peak profiles and partialities, which in turn give more accurate estimates of the structure factors. In some cases it is possible to reject outlier patterns based on the results of indexing the Bragg peaks, which may or may not improve the final dataset [63]. It should be possible to separate distinct phases of materials in the beam that have different unit cell dimensions [64], or perhaps to carry out a cluster analysis based on the intensities or unit cell dimensions [65].

4 Sample Delivery Methods

There are almost as many methods to rapidly deliver protein crystals and small particles to the beam as there are groups carrying out serial diffraction experiments; such is the vigor and diversity of this young developing field. Some of these methods have been described in reviews [66–68] and they can be grouped into methods of continuously or repetitively flowing samples across the X-ray beam, or rastering through the beam of a two-dimensional matrix in which specimens are mounted or embedded, referred to, respectively, as "jetting" or "fixed targets," as illustrated in Fig. 1. As is obvious from the discussion above, serial diffraction measurements are as much about acquiring diffraction as they are about reducing background. As emphasized by Gruner and Lattman [69], there are many sources of background in conventional crystallography experiments and thus the common practices of mounting crystals and using protecting foils to prevent crystal dehydration, for example, must be modified or abandoned when crystals approach volumes of 1 μm^3. Sample supports and foils, surrounding amorphous ice, and air present a scattering cross section (integrated over the path of the X-ray beam) that may surpass that of a small crystal by many orders of magnitude (i.e., along the beam, there are more atoms of these objects than in the crystal) and thus they contribute many orders of magnitude more photons on the detector than the Bragg diffraction of the crystal. The concentration of crystal diffraction into Bragg peaks enables this signal to be detected even when more photons contribute to the diffuse background. This ratio of total signal photons to background is independent of the X-ray fluence, so precautions are universally needed. The microdiffraction beamlines at XFELs were designed for experiments to be carried out in vacuum, with samples held on thin membranes or delivered as an aerosol jet [70–72]. The first serial crystallography experiments at LCLS [73] were carried out using a liquid jet of several micrometers diameter of a suspension of submicrometer crystals in their mother liquor. The thinness of the jet, and the fact that it could be sustained in the vacuum environment, were achieved using a gas-focusing nozzle which was one of the enabling inventions for the method [4, 5, 74].

Besides not generating too much background scatter, the delivery method should ideally not consume too much sample and should be able to replenish a new sample to the beam on each shot (possibly just before photoactivation), with the possibility of mixing or some other method of initiating a reaction. The properties of the delivery device therefore depend strongly on the repetition rate of the source, which spans 30 Hz at SACLA, 120 Hz at LCLS, to 4.5 MHz at the European XFEL (in bursts). For a flowing sample, efficiency can be parameterized by the "hit fraction," or the proportion of pulses that generate diffraction from a particle or crystal, sometimes referred to as the "hit rate." For short femtosecond pulses this can be approximated as $H = fA/(vw)$ for f particles per second injected at a velocity v as a stream of width w moving in a direction perpendicular to an X-ray beam of cross section A [75]. At any instant of time (such as when the X-ray pulse arrives) the areal density of particles or crystals as seen by the X-ray beam is $f/(vw)$. For a given consumption f, the density and hence the hit fraction are increased by slowing down the particles, which however must be travelling at a high enough speed so that the sample (and any expanding volume of destruction) clears the beam by the next pulse [76]. In this regard, in-air or in-vacuum extrusion injectors that flow crystals embedded in lipidic cubic phase [6], grease [7], or gel [8] provide speeds of several mm/s that are well matched to repetition rates of 30–120 Hz. Depositing the sample onto a moving tape (in air) also gives similar speeds [77]. The extruded pastes or moving tape are usually quite thick, however, giving rise to background counts that are many times higher than achievable with gas-focused liquid jets. Such micrometer-diameter gas-focused jets run at speeds of about 50 m/s, suitable for the MHz rates expected at the European XFEL and LCLS II. Recent developments of jetting two fluids concentrically allow for fast mixing prior to exposure [78–80] in a narrow gas-focused jet. Mixing times depend on diffusion across boundaries of liquids under laminar flow, but can be less than 1 ms, providing a temporal resolution on this order. The time delay can be continuously varied in a telescopic design or moving the nozzle position relative to the beam [79]. Faster mixing could be induced by more complicated flow-folding schemes, as used in microfluidic experiments [81].

For the mixing jet, one can also choose the sheath liquid based on its fluid properties that define the jet behaviour, such as viscosity and surface tension, giving a very reliable and stable sample delivery method [80], which may be appropriate for rapid structure determination at a dedicated station. Using the AGIPD detector, capable of reading 3520 patterns per second, 10^6 frames could be recorded in less than 5 min at high repetition rate FELs, putting greater premium on reliability of jets and the ability to automatically change sample [67]. Even at the lower pulse rates, where sample consumption per pattern is much higher, liquid jets provide a convenient method to deliver samples in liquid form and at room temperature with reasonably low background, in vacuum, or ambient atmosphere. Elongated objects become aligned along their long axis by the nonuniform fluid velocity profile across the nozzle capillary, and they tend to retain this alignment in the jet. This has a benefit for fiber diffraction, and may enable the serial diffraction methods to obtain 3D structure factors from single-fiber patterns. For most 3D crystals, however, flow alignment can lead to a missing cone of measurements in reciprocal space, requiring the ability to tilt the jet relative to the X-ray beam direction.

The lowest possible background is achieved by aerosolizing the sample and entraining it with a low-pressure gas into a beam using an aerodynamic lens. This device consists of a series of concentric apertures in a larger-diameter tube. Laminar flow of the gas through the restrictions briefly concentrates the streamlines, but the particles cannot exactly follow these lines due to their momentum and instead tend to fly to the center of the flow [82]. Aerodynamic lenses have been used successfully for single-particle diffraction experiments at FLASH [9] and LCLS [83] and are under improvement and optimization to decrease the stream width w to below 10 μm [84]. In principle, this method should be suitable for injecting small crystals, as long as the residence in the aerosol does not dehydrate them. Particle speeds are on the order of 10–50 m/s. Simpler convergent nozzles have produced jet sizes smaller than 2 μm travelling faster than 200 m/s [75]. Experiments are underway to use optical forces to further concentrate such beams [85].

Raster-scanning a structure supporting many samples can give near 100% hit fractions with very little consumption of material and minimal background. Achieving all these conditions at once calls for a careful experimental design that depends on particle or crystal size, and placement of the sample in air or vacuum. Low background requires as thin support structure as possible, such as graphene or silicon nitride, as well as ensuring that the wings of the focused X-ray beam do not interact with the supporting frame (using low-scatter clean-up slits or aperture). If crystals are large enough they can be caught in open holes in a silicon chip [12] by using a clever method of pipetting a liquid suspension onto the chip and blotting from the rear. This wicks away most liquid, for low background. When the holes are arranged in a regular array, the chip can be rapidly scanned so that a fresh sample position is probed on each shot. With a spacing between the windows of ~50 μm, scan speeds of 20 mm/s allow measurements at 120 Hz [86] plus some overhead for reversing the scan direction. Unless cryogenically cooled, these chips must be used in a humid atmosphere to prevent sample dehydration, which requires other precautions to minimize air scatter. It may be possible to sandwich the sample between graphene layers [87] to prevent dehydration in a vacuum environment. As with liquid jets, supporting surfaces might give rise to preferred orientation of crystals, which can be managed by the ability to tilt the chip. For serial diffraction at a low repetition-rate XFEL or synchrotron radiation source, these "fixed target" methods give optimum efficiencies and highest quality diffraction, although they present greater challenges for time-resolved measurements than the flowing methods, especially for irreversible reactions, where it must be ensured that only sample at one position on the chip is activated at a time.

5 Diffractive Imaging and Crystallography

The use of XFEL pulses has opened the way to structure determination not only for crystallites that are smaller than required by conventional means, but also for single non-periodic particles, two-dimensional crystals [16], fibers, and oriented molecules in the gas phase [17]. Diffraction measurements from micrometer-sized 3D crystals or smaller has also opened up several new possibilities to experimentally phase the diffraction patterns (that is, obtain phases without the use of a structural model), such as using measurable intensities between Bragg peaks that occur due to the finite extent of the crystal [88, 89], or using continuous diffraction that occurs due to deviation of the crystal structure from a perfect periodicity [90]. These ideas emerged primarily from investigations of diffraction of single (or non-periodic) objects and while that field of coherent diffractive imaging is very closely related to crystallography [91], it is worth making a brief digression to establish the core concepts in a common language.

The wavelength of X-rays is short enough to resolve atoms in a molecule. This means that the scattered waves from two neighboring atoms, illuminated coherently, can interfere at the detector to give a diffraction pattern consisting of fringes, first understood (at much longer wavelengths) in the famous Young's double-slit experiment. A molecule of more than two atoms will consist of many such pairs, each producing a fringe pattern that contributes to the overall diffraction and which encodes the spacing between the pair of atoms and direction between them. This composite fringe pattern, termed the diffraction pattern, is proportional to the square modulus of the Fourier transform of the electron density (the molecular transform). When inverse-Fourier-transformed, the diffraction pattern reveals a map, called the autocorrelation function, of the distribution of all the atom pairs in the object. By the wonderful reciprocity between real space and diffraction space, points in the diffraction pattern correspond to single spatial frequencies (that is, fringes of electron density) in the object. A single snapshot of the diffraction pattern is two dimensional, so only a 2D selection of spatial frequencies of the three dimensional structure is recorded in a single snapshot pattern, on the surface of the Ewald sphere. Even though a single pattern contains depth information (due to the Ewald sphere curvature), full structural information requires patterns measured in many directions to fill out 3D space (which, in the scheme of diffraction before destruction, must arise from a supply of reproducible objects). From such measurements, a 3D image of the molecule's electron density can be synthesized by Fourier analysis, but to go beyond the 3D map of interatomic vectors between pairs of atoms to the actual map of the positions of those atoms, requires assigning phases to the measured diffraction intensities, which in turn assigns positions to the spatial frequencies that together generate the electron density image.

The scattering strengths of such single molecules are exceedingly weak, as seen in Fig. 2. Macromolecular structures are primarily obtained using a different strategy, in which the diffraction pattern is amplified by virtue of the arrangement of molecules in a periodic lattice of a crystal. To the degree that the molecules are identical in structure and orientation, each molecule in a crystal gives rise to the same diffracted wavefield, but originating from a different place in the lattice on which the relative phase of each wavefield depends. There are so many of these waves, diffracting from so many molecules in the crystal, that they mostly cancel out (for each wave there is likely to be another with opposite phase) except in those quite sparse directions that correspond to Bragg angles. These are the directions where every wave arrives at the detector after travelling from the source via the sample by exactly an integer multiple of the wavelength and thus constructively interferes with all others. The arrangement of Bragg peaks follows the Fourier transform of the crystal lattice, known as the reciprocal lattice. The constructive interference of the diffracted wavefields in the Bragg peaks gives a huge "coherency gain" [92], amplifying the strength of the single-molecule diffraction pattern by the number of molecules in the crystal, which can give a strong enough diffraction pattern within limited tolerable dose limits [33]. Unfortunately the sparsity of the Bragg peaks comes at a high cost in the ability to assign the phases needed to reconstruct the structure. By being able to measure the diffraction only at the discrete points of the reciprocal lattice rather than to observe the continuous molecular transform, the information content of the diffraction pattern is significantly reduced [93]. This information loss usually prevents the possibility to derive the phases from the intensities alone, unless extremely high resolution data is available (for the application of direct methods and algorithms such as charge flipping [94]). This so-called "phase problem" is the familiar state of affairs in crystallography, requiring additional measurements such as multiple wavelength anomalous diffraction or isomorphous replacement to provide the needed missing information.

The continuous diffraction from a single non-periodic compact object does not suffer from the phase problem since there are generally more independent measurements in the diffraction intensities than needed to describe the object. The greatest distance between any pair of atoms in the object are those at opposite extremes of the object; these give the largest extent of the autocorrelation map. Since this map is just another representation of the diffraction data (obtained directly from the measured intensities by Fourier transformation) the number of independent measurements is equal to the number of independent points in the autocorrelation map. This itself has a much larger non-zero volume than the original object (the volume of all possible connections between atoms is larger than the actual distribution of atoms). The volume of the autocorrelation map of a spherically shaped molecule is eight times that of the object itself. Accounting for the centrosymmetry of the autocorrelation map, this still gives a constraint ratio [95] of four, i.e., fourfold surplus of the measured information content over what is needed to describe the object. This overdetermination factor depends on the shape of the object (and of its autocorrelation function) but not the resolution—that is, atomic resolution is not required. A successful approach to determine the phases is to use one of a class of algorithms that iteratively constrains the solution to be consistent with the measured diffraction and a priori information about the object's structure [96]. This additional information need not be very detailed, and may simply be that the object fits within a certain rectangular box that is smaller than the extent of the autocorrelation function [97]; that the electron density is positive; or that the histogram of electron densities follows a certain profile, common to related proteins.

For crystals, the number of independent Bragg intensities is usually smaller than what is needed to describe the object, unless atomic resolution is reached where the number of measurements comfortably exceeds the number of parameters needed to describe the atoms (i.e., their positions and amplitudes of vibrations) (*see* Chapter 22 by Jaskolski in this volume). At the usual resolutions obtained with protein crystals, an ambiguity arises because of the crystal periodicity. The autocorrelation map of the crystal repeats with the same periodicity as the crystal lattice, and so the unique volume is restricted to at most one half the volume of the unit cell (due to centrosymmetry). It is not possible to distinguish points in the correlation map as arising from the intramolecular (within the same molecule) or intermolecular context (between neighboring molecules). If there is no non-crystallographic symmetry and if the object fills the volume of the unit cell, then the measurements would only account for half of the information needed to describe the molecule (at whatever resolution the diffraction extended to). This deficiency of information has long been recognized, and the earliest (unsuccessful) attempts of phasing protein crystal diffraction by Bragg, Perutz, and others used crystals of various states of dehydration, and thus different unit cell dimensions, to obtain measurements of the molecular transform at a higher density than possible from a single crystal [98]. When the solvent content exceeds 50% of the crystal volume, the information obtainable from the Bragg peaks should be higher than that of the unknown structure, allowing iterative phasing [91, 99].

A recent method that merges the approaches of crystallography and coherent diffractive imaging utilizes the continuous diffraction from crystals exhibiting translational disorder. Disorder of any kind in a crystal is a bane to the formation of Bragg peaks, which only form when there are correlations over many unit cells. If a molecule is displaced by a vector $\vec{\sigma}$ from its ideal position in the crystal lattice then the diffracted wavefield from that molecule is modulated by a phase ramp $\exp(-2\pi i\vec{\sigma}\cdot\vec{q})$. Here the magnitude of the wave-vector transfer $\vec{q}$ is equal to $2\sin\theta/\lambda$ for a scattering angle 2θ and wavelength λ. At a Bragg peak corresponding to a particular resolution length $d = 1/q$, the wavefield of the displaced molecule will combine with those of others with a phase error of $2\pi\sigma/d$ if that displacement is in the direction corresponding to the Bragg peak. For example, a displacement of 1.5 Å would cause destructive interference (a phase shift of π) at a Bragg peak corresponding to 3 Å resolution. Small random displacements of all molecules in the crystal with a mean square displacement of $<\sigma^2>$ will lead to random phases for scattering angles at high enough resolution. Due to that randomness, for every phase shift there is likely to be an opposite phase shift, with the result that the constructive interference that gives rise to the formation of Bragg peaks will not occur. Instead, the diffracted wavefields of each molecule will sum incoherently, giving rise to the continuous diffraction pattern of a single molecule, multiplied by the number of molecules. At low resolutions ($d \gg \sigma$) the phase errors from the displacements will be small, and in that case the constructive interference of Bragg peaks will still occur, and there will be little or no continuous diffraction. In general, the Bragg intensities will be modulated by the well-known Debye–Waller factor, $\exp(-4\pi^2\sigma^2q^2)$ whereas the continuous diffraction will arise contrariwise with q as $1 - \exp(-4\pi^2\sigma^2q^2)$. The Bragg intensities are reduced by a factor of $1/e = 0.368$ at a resolution of $d = 2\pi\sigma$, and an RMS displacement of 1.5 Å would reduce the Bragg intensities by this amount at 9 Å resolution. The effects of the translational arrangement of identical objects on their diffraction patterns are illustrated in Fig. 3.

There are two important implications of translational disorder in a crystal. One is that the continuous diffraction of a translationally disordered crystal may extend to resolutions far beyond Bragg

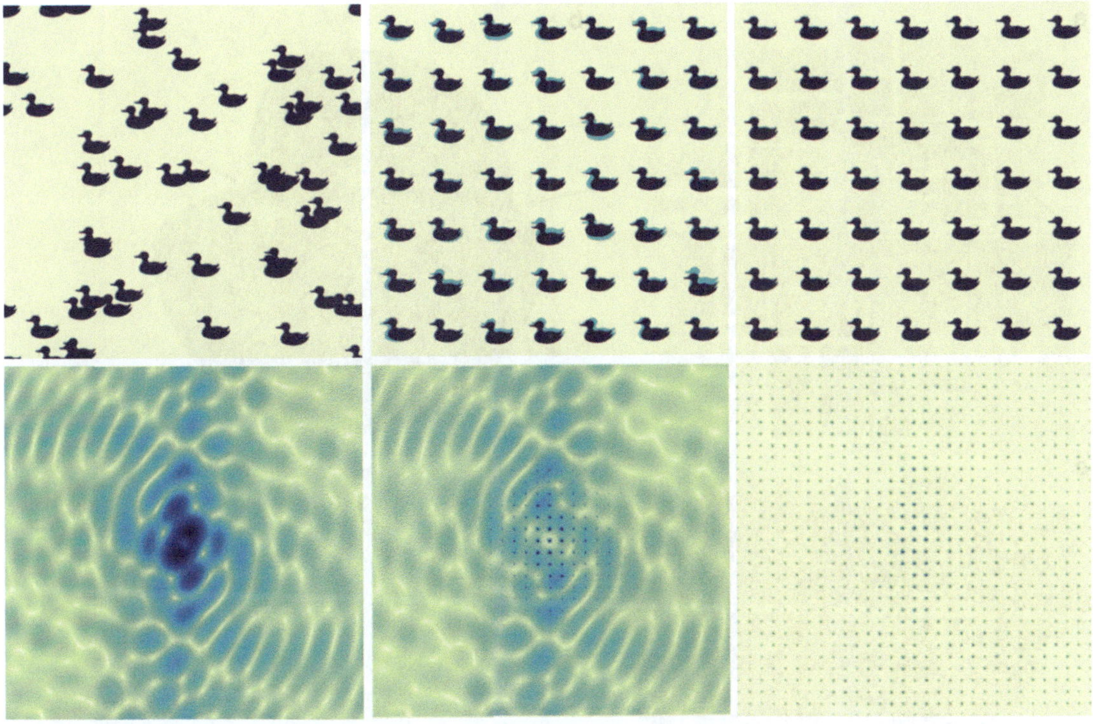

Fig. 3 Diffraction from an ensemble of similarly oriented objects depends on correlations between their positions. (**a**) A random arrangement of objects gives rise to the incoherent sum of the continuous diffraction pattern of each object (that is, *N* times the strength of the diffraction of a single object for *N* illuminated objects). (**b**) A crystal with a degree of translational disorder consists of Bragg peaks formed from the coherent sum of diffraction from all objects, modulated by a Debye–Waller factor that describes the suppression of Bragg peaks at resolutions greater than the disorder length divided by 2π. At those resolutions the incoherent sum of single-object diffraction occurs. (**c**) A perfect crystal produces solely the coherent sum of diffraction from the periodic array of scatterers

peaks—the resolution of useful information for structure determination is not necessarily limited by the extent (resolution) of the Bragg peaks. The second is that the continuous diffraction may be overdetermined by a significant factor, allowing iterative phasing methods to be used to obtain a 3D image of the molecule, without the need for a structural model [41]. The method was recently demonstrated on microcrystals of photosystem II which gave measurable Bragg peaks to a resolution of about 4.5 Å (Fig. 4). Continuous diffraction was observed to a resolution of 3.5 Å, limited by the detector extent and the number of patterns recorded. Several tests confirmed the origin of the continuous diffraction as the incoherent sum of the molecular diffraction from photosystem II dimers: the autocorrelation map computed directly from the continuous intensities was of finite extent with a boundary of the correct width and shape as corresponding to photosystem II molecules; the distribution of the intensities of the continuous

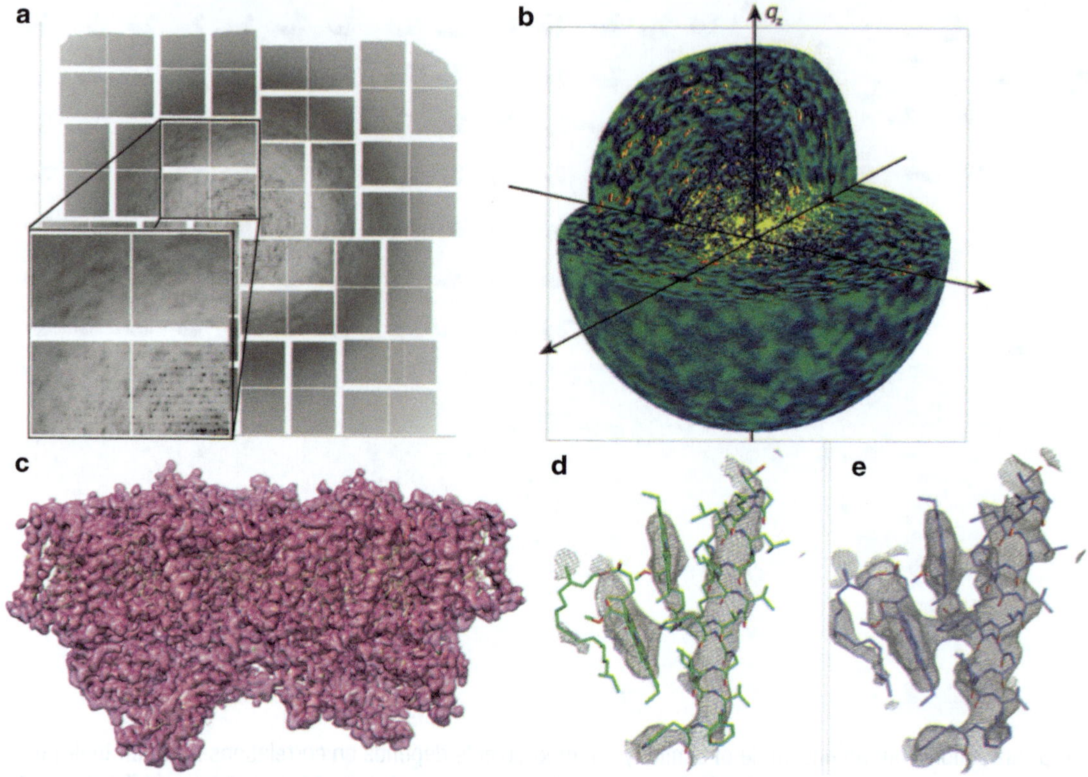

Fig. 4 Weak continuous diffraction (**a**) was observed in individual snapshot diffraction patterns recorded from photosystem II crystals at the LCLS [90]. When 2885 patterns were merged in a common frame of reference (defined by the indexed lattice) in 3D reciprocal space, the signal to noise of the continuous diffraction markedly improved, and extended significantly beyond the Bragg peaks (**b**). (**c**) The continuous diffraction could be phased using an iterative phasing algorithm to obtain a 3D image of the electron density of the photosystem II dimer. A detail of two chlorophylls of the dimer shows the improvement obtained from performing a structural refinement using the Bragg data only (to a resolution of 4.5 Å) (**d**) as compared with using together the Bragg and continuous diffraction (to a resolution of 3.5 Å) (**e**). Reprinted from [90]

diffraction followed Wilson statistics; and the diffraction could be phased by using a fixed support volume created by blurring out an initial electron density map obtained by refining a model using the Bragg intensities. The 3D image obtained by the continuous-transform phasing showed much clearer definition of structural elements such as α helices, even though the phasing algorithm had no knowledge of such structures (and the support, blurred to 8.9 Å resolution, showed no indication of these structures).

Translational disorder has not generally been expected in macromolecular crystals, although studies have been made on continuous diffraction from crystals of systems that undergo conformational dynamics [100]. It is common experience that many protein crystals only give Bragg diffraction to limited resolutions, and it is not unreasonable that the most dominant modes of disorder in a crystal with high solvent content (and few crystal contacts) would be rigid-body translations and rotations of the biological structure, followed by internal displacements. Such may be the case for membrane proteins and other large complexes which form very delicate crystals that are easily disrupted. Whether the displacements are static or in motion during the exposure does not affect the obtained diffraction, which is an average across the illuminated volume of the crystal and time. Thus, it may be possible to induce angstrom-scale acoustic modes in the crystal to achieve or enhance the continuous diffraction. At resolutions higher than the inverse of the disorder length, one can treat the crystal simply as a convenient way to place many aligned molecules in the X-ray beam, just like a gas of aligned molecules, to obtain the incoherent sum of the aligned-molecule diffraction (Fig. 3). Just as in aligned-molecule diffraction, random rigid-body rotations of the molecules give rise to a blurring of the continuous diffraction intensities that gets worse with increasing scattering angle. To avoid smearing out an individual speckle at the highest resolution, the width of the distribution of the rotations should be $\Delta\phi < d/w$, for a molecule of width w, equivalent to the Crowther condition in tomography [101]. The condition for this incoherent blurring is less stringent than the translations that disrupt the coherent interference at the Bragg peaks. The ultimate resolution of the continuous diffraction, therefore, will depend on the degree of rotational disorder and internal variabilities of the molecules.

The molecules in a crystal are aligned in several discrete orientations following the point group symmetry of the crystal. For example, photosystem II crystals have the space group symmetry $P2_12_12_1$ consisting of four dimers in unique orientations found by rotating any one of them by 180° about each of the three orthogonal axes of the orthorhombic cell (point group 222). Assuming no correlation between the translations and the orientation of molecules, the continuous diffraction is proportional to the incoherent sum of the dimer diffraction in these four orientations. For a spherically shaped molecule this overlap of orientations would reduce the information content of the continuous diffraction by a factor of four; less, if the object is non-spherical, since the summed autocorrelation functions of the various orientations will not completely overlap and hence will be partially distinguishable. In the case of photosystem II, the information content of the continuous diffraction exceeded that required to describe the dimer by a factor of 2.1 (compared with a factor of 0.86 for the Bragg peaks when accounting for the solvent fraction of the crystal) [41].

Periodicity concentrates the diffracting photons into narrow Bragg peaks that can be delineated from the background. The continuous diffraction of a disordered crystal contains as many diffracted photons as in the Bragg peaks of the same resolution from an ordered crystal—the scattering strength of atoms does not depend on their location or relation to each other. However, the continuous diffraction signal is much lower than that of Bragg peaks since the photons are spread out over many more pixels. For an average spacing of 10 pixels between Bragg peaks, for example, the continuous diffraction signal will be about 1% of that of the ordered crystal. It is also more difficult to separate the continuous diffraction from the continuous background. Its measurement requires the precautions described above to create experimental conditions with the lowest possible background. By perfecting these experiments it should be possible to further extend the achievable resolution, and to use the additional diffraction information to obtain direct information of conformational variability. This will be helped by reducing the crystal size, perhaps down to single molecules.

6 Conclusions

Free electron lasers have enabled some new paradigms for structure determination of macromolecules, and opened up new capabilities for time-resolved imaging and obtaining images from samples too small for conventional X-ray analysis. One of the main methodological innovations to use XFEL pulses, serial crystallography, and diffraction has also been shown to provide benefits when used with synchrotron radiation by being able to gather high-resolution structural information at room temperature without having to expose samples to the limits of their tolerable doses. Serial crystallography can be thought of as powder diffraction, measured one grain at a time, allowing the ensemble to be consolidated in the frame of reference of the lattice, instead of simply summing all patterns in the laboratory frame of the detector. As such, the only requirement from a single pattern is that it gives enough information to place it, in three-dimensional reciprocal space, in relation to other patterns. It has been demonstrated that millions of extremely weak patterns can be treated this way, as long as background scatter is small. Thus, the ingredients for this paradigm are a tightly focused intense X-ray beam of bandwidth up to a few percent; a method to rapidly replenish the sample into the beam that generate the lowest possible background; a high frame-rate integrating pixellated detector to record patterns of individual objects fed through the beam; and software to consolidate all diffraction data in a common frame of reference. At free-electron lasers, conventional dose limits are overcome by using femtosecond-duration pulses, which may enable exposures with such high fluences to give as many elastically scattered photons into the diffraction pattern as there are atoms in the sample, which should be enough to carry out single-molecule diffraction. The number of required patterns decreases sharply with the size of the crystalline samples, but low background and strong exposures from small crystals have enabled time-resolved measurements over timescales from below 1 ps to several seconds, as well as new phasing methods. In particular, strong exposures and low backgrounds let us measure diffraction beyond the highest scattering angles of visible Bragg peaks to acquire the continuous diffraction from single molecules in a translationally disordered crystal. Serial crystallography requires sources of high average brightness and high peak brightness (or peak power) and schemes to reach peak powers of 1 TW (e.g., giving 10^{12} photons in 1 fs) and repetition rates approaching or even exceeding 1 MHz, will enable dramatic increases in capabilities, surpassing those we have witnessed in the initial experiments at X-ray free electron lasers.

References

1. McNeil BWJ, Thompson NR (2010) X-ray free-electron lasers. Nat Photon 4:814–821
2. Liu W, Wacker D, Gati C et al (2013) Serial femtosecond crystallography of G protein-coupled receptors. Science 342:1521–1524
3. Zhou Q, Lai Y, Bacaj T et al (2015) Architecture of the synaptotagmin-snare machinery for neuronal exocytosis. Nature 525:62–67
4. DePonte DP, Weierstall U, Schmidt K et al (2008) Gas dynamic virtual nozzle for generation of microscopic droplet streams. J Phys D41:195505
5. Weierstall U, Spence JCH, Doak RB (2012) Injector for scattering measurements on fully solvated biospecies. Rev Sci Instrum 83:035108
6. Weierstall U, James D, Wang C et al (2014) Lipidic cubic phase injector facilitates membrane protein serial femtosecond crystallography. Nat Commun 5:3309
7. Sugahara M, Mizohata E, Nango E et al (2015) Grease matrix as a versatile carrier of proteins for serial crystallography. Nat Methods 12:61–63
8. Conrad CE, Basu S, James D et al (2015) A novel inert crystal delivery medium for serial femtosecond crystallography. IUCrJ 2:421–430
9. Bogan M, Benner W, Boutet S et al (2008) Single particle X-ray diffractive imaging. Nano Lett 8:310–316
10. Zarrine-Afsar A, Müller C, Talbot FO et al (2011) Self-localizing stabilizing mega-pixel picoliter arrays with size-excluding sorting capabilities. Anal Chem 83:767–773
11. Hunter MS, Segelke B, Messerschmidt M et al (2014) Fixed-target protein serial microcrystallography with an X-ray free electron laser. Sci Rep 4:6026
12. Roedig P, Vartiainen I, Duman R et al (2015) A micro-patterned silicon chip as sample holder for macromolecular crystallography experiments with minimal background scattering. Sci Rep 5:10451
13. Kirian RA, White TA, Holton JM et al (2011) Structure-factor analysis of femtosecond microdiffraction patterns from protein nanocrystals. Acta Crystallogr A 67:131–140
14. Neutze R, Wouts R, van der Spoel D et al (2000) Potential for biomolecular imaging with femtosecond X-ray pulses. Nature 406:753–757
15. Huldt G, Szoke A, Hajdu J (2003) Diffraction imaging of single particles and biomolecules. J Struct Biol 144:219–227
16. Frank M, Carlson DB, Hunter MS et al (2014) Femtosecond X-ray diffraction from two-dimensional protein crystals. IUCrJ 1:95–100
17. Küpper J, Stern S, Holmegaard L et al (2014) X-ray diffraction from isolated and strongly aligned gas-phase molecules with a free-electron laser. Phys Rev Lett 112:083002
18. Barends TRM, Foucar L, Ardevol A et al (2015) Direct observation of ultrafast collective motions in co myoglobin upon ligand dissociation. Science 350:445–450
19. Pande K, Hutchison CDM, Groenhof G et al (2016) Femtosecond structural dynamics drives the trans/cis isomerization in photoactive yellow protein. Science 352:725–729

20. Schmidt M (2013) Mix and inject: reaction initiation by diffusion for time-resolved macromolecular crystallography. Adv Cond Matter Phys 10:167276
21. Henderson R (1995) The potential and limitations of neutrons, electrons and X-rays for atomic resolution microscopy of unstained biological molecules. Q Rev Biophys 28:171–193
22. Chapman HN, Caleman C, Timneanu N (2014) Diffraction before destruction. Philos Trans R Soc B369:20130313
23. Creagh DC, Hubbell JH (2006) X-ray absorption (or attenuation) coefficients. Int Tables Crystallogr C:220–229
24. Son S-K, Young L, Santra R (2011) Impact of hollow-atom formation on coherent X-ray scattering at high intensity. Phys Rev A83:033402
25. Caleman C, Ortiz C, Marklund E et al (2009) Radiation damage in biological material: electronic properties and electron impact ionization in urea. Europhys Lett 85:18005, erratum 88: 29901
26. Barty A, Caleman C, Aquila A et al (2012) Self-terminating diffraction gates femtosecond X-ray nanocrystallography measurements. Nat Photon 6:35–40
27. Nass K, Foucar L, Barends TRM et al (2015) Indications of radiation damage in ferredoxin microcrystals using high-intensity X-FEL beams. J Synchrotron Radiat 22:225–238
28. Son S-K, Chapman HN, Santra R (2011) Multiwavelength anomalous diffraction at high X-ray intensity. Phys Rev Lett 107:218102
29. Son S-K, Chapman HN, Santra R (2013) Determination of multiwavelength anomalous diffraction coefficients at high X-ray intensity. J Phys B46:164015
30. Galli L, Son S-K, Barends TRM et al (2015) Towards phasing using high X-ray intensity. IUCrJ 2:627–634
31. Serkez S, Kocharyan V, Saldin E et al (2013) Proposal for a scheme to generate 10 TW-level femtosecond X-ray pulses for imaging single protein molecules at the European XFEL. arXiv.org:1306.0804
32. Davis KM, Kosheleva I, Henning RW et al (2013) Kinetic modeling of the X-ray-induced damage to a metalloprotein. J Phys Chem B117:9161–9169
33. Owen RL, Rudino-Pinera E, Garman EF (2006) Experimental determination of the radiation dose limit for cryocooled protein crystals. Proc Natl Acad Sci U S A 103:4912–4917
34. Garman EF, Weik M (2015) Radiation damage to macromolecules: kill or cure? J Synchrotron Radiat 22:195–200
35. Cowan A, Nave C (2008) The optimum conditions to collect X-ray data from very small samples. J Synchrotron Radiat 15:458–462
36. Coquelle N, Brewster AS, Kapp U et al (2015) Raster-scanning serial protein crystallography using micro- and nano-focused synchrotron beams. Acta Crystallogr D Biol Crystallogr 71:1184–1196
37. Koopmann R, Cupelli K, Redecke L et al (2012) In vivo protein crystallization opens new routes in structural biology. Nat Methods 9:259–262
38. Redecke L, Nass K, DePonte DP et al (2013) Natively inhibited Trypanosoma brucei cathepsin B structure determined by using an X-ray laser. Science 339:227–230
39. Jakobi AJ, Passon DM, Knoops K et al (2016) In cellulo serial crystallography of alcohol oxidase crystals inside yeast cells. IUCrJ 3: 88–95
40. Sawaya MR, Cascio D, Gingery M et al (2014) Protein crystal structure obtained at 2.9 Å resolution from injecting bacterial cells into an X-ray free-electron laser beam. Proc Natl Acad Sci U S A 111:12769–12774
41. Ayyer K, Philipp HT, Tate MW et al (2015) Determination of crystallographic intensities from sparse data. IUCrJ 2:29–34
42. Wierman JL, Lan T-Y, Tate MW et al (2016) Protein crystal structure from non-oriented, single-axis sparse X-ray data. IUCrJ 3: 43–50
43. Loh NTD, Elser V (2009) Reconstruction algorithm for single-particle diffraction imaging experiments. Phys Rev E80:026705
44. Fung R, Shneerson V, Saldin DK et al (2009) Structure from fleeting illumination of faint spinning objects in flight. Nat Phys 5:64–67
45. White TA (2014) Post-refinement method for snapshot serial crystallography. Philos Trans R Soc B369:20130330
46. Gati C, Oberthuer D, Yefanov O et al (2017) Atomic structure of granulin determined from native nanocrystalline granulovirus using an X-ray free-electron laser. Proc Natl Acad Sci U S A 114:2247–2252
47. Galli L, Metcalf P, Chapman HN (2015) Implications of the focal beam profile in serial femtosecond crystallography. Proc SPIE 9511:95110H
48. Liang M, Williams GJ, Messerschmidt M et al (2015) The coherent X-ray imaging instrument at the linac coherent light source. J Synchrotron Radiat 22:514–519

49. Hart P, Boutet S, Carini G et al (2012) The CSPAD megapixel X-ray camera at LCLS. Proc SPIE 8504:85040C–850411
50. Kärtner F, Ahr F, Calendron A-L et al (2016) AXSIS: exploring the frontiers in attosecond X-ray science, imaging and spectroscopy. Nucl Instrum Methods Phys Res A829:24–29
51. Allahgholi A, Becker J, Bianco L et al (2015) AGIPD, a high dynamic range fast detector for the European XFEL. J Instrum 10:C01023
52. Brehm W, Diederichs K (2014) Breaking the indexing ambiguity in serial crystallography. Acta Crystallogr D Biol Crystallogr 70:101–109
53. Ginn HM, Messerschmidt M, Ji X et al (2015) Structure of CPV17 polyhedrin determined by the improved analysis of serial femtosecond crystallographic data. Nat Commun 6:6435
54. Barends TRM, Foucar L, Botha S et al (2014) *De novo* protein crystal structure determination from X-ray free-electron laser data. Nature 505:244–247
55. Nakane T, Song C, Suzuki M (2015) Native sulfur/chlorine SAD phasing for serial femtosecond crystallography. Acta Crystallogr D Biol Crystallogr 71:2519–2525
56. Nass K, Meinhart A, Barends TRM et al (2016) Protein structure determination by single-wavelength anomalous diffraction phasing of X-ray free-electron laser data. IUCrJ 3:180–191
57. Schmidt S (2014) GrainSpotter: a fast and robust polycrystalline indexing algorithm. J Appl Crystallogr 47:276–284
58. Gildea RJ, Waterman DG, Parkhurst JM et al (2014) New methods for indexing multi-lattice diffraction data. Acta Crystallogr D Biol Crystallogr 70:2652–2666
59. Ginn HM, Roedig P, Kuo A et al (2016) TakeTwo: an indexing algorithm suited to still images with known crystal parameters. Acta Crystallogr D Biol Crystallogr 72:956–965
60. White TA, Barty A, Stellato F et al (2013) Crystallographic data processing for free-electron laser sources. Acta Crystallogr D Biol Crystallogr 69:1231–1240
61. Hirata K, Shinzawa-Itoh K, Yano N et al (2014) Determination of damage-free crystal structure of an X-ray-sensitive protein using an XFEL. Nat Methods 11:734–736
62. Cohen AE, Soltis SM, Gonzalez A et al (2014) Goniometer-based femtosecond crystallography with X-ray free electron lasers. Proc Natl Acad Sci U S A 111:17122–17127
63. Diederichs K, Karplus PA (2013) Better models by discarding data? Acta Crystallogr D Biol Crystallogr 69:1215–1222
64. Zhang T, Jin S, Gu Y et al (2015) SFX analysis of non-biological polycrystalline samples. IUCrJ 2:322–326
65. Liu Q, Dahmane T, Zhang Z et al (2012) Structures from anomalous diffraction of native biological macromolecules. Science 336:1033–1037
66. Schlichting I (2015) Serial femtosecond crystallography: the first five years. IUCrJ 2:246–255
67. Chavas LMG, Gumprecht L, Chapman HN (2015) Possibilities for serial femtosecond crystallography sample delivery at future light sources. Struct Dyn 2:041709
68. Chapman HN (2015) Serial femtosecond crystallography. Synchrotron Radiat News 28:20–24
69. Gruner SM, Lattman EE (2015) Biostructural science inspired by next-generation X-ray sources. Annu Rev Biophys 44:33–51
70. Boutet S, Williams SG (2010) The coherent X-ray imaging (CXI) instrument at the linac coherent light source (LCLS). New J Phys 12:035024
71. Bozek JD (2009) AMO instrumentation for the LCLS X-ray FEL. Eur Phys J 169:129–132
72. Song C, Tono K, Park J et al (2014) Multiple application X-ray imaging chamber for single-shot diffraction experiments with femtosecond X-ray laser pulses. J Appl Crystallogr 47:188–197
73. Chapman HN, Fromme P, Barty A et al (2011) Femtosecond X-ray protein nanocrystallography. Nature 470:73–77
74. Gañan Calvo AM (1998) Generation of steady liquid microthreads and micron-sized monodisperse sprays in gas streams. Phys Rev Lett 80:285–288
75. Awel S, Kirian RA, Eckerskorn N et al (2016) Visualizing aerosol-particle injection for diffractive-imaging experiments. Opt Express 24:6507–6521
76. Stan CA, Milathianaki D, Laksmono H et al (2016) Liquid explosions induced by X-ray laser pulses. Nat Phys 12:966–971
77. Roessler CG, Kuczewski A, Stearns R et al (2013) Acoustic methods for high-throughput protein crystal mounting at next-generation macromolecular crystallographic beamlines. J Synchrotron Radiat 20:805–808
78. Ganan-Calvo AM, Gonzalez-Prieto R, Riesco-Chueca P (2007) Focusing capillary jets close to the continuum limit. Nat Phys 3:737–742
79. Wang D, Weierstall U, Pollack L et al (2014) Double-focusing mixing jet for XFEL study

of chemical kinetics. J Synchrotron Radiat 21:1364–1366

80. Oberhuer D et al (2017) Room-temperature structure determination of RNA polymerase II enabled by double-flow focusing injection. Sci Rep 7:44628
81. Lee C-Y, Chang C-L, Wang Y-N et al (2011) Microfluidic mixing: a review. Int J Mol Sci 12:3263
82. Liu P, Ziemann PJ, Kittleson DB et al (1995) Generating particle beams of controlled dimensions and divergence: I. Theory of particle motion in aerodynamic lenses and nozzle expansions. Aerosol Sci Technol 22:314–324
83. Seibert MM, Ekeberg T, Maia FRNC et al (2011) Single mimivirus particles intercepted and imaged with an X-ray laser. Nature 470:78–81
84. Aquila A, Barty A, Bostedt C et al (2015) The linac coherent light source single particle imaging road map. Struct Dyn 2:041701
85. Eckerskorn N, Bowman R, Kirian RA et al (2015) Optically induced forces imposed in an optical funnel on a stream of particles in air or vacuum. Phys Rev Appl 4:064001
86. Sherrell DA, Foster AJ, Hudson L et al (2015) A modular and compact portable mini-endstation for high-precision, high-speed fixed target serial crystallography at FEL and synchrotron sources. J Synchrotron Radiat 22:1372–1378
87. Yuk JM, Park J, Ercius P et al (2012) High-resolution EM of colloidal nanocrystal growth using graphene liquid cells. Science 336:61–64
88. Spence JCH, Kirian RA, Wang X et al (2011) Phasing of coherent femtosecond X-ray diffraction from size- varying nanocrystals. Opt Express 19:2866–2873
89. Kirian RA, Bean RJ, Beyerlein KR et al (2015) Direct phasing of finite crystals illuminated with a free-electron laser. Phys Rev X 5:011015
90. Ayyer K, Yefanov OM, Oberthür D (2016) Macromolecular diffractive imaging using imperfect crystals. Nature 530:202–206
91. Millane RP (1990) Phase retrieval in crystallography and optics. J Opt Soc Am A 7:394–411
92. Sayre D, Chapman HN (1995) X-ray microscopy. Acta Crystallogr A 51:237–252
93. Sayre D (1952) Some implications of a theorem due to Shannon. Acta Crystallogr 5:843
94. Oszlányi G, Süto A (2008) The charge flipping algorithm. Acta Crystallogr A 64:123–134
95. Elser V, Millane RP (2008) Reconstruction of an object from its symmetry-averaged diffraction pattern. Acta Crystallogr A 64:273–279
96. Thibault P, Elser V (2010) X-ray diffraction microscopy. Annu Rev Cond Matter Phys 1:237–255
97. Fienup JR (1982) Phase retrieval algorithms: a comparison. Appl Opt 21:2758–2769
98. Bragg L, Perutz MF (1952) The structure of Haemoglobin. Proc R Soc Lond 213:425–435
99. He H, Su W-P (2015) Direct phasing of protein crystals with high solvent content. Acta Crystallogr A 71:92–98
100. Wall ME, Adams PD, Fraser JS et al (2014) Diffuse X-ray scattering to model protein motions. Structure 22:182–184
101. Crowther R, DeRosier D, Klug A (1970) The reconstruction of a three-dimensional structure from its projections and its applications to electron microscopy. Proc R Soc Lond 317:319–340
102. Roedig P, Duman R, Sanchez-Weatherby J et al (2016) Room-temperature macromolecular crystallography using a micro-patterned silicon chip with minimal background scattering. J Appl Crystallogr 49:968–975
103. Stellato F, Oberthür D, Liang M et al (2014) Room-temperature macromolecular serial crystallography using synchrotron radiation. IUCrJ 1:204–212
104. Nogly P, James D, Wang D et al (2015) Lipidic cubic phase serial millisecond crystallography using synchrotron radiation. IUCrJ 2:168–176
105. Botha S, Nass K, Barends TRM et al (2015) Room-temperature serial crystallography at synchrotron X-ray sources using slowly flowing free-standing high-viscosity microstreams. Acta Crystallogr D Biol Crystallogr 71:387–397
106. Boutet S, Lomb L, Williams GJ et al (2012) High-resolution protein structure determination by serial femtosecond crystallography. Science 337:362–364
107. Gati C, Bourenkov G, Klinge M et al (2014) Serial crystallography on in vivo grown microcrystals using synchrotron radiation. IUCrJ 1:87–94
108. Kupitz C, Basu S, Grotjohann I et al (2014) Serial time-resolved crystallography of photosystem II using a femtosecond X-ray laser. Nature 513:261–265
109. Pedrini B, Tsai C-J, Capitani G et al (2014) 7 Å resolution in protein two-dimensional-crystal X-ray diffraction at linac coherent light source. Philos Trans R Soc Lond B Biol Sci B369:20130500
110. Holton JM (2009) A beginner's guide to radiation damage. J Synchrotron Radiat 16:133–142

Chapter 13

Processing of XFEL Data

Thomas A. White

Abstract

The introduction of the X-ray laser to crystallography, and its impact on the types of crystallographic experiments being performed as described in the previous chapter, has meant that new data processing strategies had to be found. While some XFEL crystallography experiments approach the conventional methods quite closely, even those are not without special considerations relating to data processing. Serial femtosecond crystallography (SFX) introduces several additional problems, many of which have been solved recently, and there has been great progress towards resolving the remaining ones. Recent developments into the use of continuous scattering between the Bragg peaks will need even greater changes to the conventional data processing methods. This chapter describes the special characteristics of XFEL data and introduces the range of processing methods which are currently under development.

Key words XFEL, Data processing, Indexing, Integration, Scaling, Merging, Serial femtosecond crystallography

1 Introduction: The Unique Features of Crystallographic Data from an XFEL

The special data processing methods applied to X-ray Free-Electron Laser (XFEL) data arise because certain features of the data differentiate it from data acquired by the conventional rotation method using a synchrotron or laboratory X-ray source. To describe these features, it is important to clearly define certain terms. The term *serial crystallography*, or SX, refers to the limiting case of multi-crystal diffraction data collection where only one diffraction pattern, or at most a small handful of patterns, are acquired from each crystal. The term initially referred to a type of diffraction experiment using electron diffraction [1, 2], but the technique was applied much more successfully to XFELs, and it became the dominant application. In SX, crystals may be delivered into the path of the X-ray beam by an injection device [3] or by scanning a surface coated with crystals [4, 5]. Many other sample delivery methods can be envisaged. The SX methodology is now being applied at synchrotron light sources [5–7], where it might offer a useful data

Alexander Wlodawer et al. (eds.), *Protein Crystallography: Methods and Protocols*, Methods in Molecular Biology, vol. 1607,
DOI 10.1007/978-1-4939-7000-1_13,

acquisition option for the upcoming new generation of high-brightness diffraction-limited storage rings [8]. When serial crystallography is performed using an XFEL, it becomes *serial femtosecond crystallography*, or SFX. However, SFX is by no means the only way to acquire crystallographic data using an XFEL. Data acquisition by the rotation method, in which a crystal is moved in the X-ray path in a controlled manner, has also been performed using XFEL sources and is a rapidly growing application [9, 10]. In experiments using laboratory X-ray sources or synchrotron sources, the sample is usually rotated *during the X-ray exposure* so as to move reflections through the Bragg condition (*see* Subheading 1.2). The case where no rotation is performed during the X-ray exposure is referred to as *snapshot diffraction*. Because XFEL pulses are so short, XFEL diffraction always consists of snapshots, even if the sample is rotated between exposures: "wedge" data acquisition with an XFEL source could be approximated by strongly attenuating the X-ray pulses and then summing many acquired snapshots with very small rotations of the sample between them, but the same effect could be achieved by using a synchrotron or laboratory source, which would have a strong stability advantage (*see* Subheading 1.4), rather than an XFEL. The ability to acquire useful levels of diffraction during a single short X-ray pulse is the main advantage of XFELs for crystallography because it allows for high time resolution when studying fast processes in time-resolved experiments and allows radiation damage processes to be circumvented (*see* **Section 2 in Chapter 12 by Chapman**). Serial methods also allow nonreversible reactions to be studied, since a fresh crystal is used for each snapshot.

1.1 Random Occurrence of Crystal Diffraction "Hits"

One feature unites all of the SX applications, which is that the detector is read out repeatedly as the crystals are delivered to the beam. The detector is read out regardless of whether or not the X-rays actually hit a crystal, and so the frames containing crystal diffraction data (or "hits") occur randomly amongst many other frames, potentially much larger in number, containing no crystal diffraction (the "non-hits").

Research is underway to improve this situation by using additional sensors, such as ion time-of-flight spectrometers, to select events when the X-rays met an object [11]. However, the main method for detecting the "hits" is currently to examine the readout from the X-ray diffraction detector itself. By detecting and counting Bragg peaks on the detector, a classification into "hit" or "non-hit" can be made without much computational effort. For SX, only the frames containing more than a certain number of peaks, usually 10–30, are taken for further processing.

Despite its apparent simplicity, there are several problems to be overcome in a "hit finding" system like this. The first is that not all peaks appearing on the detector are in fact Bragg peaks. "Hot" pixels can arise from the detector and may be difficult to distinguish from real Bragg peaks (*see* Subheading 1.3). Not all Bragg peaks come from the crystals of the intended sample: crystallites of salt or ice can also appear depending on the sample delivery method, or there may be Bragg peaks from apertures or filters in the experimental station. Slowly varying background scattering from the sample delivery medium usually means that a simple intensity threshold cannot be used for locating peaks, so a more advanced method based on the gradient of the intensity must be used as a minimum. As well as all this, each crystalline system being studied has different characteristics, in particular different unit cell parameters, Bragg peak intensities and upper resolution limit of visible diffraction. A suitable minimum number of peaks per detector frame for one sample may be too high for another one which has a smaller unit cell (hence fewer reflections per pattern) or diffracts less strongly.

The "hit rate" in a serial crystallography experiment is defined as the fraction of detector frames which contain interpretable crystal diffraction. As the density of crystals in the sample delivery medium is increased from zero, so too will the hit rate. If the density is increased further, eventually there will be a significant number of detector frames containing diffraction from two or more crystals. The relationship between crystal density and hit rate is given by Poisson statistics, and it has been calculated that the rate of "single hit" acquisition (only one diffraction pattern per detector frame) is maximal when the "hit rate" is 63.2%. At this density, 36.8% of the frames will contain single hits, 26.4% will contain more than one diffraction pattern, and 36.8% will be blank [4, 12]. In practice, most serial crystallography experiments have hit rates much lower than this, which slows the data acquisition but eases sample handling and data processing (*see* Subheading 2.2).

1.2 Single-Shot Diffraction Patterns and Partially Recorded Reflections

In a serial crystallography experiment, the number of diffraction frames recorded per crystal is close to 1. This is in contrast to rotation data acquisition where an entire data set may be acquired from one crystal. Modern experiments, particularly those in macromolecular crystallography, tend to use several crystals and form a full data set by merging the intensity measurements together. SX is at the extreme end of this trend, and SFX is even more extreme because a single XFEL pulse destroys the entire crystal so there is absolutely no possibility of a second diffraction frame being recorded from it.

The first problem caused by the single-shot nature of SFX is that the orientation of the crystal, and the unit cell parameters of the crystal, if they are not already known, must be determined from a single snapshot which covers a much smaller region of reciprocal space than an angular sweep provided by a rotation experiment. Our ability to do this relies on the curvature of the Ewald sphere, which means that a single snapshot needs to contain three-dimensional information. In practice, conventional indexing algorithms such as those embodied in the programs MOSFLM [13], DirAx [14], and XDS [15] have proved sufficient even though they were not originally devised for this scenario. However, algorithms have also been devised specifically for the case of snapshot diffraction and especially the case of multiple crystals per frame (*see* Subheading 2.2).

A larger problem is that of *partiality*. This term refers to the relationship between the structure factor of a reflection and the intensity of the Bragg peak which arises from it in a certain diffraction pattern, after accounting for factors such as the intensity of the X-ray pulse (which varies from shot to shot, as described in Subheading 1.4). In a rotation experiment, each reciprocal lattice point approaches the Bragg condition, passes through it then moves away from it, and the sum of the scattered intensity is measured for the whole process. A reflection which does not pass through this entire process is called *partially recorded*, and can be corrected to give the *full* intensity if the geometry of diffraction is known accurately enough. This analysis method is common in rotation crystallography [16]. In a snapshot diffraction pattern, the intensity of a reflection depends on the distance between the corresponding reciprocal lattice point and the Ewald sphere (*see* Fig. 1). An unknown proportion of the energy in the X-ray beam is scattered into the corresponding peak, and the proportion is different for each reflection in every diffraction pattern. Determining and compensating for these proportions has been a major theme in XFEL data analysis (*see* Subheading 2.6).

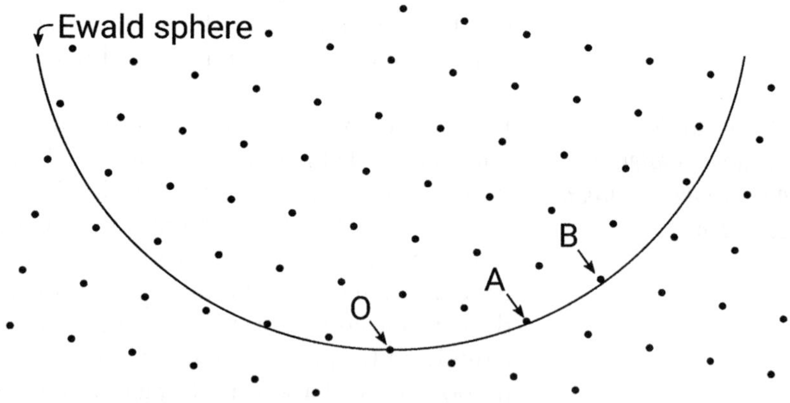

Fig. 1 Schematic diagram representing a flat cross section through reciprocal space. The spots represent reciprocal lattice points. Point O, the origin of reciprocal space, corresponds to the undiffracted X-ray beam and is always at the exact Bragg condition. Point A is close to the Ewald sphere, therefore close to the exact Bragg condition and therefore has a large partiality. Point *B* is further from the Ewald sphere, therefore further from the exact Bragg condition so has a smaller partiality. The measured intensity from reflection B would need to be scaled up by a larger factor than reflection A to correct for the partiality

Partiality can be thought of as a generalization of the problem of determining which reflections are excited in a given diffraction snapshot. In an SX experiment, without the advantage of surrounding frames in a rotation series to help, this determination must be performed using only the information from the snapshot itself. It might seem helpful to assign indices to all the visible peaks in the diffraction pattern and integrate only those, but this would neglect weak reflections. To accurately measure the intensity of a reflection, all intensity measurements from different frames should be combined, some below and some even slightly above the true value, according to the limited precision of an individual measurement. For a very weak reflection, some or perhaps all of these measurements may not correspond to obviously identifiable peaks in the diffraction patterns, and some of the estimations may even be negative after subtracting the background. Consider a hypothetical experiment in which indexing indicates that we made five measurements of a certain reflection in different snapshots and found that only one of the measurements indicated a high intensity, the rest giving very low values below their estimated error. Did we measure a strong reflection which we incorrectly determined to be excited in four of the snapshots, or did we in fact determine the excitation of the reflection correctly and measured a weak reflection with one outlier due to some other influence? By correcting for partialities, we extend the simple yes/no decision about the excitation of each reflection to a smoothly varying factor describing how excited the reflection is, in turn describing how precisely the reflection can give information about the underlying structure factor modulus.

Partiality is also a consideration in rotation experiments performed using an XFEL. Hirata et al. [10] investigated the effect of the size of the rotation step between snapshots in such an experiment, and found the optimal step size to be one third of the mosaicity of the crystal. The snapshot diffraction patterns were processed using a standard program for rotation data analysis, which was not aware that the patterns were snapshots rather than rotation wedges. Nevertheless, the final data quality compared favourably with a reference experiment using a synchrotron X-ray source.

1.3 Detectors

Detector technology for crystallographic data acquisition is a rapidly moving field. Photon counting detectors are very popular for use at synchrotron light sources. These detectors work by literally counting the individual X-ray photons as they arrive at each pixel. Unfortunately, this type of detector is of almost no use for XFEL experiments, because all the X-ray photons arrive within too short a time period and any number of photons above zero would be measured as one photon. The detectors used for XFEL experiments are of the integrating type, where the intensity from multiple photons is accumulated during the X-ray pulse and read out afterwards. Making an integrating detector which can be read out at a rate comparable to the pulse repetition rate of even the first XFEL sources (30 Hz for SACLA and 120 Hz for LCLS) is a major technological challenge, and the difficulties become even greater for the next generation of XFELs based on superconducting electron accelerators (27,000 Hz for the European XFEL, with pulses arriving with a separation of only 220 ns). Close to the focus, the peak power of an XFEL is high enough to ablate the surface and quickly cut a hole through a beam stop [17], therefore the direct beam must pass through a hole or gap in the detector to a beam stop far behind, if its full power is to be used. Far behind the detector, the divergence of the beam makes the irradiated region much larger and easier to absorb. The requirement for a hole through the middle of the detector imposes additional design constraints and precludes the use of many common detector types.

New detectors have been designed to address these unique requirements, including the pnCCD [18], CSPAD [19], MPCCD [20], and AGIPD [21]. A common feature of most XFEL detectors is that they are made up of multiple small detector "tiles." For example, the CSPAD is made up of 32 separate tiles, as shown in Fig. 2. In all XFEL detectors to date, the panels are not aligned with one another such as would allow them to be accurately mapped onto a single grid of pixels (without interpolation of pixel values). Therefore, although it is not technically a fundamental attribute of XFEL data, XFEL data processing software has by necessity been developed to handle multi-panel detector geometry, a capability which was not found in any of the popular crystallographic data processing software packages until very recently.

Fig. 2 Photograph of the CSPAD detector used at LCLS. Image courtesy of Sebastien Boutet/SLAC National Accelerator Laboratory

Detector technology for XFELs is an active field of research in its own right, and the properties of detectors are improving as time passes. Many of the detectors used to date have been first generation devices, closer to prototypes than refined commercial products, and are manufactured individually rather than in large numbers—each detector is unique. Several XFEL detectors are subject to certain artifacts which complicate data analysis, such as nonlinear response which depends on the incident intensity [22, 23], patterns of unresponsive pixels which change more rapidly than in more familiar detectors [17], and relatively small dynamic range [22]. Detector artifacts are sometimes due to damage done to the detector by strong Bragg diffraction from crystals of ice or salt in the injection media, and complete failure of whole tiles caused by one errant reflection has been known to occur.

1.4 SASE Noise and Crystal Variability

An XFEL can be thought of as an amplifier which takes a random fluctuation in the electric field of the electron bunch and amplifies it by many orders of magnitude to produce the final X-ray pulse. This process is called self-amplified spontaneous emission (SASE). The "noisy startup" of SASE means that the final X-ray pulses exhibit random fluctuations in almost all parameters. Most notably for crystallography, the intensity and spectrum of each X-ray pulse is different [24]. Typical fluctuations in intensity are about 10–30% of the mean. In data processing, all the reflection intensity measurements much be put on a common scale so that they can be merged, a process which is made significantly more practical for synchrotron and laboratory X-ray sources by adding the restraint that the scaling factors are similar between exposures on the same crystal [16]. With a randomly varying incident intensity, only a much weaker restraint can be used.

The spectral characteristics of the X-ray pulses also fluctuate because of the SASE process, and this is also of importance for crystallography. Not only does the distribution of intensity over wavelength vary from pulse to pulse, but the mean and modal wavelength also vary by a few tens of electron-Volts. When processing the data, the software should be aware of this fluctuation and be provided with an accurate estimate of the wavelength for each frame. This can be avoided by using a monochromator to select a small range of wavelengths, which will then be consistent from shot to shot. In this case, the spectral fluctuations will be reduced, but the intensity fluctuations of the pulses after the monochromator will depend on the spectrum and therefore will be greatly increased.

The fluctuations from the SASE process can be greatly reduced by *seeding* the FEL, which means to introduce an intentional perturbation to the distribution of electrons in the FEL which is then amplified, instead of starting from noise. This has been achieved at X-ray wavelengths at LCLS using a self-seeding method involving two groups of undulator segments, the first for SASE beam generation and the second for seeded beam amplification, with a diamond monochromator and electron chicane between the two groups [25]. The resulting X-ray pulses have a much narrower wavelength spectrum, around 0.5 eV, and consistent mean wavelength. The use of seeded X-ray pulses in SFX has been investigated, and it was found that, contrary to expectations, moving from SASE to seeded pulses did not increase the quality of the resulting merged data [26]. This finding is interesting because it indicates that the data quality is dominated by other sources of error aside from those arising from wavelength variation.

Apart from the fluctuations of the X-ray beam, the crystals themselves exhibit variability, which may cause difficulty in an SX experiment where a new crystal is used for each exposure. The sizes, shapes, quality, and structure of the crystals may all vary. Differences in the sizes of the crystals can be taken into account in data processing by the scaling factors, just like differences in the intensities of the X-ray pulses, and could actually be expected to have a much larger influence on the scaling factor. The shape of the crystal affects the shapes of the peaks, as seen in the previous chapter, and this may affect the determination of which reflections appear in a given frame (*see* Subheadings 1.2 and 2.2). The degree of crystalline order affects the overall strength of the Bragg peaks and also the rate at which the intensity decreases with resolution, and therefore affects the resolution limit of visible diffraction. These effects can be accounted for by a scaling algorithm which incorporates per-crystal Debye–Waller parameters (*see* Subheading 2.6). Different crystals may have large or small differences in crystal packing: large differences could be detected by examining the unit cell parameters and checking for significant numbers of patterns indexed using incompatible parameters (*see* Subheading 2.2), but smaller differences (non-isomorphism) may go unnoticed if they change the structure factors but not the unit cell parameters. Furthermore, it is not uncommon to see differences between crystallization batches, requiring care when merging data between them.

2 Algorithms and Software in XFEL Data Processing

Since the first XFEL experiments, new algorithms have been developed to address the considerations identified in the previous section. These algorithms have been embodied in new software.

The most widely used software packages for finding crystal hits are Cheetah [27] and CASS [28], both of which have much wider capabilities than crystallography. Cheetah incorporates a wide selection of options for finding single-particle diffraction patterns as well as crystal diffraction, alongside a range of options for handling detector artifacts (*see* Subheading 1.3) and secondary detectors such as spectrometers for examining the spectrum of the X-ray pulses (*see* Subheading 1.4). The scope of CASS is even wider still: it allows the user to build complete analysis pipelines according to the specific requirements of the experiment, out of preprepared modules which perform smaller tasks such as searching for peaks in a detector frame, summing the pixel values in a detector frame and calculating running averages from arrays of values. These pipelines can be applied to data stored in files from previous experiments, or "online" using a data stream direct from the facility's data acquisition system.

The process of hit finding is very data-intensive, since every frame read out from the detector must be examined to see whether it contains crystal diffraction. The volume of data once only the hits have been selected is usually much smaller, in proportion to the hit rate. Most, but not all, crystallographic data processing software for XFELs operates on the hits stored in intermediate data files after they have been found. This allows the hit-finding process to be performed once, and different processing scenarios to be investigated for the later stages without having to reexamine the potentially large number of non-hits. Additional benefits of storing the intermediate data are that the reduced data volume can more easily be brought to a home institution for further processing and archiving, and that the later processing stages will be much more efficient because they handle a smaller amount of input data.

After the hit finding stage, the most widely used software for processing diffraction patterns in SX is CrystFEL [29]. CrystFEL is designed for processing diffraction snapshots from any type of serial crystallography experiment, whether SFX or SX using other X-ray sources. Its main task is to index each diffraction pattern, integrate the reflections and merge the individual intensity measurements into a final combined dataset. Alongside this, it offers tools for other tasks such as simulating diffraction data for test purposes, calculating figures of merit, visualizing data and refining detector geometry. The pipeline for indexing and integration of patterns is shown in Fig. 3.

The other dominant piece of processing software for XFELs is cctbx.xfel [30]. This software incorporates tools for both the hit finding and crystallographic data processing stages. Recent versions of cctbx.xfel have made use of the capabilities of DIALS (Data Integration for Advanced Light Sources), a software project to create tools for the challenges posed by new X-ray sources—not just XFELs but also high-brightness synchrotrons [31].

A version of the very popular XDS package for conventional crystallographic X-ray data processing [15] has been created for SX, which is called nXDS [32]. nXDS has performed excellently in tests using data from synchrotrons, first with data from a single crystal rotation dataset processing without making use of the knowledge that the frames actually formed a rotation series, and

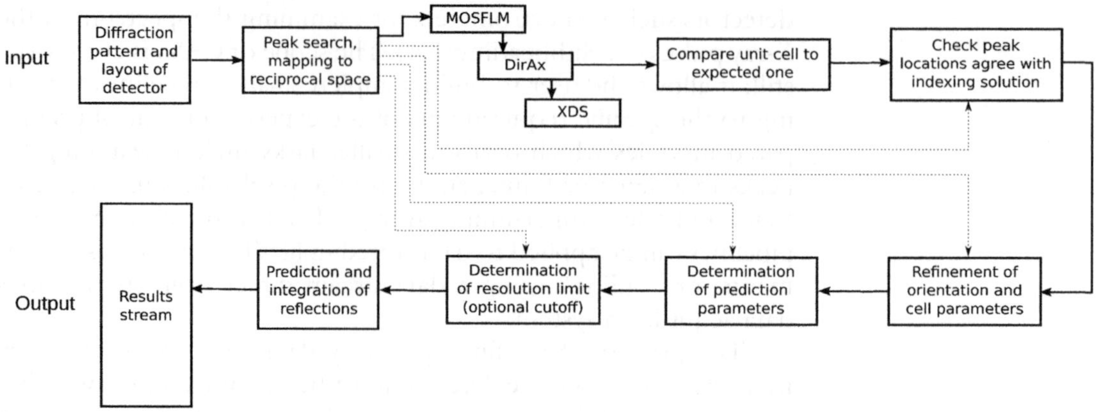

Fig. 3 Indexing and integration pipeline in CrystFEL. The *boxes* represent the different stages of processing, and the *arrows* indicate the order in which they are executed for each diffraction pattern. The indexing programs (here, MOSFLM, DirAx and XDS are shown, but the programs and their order can be chosen by the user) are tried in order until one of them produces an indexing solution which allows the later stages to be successfully performed. Reproduced from [37], © The Author licensed under CC-BY-2.0

later with data from a serial crystallography experiment performed using a viscous extrusion device at a synchrotron [33].

The software package cppxfel [34] has recently been published, and aims to be a "showcase" for new data processing methods which can then be incorporated into the more widely used packages. These new methods have already shown success with real SFX datasets [35, 36] and some of them have been incorporated into packages such as CrystFEL [37].

With the growing importance of SFX to XFEL facilities, both existing and forthcoming facilities have aimed to incorporate SFX data processing in their own frameworks, thereby providing a smooth experience for users. In the best possible case, users would need little knowledge of the details of data processing and be able to leave the facility with a merged dataset which they can use directly for further analysis at their home institution. An intermediate step is for users to take only the hits home, with all facility-specific conversions and corrections applied. Since a large amount of computing power is needed for processing diffraction patterns even after hit-finding, this can free users from having to find suitably large computer facilities. At SACLA, a combined online and offline SFX data analysis pipeline has been built based on Cheetah and CrystFEL [38], which provides immediate feedback on hit rates, unit cell parameters and resolution limits as well as automatically providing the hits for further processing.

2.1 Hit Finding

Hit finding is conceptually simple, but in practice is complicated by experimental factors and the large data volumes which require full automation. Hit finding for serial crystallography is usually based on locating Bragg peaks in the detector frames. There are many simple algorithms for finding peaks, for example using a minimum value of the pixel intensity or the gradient of the intensities [39]. Absolute threshold values are usually not sufficient unless the Bragg peaks are very strong compared to the background, because the background varies with position in the image, usually having a radial variation with rings of strong background intensity from the sample delivery medium. Aqueous media gives a diffuse ring of background scattering at around 3.5 Å, and lipidic cubic phase media gives a somewhat less diffuse ring near 5 Å. The exact size and position of the ring on the detector depends on the X-ray wavelength, which varies as has been previously mentioned (Subheading 1.4). In Cheetah, this is dealt with by performing a radius-dependent background subtraction, where the image data is examined in rings and the mean value of the background subtracted for each ring. An even more severe option for reducing the influence of background is to apply a filter where the pixels are taken in small squares and the median value of the pixels in each square subtracted from the pixel in the center of the square, repeating the process with the square sequentially centered on each pixel. This filter can remove almost any kind of background, but is computationally expensive. In all cases, if the frame contains a sufficient number of peaks, the frame should be stored *without* background subtraction. This is because the background scattering may contain important information, perhaps concerning the thickness of the sample delivery medium or even "single molecule" diffraction [40], and should therefore not be removed except as a temporary measure for simplifying hit finding. Data is stored "raw," without background subtraction.

Although simple in principle, the task of finding peaks is complicated by many factors. Several of these arise from the detector itself. Usually, a "dark calibration" must be made which consists of the values given for each pixel when no X-rays are incident on the detector. These values must be subtracted from each pixel in the real data to yield the intensity from the X-rays. The dark calibration usually varies over a timescale of several hours, or faster if the detector has been recently reconfigured (for example, a change of readout rate or gain mode), meaning that the process of acquiring and analyzing the "dark run" must be repeated rather frequently. This process has been made particularly convenient in the data acquisition system at SACLA, where a suitable number of dark frames are recorded automatically each time the users request a new acquisition run. After subtraction of the dark calibration, there is usually still a frame-to-frame "common mode" variation of the detector's background which affects entire regions of pixels at once. For the

CSPAD detector, this is handled by the presence of a small number of pixels where the detector electronics are deliberately not connected to the X-ray sensitive silicon wafer. Any variation of these pixels must therefore be due to the common mode offset, and the variations can be subtracted from all pixels in that panel of the detector.

At any one time, certain pixels of the detector, hopefully small in number, will not respond to X-rays at all, instead having a constant high or low value. These pixels must be identified and masked out. Cheetah incorporates an algorithm for this where the statistical properties of each pixel are monitored, and if they are implausible for real X-ray intensities, the pixel is automatically masked for the subsequent frames. Sometimes, entire regions of the detector may be defective (for example, two panels of the detector in the first SFX experiment [41]), and these must be masked out completely. In addition to all this, there may be "parasitic" diffraction from ice or salt crystals in the sample. Because the intensity of the incident X-ray beam is so high and ice and salt crystals scatter X-rays very strongly compared to protein crystals, this can present a serious risk of physical damage to the detector. Even if not strong enough to damage the detector, Bragg peaks from ice or salt need to be excluded from the hit finding process so that the diffraction from the sample can be identified.

Once the hits have been found amongst all the detector frames, they are often stored separately. Because the hit rate is usually much less than 100%, typically 1–10%, the amount of data storage taken for this is small compared to the raw data, and can often be handled without dedicated high-performance computing or data archiving facilities. In Cheetah, HDF5 is used as a file format for this purpose. HDF5 allows for a flexible definition of the data to be stored in the file [37], for example including information about which pixels where masked out or the locations of the peaks found in the peak search. A single file can be used for each hit, or multiple hits can be grouped together into one larger file, which reduces demand on computer filesystems somewhat. In cctbx.xfel, the Python serialization format ("Pickle") is used for a similar purpose, although the hits need not be saved separately at all.

When performing an experiment, particularly an SX experiment, it is very helpful to have near-immediate feedback about the hit rate. For example, a drop in hit rate may indicate that the sample injector has moved out of alignment with the X-ray beam, or that the crystals have settled in the sample reservoir [42]. This functionality is offered by both CASS and cctbx.xfel [43], and also by a separate program called OnDA [44] which makes use of the peak search algorithms from Cheetah via a shared library of routines.

2.2 Indexing Snapshot Diffraction Patterns

Indexing is the process of assigning indices to the Bragg peaks in the diffraction pattern, and can also be thought of as the process of determining the orientation of the crystal. As has been mentioned earlier (Subheading 1.2), this task is undoubtedly more difficult for snapshot diffraction patterns in serial crystallography than for rotation crystallography. Both CrystFEL and cctbx.xfel make use of existing software for indexing, LABELIT [45] in the case of cctbx.xfel [30], and a choice of programs including MOSFLM [46] and DirAx [14] in the case of CrystFEL. Since the underlying indexing programs cannot, in most cases, handle the multi-panel detector geometry from many XFEL facilities (*see* Subheading 1.3), some interface code is necessary to "hide" the true experimental details from the indexing engine. In CrystFEL, the spot positions are projected onto a fictional single-panel detector, and the spot coordinates on the fictional detector given to MOSFLM, whereas DirAx can use reciprocal space coordinates directly.

Indexing algorithms specific to serial crystallography are also under development. An algorithm has been developed for small unit cell crystals, which give widely spaced Bragg peaks, and specifically for serial crystallography [47]. This algorithm, embodied in the program cctbx.small_cell, works very differently to most other indexing algorithms. Rather than searching reciprocal space for directions perpendicular to clear sets of planes (these directions are the direct space vectors of the crystal lattice), it assigns indices to spots based on the moduli of their scattering vectors, and then resolves ambiguous assignments using approaches from graph theory.

The lattice type and unit cell parameters are sometimes known in advance of an SX experiment, but often are not. It may be possible to determine these by indexing a single snapshot in isolation, but a better method is to process all the frames and examine histograms of the unit cell parameters to find the most common sets of parameters. In CrystFEL, this is facilitated by the "Unit Cell Explorer" tool, shown in Fig. 4. Once the parameters have been determined, the indexing process can be repeated using them as prior information to the indexing algorithm. Different indexing algorithms are able to make use of different amounts of prior information. For example, MOSFLM can use information about the Bravais lattice type to help it choose the correct indexing solution, and recent versions can use the unit cell axis lengths and angles as well.

Recently, indexing algorithms capable of handling multiple lattices at once have emerged. Widespread adoption of such algorithms in SX data processing software would allow crystal concentrations to be increased beyond the theoretical optimum hit rate of 63%, to a point where many "multiple hit" diffraction patterns are acquired, thereby increasing the efficiency of data collection time. The most obvious way to handle a multiple hit is to index the pattern as usual, in the hope that the indexing algorithm will find one of the lattices. The peaks accounted for by that lattice

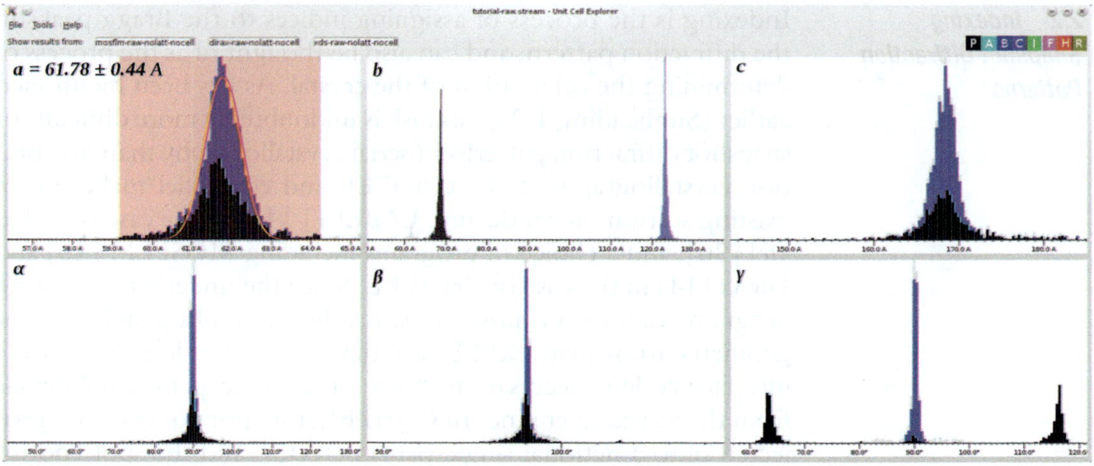

Fig. 4 Determination of unit cell parameters after an initial round of indexing without enforcing unit cell parameters or lattice type. The image shows a screenshot from the "Unit Cell Explorer" tool of CrystFEL, which displays histograms of unit cell parameters across the entire dataset, colored according to lattice centering. The most common sets of parameters can be isolated and a curve fitted to their distribution. Reproduced from [37],

are then eliminated, and the indexing algorithm rerun. This method has been demonstrated for rotation and snapshot data [13, 30]. It was later shown that by making use of prior knowledge about the unit cell parameters, this method could be made more effective, particularly on very small wedges of rotation data [48]. A totally different algorithm is that of GrainSpotter [49]. This algorithm operates in Rodrigues space, in which points correspond to orientations and straight lines correspond to continuous rotations around an axis, such as can be constructed by rotating a crystal around the axis from the origin of reciprocal space to a reciprocal lattice point, having assigned its indices. The intersection of the lines in Rodrigues space for several reflections corresponds to the orientation of the crystal, although there is again a combinatorial problem in separating ambiguous index assignments. This algorithm has been demonstrated on rotation datasets with up to seven crystals of lysozyme [50].

2.3 Refinement of Crystal Parameters and Detector Geometry

All current SX processing software incorporates a refinement stage in which at least the orientation of the crystal is refined to give the best achievable agreement between the observed spot positions and the calculated Bragg peak locations for the diffraction pattern. Simultaneously, the strongest Bragg peaks in the pattern are moved towards the exact Bragg condition. This is usually done by a standard nonlinear least squares refinement algorithm incorporating, in the residual, terms for the spot position deviation and deviation from the Bragg condition weighted according to the intensity of the reflection. The motivation for minimizing the deviation between observed and calculated spot positions is obvious, but the reason for minimizing deviation from the Bragg condition is less so. The idea is that stronger reflections are more likely to arise from reflections very close to the Bragg condition, and therefore the orientation should be chosen such that they are proportionally closer to it than weaker reflections. However, as described earlier (Subheading 1.2), this need not always be the case: a strong peak can arise from a reflection with a large structure factor far from the Bragg condition, or from one with a smaller structure factor close to the Bragg condition. Later stages of the processing pipeline attempt to mitigate the bias resulting from this (*see* Subheading 2.6).

In CrystFEL, this task is referred to as "prediction refinement," because it consists of refining the orientation and unit cell parameters of the crystal prior to "prediction" of the reflection locations and integration at those locations [37]. The algorithms employed in nXDS, cctbx.xfel, and CrystFEL are all broadly similar, being based on nonlinear least squares and a residual containing terms for spot position and distance from the exact Bragg condition.

There are parameters besides crystal orientation and unit cell parameters which contribute to the calculated spot positions on the detector. Most obvious is the geometry of the detector itself. Unfortunately, the detector geometry is usually not known to sufficient accuracy in advance, and some refinement is necessary. The capability to do this is included in cctbx.xfel [30], CrystFEL [51], cppxfel [34], and nXDS [32]. In CrystFEL, the detector geometry refinement program *geoptimiser* refines the individual panel positions and camera length using the entire data set, complemented by ability of the prediction refinement stage to refine the position of the central beam on the detector for each individual frame.

When calculating the positions of reflections for integration, it is important that not only the calculated positions are accurate, but also that the number of reflections integrated is correct. If parameters such as the X-ray spectral bandwidth, beam convergence or mosaicity of the crystal are assigned values which are too large, integration will be performed at many places in the diffraction pattern where no true Bragg peaks exist. These readings will contribute noise to the data set and complicate the later stages of scaling and merging. To avoid this, the parameters affecting the number of predicted reflections are optimized in most recent SX data processing software. The optimization is based on the clearly visible reflections in each diffraction pattern, and accurate values mean that the reflections can be integrated with confidence whether or not they are clearly visible, so that weak reflections are not neglected (*see* Subheading 1.2). In CrystFEL, the X-ray bandwidth and convergence as well as the crystal mosaicity are kept constant, while a value is determined for the notional "size" of a sphere around each reciprocal lattice point which determines how far away from the Bragg condition diffraction will still be seen, and models the Fourier truncation effects of the crystal size [37, 52, 53]. In cctbx.xfel, a similar approach is used; however, the mosaicity of the crystal is also adjusted [54]. A slightly different approach was proposed by Ginn [35], in which the crystal orientations are refined by searching for the orientation which allows all the reflections in the pattern to be predicted with the narrowest possible spectral bandwidth. This approach hence incorporates both stages of refinement in one.

2.4 Integration

Measurement of the actual spot intensities in each diffraction pattern has so far been performed both using conventional and specialized methods. CrystFEL offers a choice between "shoebox" integration, where the pixel intensities close to the predicted location are simply summed after subtracting the background level estimated from pixels further out [53], and two-dimensional profile fitting in which an average reflection profile is constructed from the strong reflections and used to fit the shape of the weaker peaks [55]. The simpler method of "shoebox" integration is much more commonly used in practice. nXDS, like the original XDS program, offers three-dimensional profile fitting where the shape of the intensity profile of each reflection in directions out of the plane of the diffraction pattern is also considered [32]. The program EVAL15 goes further still, by calculating the expected shapes of the peaks ab initio from the crystal and experimental parameters, and fitting those shapes to the observed shapes [56]. The approach in cctbx.xfel involves calculating which specific pixels in the diffraction pattern should contain signal, and hence aims to avoid including regions with no signal to an even greater extent than that given by refining the crystal parameters [30].

In XFEL experiments with very small crystals, Fourier truncation fringes become visible in the diffraction patterns (*see* **Subheading 5 in Chapter 4a**). The size and number of fringes depend on the crystal size. The first proposed method for integrating SFX data involved selecting pixels in the diffraction pattern based on their proximity to the exact Bragg condition, to select only the central maximum of the fringe structure [57]. Extending this idea, a method has been proposed for "dividing out" the fringe structure to remove the influence of varying crystal size on the data quality [58], or even as a method for solving the phase problem [59].

2.5 Resolving Indexing Ambiguities

All the indexing algorithms mentioned in Subheading 2.2 are geometrical methods which use the spot positions in each diffraction pattern to deduce the orientation and unit cell parameters of the crystal which produced it. There are obviously at least as many possible indexing solutions as there are rotational symmetry operators in the point group of the crystal. All of these indexing solutions are equivalent and therefore cause no problem. However, some symmetry classes have less rotational symmetry than their lattices. The indexing algorithms use the spot positions, which have the symmetry of the lattice, and ignore the spot intensities, which reflect the true symmetry of the structure. In cases where there is a rotation operation in the lattice symmetry which is not present in the true symmetry, the possible indexing solutions are no longer all equivalent. The conditions where this can happen are closely related to the conditions where crystal twinning can occur. Excluding all point groups containing centers of symmetry or mirror operations, exact ambiguities occur for crystals with point groups 3, 4, 6, 312, 321, and 23. There are usually only two indexing assignments to be distinguished between, but structures with hexagonal lattices and point group 3 exhibit a double ambiguity where there are four nonequivalent indexing solutions. In addition to all of these, there may be accidental ambiguities due to the unit cell parameters [60].

Indexing ambiguities can cause difficulties anywhere in crystallography where reflection data or structural models need to be compared or merged. In SX, they are a particular problem because very many sets of reflection data need to be merged together. If an ambiguity is not resolved by selecting the correct indexing assignment for each crystal, the final merged dataset will exhibit higher symmetry than the true symmetry of the structure—as if it had been acquired from a perfectly twinned crystal. The first published method for resolving the ambiguity successfully on experimental data was given by Brehm and Diederichs [61], and uses an approach of clustering the individual datasets as points in two or three dimensions, where the distances between the points depend on the similarity between the datasets. This method was shown to be effective on the very first SFX dataset [41], as processed with the earliest version of CrystFEL, so its success is not due to the other enhancements in SFX data acquisition and processing since then. A slightly simpler algorithm, based on clustering in one dimension, was subsequently developed [37], similar to a method proposed by Kabsch [32]. The key feature is that correlation coefficients are measured between many pairs of crystals and mean values taken, reducing the effect of variability between snapshots (Subheading 1.4) in a similar way to how they must be reduced for the intensity measurements themselves by taking mean values of many measurements (Subheading 2.6). A similar algorithm has also been independently developed where the correlation coefficients are calculated between an individual crystal and a merged dataset which is iteratively updated [62], also showing success with experimental data. Other approaches have been proposed based on first compensating for the effect of Fourier truncation fringes (*see* Subheading 2.4), but do not appear to have yet been demonstrated on experimental data [63, 64].

2.6 Scaling, Merging, and Post-Refinement

As has been described above, many measurements of each symmetrically unique reflection are made in an SX experiment. The measurements for each symmetrically equivalent reflection must be merged to produce the final estimate of the intensity, which is used for the later stages of structural analysis. For the first experiments, merging was performed by simply taking the mean of all the measurements. This approach was called "Monte Carlo Integration," because it achieves integration over all the confounding parameters in the experiment (*see* Subheading 1.4) by randomly sampling from their probability distributions.

Far from being a question of just merging the individual measurements, several things can be done at this stage to model diffraction. The key feature of all of these things is that they operate on the intensities after they have been measured, although the techniques can also be combined with elements of the previous processing, as will be described shortly. The first and most obvious of these is to scale the intensities. A single linear scale factor for each crystal essentially models the combined effects of incident beam intensity and crystal size. Introducing a Debye–Waller term for each crystal models variations in crystal order and temperature. This alone has been shown to improve the self-consistency of data in SFX [37].

The most widely discussed enhancement to data processing overall is that of post-refinement to correct partialities (*see* Subheading 1.2). As has been described earlier, with a sufficiently accurate geometrical model of the diffraction condition, the partialities of the reflections could be corrected, in principle removing a significant source of variance from their estimations. In post-refinement, the orientations (and possibly the unit cell parameters as well) are varied in order to maximize the agreement between the corrected partial intensities from each pattern. Rather than using spot positions, post-refinement uses the intensity measurements themselves. The name post-refinement was chosen because it is performed after all other processing is complete for each diffraction pattern.

Post-refinement is a feature of almost all data processing software for rotation crystallography. In this case, there is often a fully recorded equivalent for each partially recorded reflection, which greatly aids the process. For SX experiments, the post-refinement procedure needs to work on datasets consisting entirely of partially recorded reflections. This was first demonstrated, albeit with simulated data, in 2014 [52]. Shortly afterwards, it was demonstrated with very small rotation wedge diffraction patterns from a synchrotron source using nXDS [32] and then on SFX data using cctbx.xfel [65]. The most successful results on SFX data so far have been shown by Uervirojnangkoorn et al. [66] and Ginn et al. [36], the first of which included simultaneous refinement according to spot positions and also scaling with per-crystal linear and Debye–Waller factors.

A final consideration which appears to be very important is to exclude individual reflections or crystals which appear totally inconsistent with the dataset as a whole. Outlier rejection techniques have been elaborated by Ginn et al. [36] and Assmann et al. [67].

2.7 Data Quality Metrics

Some new metrics have been created for quantitatively assessing the quality of SX data. These figures of merit are all based on the degree of self-consistency of the dataset: if we were to hypothetically repeat the experiment under the same conditions, how closely would the results be reproduced? To estimate this, the entire dataset can be split in two, and each half merged separately. The figure of merit R_{split} is an R-factor, similar to R_{merge}, proportional to the ratio of the sum of differences between the intensity measurements in the two merged half-datasets and the sum of the mean intensities. This ratio is divided by $\sqrt{2}$ to estimate the correlation between two hypothetical experiments with the same number of crystals, rather than the correlation between the two half-datasets.

A similar figure of merit is $CC_{1/2}$, which is the Pearson correlation coefficient between the intensities in each half data set. Whereas R_{split} is specific to SX, $CC_{1/2}$ originates in conventional crystallography, enabling closer comparisons between the two techniques. Another figure of merit, CC^*, estimates the correlation coefficient between the full dataset and the hypothetical "true" dataset. CC^* is derived from $CC_{1/2}$ by a simple formula [68].

Besides the intensities themselves, the self-consistency of the differences between intensities of Bijvoet pairs of reflections can be measured. Like with the intensities, this can be done using an R-factor (R_{ano}) or a correlation coefficient (CC_{ano}).

3 Future Directions

In the short to medium term, the path for further development in XFEL data processing undoubtedly consists of improving modeling of the experimental and crystal parameters. As well as improvements to all of the procedures mentioned in this chapter, this may include modeling of the fast electronic radiation damage processes which occur under XFEL irradiation (*see* **Subheading 2 in Chapter 4a**). Eventually, this should reach a stage where the intensities of the reflections measured in an individual diffraction pattern can be accounted for within a few percent of the true value so that, in principle, only one measurement need be made of each symmetrically equivalent reflection. Then, the measurement time and sample consumption can be minimized, and more time points measured in time-resolved or dynamic experiments. In practice, multiple measurements would still be necessary so that different datasets could be compared in order to determine the relevant parameters. We can now expect to be able to routinely solve a structure in an SFX experiment with a few thousand processed diffraction patterns or even less, provided that a good model is available for phasing by molecular replacement. However, experimental phasing has much more stringent requirements on the data quality, and presents a much greater challenge for data processing. Optimizing this type of experiment must be the focus of development in XFEL data processing in the future.

In the longer term, we can expect XFEL data processing software to make increased use of the Fourier truncation fringes surrounding each reflection, when sufficient intensity is used and the crystals sufficiently small. This information might come in the form of phase estimates which can directly be used in established pipelines for experimental phasing. The quasi-single-molecule diffraction patterns recently observed and analyzed for slightly disordered crystals offer a very exciting future route.

Through carefully considered use of multi-processing using multiple CPUs and cluster environments, the overall SFX data analysis pipeline has reached a speed where it can almost keep up with current data acquisition rates. That is, taking into account that there are periods of time where data is not acquired due to the need to make alterations to the experimental setup, and also that the hit rates are usually somewhat lower than optimal (*see* Subheading 1.1). In the future, XFELs using superconducting linear accelerators will come into use, such as the European XFEL and LCLS-II. This poses a challenge not only for sample delivery and detectors but also for data processing software which will have to be made as computationally efficient as is practical to keep up with the potential peak data rates. However, with the great progress over the currently short period since the first XFELs came into use, this challenge seems likely to be surmounted.

Acknowledgments

The author acknowledges financial support from the Helmholtz Association through programme oriented funds, and thanks Anton Barty and Nadia Zatsepin for reading and providing helpful suggestions on the manuscript.

References

1. Spence JCH, Doak RB (2004) Single molecule diffraction. Phys Rev Lett 92:198102
2. Shapiro DA, Chapman HN, DePonte D et al (2008) Powder diffraction from a continuous microjet of submicrometer protein crystals. J Synchrotron Rad 15:593–599
3. DePonte DP, Weierstall U, Schmidt K et al (2008) Gas dynamic virtual nozzle for generation of microscopic droplet streams. J Phys D41:195505
4. Hunter MS, Segelke B, Messerschmidt M et al (2014) Fixed-target protein serial microcrystallography with an X-ray free electron laser. Sci Rep 4:6026
5. Gati C, Bourenkov G, Klinge M et al (2014) Serial crystallography on in vivo grown microcrystals using synchrotron radiation. IUCrJ 1:87–94
6. Nogly P, James D, Wang D et al (2015) Lipidic cubic phase serial millisecond crystallography using synchrotron radiation. IUCrJ 2:168–176
7. Stellato F, Oberthür D, Liang M et al (2014) Room-temperature macromolecular serial crystallography using synchrotron radiation. IUCrJ 1:204–212

8. Eriksson M, van der Veen JF, Quitmann C (2014) Diffraction-limited storage rings—a window to the science of tomorrow. J Synchrotron Rad 21:837–842
9. Cohen AE, Soltis SM, González A et al (2014) Goniometer-based femtosecond crystallography with X-ray free electron lasers. Proc Natl Acad Sci U S A 111:17122–17127
10. Hirata K, Shinzawa-Itoh K, Yano N et al (2014) Determination of damage-free crystal structure of an X-ray-sensitive protein using an XFEL. Nat Methods 11:734–736
11. Andreasson J, Martin AV, Liang M et al (2013) Automated identification and classification of single particle serial femtosecond X-ray diffraction data. Opt Express 22:2497–2510
12. Park J, Joti Y, Ishikawa T et al (2013) Monte Carlo study for optimal conditions in single-shot imaging with femtosecond X-ray laser pulses. Appl Phys Lett 103:264101
13. Powell HR, Johnson O, Leslie AGW (2013) Acta Crystallogr D Biol Crystallogr 69:1195–1203
14. Duisenberg AJM (1992) Indexing in single-crystal diffractometry with an obstinate list of reflections. J Appl Crystallogr 25:92–96
15. Kabsch W (1988) Evaluation of single-crystal X-ray diffraction data from a position-sensitive detector. J Appl Crystallogr 21:916–924
16. Rossmann MG, van Beek CG (1999) Data processing. Acta Crystallogr D Biol Crystallogr 55:1631–1640
17. Weidenspointner G, Epp S, Hartmann A et al (2011) Practical experience from operating the imaging pnCCD detectors of the CAMP instrument at LCLS. Proc SPIE 8078:80780U
18. Strüder L, Epp S, Rolles D et al (2010) Large-format, high-speed, X-ray pnCCDs combined with electron and ion imaging spectrometers in a multipurpose chamber for experiments at 4th generation light sources. Nucl Instrum Methods Phys Res A 614:483–496
19. Philipp HT, Koerner LJ, Hromalik MS et al (2010) Femtosecond radiation experiment detector for X-ray free-electron laser (XFEL) coherent X-ray imaging. IEEE Trans Nucl Sci 57:3795–3799
20. Kameshima T, Ono S, Kudo T et al (2014) Development of an X-ray pixel detector with multi-port charge-coupled device for X-ray free-electron laser experiments. Rev Sci Instrum 85:033110
21. Allahgoli A, Becker J, Bianco L et al (2015) AGIPD, a high dynamic range fast detector for the European XFEL. J Instrum 10:C01023
22. Carini GA, Boutet S, Chollet M et al (2013) Experience with the CSPAD during dedicated detector runs at LCLS. J Phys Conf Ser 493:012011
23. Carini GA, Boutet S, Chollet M et al (2013) Measurements at sychrotrons and FELs: some differences observed with the CSPAD. IEEE Nucl Sci Symp Med Imaging Conf Rec 1:1
24. Bonifacio R, Salvo LD, Pierini P et al (1994) A study of linewidth, noise and fluctuations in a FEL operating in SASE. Nucl Instrum Methods Phys Res A 341:181–185
25. Amann J, Berg W, Blank V et al (2012) Demonstration of self-seeding in a hard-X-ray free-electron laser. Nat Photonics 6:693–698
26. Barends TRM, White TA, Barty A et al (2015) Effects of self-seeding and crystal post-selection on the quality of Monte Carlo-integrated SFX data. J Synchrotron Radiat 22:644–652
27. Barty A, Kirian R, Maia FRNC et al (2014) Cheetah: software for high-throughput reduction and analysis of serial femtosecond X-ray diffraction data. J Appl Crystallogr 47:1118–1131
28. Foucar L, Barty A, Coppola N et al (2012) CASS—CFEL-ASG software suite. Comput Phys Commun 183:2207–2213
29. White TA, Kirian RA, Martin AV et al (2012) CrystFEL: a software suite for snapshot serial crystallography. J Appl Crystallogr 45:335–341
30. Hattne J, Echols N, Tran R et al (2014) Accurate macromolecular structures using minimal measurements from X-ray free-electron lasers. Nat Methods 11:545–548
31. Waterman DG, Winter G, Parkhurst JM et al (2013) The DIALS framework for integration software. CCP4 Newsl Prot Crystallogr 49:16–19
32. Kabsch W (2014) Processing of X-ray snapshots from crystals in random orientations. Acta Crystallogr D Biol Crystallogr 70:2204–2216
33. Botha S, Nass K, Barends TRM et al (2015) Room-temperature serial crystallography at synchrotron X-ray sources using slowly flowing free-standing high-viscosity microstreams. Acta Crystallogr D Biol Crystallogr 71:387–397
34. Ginn HM, Evans G, Sauter NK et al (2016) On the release of cppxfel for processing X-ray free electron laser images. J Appl Crystallogr 49:1065–1072
35. Ginn HM, Messerschmidt M, Ji X et al (2015) Structure of CPV17 polyhedrin determined by the improved analysis of serial femtosecond crystallographic data. Nat Commun 6:6435
36. Ginn HM, Brewster AS, Hattne J et al (2015) A revised partiality model and post-refinement algorithm for X-ray free-electron laser data.

Acta Crystallogr D Biol Crystallogr 71:1400–1410
37. White TA, Mariani V, Brehm W et al (2016) Recent developments in CrystFEL. J Appl Crystallogr 49:680–689
38. Nakane T, Joti Y, Tono K et al (2016) Data processing pipeline for serial femtosecond crystallography at SACLA. J Appl Crystallogr 49:1035–1041
39. Zaefferer S (2000) New developments of computer-aided crystallographic analysis in transmission electron microscopy. J Appl Crystallogr 33:10–25
40. Ayyer K, Yefanov OM, Oberthür D et al (2016) Macromolecular diffractive imaging using imperfect crystals. Nature 530:202–206
41. Chapman HN, Fromme P, Barty A et al (2011) Femtosecond X-ray protein nanocrystallography. Nature 470:73–77
42. Lomb L, Steinbrener J, Bari S et al (2012) An anti-settling sample delivery instrument for serial femtosecond crystallography. J Appl Crystallogr 45:674–678
43. Sauter NK, Hattne J, Grosse-Kunstleve RW et al (2013) New python-based methods for data processing. Acta Crystallogr D Biol Crystallogr 69:1274–1282
44. Mariani V, Morgan A, Yoon CH et al (2016) OnDA: online data analysis and feedback for serial X-ray imaging. J Appl Crystallogr 49:1073–1080
45. Sauter NK, Grosse-Kunstleve RW, Adams PD (2004) Robust indexing for automatic data collection. J Appl Crystallogr 37:399–409
46. Powell HR (1999) The Rossmann Fourier autoindexing algorithm in *MOSFLM*. Acta Crystallogr D Biol Crystallogr 55(10):1690–1695
47. Brewster AS, Sawaya MR, Rodriguez J et al (2015) Indexing amyloid peptide diffraction from serial femtosecond crystallography: new algorithms for sparse patterns. Acta Crystallogr D Biol Crystallogr 71:357–366
48. Gildea RJ, Waterman DG, Parkhurst JM et al (2014) New methods for indexing multi-lattice diffraction data. Acta Crystallogr D Biol Crystallogr 70:2652–2666
49. Schmidt S (2014) GrainSpotter: a fast and robust polycrystalline indexing algorithm. J Appl Crystallogr 47:276–284
50. Paithankar KS, Sørensen HO, Wright JP et al (2011) Simultaneous X-ray diffraction from multiple single crystals of macromolecules. Acta Crystallogr D Biol Crystallogr 67:608–618
51. Yefanov O, Mariani V, Gati C et al (2015) Accurate determination of segmented X-ray detector geometry. Opt Express 23:28459
52. White TA (2014) Post-refinement method for snapshot serial crystallography. Philos Trans R Soc Lond B Biol Sci B369:20130330
53. White TA, Barty A, Stellato F et al (2013) Crystallographic data processing for free-electron laser sources. Acta Crystallogr D Biol Crystallogr 69:1231–1240
54. Sauter NK, Hattne J, Brewster AS et al (2014) Improved crystal orientation and physical properties from single-shot XFEL stills. Acta Crystallogr D Biol Crystallogr 70:3299–3309
55. Rossmann MG (1979) Processing oscillation diffraction data for very large unit cells with an automatic convolution technique and profile fitting. J Appl Crystallogr 12:225–238
56. Kroon-Batenburg LMJ, Schreurs AMM, Ravelli RBG et al (2015) Accounting for partiality in serial crystallography using ray-tracing principles. Acta Crystallogr D Biol Crystallogr 71:1799–1811
57. Kirian RA, Wang X, Weierstall U et al (2010) Femtosecond X-ray protein nanocrystallography—data analysis methods. Opt Express 18:5713–5723
58. Qu K, Zhou L, Dong YH (2014) An improved integration method in serial femtosecond crystallography. Acta Crystallogr D Biol Crystallogr 70:1202–1211
59. Spence JCH, Kirian RA, Wang X et al (2011) Phasing of coherent femtosecond X-ray diffraction from size-varying nanocrystals. Opt Express 19:2866–2873
60. Barends TRM, Foucar L, Ardevol A et al (2015) Direct observation of ultrafast collective motions in CO myoglobin upon ligand dissociation. Science 350:445–450
61. Brehm W, Diederichs K (2014) Breaking the indexing ambiguity in serial crystallography. Acta Crystallogr D Biol Crystallogr 70:101–109
62. Liu H, Spence JCH (2014) The indexing ambiguity in serial femtosecond crystallography (SFX) resolved using an expectation maximization algorithm. IUCr J 1:393–401
63. Zhou L, Liu P, Dong YH (2013) Solution of the effects of twinning in femtosecond X-ray protein nanocrystallography. Chinese Phys C 37:028101
64. Donatelli J, Sethian JA (2014) An algorithmic framework for X-ray nanocrystallographic reconstruction in the presence of the indexing ambiguity. Proc Natl Acad Sci U S A 11:593–598

65. Sauter NK (2015) XFEL diffraction: developing processing methods to optimize data quality. J Synchrotron Radiat 22:239–248
66. Uervirojnangkoorn M, Zeldin OB, Lyubimov AY et al (2015) Enabling X-ray free electron laser crystallography for challenging biological systems from a limited number of crystals. elife 4:e05421
67. Assmann G, Brehm W, Diederichs K (2016) Identification of rogue datasets in serial crystallography. J Appl Crystallogr 49:1021–1028
68. Karplus PA, Diederichs K (2012) Linking crystallographic model and data quality. Science 336:1030–1033

Chapter 14

Many Ways to Derivatize Macromolecules and Their Crystals for Phasing

Miroslawa Dauter and Zbigniew Dauter

Abstract

Due to the availability of many macromolecular models in the Protein Data Bank, the majority of crystal structures are currently solved by molecular replacement. However, truly novel structures can only be solved by one of the versions of the special-atom method. The special atoms such as sulfur, phosphorus or metals could be naturally present in the macromolecules, or could be intentionally introduced in a derivatization process. The isomorphous and/or anomalous scattering of X-rays by these special atoms is then utilized for phasing. There are many ways to obtain potentially useful derivatives, ranging from the introduction of special atoms to proteins or nucleic acids by genetic engineering or by chemical synthesis, to soaking native crystals in solutions of appropriate compounds with heavy and/or anomalously scattering atoms. No approach guarantees the ultimate success and derivatization remains largely a trial-and-error process. In practice, however, there is a very good chance that one of a wide variety of the available procedures will lead to successful structure solution.

Key words Derivatization of crystals, Heavy atoms, Anomalous signal, MIR phasing, MAD phasing, SAD phasing

1 Introduction

There are two principal methods of solving macromolecular crystal structures from diffraction data by the technique of macromolecular crystallography (MX). If a model sufficiently similar to the investigated macromolecule is available, the unknown structure may be determined by the method of molecular replacement. Otherwise, the only practically available approach is to use one of the versions of the approach that can be termed the "special atom method." Such a procedure utilizes some special properties of a small number of certain atoms present among a large number of "standard" elements (C, N, O, H) within the macromolecule. The first step in this approach is the location of the special atoms, and the next stage extends this special-atom "substructure" to the complete crystal structure. The special characteristics may be a

Alexander Wlodawer et al. (eds.), *Protein Crystallography: Methods and Protocols*, Methods in Molecular Biology, vol. 1607,
DOI 10.1007/978-1-4939-7000-1_14,

large number of electrons of some heavy atoms, anomalous X-ray scattering properties of the heavier or lighter atoms, or a combination of both. The heavy and/or anomalous atoms may be naturally present in the crystallized molecules (e.g., in metalloproteins), otherwise they have to be incorporated into native macromolecules before or after crystallization in a process known as derivatization.

Currently there are ~120,000 structures in the Protein Data Bank [1] that can be used as potential search models for molecular replacement and, indeed, the majority of crystal structures are nowadays solved by this technique. However, if no suitable model can be found, there is a need to resort to the special-atom approach. This method was used for solving the first X-ray structures of hemoglobin [2, 3], myoglobin [4] and in other pioneering works. A very illuminating account of the process of crystal structure determination of lysozyme in the early 1960s was presented by Blake et al. [5]. In the early days, when diffraction data were recorded on photographic films, their accuracy was only sufficient to utilize the isomorphous signal of heavy-atom derivatives, with differences between reflection intensities of the native and derivative crystals amounting sometimes to as much as 15–25%. Only after the introduction of more accurate ways of measuring reflection intensities (using multiwire, imaging plate, CCD, and pixel detectors), and with the advent of powerful and tunable synchrotron X-ray beam lines, has it become possible to achieve diffraction data accuracy (on the order of a few percent), necessary for productive exploitation of the inherently weak anomalous signal of various elements present in macromolecular crystals. Currently the anomalous signal is used to solve a great majority of novel X-ray crystal structures, mostly thanks to the efforts of Wayne Hendrickson, who pioneered the methods of multiwavelength anomalous diffraction (MAD [6]), as well as single-wavelength anomalous diffraction (SAD [7]).

Thus, to solve a novel X-ray crystal structure of a protein or nucleic acid, one has to utilize the signal, isomorphous or anomalous, of some special atoms present in the investigated structure. These atoms might be present in the native molecules, for example, in various metalloproteins containing such metals as Fe, Cu, and Zn or could be even lighter elements, such as sulfur contained in almost all proteins and phosphorus present in all nucleic acids. The suitable elements can be introduced by genetic engineering, as is the case of selenium incorporated in the form of selenomethionine (SeMet) [8], which nowadays is the workhorse of protein crystallography owing to its relatively strong anomalous signal. Of course, the classic approach is based on the introduction of heavy metals, such as Pt, Hg, Au, Os, and lanthanides by soaking crystals in buffers also containing appropriately selected compounds. The metal cations are coordinated by reactive functional groups at the protein surface, such as the sulfhydryl group of cysteines, nitrogen atoms of histidines, or carbonyl and carboxylate oxygen atoms. Very useful for phasing large crystal structures are multi-center metal complexes [9, 10] which provide large phasing signals, especially at low resolution. Some ions, such as halides, Br^- and I^-, do not form covalent or coordination bonds with the proteins, but stick to their surface by hydrogen bonds or by nonpolar interactions, often sharing their locations with solvent water molecules [11].

There is, therefore, a large palette of approaches which can be used to solve novel X-ray crystal structures of proteins and nucleic acids when the application of molecular replacement is not possible or not successful. In general, the experimental data used for special-atom phasing have to be more accurate than those for molecular replacement. Phasing based on the anomalous signal of sulfur or phosphorus requires exceptionally accurate data, since the expected phasing signal may be at the level of 1% or less [12–14]. Whereas data used for structure refinement can tolerate a certain degree of radiation damage, those used for the isomorphous and especially anomalous applications should be collected so as to avoid at all cost any radiation damage [15]. However, radiation damage can sometimes be used for phasing, in a method known as RIP [16]. It is possible to combine data collected from several crystals [17], provided they are isomorphous.

2 Incorporation of Special Atoms

The special atoms, intended as the source of the isomorphous and/or anomalous phasing signal, can be incorporated into macromolecular crystals in a variety of ways. Obviously, to exploit the signal of sulfur, which is inherently present in most proteins, or of transition metals found in metalloproteins, no additional derivatization is necessary. If the investigated macromolecule does not contain any elements suitable for phasing, it is necessary to introduce them before conducting the diffraction experiment. This can be done in several ways.

2.1 Modification of the Macromolecules Before Crystallization

The first possibility is to prepare by chemical or biochemical methods a variant of the protein or nucleic acid containing the special atom in advance of the crystallization trials. Incorporation of selenomethionine by genetic engineering [8] is a very commonly used way of obtaining a convenient vehicle for solving protein crystal structures by the MAD or, more frequently, SAD approaches. Indeed, this is the way how the majority of novel protein crystal structures are solved these days. It is also possible to treat other elements in a similar way, e.g., by incorporating in the protein sequence the unnatural amino acid p-iodophenylalanine instead of phenylalanine [18]. The iodine atom can be substituted at the aromatic rings of tyrosine by treating the protein with N-iodosuccincimide prior to crystallization [19]. Analogously to SeMet in proteins, 5-bromouracil can be used as an anomalous marker in chemically synthesized nucleic acids, also in their complexes with proteins [20]. It is also possible to introduce selenium into nucleic acids in the form of phosphoroselenates [21]. Another possibility is to utilize the RIP or "Cheshire cat" effect of isomorphous differences originating from radiation-induced disappearance of the heavy atoms, as demonstrated by the solution of the crystal structures of an Hg-derivative of a protein [22] and bromouracil-modified nucleic acid [23].

2.2 Classic Heavy-Atom Derivatization

The standard way of obtaining heavy-atom derivatives, introduced to protein crystallography as the first method of phasing macromolecular crystal structures by Perutz [3] and his followers [4, 5], is based on soaking native crystals in aqueous solutions containing salts of the appropriate metals. In a variant of this method, the protein or nucleic acid is crystallized from a solution containing the selected salt. A variety of different inorganic and organic salts have been utilized [24], but the most popular and successful reagents are the so-called magic seven: K_2PtCl_4, $KAu(CN)_2$, K_2HgI_4, $UO_2(AcO)_2$, $HgCl_2$, $K_3UO_2F_5$, and PCMBS (para-chloro mercury benzoic acid sulfonate) [25]. A compendium of various reagents and derivatives is available at the Heavy Atom Databank (http://www.sbg.bio.ic.ac.uk/had/ [26]). An important factor to keep in mind is the high toxicity of many heavy-metal compounds, especially those containing mercury or osmium, and appropriate safety procedures must be strictly observed when working with such reagents.

The ligands (Cl^-, Br^-, I^-, CN^-, etc.) in the coordination compounds of the heavy metals must hydrolyze or be substituted by chemically reactive groups of proteins for their successful derivatization. This process can be rapid (seconds), or might take a considerable time (months). Moreover, some reactions with heavy metals may sometimes induce structural rearrangements of proteins, introducing significant non-isomorphism, or even cause visible cracking of the soaked crystals. The usual practice is to soak protein crystals in diluted, millimolar solutions of heavy-atom salts for longer time (several hours or days). If the crystals visibly deteriorate when observed under a microscope, the concentration of the heavy-atom reagent should be lowered. In fact, such a behavior has a positive side, providing a confirmation of a successful, even if too vigorous, derivatization. The excess of the unbound heavy-atoms, which is not productive for phasing, may be removed by a short soak in the mother liquor devoid of derivatization agents. This is particularly important for highly absorbing salts of very heavy metals, such as osmium. On the other hand, care must be taken not to back-soak the metal from the productive protein-binding sites.

As a modification of the usual, slow heavy-atom soaking procedure, the quick soaking approach was proposed [27], where the heavy atoms bind to the protein rapidly, before they can introduce any significant non-isomorphism or structural rearrangements of the crystal contents.

The choice of the most promising derivative is difficult and often the only way is to try a number of reagents, hoping for the eventual success, which may depend on many factors, such as the contents of the crystallization buffer, its pH, concentration, temperature, etc. If the protein contains free sulfhydryl groups, not engaged in disulfide bridges, a promising approach is to attempt mercury derivatization, using one of the many Hg reagents available, such as $HgCl_2$, K_2HgI_4, $Hg(CH_3)Cl$, PCMBS, thiomersal [EMTS, ethylmercury thiosalicylate], mersalyl ({3-[2-(carboxymethoxy)benzoyl]amino-2-methoxypropyl}[hydroxyl] mercury), and others. Compounds containing Pt, Au, Os, and similar elements are often coordinated by the nitrogen atoms of the imidazole rings of histidine or the sulfur atom of methionine. Soaking of native protein crystals in solutions containing triiodides (I_3^-) may lead to iodination of tyrosine rings [28].

A separate group of reagents, especially useful for phasing very large structures of proteins and complexes, are multinuclear metal clusters, such as $Ta_6Br_{12}^{2-}$ or $P_2W_{18}O_{62}^{6-}$ [10]. They were very helpful in cracking the structure of the ribosome [9, 29, 30].

The success of derivatization becomes apparent only after collecting the diffraction data and comparing them with the native set or, in fact, after a successful structure solution. However, some other symptoms may provide useful indications also at earlier stages. For example, the mass spectra of potentially derivatized macromolecules may confirm the successful binding of heavy atoms [25, 31]. Some heavy-atom compounds distinctly colorize the transparent crystals; for example the "magic green" tantalum bromide complex $Ta_6Br_{12}^{2-}$ makes the crystals dark green after successful binding [32]. Often the whole green color is "soaked" from the crystallization solution into crystals, which indicates that the soaking drop should be supplemented with a fresh dose of the reagent.

2.3 Quick Halide Soaks

In contrast to heavy-atom reagents, the bromide and iodide ions do not form stable bonds with proteins and they penetrate into crystals very rapidly, even during a few second soaks [11]. They populate multiple sites around the protein surface, forming ion pairs with the positively charged side chains of Arg and Lys; hydrogen bonds with the amide or hydroxyl donors; or sit in hydrophobic niches. In a variant of this approach it is possible to use triiodides I_3^-, easily prepared by dissolving elemental iodine in the solution of KI [33].

Usually a large number of partially occupied Br^- or I^- sites can be identified as they share their sites with water molecules. The Br^- ions are suitable for SAD or MAD phasing since the $K\alpha$

X-ray absorption edge of Br is at 0.92 Å, easily achievable at all synchrotron beam lies. The absorption edges of I^- iodine are not easily accessible, but iodine has a significant anomalous effect, especially at wavelengths longer than 1.5 Å, and is very convenient for SAD phasing using copper radiation data collected at home laboratories.

The halide concentration in the soaking solution should be high, up to 1 M or more, although some successful results were obtained at concentrations lower than 0.2 M. It is advisable to start testing with high concentration of NaBr or NaI and observe if the crystal survives such a treatment without visible cracking or dissolving. If a high concentration of the halides quickly deteriorates the crystal quality, it should be lowered and the procedure repeated. Due to the fast diffusion of halides into protein crystals, in practice it is enough to sweep the crystal through a drop of the cryoprotecting buffer supplemented with the halides immediately prior to freezing it for data collection.

2.4 Incorporation of Noble Gases

Noble gases, such xenon or krypton, are capable of penetrating into protein crystals under increased pressure and occupy sites at hydrophobic patches on the protein surface [34]. In practice, crystals mounted in nylon loops or in capillaries are kept in high-pressure cells for up to 1 h under a noble gas pressure of several MPa and then rapidly flash-cooled in cold nitrogen gas or liquid. Since Xe or Kr are inert and do not react with proteins chemically, they usually do not introduce significant non-isomorphism. The atoms of Xe and Kr are isoelectronic with, respectively, the I^- and Br^- ions and have analogous X-ray scattering properties. Thus, xenon derivatives display significant isomorphous and anomalous signals at longer wavelengths and krypton is suitable for MAD phasing. Pressure cells of different construction for noble gases derivatization are available commercially and can be found at many synchrotron beam lines.

3 Conclusions

A large palette of techniques exists for obtaining useful derivatives of macromolecular crystals with the incorporation of a wide selection of special atoms suitable for phasing by the MIR, MIRAS, MAD, or SAD methods. However, no technique guarantees a successful structure solution. Currently, the most popular and most successful is phasing of novel crystal structures using the anomalous signal of Se, introduced to proteins as SeMet by genetic engineering, but even this universal approach is not always applicable.

References

1. Berman HM, Westbrook J, Feng Z et al (2000) The Protein Data Bank. Nucleic Acids Res 28:235–242
2. Green DW, Ingram VM, Perutz MF (1954) The structure of haemoglobin. IV. Sign determination by the isomorphous replacement method. Proc R Soc Lond 225:287–307
3. Perutz MF, Rossmann MG, Cullis AF et al (1960) Structure of haemoglobin: a three-dimensional Fourier synthesis at 5.5-Å resolution, obtained by X-ray analysis. Nature 185: 416–421
4. Kendrew JC, Bodo G, Dintzis HM et al (1958) A three-dimensional model of the myoglobin molecule obtained by X-ray analysis. Nature 181:662–666
5. Blake CCF, Fenn RH, Johnson LN et al (2001) A historical perspective: how the structure of lysozyme was actually determined. In: International tables for crystallography, vol F. Kluwer Academic Publishers, Dordrecht, pp 745–772
6. Hendrickson WA (1991) Determination of macromolecular structures from anomalous diffraction of synchrotron radiation. Science 254:51–58
7. Hendrickson WA, Teeter MM (1981) Structure of the hydrophobic protein crambin determined directly from the anomalous scattering of sulfur. Nature 290:107–113
8. Hendrickson WA, Horton JR, LeMaster DM (1990) Selenomethionyl proteins produced for analysis by multiwavelength anomalous diffraction (MAD): a vehicle for direct determination of three dimensional structure. EMBO J 9:1665–1672
9. Gluehmann M, Zarivach R, Bashan A et al (2001) Ribosomal crystallography: from poorly diffracting microcrystals to high-resolution structures. Methods 25:292
10. Dauter Z (2005) Use of polynuclear metal clusters in protein crystallography. Compt Rend Chimie 8:1808–1181
11. Dauter Z, Dauter M, Rajashankar KR (2000) Novel approach to phasing proteins: derivatization by short cryo-soaking with halides. Acta Crystallogr D Biol Crystallogr 56:232–237
12. Wang BC (1985) Resolution of phase ambiguity in macromolecular crystallography. Methods Enzymol 115:90–112
13. Ramagopal UA, Dauter M, Dauter Z (2003) Phasing on anomalous signal of sulfurs: what is the limit? Acta Crystallogr D59:1020–1027
14. Dauter Z, Adamiak DA (2001) Anomalous signal of phosphorus used for phasing DNA oligomer: importance of data redundancy. Acta Crystallogr D57:990–995
15. Garman EF (2010) Radiation damage in macromolecular crystallography: what is it and why should we care? Acta Crystallogr D Biol Crystallogr 66:339–351
16. Ravelli RBG, Leiros HK, Pan B et al (2003) Specific radiation damage can be used to solve macromolecular crystal structures. Structure 11:217–224
17. Liu Q, Liu Q, Hendrickson WA (2013) Robust structural analysis of native biological macromolecules from multi-crystal anomalous diffraction data. Acta Crystallogr D Biol Crystallogr 69:1314–1332
18. Xie J, Wang L, Brock A et al (2004) The site-specific incorporation of p-iodo-L-phenylalanine into proteins for structure determination. Nature Biotechnol 22:1297–1301
19. Brzozowski AM, Derewenda U, Derewenda ZS et al (1991) A model for interfacial activation in lipases from the structure of a fungal lipase-inhibitor complex. Nature 351:491–494
20. Hendrickson WA, Ogata CM (1997) Phase determination from multiwavelength anomalous diffraction measurements. Methods Enzymol 276:494–513
21. Wilds CJ, Pattanayek R, Pan C et al (2002) Selenium-assisted nucleic acid crystallography: use of phosphoroselenates for MAD phasing of a DNA structure. J Am Chem Soc 124: 14910–14916
22. Ramagopal UA, Dauter Z, Thirumuruhan R et al (2005) Radiation-induced site-specific damage of mercury derivatives: phasing and implications. Acta Crystallogr D Biol Crystallogr 61:1289–1298
23. Ennifar E, Carpentier P, Ferrer JL et al (2002) X-ray induced debromination of nucleic acids at the Br K absorption edge and implications for MAD phasing. Acta Crystallogr D Biol Crystallogr 58:1262–1268
24. Agniswamy J, Joyce MG, Hammer CH et al (2008) Towards a rational approach for heavy-atom derivative screening in protein crystallography. Acta Crystallogr D Biol Crystallogr 64:354–367
25. Boggon TJ, Shapiro L (2000) Screening for phasing atoms in protein crystallography. Structure 8:R143–R149
26. Carvin D, Islam SA, Sternberg MJE et al (1998) A databank of heavy-atom binding sites in protein crystals: a resource in use for multiple isomorphous replacement and anomalous

scattering. Acta Crystallogr D Biol Crystallogr 54:1199–1206
27. Sun PD, Radaev S, Kattah M (2002) Generating isomprphous heavy-atom derivatives by a quick-soak method. Part I. Test cases. Acta Crystallogr D Biol Crystallogr 58: 1092–1098
28. Kretsinger RH (1968) A crystallographic study of iodinated sperm whale metmyoglobin. J Mol Biol 31:315–318
29. Ban N, Freeborn B, Nissen P et al (1998) A 9 Å resolution X-ray crystallographic map of the large ribosomal subunit. Cell 93:1105–1115
30. Clemons WM, May JLC, Wimberly BT et al (1999) Structure of a bacterial 30S ribosomal subunit at 5.5 Å resolution. Nature 400: 833–840
31. Joyce MG, Radaev S, Sun PD (2010) A rational approach to heavy-atom derivative screening. Acta Crystallogr D Biol Crystallogr 66:358–366
32. Pasternak O, Bujacz A, Biesiadka J et al (2008) MAD phasing using the $(Ta_6Br_{12})^{2+}$ cluster: a retrospective study. Acta Crystallogr D Biol Crystallogr 64:595–606
33. Evans G, Bricogne G (2003) Triiodide derivatization in protein crystallography. Acta Crystallogr D Biol Crystallogr 59:1923–1929
34. Prangé T, Schiltz M, Pernot L et al (1998) Exploring hydrophobic sites in proteins with xenon or krypton. Proteins 30:61–73

Chapter 15

Experimental Phasing: Substructure Solution and Density Modification as Implemented in SHELX

Andrea Thorn

Abstract

This chapter describes experimental phasing methods as implemented in SHELX. After introducing fundamental concepts underlying all experimental phasing approaches, the methods used by SHELXC/D/E are described in greater detail, such as dual-space direct methods, Patterson seeding and density modification with the sphere of influence algorithm. Intensity differences from data for experimental phasing can also be used for the generation and usage of difference maps with ANODE for validation and phasing purposes. A short section describes how molecular replacement can be combined with experimental phasing methods. The second half covers practical challenges, such as prerequisites for successful experimental phasing, evaluation of potential solutions, and what to do if substructure search or density modification fails. It is also shown how auto-tracing in SHELXE can improve automation and how it ties in with automatic model building after phasing.

Key words Experimental phasing, Substructure search, Density modification, Direct methods, Heavy atoms, Anomalous diffraction, Anomalous difference map, MAD, SAD, MR-SAD, RIP phasing

1 Introduction: What Do All Experimental Phasing Methods Have in Common?

For every Bragg reflection recorded in X-ray diffraction, there is a structure factor with an amplitude and a phase. From amplitudes and phases an electron density map can be calculated via Fourier summation.

The amplitude corresponding to a certain reflection is proportional to the square root of the intensity of said reflection. Its phase is lost. This constitutes the central problem of macromolecular crystallography: the phase problem.

As the phase cannot be obtained directly from the data, it has to be determined from a model and/or indirectly from the measured data. Experimental phasing utilizes information from one or several data sets, typically measured for this purpose to determine the phases.

Alexander Wlodawer et al. (eds.), *Protein Crystallography: Methods and Protocols*, Methods in Molecular Biology, vol. 1607,
DOI 10.1007/978-1-4939-7000-1_15,

Table 1
Experimental phasing methods can be grouped into two categories: Those relying on anomalous differences such as SAD and MAD, and those relying on the differences between crystals, such as SIR and MIR. RIP, where the chemical differences created by radiation damage in a crystal are exploited, constitutes a special case of SIR. Methods from both categories can be combined, giving rise to SIRAS, RIPAS and MIRAS

Abbreviation	Method
SAD	Single wavelength anomalous diffraction
S-SAD	SAD based on native sulfurs (special case of SAD)
MAD	Multiple wavelength anomalous diffraction
SIR	Single isomorphous replacement
RIP	Radiation damage induced phasing (special case of SIR)
MIR	Multiple isomorphous replacement
SIRAS	Single isomorphous replacement with anomalous scattering
MIRAS	Multiple isomorphous replacement with anomalous scattering

All experimental phasing methods rely on differences in intensities [1]. These can originate from small chemical differences between two or more isomorphous crystals (meaning that the atomic positions of these crystals are mostly the same, and hence they have the same space group and unit cell). They can also originate from anomalous scattering which affects otherwise equal intensities. In any case, only a small number of atoms in the unit cell cause these intensity differences, and these atoms are referred to as "marker atoms," "substructure," or sometimes "heavy atoms." Table 1 shows the most common experimental phasing methods.

2 Anomalous Scattering

Anomalous scattering is an energy-dependent change of atomic form factors f due to absorption. Each structure factor is composed of form factor contributions f from each atom in the unit cell [1, 2]:

$$f = f_0 + f' + if''.$$

f_0 depends solely on $sin(\theta)/\lambda$ for the reflection in question and the atom's element. f' and f'' are the real (dispersive) and imaginary (anomalous) component of the anomalous scattering, dependent on the wavelength of the X-rays, and are typically very small. They can be observed near the so-called absorption edge (*see* Fig. 1), where their contribution to the diffraction can be observed as a violation of Friedel's law (*see* Fig. 2).

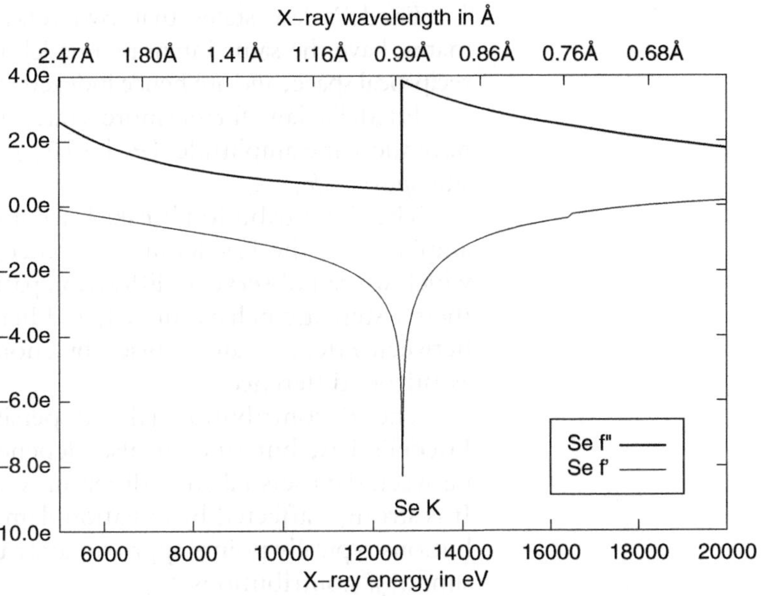

Fig. 1 Anomalous scattering coefficients for Selenium: f' and f'' against energy/wavelength. A plot like this can be obtained for any element from Ethan Merritt's homepage http://www.bmsc.washington.edu/scatter

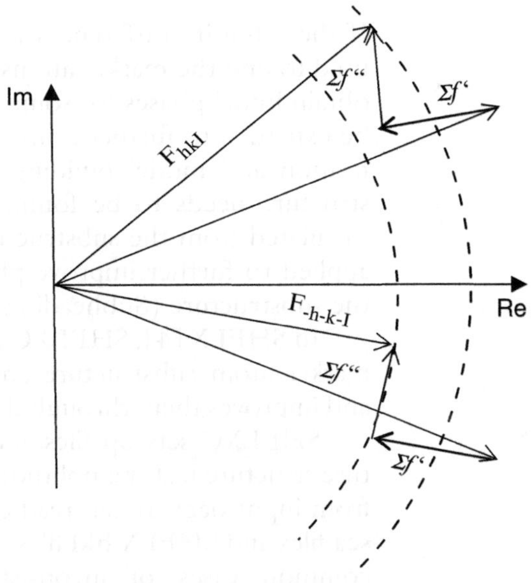

Fig. 2 Breaking of Friedel's law. F_{hkl} and F_{-h-k-l} are the structure factors for two Friedelmates; $\Sigma f'$ are the dispersive contributions from all atoms combined and $\Sigma f''$ are the anomalous contributions from all atoms combined. The $\Sigma f''$ contribution breaks Friedel's law, resulting in different amplitudes (shown by dashed circles) and thus, different intensities for Friedel mates

Friedel's law states that two reflections, the so-called Friedel mates, have the same intensity. Friedel mates are centrosymmetric in reciprocal space, and are hence indexed with *h*, *k*, *l* and −*h*, −*k*, and −*l*.

Friedel's law furthermore states that their structure factors have the same amplitude: $|F_{hkl}| = |F_{-h-k-l}|$ and their phases are opposite: $\phi_{hkl} = -\phi_{-h-k-l}$.

The f'' contribution breaks Friedel's law, altering the phase and amplitude of the Friedel mates, and consequently, their intensities which we can observe in diffraction patterns. Bijvoet first exploited their systematic behaviour [3], and hence, the intensity difference between Friedel mates, caused by anomalous scattering, is known as Bijvoet difference.

The f' contribution (the dispersive signal) does not break Friedel's law, but since it also depends on wavelength, it varies between data sets taken at different wavelengths in MAD phasing. It is strongly affected by radiation damage, as the sum of f' contributions typically points approximately towards the direction of the sum of f_0 contributions.

In isomorphous replacement methods (*see* Table 1), the difference in intensity is caused by a difference in atoms contributing to every structure factor, as some atoms are present in one data set), but not in another.

3 The Workflow in Experimental Phasing

If the intensity differences are sufficiently accurate, they may be used to find the marker atoms and then using the marker atoms to obtain initial phases for some of the reflections. These phases may be extended to further reflections and improved by density modification and model building. After data measurement, the substructure needs to be found (Subheading 4). Initial phases are calculated from the substructure and then density modification is applied to further improve phases and resolve the handedness of the substructure (Subheading 5).

In SHELX [4], SHELXC prepares the data, SHELXD finds the marker atom substructure and SHELXE calculates initial phases and improves them through density modification (compare Fig. 3).

SHELXC sets up files and calculates the estimated substructure structure factor amplitude $|F_A|$ and the phase shift α (*see* below) from input data. It can read data from XDS (**`XDS_ASCII.HKL`**), **sca** files and SHELX **hkl** files and if necessary corrects for the most common cases of inconsistent indexing between data sets. Unmerged data are preferable as input. If only an **mtz** file is available, an **hkl** file can be generated by MTZ2VARIOUS in CCP4 [6], MTZ2HKL [7]. If AIMLESS [8] is used, a **sca** file should be written out in addition to the default **mtz** file at the scaling step

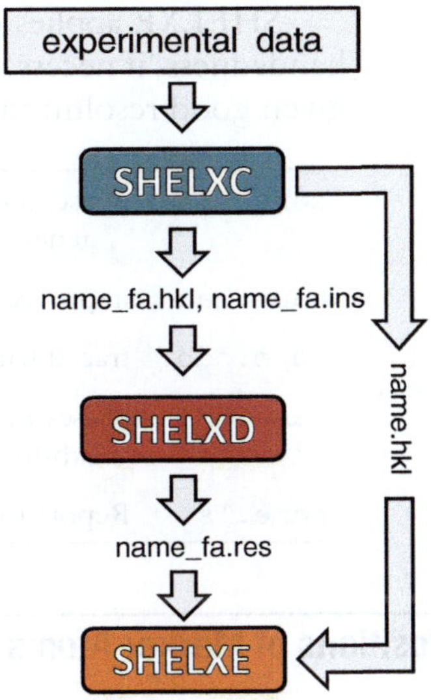

Fig. 3 Typical SHELXC/D/E workflow for experimental phasing. Picture from [5]

before phasing is attempted. If at all possible, **unmerged data** should be input to SHELXC.

In addition to the data, SHELXC requires some input parameters, such as the cell, space group, and expected number of marker atoms, typically supplied the form of a text file. Instructions on how to set up this text file are given along with other helpful information if the program SHELXC is executed without an input file. However, there are a number of graphical user interfaces for SHELX which manage the input and output automatically, such as for example CCP4i2 [9], HKL2MAP [10] and HKL3000 [11].

In the context of this chapter, '**name**' will always refer to the project file name chosen by the user. Three files are written out from input data:

`name_fa.ins`	cell, symmetry and instructions for SHELXD
`name_fa.hkl`	$h, k, l, \|F_A\|, \sigma(\|F_A\|)$ and the estimated α angles
`name.hkl`	$h, k, l, \|I_{obs}\|, \sigma(\|I_{obs}\|)$

SHELXD finds the marker atom substructure and puts out the following file:

`name_fa.res`	potential marker atom positions (best substructure result)
`name.lst`	report on substructure search

SHELXE applies density modification to establish the correct handedness, if necessary, and can auto trace the poly-Ala backbone given good resolution native data; its output files are:

name.phs	phases and amplitudes from the data, a map can be generated from this file, for example in COOT [12]
name.hat	improved marker atom positions
name.pdb	traced backbone, if auto-tracing has been used
name.pha	phases and amplitude differences, corresponding to the substructure, a map can be generated from this information
name.lst	Report for evaluation purposes

4 How Are the Positions of Marker Atoms Found?

4.1 What Information Is Actually Contained in Intensity Differences?

In order to understand how the marker atom positions can be found, it helps to look at the information content in the intensities: An electron density map is calculated from amplitudes (which are obtained from the measured intensities) and phases. Before phasing, a map can be calculated from the intensities only, with all phases set to zero. This map is called a "Patterson map," and it contains information about the length and direction of interatomic vectors, but not of their location in the unit cell. However, in the context of macromolecular crystallography, a Patterson map is not helpful to locate atoms, as there are simply too many interatomic vectors—the Patterson map is too populated, and the non-atomic resolution typical for macromolecular data further aggravates the problem.

If, however, a map is calculated from intensity differences, for example caused by anomalous scattering, and a set of good phases, this map shows only peaks where anomalously scattering atoms are located. Such a map is referred to as "anomalous map"—or "heavy atom map." It can be used for validation purposes, to identify reactions within the crystal or to confirm the element of an atom [13, 14]. The peak heights also can give information about the solvability of the substructure and the best resolution cutoff [15]. Within the SHELX framework, such a map can be obtained from a **pha** file, which can be generated by SHELXE (using phases from density modification) or ANODE (using phases from a PDB structure, *see* Fig. 4). Most marker atoms will have separate density peaks in these maps, as they are far apart. Tantalum bromide clusters or disulfide bridges may fuse into 'blobs'.

With the phases being of course initially unavailable, a map can be calculated with the intensity differences and all phases set to zero, the result is an "anomalous Patterson map"—giving only the interatomic vectors between marker atoms, but not others. These

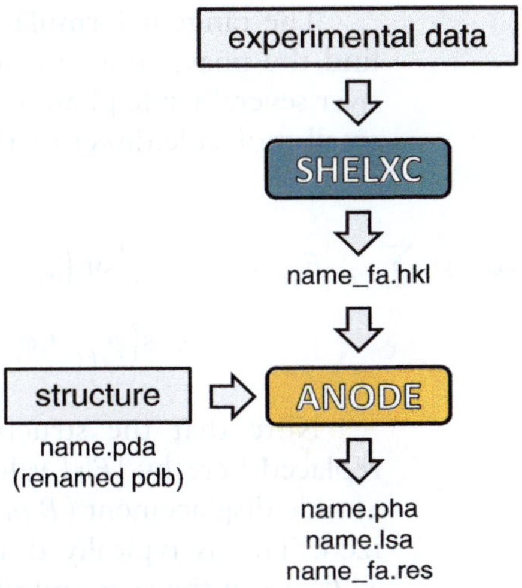

Fig. 4 Usage of ANODE within the SHELX framework to generate difference maps. The command `anode name` reads in a `name_fa.hkl` file with differences from any experimental phasing method and a `name.pdb` file and generates a difference map for validation purposes. To obtain negative and positive density (for example for RIP experiments), anode name `-n3` has to be used. The negative density corresponds to the atomic positions after the radiation damage. Picture from [5]

maps can be utilized to find marker atom positions in the same way atoms are found in small molecule crystallography.

4.2 Methods Borrowed From Small Molecule Crystallography

The most commonly used phasing procedure for small molecules is direct methods, which are essentially a search method. Direct methods exploit inherent features of electron density—it is, for example, never negative and atoms are well separated from each other—so as to establish relationships between the phases of structure factors. In order to use direct methods to find marker atoms in a macromolecular structure even at low resolutions, intensity differences have to be used instead of intensities, as the marker atom densities are well separated from each other, while electron density of a native protein tends to be continuous except at very high resolution.

One of the fundamental equations used in direct methods is the triplet equation, which enables the phase of a structure factor to be estimated from two other phases. It is, however, not an exact equation, but subject to statistical fluctuations:

$$\Phi_{hkl} = \Phi_{h'k'l'} + \Phi_{h-h'\,k-k'\,l-l'} (modulo\ 360°).$$

The tangent formula, which is simply the general formula to find the phase of a sum of complex numbers, is a weighted sum over several triple phase relations. It has played a dominant role in small-molecule direct methods of structure solution:

$$\tan(\phi_{hkl}) = \sum_{h'k'l'} \left|E_{h'k'l'} E_{h-h'\,k-k'\,l-l'}\right| \sin\left(\phi_{h'k'l'} + \phi_{h-h'\,k-k'\,l-l'}\right) / \sum_{h'k'l'} \left|E_{h'k'l'} E_{h-h'\,k-k'\,l-l'}\right| \cos\left(\phi_{h'k'l'} + \phi_{h-h'\,k-k'\,l-l'}\right).$$

Note that the structure factor amplitudes $|F_{hkl}|$ have been replaced here by $|E_{hkl}|$ values. In order to eliminate the effects of atomic displacement (*B factor*), the $|F_{hkl}|$ values have been normalized. This is typically done by dividing $|F_{hkl}^2|$ with the average $<|F_{hkl}^2|>$ in the corresponding resolution bin. *E*-values correspond to the structure factors of a point atom structure. This is of particular relevance in macromolecular phasing, as the displacement of marker atoms can be high, and *E*-values can compensate for this to a certain degree, enabling the crystallographer to obtain the substructure by direct methods.

With these equations, a set of phases for the substructure can be calculated, but a starting point is needed. This starting point can be random phases, such as in the program RANTAN [16], or it can be consistent with the interatomic distances found in the Patterson map (*see* above), which then is called Patterson seeding, such as in the SHELXD, where it is used if the keyword **PATS** is given.

In order to make the whole search converge faster, dual-space methods are used to iterate between real and reciprocal space using Fast Fourier Transforms. After Patterson Seeding, the reflections with the highest *E*-values (about 10% in SHELXD) are extended by the tangent formula followed by a peak search in real space.

It is remarkable that SHELXD was originally written for the phasing of small molecules and has been repurposed later [17].

4.3 Improving the Substructure Before Proceeding

In particular with a large amount of marker atoms or borderline cases, it is important to improve the substructure with maximum likelihood methods, such as SHARP [18] or PHASER_EP [19]. SHELXE also includes an option (**−z**) to improve the substructure before using it for density modification. In particular with large substructures containing many atoms, this "tweaking" of the substructure can be very helpful to get a usable map after density modification.

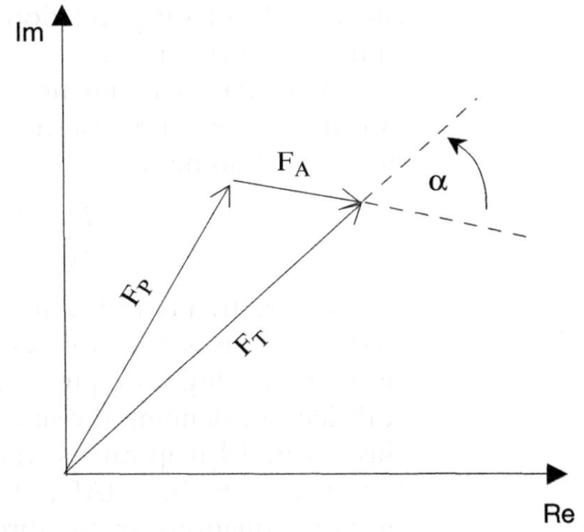

Fig. 5 Definition of the α angle: F_T is the total structure factor composed of all contributions from marker atoms F_A and the contributions from everything else (mainly Protein) F_P: $\mathbf{F_T} = \mathbf{F_P} + \mathbf{F_A}$. The angle between F_A and F_T is α

5 How Can Initial Phases Be Obtained From the Substructure?

We can think of each structure factor F_T composed of all contributions from marker atoms F_A and the contributions from everything else (mainly protein) F_P. Let the angle between F_A and F_T be α (*see* Fig. 5):

$$\alpha = \varphi_T - \varphi_A.$$

If we already know the substructure, we can calculate F_A, including its phase contribution φ_A. If we can estimate α, we can calculate the overall phase φ_T:

$$\varphi_A + \alpha = \varphi_T.$$

5.1 Calculating the α Angle

The alpha angle can be calculated using the following equations which assume that there is only one type of anomalous scatterer:

$$|F_{hkl}|^2 = |F_T|^2 + a|F_A|^2 + b|F_T||F_A|\cos\alpha + c|F_T||F_A|\sin\alpha$$

$$|F_{-h-k-l}|^2 = |F_T|^2 + a|F_A|^2 + b|F_T||F_A|\cos\alpha - c|F_T||F_A|\sin\alpha$$

$$a = \frac{f''^2 + f'^2}{f_0^2} \quad b = \frac{2f'}{f_0} \quad c = \frac{2f''}{f_0}$$

In the case of MAD phasing, there would be different a, b, c and two observations for each wavelength. $|F_A|$, $|F_T|$ and α are unknown. So given good data from at least two wavelengths, the system of equations can be solved. This works of course best if the f' differences and the sum of f'' values are large, and errors are small, as the phasing equations are only strictly true in the absence of measurement errors.

In a SAD experiment, however, only two observables are available, as only one wavelength was measured. Assumptions are necessary, leading to:

$$|F_T| = 0.5\ (|F_{hkl}| + |F_{-h-k-l}|).$$
$$|F_{hkl}| - |F_{-h-k-l}| = c|F_A|\ \sin\alpha.$$

If we restrict ourselves to the largest positive or negative anomalous differences, they will tend to have $\sin(\alpha)$ of +1 or −1 respectively. Surprisingly, despite being a very rough estimation, this is sufficient for defining the substructure and estimation of φ_T. From this, an initial map can be calculated, but the phases will still be very inaccurate. For MAD, MIR and SIRAS phasing we have three or more equations for the three unknowns F_T, F_A and α, so fewer approximations are required and the initial phases may yield an interpretable map. SAD phases alone cannot do this until they have been extended and improved by density modification.

It is very important to recognize that at this stage, the handedness of the substructure has not yet been established—this needs to be discerned at the next stage: density modification.

6 How Are Initial Phases Improved?

At this stage, especially SAD phases are still ambiguous as well as inaccurate. Density modification dramatically improves the initial phases and thus the electron density; it also resolves the handedness of the substructure. In order to do so, the substructure and its enantiomer are employed as starting points. The map which after density modification has more protein-like features, meaning a better connectivity among other things, has started from the correct substructure. The other map has less connectivity and looks 'ragged'.

In general, density modification generally follows this procedure: An electron density map is calculated with the initial phases, and then a modification of some sort is applied. The modifications make the map more resemble that of a typical macromolecular crystal structure. Fourier inversion of the modified density should then result in improved phases.

These modifications include:

Solvent Flattening [20]: Disordered solvent regions of the map are flat and has fewer features than ordered regions, i.e., less variation. Solvent flattening is an iterative method to improve the phases by setting the electron density in the solvent region to a constant value [1]. Finding the boundary between the ordered macro molecule and the disordered solvent is not trivial: The local variation of the map can be used. SHELXE uses the Sphere of influence algorithm as an alternative approach (*see* below).

Histogram Matching: In badly phased maps the electron density values have a Gaussian histogram due to noise, while well phased maps have a skewed distribution. By sharpening the density the histogram is made to resemble that of a well phased map better [21].

Non-crystallographic Symmetry: It might be possible to establish, from the number of monomers, from the substructure or even from the Patterson map non-crystallographic symmetry, which can also be used as prior knowledge in density modification. In SHELXE, the command line option `-n <N>` can be used to specify the expected number of copies in the asymmetric unit.

SHELXE employs the sphere of influence algorithm (*see* Fig. 6) for density modification to improve these phases [22]. The sphere of influence is a shell of radius 2.42 Å about a grid point in the map. If there is a large variance in the density values in this spherical shell, the grid point is more likely to correspond to an ordered atomic position as 2.42 Å is a common 1,3-interatomic distance in biological macromolecules. After several cycles, this results in flattening of the solvent regions without ever needing to define an explicit solvent boundary.

If density modification resulted in an interpretable map which can be used for structure refinement, the structure is considered solved.

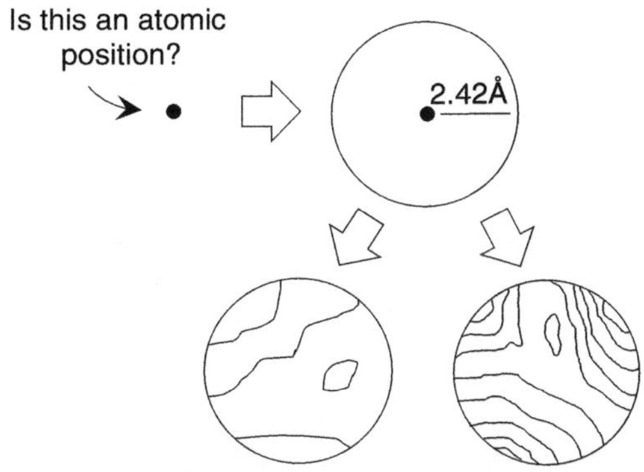

Fig. 6 Sphere-of-influence algorithm, as used in SHELXE [22]. To identify if a given peak in the initial map is an atomic position inside the protein, the electron density on a sphere with a radius of 2.42 Å around that point is calculated and the variance of the electron density evaluated: if it does not vary much (*left*), it is flipped [23], which combats model bias. As the density is later combined with a newly calculated map [24] the effect is similar to solvent flattening. In addition to the two shown options, for intermediate variance values, a weighted mean of the two operations is performed, resulting in a "fuzzy" solvent boundary. This helps to avoid a sharp boundary, locking the program into a false solution. If it varies a lot (*right*), the density is further sharpened, as the middle of the sphere likely corresponds to a position inside the ordered macromolecule

7 Can Experimental Phasing Methods Be Combined with Other Methods?

In recent years, MR-SAD [25] has become popular, which is a combination of phase information from molecular replacement (MR) and SAD phasing. The most common way to do this is to determine the substructure from a partial molecular replacement solution, i.e., to "bootstrap" the substructure with the molecular replacement phases. Then substructure optimization and density modification are applied. This in particular removes model bias from the molecular replacement and can dramatically improve the density not covered by the search model.

In SHELXE, there are two ways to combine an existing partial MR solution with the phase information in experimental phasing: The **pdb** file with the partial solution can be renamed to `name.pda` (so as to not to be overwritten by the output of auto-tracing, if used) and then this is given instead of a substructure file `name_fa.res` (*see* Fig. 7). The other way is to use ANODE to obtain the substructure from experimental data and a partial MR solution: ANODE is run with the experimental data and the partial MR solution as structure, and the resulting file `name_fa.res` (*see* Fig. 4) is then used like any other substructure file in density modification.

Density modification can even be used to improve molecular replacement phases without experimental phase information

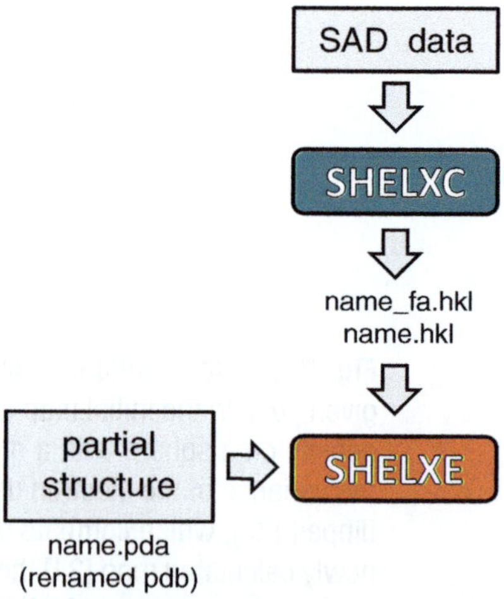

Fig. 7 Workflow for MR-SAD in SHELXE. Note how the substructure search is skipped, as initial phase information is obtained from molecular replacement. Picture from [5]

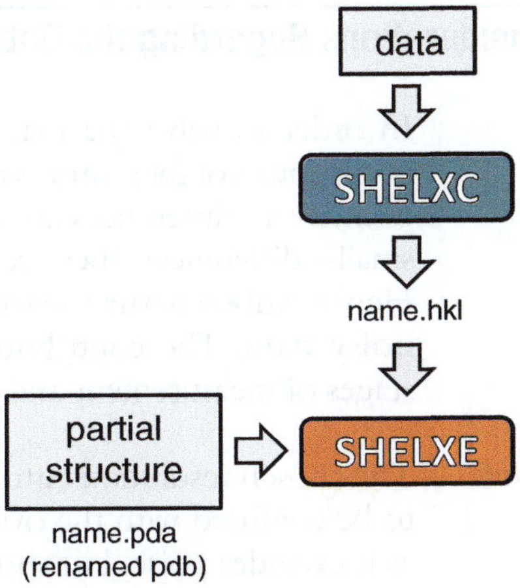

Fig. 8 Workflow for density modification on a partial molecular replacement solution in SHELXE. Input phases for SHELXE can be from a molecular replacement model, renamed from `name.pdb` to `name.pda` (so as not to be overwritten by an eventually traced backbone). The command line to use is `shelxe name.pda <options>`. Picture from [5]

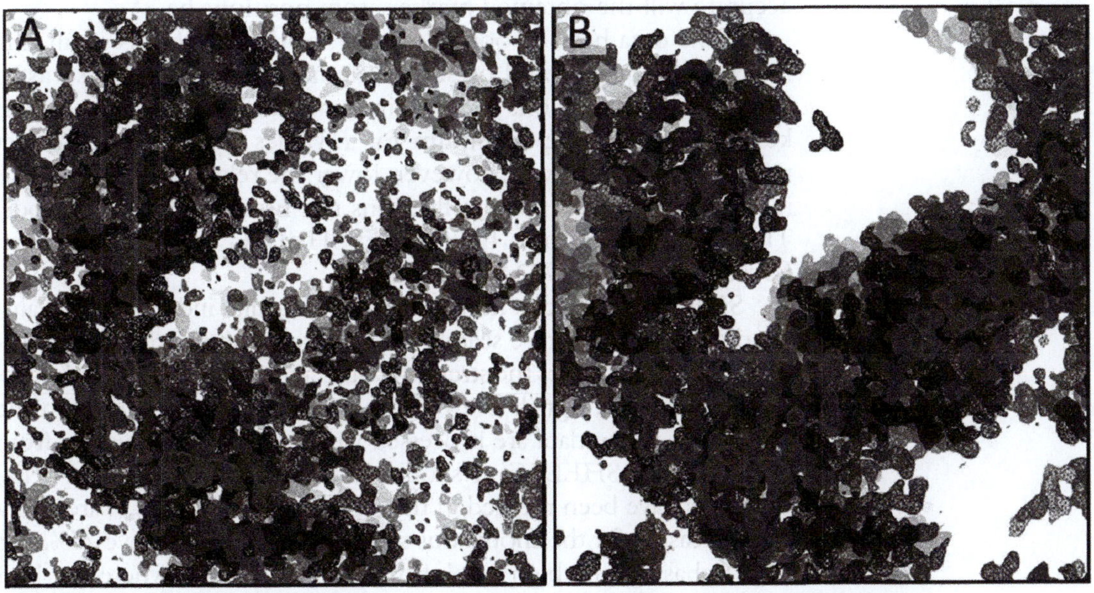

Fig. 9 Molecular Replacement electron density [26] before and after density modification

[23]. This is only feasible at very good resolutions (2.3 Å or better, depending on the structure), but it improves phases, expands the map and removes bias from the search model used for molecular replacement, as the positioned search is "thrown away" once the density modification starts (*see* Figs. 8 and 9). However, density modification will be more effective when phase information comes from an independent source, such as experimental data.

8 Practical Considerations Regarding the Data

In order to solve the marker atom substructure, the positions and occupancies of these atoms need to be determined. For this, thousands of intensity differences are available, but since they are—often rather small—differences, they are subject to measurement and other errors. Hence, **substructure search is an overdetermined problem with noisy data**. The contribution of errors should be minimized at all stages of measurement and data processing.

8.1 Resolution Cutoff

The chosen resolution cutoff for the substructure search, which is not to be confused with the overall resolution limit of the data, is crucial, as it excludes noise, but should include the signal. If the substructure search is not immediately successful, the resolution cutoff should be varied by using SHEL <lower limit> <higher limit> in SHELXC, or by simply changing the SHEL command in the file name_fa.ins, which contains the instructions for SHELXD [27].

Generally speaking, the higher the resolution, the longer SHELXD will need for a try of substructure solution and the more accurate will the obtained marker atom positions be.

If the data have a poor resolution, or the unit cell is small, the ESEL parameter should be set lower than its default 1.5 so as to have enough reflections for the direct methods calculations, as this keyword defines the minimum for *E*-values to be used in direct methods. Doing so will, however, lower the reliability of CC_{weak}, as it is calculated with those reflections not used for direct methods. This keyword may also be input to SHELXC, and is then transferred to the SHELXD input file.

SHELXC supplies a number of useful quality indicators to evaluate data for substructure search, and to decide on the resolution cutoff. ***<d''/sig>*** indicates the strength of the anomalous signal –***<d'/sig>*** for the dispersive signal. Both values should asymptote to 0.8 in the outer shell, if the data are processed well. If this line is missing from the output of SHELXC, it should be checked whether the Friedel opposites have been merged in the input data. $CC(1/2)_{anom}$ are also a good measure for the anomalous signal and should be above 25%. For MAD phasing, the high resolution cutoff should be where the best correlation coefficient between two of the data sets drops below 25%. It is also important to keep in mind that if quality indicators look suspicious, scaling and data processing should be reevaluated.

8.2 Multiplicity

High multiplicity ensures small errors and should be favoured compared to the highest possible resolution. An extra goniometer circle, for example using a mini-kappa device, can be helpful. However, as density modification and model building benefit from high resolution, an extra low-multiplicity high resolution native data set is also desirable.

8.3 Scaling

Data should have the best possible quality. Special attention should be paid to proper measurement setup, processing and to scaling statistics. Anisotropic scaling is applied by virtually every experimental phasing program today, as it often can make the difference between solving and not solving a phase problem. Anisotropic scaling of the native data is applied automatically in SHELXE.

8.4 Intensity Outliers and Low Resolution Completeness

As outliers result in very high intensity differences, properly masking the beam stop, etc. is important, as is outlier rejection. For example, a reflection which is partly obscured by the beam stop will give rise to a huge Bijvoet difference, and hence a very high E-value which is likely to dominate the substructure search and make it fail. Because of their high intensities, the low resolution reflections are very important for all kinds of experimental phasing and care should be taken to measure as many of them as possible – this can also be done by combining a higher resolution data collection pass with a lower resolution pass.

8.5 Multi-crystal Averaging

Better than averaging measurements from one crystal is to average measurements from several crystals [28–30]. However, isomorphism must be established, ideally by the correlation coefficient between data sets or the anomalous correlation coefficient. Software to do this automatically is available, such as BLEND [31] and phenix.multi_crystal_average [32].

8.6 Radiation Damage

Radiation damage is often bad, unless it is used to obtain data sets for RIP phasing. *E*-values have to be weighted, given that all reflections get weaker from radiation damage, which effectively heightens disorder in the crystal. This weighting is done by using the commands `RIPW` and `DSCA` in SHELXD. `RIPW <weight>` gives the weight of the anomalous contribution from the "before" data set as opposed to the "after" data set in a RIP-AS experiment. `DSCA` can be used to scale the "before" data set to the "after" data set in a RIP experiment. Typical values are between `1.00` (no scaling) and `0.95`. It can also be used for RIP-AS, SIR-AS, and SIR, where it is applied to the native data.

If significant radiation damage is encountered in a MAD experiment, the dispersive term can be set to zero by using the command **SMAD** in SHELXC. This is also useful if peak and inflection point data sets have been confused. In such a case, the marker atom solution from SHELXD will have good quality indicators, but the resulting map will be bad. Much phase information will be lost, but the structure can be solved in this way in some cases by giving this keyword together with the regular **MAD** data input keywords.

8.7 Fluorescence Scan

A fluorescence scan can give important information: It can prove or disprove the presence of anomalous scatterers in the crystal. However, even if it proves their incorporation, they may not form an ordered substructure.

9 Did It Work?

9.1 Substructure Search in SHELXD

The most important indicator of a successful substructure search in SHELXD is the CFOM (combined figure of merit) which is $CC_{All} + CC_{Weak}$.

CC_{All} is the correlation coefficient between the normalized structure factor differences in measured data and those calculated from a given substructure solution; CC_{Weak} is the same, but only for weak reflections which have not been used for direct methods, and hence are a somewhat independent criterion.

Unfortunately, CFOM for a correct substructure varies depending on the phasing method, resolution and other criteria—its distribution gives away the correct solution: Typically, CC_{All} and CC_{Weak} are considerably higher for full or partial solutions than for non-solutions. Almost all SHELX GUIs plot CC_{All} against CC_{Weak} to show distinct clusters for solutions and non-solutions.

Consequently, CFOM for a successful substructure solution is higher than the values for non-solutions. It can also be useful to compare the values of potential solutions to the values in alternative space groups with the same number of general positions.

An additional criterion is a clear drop in marker atom occupancies. If this is observed, it is a good indicator that the substructure is solved—however, it absence does not indicate that the solution is wrong—the number of atoms to search for (specified by the `FIND` command) may have been too low. Together with the resolution cutoff, this parameter may be critical for a successful marker atom search and should be within 20% of the true number of marker atoms in the asymmetric unit. For iodine soaks, the number of amino acid residues in the asymmetric unit divided by 15 is a good starting point. For RIP, there will be no sharp drop of occupancy values, as radiation damage may affect a great number of atoms to a varying degree. The number of disulfide bridges if present is a good starting value. For S-SAD, the number of sulfur atoms is a good starting point. Note that if atoms are closer than the high resolution limit specified in `SHEL`, they should be given as one peak, for example, when searching for the elongated peaks of a disulfide bridge: disulfide bridges are 2.03 Å long, so at a resolution of 2.1 Å or lower, the sulfur peaks fuse into one. It should also be mentioned that using the command `DSUL <number of disulphide bridges>` in SHELXD can be vital for a successful search of disulfide bridges.

If the substructure has a known geometry—such as disulfides at high resolution or the "magic triangle" [33], the geometry should be reflected in the found marker atom positions and their symmetry equivalents which can be found in the output **name_fa.res** file. For automatic substructure comparison, the programs SITCOM [34] or phenix.emma [35] are available.

The ultimate indication however is only a successful density modification resulting in a map interpretable in terms of a molecular model which can be used for refinement. So if it is unclear whether a SHELXD run worked, it is always worthwhile to use the substructure for density modification, or to subject it to substructure optimization.

If the substructure cannot be found, it is also worthwhile to increase the number of tries (`NTRY <number>` in SHELXD or SHELXC). How many tries are used depends on the computing time available, but as a rule of thumb 60,000 is a good starting point, and up to 500,000 have been necessary in some cases to find a correct solution. If the "best CFOM" is satisfactory, a SHELXD run can be interrupted by creating a file `name.fin` in the directory SHELXD is running in.

Another thing to try is to vary the resolution cutoff used for the substructure search (`SHEL` command in SHELXC or SHELXD, *see* above), in 0.2 Å steps. The *CFOM* values, however, will increase with lower resolution, as fewer data points are fitted.

Varying all these parameters, a number of different runs in SHELXD can be set up. In addition, phase information from partial MR solutions can be used (*see* Subheading 7).

9.2 Density Modification in SHELXE

The best indication of successful density modification is a map that looks like protein and in which the macromolecule can be built easily.

Connectivity (continuous stretches of electron density representing the connectivity of the macro molecule in question) and map contrast (clear delineation between ordered and disordered regions of the crystal) should be high.

By default, 20 cycles of density modification are applied, and it can be useful to increase this value (using the command line option `-m` in SHELXE if the solvent content is very high.

If high resolution data are available, a backbone trace may give away successful phasing in the form of a trace that is compact and recognizable as a protein fold, and in the form of *CC of trace against native data* (*see* Subheading 10). Up to 25 cycles of auto tracing (option `-a`) may be needed to improve the phases enough. Particularly for DNA or RNA structures, where auto-tracing is not available, using the "free lunch" algorithm [36] can be useful (`-e <resolution>`) to extend the phases to a higher resolution than data were originally measured. Like backbone tracing, this works best at high resolutions.

If density modification does not work, but the substructure seems reliable, one may want to raise the solvent content (given as input to SHELXE with the option `-s <fraction>`) a bit. The solvent content should generally be correct within 10%, as it is a crucial parameter for successful density modification. The number of density modification cycles can also be varied.

It should be ensured that both enantiomorphs of the substructure have been tried (`-i` option in SHELXE) unless the input phase information was in the form of a partial MR solution. The correct hand of the substructure is the one with the higher map contrast. The correct map should look less ragged after some cycles of density modification. Clear side chains protruding from a traced backbone are an unambiguous sign that the correct substructure enantiomorph has been chosen.

It may also be advisable to optimize the substructure (*see* Subheading 4.3 for options to do so) before density modification.

If nothing else works, scaling and data processing should be revisited, in particular with regard to data pathology, outliers, and correct parameters for integration.

10 Auto-Tracing and Automation

Today, all stages of experimental phasing are highly automatized. The available computing power usually allows for parallelized substructure search and parallel testing of different parameter sets. In addition, auto tracing of the backbone is used together with density modification, which significantly heightens its power to improve phases. Moreover, a structure that can be traced is a structure solved. In SHELXE [37] this is evaluated with a single value, the correlation coefficient of the traced backbone against the native data. At resolutions of 2.5 Å or better, a value over 25% means that the structure is solved.

This is exploited in a number of pipelines, for example ARCIMBOLDO [38], AMPLE [39] or AUTORICKSHAW [40]. It is important to use, however, the **name.phs** file for further structure building as well as the trace, as SHELXE uses the auto tracing to improve the phases, and the traced backbone does not contain as much information as the phases themselves, contained in the **phs** file.

Chemical knowledge can be further employed for a more complete auto building, which enhances the phases further, for example with programs like AUTOSOL [41], ARP/wARP [42], BUCCANEER [43].

Acknowledgments

I would like to thank Airlie McCoy and George M. Sheldrick for fruitful discussions. This work was supported by the European Union FP7 Marie-Curie IEF grant "SOUPINMYCRYSTAL" (grant No. 330033).

References

1. Rupp B (2009) Biomolecular crystallography. In: Principles, practice, and application to structural biology. Garland Science, New York
2. Drenth J (1994) Principles of protein X-ray crystallography. Springer, New York
3. Bijvoet JM (1945) Phase determination in direct Fourier synthesis of crystal structures. Koninkl Nederland Akad Wetenschap Proc 52:313–314
4. Sheldrick GM (2008) A short history of SHELX. Acta Crystallogr A 64:112–122
5. Thorn A. Lecture notes "Crystallographic Masterclass", Diamond Lightsource/University of Oxford. http://shelx.uni-ac.gwdg.de/~athorn
6. Winn MD, Ballard CC, Cowtan KD, Dodson EJ, Emsley P, Evans PR, Keegan RM, Krissinel EB, Leslie AGW, McCoy A, McNicholas SJ, Murshudov GN, Pannu NS, Potterton EA, Powell HR, Read RJ, Vagin A, Wilson KS (2011) Overview of the CCP4 suite and current developments. Acta Crystallogr D Biol Crystallogr 67:235–242
7. Gruene T (2008) mtz2sca and mtz2hkl: facilitated transition from CCP4 to the SHELX program suite. J Appl Crystallogr 41:217–218
8. Evans PR (2011) An introduction to data reduction: space-group determination, scaling and intensity statistics. Acta Crystallogr D Biol Crystallogr 67:282–292
9. Liz Potterton L (2012) Introducing CCP4i2. CCP4 Newsletter 48
10. Pape T, Schneider TR (2004) HKL2MAP: a graphical user interface for macromolecular phasing with SHELX programs. J Appl Crystallogr 37:843–844
11. Minor W, Cymborowski M, Otwinowski Z, Chruszcz M (2006) HKL-3000: the integration of data reduction and structure solution—from diffraction images to an initial model in minutes. Acta Crystallogr D Biol Crystallogr 62:859–866
12. Emsley P, Lohkamp B, Scott W, Cowtan K (2010) Features and development of coot. Acta Crystallogr D Biol Crystallogr 66:486–501
13. Thorn A, Sheldrick GM (2011) ANODE: anomalous and heavy-atom density calculation. J Appl Crystallogr 44:1285–1287
14. Terwilliger TC, Bunkóczi G, Hung L-W, Zwart PH, Smith JL, Akey DL, Adams PD (2016) Can I solve my structure by SAD phasing? Planning an experiment, scaling data and evaluating the useful anomalous correlation and anomalous signal. Acta Crystallogr D Biol Crystallogr 72:359–374
15. Thorn A (2011) Practical approaches to macromolecular X-ray structure determination. Dissertation, Georg-August University Goettingen. http://hdl.handle.net/11858/00-1735-0000-0006-B072-8
16. Yao J-X (1981) On the application of phase relationships to complex structures. XVIII. *RANTAN*—random *MULTAN*. Acta Cryst A37:642–644
17. Usón I, Sheldrick GM (1999) Advances in direct methods for protein crystallography. Curr Opin Struct Biol 9:643–648
18. Bricogne G, Vonrhein C, Flensburg C, Schiltz M, Paciorek W (2003) Generation, representation and flow of phase information in structure determination: recent developments in and around SHARP 2.0. Acta Crystallogr D Biol Crystallogr 59:2023–2030
19. McCoy AJ, Storoni LC, Read RJ (2004) Simple algorithm for a maximum-likelihood SAD function. Acta Crystallogr D Biol Crystallogr 60:1220–1228
20. Wang BC (1985) Resolution of phase ambiguity in macromolecular crystallography. Methods Enzymol 115:90–112
21. Lunin VY (1988) Use of the information on electron density modification technique for phase refinement and extension of macromolecules. Acta Crystallogr A 44:144–150
22. Abrahams JP (1997) Acta Crystallogr D Biol Crystallogr 53:371–376
23. Thorn A, Sheldrick GM (2013) Extending molecular-replacement solutions with SHELXE. Acta Crystallogr D Biol Crystallogr 69:2251–2256
24. Sheldrick GM (2002) Macromolecular phasing with SHELXE. Z Kristallogr 217:644–650
25. Schuermann JP, Tanner JJ (2003) MRSAD: using anomalous dispersion from S atoms collected at Cu K[alpha] wavelength in molecular-replacement structure determination. Acta Crystallogr D Biol Crystallogr 59:1731–1736
26. Galej WP, Oubridge C, Newman AJ, Nagai K (2013) Crystal structure of Prp8 reveals active site cavity of the spliceosome. Nature 493:638–643
27. Schneider TR, Sheldrick GM (2002) Substructure solution with SHELXD. Acta Crystallogr D Biol Crystallogr 58:1772–1779
28. Evans P (2006) Scaling and assessment of data quality. Acta Crystallogr D Biol Crystallogr 62:72–82
29. Liu Q, Dahmane T, Zhang Z, Assur Z, Brasch J, Shapiro L, Mancia F, Hendrickson WA

(2012) Structures from anomalous diffraction of native biological macromolecules. Science 336:1033–1037

30. Giordano R, Leal RMF, Bourenkov GP, McSweeney S, Popov AN (2012) The application of hierarchical cluster analysis to the selection of isomorphous crystals. Acta Crystallogr D Biol Crystallogr 68:649–658
31. Foadi J, Aller P, Alguel Y, Cameron A, Axford D, Owen RL, Armour W, Waterman DG, Iwata S, Evans G (2013) Clustering procedures for the optimal selection of data sets from multiple crystals in macromolecular crystallography. Acta Crystallogr D Biol Crystallogr 69:1617–1632
32. Terwilliger TC, Bunkoczi G, Hung L-W, Zwart PH, Smith JL, Akey DL, Adams PD (2015) Can I solve my structure by SAD phasing? Anomalous signal in SAD phasing. Acta Crystallogr D Biol Crystallogr 72:346–358
33. Beck T, Krasauskas A, Gruene T, Sheldrick GM (2008) A magic triangle for experimental phasing of macromolecules. Acta Crystallogr D Biol Crystallogr 64:1179–1182
34. Dall'Antonia F, Schneider TR (2006) SITCOM: a program for comparing sites in macromolecular substructures. J Appl Crystallogr 39:618–619
35. Adams PD, Afonine PV, Bunkoczi G, Chen VB, Davis IW, Echols N, Headd JJ, Hung LW, Kapral GJ, Grosse-Kunstleve RW, McCoy AJ, Moriarty NW, Oeffner R, Read RJ, Richardson DC, Richardson JS, Terwilliger TC, Zwart PH (2010) PHENIX: a comprehensive Python-based system for macromolecular structure solution. Acta Crystallogr D Biol Crystallogr 66:213–221
36. Caliandro R, Carrozzini B, Cascarano GL, De Caro L, Giacovazzo C, Siliqi D (2007) Advances in the free lunch method. J Appl Crystallogr 40:931–937
37. Sheldrick GM (2010) Experimental phasing with SHELXC/D/E: combining chain tracing with density modification. Acta Crystallogr D Biol Crystallogr 66:479–485
38. Rodríguez DD, Grosse C, Himmel S, González C, de Ilarduya IM, Becker S, Sheldrick GM, Usón I (2009) Crystallographic ab initio protein structure solution below atomic resolution. Nat Methods 6:651–653
39. Bibby J, Keegan R, Rigden DJ, Wynn M, Mayans O (2012) AMPLE—using ab initio modelling to tackle difficult molecular replacement cases. CCP4 Newslett 48
40. Panjikar S, Parthasarathy V, Lamzin VS, Weiss MS, Tucker PA (2005) Auto-rickshaw: an automated crystal structure determination platform as an efficient tool for the validation of an X-ray diffraction experiment. Acta Crystallogr D Biol Crystallogr 61:449–457
41. Terwilliger TC, Adams PD, Read RJ, McCoy AJ, Moriarty NW, Grosse-Kunstleve RW, Afonine PV, Zwart PH, Hung LW (2009) Decision-making in structure solution using Bayesian estimates of map quality: the PHENIX AutoSol wizard. Acta Crystallogr D Biol Crystallogr 65:582–601
42. Langer G, Cohen SX, Lamzin VS, Perrakis A (2008) Automated macromolecular model building for X-ray crystallography using ARP/wARP version 7. Nat Protoc 3:1171–1179
43. Cowtan K (2006) The buccaneer software for automated model building. Acta Crystallogr D Biol Crystallogr 62:1002–1011

Chapter 16

Contemporary Use of Anomalous Diffraction in Biomolecular Structure Analysis

Qun Liu and Wayne A. Hendrickson

Abstract

The normal elastic X-ray scattering that depends only on electron density can be modulated by an "anomalous" component due to resonance between X-rays and electronic orbitals. Anomalous scattering thereby precisely identifies atomic species, since orbitals distinguish atomic elements, which enables the multi- and single-wavelength anomalous diffraction (MAD and SAD) methods. SAD now predominates in de novo structure determination of biological macromolecules, and we focus here on the prevailing SAD method. We describe the anomalous phasing theory and the periodic table of phasing elements that are available for SAD experiments, differentiating between those readily accessible for at-resonance experiments and those that can be effective away from an edge. We describe procedures for present-day SAD phasing experiments and we discuss optimization of anomalous signals for challenging applications. We also describe methods for using anomalous signals as molecular markers for tracing and element identification. Emerging developments and perspectives are discussed in brief.

Key words Anomalous scattering, Crystal structure, Phasing problem, Native SAD, Multiple crystals, De novo structure determination

1 Introduction and Theoretical Background

X-ray diffraction analysis is very effective for determining atomic structures of biological macromolecules. It does not produce images directly, however, rather the image is synthesized computationally from the diffracted waves, for which we can record directly only the amplitudes and need to evaluate the phases by other means. Anomalous diffraction has become the method of choice for de novo structure determination of biomolecules. In this chapter, we summarize the theory, approaches, and applications that are currently most effective for using anomalous diffraction in structure analysis.

X-rays are scattered from electrons in the atoms from which molecules are built; and, when these molecules are arrayed into a crystal, the coherent component of the scattering is restricted to discretely directed and highly amplified beams. Each diffracted

Alexander Wlodawer et al. (eds.), *Protein Crystallography: Methods and Protocols*, Methods in Molecular Biology, vol. 1607, DOI 10.1007/978-1-4939-7000-1_16, © Springer Science+Business Media LLC 2017

X-ray beam (Bragg reflection) is characterized by its direction, encoded by Miller indices $\boldsymbol{h}(h,k,l)$; its amplitude, for which the structure-dependent factor is designated $|F(\boldsymbol{h})|$; and its phase $\phi(\boldsymbol{h})$. The theory of diffraction from a crystal has the form of a Fourier transformation of the distribution of electron density $\rho(\boldsymbol{r})$ for all positions $\boldsymbol{r}$ within the unit cell of the crystal: $F(\boldsymbol{h}) = |F(\boldsymbol{h})| \exp(i\phi(\boldsymbol{h})) = \mathcal{F}[\rho(\boldsymbol{r})]$. By Fourier theory, the electron-density distribution can be reconstituted by Fourier inversion of the comprehensive set of diffracted waves: $\rho(\boldsymbol{r}) = \mathcal{F}^{-1}[\{F(\boldsymbol{h})\}]$. Since X-ray experiments record only the amplitudes of diffracted X-ray waves, this poses the phase problem—what is $\phi(\boldsymbol{h})$ for each of the many thousands of Bragg reflections from a biomolecular crystal?

Several ingenious methods have been invented for solving the phase problem, and the one that took hold initially for proteins was that of multiple isomorphous replacement (MIR), which takes advantage of the distinctively strong scattering of heavy metals that can be added to the natural macromolecule. Scattering strength from matter is defined by the atomic scattering factor, f, which is measured relative to the inelastic scattering expected from a single electron and is proportional to the number of electrons in an atom (atomic number Z). Hence, as an example, mercury at $Z = 80$ is highly potent as a scatterer when inserted into biomolecules, which are largely made of carbon ($Z = 6$), nitrogen ($Z = 7$), and oxygen ($Z = 8$). Scattering from atoms includes not only this "normal" component proportional to electron density, f°, but also an "anomalous" increment, f^{Δ}, due to resonance between the incident X-ray waves and electronic orbitals. A 90° phase shift accompanies anomalous (resonant) scattering, which resolves into real and imaginary parts, f' and f''. Thus,

$$f = f^{\circ} + f^{\Delta} = f^{\circ} + f' + if''. \quad (1)$$

Anomalous scattering factors are usually small relative to normal scattering factors; nevertheless, anomalous scattering proved effective from the earliest days of protein crystallography for enhancing MIR (MIRAS) or for making single derivative analyses possible (SIRAS).

Eventually, it became clear that anomalous scattering on its own could suffice to solve the phase problem for macromolecular crystals. We have thoroughly reviewed the ensuing development of multi- and single-wavelength anomalous diffraction (MAD and SAD) [1]. Here, we simply summarize the theoretical underpinnings. Phase evaluation by MAD or SAD begins by measuring complete diffraction data $\{|{}^{\lambda}F(\pm\boldsymbol{h})|^2\}$ at an appropriate set of wavelengths λ (only one for SAD) and usually at $\pm\boldsymbol{h}$ (Friedel mates or symmetry equivalents, i.e., the Bijvoet mates); atomic positions for the anomalous scatterers are then determined, usually from an analysis of Bijvoet differences; next, contributions to the diffraction from the normal scattering, f°, of this "anomalous" substructure can be calculated, ${}^{\circ}F_A(\boldsymbol{h}) = |{}^{\circ}F_A(\boldsymbol{h})| \exp(i\, {}^{\circ}\phi_A)$; and, ultimately, these $°F_A(\boldsymbol{h})$ components serve as reference waves for evaluating structure factors, $°F_T(\boldsymbol{h}) = |°F_T(\boldsymbol{h})| \exp(i\ °\phi_T)$, that correspond to the actual electron density for the entire structure (T for total).

Such structure analyses can be made for arbitrarily complex situations, but the formulation simplifies for the case of only one kind of anomalous scatterer (e.g., Se atoms in a selenomethionyl protein). Then

$$\begin{aligned} \left|{}^{\lambda}F_T(\boldsymbol{h})\right|^2 = \left|{}^{\circ}F_T(\boldsymbol{h})\right|^2 &+ a(\lambda)\left|{}^{\circ}F_A(\boldsymbol{h})\right|^2 \\ &+ b(\lambda)\left|{}^{\circ}F_T(\boldsymbol{h})\right|\left|{}^{\circ}F_A(\boldsymbol{h})\right|\cos\left({}^{\circ}\phi_T - {}^{\circ}\phi_A\right) \\ &\pm c(\lambda)\left|{}^{\circ}F_T(\boldsymbol{h})\right|\left|{}^{\circ}F_A(\boldsymbol{h})\right|\sin\left({}^{\circ}\phi_T - {}^{\circ}\phi_A\right), \quad (2) \end{aligned}$$

where all wavelength dependence is in the factors a, b, and c:

$$a(\lambda) = \left(\left|f^{\Delta}\right|/f^{\circ}\right)^2; b(\lambda) = 2\left(f'/f^{\circ}\right);\ and\ c(\lambda) = 2\left(f''/f^{\circ}\right).$$

This system of equations from multiple wavelengths and Friedel mates ($\pm\boldsymbol{h}$) provides a basis for definitive phase evaluation by MAD [2, 3]. The definitive character of MAD is seen from the orthogonality of phase information in appropriate diffraction differences. By differencing between Friedel mates, it follows from Eq. (2) that

$$\left|{}^{\lambda}F(\boldsymbol{h})\right|^2 - \left|{}^{\lambda}F(-\boldsymbol{h})\right|^2 = 2c(\lambda)\left|{}^{\circ}F_T(\boldsymbol{h})\right|\left|{}^{\circ}F_A(\boldsymbol{h})\right|\sin({}^{\circ}\phi_T - {}^{\circ}\phi_A). \quad (3)$$

Similarly, after defining $<|{}^{\lambda}F(\boldsymbol{h})|^2> = (|{}^{\lambda}F(\boldsymbol{h})|^2 + |{}^{\lambda}F(-\boldsymbol{h})|^2)/2$, one can obtain the dispersive differences between measurements made at two wavelengths, λ_i and λ_j:

$$\begin{aligned} \left\langle\left|{}^{\lambda i}F(\boldsymbol{h})\right|^2\right\rangle - \left\langle\left|{}^{\lambda j}F(\boldsymbol{h})\right|^2\right\rangle &= \left[a(\lambda_i) - a(\lambda_j)\right]\left|{}^{\circ}F_A(\boldsymbol{h})\right|^2 \\ &+ \left[b(\lambda_i) - b(\lambda_j)\right]\left|{}^{\circ}F_T(\boldsymbol{h})\right|\left|{}^{\circ}F_A(\boldsymbol{h})\right|\cos\left({}^{\circ}\phi_T - {}^{\circ}\phi_A\right). \quad (4) \end{aligned}$$

Moreover, since $|°F_T(\boldsymbol{h})| \approx (|{}^{\lambda}F(\boldsymbol{h})| + |{}^{\lambda}F(-\boldsymbol{h})|)/2$ for typical cases where anomalous scattering is relatively weak, Eq. (3) reduces to the Bijvoet-difference equation for the desired $°\phi_T$ phase information in terms of the $°F_A(\boldsymbol{h})$ reference wave:

$$\begin{aligned} {}^{\lambda}\Delta F_{\pm h} = \left|{}^{\lambda}F(\boldsymbol{h})\right| - \left|{}^{\lambda}F(-\boldsymbol{h})\right| &\approx c(\lambda)\left|{}^{\circ}F_A(\boldsymbol{h})\right|\sin\left({}^{\circ}\phi_T - {}^{\circ}\phi_A\right) \\ &= 2\left(f''/f^{\circ}\right)\left|{}^{\circ}F_A(\boldsymbol{h})\right|\sin\left({}^{\circ}\phi_T - {}^{\circ}\phi_A\right). \quad (5) \end{aligned}$$

Equation (5) was used as the basis for determining the structure of crambin from the anomalous scattering of its intrinsic sulfur atoms [4], a method that would now be known as native SAD.

The analysis of crambin confronted the complication of phase ambiguity—from Eq. (3) we obtain the sine of an angle but need the angle itself. The partial structure of sulfur atoms was used for ambiguity resolution for the crambin analysis, but it became clear that this approach would not be powerful enough for structure determinations in general. It was the motivation to find a better alternative that led to MAD, where definitive phase evaluation is manifestly feasible—mathematically, Eq. (4) provides cosine values to complement the sine values obtained from Eqs. (3) and (5). MAD analysis developed very effectively, and its varied implementations have been reviewed by us and others [1, 3, 5–7]. By the year 2000, MAD had surpassed MIR for de novo determination of biomolecular structures [1]. At about that time, a more efficient alternative for resolving phase ambiguities emerged with density modification procedures. Density modification originated with solvent flattening as devised by Wang [8], but it was with systematic incorporation of molecular averaging and other features into the program DM [9] that its effectiveness for SAD grew. By 2006, SAD had overtaken MAD and it now predominates overall for de novo phasing of biomolecules [1].

In this chapter we summarize practical aspects of structure analysis from anomalous diffraction measurements with emphasis on SAD phasing procedures as currently practiced.

2 Phasing Elements and Anomalous Scattering Factors

MAD phasing experiments rely on the sharp variation of anomalous scattering that occurs near the resonance energy for a suitable electronic transition. The spectrum of anomalous scattering factors at the Se K edge of a selenomethionyl (SeMet) protein is shown in Fig. 1a as an example. Definitive phase evaluations can be made with a judicious selection of wavelengths, chosen as indicated to optimize the complementarity from the f' and f'' contributions defined by Eqs. (3) and (4). To a first approximation, all K edges are alike except that their resonant energies progress systematically with atomic number; likewise for L and M edges. K, L, and M resonances do give rise to successively larger anomalous scattering factors, however, as they respectively engage more electrons. Moreover, electrons in molecular orbitals may give rise to especially sharp edge variations (white lines), as seen in the f'' spectra (Fig. 1b). Peak f'' values vary from a few electrons for Se at its K edge ($E = 12.66$ keV), to some 30 electrons for Yb at its L_{III} edge ($E = 8.94$ keV), to over 100 electrons for U at its M_V edge ($E = 3.55$ keV).

SAD phasing depends only on f'', and it can be highly effective when freed by density modification from the tyranny of phase ambiguity. The strength of anomalous diffraction for SAD can be estimated as the Bijvoet diffraction ratio [4]. For the case of one kind of anomalous scatterer, this ratio approximates to

$$rms(\Delta F_{\pm h}) / rms(F_P) \approx \sqrt{2}(\sqrt{N_A} f'') / (\sqrt{N_P} Z_{eff}) \qquad (6)$$

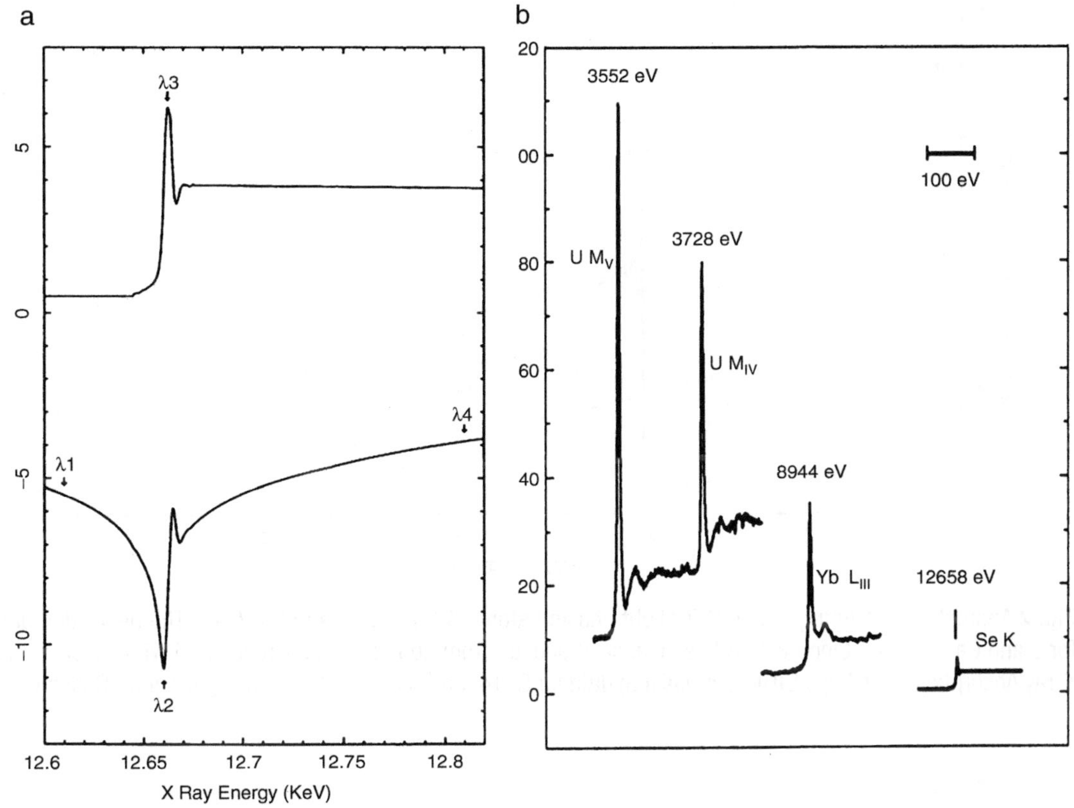

Fig. 1 X-ray anomalous scattering absorption edges. (**a**) Anomalous scattering factor real component *f′* (*bottom*) and imaginary component *f″* (*top*) from Se. (**b**) Anomalous diffraction imaginary component *f″* for Se K edge, Yb L_{III} edge, U M_{IV} and U M_V edges. The Se K edge is from a crystal of selenomethionyl human chorionic gonadotropin [10], the U M_{IV} and U M_V edges are from uranyl nitrate [11], and the Yb L_{III} edge is from a ytterbium-derivatized crystal of N-cadherin [12]. Reproduced with permission from Elsevier Ltd. for (**a**) and the Proceedings of the National Academy of Sciences of the United States of America for (**b**)

where N_A and N_P are the numbers of anomalous scatterers and total non-hydrogen atoms, respectively, and Z_{eff} is the effective atomic number (~6.7 for proteins). Thus, the Bijvoet signals needed for SAD phasing are proportional to the strength of *f″*, the imaginary component of anomalous scattering for the phasing element in a specific subject of interest, and to the relative abundance of ordered atoms of this element. For a given crystal, the optimization of *f″* is of paramount importance. For an accessible edge, the X-ray energy needs to be tuned precisely to obtain the highest possible *f″*. For lighter atoms, the resonance energy is too low to be accessible at many synchrotron beamlines, as for the sulfur and calcium K edges at 2.47 keV and 4.04 keV, respectively (Fig. 2). Nevertheless, these atoms can produce weak but measurable anomalous signals at off-resonance energies; for sulfur, *f″* is 1.31 electrons at 5 keV, 0.96 electrons at 6 keV and 0.72 electrons at 7 keV. The anomalous scattering signals from sulfur can be harvested for phasing at most crystallographic beamlines. Figure 3 has

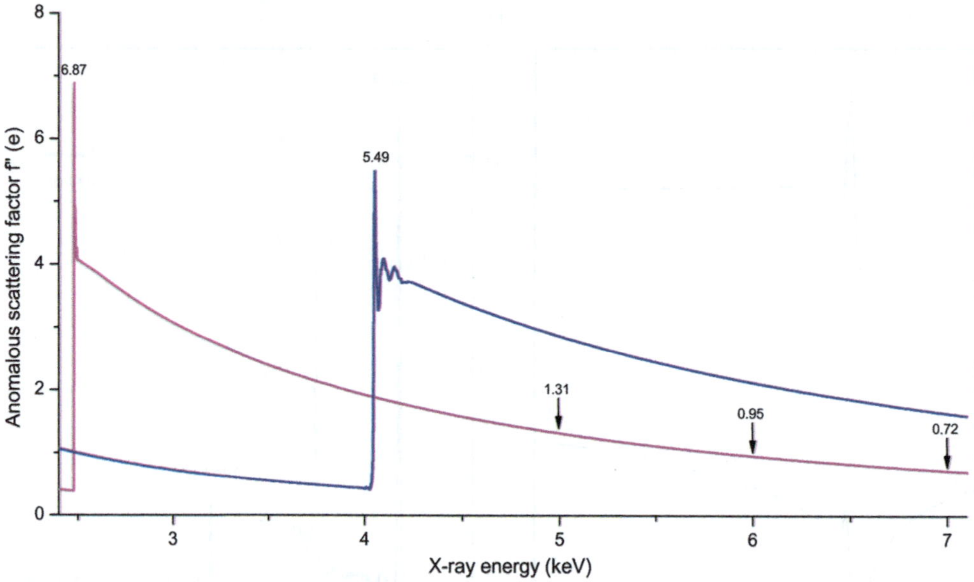

Fig. 2 Anomalous scattering factor *f*″ for light phasing atoms S (*magenta*) and Ca (*blue*). The near-edge data for S and Ca were combined with off-resonance *f*″ spectra from quantum calculations [13] with experimental X-ray Absorption Near-Edge Structure (XANES) data for S [14] and Ca [15] fitted by using program Chooch [16]

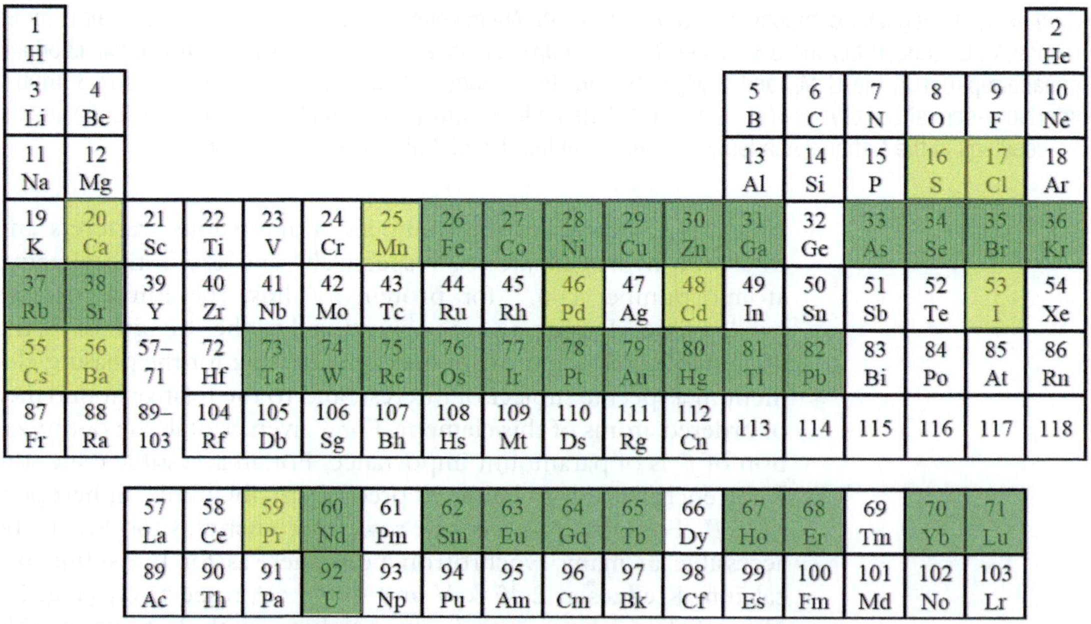

Fig. 3 Periodic table of phasing elements. Elements currently used in at-resonance experiments are highlighted in *green*; and elements used in off-resonance experiments are highlighted in *yellow*. Reproduced from [1] with permission from Cambridge University Press

the periodic table colored for phasing elements that have been used for MAD and SAD experiments, at-resonance (green) or off-resonance (yellow).

Appropriate phasing atoms have to be incorporated into biomolecules for analysis by anomalous diffraction. For proteins that do not contain desired phasing atoms natively, co-crystallization or soaking are two popular ways of introducing phasing atoms [17]. Derivatization of heavy atoms to proteins may be screened in solution ahead of crystallization experiments [18]. However, it is hard to predict the results and anomalous data have to be measured for screening suitable phasing atoms. This is a trial-and-error process with significant overhead in time and cost in dealing with toxic heavy atoms. To facilitate phasing atom incorporation, selenomethionyl (SeMet) substitution method was invented by biochemical incorporation in vivo [19]. The substituted SeMet gave reliable incorporation and robust anomalous signals and is now the most popular method for introducing phasing atoms. Of course, transition metals such as iron, zinc, and copper are present naturally in about 30% native proteins. These metals are suitable phasing atoms for at-resonance experiments (K edges are at 7.11 keV for Fe, 8.98 keV for Cu and 9.66 keV for Zn). Beyond these heavier phasing atoms, most proteins contain sulfur in methionine and cysteine residues and all nucleic acids contain phosphorus. With no need of heavy atom derivatization, native-SAD phasing is a very attractive approach.

3 Routine Procedures for SAD Phasing

SAD phasing depends on what are often relatively delicate anomalous signals embodied in the Bijvoet differences (Eq. (5)). Nevertheless, SAD structure determinations are often quite routine for metalloproteins or SeMet proteins and now even for native, only-light-atom biomolecules. SAD phasing procedures include preparation of suitable cryogenic samples, anomalous data collection and analysis, substructure and phase determination, model building and refinement. We discuss these individual procedures with our recently solved DnaK structure in ATP state (DnaK-ATP) by native SAD [20, 21].

3.1 Cryogenic Sample Preparation

Cryocooling is the most efficient way to stretch the lifetime of biomolecular crystals under X-ray exposure [22]. The standard procedure is to transfer a crystal on a micromount into cryoprotectant for a short time soaking before immersion into liquid nitrogen at 100 K. The purpose of using cryoprotectant is to slow the rate of ice nucleation so that flash cooling produces a rigid glass instead of crystalline ice. For most crystals, a few seconds of soaking suffice for the exchange process. If the standard protocol does not work, for example resulting in cracked crystals or deteriorated diffraction, stepwise cryoprotectant exchange may be necessary to minimize the osmotic and surface stress shock. To find a suitable cryoprotectant, a screening kit such as CryoPro from Hampton Research may be used. For challenging samples that could not survive externally added cryoprotectant, dehydration could be used to increase the precipitant concentration; and crystals might then be frozen directly. The size of crystals and the amount of solvents around crystals are also important factors for cryocooling [23]. For optimized anomalous data collection, one should minimize solvents around crystals as much as possible before cryocooling. This can be realized by matching micromount size to crystal size and by removing solvents with a filter paper. To prevent over-dehydration, cryogenic sample preparation is better performed at lower temperature, for example in a cold room or a temperature-controllable glove box or cabinet.

3.2 Anomalous Data Measurement

Contemporary anomalous data are preferably measured at sophisticated modern synchrotron beamlines with X-ray energy tunable to desired values. A list of synchrotron beamlines for macromolecular crystallography may be found at BioSync (biosync.sbkb.org). Most beamlines can either be tuned to cover the anomalous diffraction spectrum for multiple phasing atoms or fixed to specific energies for popular phasing atoms, such as 12.67 keV for SeMet crystals.

Prior to anomalous data collection at an energy-tunable beamline, the X-ray energy needs first to be calibrated. For at-resonance experiment, a two-stage fluorescence scanning protocol is used. With SeMet K edge resonance as an example, first the Se foil standard is used for an EXAFS scanning to calibrate the energy to Se K edge at 12.67 keV. Then a SeMet crystal is used for a second scanning from which the resonant X-ray energy is determined for anomalous data collection. For off-resonance anomalous data collection, only the foil scanning is needed for energy calibration. For sulfur off-resonance anomalous data collection, X-ray energy at around 7 keV or lower is desirable. To collect anomalous data at 7 keV, Fe K edge ($E = 7.11$ keV) is used for calibration. Similarly Cr K edge ($E = 5.99$ keV) is used for 6 keV; and Cs L_{III} edge ($E = 5.01$ keV) is used for 5 keV energy calibration.

Expected Bijvoet-difference signals are relatively weak for many SAD phasing problems, certainly so for native SAD structures and often also for low-resolution SeMet SAD cases. It then becomes imperative to take special considerations in reducing errors when making the diffraction measurements. Errors may be random, systematic, or sporadic (i.e., inexplicable). Random errors might be overcome by increasing the average measurement time for reflections, or alternatively by increasing the redundancy in measurements at a given dose rate. Radiation damage is, of course, detrimental to the purpose of reducing random errors by increased exposure [24], and this may introduce added systematic error if excessive. If multiplicity is achieved from different samples or different crystal orientations, systematic components of error tend to be randomized which drives toward accuracy. High multiplicity achieved at reduced dose in multiple orientation has proved very effective for achieving satisfactory signal-to-noise [25]. This approach requires use of a multi-axis goniometer such as PRIGo [26]. When radiation damage is a concern, data from different crystals may be used to improve data quality [27, 28]. Systematic errors that may arise from various effects, such as sample absorption, may cancel in differences obtained by inverse-beam data collection [29]. A common inverse-beam procedure is first to collect a wedge of data (5–10° rotation), and then to repeat this wedge with the crystal rotated by 180° about an axis perpendicular to the X-ray beam. By this strategy, Friedel mates recorded in the two wedges suffer similar systematic errors, including those from the similar prior radiation doses, which then tend to cancel on differencing. Sporadic errors can be eliminated by outlier rejection procedures that are integrated into most data reduction packages such as those noted below.

For a good start in SAD phasing, diffraction data better than 6 Å spacings for at-resonance experiments and better than 3.5 Å for off-resonance native SAD experiments are recommended.

3.3 Anomalous Signal Analysis

Diffraction data processing packages HKL2000 [30], d*TREK [31], XDS [32], and MOSLFM [33] may be used conveniently for anomalous data processing. All these packages index diffraction patterns to obtain the lattice information, from which reflections can be integrated and used for subsequent data reduction process either internally or through external programs. Procedures for using individual programs may be found from their website documentation and published literature. Here we use XDS [32, 34] to illustrate the deduction of anomalous signals. With XDS, the indexing may be performed from a single pattern, wedge data or the entire data. For diffraction data to 3 Å spacings or beyond, default refinement parameters may be used; while for low resolution data, worse than 3.5 Å, parameters of beam center and the sample-to-detector distance may be better fixed for reliable indexing and integration. The two parameters may be refined during the optimization steps after the orientation matrix and unit cell parameters have been well determined. To accommodate radiation damage, it is beneficial to use corrections for "ALL" factors with "STRICT_ABSORPTION_CORRECTION=TRUE" for Bijvoet mates. During integration, Bijvoet mates are separated and are not used for calculation of statistics. After integration, XSCALE and XDSCONV within the XDS package can be used for obtaining reduced data. Alternatively, XDS output can be used for downstream data processing by external packages such as CCP4 [35] or PHENIX [36]. CCP4 programs POINTLESS and AIMLESS (a new version of SCALA) [37, 38] can be used for data analysis and reduction. The same as in XDS,

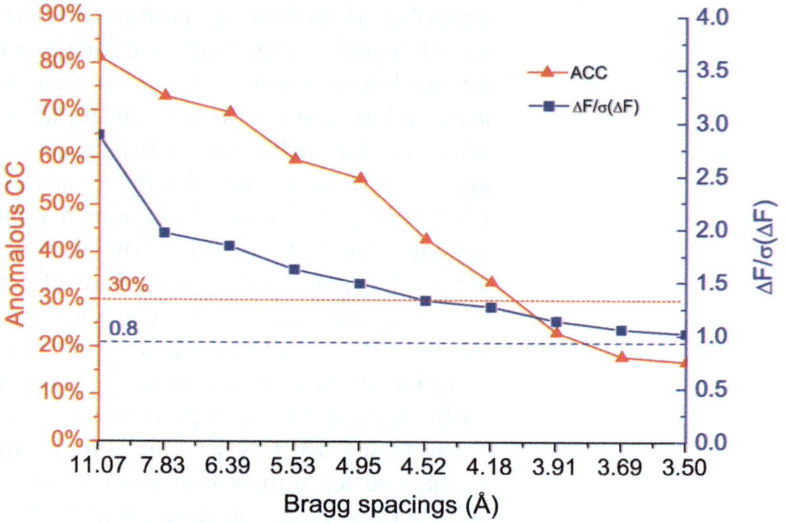

Fig. 4 Anomalous signal indicators. Anomalous correlation coefficient (ACC) (*red*) and $\Delta F/\sigma(\Delta F)$ (*blue*) for DnaK-ATP [20, 21] were shown. *Dashed red line* at 30% and *dashed blue line* at 0.8 are cutoff values for evaluation of ACC and $\Delta F/\sigma(\Delta F)$, respectively

Friedel mates are treated as two separate reflections during merging in AIMLESS or SCALA.

Data quality indicators for anomalous signals have been reviewed thoroughly [39, 40]. Among these we prefer to use anomalous CC (ACC) and $\Delta F/\sigma(\Delta F)$ to quantify anomalous signals (Fig. 4). ACC calculates the correlation coefficients between anomalous differences in randomly split halves of data. The plot of ACC with respect to Bragg spacings gives an indication of meaningful anomalous signals cutoff for substructure determination and phasing. Due to increased measurement noise, anomalous signals drop with decreasing Bragg spacings. The suggested cutoff value for ACC is 25–30% [41]; nevertheless, in our practice, data at lower ACC may still be useful for substructure determination and phasing. A second useful measure is the experimental anomalous signal-to-noise ratio, $<|\Delta F|>/\sigma(<|\Delta F|>)$, which is calculated by SHELXC or CCP4 programs and denoted $\Delta F/\sigma(\Delta F)$ for short. As for ACC, the plot of $\Delta F/\sigma(\Delta F)$ with respect to Bragg spacings also indicates the strength of anomalous signals. The expected value of $\Delta F/\sigma(\Delta F)$ for random data is $(2/\pi)^{1/2}$ [SHELX manual, http://shelx.uni-ac.gwdg.de/SHELX/]; therefore values over 0.8 are associated with meaningful anomalous signals provided that σ values are properly estimated. Anomalous signals in DnaK-ATP are significant as shown by the two dashed lines, red for ACC and blue for $\Delta F/\sigma(\Delta F)$ (Fig. 4). Based on the ACC and $\Delta F/\sigma(\Delta F)$ analyses, anomalous signals at 3.8 Å may be used for substructure determination where ACC and $\Delta F/\sigma(\Delta F)$ are 20.5% and 1.1, respectively.

3.4 Substructure Determination

To determine the phases for the overall structure, first the anomalous substructure has to be determined, which is done from $|\Delta F_{\pm h}|$ data with reference to Eq. (5) for relatedness to the $|{}^{\circ}F_A|$ coefficients for the substructure. CCP4, PHENIX, SHELXD [42], and SnB [43] packages can be used to determine the substructure of phasing atoms by direct methods. SHELXD uses correlation coefficients (CC) between normalized structure factors of observed $|\Delta F_{\pm h}|$ data and those calculated from trial models as criteria to evaluate the validity of substructure solutions. For each trial structure, CC_{all} and CC_{weak} are calculated based on all data and 30% of the weak data, respectively (Fig. 5) [41]. For DnaK-ATP substructure determinations, we used 3.8 Å data for search of 32 sites for either 10,000 or 1000 tries, both yielding clear separation of correct solutions (red cluster) and random candidates (blue cluster). Although clear separation almost certainly indicates correctness of solutions, candidates separated even marginally from the random CC_{all}/CC_{weak} cluster, as in Fig. 5b) might be useful. Such candidates may contain a partial substructure, which could be refined and expanded to a complete structure during the phasing procedure (*see* Subheading 3.5). For substructure determination by SHELXD, a few parameters have to be explored to enhance the success structure determination practice. The first parameter is the number of tries. More tries will give a high probability of finding correct solutions. For DnaK-ATP, we could not find substructures with 100 tries; but we found 5 solutions from 1000 tries and 37 from 10,000 tries. It is advisable to have 10,000 tries for routine substructure determination. The second parameter is the resolution cutoff as shown in Fig. 4. Including noisy high angle data is detrimental for substructure determination. In general, a series of resolution cutoffs for ACC between 30 and 10% may be screened for substructure

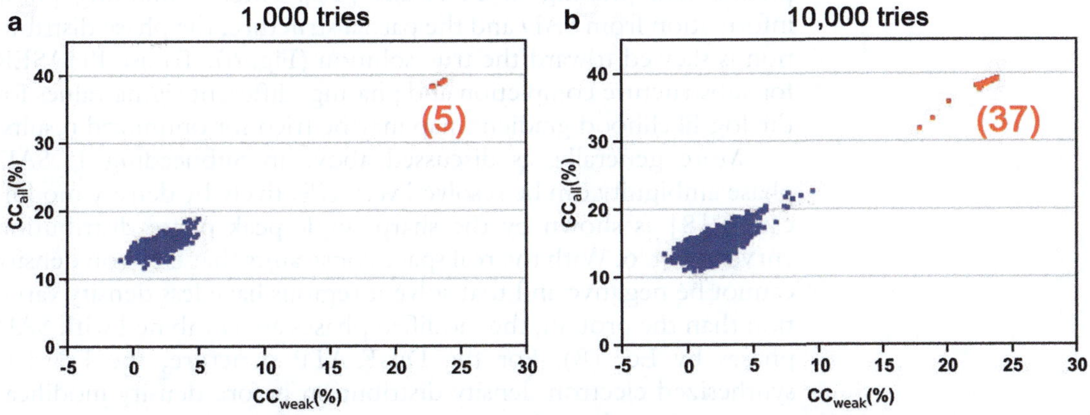

Fig. 5 CC_{weak}/CC_{all} plots for substructure determination by SHELXD. *Red* and *blue* clusters show the correct and random solutions, respectively. The numbers of correct solutions from 1000 and 10,000 tries were indicated. The native-SAD data for DnaK-ATP were used for the plots

determination. The third parameter is the number of substructure phasing atoms for the searches. The exact number of phasing atoms is often uncertain or unknown, for example, because of an ambiguous number of molecules per asymmetric unit, unclear stoichiometry of heavy-atom derivatization, or uncertainty in site flexibility and solvent ions for native SAD. In general, it is wise to search for atoms from as few as 2 to as many as 100 to best cover the possibilities.

3.5 Phasing and Density Modification

Prior to phase calculation, the coordinate, occupancy and temperature factor parameters for the deduced substructure are refined based on $\Delta F_{\pm h}$ data. Then, the refined substructure is used to calculate $|{}^{\circ}F_A(\boldsymbol{h})|$ and ${}^{\circ}\phi_A$, from which in principle ${}^{\circ}\phi_T$ for the whole structure may be evaluated algebraically from Eq. (5) as

$$ {}^{\circ}\phi_T = {}^{\circ}\phi_A + \pi/2 \pm \cos^{-1}\left[\Delta F_{\pm h} / 2\left(f''/f^{\circ}\right){}^{\circ}F_A(\boldsymbol{h})\right] \quad (7) $$

Clearly, for strict SAD phasing, with ${}^{\circ}\phi_A$ known from the substructure, ${}^{\circ}\phi_T$ has two equally possible solutions, thus posing the phase ambiguity problem. In actual practice, one uses phase probabilities rather than such an algebraic approach. The phase probability distribution, $P({}^{\circ}\phi_T)$, for this situation can be described by the form of

$$ P\left({}^{\circ}\phi_T\right) = N\left[\exp A\sin\left({}^{\circ}\phi_T\right) + B\cos\left({}^{\circ}\phi_T\right) + C\sin\left(2{}^{\circ}\phi_T\right) + D\cos\left(2{}^{\circ}\phi_T\right)\right] \quad (8) $$

with Hendrickson–Lattman coefficients A, B, C, D as defined for anomalous diffraction based on Eq. (2) [44] or as deduced from another phase probability analysis. Moreover, the substructure itself provides information for resolving the phase ambiguity intrinsic to SAD, and such partial structure information was used for solving the crambin structure [4]. Substructure refinement by maximum likelihood methods [45] allows for simultaneous substructure completion and phasing in PHASER [46]. After combining phase information from SAD and the partial structure, the phase distribution is skewed toward the true solution (Fig. 6). To use PHASER for substructure completion and phasing, different sigma values for the log-likelihood gradient map may be tried for optimized results.

More generally, as discussed above in Subheading 1, SAD phase ambiguity can be resolved very effectively by density modification [8] as shown by the sharp single-peak phase distribution curve in Fig. 6. With the real space constraints that electron density cannot be negative and that solvent regions have less density variation than the protein, the modified phases are combined with SAD phases by Eq. (8). For the DnaK-ATP structure, the Fourier-synthesized electron density distribution before density modification poorly defines the protein structure (shown as the magenta Cα traces); whereas after density modification, the boundary of the protein region is very well defined with β-stand and α-helix features clearly resolved (Fig. 7). Multiple density modification

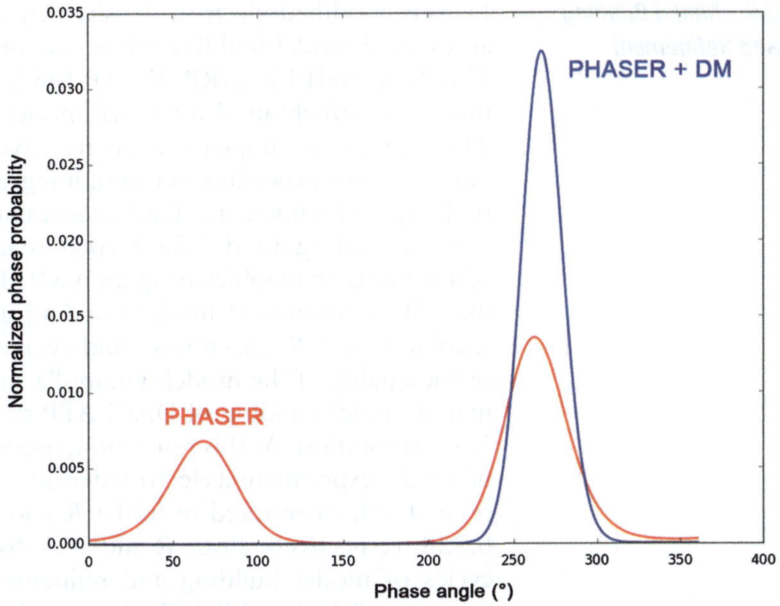

Fig. 6 Anomalous phasing ambiguity resolved by density modification. This is the phase probability distribution for a reflection in the DnaK-ATP data. For this reflection, the phase ambiguity was partially resolved (*red line*) by maximum likelihood refined substructure in PHASER [46] and fully resolved (*blue line*) after density modification in DM [9]

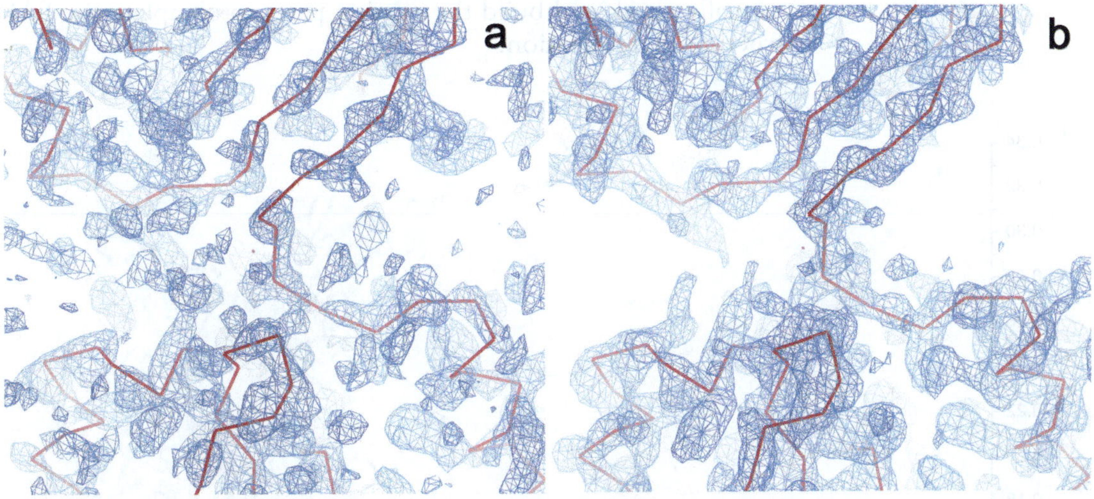

Fig. 7 Electron density maps of SAD phasing. (**a**) Electron density distribution before density modification. (**b**) Electron density distribution after density modification. The Cα tracings of the built structure are shown as *magenta lines*. The native-SAD data for DnaK-ATP were used for the figures

techniques, notably solvent flattening, solvent flipping, histogram matching, and molecular averaging, were developed and implemented in CCP4 programs DM and SOLOMON [9, 47].

3.6 Model Building and Refinement

Density-modified electron density maps may be used directly for automated model building when resolution is better than ~3.2 Å. PHENIX, SHELX, ARP/WARP [48], and BUCCANEER [49] may be routinely used for initial model building and refinement. These programs implement an iterative phase combination and improvement procedure through integration with cycles of model building and refinement. Under favorable cases, they may generate a good starting model for further manual model building and adjustments in graphics program COOT [50]. With the cycles of manual or automatic model building and refinement, crystallographic R and R_{free} factors should decrease, which is an indication of the quality of the model. Figure 8a shows progress of the automated model building of DnaK-ATP structure by ARP/WARP at 2.3 Å resolution. At this resolution, pseudoatoms were first used to fill up the experimental electron density map and then subsequently refined, which resulted in initial R and R_{free} values of 0.327 and 0.328, respectively. Both R and R_{free} dropped persistently during cycles of model building and refinement, and they finally converged to 0.178 and 0.237, respectively, after 40 cycles. We attribute this progression to the high quality of the density-modified experimental phases. With this automated process, ARP/WARP built 1110 ordered residues out of 1200 in the finally refined structure, with an estimated correctness of 97.3%. Figure 8b shows the DnaK-ATP dimer built automatically by ARP/WARP. Although programs can save a lot of work in model building, it is critical to manually check and build the missing parts to complete the SAD structure determination.

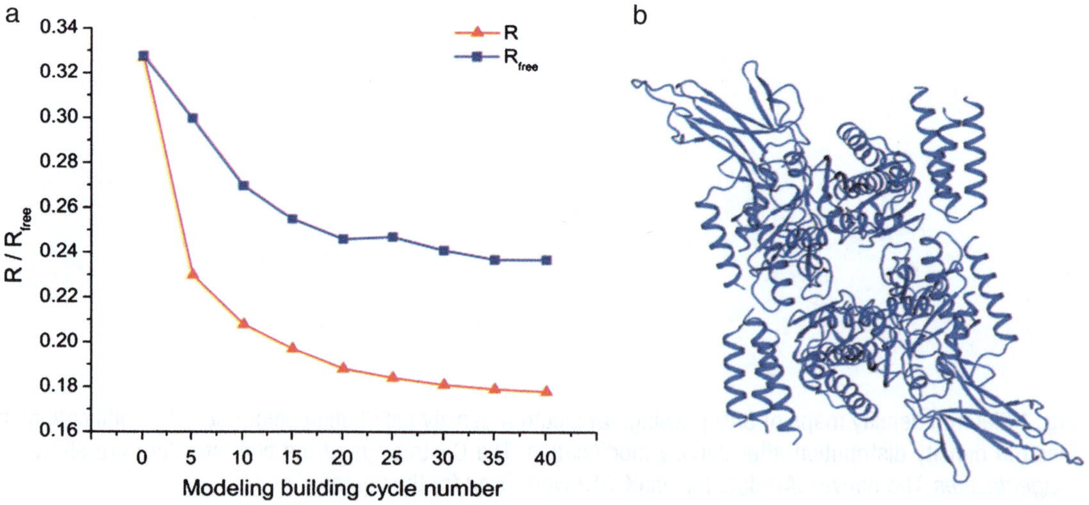

Fig. 8 Automated model building and refinements. (**a**) Progression of R and R_{free} factors during cycles of model building and refinement of DnaK-ATP in ARP/WARP. (**b**) A ribbon diagram of the auto-built DnaK structure as a dimer

4 Optimization of Anomalous Signals for Challenging Applications

SAD phasing has become sufficiently routine that it now dominates de novo crystallographic structure determination [1]. Nevertheless, complications do arise that can stymie routine analysis. These include inadequate anomalous scattering strength, limited diffraction due to poor intrinsic order, small crystals, and radiation damage. Such effects especially afflict state-of-the-art investigations such as on membrane proteins and large macromolecular complexes. Two classes of problems that have remained particularly challenging are low-resolution SAD analyses ($d_{\min} \geq 3.5$ Å) and only-light-atom native SAD analyses (anomalous scatterer $Z \leq 20$), as is manifest in the under representation of such structures in the Protein Data Bank [51]. When Bijvoet diffraction signals are relatively small (typically <1% for S-SAD at 8 keV and ~4% for SeMet-SAD at the Se K-edge) and noise-causing factors are present, it becomes challenging to achieve adequate signal-to-noise ratio in diffraction measurements (Fig. 4). One can strive to enhance the signal-to-noise ratio by increasing the strength of anomalous scattering or by reducing the level of noise.

4.1 Optimization of Anomalous Scattering Strength

Anomalous scattering strength as measured by the Bijvoet diffraction ratio (Eq. (6)) depends on N_A, the number of anomalous scatterers, and on f'', the imaginary (absorptive) component of the anomalous scattering factor, all relative to the diffraction of the entire structure. Site occupancies, not considered in Eq. (6), are additional factors of concern, typically so for heavy-atom derivatives and for SeMet proteins from eukaryotic sources. These are helpful considerations in the design of experiments, but for a particular crystal the composition is set. One then only has experimental control over f'' and that control is limited by synchrotron beamline properties. Opportunities for optimization of f'' differ for at-resonance versus off-resonance SAD experiments.

When an appropriate resonance edge is accessible for an anomalous phasing element at issue, then clearly it is best to tune to the resonance peak of f'', which can be ascertained from a fluorescence scan of the sample. The edge features for many resonances of interest are exquisitely sharp (Fig. 1), so this tuning must be done with care. Moreover, because of the sharpness, the peak value can readily be spoiled if the energy resolution of the particular beamline is not adequate. For example, whereas the peak features for Se in SeMet proteins are intrinsically very sharp [19], the focusing optics for divergent beams can blur these features and reduce the maximal achievable value of f'' [1]. Beam-defining slits can adjust beam divergence and thereby improve energy resolution. For the future, beamlines at lower emittance undulator sources, such as NYX at NSLS-II, promise to preserve the inherent fine structure at resonant edges.

The resonant edge for an element of interest may be out of reach at an available beamline; however, one might move to the lowest energy achievable to maximize f''. For example, the K (33.17 keV) and L_I (5.19 keV) edges of iodine are both inaccessible at most beamlines, but iodine-SAD experiments are highly practical even with CuKα (8.04 keV) radiation and are made even better at lower energy. Moreover, the resonances of S (2.47 keV) and Ca (4.04 keV), which are important native-SAD elements, are out of reach for most beamlines. Nevertheless, because f'' values steadily increase as the X-ray energy is lowered toward these resonance energies (Fig. 2), highly effective native-SAD experiments can be conducted at 6–7 keV on many beamlines. Practical considerations of diffraction geometry as well as parasitic X-ray absorption and background scattering complicate experiments at lower energy [20], but these are now being explored at Diamond I-23 [52] and Photon Factory beamline 1A [53], and are planned for LAX at NSLS-II.

4.2 Enhancement of Signal-to-Noise in Anomalous Diffraction

By arguments from Poisson statistics, if a reflected intensity records N counts, its standard deviation, $\sigma(N) = \sqrt{N}$, provides a measure of the noise in this intensity measurement. If the X-ray dose is increased by a factor T, for example by recording T-times longer, the signal-to-noise ratio, $I/\sigma(I)$, is expected to increase by $\sqrt{T}$. Similarly, if a given reflection is independently measured M times, the signal-to-noise ratio for these M measurements, $<I>/\sigma(<I>)$, is expected to be $\sqrt{M}$ times that of an individual measurement. The effectiveness of increasing multiplicity to improve anomalous diffraction analysis has been demonstrated in several studies [54, 55]; however, such effectiveness is limited by the radiation sensitivity of the sample [56]. In principle, by using multiple crystals the limitation from radiation damage can be circumvented provided that the crystals are statistically equivalent.

We demonstrated the effectiveness of multi-crystal SAD first by solving a relatively large histidine kinase sensor domain (1456 ordered residues) from eight SeMet crystals [27] and a membrane transporter from three SeMet crystals [57], both at the low resolution of 3.5 Å. The method is of general utility and simplicity and can be used robustly to enhance weak anomalous signals from sulfur for native-SAD phasing [20, 58]. Similar procedures have been implemented in CCP4 program BLEND [59] and PHENIX program phenix.scale_and_merge [60] for combing multi-crystal diffraction data.

To make sure that diffraction data from different crystals are indeed equivalent, we devised three statistical metrics for outlier rejection; and their effectiveness in multi-crystal native-SAD phasing has been demonstrated [20, 58]. Unit cell variation defines the combined difference of unit cell parameters (both dimensions and angles of the reduced cell). Only crystals with unit cell variation of less than 3σ may be merged together; which may be conveniently calculated by clustering analysis. In addition, the diffraction intensities may be used for precise analysis by diffraction dissimilarity analysis in which the intensities of two crystals are correlated and only compatible data sets (diffraction dissimilarity <5%) are merged. To quantify anomalous contribution from individual crystals, relative anomalous correlation coefficient (RACC) is used for data correlation analysis. The RACC analysis compares anomalous signals from individual crystals to the merged one; and checks for their relative contribution to the overall anomalous signals. If the contribution from a single crystal is too small or even negative, e.g., reducing the overall ACC, the crystal may be rejected from further use. Through the combination of these three metrics, reliable anomalous signals may be obtained for robust de novo SAD phasing. There are, however, no clear guidelines on exact rejection parameters. For real-life applications, diffraction dissimilarity and RACC are resolution dependent, and it may be worth trying different cutoff combinations. It is also worth noting that improvement from multiple crystals seems to be asymptotic, which is likely crystal and experiment dependent [58].

We again illustrate the enhanced anomalous signals with our structure determination of DnaK-ATP by native-SAD phasing [20]. This structure determination was not trivial, which is not surprising because ACC in any single data set is far inferior to the merged one (Fig. 9a). With the progressive inclusion of statistically compatible crystals as judged by the three metrics [20], ACC increased gradually and made the substructure determination and subsequent phasing successful from the merged data (Figs. 5 and 9b).

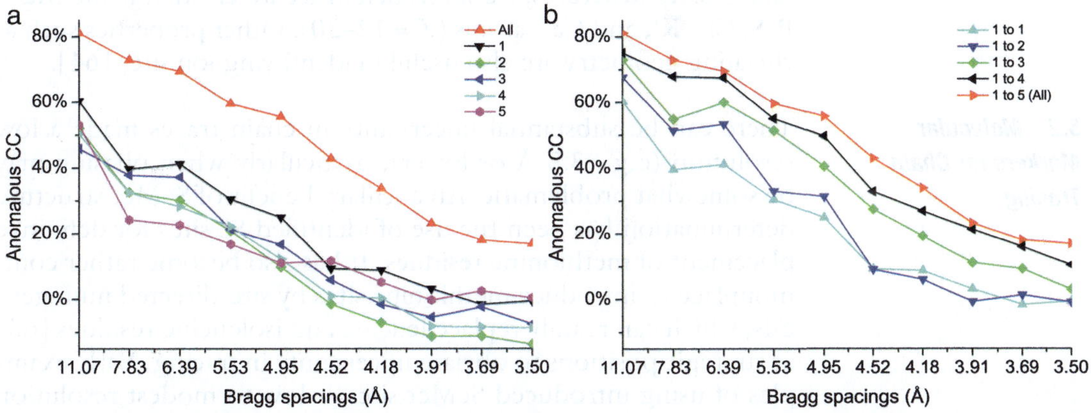

Fig. 9 Enhanced anomalous signals from using multiple crystals. (**a**) Anomalous CC of five DnaK-ATP data sets and the merged (All). (**b**) Anomalous CC of the progressive accumulation of the five DnaK-ATP data sets, added one by one

5 Anomalous Scattering in Chemical and Molecular Identification

Because anomalous scattering is a resonant phenomenon, dependent on chemical-specific orbitals, the Bijvoet differences and associated Bijvoet-difference Fourier syntheses can provide exquisitely sensitive indicators of chemical species. Such identifications from anomalous scattering can be very helpful in biomolecular structure analysis.

5.1 Element and Chemical State Identification

The identity and chemical state of metals in metalloproteins can be evident in associated X-ray absorption spectroscopy, and anomalous diffraction analyses can associate these properties with individual sites. Thus, one can readily associate elemental identity with specific sites in metalloproteins by preparing Bijvoet-difference Fourier syntheses at the peak energies of candidate ions. It is also possible to go beyond element identification to site-selective state identification by the procedure of spatially resolved anomalous dispersion refinement [61]. As applied to nitrogenase, where diffraction-derived spectra were obtained from refinements of f'' at 17 energies for 14 Fe sites in the molecule, reduced-state sites were distinguished from oxidized sites [62].

Another common use of anomalous diffraction in element identification concerns biologically relevant low-Z ions, such as Na^{+}, Mg^{2+}, K^{+}, Cl^{-}, or Ca^{2+}, which are prevalent constituents in channels, transporters and other biomolecules. Since the resonance edges of these light elements are typically inaccessible, identification are often made indirectly through substitutions such as of Na^{+} and K^{+} by Rb^{+} or Tl^{+} or of Mg^{2+} and Ca^{2+} by Sr^{2+} or a lanthanide [63]. This, of course, introduces questions of ion compatibilities. We recently introduced the effective alternative of identifying sites in multi-crystal-enhanced Bijvoet-difference syntheses and then performing f'' refinements for each candidate [20]. Using 7 keV X-rays for five native-SAD structures, we succeeded in accurately identifying Mg^{2+}, P, S, Cl^{-}, K^{+}, and Ca^{2+} atoms ($Z = 12$–20). Other properties such as chelating geometry are also useful in identifying ion sites [64].

5.2 Molecular Markers for Chain Tracing

There can be substantial uncertainty in chain traces made a low resolution (e.g., 3.5 Å or lower), particularly when phasing may be somewhat problematic. An ancillary benefit of SeMet structure determination has been the use of identified Se sites for definitive placement of methionine residues. It has also become rather commonplace to introduce methionine sites by site-directed mutagenesis, which can readily replace leucine and isoleucine residues [65] at strategic positions to obviate uncertainty in tracing. Early examples of using introduced SeMet sites at low to modest resolution include a domain-positioning analysis of a spliceosomal snRNP [66] and disambiguation of chain tracing for a CLC chloride transporter [67] and for P-glycoprotein [68]. Recently, more in the category of hypothesis testing than chain tracing, an introduced SeMet residue was used to identify a putative gating site in a TRPV6 channel structure [69]. Increasingly, as native SAD has taken hold and whenever data are measured at lower energy, the positions of sulfur atoms serve as comprehensive natural markers for methionine and cysteine residues. A procedure has been developed expressly for the purpose of defining such weak anomalous sites [70].

5.3 Localization of Ligands

Anomalous scattering can be used to locate ligands that contain identifiable anomalous scatterers, such as Mg-ATP for which the phosphorus and magnesium atoms can be located [20, 21]. Another important area of expanding application is in fragment-based drug development. Brominated or iodinated compounds are featured in fragment libraries that are used to identify weakly binding compounds that have substantial potential for chemical expansion into drug-development leads [71, 72]. In addition, these halogen atoms can even be used in structure determination by SAD phasing [73].

6 Emerging Developments and Future Perspectives

Recent developments greatly accelerated the SAD phasing which is now dominating de novo structure determination practice. With the fast read-out pixel array detectors such as PILATUS 6M, ADSC HF 4M, and EIGER 16M [74, 75], it is now routine to collect complete data sets in a few minutes or less. These detectors enable the use of raster scanning technique for identifying diffraction hot spots without visually seeing crystals as common for frozen crystals embedded in lipid cubic phase. Pixel array detectors are also ideal for collecting fine-slicing data for obtaining improved statistics and data quality [31, 76]. If radiation damage is not an issue, multiple data sets and multiple orientations from a single crystal could be used to improve diffraction statistics [55, 77]. In addition, the pixel array detectors may permit energy discrimination whereby parasitic fluorescence X-rays can be filtered out.

The integration of substructure determination and phasing are pushing the limit of SAD phasing to allow for use of very weak anomalous signals [78, 79]. The iterative model building and refinements procedures as implemented in PHENIX, SHELX, ARP/WARP and BUCCANEER have been greatly useful for automated structure determination. However these programs are most useful at resolutions of about 3.2 Å or higher. At low resolution, due to insufficient number of unique reflections for refinement and less atomic features for chain tracing and side chain docking, new algorithms are needed for automated low-resolution phasing and model building. The incorporation of chemical and bioinformatics knowledge into crystallographic model building cycles may improve geometry and reliability. Better treatment of anisotropy, disorder, and radiation damage at low resolutions are also aspects of consideration for future development.

Contemporary SAD phasing uses crystals with sizes of about 20 μm or larger. For smaller crystals, radiation damage may kill the crystal before useful anomalous signals can be obtained. X-ray free electron lasers have been promising to overcome radiation damage to micron-sized crystals for both Gd-SAD and native-SAD phasing [80–82]. However, X-ray free electron lasers require huge numbers of crystals and are not available for most crystallographers. To make microcrystal SAD routinely accessible, synchrotron beamlines need to be optimized for focused microbeam with high-accuracy goniometers for precise delivery of microcrystals into the beam. New methods for harvesting and cryocooling microcrystals also need to be developed and optimized.

Acknowledgments

This work was supported in part by NIH grants R01GM107462 and P41GM116799 and by Brookhaven National Laboratory LDRD 15-034.

References

1. Hendrickson WA (2014) Anomalous diffraction in crystallographic phase evaluation. Q Rev Biophys 47:49–93
2. Hendrickson WA (1985) Analysis of protein structure from diffraction measurements at multiple wavelengths. Trans Am Cryst Assn 21:11–21
3. Hendrickson WA (1991) Determination of macromolecular structures from anomalous diffraction of synchrotron radiation. Science 254:51–58
4. Hendrickson WA, Teeter MM (1981) Structure of the hydrophobic protein crambin determined directly from the anomalous scattering of sulfur. Nature 290:107–113
5. Hendrickson WA, Ogata CM (1997) Phase determination from multiwavelength anomalous diffraction measurements. Methods Enzymol 276:494–523
6. Walsh MA, Evans G, Sanishvili R, Dementieva I, Joachimiak A (1999) MAD data collection—current trends. Acta Crystallogr D Biol Crystallogr 55:1726–1732
7. Blow DM (2003) How Bijvoet made the difference: the growing power of anomalous scattering. Methods Enzymol 374:3–22
8. Wang BC (1985) Resolution of phase ambiguity in macromolecular crystallography. Methods Enzymol 115:90–112
9. Cowtan KD, Zhang KYJ (1999) Density modification for macromolecular phase improvement. Prog Biophys Mol Biol 72:245–270
10. Wu H, Lustbader JW, Liu Y et al (1994) Structure of human chorionic gonadotropin at 2.6 Å resolution from MAD analysis of the selenomethionyl protein. Structure 2:545–558
11. Liu Y, Ogata CM, Hendrickson WA (2001) Multiwavelength anomalous diffraction analysis at the M absorption edges of uranium. Proc Natl Acad Sci U S A 98:10648–10653
12. Shapiro L, Fannon AM, Kwong PD et al (1995) Structural basis of cell-cell adhesion by cadherins. Nature 374:327–337
13. Cromer DT, Liberman DA (1981) Anomalous dispersion calculations near to and on the long-wavelength side of an absorption edge. Acta Crystallogr A 37:267–268
14. Bovenkamp GL, Zanzen U, Krishna KS et al (2013) X-ray absorption near-edge structure (XANES) spectroscopy study of the interaction of silver ions with *Staphylococcus aureus*, *Listeria monocytogenes*, and *Escherichia coli*. Appl Environ Microbiol 79:6385–6390

15. Sepulcre F, Proietti MG, Benfatto M et al (2004) A quantitative XANES analysis of the calcium high-affinity binding site of the purple membrane. Biophys J 87:513–520
16. Evans G, Pettifer RF (2001) CHOOCH: a program for deriving anomalous-scattering factors from X-ray fluorescence spectra. J Appl Crystallogr 34:82–86
17. Pike AC, Garman EF, Krojer T et al (2016) An overview of heavy-atom derivatization of protein crystals. Acta Crystallogr D Biol Crystallogr 72:303–318
18. Boggon TJ, Shapiro L (2000) Screening for phasing atoms in protein crystallography. Struct Fold Des 8:R143–R149
19. Hendrickson WA, Horton JR, Lemaster DM (1990) Selenomethionyl proteins produced for analysis by multiwavelength anomalous diffraction (MAD)—a vehicle for direct determination of three-dimensional structure. EMBO J 9:1665–1672
20. Liu Q, Liu Q, Hendrickson WA (2013) Robust structural analysis of native biological macromolecules from multi-crystal anomalous diffraction data. Acta Crystallogr D Biol Crystallogr 69:1314–1332
21. Qi R, Sarbeng EB, Liu Q et al (2013) Allosteric opening of the polypeptide-binding site when an Hsp70 binds ATP. Nat Struct Mol Biol 20:900–907
22. Pflugrath JW (2015) Practical macromolecular cryocrystallography. Acta Crystallogr F Struct Biol Commun F71:622–642
23. Pellegrini E, Piano D, Bowler MW (2011) Direct cryocooling of naked crystals: are cryoprotection agents always necessary? Acta Crystallogr D Biol Crystallogr 67:902–906
24. Garman EF, Weik M (2015) Radiation damage to macromolecules: kill or cure? J Synchrotron Radiat 22:195–200
25. Weinert T, Olieric V, Waltersperger S et al (2015) Fast native-SAD phasing for routine macromolecular structure determination. Nat Methods 12:131–133
26. Waltersperger S, Olieric V, Pradervand C et al (2015) PRIGo: a new multi-axis goniometer for macromolecular crystallography. J Synchrotron Radiat 22:895–900
27. Liu Q, Zhang Z, Hendrickson WA (2011) Multi-crystal anomalous diffraction for low-resolution macromolecular phasing. Acta Crystallogr D Biol Crystallogr 67:45–59
28. Olieric V, Weinert T, Finke AD et al (2016) Data-collection strategy for challenging native SAD phasing. Acta Crystallogr D Biol Crystallogr 72:421–429
29. Hendrickson WA, Smith JL, Sheriff S (1985) Direct phase determination based on anomalous scattering. Methods Enzymol 115:41–55
30. Otwinowski Z, Minor W (1997) Processing of X-ray diffraction data collected in oscillation mode. Methods Enzymol 276:307–326
31. Pflugrath JW (1999) The finer things in X-ray diffraction data collection. Acta Crystallogr D Biol Crystallogr 55:1718–1725
32. Kabsch W (2010) XDS. Acta Crystallogr D Biol Crystallogr 66:125–132
33. Leslie AGW (2006) The integration of macromolecular diffraction data. Acta Crystallogr D Biol Crystallogr 62:48–57
34. Kabsch W (2010) Integration, scaling, space-group assignment and post-refinement. Acta Crystallogr D Biol Crystallogr 66:133–144
35. Winn MD, Ballard CC, Cowtan KD et al (2011) Overview of the CCP4 suite and current developments. Acta Crystallogr D Biol Crystallogr 67:235–242
36. Adams PD, Afonine PV, Bunkoczi G et al (2010) PHENIX: a comprehensive python-based system for macromolecular structure solution. Acta Crystallogr D Biol Crystallogr 66:213–221
37. Evans PR (2011) An introduction to data reduction: space-group determination, scaling and intensity statistics. Acta Crystallogr D Biol Crystallogr 67:282–292
38. Evans PR, Murshudov GN (2013) How good are my data and what is the resolution? Acta Crystallogr D Biol Crystallogr 69:1204–1214
39. Dauter Z (2006) Estimation of anomalous signal in diffraction data. Acta Crystallogr D Biol Crystallogr 62:867–876
40. Evans P (2006) Scaling and assessment of data quality. Acta Crystallogr D Biol Crystallogr 62:72–82
41. Schneider TR, Sheldrick GM (2002) Substructure solution with SHELXD. Acta Crystallogr D Biol Crystallogr 58:1772–1779
42. Sheldrick GM (2010) Experimental phasing with SHELXC/D/E: combining chain tracing with density modification. Acta Crystallogr D Biol Crystallogr 66:479–485
43. Weeks CM, Miller R (1999) The design and implementation of SnB version 2.0. J Appl Crystallogr 32:120–124
44. Pahler A, Smith JL, Hendrickson WA (1990) A probability representation for phase information from multiwavelength anomalous dispersion. Acta Crystallogr A 46:537–540
45. Fortelle E, Bricogne G (1997) Maximum-likelihood heavy-atom parameter refinement

for multiple isomorphous replacement and multiwavelength anomalous diffraction methods. Methods Enzymol 276:472–494
46. Read RJ, McCoy AJ (2011) Using SAD data in Phaser. Acta Crystallogr D Biol Crystallogr 67:338–344
47. Abrahams J, Leslie A (1996) Methods used in the structure determination of bovine mitochondrial F1 ATPase. Acta Crystallogr D Biol Crystallogr 52:30–42
48. Langer G, Cohen SX, Lamzin VS et al (2008) Automated macromolecular model building for X-ray crystallography using ARP/wARP version 7. Nat Protoc 3:1171–1179
49. Cowtan K (2006) The buccaneer software for automated model building. 1. Tracing protein chains. Acta Crystallogr D Biol Crystallogr 62:1002–1011
50. Emsley P, Lohkamp B, Scott WG et al (2010) Features and development of coot. Acta Crystallogr D Biol Crystallogr 66:486–501
51. Liu Q, Hendrickson WA (2015) Crystallographic phasing from weak anomalous signals. Curr Opin Struct Biol 34:99–107
52. Wagner A, Duman R, Henderson K et al (2016) In-vacuum long-wavelength macromolecular crystallography. Acta Crystallogr D Biol Crystallogr 72:430–439
53. Ru H, Zhao L, Ding W et al (2012) S-SAD phasing study of death receptor 6 and its solution conformation revealed by SAXS. Acta Crystallogr D Biol Crystallogr 68:521–530
54. Dauter Z, Adamiak DA (2001) Anomalous signal of phosphorus used for phasing DNA oligomer: importance of data redundancy. Acta Crystallogr D Biol Crystallogr 57: 990–995
55. Liu ZJ, Chen L, Wu D et al (2011) A multidataset data-collection strategy produces better diffraction data. Acta Crystallogr A 67:544–549
56. Garman EF (2010) Radiation damage in macromolecular crystallography: what is it and why should we care? Acta Crystallogr D Biol Crystallogr 66:339–351
57. Mancusso R, Gregorio GG, Liu Q, Wang DN (2012) Structure and mechanism of a bacterial sodium-dependent dicarboxylate transporter. Nature 491:622–626
58. Liu Q, Dahmane T, Zhang Z et al (2012) Structures from anomalous diffraction of native biological macromolecules. Science 336:1033–1037
59. Foadi J, Aller P, Alguel Y et al (2013) Clustering procedures for the optimal selection of data sets from multiple crystals in macromolecular crystallography. Acta Crystallogr D Biol Crystallogr 69:1617–1632
60. Akey DL, Terwilliger TC, Smith JL (2016) Efficient merging of data from multiple samples for determination of anomalous substructure. Acta Crystallogr D Biol Crystallogr 72:296–302
61. Einsle O, Andrade SL, Dobbek H et al (2007) Assignment of individual metal redox states in a metalloprotein by crystallographic refinement at multiple X-ray wavelengths. J Am Chem Soc 129:2210–2211
62. Spatzal T, Schlesier J, Burger EM et al (2016) Nitrogenase FeMoco investigated by spatially resolved anomalous dispersion refinement. Nat Commun 7:10902
63. Zhou Y, MacKinnon R (2003) The occupancy of ions in the K+ selectivity filter: charge balance and coupling of ion binding to a protein conformational change underlie high conduction rates. J Mol Biol 333:965–975
64. Echols N, Morshed N, Afonine PV et al (2014) Automated identification of elemental ions in macromolecular crystal structures. Acta Crystallogr D Biol Crystallogr 70:1104–1114
65. Leahy DJ, Erickson HP, Aukhil I et al (1994) Crystallization of a fragment of human fibronectin: introduction of methionine by site-directed mutagenesis to allow phasing via selenomethionine. Proteins 19:48–54
66. Oubridge C, Krummel DA, Leung AK et al (2009) Interpreting a low resolution map of human U1 snRNP using anomalous scatterers. Structure 17:930–938
67. Feng L, Campbell EB, Hsiung Y et al (2010) Structure of a eukaryotic CLC transporter defines an intermediate state in the transport cycle. Science 330:635–641
68. Jin MS, Oldham ML, Zhang Q et al (2012) Crystal structure of the multidrug transporter P-glycoprotein from *Caenorhabditis elegans*. Nature 490:566–569
69. Saotome K, Singh AK, Yelshanskaya MV et al (2016) Crystal structure of the epithelial calcium channel TRPV6. Nature 534:506–511
70. Thorn A, Sheldrick GM (2011) ANODE: anomalous and heavy-atom density calculation. J Appl Crystallogr 44:1285–1287
71. Groftehauge MK, Therkelsen MO, Taaning R et al (2013) Identifying ligand-binding hot spots in proteins using brominated fragments. Acta Crystallogr Sect F Struct Biol Cryst Commun 69:1060–1065
72. Tiefenbrunn T, Forli S, Happer M et al (2014) Crystallographic fragment-based drug discovery: use of a brominated fragment library targeting HIV protease. Chem Biol Drug Des 83:141–148
73. Bauman JD, Harrison JJ, Arnold E (2016) Rapid experimental SAD phasing and hot-spot

identification with halogenated fragments. IUCrJ 3:51–60
74. Loeliger T, Bronnimann C, Donath T et al (2012) The new PILATUS3 ASIC with instant retrigger capability. In: Proceedings of IEEE Nuclear Science Symposium and Medical Imaging Conference (NSS/MIC), IEEE, pp 610–615. doi:10.1109/NSSMIC.2012.6551180
75. Tinti G, Bergamaschi A, Cartier S et al (2015) Performance of the EIGER single photon counting detector. J Instrum 10:C03011
76. Mueller M, Wang M, Schulze-Briese C (2012) Optimal fine phi-slicing for single-photon-counting pixel detectors. Acta Crystallogr D Biol Crystallogr 68:42–56
77. Brockhauser S, Ravelli RB, McCarthy AA (2013) The use of a mini-kappa goniometer head in macromolecular crystallography diffraction experiments. Acta Crystallogr D Biol Crystallogr 69:1241–1251
78. Skubak P, Pannu NS (2013) Automatic protein structure solution from weak X-ray data. Nat Commun 4:2777
79. Bunkóczi G, McCoy AJ, Echols N et al (2014) Macromolecular X-ray structure determination using weak, single-wavelength anomalous data. Nat Methods 12:127–130
80. Barends TR, Foucar L, Botha S et al (2014) De novo protein crystal structure determination from X-ray free-electron laser data. Nature 505:244–247
81. Nakane T, Song C, Suzuki M et al (2015) Native sulfur/chlorine SAD phasing for serial femtosecond crystallography. Acta Crystallogr D Biol Crystallogr 71:2519–2525
82. Nass K, Meinhart A, Barends TR et al (2016) Protein structure determination by single-wavelength anomalous diffraction phasing of X-ray free-electron laser data. IUCrJ 3: 180–191

Chapter 17

Long-Wavelength X-Ray Diffraction and Its Applications in Macromolecular Crystallography

Manfred S. Weiss

Abstract

For many years, diffraction experiments in macromolecular crystallography at X-ray wavelengths longer than that of Cu-K_α (1.54 Å) have been largely underappreciated. Effects caused by increased X-ray absorption result in the fact that these experiments are more difficult than the standard diffraction experiments at short wavelengths. However, due to the also increased anomalous scattering of many biologically relevant atoms, important additional structural information can be obtained. This information, in turn, can be used for phase determination, for substructure identification, in molecular replacement approaches, as well as in structure refinement. This chapter reviews the possibilities and the difficulties associated with such experiments, and it provides a short description of two macromolecular crystallography synchrotron beam lines dedicated to long-wavelength X-ray diffraction experiments.

Key words Macromolecular crystallography, Long-wavelength experiments, Soft X-rays, Diffraction efficiency, X-ray absorption, Anomalous scattering, Native SAD, S-SAD phasing, Substructure identification, Molecular replacement

1 Introduction

Most diffraction experiments in Macromolecular Crystallography (MX) are nowadays performed at X-ray wavelengths at or close to λ = 1.0 Å (E = 12.4 keV). While in the early days of MX home sources using Cu-targets and the K_α-emission line of Cu (λ = 1.54 Å) were the most common means for measuring diffraction data, the world-wide development of synchrotron beam lines for MX and the increased possibilities of obtaining access to them, made wavelengths shorter than the Cu-K_α progressively more popular. The much higher intensity X-ray beams at synchrotron beam lines compared to the ones from home sources compensated for the reduced scattering at the shorter wavelengths and thus allowed to conduct diffraction experiments that are significantly less affected by X-ray absorption. Furthermore, the possibility to accessing the L-absorption edges of heavy atoms such as Hg, Pt, and Au, etc.,

Alexander Wlodawer et al. (eds.), *Protein Crystallography: Methods and Protocols*, Methods in Molecular Biology, vol. 1607, DOI 10.1007/978-1-4939-7000-1_17, © Springer Science+Business Media LLC 2017

which were at that time often used for phase determination by isomorphous replacement methods, amplified the phasing possibilities dramatically. In particular, the development of easy replacement of the amino acid Met in proteins by the isosteric artificial amino acid Se-Met and the concomitant potential for phase determination by multiple wavelength anomalous dispersion (or multiple wavelength anomalous diffraction, MAD) revolutionized the phasing step in MX. Nowadays, the majority of phase determination experiments are conducted at the Se-K_α-absorption edge (λ = 0.98 Å) using a Se-Met derivative of the protein of interest.

Consequently, little room seems to be left for X-ray diffraction experiments at other wavelengths. Nonetheless, apart from this mainstream in MX, there have always been some researchers who were interested in widening the spectrum of wavelengths for MX. Their experiments were for the most part focused on reaching absorption edges of elements outside the MX comfort zone, which for the purpose of this chapter may be defined as the wavelength range from 0.8 to 1.6 Å. The groups of Helliwell [1], Fourme [2] and Tucker (personal communication) explored the space on the short wavelength side down to 0.3 Å, while Stuhrmann and his colleagues were the main protagonists on the long wavelength side up to 6.0 Å [3–7]. Mainly as a result of the experimental difficulties associated with such experiments and the rather limited scientific justification, they have not been picked up by a larger group of experimental structural biologists. A recent survey of all entries in the Protein Data Bank [8] revealed that only about 0.7% of all diffraction experiments leading to a PDB entry were conducted at a wavelength shorter than 0.8 Å and only about 0.9% at a wavelength longer than 1.6 Å.

More recently however, diffraction experiments conducted at wavelengths somewhat longer than the Cu-K_α wavelength (λ = 1.7–3.0 Å) have gained some attention [9–11]. The main rationale for using such wavelengths for diffraction experiments is of course to increase the measurable anomalous scattering. Particularly the light atoms, which are very relevant in biology (Na, Mg, S, P, Cl, K, Ca) exhibit significantly increased anomalous scattering properties at such wavelengths compared to the typically used wavelength of around 1.0 Å. Apart from this, increased anomalous scattering can also be observed for the medium heavy elements (I, Xe, Cs) and for the very heavy elements (Hg, Pb, etc.). A further important point to make is that diffraction experiments in this wavelength range are much easier to conduct than the very long-wavelength experiments mentioned above and that they can be carried out at many standard MX beam lines around the world.

With respect to MX applications using home sources, David Blow had proposed a Cr-target for home X-ray laboratories (λ = 2.29 Å) as early as 1958 [12]. However, it was not before 2003 when the company Molecular Structure Corporation (which later became Rigaku Americas Corporation) introduced a Cr-anode into the market [13], even though some exploratory experiments [14, 15] had clearly hinted at the potential of such devices.

In this chapter, I will give an overview of what is possible with longer X-ray wavelengths in MX. I will also discuss some of the difficulties associated with such experiments and I will outline and discuss possible solutions.

2 Theoretical Background

2.1 Normal Scattering

As was noted by Arndt [16], Polikarpov and colleagues [17, 18], Murray et al. [19] and others and further elaborated on by Djinovic Carugo et al. [11], the total integrated scattered intensity I of a crystal with linear dimension x is approximately proportional to the square of the incident X-ray wavelength λ at small scattering angles (Eq. 1).

$$I \propto \lambda^2 x^3 \exp(-\mu x) \tag{1}$$

At the same time the linear absorption coefficient μ is proportional to the third power of the incident X-ray wavelength (Eq. 2).

$$\mu \propto \lambda^3 \tag{2}$$

It has therefore been argued by Helliwell [20], by Teplyakov and colleagues [21] and by others, that the wavelength at which the scattering efficiency is at a maximum is a function of the size of the sample (Eq. 3).

$$\lambda[\text{Å}] \approx (300 / x[\mu m])^{1/3} \tag{3}$$

In Eqs. 1–3, λ is the incident X-ray wavelength and x is the approximate diameter of the crystalline sample. This derivation is of course only valid in the absence of any X-ray absorption edges nearby. It is also neglecting further instrumental effects such as, for instance, the detector efficiency. A graphical representation of the scattering efficiency (defined as the ratio between integrated scattered intensity and absorption) for different crystal sizes is shown in Fig. 1. The curves clearly suggest that for smaller samples the scattering efficiency is higher at longer wavelengths.

2.2 Anomalous Scattering

The situation is slightly different when anomalous scattering is considered. Far from an elemental absorption edge, the anomalous scattering length $\Delta f''$ given in units of electrons is proportional to the square of the incident X-ray wavelength λ. Consequently, the anomalous part of the structure factor increases with λ^2 and the anomalous intensity differences with λ^4. Taking the λ^3-dependance of X-ray absorption into account (Eq. 2) means that anomalous

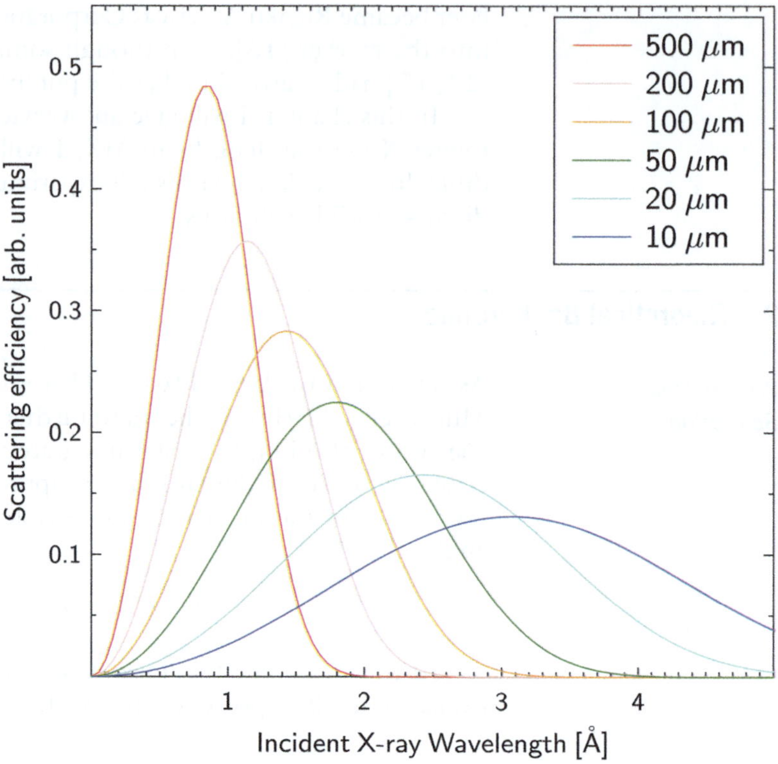

Fig. 1 Scattering efficiency in arbitrary units of a protein crystal of average composition (9% H, 27% C, 8% N, 55% O and 1% S, see also [22]) as a function of the linear crystal dimension x and incident X-ray wavelength λ. The curve was calculated for six different crystal sizes ranging from 10 to 500 μm. In order to place the six curves on the same scale, they were normalized by the sample cross section x^2. The mass energy absorption coefficient was calculated using the NIST (Gaithersburg, MD, USA) XCOM applet, which can be accessed at http://physics.nist.gov/PhysRefData/Xcom/html/xcom1-t.html. A standard value for the protein density of 1.35 g/cm^3 was assumed

intensity differences should become more pronounced at longer X-ray wavelengths. However, this argument is only true in the absence of experimental errors. Increased absorption makes data collection and processing more difficult, and since in MX absorption effects are only treated implicitly during the scaling stage, the larger absorption effects at longer wavelengths will at some point outweigh the gain in signal that can be achieved. In a large systematic study Mueller-Dieckmann and colleagues observed that the maximum anomalous signal-to-noise ratio can be obtained at wavelengths around 2.0 Å nearly irrespective of the nature and the composition of the sample [23].

3 Possibilities with Longer-Wavelength X-Ray Diffraction

3.1 Light-Atom Based Phase Determination (Native SAD)

Probably the most important and the most widely known application of longer-wavelength diffraction experiments in MX is phase determination by the single wavelength anomalous diffraction (SAD) approach. As outlined above in the introductory paragraph, the anomalous scattering length of light atoms (Na, Mg, P, S, Cl, K, Ca, etc.) is significantly enhanced at longer wavelengths compared to the more typically used wavelengths in MX. This leads to appreciable and measurable anomalous intensity differences. Although these differences are still small, it is possible to measure them accurately enough to allow the determination of the anomalously scattering substructure. Once this substructure is known, it may serve as a reference point for phase determination. This structure determination approach is called sulfur-SAD or S-SAD. More recently, many researchers refer to it as native SAD, because sulfur is of course not the only relevant element in this respect.

The first successful experiment employing this method was the determination of the structure of the small protein crambin in 1981 by Wayne Hendrickson and Martha Teeter [24]. After that it took almost 20 years until Zbigniew Dauter and colleagues demonstrated that the structure of the model protein hen egg-white lysozyme could be solved using anomalous intensity differences [25]. This paper quickly became an eye-opener for many researchers interested in this field. Consequently, the following decade witnessed a number of successful real case native SAD structure determinations as well as some important methodological studies. In the following, some landmark native SAD structure determination experiments will be mentioned and briefly discussed. This list is necessarily subjective and thus incomplete, and the author would like to apologize to all scientists who carried out all the other important experiments, which are not mentioned explicitly here.

The first native SAD structure for which longer wavelength data were employed was the photoprotein obelin, solved by the group of B.-C. Wang [26]. α-Crustacyanin C1 was the first structure determined by long-wavelength native SAD, in which non-crystallographic symmetry was present in the crystal [27]. Then, the hyperthermophile protein Sso10a, a dimeric winged helix DNA binding domain, was the first novel structure to be solved from data collected using a Cr-anode [28]. The structure of CIB-1, solved based on data also collected using a Cr-anode, was the largest structure at that time [29], and the structure of Dsrc was the first one for which a combination of radiation damage induced phasing (RIP) and native SAD was employed [30]. The next record for the largest structure determined by native SAD was set in 2005 (although the corresponding publication only appeared in 2008) with the structure determination of SusB, an 84-kDa α-glucoside hydrolase involved in the starch utilization system [31]. Two times 738 amino acids made up the asymmetric unit of the monoclinic crystal. This constituted a new size record, which remained unchallenged for quite some time. Between about 2005 and 2010, the field consolidated with few further breakthroughs. Also, among many researchers the notion set in that native SAD would only work for small proteins that crystallized in high symmetry space groups and diffracted to high resolution. This period of disillusionment came to a sudden end in 2012, when Wayne Hendrickson and colleagues introduced the multi-crystal native SAD approach and demonstrated that native SAD was by no means limited in any way [32, 33]. Finally in 2014, the spectacular structure determination of the tubulin-stathmin-TTL complex was reported [34]. With a total molecular weight of about 260 kDa in the asymmetric unit and an anomalously scattering substructure consisting of 136 light atoms (118 S, 13 P, 3 Ca, 2 Cl), this structure determination obliterated the previous size record both in terms of total molecular weight as well as in terms of the number of light atom anomalous scatterers. It also made it clear to everyone in the field that no apparent limit for native SAD exists. With respect to data resolution, an important study was published by El Omari and colleagues on the structure determination of the ectodomain of HCV E1 [35]. They were able to show that even for crystals diffracting to lower than 4 Å resolution, a native SAD approach is feasible. With a data set averaged from 32 crystals, a multiplicity of more than 120 was obtained, which allowed the calculation of an interpretable electron density map. As of 2015, a total of about 150 structures have been solved using native SAD (Fig. 2). Recent overviews, to which the reader may be referred to, have been published as supplementary material in Gorgel et al. [36] and Rose et al. [37]. Given this still rather small number of reported successful cases, the method appears to be of little significance compared to the total number of PDB entries, but the number is rising. More and more researchers are becoming aware of the possibilities offered by this method and it may be anticipated that significantly more native SAD structures will appear in the years to come.

3.2 Substructure Identification

A largely underappreciated by-product of a long-wavelength X-ray diffraction experiment is that it is able to provide additional and orthogonal information with respect to atom identities. Unless very high resolution data have been collected or additional biochemical data are available, the usual procedure in MX is to assign any spherical electron density, which is left over after the macromolecule structure has been built, to water molecules. Given the fact that most macromolecule crystals grow from solutions containing a complex mixture of chemicals, this approach seems too simplistic. A recent survey of the Protein Data Bank [8] showed that only about 4% of all entries determined by X-ray diffraction

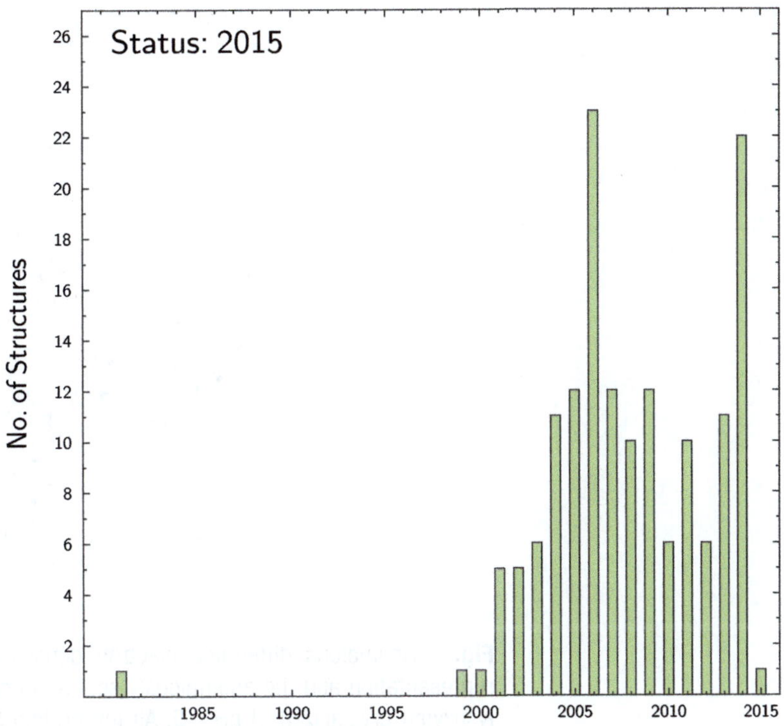

Fig. 2 Number of reported native SAD structures per year

methods contained phosphate, 14% contained sulfate, 9% contained chloride, 2% contained potassium and 8% contained calcium. These numbers seem awfully low given that chemicals such NaCl are frequently used in protein buffers and that ammonium sulfate is one of the most popular precipitants in protein crystallization. A few studies reported in the literature suggested that an anomalous difference electron density map [38] can be of help in assigning atom identities and in distinguishing water molecules from other entities bound to the macromolecule. Some interesting examples for this have been reported by Einspahr et al. [39], Weiss et al. [9, 40], Kuettner et al. [41], Ferreira et al. [42], Sekar et al. [43] and others. In 2007, Mueller-Dieckmann and colleagues published a thorough analysis of 23 protein structures based on high-resolution X-ray diffraction data, all collected at a wavelength of 2.0 Å [44]. The somewhat surprising finding was that in about 90% of all macromolecular structures, something else than just water was found to bind to the protein surface (Fig. 3). Since such binding sites may not only be fortuitous and since they may point to some hitherto undetected biological function, Mueller-Dieckmann and colleagues argued that any macromolecular structure determination ought to be augmented with a long-wavelength diffraction data set. Yet another aspect in this context is the possibility to

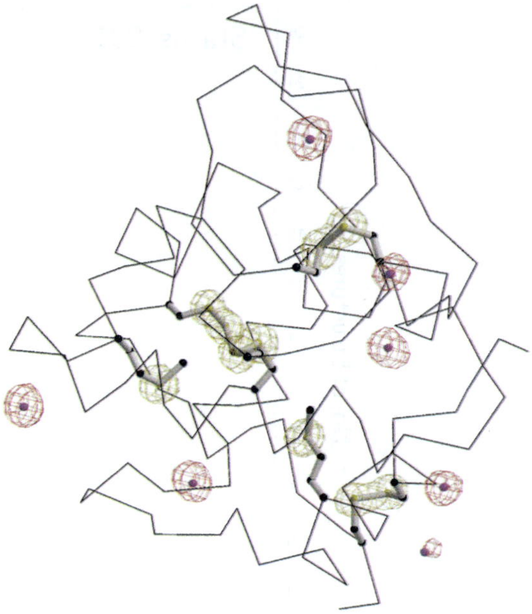

Fig. 3 Anomalous difference electron density map superimposed onto C_α-representation and the anomalously scattering substructure for hen egg-white lysozyme crystallized at pH 4.5. All ten protein S atoms (Cys6 SG, Met12 SD, Cys30 SG, Cys64 SG, Cys76 SG, Cys80 SG, Cys94 SG, Met105 SD, Cys115 SG, and Cys127 SG) are visible (*yellow*) as well as seven surface-bound chloride ions (*red*). The figure has been reproduced with permission from Mueller-Dieckmann et al. [44]

unambiguously place a ligand, which contains P, S, Cl or heavier atoms in protein-ligand complex structures. A very nice example for this has been provided by Raaf et al. [45].

3.3 Molecular Replacement

About two thirds of all macromolecular structures are nowadays determined using the molecular replacement approach. Since the number of protein folds seems to be limited [46, 47] and since more and more of the folds are discovered, it can be safely predicted that the importance of molecular replacement in MX will continue to rise. The method of molecular replacement requires that a structure (also called the search structure) which bears some similarity to the structure to be determined (also called the target structure) is known and available. Structural similarity may, for instance, be inferred by comparing amino acid sequences. However, structural similarity is not evenly distributed throughout the structure. When related structures are compared to each other, it is mostly the core of the protein which exhibits the highest degree of similarity, while surface residues and loops are often more diverse. Since the sulfur containing amino acids Cys and Met are more conserved than most other amino acids, it has been proposed by Unge et al. [48] that the sensitivity of molecular replacement may be increased by combining the traditional approach with one that is based on anomalous differences and the anomalously scattering substructure. Unge and colleagues calculated a combined rotation function RF_{comb}, which is a linear combination of the rotation functions based on the real structure factors (RF_{prot}) and on the anomalous differences (RF_{anom}) (Fig. 4). Using the 23 long-wavelength data sets described in Mueller-Dieckmann et al. [44], Unge et al. convincingly demonstrated that factoring in the anomalous differences leads in most cases to clearer signals and higher peaks than in the traditional molecular replacement approach relying on intensities alone.

Other researchers have also used anomalous differences in a molecular replacement type approach. In the MRSAD approach introduced by Schuermann and Tanner [49] anomalous differences were used to distinguish the correct molecular solutions from the incorrect ones. In another approach the anomalous substructure and anomalous differences are used to calculate an electron density map which is unbiased by the search model [50].

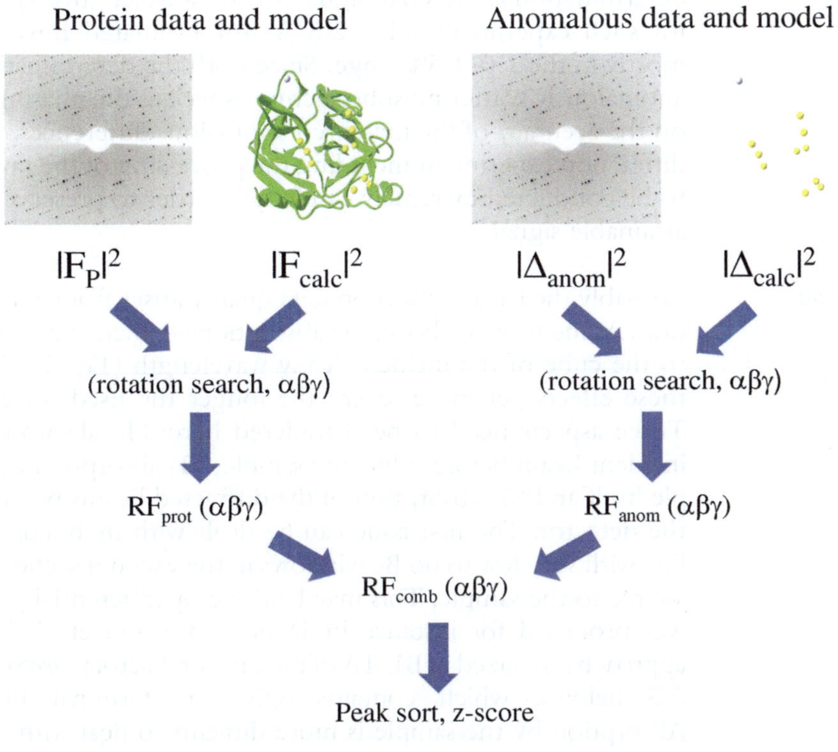

Fig. 4 Schematic illustration of how anomalous differences may be used in a molecular replacement calculation. The rotation function RF_{prot} is calculated from the observed structure-factor amplitudes and the calculated structure-factor amplitudes of the corresponding protein model. The anomalous rotation function RF_{anom} is calculated based on the experimental anomalous differences and the corresponding anomalous substructure of the model. The figure has been reproduced with permission from Unge et al. [48]

Thus two independent phase sets can be obtained, which may be combined for more efficient structure determination. This idea has been automated and implemented for instance in the Auto-Rickshaw structure determination pipeline [51].

3.4 Structure Refinement

Very few studies have been conducted taking anomalous differences into account in refinement, although it seems quite natural to use any kind of measured signals in the diffraction data for optimal definition of the coordinate set, which fits best the observed data. Initial studies using the 23 long-wavelength data sets described in Mueller-Dieckmann et al. [44] indicated that refinement statistics can indeed be improved to some extent when anomalous differences are taken into account explicitly (data not shown). However, at present it is not clear whether these results can be generalized.

4 A Brief Survey of Experimental Difficulties and Possible Solutions

4.1 Small Signals

The root of all the problems associated with a native SAD structure determination is the small signal. At the wavelengths typically used for such experiments (1.7–2.3 Å) the estimated Bijvoet ratio is mostly in the 1.0–1.5% range. Since both the determination of the anomalously scattering substructure as well as the phasing step rely on the accuracy of the measured anomalous differences, the actual diffraction experiment including the processing of the raw data has to be conducted extremely carefully in order to preserve the small attainable signal.

4.2 Absorption

Probably the largest effect on data quality arises from X-ray absorption. As mentioned above, the absorption coefficient is proportional to the cube of the incident X-ray wavelength (Eq. 2). Therefore, these effects get more severe the longer the used wavelength is. Three aspects need to be considered here: (1) absorption of the incident beam before it hits the sample, (2) absorption by the sample itself and (3) absorption of the diffracted beams before they hit the detector. The first issue can be dealt with by building a beam line with very few to no Be windows in the vacuum section from the source to the sample. This may be done by differential pumping as was proposed for instance in Djinovic Carugo et al. [11]. This approach is realized in BL-1A of the Photon Factory (*see* Subheading 5.3 below), which contains only one terminal Be-window. Absorption by the sample is more difficult to deal with. Unlike in small molecule crystallography, MX diffraction data are usually not corrected for absorption effects by employing an analytical absorption correction. Instead, absorption is factored in implicitly during the scaling step. This approach relies on high data multiplicity and will work well up to a certain point [23], but not anymore if the absorption effects become too large. As a consequence Mueller-Dieckmann and colleagues suggested an optimal wavelength of about 2.0 Å irrespective of the sample composition, at which the maximum anomalous signal-to-noise ratio can be obtained [23]. More recently, data collection strategies are employed that rely on multiple crystal orientations [34], thus making the implicit absorption correction by scaling more effective. The most elegant approach in this respect appears to be tomographic sample reconstruction [52]. Having the exact size and shape of the entire sample (crystal, surrounding liquid, sample holder) available should make an analytical treatment of absorption feasible. However, its beneficial use in practice still has to be established. Finally, absorption of the scattered beam can be dealt with by either inserting a He path between the sample and the detector [53] or by having the whole end station immersed in a He atmosphere (for instance at BL-1A, *see* Subheading 5.3 below) or in vacuum as it is realized at beam line I23 (*see* Subheading 5.4 below).

4.3 Large Scattering Angles

Another difficulty associated with long-wavelength data collection is due to the increased scattering angles. At a wavelength of $\lambda = 2$ Å, a scattering angle 2θ of 60° is needed to obtain a resolution of 2.0 Å. This can only be realized with a large flat detector by making provisions for a very small sample-to-detector distance, or by mounting the detector onto a 2θ-stage. Both approaches have their disadvantages for diffraction data collection. Small sample-to-detector distances lead to overlapping reflections on the detector and employing a 2θ-stage makes it much more cumbersome and more difficult to obtain a complete, high multiplicity data set. An alternative solution is to work with a curved detector, which is realized at the dedicated long-wavelength beam line I23 at the Diamond Light Source (*see* Subheading 5.4).

4.4 Third Harmonic Contamination

Most tuneable MX beam lines around the world utilize a Si double-crystal monochromator in Si(111)-geometry. This has the inherent disadvantage that, in addition to the diffraction arising from the incident wavelength λ coming through the Si(111)-reflection, diffraction arising from the wavelength $\lambda/3$ and the Si(333)-reflection will be observed. This is typically referred to as third harmonic contamination. The corresponding diffraction images will then show two lattices, which overlap partially. If such a situation is encountered, the anomalous difference cannot be measured accurately enough for structure determination. Consequently, at beam lines which exhibit a spectrum with appreciable intensity at the $\lambda/3$ wavelength, additional mirrors have to be inserted in order to get rid of this effect. Alternatively, if the two monochromator crystals are separate from each other, a slight detuning of the second crystal can efficiently suppress the $\lambda/3$ wavelength although at the cost of concomitant loss of X-ray intensity which can be as large as 30–50%.

4.5 Sample Mounting

A very important point to consider is sample mounting. It has been convincingly demonstrated and reported by Alkire et al. [54, 55] that the choice of the sample mount can greatly affect data quality. Lithographic structures seem to be preferred here over the simpler nylon loops because of the tendency of nylon loops to vibrate in the gaseous nitrogen stream. Another aspect is the sample itself. Since the sample generally consists of the crystal, the surrounding mother liquor and the sample holder structure, it makes sense to think of ways to reduce everything as much as possible to the crystal alone. Kitago and colleagues developed the so-called loop-less mounting method [56, 57], which involves the removal of all mother liquor and the loop from around the crystal. Since this method is not easy to master, other approaches have been developed which achieve similar improvements. One such approach is to use a dehydration device to remove all mother liquor from around the crystal [58] and another one is to wrap the crystals in graphene sheaths [59, 60].

4.6 Data Collection Strategies

As mentioned in Subheading 4.2, the correction of the raw intensities obtained from an MX diffraction experiment for absorption is only dealt with implicitly during the scaling stage. Efficient scaling requires that the multiplicity of the data set be high. This has been noted a number of times in the literature [10, 61, 62]. It is important to mention that high multiplicity does not mean to repeat the same measurement over and over again. Slight variations in the conditions for data collection help to iron out systematic errors by adding a random component to them. This is the basis for the approach to use multiple orientations of the same crystal for data collection [34, 63]. Alternatively, it is also possible to assemble a high multiplicity data set from diffraction data measured from multiple crystals in random orientations [32, 33]. In such cases, some sort of a clustering approach is often necessary in order to identify individual data sets which do not fit well to the remaining ones and which need to be left out from the merging process. Since a whole chapter in this book is devoted to multi-crystal approaches [64], it will not be further elaborated on here. Finally, one can also resort to more traditional, superior data collection schemes such as the inverse beam method [65] or to align the crystal using a suitable goniometer [66, 67] in order to record Bijvoet pairs on the same image.

4.7 Detection of Soft X-Rays

In the recent years, the quality of X-ray detectors has improved greatly. Hybrid photon counting detectors [68] which are now available at MX beam lines around the world have had a significant impact on the data quality achieved. On the HZB-MX beam line BL14.1 [69, 70], merely the replacement of a CCD-detector by a PILATUS 6M in 2013 has led to an improvement of the asymptotic $I/\sigma(I)$-values (ISa-values [71]) of the collected diffraction data sets on average by 10–20 units (data not shown). These detectors are also particularly attractive for diffraction experiments at longer wavelengths since they approach a detected quantum efficiency (DQE) of 1 at energies of about 7.5 keV, which is equivalent to wavelengths around 1.65 Å (https://www.dectris.com/sensor_details.html). Coupled with the absence of detector readout noise and the possibility to use fine-slicing in combination with shutterless data collection these new detectors enable the collection of very high quality data. Nevertheless, Holton and colleagues argue that a perfect detector should be able to deliver data sets with ISa values exceeding 100 and they identify detector calibration deficiencies as the main reason why this is still not possible [72]. It seems therefore not unrealistic to believe that further improvements of detector technology are achievable and that there will be even more efficient detectors available in the not too distant future.

4.8 Data Processing: Integration and Scaling

Processing the raw diffraction data obtained in longer-wavelength diffraction experiments is no different from processing diffraction data collected at other wavelengths. The very same software packages, e.g., *iMOSFLM* [73], *HKL2000* [74], *XDS* [75–77] and *d*TREK* [78] can be used. Of these *XDS* appears to be particularly suitable to process the fine-sliced data obtained from the shutterless data collections using a PILATUS detector (*see* paragraph Subheading 4.7) due to its 3D-profile fitting algorithm.

When it comes to scaling, the situation is a little different. It has been shown by Mueller-Dieckmann and colleagues [79] that the most accurate anomalous differences can be obtained when a 3D-scaling protocol is employed. The reason for this is that absorption effects, which are certainly more serious at longer wavelengths than at short wavelengths, are usually dealt with implicitly at the scaling stage (*see* also Subheading 4.2). Since the sample is a three-dimensional object it seems obvious that only a 3D-scaling model can properly account for these effects. A further slight improvement may be obtained when a data set collected at a short wavelength is available, which can be employed as a reference data set for scaling. This has also been demonstrated by Mueller-Dieckmann et al. [79], but it has rather little been used in the past few years. Nowadays, all relevant scaling programs (e.g., AIMLESS [80], SCALEPACK [74], XSCALE [77], etc.) do employ a 3D-scaling model and it probably makes little difference which programs are used, provided that the 3D-scaling option is switched on.

4.9 Radiation Damage

A recurring problem in MX, and in particular for native SAD approaches, is radiation damage [81, 82]. Even before radiation damage leads to a visible loss of diffraction power of a crystal, specific radiation damage effects may have set in. In proteins, the primary sites where specific damage occurs are metal centers and disulfide bridges. Since a reduction of a disulfide bridge inevitably leads to a movement of the associated sulfur atoms and thus to a change in the anomalously scattering substructure, the anomalous scattering contribution to the total structure factor at the end of the diffraction experiment will not be the same as at the beginning of the experiment. This puts a limit on the accuracy of anomalous intensity differences which can be collected from a crystal and it calls for an inclusion of radiation damage considerations into data collection strategy [82]. Sometimes, however, effects from radiation damage can even be turned into an advantage, as has for instance been demonstrated by the structure determination of Dsrc [30]. Here, the alteration of the anomalously scattering substructure during the experiment was taken into account during the phasing step, thus leading to better phases.

5 Beam Lines for Long-Wavelength X-Ray Diffraction

5.1 A Little Bit of History

In the early 2000s, when the interest in long-wavelength diffraction experiments began to rise, there was hardly a synchrotron beam line set up to routinely carry out diffraction experiments at such wavelengths. A survey of MX beam lines at synchrotron facilities worldwide conducted in 2005 by Djinovic Carugo and colleagues showed that although longer wavelengths were indeed accessible at several beam lines, which were operational at that time [11], most of the early experiments were conducted on very few beam lines. In Europe, early experiments were carried out mainly at the XRD1 beam line [83] of the ELETTRA synchrotron (Trieste, Italy) and at BM14 of the ESRF (Grenoble, France), while in the USA the SERCAT team (http://www.ser-cat.org) on the beam lines 22-ID and 22-BM at the APS (Argonne, USA) led the field. In Japan, researchers at beam line BL-1A of the Photon Factory (Tsukuba, Japan) were most active. Some time later, when the general user community as well as beam line staff started to appreciate the advantages offered by longer-wavelength diffraction experiments for MX, other beam lines expanded their wavelength spectrum and made such experiments possible as well. These included, among many others, beam line ID29 at the ESRF [84], the newly built X12 at the DORIS ring at DESY (Hamburg, Germany) operated by the EMBL Hamburg Outstation, the MX beam lines at the BESSY II storage ring in Berlin [69, 70], and beam line 10 [85] at the Daresbury synchrotron radiation source SRS (Daresbury, UK). Due to the closure of the DORIS ring in Hamburg and the SRS in Daresbury, the latter two beam lines are not in operation anymore.

5.2 Status Today

Nowadays, the situation has changed appreciably. Pretty much every newly built tuneable synchrotron beam line for MX is able to provide access to wavelengths up to 2.5 Å or even beyond.

Furthermore, beam line staff is increasingly aware of the requirements for longer-wavelength diffraction experiments and provisions have been made to make such experiments possible and to optimally assist users during such experiments. In the following two paragraphs, two new synchrotron beam lines for MX, which are specifically designed and built for long-wavelength diffraction experiments, will be described in greater detail. These may be called dedicated long-wavelength diffraction beam lines, even though they do provide access to shorter wavelengths as well.

5.3 Beam Line BL-1A at the Photon Factory

BL-1A at the Photon Factory (Tsukuba, Japan) is dedicated to micro-crystallography and long-wavelength experiments. It has two operation modes, one in the wavelength range 0.95–1.1 Å and the other one in the range 2.7–3.3 Å. Fed from an *in-vacuum* short gap undulator as the X-ray source, the vacuum section of the beam line has only one terminal Be window. It contains a cryo-cooled channel-cut Si(111) double-crystal monochromator followed by a bimorph KB pair of focussing mirrors. In the end station, the diffractometer is equipped with a He cryostream and a specially designed He chamber to minimize the attenuation of the X-ray beam at longer wavelengths and the corresponding background. More information on this beam line can be found on the beam line web page http://pfweis.kek.jp/protein/BeamLine/BL1/bl1.html and in Liebschner et al. [86] and Hiraki et al. [87].

5.4 Beam Line I23 at Diamond

At the Diamond Light Source in the UK, researchers are presently commissioning beam line I23. This beam line constitutes one of most ambitious MX beam line projects of the last decade. I23 will be a unique facility specifically designed and built for long-wavelength MX experiments. Fed from a 2 m long *in-vacuum* undulator, the beam line comprises a double-crystal Si(111) monochromator and a four-mirror system: a cylindrical and an elliptical mirror for vertical and horizontal focussing respectively and two flat mirrors for harmonic rejection. I23 will provide X-rays over a large wavelength range from 1.0 to 5.9 Å, with the optimum between 1.5 and 4.0 Å. The complete end station (sample, goniometer and detector) will be completely housed in a large vacuum chamber. A spectacular custom-made curved PILATUS 12 M detector will make diffraction angles up to $2\theta = 100°$ accessible. Furthermore, an X-ray tomography setup is planned that completes the experimental end station. With this it will be possible to accurately define the sample dimensions for analytical absorption correction [52]. The beam line has received its first users in early 2016 and is currently in the commissioning phase. First results from this beam line are eagerly awaited. More information on this beam line can be found on the beam line web page http://www.diamond.ac.uk/Beamlines/Mx/I23.html and in Wagner et al. [88].

6 Conclusions and Outlook

The developments in the recent past clearly indicate that longer-wavelength applications are now close to entering the mainstream of MX. Many researchers are nowadays aware of the possibilities of such experiments. Also, many of the early notions about the difficulties of such experiments, in particular the difficulties associated with native SAD, for instance that only small proteins, crystallized in highly symmetric space groups and diffracting to high resolution, would be amenable to this method, have now been proven incorrect. It remains to be seen, however, whether this will translate into many more de novo structure determinations being carried out by native SAD. It is also clear that methodological improvements, both on the hardware and on the software sides, will certainly come to contribute to the success of the method. Even at the upcoming X-ray free electron lasers (XFELs) native SAD structure determinations are now being carried out [89, 90]. This will further help to increase the awareness that such experiments are feasible and that they are able to deliver important information that can otherwise not be easily obtained.

Acknowledgements

I would like to thank all of my friends and colleagues who have worked with me in this exciting field of longer-wavelength MX over many years and contributed many of the ideas and hypotheses presented in this chapter. I would also like to thank Rachel Kramer Green from the PDB for providing the statistics on the wavelength data in the PDB entries.

References

1. Helliwell JR, Ealick S, Doing P et al (1993) Towards the measurement of ideal data for macromolecular crystallography using synchrotron sources. Acta Crystallogr D Biol Crystallogr 49:120–128
2. Schiltz M, Kvick A, Svensson OS et al (1997) Protein crystallography at ultra-short wavelengths: feasibility study of anomalous-dispersion experiments at the xenon *K*-edge. J Synchrotron Radiat 4:287–297
3. Lehmann MS, Müller HH, Stuhrmann HB (1993) Protein single-crystal diffraction with 5 Å synchrotron X-rays at the sulfur *K*-absorption edge. Acta Crystallogr D Biol Crystallogr 49:308–310
4. Stuhrmann S, Hütsch M, Trame C et al (1995) Anomalous dispersion with edges in the soft X-ray region: first results of diffraction from single crystals of ribosomes near the *K*-absorption edge of phosphorus. J Synchrotron Radiat 2:83–86
5. Stuhrmann S, Bartels KS, Braunwarth W et al (1997) Anomalous dispersion with edges in the soft X-ray region: first results of diffraction from single crystals of trypsin near the *K*-absorption edge of sulfur. J Synchrotron Radiat 4:298–310
6. Behrens W, Otto H, Stuhrmann HB et al (1998) Sulfur distribution in bacteriorhodopsin from multiple wavelength anomalous diffraction near

the sulfur *K*-edge with synchrotron X-ray radiation. Biophys J 75:255–263

7. Carpentier P, Berthet-Colominas C, Capitan M et al (2000) Anomalous X-ray diffraction with soft X-ray synchrotron radiation. Cell Mol Biol 46:915–935
8. Berman HM, Westbrook J, Feng Z et al (2000) The protein data bank. Nucleic Acids Res 28:235–242
9. Weiss MS, Sicker T, Djinović Carugo K et al (2001) On the routine use of soft X-rays in macromolecular crystallography. Acta Crystallogr D Biol Crystallogr 57:689–695
10. Weiss MS, Sicker T, Hilgenfeld R (2001) Soft X-rays, high redundancy and proper scaling: a new procedure for automated structure determination via SAS. Structure 9:771–777
11. Djinovic Carugo K, Helliwell JR, Stuhrmann H et al (2005) Softer and soft X-rays in macromolecular crystallography. J Synchrotron Radiat 12:410–419
12. Blow DM (1958) The structure of haemoglobin. VII. Determination of phase angles in the non-centrosymmetric [100] zone. Proc R Soc A247:302–336
13. Yang C, Pflugrath JW, Courville DA et al (2003) Away from the edge: SAD phasing from the sulfur anomalous signal measured in-house with chromium radiation. Acta Crystallogr D Biol Crystallogr 59:1943–1957
14. Anderson DH, Weiss MS, Eisenberg D (1996) A challenging case for protein crystal structure determination: the mating pheromone E*r*-1 from *Euplotes raikovi*. Acta Crystallogr D Biol Crystallogr 52:469–480
15. Kwiatkowski W, Noel JP, Choe S (2000) Use of Cr K_α radiation to enhance the signal from anomalous scatterers including sulphur. J Appl Cryst 33:876–881
16. Arndt UW (1984) Optimum X-ray wavelength for protein crystallography. J Appl Cryst 17:118–119
17. Polikarpov I (1997) Protein crystallography in the soft X-ray region: crystal lifetime and diffraction efficiency. J Synchrotron Radiat 4:17–20
18. Polikarpov I, Teplyakov A, Oliva G (1997) The ultimate wavelength for protein crystallography? Acta Crystallogr D Biol Crystallogr 53:734–737
19. Murray JW, Garman EF, Ravelli RBG (2004) X-ray absorption by macromolecular crystals: the effects of wavelength and crystal composition on absorbed dose. J Appl Cryst 37:513–522
20. Helliwell JR (1993) The choice of X-ray wavelength in macromolecular crystallography. Daresbury CCP4 study weekend Proceedings DL/SCI/R34. CCLRC Daresbury Laboratory, Warrington, UK, pp 80–88
21. Teplyakov A, Oliva G, Polikarpov I (1998) On the choice of an optimal wavelength in macromolecular crystallography. Acta Crystallogr D Biol Crystallogr 54:610–614
22. Weiss MS, Panjikar S, Mueller-Dieckmann C et al (2005) On the influence of the incident photon energy on the radiation damage in crystalline biological samples. J Synchrotron Radiat 12:304–309
23. Mueller-Dieckmann C, Panjikar S, Tucker PA et al (2005) On the routine use of soft X-rays in macromolecular crystallography, part III—the optimal data collection wavelength. Acta Crystallogr D Biol Crystallogr 61:1263–1272
24. Hendrickson WA, Teeter MM (1981) Structure of the hydrophobic protein crambin determined directly from the anomalous scattering of sulphur. Nature 290:107–113
25. Dauter Z, Dauter M, de La Fortelle E et al (1999) Can anomalous signal of sulfur become a tool for solving protein crystal structures? J Mol Biol 289:83–92
26. Liu ZJ, Vysotski ES, Chen CJ et al (2000) Structure of the Ca^{2+}-regulated photoprotein obelin at 1.7 Å resolution determined directly from its sulfur substructure. Protein Sci 9:2085–2093
27. Gordon EJ, Leonard GA, McSweeney S et al (2001) The C1 subunit of α-crustacyanin: the de novo phasing of the crystal structure of a 40 kDa homodimeric protein using the anomalous scattering from S atoms combined with direct methods. Acta Crystallogr D Biol Crystallogr 57:1230–1237
28. Chen L, Chen LR, Zhou XE et al (2004) The hyperthermophile protein Sso10a is a dimer of winged helix DNA-binding domains linked by an antiparallel coiled coil rod. J Mol Biol 341:73–91
29. Gentry HR, Singer AU, Betts L et al (2005) Structural and biochemical characterization of CIB1 delineates a new family of EF-hand-containing proteins. J Biol Chem 280:8407–8415
30. Weiss MS, Mander G, Hedderich R et al (2004) Determination of a novel structure by a combination of long wavelength sulfur phasing and radiation damage induced phasing. Acta Crystallogr D Biol Crystallogr 60:686–695

31. Kitamura M, Okuyama M, Tanzawa F et al (2008) Structural and functional analysis of a glycoside hydrolase family 97 enzyme from *Bacteroides thetaiotaomicron*. J Biol Chem 283:36326–36337
32. Liu Q, Dahmane T, Zhang Z et al (2012) Structures from anomalous diffraction of native biological macromolecules. Science 336:1033–1037
33. Liu Q, Liu Q, Hendrickson WA (2013) Robust structural analysis of native biological macromolecules from multi-crystal anomalous diffraction data. Acta Crystallogr D Biol Crystallogr 69:1314–1332
34. Weinert T, Olieric V, Waltersperger S et al (2015) Fast native-SAD phasing for routine macromolecular structure determination. Nat Methods 12:131–133
35. El Omari K, Lourin O, Kadlec J et al (2014) Pushing the limits of sulfur SAD phasing: de novo structure solution of the N-terminal domain of the ectodomain of HCV E1. Acta Crystallogr D Biol Crystallogr 70:2197–2203
36. Gorgel M, Bøggild A, Ulstrup JJ et al (2015) Against the odds? De-novo structure determination of a type IV pilin with two cysteine residues by sulfur SAD. Acta Crystallogr D Biol Crystallogr 71:1095–1101
37. Rose JP, Wang BC, Weiss MS (2015) Native SAD is maturing. IUCr J 2:431–440
38. Strahs G, Kraut J (1968) Low-resolution electron-density and anomalous-scattering-density maps of Chromatium high-potential iron protein. J Mol Biol 35:503–512
39. Einspahr H, Suguna K, Suddath FL et al (1985) The location of manganese and calcium ion cofactors in pea lectin crystals by use of anomalous dispersion and tuneable synchrotron X-radiation. Acta Crystallogr B 41:336–341
40. Weiss MS, Panjikar S, Nowak E et al (2002) Metal binding to porcine pancreatic elastase: calcium or not calcium. Acta Crystallogr D Biol Crystallogr 58:1407–1412
41. Kuettner EB, Hilgenfeld R, Weiss MS (2002) The active principle of garlic at atomic resolution. J Biol Chem 277:46402–46407
42. Ferreira KN, Iverson TM, Maghlaoui K et al (2004) Architecture of the photosynthetic oxygen-evolving center. Science 303:1831–1838
43. Sekar K, Rajakannan V, Velmurugan D et al (2004) A redetermination of the structure of the triple mutant (K53,56,120M) of phospholipase A2 at 1.6 Å resolution using sulfur-SAS at 1.54 Å wavelength. Acta Crystallogr D Biol Crystallogr 60:1586–1590
44. Mueller-Dieckmann C, Panjikar S, Schmidt A et al (2007) On the routine use of soft X-rays in macromolecular crystallography, part IV—efficient determination of anomalous substructures in bio-macromolecules using longer X-ray wavelengths. Acta Crystallogr D Biol Crystallogr 63:366–380
45. Raaf J, Issinger OG, Niefind K (2008) Insights from soft X-rays: the chlorine and sulfur substructures of a CK2alpha/DRB complex. Mol Cell Biochem 316:15–23
46. Chothia C (1992) Proteins. One thousand families for the molecular biologist. Nature 357:543–544
47. Liu X, Fan K, Wang W (2004) The number of protein folds and their distribution over families in nature. Proteins 54:491–499
48. Unge J, Mueller-Dieckmann C, Panjikar S et al (2011) On the routine use of soft X-rays in macromolecular crystallography, part V – molecular replacement and anomalous scattering. Acta Crystallogr D Biol Crystallogr 67:729–738
49. Schuermann JP, Tanner JJ (2003) MRSAD: using anomalous dispersion from S atoms collected at Cu K_α wavelength in molecular-replacement structure determination. Acta Crystallogr D Biol Crystallogr 59:1731–1736
50. Baker EN, Anderson BF, Dobbs AJ et al (1995) Use of iron anomalous scattering with multiple models and data sets to identify and refine a weak molecular replacement solution: structure analysis of cytochrome c' from two bacterial species. Acta Crystallogr D Biol Crystallogr 51:282–289
51. Panjikar S, Parthasarathy V, Lamzin VS et al (2009) On the combination of molecular replacement and single-wavelength anomalous diffraction phasing for automated structure determination. Acta Crystallogr D Biol Crystallogr 65:1089–1097
52. Brockhauser S, Di Michiel M, McGeehan JE et al (2008) X-ray tomographic reconstruction of macromolecular samples. J Appl Cryst 41:1057–1066
53. Polentarutti M, Glazer R, Djinovic Carugo K (2004) A helium-purged beam path to improve soft and softer X-ray data quality. J Appl Cryst 37:319–324
54. Alkire RW, Duke NEC, Rotella FJ (2008) Is your cold-stream working for you or against you? An in-depth look at temperature and sample motion. J Appl Cryst 41:1122–1133
55. Alkire RW, Rotella FJ, Duke NEC (2013) Testing commercial protein crystallography sample mounting loops for movement in a cold-stream. J Appl Cryst 46:525–536
56. Kitago Y, Watanabe N, Tanaka I (2005) Structure determination of a novel protein by sulfur SAD using chromium radiation in combination with a new crystal-mounting method. Acta Crystallogr D Biol Crystallogr 61:1013–1021

57. Kitago Y, Watanabe N, Tanaka I (2010) Semi-automated protein crystal mounting device for the sulphur single-wavelength anomalous diffraction method. J Appl Cryst 43:341–346
58. Bowler MW, Mueller U, Weiss MS et al (2015) Automation and experience of controlled crystal dehydration: results from the European synchrotron HC1 collaboration. Cryst Growth Des 15:1043–1054
59. Wierman JL, Alden JS, Kim CU et al (2013) Graphene as a protein crystal mounting material to reduce background scatter. J Appl Cryst 46:1501–1507
60. Warren AJ, Crawshaw AD, Trincao J et al (2015) *In vacuo* X-ray data collection from graphene wrapped protein crystals. Acta Crystallogr D Biol Crystallogr 71:2079–2088
61. Weiss MS (2001) Global indicators of X-ray data quality. J Appl Cryst 34:130–135
62. Cianci M, Helliwell JR, Suzuki A (2008) The interdependence of wavelength, redundancy and dose in sulfur SAD experiments. Acta Crystallogr D Biol Crystallogr 64:1196–1209
63. Olieric V, Weinert T, Finke AD et al (2016) Data-collection strategy for challenging native SAD phasing. Acta Crystallogr D Biol Crystallogr 72:421–429
64. Liu Q, Hendrickson WA (2017) Contemporary use of anomalous diffraction in biomolecular structure analysis. In: Wlodawer A, Dauter Z, Jaskolski M (eds) Protein crystallography. Springer, New York
65. Hendrickson WA, Pähler A, Smith JL et al (1989) Crystal structure of core streptavidin determined from multiwavelength anomalous diffraction of synchrotron radiation. Proc Natl Acad Sci U S A 86:2190–2194
66. Brockhauser S, Ravelli RBG, McCarthy AA (2013) The use of a mini-κ goniometer head in macromolecular crystallography diffraction experiments. Acta Crystallogr D Biol Crystallogr 69:1241–1251
67. Waltersperger S, Olieric V, Pradervand C et al (2015) PRIGo: a new multi-axis goniometer for macromolecular crystallography. J Synchrotron Radiat 22:895–900
68. Broennimann C, Eikenberry EF, Henrich B et al (2006) The PILATUS 1M detector. J Synchrotron Radiat 13:120–130
69. Mueller U, Darowski N, Fuchs MR et al (2012) Facilities for macromolecular crystallography at the Helmholtz-Zentrum Berlin. J Synchrotron Radiat 19:442–449
70. Mueller U, Förster R, Hellmig M et al (2015) The macromolecular crystallography beamlines at BESSY II of the Helmholtz-Zentrum Berlin: current status and perspectives. Eur Phys J Plus 130:141–150
71. Diederichs K (2010) Quantifying instrument errors in macromolecular X-ray data sets. Acta Crystallogr D Biol Crystallogr 66:733–740
72. Holton JM, Classen S, Frankel KA et al (2014) The R-factor gap in macromolecular crystallography: an untapped potential for insights on accurate structures. FEBS J 281:4046–4060
73. Battye TGG, Kontogiannis L, Johnson O et al (2011) *iMOSFLM*: a new graphical interface for diffraction-image processing with *MOSFLM*. Acta Crystallogr D Biol Crystallogr 67:271–281
74. Otwinowski Z, Minor W (1997) Processing of X-ray diffraction data collected in oscillation mode. Methods Enzymol 276:307–326
75. Kabsch W (1993) Automatic processing of rotation diffraction data from crystals of initially unknown symmetry and cell constants. J Appl Cryst 26:795–800
76. Kabsch W (2010) XDS. Acta Crystallogr D Biol Crystallogr 66:125–132
77. Kabsch W (2010) Integration, scaling, space-group assignment and post-refinement. Acta Crystallogr D Biol Crystallogr 66:133–144
78. Pflugrath JW (1999) The finer things in X-ray diffraction data collection. Acta Crystallogr D Biol Crystallogr 55:1718–1725
79. Mueller-Dieckmann C, Polentarutti M, Djinović-Carugo K et al (2004) On the routine use of soft X-rays in macromolecular crystallography, part II: data collection wavelength and scaling models. Acta Crystallogr D Biol Crystallogr 60:28–38
80. Evans PR, Murshudov GN (2013) How good are my data and what is the resolution? Acta Crystallogr D Biol Crystallogr 69:1204–1214
81. Garman EF (2013) Radiation damage in macromolecular crystallography: what is it and why do we care? In: Read R, Urzhumtsev AG, Lunin VY (eds) Advancing methods for biomolecular crystallography. Springer, Dordrecht, pp 69–77
82. Zeldin OB, Brockhauser S, Bremridge J et al (2013) Predicting the X-ray lifetime of protein crystals. Proc Natl Acad Sci U S A 110:20551–20556
83. Lausi A, Polentarutti M, Onesti S et al (2015) Status of the crystallography beamlines at Elettra. Eur Phys J Plus 130:43
84. De Sanctis D, Beteva A, Caserotto H et al (2012) ID29: a high-intensity highly automated ESRF beamline for macromolecular crystallography experiments exploiting anomalous scattering. J Synchrotron Radiat 19:455–461
85. Cianci M, Antonyuk S, Bliss N et al (2005) A high-throughput structural biology/proteomics beamline at the SRS on a new multi-pole wiggler. J Synchrotron Radiat 12:455–466

86. Liebschner D, Yamada Y, Matsugaki N et al (2016) On the influence of crystal size and wavelength on native SAD phasing. Acta Crystallogr D Biol Crystallogr 72:728–741
87. Hiraki M, Matsugaki N, Yamada Y et al (2016) Development of sample exchange robot PAM-HC for beamline BL-1A at the photon factory. AIP Conf Proc 1741:030029
88. Wagner A, Duman R, Henderson K et al (2016) *In vacuum* long-wavelength macromolecular crystallography. Acta Crystallogr D Biol Crystallogr 72:430–439
89. Nakane T, Song C, Suzuki M et al (2015) Native sulfur/chlorine SAD phasing for serial femtosecond crystallography. Acta Crystallogr D Biol Crystallogr 71:2519–2525
90. Nass K, Meinhart A, Barends TRM et al (2016) Protein structure determination by single-wavelength anomalous diffraction phasing of X-ray free-electron laser data. IUCr J 3:180–191

Chapter 18

Acknowledging Errors: Advanced Molecular Replacement with Phaser

Airlie J. McCoy

Abstract

Molecular replacement is a method for solving the crystallographic phase problem using an atomic model for the target structure. State-of-the-art methods have moved the field significantly from when it was first envisaged as a method for solving cases of high homology and completeness between a model and target structure. Improvements brought about by application of maximum likelihood statistics mean that various errors in the model and pathologies in the data can be accounted for, so that cases hitherto thought to be intractable are standardly solvable. As a result, molecular replacement phasing now accounts for the lion's share of structures deposited in the Protein Data Bank. However, there will always be cases at the fringes of solvability. I discuss here the approaches that will help tackle challenging molecular replacement cases.

Key words Molecular replacement, Maximum likelihood, LLGI

1 Introduction

As originally conceived [1–3], the aim of molecular replacement (MR [4]; *see* **Note 1**) was to correctly orient and place a model that had high homology to the target and represented the bulk of the scattering, for the purpose of phasing. It has since been generalized to cases of targets being modeled by any number of components with any homology to the target, and each component representing any fraction of the scattering in an asymmetric unit [5]. The central problem of MR is to identify the correct placement (where *placement* refers to the three orientation angles and the three translation coordinates) of all model components in the asymmetric unit, with the hope that the resulting phases will be good enough to see novel features of the target structure and for iterative cycles of model building and refinement to commence [6].

MR consists of two aspects: a search procedure, for sampling orientations and translations of the model(s) in the crystal asymmetric unit; and a scoring function, for determining the (best) match of the structure factors calculated from the oriented and

Alexander Wlodawer et al. (eds.), *Protein Crystallography: Methods and Protocols*, Methods in Molecular Biology, vol. 1607, DOI 10.1007/978-1-4939-7000-1_18,

positioned model(s) to the observed structure factors, and hence the correct placement of the components. If the model is good and the data are free of pathologies, MR is likely to be successful with many, or all, of the implemented search strategies and scoring functions (X-PLOR [7], CNS [8], AMoRe [9], MOLREP [10], EPMR [11], Qs [12], SOMoRe [13], COMO [14], and Phaser [15]), each with their own strengths [16].

When it works, the speed and automation of MR rivals that of the direct methods used for small molecule crystallography, but it has a dark side. Because it is a search procedure, the success or failure of the method depends on the signal-to-noise of the correct placement, which depends on the quality of the model and data. Quick when it works with the first model and dataset input, it can be prohibitively slow if it does not, leading to an ever-increasing drain of computational resources. Paradoxically, successful MR strategies include knowing when to stop searching and attempt other structure solution methods.

With the extension of the Protein Data Bank ([17] PDB) to cover much of fold space, the chances are good that there will be a structure already in the PDB with the same fold as the target protein [18]. Despite this, it is still common for MR models to have very low or even barely detectable sequence identity with the target (*see* **Note 2**). Statistically, this is not a surprise, given the uncountable number of ways proteins can diverge in sequence from one another. It is also natural that researchers choose to crystallize proteins only when they require novel structural information.

Although much smaller, the database of nucleic acid and nucleic acid-protein complexes also offers a wealth of opportunity for MR phasing, partly because nucleic acid helices can adopt similar conformations with drastically different sequences, and because it is now recognized that there are nucleic acid structural building blocks [19].

2 Protocols

The aim of this review is not to provide a set of prescriptive protocols for MR with Phaser [15], and indeed some of the approaches can be taken with other software. I assume that the reader is familiar with basic MR theory and practice. When MR is nontrivial, no two pathways to structure solution will be identical. Apart from the crystal-specific differences, there is the constantly changing background of instrumentation and software. Therefore, I aim to describe approaches to difficult cases that can be flexibly adapted to the problem at hand.

3 Overview

This review is directed at maximum likelihood MR (MLMR), and specifically the use of the LLGI (Log-Likelihood Gain on Intensity) target [20]. MLMR scoring methods are superior in discriminating correctly from incorrectly placed models than Patterson methods [21]. LLGI adds the ability to account for experimental error in the data to the well-established ability of MLMR to account for errors in the models. It removes biases in MLMR targets formulated in terms of structure factor amplitudes, where the very poorly measured reflections are not appropriately down-weighted. LLGI has the correct asymptotic behaviour for data with infinite experimental error: these data have no contribution to the total LLGI. LLGI abolishes the need for the conversion of intensity data to amplitudes (usually performed with the French and Wilson method [22]) before MR.

Most of the problems with MR arise when there is a need to place a large number of components in the asymmetric unit, particularly if there is also low structural homology between models and targets. These situations may be engendered by the choice of crystallization target, for example, a macromolecular complex for which the structures for individual components, in isolation, are known, but not the complex in its entirety; or it may come about because the crystal happens to grow with many copies of the macromolecule in the asymmetric unit; or it may arise because the crystallographer chooses to attempt MR with small, generic, structural elements. Large errors are intrinsic to these problems, which is why MLMR targets are well suited to tackling them.

It is also possible for cases that seemed likely to be trivial at the outset to turn out to be fiendishly difficult, due to particular pathologies. MR is increasingly being attempted with crystals that are inherently twinned, show highly anisotropic diffraction and/or have translational non-crystallographic symmetry. MLMR approaches account for the intensity modulations arising from anisotropy and translational non-crystallographic symmetry, and the use of the LLGI target correctly weights the weak data with high error that are intrinsic to these data.

The most critical difference between MLMR approaches and Patterson approaches to MR is that MLMR is optimized when both the mean of the distribution *and the standard deviation of the distribution* are closest to the real values used to generate the data. The standard deviation is a fully fledged parameter, and can be refined along with the mean in minimization (optimization) algorithms. If the errors are low, optimizing the parameters contributing to the standard deviation will make little difference to the outcome of MR. However, successful MLMR in borderline cases is not simply about good estimates of structure factors; it is also about good estimates of the *errors in the structure factors* (Fig. 1).

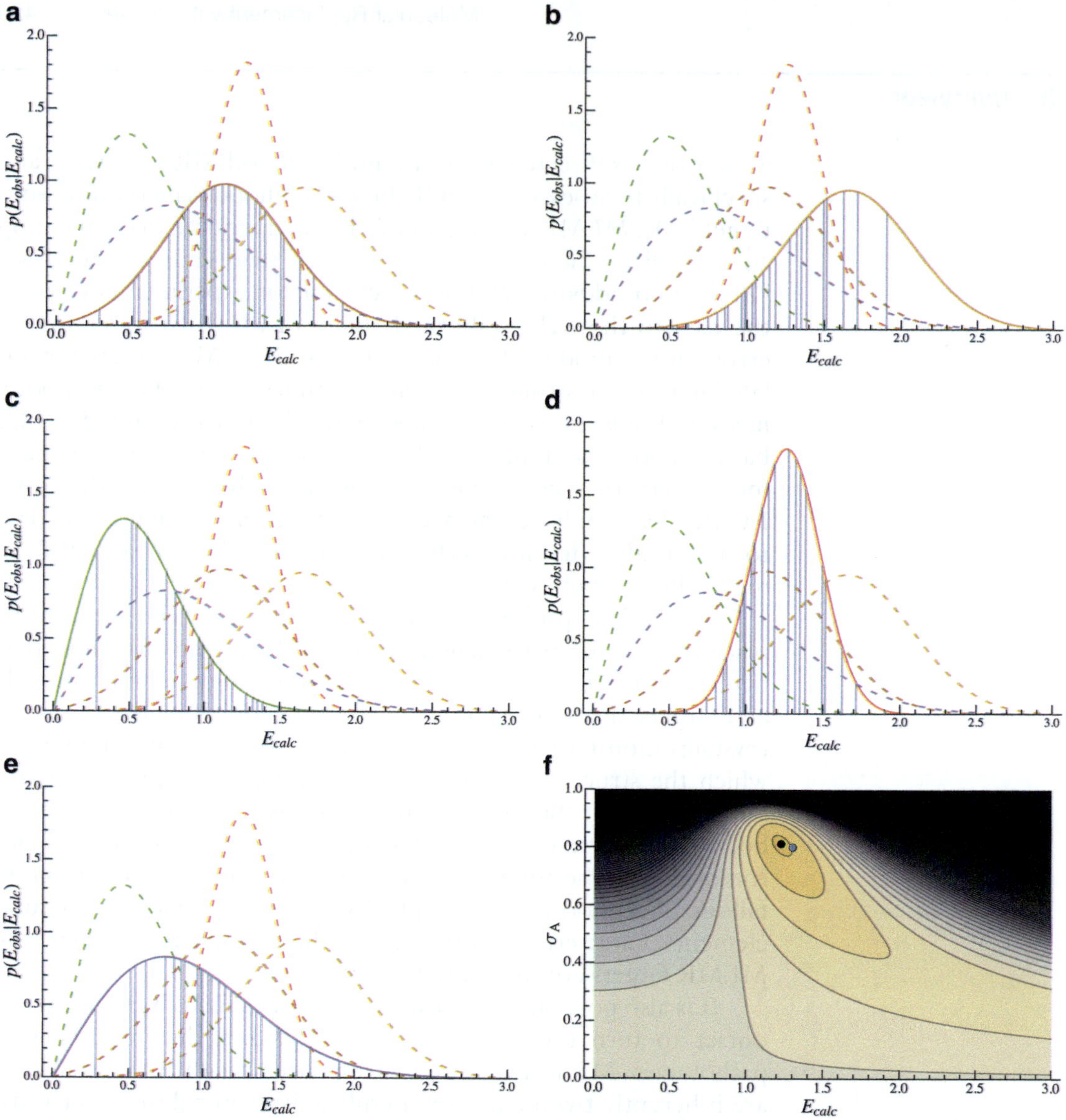

Fig. 1 Deducing E_{calc} and σ_A from a set of E_{obs}. The likelihood of E_{obs} given E_{calc} is given by the Rice distribution [15, 117]. Twenty-five E_{obs} were randomly generated from a Rice distribution with $E_{calc} = 1.3$ and $\sigma_A = 0.8$. The vertical bars correspond to the E_{obs}. The height of each bar represents the probability of E_{obs}, given the E_{calc} and σ_A of the Rice distribution shown. Dotted curves show probabilities from other panels, for comparison. The log-likelihood is the sum of the log-likelihoods for each E_{obs}. (**a**) Twenty-five E_{obs} shown with the Rice function that was used to generate them. The centre of the distribution is most heavily populated by the data, and none of the probabilities is very low. The total log-likelihood is −11.9. (**b**) Change the E_{calc} of the Rice distribution to 2. The E_{obs} on the low end of the Rice distribution become very improbable, which will reduce the likelihood. Fewer of the data points are now in the peak region. The total log-likelihood is −35.7. (**c**) Change the E_{calc} of the Rice distribution to 0.3. The total log-likelihood −40.9. (**d**) Change the σ_A of the Rice distribution to 0.95. In the heavily populated centre, the probability values go up, but the values in the two tails go down even more, so that the overall value of the likelihood is reduced. The total log-likelihood is −32.1. (**e**) Change the σ_A of the Rice distribution to 0.3. The probabilities in the tails go up, but decrease in the heavily populated peak. The total log-likelihood is −15.3. (**f**) Contour plot of the log-likelihood for pairs of E_{calc} and σ_A. The peak in this distribution (*black dot*) is close to the E_{calc} and σ_A that were used in generating the data (*blue dot*). With the correct E_{calc} the likelihood function will balance out the influence of the sparsely populated tails and the heavily populated centre to give the correct σ_A

When the errors are high it is important to understand the sources of error so that they can be reduced and/or correct estimates optimally incorporated in the likelihood functions, so that the LLGI is maximized.

The contribution to the total LLGI from any individual reflection depends on the variables E_{calc}, σ_A, I_{obs} and σ_{Iobs} [20]. The total LLGI is the sum of the reflection LLGI values. These principles are the basis for the discussion of optimization of the signal in MLMR (Fig. 2).

4 Methods

MLMR targets and target specific search strategies for MLMR are implemented in Phaser [15] (and previously BEAST [21]). Phaser is distributed through the CCP4 [23] and phenix [24] software suites. The software can be run from the command line, from python scripts, or through the ccp4i [25] or phenix interfaces [26]. Phaser is used in MR pipelines including MrBump [27] from the CCP4 suite, MRage [28] from the phenix suite and the WS-MR SBGrid [29]. It is also the basis of the anisotropy server [30]. Phaser has been incorporated into the development of ab initio phasing via MR in Arcimboldo [31] and Ample [32]. Many of the methods discussed here are relevant to all versions of Phaser, but some require Phaser-2.7.12 and above.

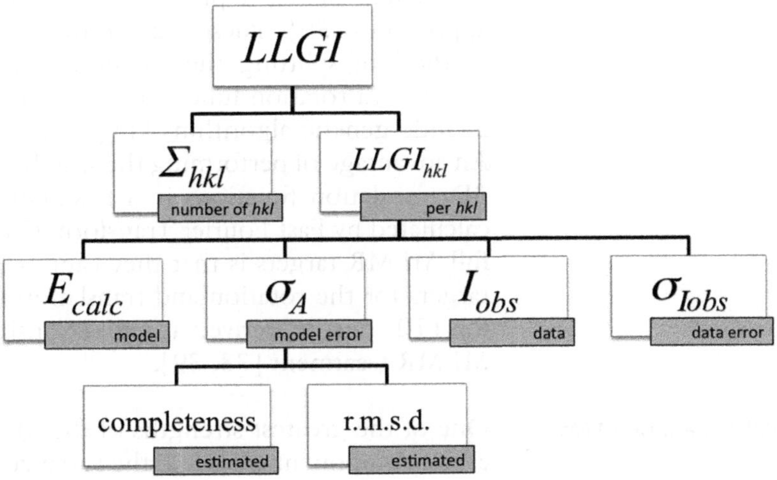

Fig. 2 Dependence of LLGI on parameters of the data and the model. The total LLGI is the sum of the LLGI for each reflection, so the more reflections the higher the LLGI. The LLGI per reflection depends on E_{calc}, σ_A, I_{obs} and σ_{Iobs}. The σ_A values are estimated from the fraction scattering of the model and the expected rmsd

4.1 Target Function

Phaser's LLGI target is the log of the likelihood of the MLMR hypothesis minus the log of the likelihood of the null hypothesis, where the hypothesis is formulated in terms of intensities [20]. The MLMR hypothesis is the current orientation and placement (translation function) or just orientation (rotation function) of the search component, within the background of the orientation and placement of other components under consideration. The null hypothesis is the Wilson distribution [33] of intensities, arising from a random distribution of isotropically scattering atoms in the asymmetric unit.

4.2 Search Strategies

Phaser implements automated search strategies for finding multiple components. In the default search strategy, data are corrected for anisotropy and translational non-crystallographic symmetry; rotation and translation functions run with automated selection of potentially correct orientations and translations; packing checks performed; the partial solutions rigid body refined; and these steps iterated over the number of components. The resolution is optimized for speed [34]. Steps are automatically repeated with altered parameters if the first set of parameters fails to yield a solution. The default search strategy is likely to find a solution if at all possible, but is also highly configurable (see documentation for details [35]).

4.2.1 Fast Fourier Transform

Conceptually, the full MR search space is $6N$ dimensional, i.e., the three rotational dimensions plus the three translational dimensions multiplied by "N," the number of models to be placed. An exhaustive $6N$ dimensional search becomes infeasible even with N in the low single digits, a problem that has spurred sparse sampling approaches [36]. These include the standard divide-and-conquer method of splitting the rotation and translation into two 3D searches (a rotation function and a translation function), but also include genetic algorithms [11] and *Monte Carlo* methods [12]. An advantage of performing the search as separate 3D rotation and 3D translation functions is that suitable target functions can be calculated by Fast Fourier Transform (FFT) [37]. The drawback of full MLMR targets is that they cannot be calculated by FFT, but targets for the rotation and translation functions that are suitable for FFT can be derived using the insights gained from the full MLMR treatment [38, 39].

4.2.2 Sequential Addition

One of the greatest strengths of the MLMR targets is that whenever a component is placed, the variances representing the remaining uncertainty in explaining the structure factors are reduced, thus increasing the signal-to-noise of the search for the next component. Thus, the natural way to build an MR solution of multiple components using MLMR is by sequential addition (*see* **Note 3**). Phaser's default search strategy is to run consecutive rotation and translation functions, iterating the two 3D searches over the number of models N, and using the structure factors calculated from the known (already placed) components to leverage the search for the next component.

4.2.3 Peak Selection

Since the Phaser search strategy is to iterate the rotation and translation function searches over the number of components, it is necessary to set the selection criteria for the rotation and translation function partial solutions so that the correct rotation(s)/translation(s) are included in the list of rotation(s)/translation(s) carried through to the next step or iteration. The process is to sort the rotation(s)/translation(s) at the rotation/translation step in LLGI order and to select the highest. Selection criteria rely on there being at least some signal at the partial solution stage, so that the correct rotation(s)/translation(s) will be sorted toward the top of the list. The ideal place to prune the list is just below the correct rotation(s)/translation(s), but of course this is not known. Rather, (by default) solutions are selected if they have a LLGI that is over 75% of the difference between the top and the mean of the search. This has the advantage that if the signal is high, then only the single top (correct) solution will be carried through, but many will be carried through if the signal is low. Other selection criteria are possible (see documentation for details [35]).

5 Identifying MR Solutions

Since MR is a search procedure, the correct solution is identified by the signal-to-noise of the correct placement. The correct placement is obvious when a single point in the $6N$ dimensional search has a high LLGI value that is clearly discriminated from all others. As previously discussed, this point is not normally found with a $6N$ dimensional search. Instead, search methods rely on intermediate steps systematically eliminating regions of search space as they home in on the correct placement. The correct placement is found as long as intermediate steps do not eliminate it from the search space along the way. It is not necessary for each step in the search procedure to have high signal-to-noise in and of itself.

If the signal in the rotation or translation function is low, the default peak selection criteria may be eliminating the correct orientation or position from the search space. Since the signal from the rotation function is generally lower than that of the translation function, an obvious first parameter to change is the number of rotation function peaks being carried through to the translation function. The default *FAST* search strategy in Phaser automatically reserves a second tranche of rotation function peaks to pass to the translation function if the first (upper) tranche fails to yield a placement with a translation function Z-score (TFZ; see documentation for details [35]) over 8. If the signal in the translation function is also low, it may be necessary to change the number of translation function peaks being carried through at intermediate stages of the search (see documentation for details [35]).

5.1 Success

The LLGI is a direct measure of the probability of a placement being correct [34]. There is also a direct relationship between the absolute value of the LLGI and its discrimination from the noise; the higher the LLGI for the correct placement, the higher its TFZ [34]. LLGI values for a model of the whole asymmetric unit greater than 60 generally have a TFZ of 8 and almost definitely represent a true solution [40]. The LLGI can be lower (50; TFZ ~ 7) and still indicate the correct placement of the first molecule in polar space groups, where there is one less degree of freedom [34]. Clear discrimination from the noise is an excellent secondary indicator that a solution is correct.

Many (different and wrong) placements with an LLGI over 60 (TFZ over 8) indicate some unexpected pathology in the data that breaks the assumptions of the likelihood hypothesis. Common pathologies include twinning (possibly complicated by pseudosymmetry) and errors in the space group determination.

5.2 Failure

Assuming that a MR solution exists using the models provided, if the correct placements of the components, as determined by superposition after structure solution, is not indicated by a LLGI clearly discriminated from the noise, then the MR with that set of components will never be conclusive, whatever search strategy is used. If many MR trials (*see* **Note 4**) do not produce a peak in the LLGI, then the crystallographer is justified in considering MR to have failed. However, putative model structures that may be somewhat superimposed on the target after structure solution, but whose placement is not indicated by a signal in the LLGI, will have very high phase errors, so that, had the placement been identified prior to structure solution, taking the MR solution forward to refinement would be extremely problematic.

5.3 Enrichment

If there are a small number of solutions with LLGI approaching 60 (TFZ ~ 8), it is likely that one of these represents the true placement. Approximately half of the solutions with an LLGI around 30 (TFZ ~ 5.5) are correct [40]. If the list of potential placements is small, then it is likely that the signal-to-noise of these possible solutions is also relatively high, and that the low likelihood is not due to pathology. The solution list is *enriched*, even though MR is not conclusive.

It may be possible to distinguish the correct MR solution in an enriched list by taking each potential solution through to refinement. This is the approach taken in Balbes [41]. Rosetta software [42] incorporates a wide convergence radius refinement method using approaches from ab initio modeling, and a pipeline for examining an enriched solution list from Phaser is implemented in phenix.mr_rosetta [43].

Anomalous data (e.g., from S-SAD, Se-SAD or a heavy metal soak) may also help to find the correct MR solution in an enriched list, even if the anomalous substructure has not been determined. In the MRPM (MR parameter matrix) search, the putative MR solutions from an enriched list are used to phase an anomalous difference Fourier and the MR solutions scored with respect to the peak heights in the resulting map [44]. If an anomalous substructure has been determined independently of a MR solution, but the resulting phases do not yield interpretable electron density, then the phases may still be good enough to identify the correct MR placement, simply by calculating the phase correlation between the experimental and (putative) MR phases. If both experimental and MR phase information is available, then phase combination will help bootstrap structure solution (*see* **Note 5**).

5.4 Persistence

If MR is failing, the crystallographer will have many MR trials from which to draw additional information. If the correct solution is somewhere at the top of the LLGI list in a number of trials, then considering the results of many trials in totality can identify the correct solution by the *persistence* of a solution across trials. This process was first introduced to MR in the context of looking across multiple rotation functions for the correct orientation [45], and later for translations [46] using AMoRe. The identification of similar placements can be done in real space, by clustering rotations and translations, or, for translations, in reciprocal space, by looking for phase correlations [47]. If done in real space, it is advisable to pre-align all homologous model structures so that high-scoring orientations and translations from different models can be compared easily.

5.5 R-Value

When all components of the asymmetric unit have been placed, it is usual to calculate other scores to test their validity, particularly the R-value ($\Sigma||F_{obs}|-|F_{calc}||/\Sigma|F_{obs}|$). The theoretical R-value for a random distribution of atoms, i.e., a maximally incorrect solution, is 0.586 [48]. In practice, wrong solutions have R-values somewhat lower because the absolute scale of F_{obs} is not known, and $|F_{obs}|$ is scaled to $|F_{calc}|$. However, the R-value need not be lower than 0.586 immediately after MR. A notion that the R-value should be low after MR is equivalent to a notion that the R-value should show a signal for MR, and therefore would make a good target function, an idea abandoned with the introduction of the correlation coefficient [49, 50] even before the introduction of MLMR. The R-value takes no account of the errors in F_{obs} and F_{calc}, and is only a useful indicator when the phase error is small. Most models, even when correctly placed by MR, will still have very considerable errors. However, the R-value does give an indication of how straightforward will be the progression into model building and refinement. At values less than 0.40, the R-value becomes a reliable indicator of good phases. MR solutions giving high *R*-values will require advanced techniques to refine. For a detailed discussion of refinement, *see* Chapter 23 by Nicholls, Kovalevskiy and Murshudov.

5.6 Termination

If the composition of the asymmetric unit is uncertain, it can be difficult to know when all components have been placed and the MR search can be terminated. The termination problem is usually solved when the signal-to-noise for adding components, which should increase with each additional component added, suddenly disappears and/or there ceases to be space for additional components in the asymmetric unit. A necessary but not sufficient condition for an MR solution is that the components form a connected lattice. In the end though, the MR search is only definitively terminated after the structure has been refined and passed validation tests.

6 Models

The important criteria for MR searches have been the subject of rules-of-thumb about sequence identity between model and target, editing of the model, and required size of the model [18, 51–54]. Only some of these have been systematically studied (e.g., editing of the model [55]). The properties of the LLGI clearly indicate the veracity or otherwise of these rules-of-thumb for MLMR [34]. Contrary to these traditions [18, 51, 52], there are no generally applicable cutoffs in sequence identity or root-mean-square deviation (rmsd) of a model to the target for successful MR. The sequence identity per se is irrelevant, except in that it allows homologs to be identified and an initial estimate to be made of the rmsd. Exactly how low the rmsd between the model and the target needs to be for success depends on the other parameters, particularly the model completeness (fraction of the scattering) and the number of reflections. High rmsd can be compensated by high model completeness. Low completeness can be compensated by low rmsd between model and target, and a large number of reflections (Fig. 2). Of course, the rmsd of the model to the target cannot be predicted reliably before MR. Strategies described here are designed to minimize the rmsd prior to MR.

6.1 Model Improvement

Because the rmsd of the model to the target cannot be predicted reliably before MR, the best model of all the alternatives cannot be chosen reliably either. Having the best possible starting model will not only facilitate MR, but will also make subsequent refinement and rebuilding much easier, so it is well worth spending some time evaluating a variety of alternative models, especially in difficult cases. MR models can be derived from different template structures in the PDB, processed in different ways by pruning, remodeling, or collected into ensembles, with or without trimming to a conserved core structure. In testing many models, it can be very helpful to use MR pipeline software such as MRage [28], which compares and combines results from many models in parallel.

6.1.1 Modeling

Techniques developed for ab initio modeling of protein structures have come of age for improving structures for MR. Application of chemical force fields can improve the structure to the point where the rmsd is low enough to find the solution [56]. Modeling specifically for MR is implemented in Ample [32]. For a detailed discussion of ab initio methods, *see* Chapter 19 by DiMaio.

6.1.2 Normal Mode Analysis

Conformational change in proteins is particularly problematic for MR. The crystallographer may expect conformational change—even be hoping to probe it—from previous studies of the macromolecule or macromolecular complex. Conformational change in proteins has been shown to be modeled by normal-mode analysis of the elastic network model [57–61]. One or more normal modes may contribute to a given conformational change [60, 61]. Perturbations along normal modes were first used successfully to find MR solutions with *AMoRe* [62]. Neither the normal modes that model the conformational change nor the perturbation distance along the modes are known in advance; multiple perturbed models need to be generated with different normal mode combinations and perturbation distances. By chance, one of the conformations generated may have a lower rmsd to the target structure than the original model, and hence yield a signal in MR. However, it is necessary to sample hundreds or even thousands of possible perturbations in order to sample conformational space finely enough to generate a good model. Normal mode perturbation of protein structures in rmsd increments can be performed with Phaser (see documentation for details [35]).

6.1.3 Conformational Sampling

Some families of proteins have been intensively studied and are present in the Protein Data Bank in many different conformations. Kinases are a prominent class of protein for which the structure with the highest sequence identity in the PDB is not likely to be the one with the lowest rmsd to the target. Kinases undergo a conformational change upon NTP binding, but the changes are not well-modeled as a simple change in disposition of domains [63]. There are thousands of kinase structures in the PDB (including serine/threonine, tyrosine, histidine, receptor and non-receptor types), and these represent many different kinase conformations. Although not all are unique, the conformational sampling they represent can be used to solve MR problems by trying all kinase structures regardless of sequence identity.

6.1.4 Wide Search

Wide Search MR (WS-MR) [29] is the extension of the database of search models to the entire PDB. The CPU intensive search becomes tractable through the use of national supercomputer grids. The approach allows optimization of the MR search model by brute force: it does not rely on sequence identity to identify models. As a consequence, MR solutions can be found with very low sequence identity and/or sequence coverage. As implemented through the SBGrid [64], the LLG and TFZ scores from Phaser are initially used to filter possible solutions, and then the structures that generate these solutions clustered by fold to find folds that persist in the solution list (*see* **Note 6**).

6.1.5 Ensembling

A reduction in model errors can be achieved using an ensemble of superimposed structures that are similar. The result of the *ensembling* [21] procedure is a set of E_{calc}, which are used *in lieu* of structure factors from a single model. The errors in the ensemble E_{calc} are lower than those of each model individually. The assumption is that some parts of the structures will be systematically closer to the target than others. The scattering from these sections will be reinforced, while regions that differ will be down-weighted. If all the structures were weighted identically, ensemble E_{calc} would be equivalent to summing structure factors from the components and dividing by the number of components, or equivalently, taking the N superimposed structures, each with the fractional occupancy $1/N$. A more sophisticated approach is to weight the structures according to the expected rmsd to the target structure. An ensemble of structures has been shown to be particularly effective when there are a number of low sequence identity models available for the target structure [21, 65].

6.1.6 Bulk Solvent

Ordered atoms are only part of the scattering matter present in the crystal. Also present are disordered atoms in the bulk solvent. Solvent corrections to the structure factors were originally developed for refinement [8, 66] where they clearly improve map quality. A mask-based solvent correction has been successfully applied to fast translation searches in MR [67] with *AMoRe* [49] and with *CNS* [8]. Phaser has the option of applying a mask bulk solvent correction throughout MR. The model structure factors are calculated for structure factor interpolation, by placing the model in a large *P*1 cell with less contrast to the surrounding solvent as compared to the default calculation, which places the model in vacuo. Models that come from electron microscopy reconstructions may already include the average contribution of vitreous ice [68]. When bulk solvent is included in the model, the error in the calculated structure factors at low resolution is reduced, and hence the σ_A at low resolution should be increased. Phaser's bulk solvent *siga fsol* parameter should be decreased from the default (see documentation for details [35]). Altering the bulk solvent terms can rescue failed MR in some cases [68].

6.2 Model Completeness

There is a penalty to the LLGI associated with reducing the size of the model. However, reducing the size of a model can be advantageous if the atoms in the wrong (relative) positions can be removed prior to the search.

6.2.1 Pruning

The longest standing method for removing atoms in the wrong (relative) positions is pruning the amino acid side chains of the model. Amino acids that are not conserved between model and target should be trimmed back to a common core. In the simplest analysis, this is the C_β atom (polyalanine), but more complex analysis can add one or two atoms further along the amino acid side chain where there are conserved atoms between common rotamers of spatially equivalent model and target amino acids. Pruning of the model has been shown to be decisive in the solution of MR problems [55]. Pruning can be performed with CHAINSAW [69] from the CCP4 suite or *phenix.sculptor* [70] from the phenix suite.

6.2.2 Domain Analysis

Protein domains are variously defined, for example in terms of sequence motifs, functional elements or evolutionary modules. For the purposes of MR, a domain is a structural element within which the atoms are fixed relative to one another between model and target, and hence are suitable sub-structures for MR. There are often changes in the disposition of domains in multi-domain proteins sometimes related to function, but also simply due to flexibility and crystal packing forces. When the protein (or a homolog) has been solved in two or more conformers, the structurally invariant regions can easily be identified [71–74]. It is more challenging to identify domains for MR when the structure of only one conformer has been solved. In simple cases, visual inspection may be sufficient to identify potential rigid domains. Various automated approaches have been taken including considerations of surface area [75], molecular dynamics simulations [72], TLS group analysis [76], and normal-mode analysis [77], among others. Phaser implements the SCEDS procedure [74].

If a model is split into N domains, the search for that component becomes $6N$ dimensional to allow each set of atoms with correct (relative) positions to be optimally positioned. The signal for the correct placement of all the domains may not be discriminated from noise until the final component is placed. In difficult cases, it is therefore advisable to search for all components in one run of Phaser, allowing the software to build a complete solution component by component, and optimizing the signal from each component as it progresses.

6.2.3 Oligomers

If the protein or proteins in the crystal are known to form oligomers, either hetero-oligomers or homo-oligomers, then searching with models that have the target's oligomeric arrangement will increase the signal. Dimers, trimers, tetramers, and hexamers with point group symmetry are able to crystallize with one unit (which may itself be made up of a protein assembly) in the asymmetric unit of the crystal in space groups with the same two-, three-, four-, or sixfold point group symmetry. Fibres, which are infinite chains of proteins with screw symmetry, must crystallize so that the crystal screw symmetry generates the infinite chain. Searching with an oligomer that has more scattering matter than that present in one asymmetric unit, where the oligomer is placed on a special position with respect to the crystal symmetry, is supported in Phaser.

6.2.4 Brute Searches

In difficult cases, the full MLMR targets can be calculated point by point on rotational and translational grids [21], rather than using the likelihood enhanced fast rotation/translation functions and FFT [38, 39]. This is termed a "brute" search. Since the full likelihood functions are slow to compute, brute searches are most useful when the search space can be restricted to a particular set of angles/coordinates near a particular placement. Such a scenario occurs when searching for a multi-domain, flexible protein for which a model of the entire target exists. It is often possible to place the large domain(s) but not the small domain(s). The approximate placement of the small domain(s) can be inferred from the placement of the large domain(s). Performing a brute rotation/translation searches restricting the orientation/position to angles/coordinates within a few tens of degrees/Ångstroms of the position relative to the (large) placed domain(s) often finds the correct placement of small domain(s) with high signal-to-noise, using the power of MLMR. In practice, it is usually sufficient to carry out a brute-force limited search of orientations combined with a fast translation search over the entire volume, because the signal is much stronger for the translation search than the rotation search. Obtaining a solution consistent with connectivity between the domains increases confidence in the correctness of that solution. The brute search method can be thought of as a wide-convergence-radius rigid-body minimization.

6.2.5 Fragments

If the number of reflections is high then it becomes feasible to use very small but accurate (low rmsd) search fragments for MR. Elements of secondary structure can prove useful generic models. Helices are particularly suitable as they are very regular over lengths of several turns; beta sheets have twists that distort the disposition of atoms within a short stretch of amino acids. This approach is particularly effective in solving coiled-coil structures [78], where MR with Phaser often fails to dock the sequence onto the helix, probably due to the strong helical modulations of the diffraction pattern. That small accurate fragments can be used to solve MR problems when whole accurate models are not available is the basis for Arcimboldo [31], Ample [32], Arcimboldo-Borges [79], and Arcimboldo-Shredder [80].

6.2.6 Search B-Factor

Model components differ not only in the rmsd to the target, and model completeness, but also the relative B-factor. The components with low B-factors are generally found first in any search. The high B-factor components can be very hard to place, because these contribute less to scattering at high resolution than other components. The relative B-factors of components are not known before structure solution, but if later components in a search are proving difficult to locate, high B-factors should be suspected. This is particularly likely if one copy of a component has been found, and therefore shown to be a good model. In Phaser, the average B-factor of all ensembles (and members of an ensemble) is, in effect, set to the Wilson B-factor. Thus, by default, differences in average B-factor between models do not affect MR, but Phaser has the option to explicitly add a relative B-factor to the search for a component to down-weight the structure factors at high resolution (see documentation for details [35]).

6.3 Model Errors

Model errors are important parameters in the likelihood targets. Correct estimation will improve signal-to-noise in borderline cases. The model error, σ_A, is computed from the estimated rmsd of the coordinates between model and target and the fraction scattering that it represents.

The LLGI for the placement of a component should be positive, and should increase as components are added. If it is negative or decreasing, it means that the parameters of the likelihood function are predicting the data worse than would a collection of random atoms. The errors are underestimated, too optimistic about how well the model can predict the data: the completeness is being over estimated and/or the rmsd of the coordinates is being underestimated.

6.3.1 Sequence Identity

Although not known exactly until after structure solution, the rmsd can be estimated from the sequence identity [81] or more accurately by also taking into account the size of the protein [40]. Optimization of the estimated rmsd can be the difference between success and failure in MR trials with low signal [40]. In a database of 3375 borderline MR cases, of which 504 were not solved using the rmsd expected from the sequence identity between model and target, a third of the failed cases could be rescued by varying the rmsd from the expected value [40].

6.3.2 Composition

The fraction scattering of a given ensemble is calculated in Phaser from the atomic composition of the input ensemble and the total atomic composition of the asymmetric unit, usually entered as protein and/or nucleic acid sequence and number of copies. The asymmetric unit composition is thus an important parameter in MLMR. Increasing the composition of the asymmetric unit will decrease the fraction of the scattering accounted for by each component.

If the composition of the asymmetric unit is uncertain, then so too will be the fraction scattering of each component. If the asymmetric unit is assumed to have less scattering than actually present, then σ_A will be over estimated, and vice versa. The LLGI will be optimized when the composition is correct. In difficult cases, it will be necessary to perform MR not only altering the number of search components but also the composition.

6.3.3 Conformational Change

When modeling conformational change, the rmsd used to estimate the σ_A should be close to the rmsd expected to apply after successful structure solution, not the higher value expected between model and target before modeling the conformational change. If conformational change is being modeled by normal mode perturbations, then the rmsd between perturbations will give an estimation of the upper limit for rmsd of the best model to the target. Phaser generates normal-mode perturbed structures by rmsd increments for this purpose (see documentation for details [35]).

6.3.4 Atomic B-Factors

Although the *overall* scale of the B-factors of the model coordinates does not affect MR with Phaser (*see* Subheading 6.2.6), differences in B-factors *between* atoms in a model affect the relative contribution of the scattering of each atom to the calculated structure factors at different resolutions; scattering from regions of high B-factor are down-weighted at high resolution. The atomic B-factors should be set proportional to the expected mean-square displacement. Modeling expected coordinate errors along the polypeptide chain as B-factors, usually lowest in the core and highest on the protein surface, have been shown to dramatically improve the utility of homology models for MR [82].

6.3.5 VRMS Refinement

Phaser refines the coordinate errors (VRMS [40]) for each component in conjunction with the rotation and translation of the model. The VRMS will often refine to a lower value than the input rmsd for a correct solution. If VRMS values of a solution refine to a significantly different value than input, then repeating the search with the refined VRMS input from the start should increase the signal-to-noise of the rotation and translation functions.

6.4 Model Case Study: Antibodies

The approaches to optimizing a model are well illustrated by the long-standing MR problem of how to solve Fab antibody structures [83], with or without their protein antigens. The elbow angle, the angle between the variable (Fv) and constant (Fc) domains of the antibody, is highly variable [84]. If the data are high enough resolution (i.e., there are a large number of reflections) then Fab placement will be possible by splitting the Fab into Fv and Fc domains and searching for these consecutively, even using Fabs with low sequence identity to the target. Pruning the Fv and Fc to the core conserved with the target is always advisable. Because of the flexibility at the elbow angle, the B-factors of one of the domains may be high, causing problems for the MR. If the Fv domain is well ordered (due to binding to its well-ordered protein antigen), and hence is easily located by MR, then a partly disordered Fc may be found by increasing the B-factor in the search for Fc or by local brute search. If the data are not so numerous, then a good signal will only be obtained searching with the whole Fab and with the elbow angle of the Fab correctly modeled. The only correlation between elbow angle and sequence is via the subtype of light chain (κ or λ) [84]. The correct antibody hinge angle may be found among those Fab structures already in the PDB, or novel conformations may need to be generated with normal mode perturbations. If the data are even poorer, then the signal can be further improved by modeling the Fv. Modeling approaches for Fv domains regularly achieve an rmsd of 1 Å or better [85]. Searches may be necessary using a range of rmsd values, or an ensemble of Fv models may be useful. Placing the Fv and Fc domains correctly in the asymmetric unit can bootstrap the placement by MR of the protein antigen, or indeed phasing by other methods [86].

7 Data

Guidance about good data collection strategies becomes particularly relevant in difficult MR cases. For a detailed discussion of data collection strategies, *see* Chapter 7 by Dauter. Some problems arising from fundamentally bad data collection simply cannot be resolved by data processing and will be fatal to MR. The following discussion assumes that the data are correctly indexed, are free of overlaps and overloads, and that the σ_{Iobs} associated with an I_{obs} encapsulates the measurement error reasonably accurately.

Like model preparation, data preparation for MR has also been the subject of rules-of-thumb regarding the resolution of the data, the need for completeness of the low-resolution data, and so forth [18, 51–54]. Again, the properties of the LLGI clearly indicate the veracity or otherwise of these rules-of-thumb for MLMR [34]. If the data have no pathology, then, for a particular model, the LLGI depends only on the number of reflections, not the resolution of the data, or the completeness of the data in resolution shells (Fig. 2). This runs contrary to experiences with Patterson methods, where the completeness of the low-resolution data is critical to the success of MR [54, 87] and where high-resolution data are not essential [88].

7.1 High-Resolution Data

Although the number of reflections is a key factor in determining the LLGI, reflections with a resolution higher than 1.8 times the rmsd of the model have σ_A values so small that they contribute insignificantly to the total LLGI. Estimates of the rmsd for MR show that for sequence identities of 15%, the rmsd is estimated as

1.5 Å for small models and up to 2.5 Å for large (1500 residue) models [40], which implies that data better than 2.7 Å for small models, and 4.5 Å for large models, will only increase CPU time. However, in cases where MR is not expected to succeed based on the most likely rmsd, success will only be found for models that happen to be somewhat better than expected, so it can help to run trials with optimistic values for the rmsd. An rmsd of only one standard deviation below the expected value (0.2 times the expected value [40]) increases the useful resolution by nearly 40%. If the VRMS is lowered in refinement, it will benefit from the additional data. Deliberately truncating data, for example at 3.5 Å, can lose critical signal for marginal cases. Phaser sets the resolution limit optimally for the rmsd input, and changes the resolution limit during the course of MR depending on how much signal is present. Using the same argument as in Subheading 7.1.1, MR with small accurate fragments will benefit greatly from high-resolution data. On no account should resolution be truncated in the search for helices or other small structural motifs. The rmsd for models consisting of single atoms is zero, and hence single atom MR is possible with very high resolution data [34].

7.2 Measurement Error

At the diffraction limit of the crystal, the issue for MR becomes measurement error. Since the demonstration that useful information can be extracted from very weak diffraction data in refinement [89, 90], and the introduction of pixel counting detectors, data are now frequently integrated beyond traditional resolution limits (e.g., merged $I_{obs}/\sigma_{Iobso} > 2$ in the outer shell). The LLGI will down weight the contribution for the poorly measured reflections at the diffraction limit of the crystal. Using LLGI, adding data at the high-resolution limit with high experimental error will not bias the MLMR target in the way that amplitude-based likelihood targets do, and, at the same time, will allow all well measured reflections, regardless of the overall I_{obs}/σ_{Iobs} in their resolution shell, to contribute to structure solution. With the use of LLGI, it should not be necessary to vary the high-resolution limit for MR in an attempt to get a solution, unless there is some pathology in the data at high resolution (e.g., an ice ring near the resolution limit). However, in the extreme of integrating data well beyond any reasonable diffraction limit, e.g., 2.0 Å ($I_{obs}/\sigma_{Iobs} = 2$) data integrated to 1.0 Å, the integration and scaling programs may do a poorer job of estimating the intensities and standard deviations, and some degree of restraint should be exercised.

7.3 Low-Resolution Data

If MR is failing and the data are poor, then improving the data should be a priority. The higher the resolution of the data, the more options for attempting MR with smaller, more accurate fragments. Low-resolution data below 15 Å is disproportionately affected by the poorly modeled diffraction from the solvent, and so has lower σ_A values than do data around 6 Å. Mid-resolution data thus give more signal per reflection than do low resolution data. Unlike Patterson based MR, high completeness at low resolution is not particularly valuable for MLMR.

7.4 Completeness

Because the LLGI is dependent on the number of reflections, it is obvious that collecting complete data will maximize the number of reflections to the diffraction limit of the crystal. Missing data affects the map resolution: the electron density is convoluted with the Fourier Transform of the mask of the missing data. Randomly missing data lower the effective resolution of the map isotropically. If data are systematically missing in a wedge, then the effective resolution in the plane perpendicular to the wedge will be lower. MR orientation and position parameters will be less accurate in the direction where the effective resolution is lowest.

7.5 Intensities

Structure factor amplitudes are normally generated from intensities by the French and Wilson [22] *truncate* procedure. In some structure solution pipelines, data are subjected to the truncate procedure by default, and all subsequent steps are performed with these amplitudes, however this transformation introduces serious biases in the likelihood targets. The LLGI targets abrogate the need for any transformation to amplitudes during MR, and it is important to input the data to Phaser in terms of intensities rather than amplitudes [20].

7.6 Alternative Datasets

If data are generally poor, it is advisable to forward a number of differently processed datasets of merged intensities for MR trials. This is a good strategy in the presence of radiation damage, where it is often not clear where to cut the data with dose to balance merging R-values against multiplicity and completeness. Differently processed or merged data sets can be used to test the *persistence* of a solution (*see* Subheading 5.4).

7.7 Space Group

Patterson based likelihood targets are less effective for higher symmetry space groups, due to the presence of inter-molecular vectors in the Patterson calculated from the data. As the symmetry increases, more and more inter-molecular vectors crowd the observed Patterson, and the signal is reduced. For MLMR, higher symmetry also increases difficulty in structure solution, because greater uncertainty in adding up structure factor contributions from symmetry-related molecules with unknown relative phase increases the variance of the rotation likelihood target. The space group has no equivalent effect on the difficulty of the translation step in MLMR.

7.8 Alternative Space Groups

Enantiomorphic space groups cannot be distinguished in the data processing stage, only by structure solution. Space groups that only differ by screw symmetry can be distinguished by the presence of systematic absences, but if the axial data are weak or missing, then the assignment of screw axes is not certain, and again, the correct space group can only be distinguished by structure solution. Clear identification of space group among a list of alternatives is a good secondary indicator of the validity of a solution.

7.9 Anisotropy

There are often differences in long-range order in different directions in reciprocal space. MLMR relies on comparing structure factors computed from a model isotropically scattering atoms with the observed data. If the implicit assumption of isotropic scattering is wrong, MLMR will not score the placements correctly and structure solution will fail. The anisotropy parameters are refined by fitting the structure factor intensity to the Wilson distribution, and these parameters are used to correct the data for anisotropy and allow structure solution to proceed as for isotropic data. The anisotropy correction is applied to both the data and the experimental errors in the data. The anisotropic correction factors calculated by fitting the data to the Wilson distribution will not be as good as those that can be calculated once the atomic model is known. Anisotropically corrected structure factors used for MR should not be passed to refinement programs.

8 Non-crystallographic Symmetry

8.1 General NCS

There is nothing particularly special about the presence of general non-crystallographic symmetry in determining the solvability of the problem by MR, as compared to any other MR problem with multiple components in the asymmetric unit.

The Matthews coefficient [91], originally established from a study of protein content in protein crystals, has been reinvestigated for crystals of nucleic acid-protein complexes and nucleic acid alone [92, 93]. The most likely number of macromolecules in the asymmetric unit is the number that gives the most likely solvent content. When the most likely number of macromolecules is one or two, it is well determined, but as the number of macromolecules increases so too does the uncertainty.

Clues to the crystal composition can be gleaned from sources other than the crystal data in hand. The number of copies in the asymmetric unit may be informed by the oligomeric state of the complex in solution, combined with the presence or absence of pure rotational symmetry operators in the space group. Light scattering, native gel electrophoresis, ultracentrifugation, and electron microscopy can indicate oligomeric state. However, differences between the buffers in which these experiments are performed (such as salt and pH), physical forces, and possible proteolysis, mean that these experiments are not necessarily good indicators of the oligomeric state in the crystal.

Information about the NCS can also be gleaned from the self-rotation function (SRF [94]). The SRF is most intuitively specified with three polar angles: the azimuthal angle and the zenith angle, which specify the direction of the rotation axis; and κ, the rotation about this axis. When there are multiple copies in the asymmetric unit, the SRF is complicated and generally not interpretable, unless there is rotational symmetry. The κ section of the peak in the SRF shows the rotation order = $360/\kappa$, e.g., two folds will appear as peaks on the 180° κ section. If rotational symmetry of a given order is clearly present, then the number of copies in the asymmetric unit is likely to be a multiple of the rotation order.

If the number of copies of the macromolecule in the asymmetric unit is not well determined, a lack of certainty becomes a significant problem for MR through not knowing when to terminate the search, and not knowing the fraction scattering of the components as the search is progressing.

Although the presence of NCS can increase the difficulty of MR, it has the compensating advantage of enabling NCS averaging after MR, which will remove some of the model bias. This is especially valuable in low-resolution structure determinations.

8.2 Translational NCS

Translational non-crystallographic symmetry (tNCS) arises when two or more copies of a macromolecule or macromolecular complex are present in the asymmetric unit in the same orientation. The presence of tNCS modulates the diffraction pattern in a way that is problematic for likelihood functions, because, like anisotropy, it violates the implicit assumption behind likelihood targets that the data follow an isotropic Wilson distribution. Macromolecules related by tNCS will have an associated peak in the native Patterson. The magnitude of the Patterson peak is a measure of both how exactly the translation vector models the translation between all atoms in copies of the macromolecule and the strength of the resulting diffraction modulation. Peaks in the native Patterson more than 20% of the origin peak are a good indicator of macromolecules being present in approximately the same orientation (up to 10° rotation for an average size protein), and for the modulation being a significant hindrance to the likelihood targets.

An important aspect of accounting for tNCS with likelihood is the modeling of the errors. The tNCS is characterized not only with a vector, but also with parameters describing the deviation from simple translations of identical coordinates between the tNCS copies. The naïve, non-likelihood approach, of modeling the tNCS as a simple translation of one structure by the tNCS vector, is inadequate for structure solution in the majority of crystallographic problems with tNCS. The likelihood correction to the tNCS is performed by refining expected intensity factors for each reflection, derived from the tNCS model of tNCS vector(s) and errors. The expected intensity factors are then used in the likelihood functions as usual, and in many cases structure solution becomes straightforward [95].

When tNCS is present and can be characterized and the intensity modulations accounted for, it can be considered an advantage for MR, because there are fewer independent copies in the asymmetric unit to place versus the same asymmetric unit contents without tNCS. On the other hand, tNCS reduces the power of NCS averaging to improve phase quality [96].

8.2.1 tNCS Order

Frequently, tNCS associates *NMOL* macromolecules in the asymmetric unit in a series of vectors that are multiples of 1, 2, 3 ... (*NMOL* – 1) times a basic translation vector (*TVEC*), with *NMOL*× *TVEC* being a unit cell translation, possibly along a unit cell diagonal. In this case the tNCS represents a pseudo-cell and is known as commensurate modulation. The integer *NMOL* is the order of the tNCS. Trying to find the related set of vectors by inspection is complicated by the Patterson symmetry and cell translations. The series will not generally have all peaks the same height. Lower peaks in the vector series represent relative rotations between vector-related molecules that are larger, and may even be missed by the default 20% origin cutoff. Phaser performs a Fourier analysis of the Patterson to assist in finding the order of tNCS and the translation vector in cases of commensurate modulation.

8.2.2 Pairs of Molecules

If there is a single peak in the native Patterson, it represents macromolecules clustered into two groups (*NMOL* = 2) related by a single tNCS vector. In these cases, Phaser can refine not only an rmsd between tNCS related copies but also a specific relative orientation between the macromolecules in the two groups. Starting from the Patterson translation vector, an estimate for the rmsd between copies, and a small number of initial rotational perturbations, the parameters are refined against the Wilson distribution to optimize the expected intensity factors for use in the LLGI.

8.2.3 Complex tNCS

If there are many macromolecules in the asymmetric unit but they are not all related by tNCS, or there are sub-groups of macromolecules related by different tNCS vectors, then the modulations of the expected intensities due to the tNCS will be much less significant than for commensurate modulation or for pairs of macromolecules. In these cases it is possible that structure solution will be achieved without any tNCS correction factors being applied. Indeed, searching exclusively for tNCS-related multiples when some molecules are not related by tNCS will cause structure solution to fail.

If ignoring tNCS fails to give a solution, then the solution must be approached stepwise. Firstly, consider the highest native Patterson peak, apply the associated tNCS correction factors, and locate all the molecules with this tNCS. Then, fix these components, and take the second independent native Patterson peak, apply the correction factors associated with it, and locate the second set of molecules. Finally, turn tNCS correction off to find any orphan molecules.

8.2.4 Helices

Crystals of nucleic acid, particularly DNA duplexes, and α-helical coiled-coils, show clear helical modulation of the diffraction pattern, and have correspondingly large Patterson peaks, due to the helical repeats. The direction of the helices can be inferred from the large Patterson peaks alone. Phaser's tNCS correction should not be applied. Arcimboldo [31], which follows MR with density modification and chain tracing, solves a high proportion of coiled coil structures despite the difficulties in the MR caused by the helical modulations.

9 Twinning

In general, MR works well with twinned data. The errors in the calculated structure factors need to be only slightly lower than would be needed for untwinned data from the same crystal form. Twinning may not even be suspected [97]. For a more detailed discussion of twinning, *see* Chapter 8 by Thompson.

9.1 Merohedral

With hemihedrally twinned data, Phaser should produce two sets of solutions that are equivalent under the twin operator, although they may not be on the same origin. However, if the twin fraction α is even slightly less than 0.5, Phaser may only give one solution. For more than two twin domains Phaser may (or may not) produce more than one solution, related by the twin law(s). To test for the twinning with a particular twin operator, twin related solution(s) can be generated manually and the LLGI calculated to compare with the original solution.

Twinning is detected with a range of tests [98]. Twinning tests that rely on structure factor intensity statistics work poorly in the presence of anisotropy and tNCS, but if the anisotropic and tNCS intensity modulations are corrected as described above, these tests become reliable [99]. Phaser reports *p*-values that will suggest whether twinning is present after removing the systematic intensity modulation effects [95].

The main problem with merohedral twinning in the context of MR occurs when perfect twinning causes the space group to be misidentified and the data are merged in a higher symmetry than the true symmetry. MR will then either fail outright, give a partial solution, or the *R*-value of what appears to be a full solution may stall during refinement, with the electron density showing breaks and spurious features that cannot be corrected by model building.

It can be difficult to detect perfect twinning masquerading as crystallographic symmetry, unless the asymmetric unit volume is too small to contain even a single copy of the macromolecule. If the data are merged in too high symmetry, the twinning tests that depend on twin laws, which compare reflections that are equivalent according to a possible twin law, cannot be performed. Only the tests for twinning that consider intensity statistics, such as the moment test in Phaser, will still indicate that twinning is present.

If twinning is indicated by the intensity statistics, and MR/refinement fails, then the true symmetry is probably lower. However, any or all of the symmetry operators could correspond with the twin operator(s). Phaser reports all the subgroups of the current space group, any of which could be the true space group in the presence of twinning. Especially in higher symmetry space groups, the number of subgroups can be very considerable, as screw symmetries also need to be considered. These can be systematically investigated by merging the data in all the lower symmetry point groups. However, if the twinning is perfect, the data can simply be expanded to lower symmetry without it being necessary to remerge the data. MR pipelines can run numerous jobs simultaneously [28].

If the MR model is good, then solving the structure in *P*1 and using the symmetry of the resulting structure to determine the true space group can bypass the expansion to all subgroups. The calculated structure factors from the MR model in *P*1 are tested to see if they obey higher symmetry [100–102].

9.2 Reticular Merohedral

In reticular merohedry, the reciprocal lattices of the twin domains superimpose exactly, but for only a fraction of the reflections. A characteristic warning sign is a pattern of "bizarre" apparent systematic absences, which are not consistent with any space group [103, 104]. The problem can be one of unit cell and/or space group determination because overlapping lattices may be interpreted as a single lattice. Overlapping lattices can be interpreted as a large unit cell, or make a centred space group appear primitive. Indexing twin-related lattices as one will cause MR to fail. If the strongest twin component can be indexed and integrated independently of the others, and enough of these reflections are unaffected by twinning, MR should be possible using the unaffected reflections alone. Data may be augmented by adding the intensities of the common reflections divided by the number of twin contributors [103].

9.3 Pseudo-merohedral

Pseudo-merohedrally twinned data are equivalent to merohedral twins for the purpose of MR. The difficulty with pseudo-merohedral twins is in the data integration step. If the difference in unit cell dimensions is very small, and the reflections are overlapping one another on the detector, then the aim of integration is to mask the twinned reflections into a single reflection so that reflections of the same index are integrated as one, so as to, in effect, force the data to be merohedrally twinned.

9.4 Static Disorder

Twinning is just one of the crystal pathologies of crystal disorder. On the other end of the continuum is statistical disorder, where the mosaic blocks are small compared to the coherence length of the X-rays. MR with statistical disorder is likely to produce several solutions with high signal-to-noise with severe packing clashes. Refinement will involve setting the occupancy of overlapping components in the asymmetric unit to appropriate values.

10 Packing

Explicit checks for the presence of overlap among and between the crystallographically and non-crystallographically related components in the unit cell are powerful additional criteria for the selection of MR solutions. The problem with these overlap tests, also known as packing tests, is that any errors in the MR model will mean that the model will not fill the same molecular volume as the true structure it represents, and so there are errors in the packing tests that cannot be properly accounted for.

A measure of the packing is given by the FFT-calculated overlap function [105], which quantitates the total volume of the unit cell that is occupied. This is a continuous function, and has been used to weight the translation function score in proportion to the total volume occupied, with the effect of (potentially) reordering the translation function peaks in MOLREP [106] and AMoRe [107]. The overlap function becomes less useful as the number of components in the asymmetric unit increases. If there is only one component in the asymmetric unit, then any reduction in the total volume occupied can be fully attributed to overlap between crystallographically related copies of that component. If there are many components, then the reduction in the total volume occupied may be entirely due to overlap of one component, or some overlap of them all. The latter should be accommodated, but the former should not. The two cases can be distinguished by counting atomic (or atomically representative) contacts, and this is the basis of the packing analysis in Phaser. Solutions are excluded if the pairwise overlap between two components is more than a given percentage. The Phaser packing test is therefore pass/fail, rather than a continuous function, and does not reorder the translation function peaks.

10.1 Trace of Coordinates

It is prohibitively slow to include all atoms in the analysis unless the model has less than about 1000 atoms. Instead, "trace" atoms represent the volume of the components. These can be Cα atoms for protein and a selection of phosphate backbone and base atoms in nucleic acid. Alternatively, the trace atoms can be abstracted to a set of points filling the molecular volume, for example to points on a hexagonal grid within the van der Waals volume of the protein. The default trace used to represent a set of coordinates adjusts to the size of the macromolecule so that the volume is represented by a maximum of 1000 trace points (see documentation for details [35]).

10.2 Trace of Maps

Electron density maps can be used to define a model for MR in Phaser, using similar input to that for a coordinate-based definition of ensembles. Putative solutions from electron density maps are tested for packing in Phaser by filling the Wang volume [108] with a hexagonal grid of points and proceeding as for the packing of atomic models.

10.3 Explicit Trace

By default, the trace of an ensemble used for packing is derived from the ensemble coordinates or Wang volume. However, it is possible to input the trace to be used so that it is defined independently of the coordinates or electron density input for calculating structure factors for the likelihood targets. This can be useful if searching with small fragments, where it is possible to exclude a larger volume around the search fragment in the packing tests points (see documentation for details [35]).

10.4 High TFZ Solutions

Solutions that have high LLGI, indicating that a placement is correct, but which fail the packing test, need to be investigated more closely. A second copy of a component may be placed on top of an identical, previously placed component if the component has a B-factor significantly lower than the Wilson B-factor: the second copy attempts to model the missing scattering. Significant overlap may also be caused by the presence of static disorder. Solutions with minor overlaps may be excluded because the allowed percentage for overlap is too strict given the accuracy of the model. Although the solution may be accepted by being more accommodating of overlap, ideally the model should be edited to remove atoms that are not shared between the model and the target, which will also increase the LLGI.

10.5 Packing During Translation Function

A high LLGI solution that does not pack influences the results of the translation function if it is the top-most peak from the translation function, since, (in the default selection criteria), the top peak is taken as the reference for the cutoff LLGI value for acceptance. If the LLGI of this top peak is much higher than any others, then it may be the case that no other solutions are output from the translation function, causing structure solution to fail in the subsequent packing test due to the loss of other candidate solutions. To avoid this case, a packing test is performed on the top solution *during* the translation function and the top peak is discarded if the overlap of any component is more than 50% of the volume. Alternative placements due to static disorder will likely be lost in this process.

11 Electron Microscopy Maps

Improvements in detectors and reconstruction software now allow atomic resolution electron microscopy (EM) imaging [109]. With high-resolution images from electron microscopy now available, it is possible to bring X-ray crystallography and electron microscopy together in two ways. Structures solved by X-ray crystallography can be docked into the high resolution EM maps, in a process analogous in many ways to MR, but this is not the subject of this review. Secondly, the electron microscopy images can be used as models for MR. This is possible even if the electron microscopy imaging has not (yet) yielded an atomic resolution structure. Since the model used in the likelihood targets is represented by the calculated structure factors, it is trivial to replace the structure factors calculated from a model with the observed structure factors from EM. The likelihood functions are then deployed without modification.

An important additional consideration when using EM maps as models in MR is that the scale of the electron micrograph may be miscalibrated by several percent [68]. Miscalibration will at the very least add noise to the MR search, and will often prevent structure solution. The MR search should be done with the scale of the EM map varying +/−10%.

The resolution of the search using EM as a model is restricted by the resolution of the EM map. Phase extension utilizes NCS averaging (if present) or other density modification processes. It may be necessary to resort to experimental phasing to get high-resolution phase information; however, derivative screening and heavy atom location will be greatly facilitated by the phases to low resolution, for example using MR-SAD in Phaser.

A detailed description of the protocol for phasing with EM maps has been published [68].

12 Notes

1. The term "molecular replacement" was coined by Michael Rossmann [110] for methods that exploit non-crystallographic symmetry for phasing, whether within or between crystal forms. However, it has come to mean the case where an unknown structure is solved with a known structure [111]. Other uses of the technique are now referred to as "non-crystallographic symmetry averaging" and "cross-crystal averaging."
2. Low homology models are detected with multiple sequence alignment methods and have benefitted greatly from whole genome sequencing. For a detailed discussion of sequence database searches, *see* Chapter 19 by DiMaio.

3. The natural way to build a solution by Patterson methods is to identify the correct placement of each component independently before assembling the solution. While it is possible to account for partial structures with Patterson translation functions or the correlation coefficient, accounting for partial structure in Patterson rotation functions is much more difficult. Patterson subtraction methods for the rotation function are highly susceptible to differences in B-factors between the component placed and the component remaining to be found, as well as coordinate differences. With low signal-to-noise for the rotation function, solutions are easily lost.
4. How many is "too many"? It depends on the time and computational resources available to the crystallographer, the possibility of better data becoming available, other options for structure solution, and significance of the project.
5. There are several other ways to combine experimental phasing with MR. If experimental phases can be determined (i.e., substructures found), then spherically averaged phased translation functions [112] and phased translation functions [113] can be used to dock models into the experimentally determined electron density [10]. If a MR solution is clear, then experimental phases can be extracted even from poor derivatives by using the MR solution to determine the substructure. The MR-SAD [114] (MR-single-wavelength anomalous dispersion) version of this technique can be performed in Phaser (see documentation for details [35]).
6. Wide Search MR can be used to resolve structure solution in cases when a protein contaminant accidentally crystallizes rather than the protein of interest. MR using models with sequence identity to the intended target will obviously fail [115]. Resources specifically designed for identifying contaminant proteins by MR have also been developed [116].

13 Conclusions

Just because MR has solved a structure does not mean that refinement will be straightforward. Because of the sensitivity of the LLGI target, MR solutions can be obtained when the phase accuracy is very low. Solutions with low phase accuracy will have model bias, and will struggle to show novel features in the electron density that could move structure solution forward. Even if MR is showing a clear solution, the approaches described here in the context of improving the models prior to MR, can also be used as an additional step between MR and refinement.

Advanced MR strategies will, almost by definition, remain non-automated. However, methods continue to be developed at the boundaries of MR, and the comments here will be superseded as advances are made.

Acknowledgments

I thank Isabel Usón for content suggestions and for proposing the title, and Randy Read for critical reading of the manuscript, discussions, and for the concept for Fig. 1. This work was supported by grant BB/L006014/1 from the BBSRC, UK.

References

1. Tollin P (1969) Determination of the orientation and position of the myoglobin molecule in the crystal of seal myoglobin. J Mol Biol 45:481–490
2. Ward KB, Wishner BC, Lattman EE et al (1975) Structure of deoxyhemoglobin a crystals grown from polyethylene glycol solutions. J Mol Biol 98:161–177
3. Schmid MF, Herriott JR, Lattman EE (1974) The structure of bovine carboxypeptidase B: results at 5.5 Ångström resolution. J Mol Biol 84:97–101
4. Rossmann MG, Blow DM (1962) The detection of sub-units within the crystallographic asymmetric unit. Acta Crystallogr 15:24–31
5. McCoy AJ (2007) Solving structures of protein complexes by molecular replacement with Phaser. Acta Crystallogr D Biol Crystallogr 63:32–41
6. Rupp B (2009) Biomolecular crystallography: principles, practice and applications to structural biology. Garland Science, New York
7. Brunger AT (1992) X-PLOR: version 3.1 a system for X-ray crystallography and NMR. Yale University Press, New Haven, CT
8. Brünger AT, Adams PD, Clore GM et al (1998) Crystallography & NMR System: a new software suite for macromolecular structure determination. Acta Crystallogr D Biol Crystallogr 54:905–921
9. Navaza J (2001) Implementation of molecular replacement in AMoRe. Acta Crystallogr D Biol Crystallogr 57:1367–1372
10. Vagin A, Teplyakov A (2010) Molecular replacement with MOLREP. Acta Crystallogr D Biol Crystallogr 66:22–25
11. Kissinger CR, Gehlhaar DK, Fogel DB (1999) Rapid automated molecular replacement by evolutionary search. Acta Crystallogr D Biol Crystallogr 55:484–491
12. Glykos NM, Kokkinidis M (2001) Multidimensional molecular replacement. Acta Crystallogr D Biol Crystallogr 57:1462–1473
13. Jamrog DC, Zhang Y, Phillips GN (2003) SOMoRe: a multi-dimensional search and optimization approach to molecular replacement. Acta Crystallogr D Biol Crystallogr 59:304–314
14. Jogl G, Tao X, Xu Y, Tong L (2001) COMO: a program for combined molecular replacement. Acta Crystallogr D Biol Crystallogr 57:1127–1134
15. McCoy AJ, Grosse-Kunstleve RW, Adams PD et al (2007) Phaser crystallographic software. J Appl Cryst 40:658–674
16. Toth EA (2007) Molecular replacement. Methods Mol Biol 364:121–148
17. Berman H, Henrick K, Nakamura H (2003) Announcing the worldwide Protein Data Bank. Nat Struct Biol 10:980
18. Scapin G (2013) Molecular replacement then and now. Acta Crystallogr D Biol Crystallogr 69:2266–2275
19. Marcia M, Humphris-Narayanan E, Keating KS et al (2013) Solving nucleic acid structures by molecular replacement: examples from group II intron studies. Acta Crystallogr D Biol Crystallogr 69:2174–2185
20. Read RJ, McCoy AJ (2016) A log-likelihood-gain intensity target for crystallographic phasing that accounts for experimental error. Acta Crystallogr D Biol Crystallogr 72:375–387
21. Read RJ (2001) Pushing the boundaries of molecular replacement with maximum likelihood. Acta Crystallogr D Biol Crystallogr 57:1373–1382

22. French S, Wilson K (1978) On the treatment of negative intensity observations. Acta Crystallogr A 34:517–525
23. Winn MD, Ballard CC, Cowtan KD et al (2011) Overview of the CCP4 suite and current developments. Acta Crystallogr D Biol Crystallogr 67:235–242
24. Adams PD, Afonine PV, Bunkóczi G et al (2010) PHENIX: a comprehensive Python-based system for macromolecular structure solution. Acta Crystallogr D Biol Crystallogr 66:213–221
25. Potterton E, Briggs P, Turkenburg M, Dodson E (2003) A graphical user interface to the CCP 4 program suite. Acta Crystallogr D Biol Crystallogr 59:1131–1137
26. Echols N, Grosse-Kunstleve RW, Afonine PV et al (2012) Graphical tools for macromolecular crystallography in PHENIX. J Appl Cryst 45:581–586
27. Keegan RM, Winn MD (2008) MrBUMP: an automated pipeline for molecular replacement. Acta Crystallogr D Biol Crystallogr 64:119–124
28. Bunkóczi G, Echols N, McCoy AJ et al (2013) Phaser.MRage: Automated molecular replacement. Acta Crystallogr D Biol Crystallogr 69:2276–2286
29. Stokes-Rees I, Sliz P (2010) Protein structure determination by exhaustive search of Protein Data Bank derived databases. Proc Natl Acad Sci U S A 107:21476–21481
30. Strong M, Sawaya MR, Wang S et al (2006) Toward the structural genomics of complexes: crystal structure of a PE/PPE protein complex from Mycobacterium tuberculosis. Proc Natl Acad Sci U S A 103:8060–8065
31. Rodríguez DD, Grosse C, Himmel S et al (2009) Crystallographic ab initio protein structure solution below atomic resolution. Nat Methods 6:651–653
32. Bibby J, Keegan RM, Mayans O et al (2012) AMPLE: A cluster-and-truncate approach to solve the crystal structures of small proteins using rapidly computed ab initio models. Acta Crystallogr D Biol Crystallogr 68:1622–1631
33. Wilson AJC (1942) Determination of absolute from relative X-ray intensity data. Nature 150:152
34. McCoy AJ, Oeffner RD, Wrobel AG, Ojala JRM, Tryggvason K, Lohkamp B, Read RJ (2017) Ab initio solution of macromolecular crystal structures without direct methods. Proc Natl Acad Sci U S A 114:3637–3641
35. McCoy AJ, Read RJ, Bunkóczi G et al Phaserwiki. http://www.phaser.cimr.cam.ac.uk
36. Evans P, McCoy A (2008) An introduction to molecular replacement. Acta Crystallogr D Biol Crystallogr 64:1–10
37. Ten Eyck LF (1973) Crystallographic fast Fourier transforms. Acta Crystallogr A 29:183–191
38. Storoni LC, McCoy AJ, Read RJ (2004) Likelihood-enhanced fast rotation functions. Acta Crystallogr D Biol Crystallogr 60:432–438
39. McCoy AJ, Grosse-Kunstleve RW, Storoni LC et al (2005) Likelihood-enhanced fast translation functions. Acta Crystallogr D Biol Crystallogr 61:458–464
40. Oeffner RD, Bunkóczi G, McCoy AJ et al (2013) Improved estimates of coordinate error for molecular replacement. Acta Crystallogr D Biol Crystallogr 69:2209–2215
41. Long F, Vagin AA, Young P et al (2008) BALBES: a molecular-replacement pipeline. Acta Crystallogr D Biol Crystallogr 64:125–132
42. Rosetta Commons. https://www.rosettacommons.org/about/pubs
43. DiMaio F, Echols N, Headd JJ et al (2013) Improved low-resolution crystallographic refinement with Phenix and Rosetta. Nat Methods 10:1102–1104
44. Pedersen BP, Gourdon P, Liu X et al (2016) Initiating heavy-atom-based phasing by multidimensional molecular replacement. Acta Crystallogr D Biol Crystallogr 72:440–445
45. Urzhumtseva L, Urzhumtsev A (2002) COMPANG: automated comparison of orientations. J Appl Cryst 35:644–647
46. Buehler A, Urzhumtseva L, Lunin VY et al (2009) Cluster analysis for phasing with molecular replacement: a feasibility study. Acta Crystallogr D Biol Crystallogr 65:644–650
47. Millán C, Sammito M, Garcia-Ferrer I, Goulas T, Sheldrick GM, Usón I (2015) Combining phase information in reciprocal space for molecular replacement with partial models. Acta Crystallogr D 71:1931–1945
48. Phillips DC, Rogers D, Wilson AJC (1950) Reliability index for centrosymmetric and non-centrosymmetric structures. Acta Crystallogr 3:398–399
49. Navaza J (1994) AMoRe : an automated package for molecular replacement. Acta Crystallogr A 50:157–163
50. Fujinaga M, Read RJ (1987) Experiences with a new translation-function program. J Appl Cryst 20:517–521
51. Delarue M (2007) Molecular replacement techniques for high-throughput structure determination. In: Sanderson MR, Skelly

JV (eds) Macromolecular crystallography: conventional and high-throughput methods. Oxford University Press, Oxford

52. Abergel C (2013) Molecular replacement: tricks and treats. Acta Crystallogr D Biol Crystallogr 69:2167–2173
53. Turkenburg JP, Dodson EJ (1996) Modern developments in molecular replacement. Curr Opin Struct Biol 6:604–610
54. Dodson E (2008) The befores and afters of molecular replacement. Acta Crystallogr D Biol Crystallogr 64:17–24
55. Schwarzenbacher R, Godzik A, Grzechnik SK et al (2004) The importance of alignment accuracy for molecular replacement. Acta Crystallogr D Biol Crystallogr 60:1229–1236
56. Qian B, Raman S, Das R et al (2007) High-resolution structure prediction and the crystallographic phase problem. Nature 450:259–264
57. Bahar I, Atilgan AR, Erman B (1997) Direct evaluation of thermal fluctuations in proteins using a single-parameter harmonic potential. Fold Des 2:173–181
58. Haliloglu T, Bahar I (1999) Structure-based analysis of protein dynamics: comparison of theoretical results for hen lysozyme with X-ray diffraction and NMR relaxation data. Proteins 37:654–667
59. Tirion M (1996) Large amplitude elastic motions in proteins from a single-parameter, atomic analysis. Phys Rev Lett 77: 1905–1908
60. Tama F, Sanejouand YH (2001) Conformational change of proteins arising from normal mode calculations. Protein Eng 14:1–6
61. Krebs WG, Alexandrov V, Wilson CA et al (2002) Normal mode analysis of macromolecular motions in a database framework: developing mode concentration as a useful classifying statistic. Proteins 48:682–695
62. Suhre K, Sanejouand YH (2004) On the potential of normal-mode analysis for solving difficult molecular-replacement problems. Acta Crystallogr D Biol Crystallogr 60:796–799
63. Blaszczyk J, Li Y, Yan H et al (2001) Crystal structure of unligated guanylate kinase from yeast reveals GMP-induced conformational changes. J Mol Biol 307:247–257
64. SBGrid Science Portal. https://portal.sbgrid.org/d/apps/wsmr/docs
65. Zhou A, Carrell RW, Murphy MP et al (2010) A redox switch in angiotensinogen modulates angiotensin release. Nature 468:108–111
66. Tronrud DE (1997) TNT refinement package. Methods Enzymol 277:306–319
67. Fokine A, Capitani G, Grütter MG et al (2003) Bulk-solvent correction for fast translation search in molecular replacement: service programs for AMoRe and CNS. J Appl Cryst 36:352–355
68. Jackson RN, McCoy AJ, Terwilliger TC et al (2015) X-ray structure determination using low-resolution electron microscopy maps for molecular replacement. Nat Protoc 10:1275–1284
69. Stein N (2008) CHAINSAW : a program for mutating pdb files used as templates in molecular replacement. J Appl Cryst 41:641–643
70. Bunkóczi G, Read RJ (2011) Improvement of molecular-replacement models with Sculptor. Acta Crystallogr D Biol Crystallogr 67:303–312
71. Wriggers W, Schulten K (1997) Protein domain movements: detection of rigid domains and visualization of hinges in comparisons of atomic coordinates. Proteins 29:1–14
72. Hayward S, Berendsen HJ (1998) Systematic analysis of domain motions in proteins from conformational change: new results on citrate synthase and T4 lysozyme. Proteins 30:144–154
73. Schneider TR (2000) Objective comparison of protein structures: error-scaled difference distance matrices. Acta Crystallogr D Biol Crystallogr 56:714–721
74. McCoy AJ, Nicholls RA, Schneider TR (2013) SCEDS: protein fragments for molecular replacement in Phaser. Acta Crystallogr D Biol Crystallogr 69:2216–2225
75. Wodak SJ, Janin J (1980) Analytical approximation to the accessible surface area of proteins. Proc Natl Acad Sci U S A 77:1736–1740
76. Painter J, Merritt EA (2006) Optimal description of a protein structure in terms of multiple groups undergoing TLS motion. Acta Crystallogr D Biol Crystallogr 62:439–450
77. Hinsen K (1998) Analysis of domain motions by approximate normal mode calculations. Proteins 33:417–429
78. Thomas JMH, Keegan RM, Bibby J et al (2015) Routine phasing of coiled-coil protein crystal structures with AMPLE. IUCr J 2:198–206
79. Sammito M, Millán C, Rodríguez DD et al (2013) Exploiting tertiary structure through local folds for crystallographic phasing. Nat Methods 10:1099–1101
80. Sammito M, Meindl K, de Ilarduya IM et al (2014) Structure solution with ARCIMBOLDO using fragments derived from distant homology models. FEBS J 281:4029–4045

81. Chothia C, Lesk AM (1986) The relation between the divergence of sequence and structure in proteins. EMBO J 5:823–826
82. Bunkóczi G, Wallner B, Read RJ (2015) Local error estimates dramatically improve the utility of homology models for solving crystal structures by molecular replacement. Structure 23:397–406
83. Brünger AT (1993) Structure determination of antibodies and antibody-antigen complexes by molecular replacement. Immunomethods 3:180–190
84. Stanfield RL, Zemla A, Wilson IA et al (2006) Antibody elbow angles are influenced by their light chain class. J Mol Biol 357:1566–1574
85. Almagro JC, Beavers MP, Hernandez-Guzman F et al (2011) Antibody modeling assessment. Proteins 79:3050–3066
86. Griffin L, Lawson A (2011) Antibody fragments as tools in crystallography. Clin Exp Immunol 165:285–291
87. Tollin P, Rossmann MG (1966) A description of various rotation function programs. Acta Crystallogr 21:872–876
88. Jeffery P Molecular replacement guide. http://xray0.princeton.edu/~phil/Facility/Guides/MolecularReplacement.html
89. Ling H, Boodhoo A, Hazes B et al (1998) Structure of the shiga-like toxin I B-pentamer complexed with an analogue of its receptor Gb3. Biochemistry 37:1777–1788
90. Karplus PA, Diederichs K (2012) Linking crystallographic model and data quality. Science 336:1030–1033
91. Matthews BW (1968) Solvent content of protein crystals. J Mol Biol 33:491–497
92. Kantardjieff KA, Rupp B (2003) Matthews coefficient probabilities: Improved estimates for unit cell contents of proteins, DNA, and protein-nucleic acid complex crystals. Protein Sci 12:1865–1871
93. Weichenberger CX, Rupp B (2014) Ten years of probabilistic estimates of biocrystal solvent content: new insights via nonparametric kernel density estimate. Acta Crystallogr D Biol Crystallogr 70:1579–1588
94. Sawaya MR (2007) Characterizing a crystal from an initial native dataset. Methods Mol Biol 364:95–120
95. Read RJ, Adams PD, McCoy AJ (2013) Intensity statistics in the presence of translational noncrystallographic symmetry. Acta Crystallogr D Biol Crystallogr 69:176–183
96. Kleywegt GJ, Read RJ (1997) Not your average density. Structure 5:1557–1569
97. Lebedev AA, Vagin AA, Murshudov GN (2006) Intensity statistics in twinned crystals with examples from the PDB. Acta Crystallogr D Biol Crystallogr 62:83–95
98. Yeates TO, Fam BC (1999) Protein crystals and their evil twins. Structure 7:R25–R29
99. Sliwiak J, Jaskolski M, Dauter Z et al (2014) Likelihood-based molecular-replacement solution for a highly pathological crystal with tetartohedral twinning and sevenfold translational noncrystallographic symmetry. Acta Crystallogr D Biol Crystallogr 70:471–480
100. Evans P (2006) Scaling and assessment of data quality. Acta Crystallogr D Biol Crystallogr 62:72–82
101. Zwart PH, Grosse-Kunstleve RW, Adams PD (2005) Xtriage and Fest: automatic assessment of X-ray data and substructure structure factor estimation. CCP4 Newsl 43:27–35
102. Lebedev AA, Isupov MN (2014) Space-group and origin ambiguity in macromolecular structures with pseudo-symmetry and its treatment with the program Zanuda. Acta Crystallogr D Biol Crystallogr 70:2430–2443
103. Herbst-Irmer R, Sheldrick GM (1998) Refinement of twinned structures with SHELXL97. Acta Crystallogr B 54:443–449
104. Dauter Z (2003) Twinned crystals and anomalous phasing. Acta Crystallogr D Biol Crystallogr 59:2004–2016
105. Harada Y, Lifchitz A, Berthou J et al (1981) A translation function combining packing and diffraction information: an application to lysozyme (high-temperature form). Acta Crystallogr A 37:398–406
106. Vagin A, Teplyakov A (1997) MOLREP : an automated program for molecular replacement. J Appl Cryst 30:1022–1025
107. Navaza J, Vernoslova E (1995) On the fast translation functions for molecular replacement. Acta Crystallogr A 51:445–449
108. Wang BC (1985) Resolution of phase ambiguity in macromolecular crystallography. Methods Enzymol 115:90–112
109. Bai X, McMullan G, Scheres SH (2014) How cryo-EM is revolutionizing structural biology. Trends Biochem Sci 40:49–57
110. Rossmann MG (1972) The molecular replacement method. Gordon & Breach, New York, NY
111. Rossmann MG (2001) Molecular replacement – historical background. Acta Crystallogr D Biol Crystallogr 57:1360–1366
112. Vagin AA, Isupov MN (2001) Spherically averaged phased translation function and its

application to the search for molecules and fragments in electron-density maps. Acta Crystallogr D Biol Crystallogr 57:1451–1456

113. Colman PM, Fehlhammer H (1976) The use of rotation and translation functions in the interpretation of low resolution electron density maps. J Mol Biol 100:278–282

114. Schuermann JP, Tanner JJ (2003) MRSAD: using anomalous dispersion from S atoms collected at Cu Kα wavelength in molecular-replacement structure determination. Acta Crystallogr D Biol Crystallogr 59:1731–1736

115. Niedzialkowska E, Gasiorowska O, Handing KB et al (2016) Protein purification and crystallization artifacts: The tale usually not told. Protein Sci 25:720–733

116. Hungler A, Momin A, Diederichs K, Arold ST (2016) ContaMiner and ContaBase: a webserver and database for early identification of unwantedly crystallized protein contaminants. J Appl Cryst 46:2252–2258

117. Rice SO (1945) Mathematical analysis of random noise. Bell Syst Tech J 24:46–156

Chapter 19

Rosetta Structure Prediction as a Tool for Solving Difficult Molecular Replacement Problems

Frank DiMaio

Abstract

Molecular replacement (MR), a method for solving the crystallographic phase problem using phases derived from a model of the target structure, has proven extremely valuable, accounting for the vast majority of structures solved by X-ray crystallography. However, when the resolution of data is low, or the starting model is very dissimilar to the target protein, solving structures via molecular replacement may be very challenging. In recent years, protein structure prediction methodology has emerged as a powerful tool in model building and model refinement for difficult molecular replacement problems. This chapter describes some of the tools available in Rosetta for model building and model refinement specifically geared toward difficult molecular replacement cases.

Key words Molecular replacement, Protein structure determination, Structure refinement

1 Introduction

The phase problem in X-ray crystallography is challenging to overcome. Most structures [1] currently solved by crystallography have their phases determined by molecular replacement [2], where a model of the crystallized protein is constructed, one or more copies are correctly oriented to maximize agreement with the data, phases are computed from the placed model, and an electron density map is calculated. If the model has sufficient similarity to the target, and the data are of reasonably high resolution, the resulting map is readily interpretable, and structure determination is straightforward. However, when data are of low resolution, or models are somewhat dissimilar to the target, interpretation becomes increasingly difficult.

A wide variety of crystallographic software for performing the initial molecular replacement search is available [3, 4], as well as automated pipelines featuring these tools [5, 6]. Detailed descriptions of the search procedure and target functions are provided elsewhere in this volume by McCoy. Generally, given a good starting

Alexander Wlodawer et al. (eds.), *Protein Crystallography: Methods and Protocols*, Methods in Molecular Biology, vol. 1607,
DOI 10.1007/978-1-4939-7000-1_19, © Springer Science+Business Media LLC 2017

model and sufficiently high-resolution data free of pathologies, any of these strategies will work, and—when coupled with automatic chain-tracing software [7, 8]—interpretation is largely automatic and relatively straightforward. However, when models are not of good quality, molecular replacement may fail: either the search step may fail to unambiguously identify molecular placement, or the search step may succeed, but the resulting phases (and thus map) are of too poor quality to interpret. It is with these difficult molecular replacement cases that this chapter is concerned.

This chapter describes the use of the Rosetta structure prediction software suite [9, 10] to aid in solving difficult molecular replacement problems, where low-resolution data or poor-quality initial models make structure determination challenging. We describe two separate challenges, depending on the failure of MR: for cases where the failure is in the search step, we describe several structure prediction tools that have been successfully used to phase crystallographic datasets; when the failure is not in model placement, but rather in refinement of the resulting solution, we have several density and reciprocal-space-data guided protocols that have a larger radius of convergence than other refinement approaches, allowing for the solution of difficult cases that would otherwise be uninterpretable. Figure 1 presents a schematic overview of these procedures.

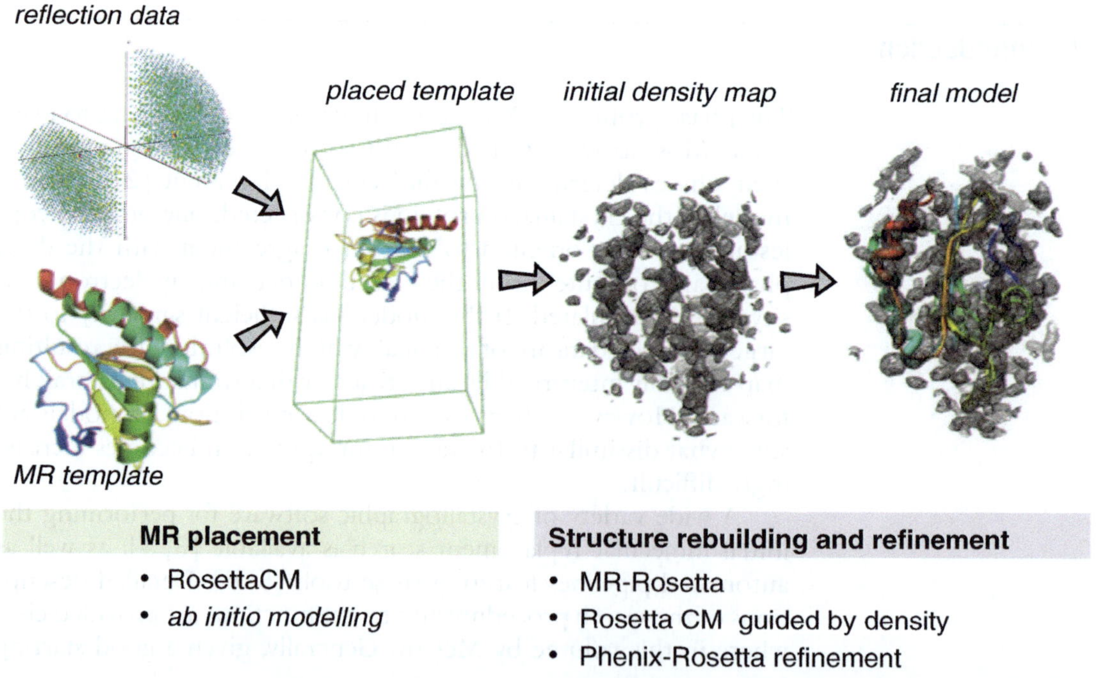

Fig. 1 A schematic overview of molecular replacement, showing the two main steps of MR search, and model rebuilding and refinement. This chapter describes structure modeling tools in Rosetta aimed at assisting in both steps

The majority of this chapter describes the wide variety of broad modeling strategies available for solving difficult molecular replacement problems. We describe tools for both homology modeling and ab initio modeling of proteins, though the former has proven more valuable in structure determination for difficult molecular replacement problems.

2 Model Building in the Absence of Experimental Data

In many cases where molecular replacement fails, the reason is that the search template is too distant from the target [11]. As sequence identity between the target and template model decreases, the structural differences between the two increase [12], leading to a "twilight zone" [13] of around 10–30% sequence identity where structural homologs may be identified, but the resulting models make molecular replacement difficult. In this regime, initial placement of the molecular replacement template is often not possible, even following trimming of surface loops and poorly aligned regions [14] or making use of multiple template structures [15].

Fortunately, methods from structure prediction may help in these cases. As results from recent structure prediction blind challenges indicate [16], state-of-the-art structure prediction methods are often able to refine homology models closer to the native conformation [16]. Furthermore, ab initio structure prediction is able, in some cases, to predict small protein structures with atomic accuracy [17]. Consequently, such models may serve as better molecular replacement templates than the homology-derived template itself.

Rosetta [9, 10] is a state-of-the-art protein structure prediction software suite that combines an all-atom forcefield with knowledge-based methods for sampling protein conformational space. The remainder of this section describes methods within Rosetta useful for solving difficult molecular replacement problems. Throughout the section, examples where Rosetta has been successfully employed are provided, as well as links to appropriate tools within Rosetta, and general strategies for using structure prediction methods in solving MR problems.

2.1 Homology Modeling

One of the most common uses of structure prediction for difficult MR problems is in solving problems from templates in the "twilight regime," where the best available high-resolution template structures are only 10–30% identical in their sequence compared to the target. While such templates are not readily identified by a simple BLAST search, a variety of sequence-profile and structural-property based methods are often capable of identifying such templates from a given target sequence. These include tools such as HHpred [18], Sparks [19], and Raptor [20], which have all been shown to perform extremely well at template identification in blind structure prediction competitions [21]. Even though these templates are readily identified, since they often differ from the target structure by more than 1.5 Å rmsd, it is challenging to use them for molecular replacement, since they often show rms deviations of more than 1.5 Å from the target structure.

It has been shown that Rosetta structure prediction may be useful in solving structures in these cases. In particular, RosettaCM (Fig. 2) combines Monte Carlo sampling of insertions and deletions, minimization with a low-resolution statistical potential, and all-atom refinement to sample an ensemble of low-energy structures within a particular topology [22]. Starting with a sequence alignment, RosettaCM first generates a "partial" threaded structure, where all unaligned residues in the template are deleted, and the residue identity of all mutations is changed to the identity of the target. Then, as illustrated in Fig. 2, Rosetta first alternates fragment-based sampling of unaligned residues and residues within eight residues of an insertion or deletion, with minimization against a statistical energy function [22]. Finally, after ~1000 iterations, the resulting model is refined against the Rosetta all-atom energy function.

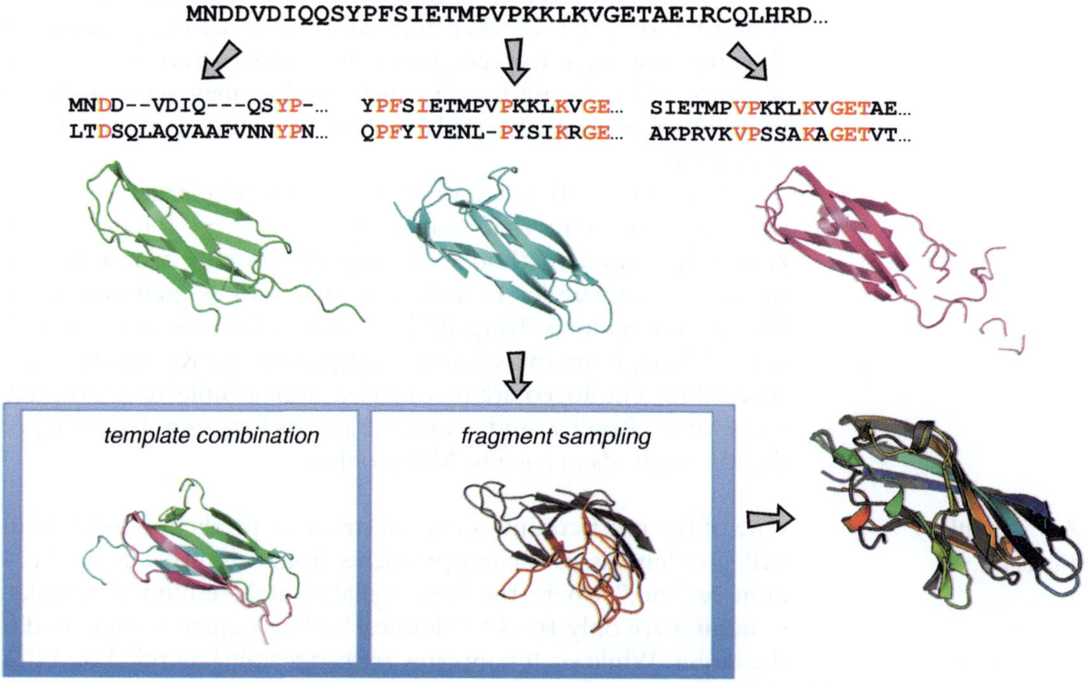

Fig. 2 An overview of RosettaCM for structure prediction guided by homologous proteins. Modeling in RosettaCM combines recombination of template structures with "fragment-based" conformational sampling in unassigned regions: borrowing short segments of backbone from proteins with similar local sequence

Additionally, in some cases, many low sequence-identity structural homologs may be known. In previous studies, multiple models have proved valuable in solving molecular replacement problems in the twilight regime [13]. In such cases, RosettaCM can make use of all such models in refinement; in these cases, the initial iteration samples—in addition to fragments in unaligned regions—segments of aligned residues from alternate templates. In this way, RosettaCM may combine regions from multiple models; this is often valuable when the best template is not the most complete one.

Generally, this protocol is run in several hundred to several thousand independent Monte Carlo trajectories depending on: (1) available CPU time, (2) the emergence of a clear molecular replacement solution, and (3) the quality of the sequence alignment. These trajectories are then each searched as independent models in MR calculations. Success criteria may be measured as in the previous chapter by McCoy, though, when making use of several thousand search models, the chances of an incorrect model scoring highly increase dramatically, and thus care must be taken.

If the number of models is extremely large, one may separate the rotational and translational search steps over all the models; for example, given a million search models, one may perform a rotational search with all models, then take the top 10^6 orientations (potentially taking more than one orientation per model), and perform the subsequent translational search. This approach proved valuable in previous work, where over a million models were generated in order to solve the structure of the Mason Pfizer Monkey Virus retroviral protease by molecular replacement [23, 24]. In this case, to initially identify the correct molecular replacement solution, all 1.7 million models underwent a rotational search. The top 1000 rotations were carried over to translational search, which quickly identified the correct solution.

Figure 3 shows an example of a crystal structure (actin-like protein Alp7) for which this method was successfully used to solve the molecular replacement problem and where numerous individual templates alone were insufficient. None of the individual templates gave reasonable hits with Phaser LLG > 33 or TF > 6. However, following extensive model building with RosettaCM—using the ten closest homologs according to HHpred [18]—the model could be placed with LLG = 40 and TFZ = 7. Subsequent rebuilding and refinement with MR-Rosetta led to successful structure determination.

RosettaCM is run through the XML interface of Rosetta, and is fully described in the Rosetta manual [25]. The XML file allows declaration of an unlimited number of template structures from which to draw, but in practice sampling becomes a limiting factor at around 20 input models; if more templates are available, multiple independent runs with subsets of 20 models are suggested. Alternatively, one may cluster the input models to maximize model

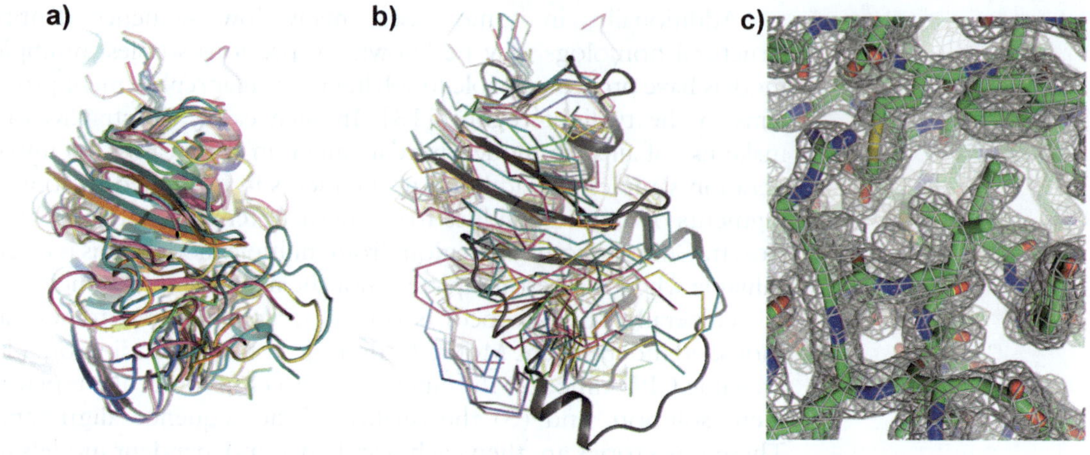

Fig. 3 An example of RosettaCM structure modeling leading to successful MR placement (in review). (**a**) A superposition of ten structural homologs used in model-building. (**b**) The resulting model, successfully used for molecular replacement. (**c**) The final structure, following MR-Rosetta and automatic chain tracing in phenix. autobuild [7]. A 2mFo-dFc map at a contour level of 1.5σ is shown

diversity. RosettaCM behaves poorly when the input models are too different from one another; for example, RosettaCM converges very poorly if templates with two different topologies are provided.

2.2 De Novo Model Building

In cases where no identifiable MR templates are available, there are still several techniques that may be employed. If the data are at very high resolution and the target structure is expected to contain α-helices, fragment-based de novo phasing has proven successful [26]. In some cases, brute force MR searches of all known protein topologies [27] can yield a successful MR solution. Finally, in cases where the target protein is small (<100 residues) it may be possible to make use of ab initio modeling in order to solve the molecular replacement problem.

Rosetta, which has previously been used to solve several molecular replacement problems via ab initio modeling, builds models in a manner similar to the way missing residues are rebuilt in RosettaCM. Predicted local structural motifs based on local structure, or "fragments," are sampled at each position in the protein in a Monte Carlo trajectory; each sampled conformation is evaluated with a statistical energy function, and accepted or rejected by the Metropolis criteria. Resulting models are refined with the Rosetta all-atom energy function following the same procedure as RosettaCM.

Despite previous successes, ab initio modeling often fails to converge on the correct topology, even when applied to soluble monomeric proteins 60–100 residues in length. However, it previously has produced models of a quality suitable for molecular replacement [17]. This work has estimated that for about one in six proteins of less than 100 residues, the procedure converges on a near-native atomic model [28]. Given all the limitations, ab initio modeling may be used as a last-ditch effort to solve difficult molecular replacement problems.

2.3 Preparing Models for Molecular Replacement

As previous work has shown, successful molecular replacement from structure prediction can be aided by proper model preparation. That includes removing surface side chains at the Cγ atom, as well as any regions poorly converging in modeling [12]. Additionally, previous studies have shown that accurate error estimates are quite important for the ability to perform MR using such models [14]. Using the resulting structures of multiple independent trajectories, one may make relatively accurate error estimates, by superposing low-energy models, calculating per-atom mean deviations μ, and converting them to B factors with $B = 8\pi^2\mu^2$. Indeed, this is done with models predicted on the Robetta structure prediction server [29], making these models more appropriate for molecular replacement.

3 Model Building and Refinement Guided by Experimental Data

As indicated in Fig. 1, model placement is not the only step where MR may fail. Indeed, it is possible to correctly (albeit ambiguously) place a molecule in the molecular replacement search, but the resulting phases, and the resulting electron density maps, may be too poor to interpret. This is particularly problematic at low to medium-low resolution (2.5 Å and worse). Rosetta has recently been augmented with real- and reciprocal-space refinement methods, and model building and refinement protocols taking these sources of data into account have been developed. The physically realistic forcefield of Rosetta better handles the sparsity and noise of low resolution and poorly phased datasets than standard crystallographic refinement software, allowing refinement in cases where standard refinement might not suffice [30, 31]. The remainder of this chapter introduces some general-purpose tools in Rosetta for improvement of borderline MR solutions, or MR solutions at low resolution.

3.1 Refinement Against Poorly Phased Data with MR-Rosetta

MR-Rosetta was developed as a tool for improving very weak molecular replacement solutions. The original hypothesis was that often when molecular replacement fails, the search is successful (though it may not be readily clear), but subsequent map interpretation is not possible. The idea behind MR-Rosetta, illustrated schematically in Fig. 4, is to perform many MR searches using different homologs and various potential trimmings of those homologs, use them to putatively phase the data, rebuild and refine each model in real space, and reevaluate the model against the

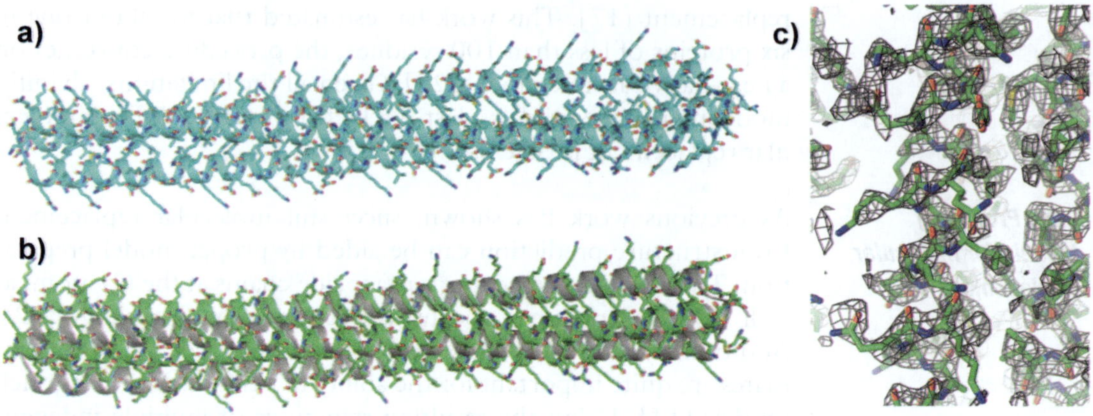

Fig. 4 A challenging molecular replacement problem solved by combination of MR-Rosetta and Phenix-Rosetta reciprocal-space modeling [32]. (**a**) A designed helical bundle was crystallized but the design model was unsuitable for MR. (**b**) A combination of MR-Rosetta and Phenix-Rosetta reciprocal-space refinement against the twinned data yielded the final structure (*green*, compare to model in *gray*). (**c**) The electron density map following this process. A 2mFo-dFc map at a contour level of 1.5σ is shown

reciprocal space data. In cases where one of the placements is correct, Rosetta refinement is often able to identify the correct solution, and often improve the phases enough to make the map readily interpretable.

Previous work has shown [30] that MR-Rosetta is successful primarily in the 15–30% sequence identity range, where it achieves a ~50% success rate, provided data resolution is better than 3.2 Å and there are fewer than four copies of the molecule in the asymmetric unit. The rebuilding and refinement strategy underlying MR-Rosetta is similar to that of RosettaCM, with two key differences. Firstly, during model building and refinement, agreement of the model with electron density is used as an additional scoring term. Secondly, only short unmodeled segments (less than nine residues in length) are rebuilt, as longer segments are less likely to be correct. Also, unlike RosettaCM, MR-Rosetta only makes use of a single initial model.

Finally, in cases where there are multiple templates available, and the molecular replacement solution is ambiguous, it is possible to use electron density and multiple templates, though the setup is a bit more complex. One may run RosettaCM with an additional term enforcing agreement with electron density. This is done by aligning all templates to the ambiguous MR hit, adding the flag "*realign_domains* = 0" to the Hybridize mover, and enabling the score term *elec_dens_fast* in stage 1, stage 2, and the full-atom stage (with weight 10, 10, and 20, respectively).

MR-Rosetta is run as a command line tool within Rosetta. A demo file, *electron_density_molecular_replacement*, shows the typical usage. Inputs include the electron density map, a placed template, and sequence alignment between template and target.

Several options are user-controllable. The primary argument that may affect results is the longest unmodeled segment option (*-max_gaplength_to_model*). This takes a default value of 8 but in some cases may benefit from either reduction (for example, if CPU is limited) or increase (for example, when segments to rebuild have high secondary structure content).

3.2 Reciprocal-Space Refinement with Phenix-Rosetta

With very low-resolution crystallographic data, automatic chain-tracing and refinement tools may perform poorly, even if the starting model is reasonably close to correct. Standard refinement tools may perform poorly because of small radius of convergence [31]. However, one may use the same idea as with MR-Rosetta—applying physically realistic structure prediction forcefield and conformational sampling tools—in order to improve the performance of refinement methods against low-resolution crystallographic data.

This approach has been implemented as a combined protocol of Rosetta and Phenix [33], with Rosetta and Phenix calling one another through a Python bridge interface. Previous work has shown that such a refinement has an improved radius of convergence with 3–4.5 Å crystallographic data. On average, better free R factors, model geometry (as assessed by Molprobity) and rmsd to the deposited model were seen, when compared to Refmac [34] with jelly-body restraints [35], phenix.refine [36], and CNS with DEN restraints [37]. Thus, this method should prove useful for real-space model rebuilding and refinement in MR-Rosetta with low-resolution data.

Indeed, we have found this combination of tools useful in practice. Figure 4 shows one sample where this approach was used in conjunction with MR-Rosetta to solve a very difficult structure. The protein in question was a helical bundle with internal pseudo-symmetry, which made MR difficult. Additionally, the data were at medium-low resolution (2.8 Å) and twinned, which made refinement difficult. However, the combination of these two tools produced a solution of this structure, as indicated in Fig. 4, with an R/R_{free} of 0.26/0.31 [32].

Like the comparative modeling tools described in the Subheading 2.1, reciprocal space refinement within Rosetta makes use of an XML interface to Rosetta. This allows a combination of standard structure prediction sampling methods with a score function, *xtal_ml*, that assesses agreement of a model with reciprocal space data, using a reciprocal space likelihood function [31]. In addition, several crystal-specific tools have been added to Rosetta including a mover *SetRefinementOptions*, which allows (1) definition of twin laws, (2) specification of map-type to use in real-space refinement, and (3) refinement with ligand molecules. Finally, a general-purpose refinement script, *refine_low_resolution.xml*, has been included as part of Rosetta and is also available through the phenix tool *phenix.rosetta_refine*.

3.3 De Novo Density-Guided Model-Building Using CryoEM Tools

Finally, in cases where phases are available but there are no recognizable homology models, there are some tools from cryoEM model-building that may prove useful, though they have yet to be used for crystallographic refinement. Such tools may be useful in cases where initial estimates of phases are available, but there is no corresponding model. This includes phases determined experimentally, as well as several methods that allow for MR without the need for an identifiable structural homolog [26, 27]. With medium- to low-resolution data, these methods may yield electron density maps with some information, but are difficult to interpret. Thus, the cryoEM modeling methods may prove useful for problems in X-ray crystallography.

In particular, a Rosetta fragment based de novo approach has been shown in some cases to automatically solve structures from ~4 Å cryoEM density [38], including two cases uninterpretable by humans [39, 40]. Such approaches might prove useful for interpreting phased low-resolution crystallographic datasets, where automatic chain tracing software tends to fail [7, 8]. However, to date, no crystallographic datasets have been tackled by this approach.

4 Discussion

We have described a set of tools, using the Rosetta structure prediction software package, that may be useful in solving difficult molecular replacement problems. These include tools for generating models that may be more useful in molecular replacement search, as well as tools for rebuilding and refining a solution, guided by noisy electron density from an initially weak molecular replacement hit. In particular, these tools have been shown as most useful in solving the phase problem for high-resolution datasets, provided template sequence identity of ~15–30%.

Recent advances in predicted coevolving residues [41, 42] seem to show additional promise in solving difficult MR cases. This information, which can provide moderately accurate predictions of residue contacts, can be used as an additional "orthogonal" constraint in conformational modeling. Previous work has shown that with such constraints, the accuracy of ab initio modeling, even for large proteins of 200–300 residues, is increased significantly [43]. As the number of known protein sequences increases, this should prove an increasingly powerful tool for solving difficult MR problems.

Finally, further improvements to this approach are likely to come from fundamental improvements to protein structure prediction, which relies on the development of improved conformational sampling algorithms and improved models of protein energetics. One of the main roles of structure prediction in solving difficult crystallographic problems is reducing the "effective degrees of freedom." Even though a protein may have 1000 rotatable bonds which, without restraints, may take any value, only a very small combination of the possible settings of these bonds leads to physically realistic protein conformations. To allow structure determination with "less" data (broadly referring to worse quality homologs and lower-resolution experimental data), fundamentally better models of protein energetics are required. These fundamental improvements will be key to further advances in structure modeling for MR.

References

1. Scalpin G (2013) Molecular replacement then and now. Acta Crystallogr D Biol Crystallogr 69:2266–2275
2. Rossmann MG, Blow DM (1962) The detection of sub-units within the crystallographic asymmetric unit. Acta Crystallogr 15:24–31
3. Vagin A, Teplyakov A (2010) Molecular replacement with MOLREP. Acta Crystallogr D Biol Crystallogr 66:22–25
4. McCoy AJ, Grosse-Kunstleve RW, Adams PD et al (2007) Phaser crystallographic software. J Appl Cryst 40:658–674
5. Keegan RM, Winn MD (2007) Automated search-model discovery and preparation for structure solution by molecular replacement. Acta Crystallogr D Biol Crystallogr 63:447–457
6. Long F, Vagin AA, Young P et al (2008) BALBES: a molecular-replacement pipeline. Acta Crystallogr D Biol Crystallogr 64:125–132
7. Terwilliger TC, Grosse-Kunstleve RW, Afonine PV et al (2008) Iterative model building, structure refinement and density modification with the PHENIX AutoBuild wizard. Acta Crystallogr D Biol Crystallogr 64:61–69
8. Cowtan K (2006) The Buccaneer software for automated model building. 1. Tracing protein chains. Acta Crystallogr D Biol Crystallogr 62:1002–1011
9. Rohl CA, Strauss CE, Misura KM et al (2004) Protein structure prediction using Rosetta. Methods Enzymol 383:66–93
10. Leaver-Fay A, Tyka M, Lewis SM et al (2011) ROSETTA3: an object-oriented software suite for the simulation and design of macromolecules. Methods Enzymol 487:545–574
11. Abergel C (2013) Molecular replacement: tricks and treats. Acta Crystallogr D Biol Crystallogr 69:2167–2173
12. Chothia C, Lesk AM (1986) The relation between the divergence of sequence and structure in proteins. EMBO J 5:823–826
13. Rost B (1999) Twilight zone of protein sequence alignments. Protein Eng 12:85–94
14. Stein N (2008) CHAINSAW: a program for mutating pdb files used as templates in molecular replacement. J Appl Cryst 41:641–643
15. Bunkóczi G, Echols N, McCoy A et al (2013) Phaser.MRage: automated molecular replacement. Acta Crystallogr D Biol Crystallogr 69:2276–2286
16. Nugent T, Cozzetto D, Jones D (2014) Evaluation of predictions in the CASP10 model refinement category. Proteins 82:98–111
17. Qian B, Raman S, Das R et al (2007) High-resolution structure prediction and the crystallographic phase problem. Nature 450:259–264
18. Söding J, Biegert A, Lupas AN (2005) The HHpred interactive server for protein homology detection and structure prediction. Nucleic Acids Res 33:W244–W248
19. Yang Y, Faraggi E, Zhao H et al (2011) Improving protein fold recognition and template-based modeling by employing probabilistic-based matching between predicted one-dimensional structural properties of the query and corresponding native properties of templates. Bioinformatics 27:2076–2082
20. Källberg M, Wang H, Wang S et al (2012) Template-based protein structure modeling using the RaptorX web server. Nat Protoc 7:1511–1522
21. Modi V, Xu Q, Adhikari S et al (2016) Assessment of template-based modeling of protein structure in CASP11. Proteins 84(Suppl. 1):200–220
22. Song Y, DiMaio F, Wang RY et al (2013) High-resolution comparative modeling with RosettaCM. Structure 21:1735–1742
23. Khatib F, DiMaio F, Foldit Contenders Group et al (2011) Crystal structure of a monomeric retroviral protease solved by protein folding game players. Nat Struct Mol Biol 18:1175–1177

24. Gilski M, Kazmierczyk M, Krzywda S et al (2011) High-resolution structure of a retroviral protease folded as a monomer. Acta Crystallogr D Biol Crystallogr 67:907–914
25. DiMaio F, Rämisch S, Adolf-Bryfogle J (2013) RosettaCM - Comparative Modeling with Rosetta. https://www.rosettacommons.org/docs/latest/application_documentation/structure_prediction/RosettaCM
26. Martínez D, Grosse C, Himmel S et al (2009) ARCIMBOLDO: crystallographic ab initio protein solution below atomic resolution. Nat Methods 6:651–653
27. Stokes-Reesa I, Sliz P (2010) Protein structure determination by exhaustive search of Protein Data Bank derived databases. Proc Natl Acad Sci U S A 107:21476–21481
28. Das R, Baker D (2009) Prospects for de novo phasing with de novo protein models. Acta Crystallogr D Biol Crystallogr 65:169–175
29. Kim D, Chivian D, Baker D (2004) Protein structure prediction and analysis using the Robetta server. Nucleic Acids Res 32:W526–W531
30. DiMaio F, Terwilliger T, Read R et al (2011) Improving molecular replacement by density- and energy- guided protein structure optimization. Nature 473:540–543
31. DiMaio F, Echols N, Headd J et al (2013) Improved protein crystal structures at low resolution by integrated refinement with Phenix and Rosetta. Nat Methods 10:1102–1104
32. Huang PS, Oberdorfer G, Xu C et al (2014) High thermodynamic stability of parametrically designed helical bundles. Science 346:481–485
33. Adams PD, Afonine PV, Bunkóczi G et al (2010) PHENIX: a comprehensive Python-based system for macromolecular structure solution. Acta Crystallogr D Biol Crystallogr 66:213–221
34. Murshudov GN, Vagin AA, Dodson EJ (1997) Refinement of macromolecular structures by the maximum-likelihood method. Acta Crystallogr D Biol Crystallogr 53:240–255
35. Nicholls RA, Long F, Murshudov GN (2012) Low-resolution refinement tools in REFMAC5. Acta Crystallogr D Biol Crystallogr 68:404–417
36. Afonine PV, Grosse-Kunstleve RW, Echols N et al (2012) Towards automated crystallographic structure refinement with phenix.refine. Acta Crystallogr D Biol Crystallogr 68:352–367
37. Schröder G, Levitt M, Brunger A (2010) Super-resolution biomolecular crystallography with low-resolution data. Nature 464:1218–1222
38. Wang R, Kudryashev M, Li X et al (2015) Accurate de novo protein structure determination from near-atomic resolution cryo-EM maps. Nat Methods 12:335–338
39. Walls AC, Tortorici MA, Bosch BJ et al (2016) Cryo-electron microscopy structure of a coronavirus spike glycoprotein trimer. Nature 531:114–117
40. Kudryashev M, Wang RYR, Brackmann M et al (2015) The structure of the type six secretion system contractile sheath solved by cryo-electron microscopy. Cell 160:952–962
41. Marks DS, Hopf TA, Sander C (2012) Protein structure prediction from sequence variation. Nat Biotechnol 30:1072–1080
42. Ovchinnikov S, Kamisetty H, Baker D (2014) Robust and accurate prediction of residue-residue interactions across protein interfaces using evolutionary information. Elife 3:e02030
43. Kim DE, DiMaio F, Wang RYR et al (2014) One contact for every twelve residues allows robust and accurate topology-level protein structure modeling. Proteins 82:208–218

Chapter 20

Radiation Damage in Macromolecular Crystallography

Elspeth F. Garman and Martin Weik

Abstract

Radiation damage inflicted on macromolecular crystals during X-ray diffraction experiments remains a limiting factor for structure solution, even when samples are cooled to cryotemperatures (~100 K). Efforts to establish mitigation strategies are ongoing and various approaches, summarized below, have been investigated over the last 15 years, resulting in a deeper understanding of the physical and chemical factors affecting damage rates. The recent advent of X-ray free electron lasers permits "diffraction-before-destruction" by providing highly brilliant and short (a few tens of fs) X-ray pulses. New fourth generation synchrotron sources now coming on line with higher X-ray flux densities than those available from third generation synchrotrons will bring the issue of radiation damage once more to the fore for structural biologists.

Key words X-ray-matter interactions, Global and specific radiation damage, Radicals and their scavengers, Absorbed dose, Radiation damage mitigation, Cryocrystallography

1 Introduction

Since the earliest days of macromolecular crystallography (MX), radiation damage to the sample during X-ray irradiation has been a limiting factor for three-dimensional structure solution. At room temperature (RT), the highest resolution reflections from diffracting protein crystals typically start to fade before a full dataset has been collected. For RT experiments, Blundell and Johnson [1] recommended that when the intensity of a monitored reflection dropped to 0.85 of its original value (I_0), the crystal should be replaced by a new one, but if the crystal supply was limited, this could be pushed to 0.7 I_0. Following the pioneering work of Hope on flash cooling of ribosome crystals [2] and the introduction of the loop mounting method for samples [3], the development of cryocrystallographic techniques in the 1990s [4–7] allowed routine flash cooling of crystals to ~100 K. At such cryotemperatures it was found that radiation damage rates could be reduced by around a factor of 70 compared with those seen at RT [8]. However, in 1994, reflection fading was again observable during

Alexander Wlodawer et al. (eds.), *Protein Crystallography: Methods and Protocols*, Methods in Molecular Biology, vol. 1607, DOI 10.1007/978-1-4939-7000-1_20, © Springer Science+Business Media LLC 2017

quantitative experiments at cryotemperatures with a white beam carried out at the Daresbury synchrotron [9]. With the advent of higher flux density X-ray beams produced by third generation synchrotrons in the mid to late 1990s, damage effects were observed at 100 K with monochromatic beams. Below we summarize some of the ongoing efforts over the last 15 years to elucidate the manifestations and origins of radiation damage and to establish mitigation strategies using various approaches. These have resulted in a deeper understanding of the physical and chemical factors affecting damage rates. However, new fourth generation synchrotron sources now coming online with even higher flux densities than hitherto utilized, will bring the issue of radiation damage into even sharper focus for structural biologists. An awareness of the effects of radiation damage on diffraction data and on the macromolecular structures derived from them, will therefore become increasingly important. The use of very short X-ray pulses (a few tens of fs) produced by X-ray free electron laser sources allows "diffraction-before-destruction" with no significant radiation damage. The new methodology associated with these experiments is covered elsewhere in Chapter 12 by Chapman.

2 Interaction of X-Ray Photons with Matter

The physics involved in radiation damage is well understood and characterized, although the same cannot be said about the chemical aspects. When an X-ray beam of the energy range usually used for MX (6–15 keV) is incident on a sample, it interacts by three main mechanisms (Fig. 1): elastic (Thomson, coherent) scattering, Compton (incoherent) scattering, and photoelectric absorption [10]. Most of the beam passes straight through the sample without interacting at all: for instance for a 12.4 keV (1 Å) beam and a 30 μm [100 μm] thick crystal, 99.2% [98%] of the photons will transit unaffected and be captured by the beamstop (this reduces to 87.3% [64%] for a 5 keV beam). Of the interacting 0.8% [2%] of the beam, 0.06% [0.16%] will undergo elastic scattering which is the desired interaction for diffraction: the photon scatters from the electrons in the crystal without leaving any energy behind. However, due to the cross sections subtended by the atoms to the X-rays, 92% [92%] of the interacting photons (~0.74% [1.84%] of the incident beam) are absorbed, and the photons lose some or all of their energy in the sample.

At 12.4 keV, Compton scattering accounts for around 9% of these absorption events, and involves the inelastic collision of an X-ray photon with the electrons in an atom to give a lower energy X-ray photon. This photon may or may not escape the sample, depending on how much energy was lost on the collision and the size and composition of the crystal. The third interaction,

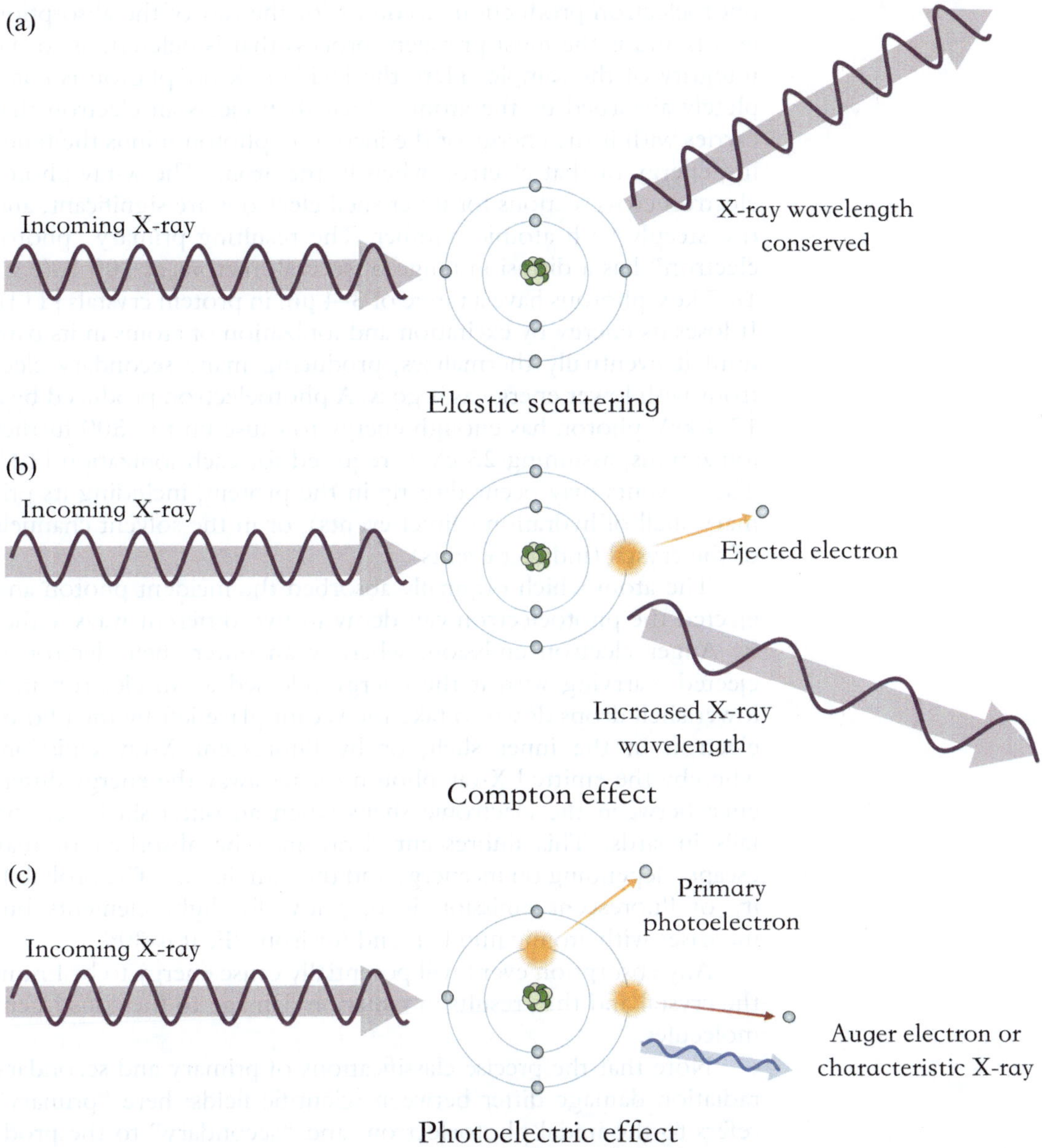

Fig. 1 Primary X-ray interaction processes with atoms of the macromolecule and solvent. (**a**) Elastic (Thomson, coherent) scattering. The waves are phase shifted by 180° on scattering and add vectorially to give the diffraction pattern. (**b**) Compton (incoherent) scattering. The X-ray transfers some energy to an atomic electron and thus has lower energy (longer wavelength) after the interaction. Energy is lost in the crystal, contributing to the absorbed dose. (**c**) Photoelectric absorption. The X-ray transfers all its energy to an atomic electron, which is then ejected and can give rise to the ionization of up to 500 other atoms. The excited atom can then emit a characteristic (fluorescent) X-ray or an Auger electron to return to its ground state

photoelectron production, accounts for the rest of the absorption events and is the most prevalent process that is deleterious to the integrity of the sample. Here the incident X-ray photon is completely absorbed by the atom, which then ejects an electron that carries with it the energy of the incoming photon minus the binding energy of that electron when in the atom. The X-ray photoelectron cross sections for inner shell electrons are significant, and rise steeply with atomic number. The resulting primary "photoelectron" has a diffusion range of several microns at 100 K (e.g., 18.7 keV photons have a range of 3–4 μm in protein crystals [11]). It loses its energy by excitation and ionization of atoms in its path until it eventually thermalizes, producing many secondary electrons with lower energy as it goes. A photoelectron produced by a 12.4 keV photon has enough energy to cause up to ~500 further ionizations, assuming 25 eV is required for each ionization [12]. These events may occur directly in the protein, including its primary shell of hydration (direct events), or in the solvent channels of the crystal (indirect events).

The atom which originally absorbed the incident photon and ejected the photoelectron can decay in two different ways: either by Auger electron emission, whereby an outer shell electron is ejected, carrying with it the energy released as an electron in a lower shell drops down to take the vacant place left by the photoelectron in the inner shell; or by fluorescent X-ray emission, whereby the emitted X-ray photon carries away the energy difference between the electronic shells when an outer-shell electron falls inwards. This fluorescent X-ray may be absorbed or may escape, depending on its energy and the sample size. The probability of fluorescent emission is very low for light elements but increases with atomic number, and for iron ^{55}Fe it is 30%.

Any absorption event will potentially cause energy to be lost in the crystal and thus results in radiation damage to the constituent molecules.

Note that the precise classifications of primary and secondary radiation damage differ between scientific fields; here "primary" refers to the initial photoelectron, and "secondary" to the products it induces. In addition, in the case of crystalline material, tertiary damage is suffered by the lattice which may be destabilized by damage to atoms involved in crystal contacts, or by gas formation within the crystal.

From a chemical viewpoint, the ionizations described above will lead to electron-gain and electron-loss (holes) centers resulting from the direct effects and giving rise to radicals. These may recombine, leading to an excited state, the deactivation of which may or may not cause damage. These recombination processes compete with charge separation through migration, which for electron and hole centers may occur by tunneling (essentially temperature independent) or hopping (temperature dependent) to given sites in the protein, where the radicals become localized.

3 Mitigation of Secondary Radiation Damage by Cryocooling

The primary absorption events (Compton and photoelectric effects) described above are a fact of physics: they cannot be avoided in the diffraction experiment and are temperature independent. However, diffusion of most of the secondary radiation induced products can be prevented by maintaining the sample at around 100 K during data collection. This temperature was originally selected as it can be reached conveniently using open flow nitrogen cryostats [13], but it was a fortuitous choice, since below 110 K, hydroxyl radicals produced by the radiolysis of water are thought to be immobile [14]. At 100 K, all larger species produced by reactions of secondary products, such as peroxide and superoxide (in the presence of oxygen), are unable to diffuse through the solvent channels of the crystal. However, even at very low temperatures, electrons and holes are both able to migrate by quantum mechanical tunneling [15], which is a temperature independent phenomenon. In addition, any thermal energy available will allow them to "hop," but the probability of this phenomenon is temperature dependent. The effect of these mobile species at 100 K is to induce specific structural damage to the protein (see below). Prompt hole transfer to the protein from primary absorption events occurring in solvent adjacent to the protein might also be expected.

The overall effect of cryocooling is to significantly improve the dose tolerance of macromolecular crystals at 100 K so that for most samples the rate of radiation damage is decreased by nearly two orders of magnitude. The use of cryocrystallographic techniques has thus allowed third generation synchrotron beams to be used to good effect to solve macromolecular crystal structures; 84% of the crystal structures deposited in the PDB with temperature of data collection information have been determined with the crystal held at or near 100 K (90% held below 160 K). In particular, the success of the Multi-wavelength Anomalous Diffraction (MAD) method of structure solution has relied on taking data at two or more different wavelengths from the same crystal, a feat possible at 100 K but not at RT.

4 Absorbed Dose, Estimating Dose and Dose Limits

To monitor the effects of radiation damage, various observables can be plotted as a function of image number or time. However, using either of these parameters as the abscissa has the disadvantage that experiments performed under different conditions cannot be compared. Thus a much more generally applicable metric is required for the study and quantitation of radiation damage effects.

The absorbed dose is such a metric: it is the energy absorbed per mass of sample, expressed in J/kg = gray (or Gy) units. This is not a directly measurable quantity, but must be estimated from both the contents of the crystal and the particular experimental conditions. For a sample with linear absorption coefficient μ_{abs}, the incident beam intensity, J_0, falls off exponentially (first order decay) with crystal thickness, x, to become J, according to the relationship:

$$J = J_0 e^{-\mu_{abs} x}$$

The energy lost in the sample is then $J_0 - J$ multiplied by the incident photon energy. The cross section subtended by the sample to the X-ray beam, μ_{abs}, is computed from its atomic composition, which includes the amino acid residues of the protein, any bound heavy atoms, ligands or nucleotides, and also the components of the buffer in the solvent channels. The cross section for the photoelectric effect rises steeply with atomic number, Z, and the presence of a few heavy atoms in the unit cell can have a large effect on the absorption of X-rays and thus the absorbed dose. Also required for the dose calculation are the beam characteristics, namely its energy, size and profile, the incident photon flux (J_0), and the exposure time. This requires regular calibration of synchrotron beamlines and straightforward methodology has been developed to expedite this step [16].

As a general dose yard stick, 1 MGy/s will be absorbed by a 100 μm cubed metal-free crystal in a 100 μm × 100 μm beam of 12.4 keV (1 Å) X-rays and a flux of 10^{13} photons/s, and will cause approximately 1 ionization per 20 amino acid residues per second [12].

In order to facilitate dose estimation in a spatially and temporally resolved way for MX, the computer program RADDOSE-3D [17] has been developed, which can model an X-ray diffraction experiment. It includes the ability to model complicated data collection protocols [18] such as helical scans [19] and translations, as well as MAD experiments. An experimentally measured beam profile can be used to gain a more realistic photon flux distribution, and buffer components can be entered as mM concentrations. The program can also use polygon crystal shapes to improve the dose estimation. Its shortcomings are that account is not taken of the escape of fluorescent X-rays (produced by heavy atoms after photoelectron ejection—see above), and also that the likely escape of the primary photoelectron [20] is neglected. Since the proportion of electrons escaping becomes significant for crystals with volumes smaller than 20 μm^3, the doses computed by RADDOSE-3D for micron-sized crystals irradiated by micron sized beams are overestimated.

The earlier RADDOSE versions 1, 2, and 3 [21–23] were developed when crystals were generally smaller than beams and thus crystal rotation could be neglected. These versions gave the maximum dose experienced by the crystal (usually at its center—see below) and this approach was widely used prior to the availability of RADDOSE-3D.

The distribution of damage in an irradiated crystal depends pivotally on the beam profile. If this is top-hat shaped, the crystal will be uniformly irradiated, and will be damaged at the same rate throughout. If, however, the flux is distributed in a rough Gaussian profile, the center of a crystal which is aligned in the beam and rotated on an axis intersecting the peak of the beam, will suffer faster damage than the parts farther away and on the tails of the flux distribution (Fig. 2). A study of the damage rates in 43

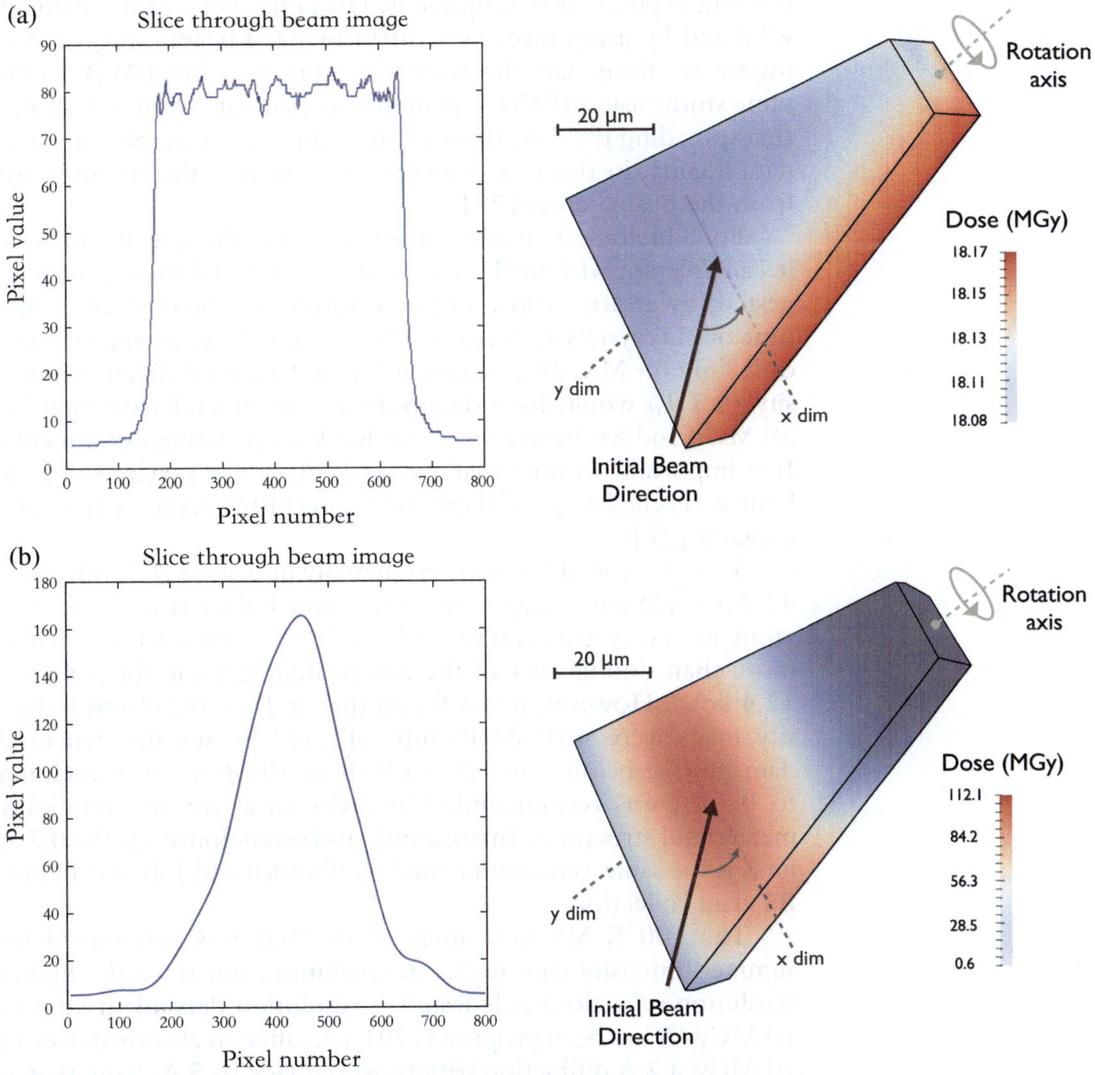

Fig. 2 RADDOSE-3D calculated dose distributions in a TRAP-RNA complex crystal rotated 180° while irradiated in two different X-ray beams, with (**a**) a top-hat profile (from EMBL beamline P14, PETRA III, DESY, Hamburg) giving dose values between 18.08 and 18.17 MGy, and (**b**) a typical Gaussian profile causing a much greater dose inhomogeneity of 0.59–112.1 MGy

datasets from 34 different crystals correlating beam profile with data quality showed that the beam profile was a vital parameter required to explain the results [24].

It is clear that it is hard to quote a representative dose for a crystal such as the one shown in Fig. 2a, where the values range from 0 up to 112 MGy. The options include the average dose for the whole crystal or the maximum dose. To address this problem, a new metric, Diffraction Weighted Dose (DWD), has been introduced, and is computed by RADDOSE-3D. It combines information from the aggregation of dose within each volume element of the crystal up to a given time, with the way the crystal is being exposed at that moment. DWD has been experimentally validated by using three very different sized beams and comparing the resulting data characteristics plotted against DWD. In the same study, using DWD to plan the experiment, it was also shown that spreading the dose through the sample more evenly improves data quality, in this case achieved by offsetting the rotation axis from the beam center [25].

For a biological sample, there is a limit to the level of radiation it can tolerate without losing its integrity. Henderson [26] suggested, by analogy and using observations of the diffraction lifetime of 2D crystals in electron microscopy at 77 K, an approximate dose limit for MX, $D_{1/2}$, after which half the total diffraction intensity (0.5 I_0) would have disappeared. The "Henderson limit" is 20 MGy, and has been a useful yardstick for planning experiments. It is important to note that a crystal might not survive until the limit is reached (e.g., if there were susceptible residues at crystal contacts [27]).

The $D_{1/2}$ for MX was experimentally measured at 100 K to be 43 MGy [28] using apo- (no iron) and holo-ferritin (one iron atom for every two amino acid residues, leading to μ_{abs} that is more than double that of the apo-protein enzyme for X-rays at 12.4 keV). However, it was found that at $D_{1/2}$, the electron density maps were very significantly affected by specific structural damage (see below), and thus a limit of 30 MGy, corresponding to 0.7 I_0, was recommended in order to arrive at biologically meaningful structures. Interestingly and serendipitously, the 0.7 I_0 limit is the same as recommended by Blundell and Johnson [1] for RT data collection.

The 100 K MX dose limit of 30 MGy was determined for summed intensity data to 2.2 Å resolution, but since the higher resolution reflections fade fastest, a resolution dependent limit of 10 MGy/Å has been proposed [29], i.e., after an absorbed dose of 10 MGy, a 2 Å diffraction pattern would fade to 3 Å. Note that at RT there appears to be a large range of doses (from a few kGy to ~1.5 MGy) for 0.7 I_0 tolerated by protein crystals, but 150 kGy has recently been suggested as a suitable RT limit [30].

5 Global Radiation Damage at Cryotemperatures

The effects of radiation damage which are observed in reciprocal space are together termed "global" damage. The most visible sign as the absorbed dose increases is the gradual fading of the intensity of the diffraction pattern, with the highest-resolution reflections being the first to disappear [9]. Thus the resolution limit of the data is degraded. Once the data are processed, other effects become clear from the inspection of the cell parameters and merging statistics: the unit cell volume increases with dose [31–33], Wilson *B*-factors increase, merging *R* factors worsen, $I/\sigma(I)$ decreases as the intensity decreases and the noise ($\sigma(I)$) concomitantly increases, and the mosaicity usually increases with dose.

The expansion of unit cell with dose was originally thought to be a possible metric that could be used at the beam line during the experiment to judge whether a crystal should be discarded due to excessive radiation damage [31]. However, systematic experimental studies of the phenomenon led to the conclusion that the rate of cell volume increase was too variable, even among fragments of the same large crystal, for it to be a robust metric [34, 35]. The volume expansion is thought to be caused by hydrogen gas produced from radiolytic reactions, which gathers at domain boundaries [36] causing the unit cell to increase in size.

There are three indications of global radiation damage that can be usefully plotted as a function of absorbed dose, *D* (Fig. 3). The first is the summed intensity of a dataset or data wedge (I_n) divided by the initial summed intensity of that dataset or wedge (I_1), I_n/I_1 [28]. This is a preferred representation of the intensity decay over $I/\sigma(I)$, since $\sigma(I)$ increases with dose. The second informative plot is of the relative scaling *B*-factor, B_{rel}, which is the difference in Wilson *B*-factor between the *n*th dataset or wedge and that of the first dataset or wedge. This function has been observed to rise linearly with dose and its gradient gives an indication of the radiation sensitivity of the crystal. This coefficient of sensitivity is defined as $s_{\text{AD}} = \Delta B_{\text{rel}}/8\pi^2\Delta D$ and has a value for hen egg white lysozyme (HEWL) crystals at 100 K of 0.012 Å^2/Gy. It does not appear to vary significantly with protein species [37]. The third indicator that is sometimes used is a pairwise *R*-factor, R_{d}, between identical and symmetry related reflections occurring on different diffraction images, plotted against the difference in dose ΔD between those images [38].

Unfortunately these three indicators can give inconsistent results for analysis of the same data (see for instance plots of I_n/I_1 and R_{d} in Figs. 3 and 4 of [39]). This is an as yet unresolved issue in systematic radiation damage studies.

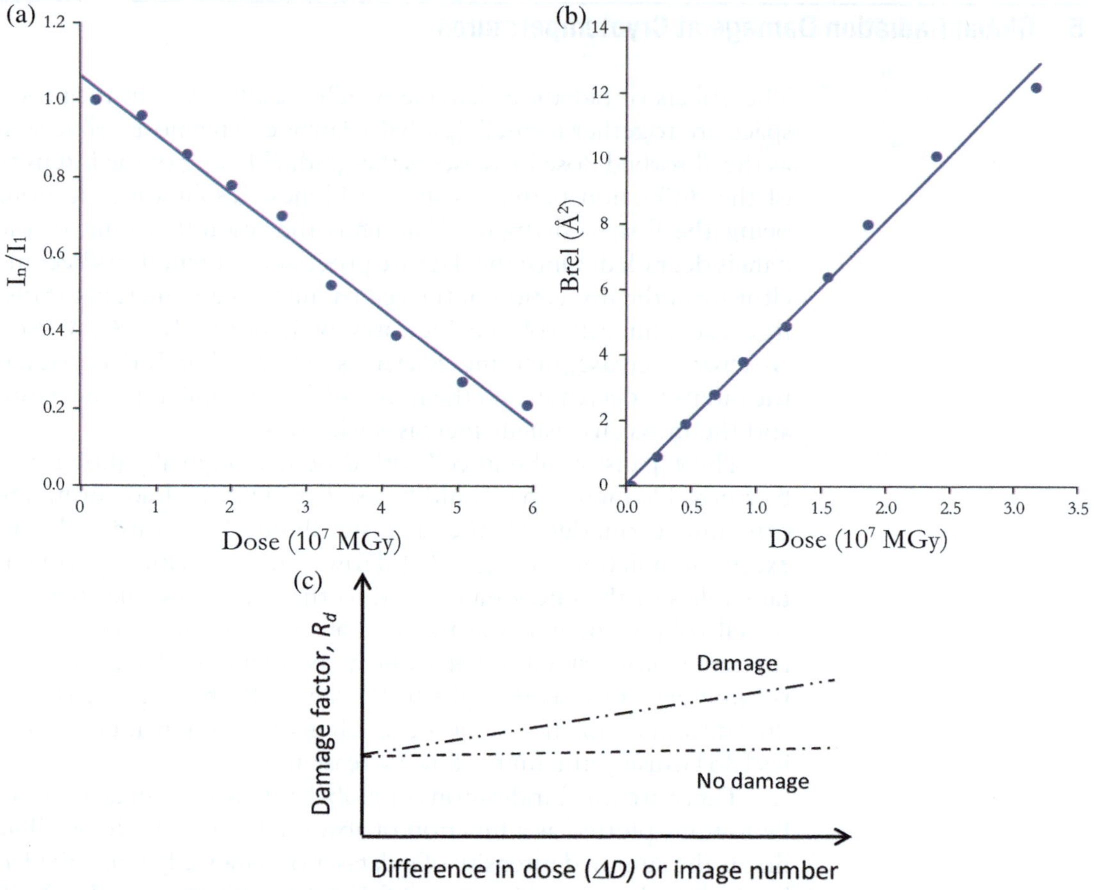

Fig. 3 Three indicators of global radiation damage. (**a**) Normalized summed intensity decay of complete consecutive datasets for a holoferritin crystal at 100 K, I_n/I_1, against dose, exhibiting approximately linear decay behaviour (fitted line shown) (**b**) Relative *B*-factor, $B_{rel} = B_n - B_1$ for the same crystal as in (**a**), showing linear increase (fitted line shown), (**c**) An idealized plot of R_d [38], the pairwise merging *R*-factor ("decay" or "damage" factor) between identical and symmetry-related reflections occurring on different diffraction images, plotted against the difference in dose, ΔD, between the images on which the reflections were collected, or against the difference in image number if dose values are not available. The plot is a straight line parallel to the *x* axis if there is no damage, but rises approximately linearly in the presence of damage

6 Specific Structural Radiation Damage at Cryotemperatures

X-ray irradiation produces specific structural and chemical damage besides the global radiation damage that affects the diffraction power described above. Disulfide bridges were already identified as being particularly radiation-sensitive at RT as early as 1988 [40]. When highly brilliant third generation synchrotron sources came online in the mid-1990s, systematic studies at cryotemperatures identified changes in the electron density of disulfide bridges and glutamic and aspartic acid residues (Fig. 4) as being the first

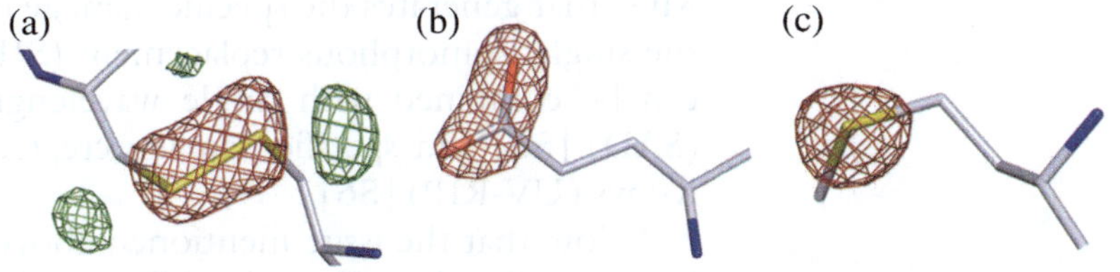

Fig. 4 Specific structural damage. (**a**) Disulfide bond cleavage, (**b**) Glu decarboxylation, and (**c**) Met sulfur disordering within myrosinase, PDB: 1DWA [32]. $F_{obs}(4)$-$F_{obs}(1)$ Fourier difference maps between dataset 1 and 4 collected on the same crystal are shown, contoured at +/−5 σ. Negative difference density (*red*) indicates disordering of the atomic positions with an accumulated dose of ~14 MGy. In (**a**), the positive difference density (*green*) indicates the repositioning of the two sulfurs upon cleavage of the disulfide bond. The maps were calculated using the phases of the structure refined from dataset 1

obvious signs of specific radiation damage, originating from reduction and decarboxylation, respectively [31, 32, 41]. Interestingly, chemically identical species appeared not to be equally affected, indicating that specific damage is not caused by X-ray absorption of the susceptible atom or group (primary damage), but is rather the consequence of interaction with secondary radicals generated by the primary event [33, 34] (the primary event typically being a result of the photoelectric effect, see above [27]). Disulfide bridges, for example, trap a secondary electron and then either elongate upon formation of a disulfide radical anion [42] or break following a complex multi-track model [43]. Among the residues suffering specific damage, those located in protein active sites have consistently been observed as being most susceptible (e.g., in acetylcholinesterase [41], bacteriorhodopsin [44], photoactive yellow protein [45], DNA photolyase [46], malate dehydrogenases [47], carbonic anhydrase [48], and fluorescent proteins [49]), possibly because they are subject to chemical and geometric strains [31, 50]. Consequently, the biological information extracted from protein crystal structures can be altered by convolution with radiation-induced structural and chemical modifications.

Although there is now a wide body of literature devoted to understanding the mechanisms behind the specific damage of proteins, far less is known regarding radiation-induced damage to crystalline nucleic acids and the wider class of nucleoprotein complexes. Recent systematic studies on a DNA-protein complex [51] and a 91 kDa RNA-protein complex [52] have clearly shown both DNA and RNA to be much more robust than protein in terms of specific structural damage.

The specific structural damage inflicted by X-rays can also be put to good use in so-called radiation-induced phasing (RIP) [53]. Two low-dose data sets are separated by an X-ray *burn* of several

MGy that generates the specific damage exploited for phasing as in the single isomorphous replacement (SIR) method [53, 54]. RIP can be combined with single wavelength anomalous diffraction (SAD) [55] and specific damage created by UV light instead of X-rays (UV-RIP) [56].

Note that the issue mentioned above concerning non-top-hat beam profiles also affects the differential specific damage rates, and these result in a mixture of molecular conformations across the crystal (e.g., disulfide bonds in various stages of breakage). Thus the resulting electron density maps will be an average over these states, and will not be as sharp as when a top-hat profile beam is used. The unit cell distribution will also be affected, those in the center expanding faster than those at the margins of the crystal. This causes non-isomorphism both within single datasets and between datasets during MAD experiments. The reflection intensity changes induced by these non-isomorphism effects can be larger than those expected between data sets recorded at different wavelengths, and thus can lead to failure of structure solution.

7 X-Ray Induced Changes in Chromophore-Containing Proteins at 100 K

The combination of X-ray crystallography with complementary *in crystallo* spectroscopic techniques, such as optical absorption spectroscopy [57], Raman spectroscopy [58], X-ray absorption spectroscopy [59, 60], and electron paramagnetic resonance [61], provided evidence for X-ray induced modifications in crystalline proteins containing chromophores such as metal complexes [62] or conjugated π-electron systems. Such modifications occur at doses over three orders of magnitude lower than the experimental limit of 30 MGy (see above) determined for global and specific radiation damage at 100 K. For the ferric (oxidized) heme iron in metmyoglobin [62], for instance, at 100 K 90% of the unreduced state remains after a dose (termed *spectroscopic lifedose* [63]) of 0.01 MGy [63]. Similarly, X-ray induced formation of a spectroscopically observed orange species in the retinal-containing membrane protein bacteriorhodopsin [44] occurs at a lifedose of 0.04 MGy [64]. Other X-ray induced modifications include heme reduction in a photosynthetic reaction center [65], in high-molecular weight cytochrome c [66] and in cytochrome c peroxidase [67], redox changes in a methylamine dehydrogenase [68], deprotonation of the bilin chromophore in a phytochrome [69], and photobleaching of a fluorescent protein [49].

Due to the particularly high radiation sensitivity of chromophore-containing proteins, it is important to identify their redox status by complementary spectroscopic methods when their structures are being solved by X-ray crystallography. As a second step, one or several strategies can be adopted to minimize or even prevent radiation-induced redox modifications (typically reduction). In a composite data-collection strategy, for instance, the X-ray dose is distributed over several locations of a large crystal [70] or over several crystals, as carried out by Aoyama et al., who collected data from 400 crystals to solve the structure of fully oxidized cytochrome c oxidase [71]. Alternatively, the data collection cryotemperature can be decreased to below the usual 100 K. Indeed, metal reduction has been reported to be reduced 30-fold when collecting data at 40 K instead of 110 K [72]. Additionally, certain scavengers can provide some protection against metal reduction [73].

Although X-ray-induced protein-chromophore reduction is generally a nuisance, it can be put to good use for studying macromolecular processes. A prominent example is the structural characterization of intermediates in the P450cam cytochrome reaction pathway, triggered by an electron generated by X-ray radiolysis of water [74]. Similarly, several redox intermediates of horseradish peroxidase have been generated by X-ray induced electrons and characterized by the aforementioned dose-dependent composite data-collection strategy [70]. More recently, enzyme catalysis in copper nitrite reductase has been elucidated by serial synchrotron crystallography (SSX) at 100 K [75] (see below). Furthermore, the catalytic cycle of the non-chromophore-containing urate oxidase, kick-started by X-ray absorption, has been studied by combining *in crystallo* Raman spectroscopy, QM/MM simulations and X-ray crystallography [76].

8 Temperature Dependence of Radiation Damage

Cryocrystallography [2] replaced RT data collection at synchrotron sources some two decades ago [77] because it increased the crystal lifetime in the X-ray beam by up to two orders of magnitude [8, 78, 79]. The main benefit of cryocooling is obtained when decreasing the temperature down to 200 K [80] where most atomic motions are quenched [81] because of vitrification (transition to a glass state) and dynamical transitions of the solvent and protein, respectively [82–84]. Decreasing the data collection temperature further to 40 K yields only a very small decrease in specific and global radiation damage compared to 100 K [36, 85, 86] (except for the large effect on metal reduction mentioned in the previous section), so that helium cooling has not replaced nitrogen cooling in standard cryocrystallographic experiments.

The combination of X-ray induced structural changes and temperature-controlled crystallography provides a means to initiate and study macromolecular functioning. In the P450cam cytochrome example mentioned above [74], the crystals were transiently thawed to RT after radiolytic electron generation to unlock motional freedom and allow the pathway to proceed. In another example, a substrate analog bound to crystalline acetylcholinesterase was radiolyzed in two dose-dependent data collection series at 100 K and at 155 K [87], respectively. Only at 155 K, but not at 100 K, was the protein flexibility high enough so that exit of the radiolytic products could be identified and followed structurally. Temperature-controlled crystallography has a long history (reviewed in [88]) and was initiated by Frauenfelder, Petsko, and Tsernoglou [89]. If radiation-induced changes are not deliberately incorporated into the experimental protocol as in the two examples above, careful control experiments must be conducted to avoid or deconvolute radiation-induced from temperature-induced changes, as illustrated by a recent multi-temperature study on crystals of the enzyme cyclophilin A [90].

9 Radiation Damage at Room Temperature

The first published study of radiation damage in MX was the seminal paper by Blake and Phillips [91] on RT myoglobin crystals. The conclusions drawn were that each 8 keV photon disrupts around 70 protein molecules and disorders a further 90 for doses up to about 0.2 MGy. The $D_{1/2}$ was reported as 0.59 MGy, around 70 times smaller than the experimental dose limit $D_{1/2}$ of 43 MGy [8]. Since the authors observed that the structure factors of some reflections increased slightly, while others decreased, they deduced that specific structural damage to particular amino acid residues must be occurring within the crystal, a hypothesis confirmed many years later [40].

Despite the high rate of damage, RT data collection is currently witnessing a renaissance for several reasons. Technically it has become possible to mount an entire crystallization tray onto a goniometer and irradiate the crystals in situ, with no crystal handling at all [92] and computational tools have been created to enable the images thus collected to be combined into complete datasets. In addition, methods have been developed for enclosing loop-mounted crystals in plastic sleeves [93], which are much less damaging to the crystals than transferring them to quartz or glass capillary tubes. Unenclosed loop-mounted crystals can be protected from drying at RT by special hydration devices [94]. From the standpoint of structural biology, is has now been recognized that the modification of conformational heterogeneity that may occur during cryocooling is avoided in RT experiments [95].

Despite the greatly increased radiation sensitivity at RT compared to 100 K, there is evidence that some of the global radiation damage can be outrun at high dose rates at (or at a lower but close to) RT [79, 96, 97]. Specific damage is very difficult to track in electron density derived from RT measurements, because of the reduced crystal lifetime hampering collection of consecutive dose-dependent datasets, which is possible at 100 K (see above). Indications of specific damage to disulfide bonds have been reported in traditional oscillation [79] and SSX [98] experiments. However, a more recent SSX study does not report any specific damage, even though the diffraction power of the crystals decreased to 20% of its original value [30]. The current rapid development of SSX will allow in-depth studies of RT radiation damage because the dose required for collecting a data set is spread over thousands of crystals or more.

10 Practical Aspects: Wavelength (In)dependence

The question of whether damage rates depend on the incident X-ray wavelength has been both theoretically and experimentally addressed for MX. Arndt [99] pointed out that both the elastic and photoelectric effect cross sections are greater at lower energy, so in fact the ratio of cross sections for diffraction to absorption events does not change significantly over the incident X-ray energy range used in MX. This was experimentally investigated by Weiss and coworkers [100], who analyzed the structures derived from elastase data collected at X-ray energies of 12.4 and 6.2 keV, and concluded that there was no difference in rates of specific damage to disulfides and cadmium ions. This result was corroborated by Shimizu and coworkers [101], who collected data at 6.5, 7.1, 8.3, 9.9, 12.4, 16.5, 20.0, 24.8, and 33.0 keV from HEWL crystals, and detected no difference in the rate of specific damage in the corresponding refined structures nor their final atomic *B*-factors. Moreover, neither were there any significant differences in any of the global damage metrics at the various energies. However, Homer and coworkers [102] found that the rate of electron density decrease (electrons/Å^3/MGy) was greater at 14 keV than at 9 keV for cysteine sulfur atoms involved in disulfide bridges in HEWL crystals, although no statistically significant differences in the decay rates were found for methionine S atoms. Also, Fourme and coworkers reported an eightfold increase in data collection efficiency at 33 keV compared with at 8 keV [103]. More recently, Liebschner and coworkers [104] took a series of diffraction images from thaumatin crystals at energies of 6.33, 12.66, and 19.00 keV at 100 K over small (2°) repetitive rotation intervals and found that for 2.45 Å resolution data, $D_{0.7}$ was 7.5 MGy for the 6.33 keV dataset and for the two higher energies it was ~11 MGy. Thus there is as yet no consensus on the question of the dependence of radiation damage rates on incident radiation wavelength.

11 Practical Aspects: Is There a Dose Rate Effect at 100 K?

Evidence from experiments with the flux densities currently utilized is that there is no significant dose rate effect at 100 K. Gonzalez and Nave [9] used two different attenuator settings on a second generation synchrotron beamline, and showed similar reflection intensity decay rates. With flux densities of up to 10^{15} photons/s/mm^2, Sliz and coworkers [105] also detected no dose rate differences in global decay metrics. Another study described similar observations when monitoring global data quality indicators, but on the basis of an analysis of difference electron density maps, concluded that there could be a second-order dose rate effect [106]. A small dose rate dependent $D_{1/2}$ decrease was observed by Owen and coworkers [28] when monitoring the summed intensity loss. Close to RT, however, significant dose rate effects have been observed (see above).

Under the experimental conditions used in all these studies, crystal heating is not thought to be significant [107]. However with the advent of fourth generation synchrotron beams for MX, this issue should be revisited, since if the beam induces temperature rises above 110 K in the sample, hydroxyl radicals will become mobile and diffuse through the crystal, thus potentially increasing the rate of damage.

12 Practical Aspects: Scavengers

A possible way to reduce the rate of radiation damage is to make use of small molecule radioprotectants, either by adding them to the cryobuffer prior to flash cooling the crystals, or by soaking the crystals before an RT experiment. These compounds either intercept free radicals (radical scavengers) preventing them from reaching the protein, or repair centers of ionization already present. A number of studies reporting conflicting results have been published, showing significant variation between crystals treated in nominally the same way with scavengers: all (46) different compounds tried prior to 2011 are summarized in a table in the supplementary material of [108]. Some scavengers have been tested by several groups but the results are not always in accord. There is thus a lack of consensus on the efficacy of scavengers that have been investigated by various means. Very few produced more than a twofold increase of $D_{1/2}$, which is a criterion suggested by Holton [109] for judging effective radiation damage mitigation strategies.

A number of scavenger studies have observed little, or even adverse effects induced by these additives. For instance, Kmetko and coworkers [110] investigated 19 putative scavengers using the B_{rel} metric, and found none were effective at 100 K. At RT, they reported that 12 were ineffective, six sensitized the crystals, and only one, sodium nitrate (which is a very efficient electron scavenger), had a protective effect. However, Barker and coworkers [111] found that 1,4-benzoquinone at RT increased the dose tolerance of HEWL crystals to global damage by a factor of 9 as monitored by intensity decay, and significantly reduced the specific damage to susceptible residues (with some disulfides remaining undamaged even at 0.62 MGy). For sodium nitrate at 100 K, de la Mora and coworkers [39] found that specific damage to disulfides was reduced by more than a factor of 5 compared to the structure determined from an HEWL crystal not soaked in nitrate, while the global damage, as monitored by diffraction intensity decay, was lessened by a factor of 2. In that study, in the soaked-crystal structure, a bound nitrate anion adjacent to a disulfide bond was seen to be reduced as a function of dose and to significantly protect the bond from cleavage (Fig. 5) compared to the non-soaked case. This observation reinforces the evidence that breakage of disulfide bonds is caused by mobile electrons which travel round the protein structure at 100 K and reduce the most electron-affinic group (in the absence of bound metal cations), and illustrates radiation chemistry in action.

Due to the fundamental disagreement within the MX literature on the general utility of scavengers, they are seldom employed to slow the progression of damage, despite some impressive anecdotal evidence that they can be effective and are worth trying. It should

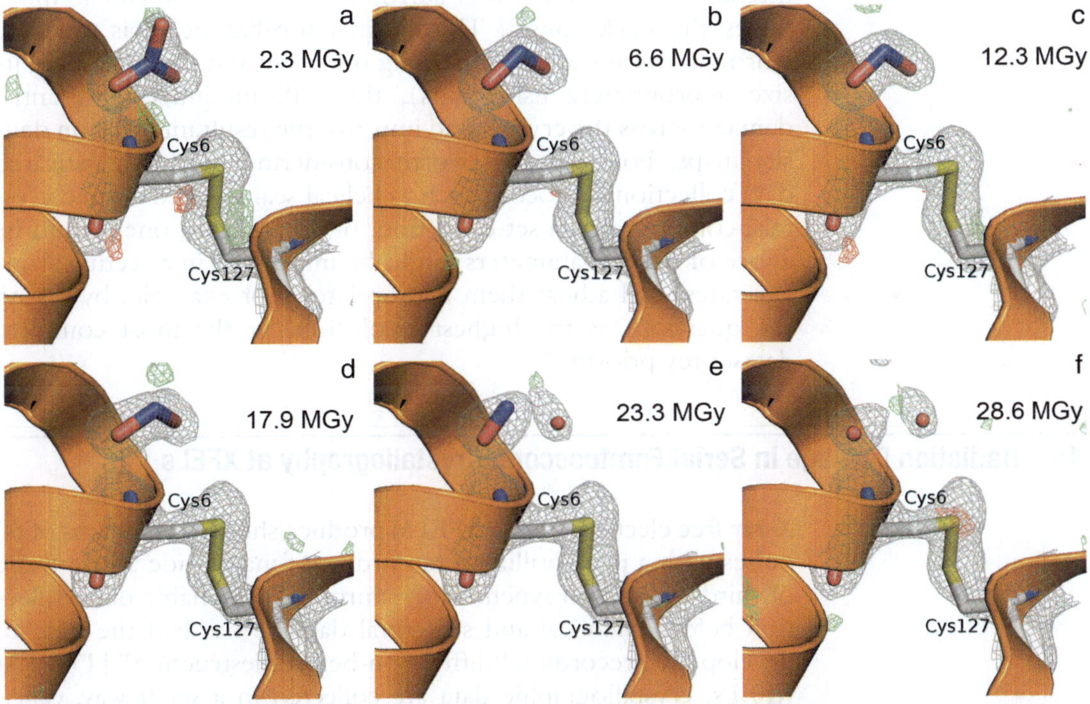

Fig. 5 Radiation chemistry of a scavenger. Dose-dependent nitrate reduction observed in an HEWL crystal soaked in 0.5 M $NaNO_3$ for 4 min prior to cryocooling and data collection. The vicinity of the Cys6–Cys127 disulfide bond is shown [panels (**a**)–(**f**) correspond to structures derived from consecutive datasets 1–6 at 2.3–28.6 MGy, respectively]. The $2F_o - F_c$ map (*grey*) is contoured at 1.0σ and the $F_o - F_c$ map is contoured at $\pm 3.0\sigma$ (green: +; red: −). A bound nitrate anion is apparently reduced to NO_2 (6.6 MGy) and then to NO (23.3 MGy) by mobile electrons produced by the X-ray beam, and the disulfide bond is protected from damage until the nitrate scavenging capacity is exhausted. The refined models corresponding to the PDB codes 2byh, 2byi, 2byj, 2byl, 2bym, 2byn are shown in panels (**a**)–(**f**), respectively

be noted that many components of crystallization buffers, and in particular cryoprotectants, are already good scavengers for some of the damage agents. For instance, ethylene glycol, PEG and glycerol have high rate constants for scavenging hydroxyl radicals.

13 What Can the Experimenter Do to Minimize Radiation Damage Rates?

Despite the systematic studies mentioned above, the pivotal question remains: what can the experimenter do to minimize radiation damage rates? There are several general strategies that can make a difference to the damage rate during a data collection. Firstly, if there are any heavy atoms in the buffer (e.g., cacodylate, which contains arsenic), back-soaking can reduce the absorption coefficients of the crystal and thus the rate of damage significantly. Holton [109] used RADDOSE to compute the "dose doubling concentrations" of various buffers, and for arsenic this is only 350 mM. Secondly, matching the beam size to the crystal reduces interactions with the buffer surrounding the crystal and so minimizes the background. Thirdly, if a top-hat beam is available (through, for instance, defocusing the beam and then reducing its size appropriately using slits), this will minimize differential damage across the crystal, and improve the resultant electron density maps. Fourthly, it is worth considering more sophisticated data collection protocols, such as helical scans of rod shaped crystals, composite data set collection, or SSX. Lastly, one should be aware of which parameters might be important in affecting damage rates, and adjust them appropriately, for example, by asking the question "is the highest resolution, or the most complete dataset my priority?"

14 Radiation Damage in Serial Femtosecond Crystallography at XFELs

X-ray free electron lasers (XFELs) produce short (several tens of fs) pulses with a peak brilliance ten orders of magnitude above those of third generation synchrotron sources. They enable data collection before chemical and structural damage has had the time to develop, i.e., recording "diffraction-before-destruction" [112]. At XFELs, crystallographic data are collected in a serial way, called serial femtosecond crystallography (SFX) [113], that leads to high-resolution protein structures [114], in most cases devoid of specific radiation damage [115, 116]. However, if SFX data are collected at very high doses (e.g., up to 3 GGy [117], but see the cautionary note above on dose calculations for small crystals), global radiation damage has been reported [117, 118], and there is evidence that specific damage might occur [117]. Such specific damage has been generated and studied in detail by deliberately exposing ferredoxin crystals to unattenuated 80 fs XFEL pulses at Stanford, leading to an absorbed dose of 30 GGy per crystal [119]. In comparison to a low-dose synchrotron data set, the SFX XFEL data showed reduced electron density for the iron atoms of the two [4Fe-4S] clusters in ferredoxin, with the effect being stronger in one of the clusters. Specific damage at sulfur sites in cathepsin B has been identified in a difference Fourier map computed from SFX data collected with a high and a low X-ray fluence [120]. Consequently, XFEL pulses need to be kept short and attenuated in order to avoid global and specific radiation damage.

Acknowledgments

We thank Ian Carmichael and Kathryn Shelley for their comments on this contribution, Charles Bury and Eugenio de la Mora for making the figures, and Markus Gerstel for carrying out a survey of the PDB regarding the temperature of data collection. We also acknowledge those many colleagues with whom we have discussed the issues related to radiation damage in MX. We especially thank the ESRF for access to beamtime since 2000 under the Radiation Damage BAG, which has allowed us to carry out systematic studies on many aspects of the topic.

References

1. Blundell TL, Johnson LN (1976) Protein crystallography. Academic Press, London
2. Hope H (1988) Cryocrystallography of biological macromolecules: a generally applicable method. Acta Crystallogr B 44:22–26
3. Teng T (1990) Mounting of crystals for macromolecular crystallography in a free-standing thin film. J Appl Cryst 23:387–391
4. Rodgers DW (1994) Cryocrystallography. Structure 2:1135–1140
5. Rodgers DW (1997) Practical cryocrystallography. Methods Enzymol 276:183–203
6. Garman EF, Schneider TR (1997) Macromolecular cryocrystallography. J Appl Cryst 30:211–237
7. Garman E (1999) Cool data: quantity AND quality. Acta Crystallogr D Biol Crystallogr 55:1641–1653
8. Nave C, Garman EF (2005) Towards an understanding of radiation damage in cryocooled macromolecular crystals. J Synchrotron Radiat 12:257–260
9. Gonzalez A, Nave C (1994) Radiation damage in protein crystals at low temperature. Acta Crystallogr D Biol Crystallogr 50:874–877
10. Nave C (1995) Applications of synchrotron X-radiation. Radiation damage in protein crystallography. Radiat Phys Chem 45:483–490
11. Sanishvili R, Yoder DW, Pothineni SB et al (2011) Radiation damage in protein crystals is reduced with a micron-sized X-ray beam. Proc Natl Acad Sci U S A 108:6127–6132
12. O'Neill P, Stevens DL, Garman EF (2002) Physical and chemical considerations of damage induced in protein crystals by synchrotron radiation: a radiation chemical perspective. J Synchrotron Radiat 9:329–332
13. Cosier J, Glazer AM (1986) A nitrogen-gas-stream cryostat for general X-ray diffraction studies. J Appl Cryst 19:105–107
14. Owen RL, Axford D, Nettleship JE et al (2012) Outrunning free radicals in room-temperature macromolecular crystallography. Acta Crystallogr D Biol Crystallogr 68:810–818
15. Jones GD, Lea JS, Symons MC et al (1987) Structure and mobility of electron gain and loss centres in proteins. Nature 330:772–773
16. Owen RL, Holton JM, Schulze-Briese C et al (2009) Determination of X-ray flux using silicon pin diodes. J Synchrotron Radiat 16:143–151

17. Zeldin OB, Gerstel M, Garman EF (2013) RADDOSE-3D: time- and space-resolved modelling of dose in macromolecular crystallography. J Appl Cryst 46:1225–1230
18. Zeldin OB, Gerstel M, Garman EF (2013) Optimizing the spatial distribution of dose in X-ray macromolecular crystallography. J Synchrotron Radiat 20:49–57
19. Flot D, Mairs T, Giraud T et al (2010) The ID23-2 structural biology microfocus beamline at the ESRF. J Synchrotron Radiat 17:107–118
20. Cowan JA, Nave C (2008) The optimum conditions to collect X-ray data from very small samples. J Synchrotron Radiat 15:458–462
21. Murray JW, Garman EF, Ravelli RBG (2004) X-ray absorption by macromolecular crystals: the effects of wavelength and crystal composition on absorbed dose. J Appl Cryst 37:513–522
22. Paithankar KS, Owen RL, Garman EF (2009) Absorbed dose calculations for macromolecular crystals: improvements to RADDOSE. J Synchrotron Radiat 16:152–162
23. Paithankar KS, Garman EF (2010) Know your dose: RADDOSE. Acta Crystallogr D Biol Crystallogr 66:381–388
24. Krojer T, von Delft F (2011) Assessment of radiation damage behaviour in a large collection of empirically optimized datasets highlights the importance of unmeasured complicating effects. J Synchrotron Radiat 18:387–397
25. Zeldin OB, Brockhauser S, Bremridge J et al (2013) Predicting the X-ray lifetime of protein crystals. Proc Natl Acad Sci U S A 110:20551–20556
26. Henderson R (1990) Cryo-protection of protein crystals against radiation damage in electron and X-ray diffraction. Proc R Soc London Ser B 241:6–8
27. Murray JW, Rudino-Pinera E, Owen RL et al (2005) Parameters affecting the X-ray dose absorbed by macromolecular crystals. J Synchrotron Radiat 12:268–275
28. Owen RL, Rudino-Pinera E, Garman EF (2006) Experimental determination of the radiation dose limit for cryocooled protein crystals. Proc Natl Acad Sci U S A 103:4912–4917
29. Howells MR, Beetz T, Chapman HN et al (2009) An assessment of the resolution limitation due to radiation-damage in x-ray diffraction microscopy. J Electron Spectrosc 170:4–12
30. Roedig P, Duman R, Sanchez-Weathertby J et al (2016) Room-temperature macromolecular crystallography using a micro-patterned silicon chip with minimal background scattering. J Appl Cryst 49:968–975
31. Ravelli RB, McSweeney SM (2000) The 'fingerprint' that X-rays can leave on structures. Structure 8:315–328
32. Burmeister WP (2000) Structural changes in a cryo-cooled protein crystal owing to radiation damage. Acta Crystallogr D Biol Crystallogr 56:328–341
33. Teng T, Moffat K (2000) Primary radiation damage of protein crystals by intense synchrotron radiation. J Synchrotron Radiat 7:313–317
34. Murray J, Garman E (2002) Investigation of possible free-radical scavengers and metrics for radiation damage in protein cryocrystallography. J Synchrotron Radiat 9:347–354
35. Ravelli RB, Theveneau P, McSweeney S et al (2002) Unit-cell volume change as a metric of radiation damage in crystals of macromolecules. J Synchrotron Radiat 9:355–360
36. Meents A, Gutmann S, Wagner A et al (2010) Origin and temperature dependence of radiation damage in biological samples at cryogenic temperatures. Proc Natl Acad Sci U S A 107:1094–1099
37. Kmetko J, Husseini NS, Naides M et al (2006) Quantifying X-ray radiation damage in protein crystals at cryogenic temperatures. Acta Crystallogr D Biol Crystallogr 62:1030–1038
38. Diederichs K, McSweeney S, Ravelli RB (2003) Zero-dose extrapolation as part of macromolecular synchrotron data reduction. Acta Crystallogr D Biol Crystallogr 59:903–909
39. De la Mora E, Carmichael I, Garman EF (2011) Effective scavenging at cryotemperatures: further increasing the dose tolerance of protein crystals. J Synchrotron Radiat 18:346–357
40. Helliwell JR (1988) Protein crystal perfection and the nature of radiation damage. J Cryst Growth 90:259–272
41. Weik M, Ravelli RB, Kryger G et al (2000) Specific chemical and structural damage to proteins produced by synchrotron radiation. Proc Natl Acad Sci U S A 97:623–628
42. Weik M, Berges J, Raves ML et al (2002) Evidence for the formation of disulfide radicals in protein crystals upon X-ray irradiation. J Synchrotron Radiat 9:342–346
43. Sutton KA, Black PJ, Mercer KR et al (2013) Insights into the mechanism of X-ray-induced disulfide-bond cleavage in lysozyme crystals based on EPR, optical absorption and X-ray diffraction studies. Acta Crystallogr D Biol Crystallogr 69:2381–2394

44. Matsui Y, Sakai K, Murakami M et al (2002) Specific damage induced by X-ray radiatioin and structural changes in the primary photoreaction of bacteriorhodopsin. J Mol Biol 324:469–481
45. Kort R, Hellingwerf KJ, Ravelli RB (2004) Initial events in the photocycle of photoactive yellow protein. J Biol Chem 279:26417–26424
46. Mees A, Klar T, Gnau P et al (2004) Crystal structure of a photolyase bound to a CPD-like DNA lesion after in situ repair. Science 306:1789–1793
47. Fioravanti E, Vellieux FM, Amara P et al (2007) Specific radiation damage to acidic residues and its relation to their chemical and structural environment. J Synchrotron Radiat 14:84–91
48. Sjoblom B, Polentarutti M, Djinovic-Carugo K (2009) Structural study of X-ray induced activation of carbonic anhydrase. Proc Natl Acad Sci U S A 106:10609–10613
49. Adam V, Carpentier P, Violot S et al (2009) Structural basis of X-ray-induced transient photobleaching in a photoactivatable green fluorescent protein. J Am Chem Soc 131:18063–18065
50. Dubnovitsky AP, Ravelli RB, Popov AN et al (2005) Strain relief at the active site of phosphoserine aminotransferase induced by radiation damage. Protein Sci 14:1498–1507
51. Bury C, Garman EF, Ginn HM et al (2015) Radiation damage to nucleoprotein complexes in macromolecular crystallography. J Synchrotron Radiat 22:213–224
52. Bury CS, McGeehan JE, Antson AA et al (2016) RNA protects a nucleoprotein complex against radiation damage. Acta Crystallogr D Biol Crystallogr 72:648–657
53. Ravelli RB, Leiros HK, Pan B et al (2003) Specific radiation damage can be used to solve macromolecular crystal structures. Structure 11:217–224
54. Banumathi S, Zwart PH, Ramagopal UA et al (2004) Structural effects of radiation damage and its potential for phasing. Acta Crystallogr D Biol Crystallogr 60:1085–1093
55. Ravelli RB, Nanao MH, Lovering A et al (2005) Phasing in the presence of radiation damage. J Synchrotron Radiat 12:276–284
56. Nanao MH, Ravelli RB (2006) Phasing macromolecular structures with UV-induced structural changes. Structure 14:791–800
57. McGeehan J, Ravelli RGB, Murray JW et al (2009) Colouring cryo-cooled crystals: online microspectrophotometry. J Synchrotron Radiat 16:163–172
58. Carpentier P, Royant A, Ohana J et al (2007) Advances in spectroscopic methods for biological crystals. 2. Raman spectroscopy. J Appl Cryst 40:1113–1122
59. Yano J, Kern J, Irrgang K-D et al (2005) X-ray damage to the Mn4Ca complex in single crystals of photosystem II: a case study for metalloprotein crystallography. Proc Natl Acad Sci U S A 102:12047–12052
60. Holton JM (2007) XANES measurements of the rate of radiation damage to selenomethionine side chains. J Synchrotron Radiat 14:51–72
61. Utschig LM, Chemerisov SD, Tiede DM et al (2008) Electron paramagnetic resonance study of radiation damage in photosynthetic reaction center crystals. Biochemistry 47:9251–9257
62. Beitlich T, Kuhnel K, Schulze-Briese C et al (2007) Cryoradiolytic reduction of crystalline heme proteins: analysis by UV-Vis spectroscopy and X-ray crystallography. J Synchrotron Radiat 14:11–23
63. Hersleth HP, Andersson KK (2011) How different oxidation states of crystalline myoglobin are influenced by X-rays. Biochim Biophys Acta 1814:785–796
64. Borshchevskiy V, Round E, Erofeev I et al (2014) Low-dose X-ray radiation induces structural alterations in proteins. Acta Crystallogr D Biol Crystallogr 70:2675–2685
65. Baxter RH, Seagle BL, Ponomarenko N et al (2004) Specific radiation damage illustrates light-induced structural changes in the photosynthetic reaction center. J Am Chem Soc 126:16728–16729
66. Sato M, Shibata N, Morimoto Y et al (2004) X-ray induced reduction of the crystal of high-molecular-weight cytochrome c revealed by microspectrophotometry. J Synchrotron Radiat 11:113–116
67. Echalier A, Goodhew CF, Pettigrew GW et al (2006) Activation and catalysis of the di-heme cytochrome c peroxidase from Paracoccus pantotrophus. Structure 14:107–117
68. Pearson AR, Pahl R, Kovaleva EG et al (2007) Tracking X-ray-derived redox changes in crystals of a methylamine dehydrogenase/amicyanin complex using single-crystal UV/Vis microspectrophotometry. J Synchrotron Radiat 14:92–98
69. Li F, Burgie ES, Yu T et al (2015) X-ray radiation induces deprotonation of the bilin chromophore in crystalline D Radiodurans phytochrome. J Am Chem Soc 137:2792–2795
70. Berglund GI, Carlsson GH, Smith AT et al (2002) The catalytic pathway of horseradish peroxidase at high resolution. Nature 417:463–468

71. Aoyama H, Muramoto K, Shinzawa-Itoh K et al (2009) A peroxide bridge between Fe and Cu ions in the O2 reduction site of fully oxidized cytochrome c oxidase could suppress the proton pump. Proc Natl Acad Sci U S A 106:2165–2169
72. Corbett MC, Latimer MJ, Poulos TL et al (2007) Photoreduction of the active site of the metalloprotein putidaredoxin by synchrotron radiation. Acta Crystallogr D Biol Crystallogr 63:951–960
73. Macedo S, Pechlaner M, Schmid W et al (2009) Can soaked-in scavengers protect metalloprotein active sites from reduction during data collection? J Synchrotron Radiat 16:191–204
74. Schlichting I, Berendzen J, Chu K et al (2000) The catalytic pathway of cytochrome p450cam at atomic resolution. Science 287:1615–1622
75. Horrell S, Antonyuk SV, Eady RR et al (2016) Serial crystallography captures enzyme catalysis in copper nitrite reductase at atomic resolution from one crystal. IUCr J 3:271–281
76. Bui S, von Stetten D, Jambrina PG et al (2014) Direct evidence for a peroxide intermediate and a reactive enzyme-substrate-dioxygen configuration in a cofactor-free oxidase. Angew Chem Int Ed 53:13710–13714
77. Garman EF, Doublie S (2003) Cryocooling of macromolecular crystals: optimization methods. Methods Enzymol 368:188–216
78. Southworth-Davies RJ, Medina MA, Carmichael I et al (2007) Observation of decreased radiation damage at higher dose rates in room temperature protein crystallography. Structure 15:1531–1541
79. Warkentin M, Hopkins JB, Badeau R et al (2013) Global radiation damage: temperature dependence, time dependence and how to outrun it. J Synchrotron Radiat 20:7–13
80. Warkentin M, Thorne RE (2010) Glass transition in thaumatin crystals revealed through temperature-dependent radiation-sensitivity measurements. Acta Crystallogr D Biol Crystallogr 66:1092–1100
81. Halle B (2004) Biomolecular cryocrystallography: structural changes during flash-cooling. Proc Natl Acad Sci U S A 101:4793–4798
82. Vitkup D, Ringe D, Petsko GA et al (2000) Solvent mobility and the protein 'glass' transition. Nat Struct Biol 7:34–38
83. Weik M, Kryger G, Schreurs AM et al (2001) Solvent behaviour in flash-cooled protein crystals at cryogenic temperatures. Acta Crystallogr D Biol Crystallogr 57:566–573
84. Schiro G, Fichou Y, Gaillat F-X et al (2015) Translational diffusion of hydration water correlates with functional motions in folded and intrinsically disordered proteins. Nat Commun 6:6490
85. Chinte U, Shah B, Chen Y-S et al (2007) Cryogenic (<20 K) helium cooling mitigates radiation damage to protein crystals. Acta Crystallogr D Biol Crystallogr 63:486–492
86. Meents A, Wagner A, Schneider R et al (2007) Reduction of X-ray-induced radiation damage of macromolecular crystals by data collection at 15 K: a systematic study. Acta Crystallogr D Biol Crystallogr 63:302–309
87. Colletier JP, Bourgeois D, Sanson B et al (2008) Shoot-and-Trap: use of specific x-ray damage to study structural protein dynamics by temperature-controlled cryo-crystallography. Proc Natl Acad Sci U S A 105:11742–11747
88. Weik M, Colletier J-P (2010) Temperature-dependent macromolecular X-ray crystallography. Acta Crystallogr D Biol Crystallogr 66:437–446
89. Frauenfelder H, Petsko GA, Tsernoglou D (1979) Temperature-dependent X-ray diffraction as a probe of protein structural dynamics. Nature 280:558–563
90. Keedy DA, Kenner LR, Warkentin M et al (2015) Mapping the conformational landscape of a dynamic enzyme by multitemperature and XFEL crystallography. Elife 4:e07574
91. Blake C, Phillips DC (1962) Effects of X-irradiation on single crystals of myoglobin. In: Proceedings of the Symposium on the Biological Effects of Ionising Radiation at the Molecular Level, Vienna, pp 183–191
92. Jacquamet L, Ohana J, Joly J et al (2004) Automated analysis of vapor diffusion crystallization drops with an X-ray beam. Structure 12:1219–1225
93. Kalinin Y, Kmetko J, Bartnik A et al (2005) A new sample mounting technique for room-temperature macromolecular crystallography. J Appl Cryst 38:333–339
94. Sanchez-Weatherby J, Bowler MW, Huet J et al (2009) Improving diffraction by humidity control: a novel device compatible with X-ray beamlines. Acta Crystallogr D Biol Crystallogr 65:1237–1246
95. Fraser JS, van den Bedem H, Samelson AJ et al (2011) Accessing protein conformational ensembles using room-temperature X-ray crystallography. Proc Natl Acad Sci U S A 108:16247–16252
96. Warkentin M, Badeau R, Hopkins JB et al (2012) Global radiation damage at 300 and 260 K with dose rates approaching 1 MGy s^{-1}. Acta Crystallogr D Biol Crystallogr 68:124–133

97. Owen RL, Paterson N, Axford D et al (2014) Exploiting fast detectors to enter a new dimension in room-temperature crystallography. Acta Crystallogr D Biol Crystallogr 70:1248–1256
98. Coquelle N, Brewster AS, Kapp U et al (2015) Raster-scanning serial protein crystallography using micro- and nano-focused synchrotron beams. Acta Crystallogr D Biol Crystallogr 71:1184–1196
99. Arndt UW (1984) Optimum X-ray wavelength for protein crystallography. J Appl Cryst 17:118–119
100. Weiss MS, Panjikar S, Mueller-Dieckmann C et al (2005) On the influence of the incident photon energy on the radiation damage in crystalline biological samples. J Synchrotron Radiat 12:304–309
101. Shimizu N, Hirata K, Hasegawa K et al (2007) Dose dependence of radiation damage for protein crystals studied at various X-ray energies. J Synchrotron Radiat 14:4–10
102. Homer C, Cooper L, Gonzalez A (2011) Energy dependence of site-specific radiation damage in protein crystals. J Synchrotron Radiat 18:338–345
103. Fourme R, Honkimäki V, Girard E et al (2012) Reduction of radiation damage and other benefits of short wavelengths for macromolecular crystallography data collection. J Appl Cryst 45:652–661
104. Liebschner D, Rosenbaum G, Dauter M et al (2015) Radiation decay of thaumatin crystals at three X-ray energies. Acta Crystallogr D Biol Crystallogr 71:772–778
105. Sliz P, Harrison SC, Rosenbaum G (2003) How does radiation damage in protein crystals depend on X-ray dose? Structure 11:13–19
106. Leiros HK, Timmins J, Ravelli RB et al (2006) Is radiation damage dependent on the dose rate used during macromolecular crystallography data collection? Acta Crystallogr D Biol Crystallogr 62:125–132
107. Mhaisekar A, Kazmierczak MJ, Banerjee R (2005) Three-dimensional numerical analysis of convection and conduction cooling of spherical biocrystals with localized heating from synchrotron X-ray beams. J Synchrotron Radiat 12:318–328
108. Allan EG, Kander MC, Carmichael I et al (2013) To scavenge or not to scavenge, that is STILL the question. J Synchrotron Radiat 20:23–36
109. Holton JM (2009) A beginner's guide to radiation damage. J Synchrotron Radiat 16:133–142
110. Kmetko J, Warkentin M, Englich U et al (2011) Can radiation damage to protein crystals be reduced using small-molecule compounds? Acta Crystallogr D Biol Crystallogr 67:881–893
111. Barker AI, Southworth-Davies RJ, Paithankar KS et al (2009) Room-temperature scavengers for macromolecular crystallography: increased lifetimes and modified dose dependence of the intensity decay. J Synchrotron Radiat 16:205–216
112. Neutze R, Wouts R, van der Spoel D et al (2000) Potential for biomolecular imaging with femtosecond X-ray pulses. Nature 406:752–757
113. Chapman HN, Fromme P, Barty A et al (2011) Femtosecond X-ray protein nanocrystallography. Nature 470:73–77
114. Boutet S, Lomb L, Williams GJ et al (2012) High-resolution protein structure determination by serial femtosecond crystallography. Science 337:362–364
115. Hirata K, Shinzawa-Itoh K, Yano N et al (2014) Determination of damage-free crystal structure of an X-ray-sensitive protein using an XFEL. Nat Methods 11:734–736
116. Chreifi G, Baxter EL, Doukov T et al (2016) Crystal structure of the pristine peroxidase ferryl center and its relevance to proton-coupled electron transfer. Proc Natl Acad Sci U S A 113:1226–1231
117. Lomb L, Barends TRM, Kassemeyer S et al (2011) Radiation damage in protein serial femtosecond crystallography using an x-ray free-electron laser. Phys Rev B Condens Matter Mater Phys B84:214111
118. Barty A, Caleman C, Aquila A et al (2012) Self-terminating diffraction gates femtosecond X-ray nanocrystallography measurements. Nat Photonics 6:35–40
119. Nass K, Foucar L, Barends TR et al (2015) Indications of radiation damage in ferredoxin microcrystals using high-intensity X-FEL beams. J Synchrotron Radiat 22:225–238
120. Galli L, Son S-K, Klinge M et al (2015) Electronic damage in S atoms in a native protein crystal induced by an intense X-ray free-electron laser pulse. Struct Dyn 2:041703

Chapter 21

Boxes of Model Building and Visualization

Dušan Turk

Abstract

Macromolecular crystallography and electron microscopy (single-particle and in situ tomography) are merging into a single approach used by the two coalescing scientific communities. The merger is a consequence of technical developments that enabled determination of atomic structures of macromolecules by electron microscopy. Technological progress in experimental methods of macromolecular structure determination, computer hardware, and software changed and continues to change the nature of model building and visualization of molecular structures. However, the increase in automation and availability of structure validation are reducing interactive manual model building to fiddling with details. On the other hand, interactive modeling tools increasingly rely on search and complex energy calculation procedures, which make manually driven changes in geometry increasingly powerful and at the same time less demanding. Thus, the need for accurate manual positioning of a model is decreasing. The user's push only needs to be sufficient to bring the model within the increasing convergence radius of the computing tools. It seems that we can now better than ever determine an average single structure. The tools work better, requirements for engagement of human brain are lowered, and the frontier of intellectual and scientific challenges has moved on. The quest for resolution of new challenges requires out-of-the-box thinking. A few issues such as model bias and correctness of structure, ongoing developments in parameters defining geometric restraints, limitations of the ideal average single structure, and limitations of Bragg spot data are discussed here, together with the challenges that lie ahead.

Key words Model building, Molecular graphics, Macromolecular crystallography, Electron microscopy, Single average model, Ensemble

1 Introduction

The scope of this book includes methods in macromolecular crystallography (MX); however, the structures recently determined at near atomic resolution by electron microscopy (EM) [1–3] require the author to consider these two approaches in combination. In particular, this area of research covers computational tools of model building, visualization, and refinement. The purpose of this review is to provide insights into model building, address its underlying concepts, and to indicate current challenges and ongoing developments.

Alexander Wlodawer et al. (eds.), *Protein Crystallography: Methods and Protocols*, Methods in Molecular Biology, vol. 1607,
DOI 10.1007/978-1-4939-7000-1_21, © Springer Science+Business Media LLC 2017

It is said that MX is a mature science, in which routine tasks deliver anticipated structures [4]. It is true that handling of the structure determination process is heavily supported by a variety of increasingly automated tools that deliver thousands of structures. This heavy support has taken away intellectual and scientific challenges. The maturity refers to the broad use of the technology by the biologically educated community that requires easy-to-use and easy-to-learn software. However, this does not equally apply to the ongoing developments and to integration of approaches and tools in structural biology.

1.1 The Difference Between a Model and a Structure

Once a molecular model is declared final, it is called the crystal, EM structure, or structure in short. This definition indicates that the molecular model is a working hypothesis being improved in the process of structure determination by its iterative verification against empirical data. We build molecular models to make them represent the empirical data in the best possible way: that is in the most objective way and often not to our liking. Molecular models reflect our understanding of molecular structure under study. Our understanding is mapped onto mathematical models that assign physical and chemical properties to amino and nucleic acids, the elementary constituents of biological polymers, and to the ligands. The mathematical model contains information about their topology (covalent-bonding network) and geometric parameters describing the bonding and nonbonding interactions of atomic constituents. Model building combines our previously gathered information such as sequences of biological polymers, chemical structure of residues and ligands, and preexisting atomic structures into a determined structure. The resulting structure is an optimized balance between the mathematical model, our understanding of structure, empirical data, and feasibility of the structure determination process. This review mostly addresses model building and analysis of an average single molecular structure solution, whereas occasionally limitations of concepts underlying its interpretation are exposed to indicate ongoing and possible developments.

1.2 Model Bias: In- and Out-of-the-Box Concepts

Molecular models are interpretations of empirical data. Interpretation of every experimental work, including structures, is intertwined with ghosts of model bias, overinterpretation, and overfitting. There is the model bias inside "the box," and there is the model bias of "the box" itself. To be aware what the concepts impose, and on the other hand, to prevent us from errors, it is important to understand them. Here, the model bias inside "the box" addresses the fulfillment of the requirements of model validation criteria, whereas the model bias of "the box" itself represents the bias imposed on the molecular model by the concepts underlying its creation and optimization. By default, we deal with the model bias inside the concept because the bias of the concept, once accepted, becomes an ideal in which the community believes and a custom that is nearly unquestionable. First, an illustration from the past: today it is almost forgotten that once it was argued that iterative cyclic refinement of structures should not be performed because the resulting maps are "hocus-pocus" based on overfitting and model bias [5]. As we know today, the iterative refinement practice prevailed [6, 7] because the goal in experimental sciences is to provide models that represent the empirical data in the best possible way, not models that contain less model bias. Now, a contemporary illustration: the model bias of the absence of a model is crystal clear to us. The absence of structure contains no model bias of the built atoms, yet it also contains no atoms. If we wanted to optimize for model bias, the absence of a model would have to be the result. Evidently, this concept is rather unhelpful; however, when only small parts of the model are omitted, this approach allows us to question validity of our working hypothesis. This can also be done systematically throughout the structure by automated model rebuilding assisted by iterative omit maps [8]. Yet structures with parts omitted are evidently not our final structure. Toward the completion of the structure determination process, we struggle to fill every uninterpreted blob in density maps. This is not at all the case for data from reciprocal space, where we leave out the same TEST subset of reflections for the sake of objectivity and to avoid overfitting during the whole course of refinement. For consistency with the TEST set model, a proportionally random number of atoms would have to be omitted from the structure too. Evidently, "the boxes" for model building and refinement and structure validation are quite different in our community. It is as if the structure factors of reciprocal space were not the projection of the real space density map by the Fourier transformation and vice versa.

1.3 Molecular Models Are Interpretations of Density Maps

Density maps precede molecular models, considering that molecular replacement is a part of the phasing procedure. In MX, none of the maps used today are direct conversion of structure factors. They are all a result of numerous, sometimes cyclic, modifications or weighting schemes that result in the most objective image of the true structure. In MX, density maps are calculated from phased structure factors of the diffraction pattern of a crystal, whereas in EM, they are obtained by 3D reconstruction of 2D projections. From the beginnings of structural biology in the 1950s, density map calculations underwent a long list of improvements. Only a few in use today are listed here: solvent flattening [9], sigma-A maps [10], the bulk solvent model [11], anisotropic correction [12], simulated annealing maps [13], average kick maps [14], density modification Schemes [15], and feature-enhanced maps [16].

Model building is a pattern recognition process during which topology and patterns of 3D structure of molecular models or their components are matched against the features of a density map. Once a match is found, models or their components are built as superimposed on the corresponding features of the density maps. Technically, model building is a process that includes creation and modification of the topology of molecules (connectivity, atoms, and residue records) and their placement into the desired positions in density maps. Tools that make this possible are computer programs that perform model building automatically, such as Auto-Rickshaw [17], PHENIX [18], Buccaneer [19], ARPwARP [20], SHELX [15], Rosetta [21], and those that assist with interactive model building via a computer graphical 3D display interface, such as O [22], VMD [23], Coot [24], Sculptor [25], and MAIN [26]. In this chapter when I make illustrations and present ideas and cases, I am mostly referring to MAIN, the software I have written, because I know it better than any other program.

2 The Essentials of Molecular Models: Patterns to be Recognized

The information on the chemical composition of molecules combined with prior knowledge of molecular structure is compiled into 3D molecular models. These models represent polymers of amino and nucleic acid residues and ligands. To build molecular models in density maps, it is mandatory to recognize their features. This knowledge enables us to build chemically reasonable models and avoid building chemically unreasonable ones.

Mathematical descriptions of models in structural biology use the concept of atoms as points in space interacting through bonding and nonbonding energy terms. The information on structure and topology of molecules is compiled into libraries of geometric parameters. Today, parameters for ideal bond length and bonding angles and forces preserving them that are used in model building and refinement are a result of statistical analysis of highly accurate geometry of small-molecule crystal structures. These libraries contain a description of amino and nucleic acid residues, based on Engh and Huber [27] and Parkinson parameter sets [28], respectively. A recent paper again strongly encourages the use of stereochemistry-dependent restraints [29]. The parameters for hetero compounds can be found in a monomer library [30] or complied from schemes, SMILES, and 3D models by such software applications as PRO-DRG [31], PURY [32], and JLigand [33]. These libraries are used by model building and refinement programs. By occasionally watching Robert Huber over his shoulder and asking him questions (while he was interpreting electron density maps using FRODO [34] during my PhD studies), I learned the elementary rule of structure correctness for density map interpretation: "Correct structure must be correct locally and globally and match electron density maps." The term *local geometry* refers to the conformation of a residue and its immediate neighbors, whereas the term *global geometry* implies folding and packing of atoms in space. The second rule that I learned was, "Do not waste time with interpretation of uninterpretable areas of electron density, make them interpretable first." To recognize chemically reasonable models corresponding to correct structure, and which map patterns correspond to them, one must be acquainted with the local and global geometry of macromolecular structures. A short description is presented below.

2.1 Local Geometry

Biological polymers discussed here are proteins and nucleic acids. Proteins are built from amino acids linked together by the peptide bond. They have N and C termini according with the terminal amino and carboxylic groups that remain unlinked. These residues have a chiral center at the CA atom to which a side group is attached. The 20 standard side groups provide them with limitless combinatorial possibilities of their sequence and hence structure. RNA and DNA are built from nucleic acid residues. They consist of two main chain components: a pentose carbohydrate ring (ribose or deoxyribose) and a phosphate group. Each phosphate group is linked to two ribose rings with two phosphoester bonds, one on each side. Their termini are called 3′ and 5′ in accordance with the first and last ribose atom in the polymer. There are five standard different nucleic acid residues, only four of them build each kind of polymer.

Amino and nucleic acid residues are built from carbon, nitrogen, oxygen, sulfur, phosphorus, and hydrogen atoms. Apart from hydrogens, these atoms appear in tetrahedral (sp^3) and planar (sp^2) hybridization states. All bonds in which tetrahedral atoms are involved share a single bond character. They are rotatable, because their rotational energy barrier is low enough to be crossed by thermal motions. The planar systems show the double bond character or the nature of aromatic systems. In biopolymers, only the peptide bond exists in two conformations, whereas the other planar systems are in rings or exhibit no internal rotations e.g., side chain COOH and CONH2 groups in ASP, GLU, ASN, and GLN or guanidine in Arg. To address the conformational space in availability criteria, we describe a conformation of fragments of macromolecular structures as favorable, allowed, and disallowed.

2.1.1 Main-Chain Geometry

Proteins

Each nonterminal residue in the chain is linked to its neighbors by two peptide bonds. The peptide is a planar fragment consisting of atoms around the amide group (CA–NH–CO–CA). Peptide bonds (NH–CO) that are in the protein are not freely rotatable. However, the single bonds (CA–N, CA–C) by which they are linked to the CA atoms are rotatable. These rotations endow proteins with the abundance of differently folded structures. The available conformation space nonetheless is limited. The limitations arise from the clashes of atoms rotating about the two bonds. The geometry of these two rotations is described by the pair of dihedral angles (also called torsions) defined by the four atoms C(−1)–N–CA–C and N–CA–C–N(+1). They are called φ and ψ, respectively. The numbers in parentheses specify that the C(−1) and N(+1) atoms belong to the previous and next subsequent residue in a sequence. The projection of φ and ψ angles on a 2D plot is called the Ramachandran plot [35], Fig. 1. Ramachandran et al. originally used an alanine residue with two peptide bonds attached to it to make the first projections and calculate their packing allowance. They wrote, "The demarcation of these regions depends on the choice of the permitted van der Waals contact distances. Two such sets, termed 'normally allowed' and 'outer limit,' were worked out from a detailed analysis of the available structural data on various organic compounds, including in particular amino acids and peptides." Nowadays, packing functions to mark available conformational space are still used, but with the abundance of protein structures available today, statistical distributions have become the prevailing form [36].

When building a protein chain, one needs to be aware that for non-glycine amino acid residues, there are three general favored areas: β (−139, 135), right-handed α-helical (−48, −57), and left-handed helical (48, 45) (Fig. 2). The values in braces are central φ and ψ angles. Areas corresponding to them are shown in green in the Ramachandran plots, Fig. 1. In Fig. 2 the β conformation, the NH and CO groups of the same residue point in the same direction, whereas in the helical conformations, they point in the opposite directions. To demonstrate the differences, the atoms O(−1), HA, and HN(+1) are connected by dashed lines. In the β conformation, the lines connect them on the same side, whereas in the cases of the helical conformations, the lines cross the peptide bond either on the N or C side of the peptide bond. Note also that the side chain positions among the three alternatives differ. Although in the extended β conformation, the side chains are positioned on alternative sides, in the helical conformations, the side chains are positioned on the external side of the helical turn.

Nucleic Acids

The side chain nucleoside is attached to the ribose. In contrast to proteins, the insight into their structure suggests that their 3D folding pattern is primarily governed by the side chain interactions. Their main-chain conformation consists of seven single bonds, and only the rotation around the C3'–C4' bond is restrained within the ribose ring. This situation enables the main chain to adopt a conformation that supports the side chain packing. The seven-dimensional space required to show backbone torsions in a Ramachandran plot-like manner is beyond human comprehension.

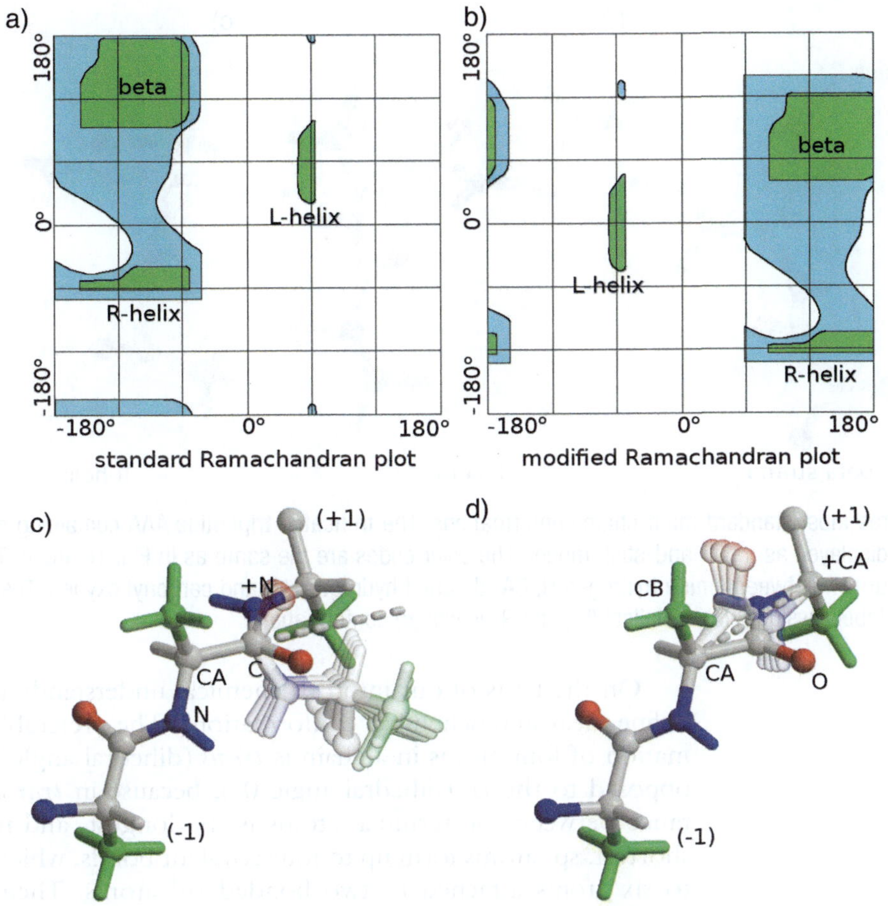

Fig. 1 Ramachandran plots. (**a**) In the upper row on the left is the standard plot used for validation of crystal structures today, where horizontal and vertical axes represent the dihedral angles C(−1)–N–CA–C and N–CA–C–N(+1), respectively. The +1 and −1 labels refer to the preceding and next residue, respectively. (**b**) On the right is the modified plot representing the dihedral angles O(−1)–CA(−1)–CA–CB and CB–CA–CA(+1)–O. The *green* and *cyan* areas indicate the favorable and allowed areas, respectively. The favorable areas represent the three most standard conformations (beta strand, R-, and L-helix) and are labeled. In the lower row, the rotation axes corresponding to the standard and modified Ramachandran plots are shown. Rotational axes about the psi dihedral angle and the CA–CA direction are shown as *dashed lines* on the left in (**c**) and on the right in (**d**), respectively. The truncated tripeptide AA containing two peptide bonds is displayed as a ball-and-stick model. Main-chain carbon, nitrogen, and oxygen atoms are colored *white*, *blue*, and *red*, respectively. The side chain carbon atoms are shown in *green*. Hydrogen atoms are colored according to the color of the atom they are attached to. The rotation axes are shown as *dashed lines*. The starting conformation is shown as opaque, whereas the transparent images indicate the rotated atoms

2.1.2 Side Chain Geometry

In side chains, rotations about single bonds are free, whereas double or aromatic systems remain rigid. A side chain rotamer is a conformation described by torsion or dihedral angles, or torsions in short. The middle pair of atoms defines the rotatable bond, whereas the terminal atoms are attached to one of the bonding atoms each.

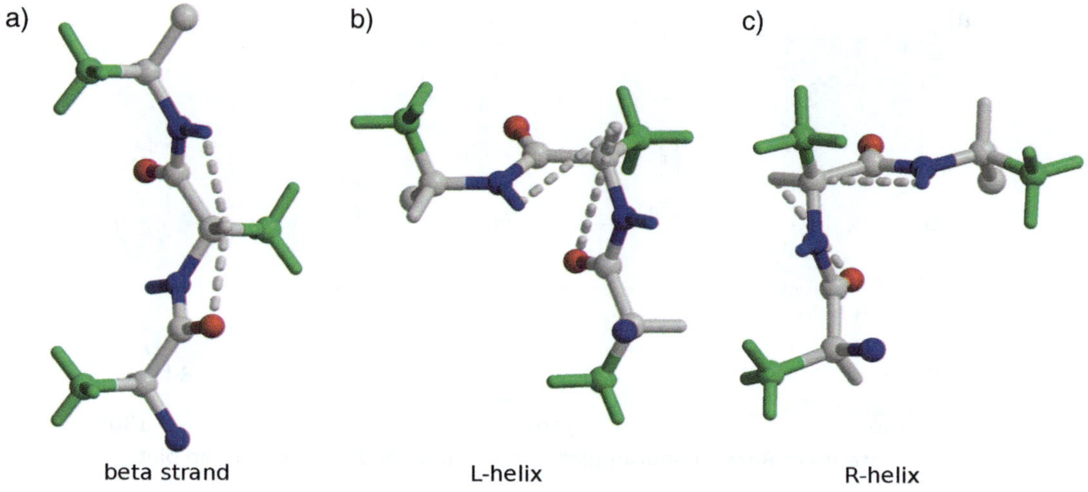

Fig. 2 Three most standard main-chain conformations. The truncated tripeptide AAA containing two peptide bonds is displayed as a ball-and-stick model. The color codes are the same as in Fig. 1c and d. The *dashed lines* are drawn between amide hydrogen H, CA-attached hydrogen HA, and carbonyl oxygen. The secondary structure labels assign β (**a**), L-helical (**b**), and R-helical (**c**) conformations

On the basis of our intuitive chemical understanding, we can define ideal stereochemical conformations. The preferable conformation of four atoms in a chain is *trans* (dihedral angle 180°), as opposed to the *cis* (dihedral angle 0°), because in *trans*, the distance between the terminal atoms is the longest, and in *cis*, the shortest. sp^3 atoms form up to four covalent bonds, which yield up to six atoms attached to two bonded sp^3 atoms. Their smallest overlap is achieved when they are ±60° and 180° apart, as demonstrated in Fig. 3, for the valine residue with displayed side chain atoms HB, CB1, and CB2 attached to CA. Atoms of planar fragments allow only *trans* and *cis* conformations; however, in amino and nucleic acid residues, the only fragment that appears in both conformations is the peptide bond. Other planar fragments are either internally constrained (rings of aromatic amino acids and nucleotides), symmetrical (guanidinium group of arginine), or too small (carboxylic and amide groups). The bond between sp^3 and sp^2 atoms is single. Their ideal positions are when a planar fragment is positioned between two neighboring groups of sp^3 atoms (Fig. 4). As shown, there can be two or four such possibilities depending on the internal symmetry of the sp^2 fragments. If one takes into account the three possibilities per rotation around the bond within each pair of sp^3 non-hydrogen atoms and four per each sp^2–sp^3 pair (and add two for proline puckers), then their numbers add up to 338 hypothetical, ideal side-chain rotamers for 20 standard amino acid residues in total. The number would be substantially reduced if overlapping conformations of longer side chains were excluded from the counting. The Arg and Lys side

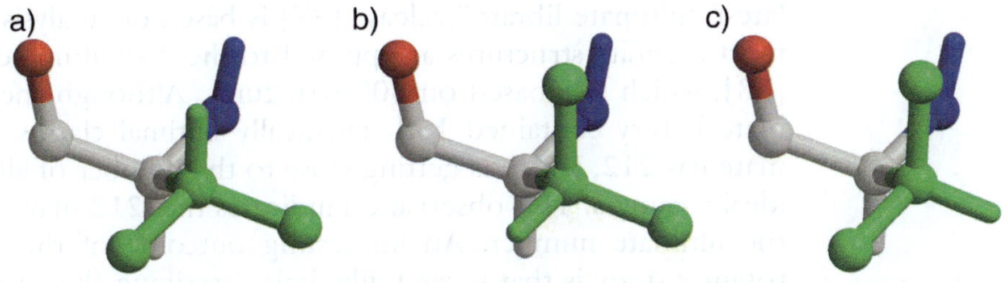

Fig. 3 Ideal sp³–sp³ rotamers of the Val side group. The valine residue is shown as a ball-and-stick model. Main-chain CA, N, and O atoms are colored *white*, *blue*, and *red*, respectively. Side chain CB atoms are colored green. Hydrogens are shown as sticks colored in the color of the attached atoms. Hydrogens attached to CG atoms are not shown. Three ideal conformations 120° apart are shown with the CB hydrogen in *trans* (180°), minus (−60°), and plus (+60°) dihedral orientation in relation to the CA hydrogen

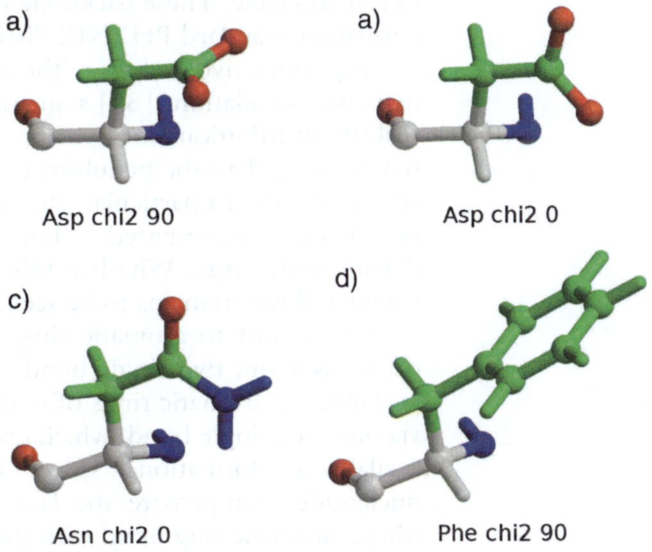

Fig. 4 Ideal sp³–sp² rotamers of Asp, Asn, and Phe side groups. Main-chain CA, N, and O atoms are colored *white*, *blue*, and *red*, respectively. Side group CB atoms are colored green. Hydrogens are shown as *sticks* with the color of the attached atoms. The carboxylic group of Asp is symmetric, and therefore its four x^2 conformations are reduced to two. Asn is shown in a conformation in which the amide hydrogens are placed farthest apart from the CB hydrogens. Phe is shown in a conformation with the least interference with the main-chain and CB atom hydrogens

chains alone, with 108 and 81 hypothetical ideal conformations, respectively, contribute to over a half of the number of ideal rotamers.

Statistical analysis of rotamers of real protein structures revealed that the side chains of amino acid residues adopt conformations in unevenly distributed populations that deviate from the above-mentioned intuitive understanding of ideal rotamers [37, 38]. The latest "ultimate library" release [37] is based on analysis of 8000 most accurate structures as opposed to the "penultimate" library [38], which was based on 500 structures. Although the penultimate library contained 153 statistically optimal classes, the ultimate has 212, which is getting close to the number of all possible ideal rotamers. This observation indicates that 212 may indeed be the ultimate number. An interesting outcome of the ultimate-rotamer study is that several dihedrals describing the geometry of sp^3–sp^2 bonds have such large standard deviations that it is difficult to assign central values to them. In particular, this is evident for the Asp and Asn [37]. Both analyses highlighted the dependence of rotamers on the backbone geometry; this state of affairs is not surprising because the peptide bond atoms on both sides of the residue may enter the space accessible to side chains and render parts of it unavailable. These backbone-dependent parameters are now a part of the standard PHENIX distribution [29].

A comparative analysis of the side chain rotamers by molecular dynamics simulations [39] suggests that such analysis yields more realistic distributions for dynamic proteins in solution at ambient temperature than the penultimate library derived from the crystal structure data. In particular, the charged surface residues seem to be better represented (library download: http://www.dynameomics.org). Whether this notion holds for the ultimate rotamer library remains to be seen.

In contrast to aromatic rings of amino acid residues, where rotations about two single bonds are available to adopt favorable packing, the aromatic rings of nucleotides are linked to the ribose via only one single bond, which dramatically reduces the region of available conformational space. Therefore, it appears that the nucleotides compensate the lack of degrees of freedom of the ribose–aromatic ring bond with the flexibility of the main chain.

2.2 Secondary Structures

These are regular repeating patterns of main chain geometry of macromolecules. Both amino and nucleic acid polymers have secondary structures, but they differ so profoundly that they have essentially nothing in common. Not even every regular structure pattern of RNA relies on hydrogen bonds.

In proteins, there are two major types of secondary structures called helices and β-sheets. They are stabilized by hydrogen bonds bringing together main-chain carbonyl and amide groups. Repetitive helical turns build helices. There are α, 3_{10}, and π helices (Fig. 5). The three helices shown contain the same number of residues. Their differences in height indicate that they differ in the number of residues per turn. In the first and last turns of the helices, the NH and CO groups are not stabilized by hydrogen bonds within the helical structure. In the most common α-helix carbonyl group forms a hydrogen bond with the amide of residue +4, whereas in

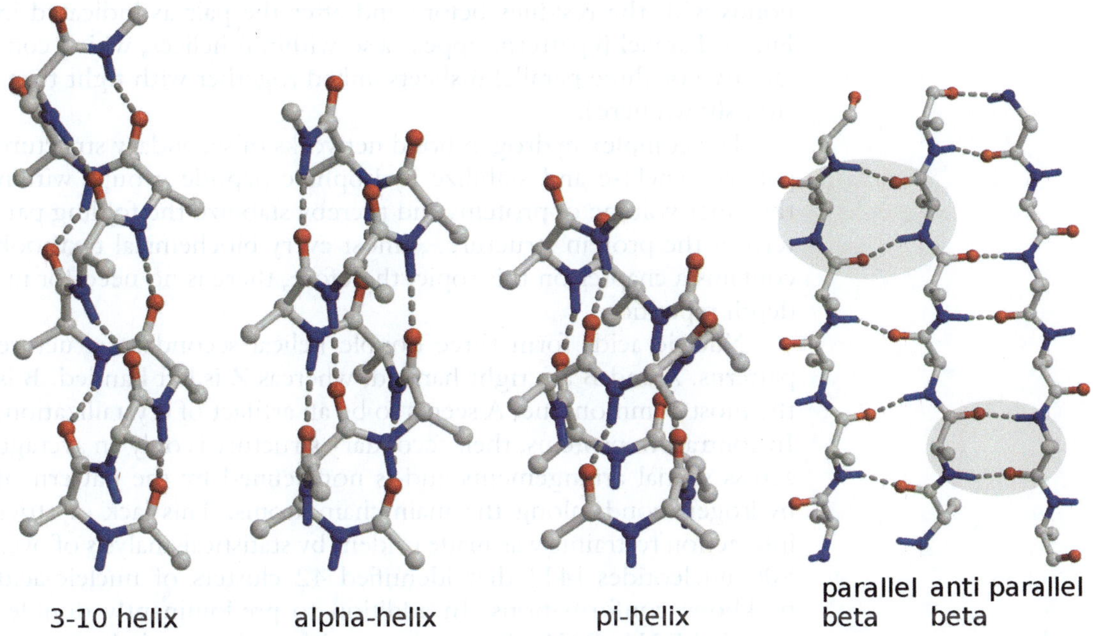

Fig. 5 Protein secondary structures. The model is shown as ball-and-stick representation. Carbons are shown in *white*, carbonyl oxygens in *red*, and nitrogen and its hydrogens in *blue*. Hydrogen-bonding patterns are shown as broken connections. The three images on the left show 3_{10}, α, and π helices composed of ten alanine residues. In the right-hand panel, hydrogen-bonding patterns in parallel and antiparallel β strand arrangements in a β-sheet are shown

the 3_{10} and π helix, the hydrogen bond is formed between residues +3 and +5, respectively. It is not uncommon for an α-helix to tighten up at the top into a 3_{10} helix. This change stabilizes the terminus by enclosing one more peptide bond in the helical hydrogen-bonding network. π helices, on the other hand, have one residue more per turn, and hence, one hydrogen bond less within the termini. Therefore, they tend to appear within an α-helix as a single turn. It has been said numerous times that the side chains in an α-helix show a "Christmas tree" pattern. This pattern can be observed when an right-handed (R) helix is displayed vertically with the N terminus at the bottom (Fig. 5). In the equivalent view on the left-handed (L) helix, the side chains would point upward (not shown here). In contrast to α-helices, which are the most abundant secondary structure elements, L-helices are rare [40].

Hydrogen bonds in helices link together peptide bonds with carbonyls and amides in parallel directions, whereas in β-sheets, they are oriented alternately along the main chain. Parallel and antiparallel strand arrangements in the β-sheet show different hydrogen-bonding patterns. In antiparallel arrangements, pairs of residues are enclosed in hydrogen bonds, whereas in parallel arrangements, one residue in a strand is involved in hydrogen bonds with the residues before and after the pair as indicated in Fig. 5. Parallel β patterns appear also within β-helices, which contain two or three parallel β-sheets linked together with tight turns (not shown here).

The complex hydrogen bond networks of secondary structure patterns enclose and stabilize hydrophilic peptide groups within the inner volume of proteins and thereby stabilize the folding pattern of the protein structure. Almost every biochemical textbook contains a chapter on this topic; therefore, there is no need for in-depth repetition.

Nucleic acids form three double helical secondary structure patterns. A and B are right handed, whereas Z is left handed. B is the most common one, A seems to be an artifact of crystallization. In contrast to proteins, their secondary structure is only an average across spatial arrangements and is not defined by the pattern of hydrogen bonds along the main-chain atoms. This lack of strict interaction restraints was made evident by statistical analysis of over 800 nucleotides [41] that identified 42 clusters of nucleic-acid backbone conformations. In addition to predominantly double-stranded DNA, RNA often appears to be single stranded.

These observations taken together indicate that the secondary structure of nucleic acids is guided by interactions between the side chains of nucleic bases. The side chains form two kinds of interactions: First, by the stacking of aromatic rings with an offset that reduces the repelling interactions between the rings of π electrons on both sides of aromatic rings, and second, by the hydrogen bonds formed by their edges. Hydrogen bonds stabilize the interactions between two strands of a double helix with the Watson–Crick (A–T, C–G) pairings (Fig. 6). An alternative hydrogen-bond pairing between nucleic bases, called Hoogsteen pairing, is at the position shown below (Fig. 6), with a different orientation of the base pairs.

An illustration of a hydrogen-bonding pattern of nucleic acids can provide a clear insight into regularity and disruption of edge pairing. However, here one can view contacts only on a one-by-one basis because, unfortunately, a view enabling the overview of the nucleic-acid hydrogen bond network is not possible. When one looks either down the helical axes or from the side, the overlapping hydrogen bonds obscure the overview and prevent elucidation of regular patterns.

2.3 Packing It All Together

The internal consistency of an atomic model can be checked when the fold is traced and the sequence is established. The fold should conform to the established folding patterns; the sequence of residues must match the density maps and side chains must find matching neighbors.

Basic insight into the packing of secondary structure elements in folding patterns can be of significant help for getting the

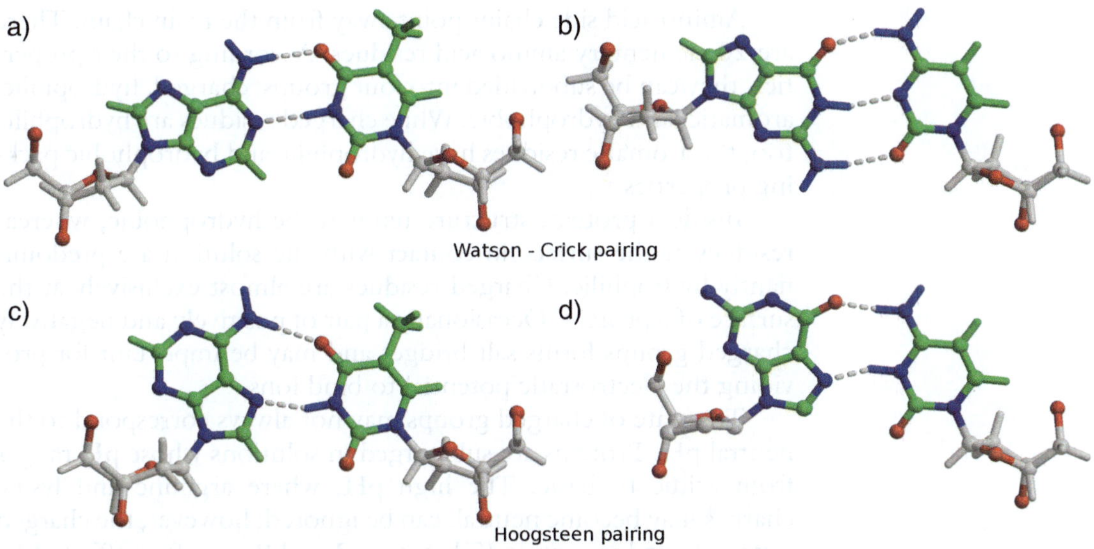

Fig. 6 Base pairing of nucleic acids. Adenine–thymine (**a**, **c**) and guanine–cytosine (**b**, **d**) with Watson–Crick (**a**, **b**) and Hoogsteen (**c**, **d**) pairing patterns of hydrogen bonds are shown as ball-and-stick models. Oxygen and nitrogen atoms are show in *red* and *blue* respectively. Nucleic bases and ribose carbon atoms are colored *green* and *white*, respectively. Hydrogen atoms share the color of the atoms they are attached to. The phosphate groups are not shown. Hydrogen bonds are shown as *dashed lines*

structure right. Nucleic acids are somewhat simpler because their secondary structures are extending over large regions. They fold differently from proteins, so that it is essentially impossible to mix them up. There is not much to add to the section above about the structure of nucleic acids. However, protein structures are more complicated. In a typical globular protein, approximately a half of the sequence is in secondary structure elements, the rest are wide and tight loops connecting the secondary structure elements. A protein fold is stabilized by numerous hydrogen bonds, which bury the hydrophilic peptide bonds inside the protein structure: in most cases, within the secondary structure elements. The secondary structures form motifs that are composed of several secondary structure elements. Combined motifs build domains: packed folding patterns. According to PDB statistics, the number of different protein folds has not increased lately. The last new fold by SCOP was added in 2008, and by CATH in 2012; however, they both give approximately the same number, close to 1400 (http://www.rcsb.org/pdb/statistics/). Unless you are like Alexei Murzin, the scientist behind the Structural Classification of Proteins (SCOP, SCOP2), you will likely not be able to perceive an imprint of all the folds in your brain [42, 43]. As an introduction to the architecture of folds, I recommend the book "Introduction to Protein Structure" by Branden and Tooze [44], which is somewhat outdated; however, the principles of protein structure have not changed.

Amino acid side chains point away from the main chain. There are 20 elementary amino acid residues. According to their properties, they can be subdivided into four groups: charged, hydrophilic, aromatic, and hydrophobic. While charged residues are hydrophilic too, the aromatic residues have hydrophilic and hydrophobic packing properties.

Inside a protein, structure tends to be hydrophobic, whereas residues at the surface in contact with the solution are predominantly hydrophilic. Charged residues are almost exclusively at the surface of a protein. Occasionally a pair of positively and negatively charged groups forms salt bridges and may be important for providing the electrostatic potential to bind ions.

The state of charged groups may not always correspond to the neutral pH. Proteins are submerged in solutions whose pH ranges from acidic to basic. The high pH, where arginine and lysine charges may become neutral, can be ignored; however, the charged states of residues with pK_a between 3 and 9 are often affected by the pH of solutions where proteins are kept in biological and "in vitro" environments. Imidazole side chains of histidines, carboxylic groups of aspartic and glutamic acids, C termini, and occasionally SH groups of cysteine residues, have pK_a values in this range. The three possible protonation states of histidines are commonly considered, whereas the carboxylic groups may appear in pairs sharing a proton. Occasionally, a local electrostatic environment plays a major role. For example, the catalytic site contains negatively charged cysteine residues in papain-like cysteine proteases. They form an ion pair with positively charged histidines, and are positioned at the N terminus of the alpha-helix. Their pK_a is ~3.5 [45], whereas in solution, it is ~8.5. When presumably charged residues get buried, they may be neutral as, for example, the E171 residue in human cathepsin B structure [46].

Uncharged hydrophilic groups such as hydroxyl or amide groups appear in roles of hydrogen-bonding donors and acceptors. Among them is also the always present ambiguity of the amide groups of asparagine and glutamine residues. Quite often, the hydrogen-bonding network and local charges suggest their orientation, but there are also many situations when this is not the case.

Aromatic residues share a dual nature. Their character enables them to pack in hydrophobic as well as in hydrophilic environments. The basis of this dual behavior is the bipolar charge distribution (a quadrupole moment) of side chains. Their π electrons at the sides of the rings make these surfaces negatively charged, whereas the hydrogens at the edges bear a partial positive charge. Therefore, when they are inside a protein, their rings tend to stack together either parallelly or perpendicularly. When they pack parallel to one another, their π electron clouds induce an offset similar to the one observed in nucleic acids, which reduces the repulsive electrostatic interactions.

Hydrophobic residues tend to form patches inside the structure. Nonetheless, their occurrence in surface regions is not uncommon.

There is a persistent desire to be able to simplify the answer and to reduce the structure correctness to a single criterion described by a single number. Commonly, *R*-free is used for this purpose. *R*-free is useful at the initial stages of structure refinement. When *R*-free does not change, but *R*-work drops substantially, it is quite likely that the structure is wrong. However, wrong structure solutions should trigger a number of other alarms anyway. When we validated the Free Kick target function, one of the test cases was based on reverse chain tracing [47]. Apart from the *R*-factor, discontinuity of density maps, and local geometry-related criteria, this model also had problems with packing of residues. The vicinity violations were mostly taken care of by the repulsive forces during refinement; however, positioning of the charged residues in unsuitable hydrophobic environments could not be dealt with. Similarly, it was shown that the *R*-free gap, as a measure of structure correctness, is not reliable when it comes to details [48]. It was shown [47] that different choices of the TEST sets deliver different phase errors that do not correlate with the *R*-free gap. In my opinion, the persistent use of *R*-free as an indicator of structure correctness is a result of the desire to simplify the reality by wishful thinking. Ironically, *R*-work and *R*-free behave independently only when refinement is not converging—that is when the progress is made in the wrong direction. In such cases, models are outside the reach of structure improvement. Because there is no improvement, these models are also called stuck or wrong. Further, of course, *R*-free should decrease when *R*-work decreases, as the structure is converging toward a consistent solution. In these cases *R*-free and *R*-work are not independent. Nonetheless, the solution delivered by the *R*-free concept is off-target, and we can do better. A longer explanation and suggestion how to avoid the TEST set concept altogether by means of the Free Kick refinement approach is outside the scope of this chapter; readers can see our article [47]. What matters here is that inspection and validation of model geometry, packing and match to density maps and to measured reflections, and consistency of biochemical data are the criteria for establishing the structure correctness.

2.4 Ligands

Structures of ligands extend over most of the space of chemical compounds. They are described in another chapter. The following, however, requires attention:

– Besides the noncovalently attached, there are also covalently attached ligands. The covalent bond changes the atom class or type of the reacting groups and thereby their stereochemical parameters. The parameters of the link require oversight because they may not be configured correctly by the automatic procedures.

- Coordination bonds between metal ions and other compounds are not described well by the current tools of trade. PURY [32] covers several interactions satisfactorily but not all of them, and their occurrence is often too small to provide statistically significant populations.

2.5 Hydrogen Atoms

It is important to address the usually ignored issue of hydrogens, the smallest atoms in a molecule. In contrast to the NMR field and neutron diffraction, where their positions and interactions are measurable, in MX, they are mostly absent from the structure files. There are good reasons for that. They obscure the view by approximately doubling the number of lines of an atomic model on the screen, and they cannot be discerned from the electron density maps apart from the crystals exposed to neutrons. There are, however, also multiple reasons why hydrogen atoms should be included and made visible during the model building stage (and be present not only during refinement and validation):

- Hydrogens are excellent indicators of hydrogen-bonding patterns, protein secondary structures, and double helix pairing of nucleic acids in particular. The use of polar hydrogens is sufficient for this purpose.
- Hydrogens provide a clearer insight into the packing of atoms and thereby make it easier to detect clashes and unfavorable conformations. There is no need to display hydrogens all the time. In various graphics programs there is a switch that turns their display on and off.
- The absence of hydrogens from model building software introduces inconsistency of real space refinement with the subsequent refinement in reciprocal space and validation that both use hydrogen atoms.
- Hydrogens play an important role in electrostatic interactions and may facilitate the ignored contribution to quadrupole moments of aromatic systems and thereby electrostatic interactions [49].
- In a protein structure, their number approximately equals the number of carbon, nitrogen, oxygen, and sulfur atoms put together. If we consider a protein structure consisting of 1000 non-hydrogen atoms, there are also 1000 hydrogen atoms. Thus, in such a structure, there are 6000–7000 electrons originating from non-hydrogen atoms and ~1000 from hydrogen atoms. Hence, to the total number of electrons, the hydrogen atoms contribute ~15% of the total diffracting power.

3 Visualization in Model Building

The role of an interactive computer graphical interface is to enable a human user to perform interpretation of electron density maps by building molecular models. During this process, the model and density patterns are matched. Success of pattern matching in large part depends on the perception of the displayed objects. Hence, for model building, we aim to provide an overview into the desired areas of space. The smaller the space we are observing, the less the care required to obtain a reasonable insight, and vice versa: the larger the area explored, the more important it is how objects are displayed. The easier the interpretation of a density map, the less we need to see, and the less we need to know, and the other way around. The harder it gets, the better the insight and the greater the expertise required. There is no optimal solution for every possible case; therefore, it is good to bear in mind that presentations can be optimized and adjusted in several ways. Another useful notion is that optimal default values of a computer program may save a lot of effort. An inappropriate combination in the choice of color, line thickness, fog parameters, and antialiasing is, in my opinion, the major obstacle affecting the perception, hence, the effectiveness when an overview of large map areas is required. In particular, this is relevant to the cases of demanding structure determination projects of large macromolecular complexes (at low resolution). Chimera [50] provides a good overview, but limited model-building capabilities, whereas with Coot it is the other way around. MAIN on the other hand provides good overview and modeling capabilities, but has steeper learning curve. Because different software developers have chosen different approaches, and users may not be aware of the relevance of available settings, the text below covers the most relevant issues of graphical presentation of molecules during model building.

3.1 Color of the Background

Usually ignored, but for me, of the utmost importance is the choice of the background color (prevailing color in the image), which has a physiological basis in the response of the eye to brightness. In a bright environment, the eye's pupil narrows to reduce the amount of incident light on the retina, whereas in the dark, pupils widen to increase the amount of light on the retina. As a consequence, we see the bright objects sharper, whereas the dark ones cannot be resolved in equivalent detail. Let us consider the extreme case first. We all know that when the sun shines on our screen or behind it, not much can be seen, no matter how hard we strain our eyes, because the screen does not produce enough light. Evidently, a combination of dark and bright objects reduces the perception of dark objects. The obvious conclusion is that the brightness of the environment has to be adjusted to that of the electronic media such as screens and projectors. However, it is not only the brightness of the environment that needs to be adjusted, the same applies also to the picture on the screen. In the RGB world, white color is composed of maximal values of red, green, and blue light, whereas all other colors have at least one of the maximal values reduced. To compensate for the loss of resolution on the bright background, the surface area of displayed darker objects has to be increased. Therefore, a white background requires surfaces and ball-and-stick models, while thin colored lines are barely resolved. This situation calls for a white (grey) background in a bright environment without lines. Nonetheless, when details of molecular models are matched against superimposed density maps, the success of the pattern recognition process requires in-depth elucidation of their complex 3D shapes. This insight is optimally conveyed by lines because other objects like surfaces obscure the view of the objects behind. Line perception is facilitated by the black background, where the colored lines stand out. There is, however, yet another reason to stick to the black background. The picture on electronic media is composed of pixels, which are separated by areas emitting no light. Regardless of the use of colors, the black raster covers a significant part of the screen surface. As a result, the combination of the dark raster with a white background reduces the perception of clarity of colors, graphical objects and thus of the image. In addition, the brain is additionally loaded by the process of masking out the spacing between pixels.

Hence, the ultimate advice is to use a black background for electronic media presentations (including wall projections) and white for paper presentations because white is the background color of the paper. When, however, the environment cannot be made dark enough—this is often the case in lecture halls and seminar rooms—then the bright color alternatives should be considered, too.

3.2 The Number of Details

In model building, it all comes down to which and how many details can be resolved and perceived on a computer screen. Brain catches features such as cross-links, edges, cavities, bumps, and sharp points. Smoother objects attract less attention, but when there are too many, they will obscure the view anyway. To optimize our perception, we need to establish the right number of displayed graphical components and the optimal art of their visual representation. For a researcher to be able to interpret complex patterns of electron density maps, the number of displayed details must be in balance with an overview. When the number of displayed details is too large, comprehension of the structure is diminished or even lost, but when it is too small, no progress toward the final structure can be made.

3.3 Performance of Graphics Cards and Interactive Devices

Good performance of a computer graphics card is a prerequisite for 3D perception of objects on the screen. 3D perception is achieved by smooth rotation and an appropriate choice of the background/foreground contrast (fog parameter in OpenGL). The essential criterion of performance is the ability to rotate graphical objects on the screen smoothly. Seamless exchange of consecutive images is achieved when the time for drawing the next one is shorter than what the eye can notice and when the change between the two consecutive images is overlapping to an extent that gives the impression of a continuous transition. Any noticeable jumps between two consecutive images may overload the processing of images by the brain and can lead to headaches. Clearly, using "crystal eyes" with quad buffer stereo may significantly shift the boundaries of the amount of information processable by the human brain; however, in the end, the kinds of limitations a researcher has to deal with are the same as on a mono display. Although satisfactory mono image processing can be achieved with a reasonably priced graphics card, quad buffer stereo remains expensive. Stereo graphics cards in combination with the use of double screens with a semitransparent mirror are the most luxurious system. Hence, the performance of graphics cards is still relevant, but to a lesser extent. Quad stereo on a single screen requires high-end graphics cards; however, the highest quality stereo with two screens and a semitransparent mirror is still outside the reach of most mortals.

Graphical performance is tightly linked to the use of interactive devices such as the mouse, keyboard, dials, and tablets. They are responsible for linking the user's perception with a computer response. The use of interactive devices is described below in Subheading 5.3.

3.4 The Art of Presentation of Electron Density Maps and Molecular Models

It is not only the number of graphical objects and screen resolution, but also the art of their presentation, that has a strong impact on our perception of 3D objects. A technique used to add greater realism to a digital image by smoothing jagged edges on lines and other objects is called antialiasing. Therefore, the antialiasing level is not a parameter to be optimized at the expense of a decrease in performance. Antialiasing should always be at the maximum. Too low or even switched-off line or scene antialiasing significantly reduces comprehension by overloading the brain with details of jagged lines. To respond to this overload of noise, which must be filtered out by the brain, a user intuitively zooms into the scene and examines smaller regions. This solution is not recommended because it backfires on perception via the diminished overview and insight.

Density maps (Fig. 7) can be presented as wire frame objects, surfaces, or their combination, while molecular objects can be presented as lines and sticks (representing bonds), spheres (for atoms' surfaces), and as ribbons.

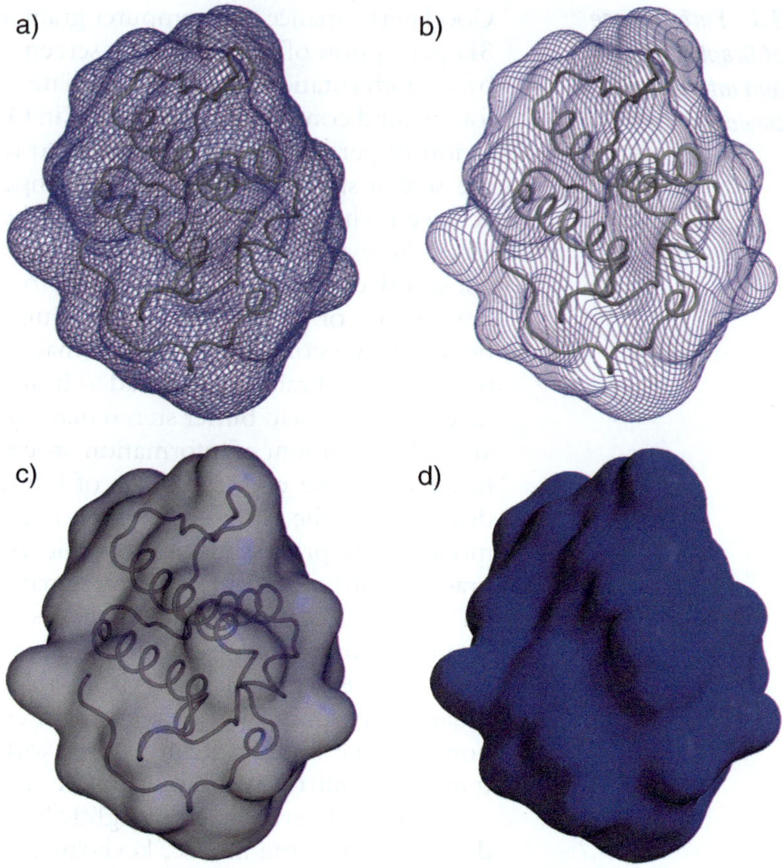

Fig. 7 Map presentations. The same map simplified to an envelope is shown around the same molecule (PDB ID: 4DIH) as a wire frame model (contoured along planes perpendicular to *X*, *Y*, and *Z* axes), contoured along planes perpendicular to the *X*-axis and as transparent and solid surface. The within-chain trace is shown in *white*

The contours of maps facilitating the understanding of atomic structure require in-depth perception of multiple objects. Among graphical objects, lines provide the most precise insight into details of maps and atomic structure. Therefore, the wireframe map presentations are the classics of MX. The line thickness does not depend on the scale of the image; therefore, a simple zoom into an overcrowded scene will separate the objects. In contrast, the thickness of polygonal objects such as sticks remains proportional to the object dimensions; thus, zooming in will also enlarge the thickness and will not result in the same object discrimination. It should also be mentioned that the use of clipping planes cannot compensate for the thickness of polygonal objects entirely.

Although higher-resolution maps contain details of molecular structure, the lower-resolution EM maps may reveal only their secondary structure patterns or envelopes of molecules. Displaying map objects as surfaces and molecules as ribbons may provide enough comprehension to establish matching patterns of molecular folds in density maps and thereby to enable positioning of the molecular models. The advantage of surfaces is their smoothness. They reveal less detail (fewer sharply resolved edges) that catches the eye and can make perception of the envelope easier to comprehend. Surfaces can also be made transparent. Nevertheless, when in-depth resolution is important, the use of a wireframe model is still recommended. If three-directional contouring results in too large a number of lines obscuring the view, in MAIN, one can present the low-resolution maps and envelopes contoured only along one projection: X, Y, or Z.

3.5 Insights into the Precision of Density Maps

Maps have yet another parameter that affects the comprehension: the spacing in the grid used for their storage. The denser the grid, the smoother, and apparently better resolved are the maps. (In addition, the increase in map smoothness is a strong indicator of reduced noise.) On the other hand, this increase in the satisfaction factor comes at the expense of the reduced overview of the map features. The grid space is cubic, meaning that the grid size of 0.5 Å in comparison with 0.8 Å size contains fourfold more points, which in turn increase the number of lines used to display a map image. Therefore, in MAIN by default, spacing for map grids is 1/3 of resolution. As for low-resolution EM maps, in Chimera [50], it is even possible to reduce the grid spacing by displaying a smaller number of "voxels" to improve comprehension in large map areas.

The simplest way of increasing the insight area (and reducing the number of displayed lines in a map) is to increase the contouring level; another method involves a shorter distance between the clipping planes, and yet another is based on the "score map" calculation as a local map averaging described below. To reduce the number of elements in a map presentation in MAIN, contouring along one of the three grid planes only was implemented. A different approach to improve the overview and to reduce in graphical primitives is map skeletonization, which, in terms of complexity, approaches the images of molecular models.

3.6 Optimal Line Attributes for Model Building

A line has two attributes: thickness and color. To balance the perception of all objects, the objects with the largest number of lines should appear less pronounced than those with a smaller number of lines. For human visual perception, green, yellow, cyan, and white appear brighter than blue, violet, and red colors. With appropriate choices, color brightness can compensate for the differences in line thickness. Therefore, in MAIN, line thickness is optimized for each combination of a screen and graphics card. Once the optimal line thickness is established (1.0–2.5 pixels is the usual range), molecular and map objects are displayed with equal line thickness but with differently bright colors. Because the number of lines in a density map is greater than that of the atoms, maps should appear darker than the model. There is not much to add here: the color choices established in the 1980s for vector graphics displays remain the best approach: blue for the basic density map and yellow (gold) for a molecular model on the black background. ("Basic density map" means any kind of experimental maps and $2F_{\mathrm{obs}} - F_{\mathrm{model}}$ maps.) This coloring scheme has the following rationale: Blue and yellow are the highest-contrast colors: blue is dark and cold, whereas yellow is bright and warm, blue is in the HUE scheme as far away as possible from yellow; in the RGB scheme, yellow is composed of equal amounts of red and green, whereas blue is absent. To lay emphasis on the difference between the main and side chain, in MAIN, the main-chain trace (N, CA, C atoms) is displayed in white as the brightest color. This blue yellow/white combination enables insight into the largest map areas and parts of the structure. When there are multiple molecules, regardless of whether they are identical copies related by noncrystallographic symmetry (NCS) or different molecules, the default MAIN coloring assumes variance from orange and yellow toward green. Green, however, is reserved for atoms with zero occupancy, whereas red and pink tones are used for displaying symmetry-related molecules. As soon as details are added, such as coloring schemes for atom types, their stick-and-ball presentations, or difference density maps of the $F_{\mathrm{obs}} - F_{\mathrm{model}}$ kind, zooming in the area is required to preserve perception of the displayed objects in an image. For ligands, coloring of atoms according to the atom type is recommended because in contrast to the amino and nucleic acid residues, users do not have enough experience to immediately recognize the atom types from the bonding network alone.

For low-resolution cases, when maps are presented as surfaces and models as ribbons, the color of the map should keep it as an object that stands out the least. White in shades of grey and commonly the transparent mode seem the best solution, whereas a ribbon presentation of models can be shown in any color as long the color makes them different from the maps. This way, color can be used to differentiate the molecules too.

3.7 Graphics Software as a Debugging Tool

Due to the complexity of the structure determination process, oversight of the numerical output of structure determination programs is cumbersome. Graphics can provide intuitive understanding of the progress of the processes and simultaneously enable debugging and analysis of computational procedures. This was always an important factor that sped up development of MAIN features and provided a complete graphical insight into every component of the structure determination process, even in such unusual data fields as derivatives and numerical output of values at grid points of maps in 3D. Graphics at the debugging level for insights into (and control of) the structure-determination process are also used in a number of other computational projects, such as a real-time plug-in of PyMol for visualization of Rosetta and PyRosetta [51] and a graphical interface as a window into the decision-making process of ARPwARP [52].

4 Models Meet Experimental Data

The practical goal of model building is to bring the atomic model into the state that can be rendered by computational tools: that is, to bring the model inside their convergence radius. Hence, the human intervention via interactive model building is a push toward the increased consistency of a model with experimental data that are represented by density maps. This approach should be accurate enough to bring the model over the recognized energy barriers, yet there is no need to exaggerate because when it comes to details and their precision, the computational tools based on energy optimization can do it better than humans. Energy optimizations of model geometry restrained by density maps are also called real space refinement. Today, restrained energy optimizations are integrated into model building to such an extent that model building without these tools is not imaginable anymore.

The resulting models reflect two general approaches to structure determination:

- The deterministic approach, which delivers the average single structural model.
- The probabilistic approach, which provides an ensemble of solutions where each model represents a possible interpretation of the data, but only collectively do they explain the dataset as a whole.

The deterministic approach is described first. In my opinion, the probabilistic approach has a right to exist only after the attempts to determine the average structure have nothing more to offer. Namely, in a structure determination process, the model errors have to be decoupled from the structural variability, uncertainty, noise, and absence of data. If the variability is introduced too early, it hinders convergence of the structure determination process. In interactive parts of model building, the role of the human operator is confined to building the average model, whereas generation of probabilistic ensemble models is exclusively a domain of noninteractive computational procedures.

The approaches used for biological structure determination depend on the number of observations. The accuracy or uncertainty, precision and details of the model as a whole, and its components are a result of empirical observations we have at hand, and the other side of the coin is the knowledge and technology we use to create and optimize them. In light of the recent changes in structural biology [2, 53], it seems more appropriate to step outside of the crystallographic definition of ranges of structures based on resolution (ultra-high, high, medium, and low resolution) to the three levels of models at which they are built today:

- The models that contain direct information that links their chemical composition (sequence) to the relative position of atoms in 3D space. Essentially, every structure determined by macromolecular crystallography and a vast majority of structures deposited in PDB belong to this group.
- The models that give clues to the recognizable patterns in the folding of macromolecules and some larger ligands without revealing the link between the chemical composition and their position in 3D space (pioneering work on hemoglobin by Max Perutz [54] and bacteriorhodopsin by Richard Henderson [55]). When available, these models enable adaptation of previously determined structures.
- The models that incorporate previously determined structures or their components on the basis of shape complementarity and other empirical data like cross-linking restraints, identification by antibodies, or knowledge about interacting partners. Integrative or hybrid model building including Small Angle X-ray Scattering (SAXS) data is in this category.

The focus of this review is the models that allow for interpretation of experimental data to position atomic structures in space, yet parts are dedicated to the other two kinds of models in order to stress that our understanding of biological structures does not solely depend on structures resolved at atomic detail.

Prior knowledge (chemical and physical parameters of macromolecular topology and 3D structure like rotamer libraries and Ramachandran plots) plays an important role in the structure determination process. Nevertheless, these are guidelines and not strict rules to be obeyed blindly. Every structure can have details that deviate from the guidelines compiled in validation reports. What matters is that these deviations are not a consequence of ignorance or sloppy work but a result of a specific reasoning based on strong experimental evidence.

4.1 Deterministic View: Average Single Structure

The deterministic view follows the premise that the structural parts that can be unambiguously resolved by the density maps are built, whereas the others are not. This approach ensures that the number of parameters describing the model stays low. The result is single average model. The average models have an advantage: they yield a simple view that is easy to present, easy to comprehend, and therefore easier linked with biological data. The vast majority of researchers in this field are examining structures this way. The manual and automated average model building practice is well established and widely used. We just love to see when all pieces of the puzzle nicely come together. I think that a big part of our love of (and devotion to) crystallography has the roots in the idea that our results are among the least ambiguous in life sciences; this feeling expresses our hesitation to embrace the solution structures obtained by NMR, which delivers an ensemble of solutions and the average structure. In contrast to NMR, the EM structures share with crystallography two elementary constituents—maps and average structures—and are therefore closer associated with MX. Nonetheless, the concept of an average single structure has its limitations. There seems to be no such MX case where a protein structure was traced from the N-terminal to the C-terminal residue unambiguously revealing only a single confirmation of every non-hydrogen atom in the structure. This observation indicates that the average structure is an ideal representation of a structure that we try to achieve through the structure determination process.

The structures built along the lines of the deterministic view are the classics of MX and single-particle EM structure determination. Statistically, the average value is the most probable answer, especially when we are averaging across a population composed of a vast number of entities. In MX, the averaging begins already in the crystal as the source of data. In a macromolecular crystal, there is a large number of molecules. Depending on the crystal size and the size of the beam, this large number of molecules may reach 10^{13} or more (estimated in a putative cubic crystal with dimensions of 0.3 mm and unit cell dimensions of 100 Å composed of several asymmetric units). When all is said and done, the number of observed molecules in XFEL structures determined from much smaller crystals is not significantly different because they are determined by means of a massive number of crystals. Averaging of data continues during processing, when equivalent and symmetry-related diffraction intensity values are merged into a unique dataset.

Slightly smaller is the number of molecules observed during single-particle analysis, whose images are used to perform the 3D reconstruction of particles in EM. Their number can vary from several tens of thousands up to hundreds of thousands [56]. One must note, however, that an EM image, even though in real space, contains the amount of information comparable to the diffraction image in MX because each image contains a projection of the object on a certain plane. Hence, in EM too, the average image obtained from the vast amount of data still gets rather close to the true picture.

By means of density maps constructed from measured data, molecular models are traced, extended, and rebuilt. The order of the following subsections is intentionally reversed, because today, the majority of initial models are automatically generated or represent the result of a molecular replacement solution in MX or positioning of an atomic model in an EM map; these methods usually require model adjustments, not building from scratch.

4.1.1 Model Rebuilding

Efficient model rebuilding is achieved by combining the automated and manual model rebuilding tools coupled with energy optimization. Once a residue or section is targeted for remodeling in MAIN, energy minimization algorithms (fragmented, kicked assisted, or simply minimized) applied in a random order will position the model in place without the need for manual interventions.

For a larger displacement, manual interventions are necessary. An interesting approach with a lot of potential in MX and EM was proposed by Croll [57], who used the VMD package for rebuilding the ectodomain structure of insulin receptor. Molecular dynamics flexible fitting (MDFF) is an approach whereby the model is pulled around, while a molecular dynamics simulation keeps shaking it in the background of a density map. The molecular dynamics have a potential to keep levels of model distortion at limited levels, while the pulling forces bring the model to the desired positions. The latest release of the VMD package appears to have all these tools included [58].

When none of the approaches works, there are two remaining possibilities: the phase errors of structure factors used for calculation of maps are still too large to provide maps that can enable exact positioning of the model, or the region is disordered. In the first case, the advice is to wait for a density map from the next cycle of density modification or refinement. A few alternatives relevant for the second case are discussed below. There is always the option to abandon model building (rephrasing George Sheldrick) and declare the model as the final structure.

Main Chain

The simplest way to detect model errors and irregularities in the secondary-structure pattern is to display the hydrogen-bonding network on the background of a main-chain trace (Fig. 8). Deviations from the expected direction and instances of absence are indicative of poor geometry. When hydrogen bonds are used as restraints in energy minimization, they may be a decisive help for building regular secondary structure patterns. At low resolution and in noisy maps, where carbonyl orientation is not held in place by the density, the use of hydrogen-bonding restraints is crucial for preserving chemically reasonable geometry. The effects on regularization of secondary structure are demonstrated with a pair of images that show a model before and after minimization with hydrogen bond distance restraints (Fig. 8).

For spotting local problems in general, the Ramachandran plot is a useful tool. The common approach seems to involve linking a

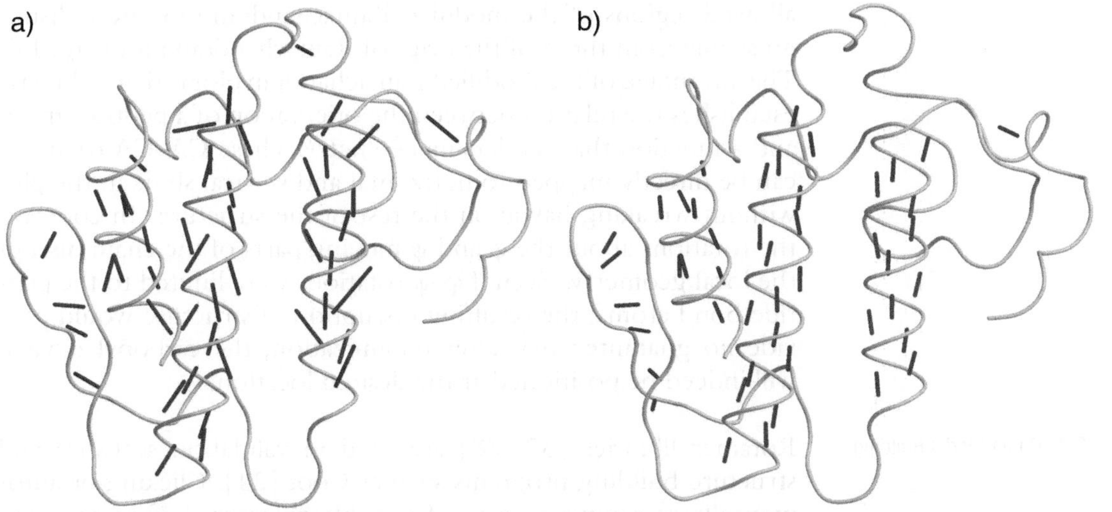

Fig. 8 Regularity of hydrogen-bonding patterns. The chain trace is shown as a *white coil* and hydrogen bonds as *pink sticks*. The figure shows a pair of models. The panel on the left shows the model of a 4DIH (PDB ID) structure after refinement of the molecular replacement model at 3 Å resolution, whereas the figure on the right shows the model after minimization with explicit hydrogen bonds used as distance restraints

Ramachandran plot to the residue of interest by providing a tool that allows a user to click on the outlier position in the plot and center the view of the residue in a 3D image. In contrast, elucidation of the problem is not directly resolved by the plot because the use of φ and ψ dihedrals as rotation axes will distort the already built model (Fig. 1c).

The solution to this problem in MAIN is not to focus on the φ and ψ angles of one residue, but on the peptide bonds shared between two neighbors (Fig. 1d). There are flip and manual rotation of the peptide bond atoms about the CA–CA axis and an automated tool that evaluates the possibilities of a range of residues and their regularity and match with the density map. When these tools do not provide a satisfactory solution, movement of residues that changes orientation of the side chain can be added.

To construct a plot that will directly help with main-chain rebuilding in MAIN, a plot can be generated that addresses not the φ and ψ angles, but the orientation of peptide bonds in relation to the side-chain CB atom of the residue (HA2 in glycine; Fig. 1b or alanine which HA2 is not marked in Fig. 1). The orientations are described by the pair of dihedral angles CB–CA–CA(−1)–O(−1) and CB–CA–CA(+1)–O, which are called modified φ (m_phi) and modified ψ (m_psi) angles. Their corresponding plot is called a modified Ramachandran plot. The relation between φ and ψ *and* m_phi and m_psi angles is very close to the shift (m_phi = φ − 120 ° and m_psi = ψ − 60 °). The shift is not mathematically identical throughout all positions (reflecting imperfection of the geometry), yet the resemblance is so close that the shape of the favorable and allowed regions of the modified Ramachandran plot are indistinguishable from those of the original Ramachandran plot (Fig. 1a). The advantage of the modified Ramachandran plot is that it directly establishes the relation between the orientation of a peptide group and a function that renders its geometry. These CA–CA rotations can be directly mapped to horizontal and vertical shifts in the plot without wreaking havoc on the rest of the structure. In contrast, the rotations about the φ and ψ moving parts of the chain disrupt the local geometry. Even if φ–ψ rotations were limited to the peptide bond atoms, the resulting disruption of structure would provide no guarantee that after minimization, the carbonyl oxygen will indeed be positioned at the desired location.

Side Chain Model Building

Rotamer libraries [37, 38] are used in validation software and structure building programs such as Coot [24]. Clicking or automatically screening a series of possible rotamers before selecting the correct one is the most common approach.

The generic approach is my preference over the list of possibilities from a precompiled library. In particular, generic approaches rely on common principles of structure and can therefore be applied to molecules of any kind of topology, including ligands, assuming that at least approximate nonbonding energy parameters are available.

For the side chain fitting, two generic approaches are implemented in MAIN:

- Click on ideal dihedral angles in an extended (all-*trans)* rotamer and the closest rotamer (dihedrals are set to the closest ideal dihedral angle: 180°, ±60°, ±90°, or 0°).
- A 5° search about every rotatable bond in the side chain taking into account the nonbonding energy of the side chain conformation and its fit to density. The number of searched conformations is optimized using the principle of dead-end elimination by skipping the impossible branches either because of a too short nonbonding contact or an electron density violation.

When combined with energy minimizations assisted by kicking and fragment rigid-body minimization, these tools will in most cases successfully fit the residues to the density maps. Thus, manual intervention is mostly confined to selecting and clicking the appropriate tool. This approach does not work when the side chain is pointing in a completely wrong direction. In such cases, the whole residue has to be rotated to redirect the CA–CB bond. Tristan Croll indicated that the side-chain fitting can be facilitated by molecular dynamics, using iMDFF [58].

4.1.2 Extension of the Polymer Model and Its Connectivity

Molecular model building enters the extension phase when the basic architecture of the structure was revealed, but the model is not completed. In MX, the task at this stage is to add as many atoms as possible to improve the scattering power of the structure. The model can grow by interpretation of the side and main-chain atoms. In most cases, the initial at least partial sequence assignment is performed by automated programs. When the solution is not provided or incomplete, interactive model-building software applications have to be used. It is important that after each extension cycle, the whole model is refined to maximize the improvement of the map. Refinement can be performed either in reciprocal or real space.

Adding Side Chains and Initial Sequence Recognition

The first extension of the model is addition of side chains. Novices in the field often argue that the sequence is the most reliable source of data. However, the structure of the side chain atoms turns from a hypothesis into hard data only after they are placed at the correct position. Too early an introduction of a partial macromolecule sequence may cause errors, which may be difficult to correct due to the bias they introduce. The exceptions are methionines, when their location can be identified by peaks in anomalous maps or when they were found during the phasing process, and occasionally tryptophans, which are the largest amino acid residues. The following paragraphs describe approaches in the order that is likely to result in the fastest building of the correct sequence.

When homologous structures are available, it is advisable to superimpose them on the model and use their homology to facilitate correct sequence identification.

The separation between selenomethionine residues might help identify a stretch in the protein sequence. For the rest of residues, the suggestion is to make the best possible guess of side chains by screening the density fits of the list of topologically different residues. Such screening can be automated in model building programs like MAIN, which can directly continue from the chain trace and map produced by phasing and density modification software like SHELXC/D/E [15]. This procedure is built on the assumption that shapes in density maps indicate side chains that contain a smaller or correct size of atoms. After the guessed sequence is introduced into the model, it can be matched against the real polymer sequence. MAIN has its own mapping function, otherwise Clustal [59] or similar alignment program can produce the desired result too. Especially when making an attempt to identify the first (and single) stretch of the guessed sequence, the stretch should be long enough, with a clearly identified number of residues building it (secondary structure elements are most suitable for ensuring the correctness of length) and should contain large residues to increase the signal in the scoring function. The length of ~10 amino acid residues will usually work. In nucleic acid sequences, the guessing is easier because there are only four nucleotides, which have only two distinctly different shapes.

All software packages enable residues to be changed/mutated manually. Once a sufficient portion of the sequence is recognized, the sequence can be automatically changed in most programs to the polymer sequence.

Main-Chain Extension and Connectivity and Loop Structures

When the stretches of identified sequence are substantial, they are our best guide to establish proper connections between the fragments. The alternative approach is to use homologous structures, which (already at an early stage of model building) can identify the architecture and fold of the structure. If none of these is available, it is best to resolve other issues and return to this place when sufficient progress elsewhere enabled one or the other approach. There is also another possibility: there are still a few structural biologists around who have (in spite of automated model building) not forgotten how proteins and nucleic acids are folded and can be of great help, when chain connectivity needs to be determined.

4.1.3 Model Tracing

When density maps provide a clear insight into the structure, residues can be built from a starting position by addition of one or two at a time. This approach is made possible in essentially every model building program. There are either rotations about the main-chain bonds or the most probable choices such as the "baton" building first implemented in O [60] and later in Coot [24]. The problem with building of a residue or two at a time is that building a structure on a residue basis requires a closer view into the density map. Nonetheless, this approach does not always work. When one cannot be sure where to build the side chain and where the main chain, it is clear that the baton extension will likely not lead toward the completion of the structure. The reason is either a poor initial density map or also disorder in the structure. In such cases, patterns of stretches of residues must be recognized.

It is not a coincidence that secondary structures are the first structural features that can be recognized in density maps. These are the only parts that show regularity (a repeating pattern) that can be recognized by a trained eye against a noisy background of the density map. Namely, noisy density maps are often discontinuous and one needs to extrapolate over the uninterpretable parts bearing in mind that α-helices are rather long sausages surrounded by less density and that β-sheets contain mostly uninterrupted strands in β conformation laid out in sheets at specific distances, with a side chain positioned alternately along the strands, but aligned when being viewed from a side. Ways to recognize them are well described in literature, e.g., by Richardsons [61]; thus, there is no need to provided details here.

An overview of a map can be increased by displaying the map skeletons; however, skeletonization is still only a density map

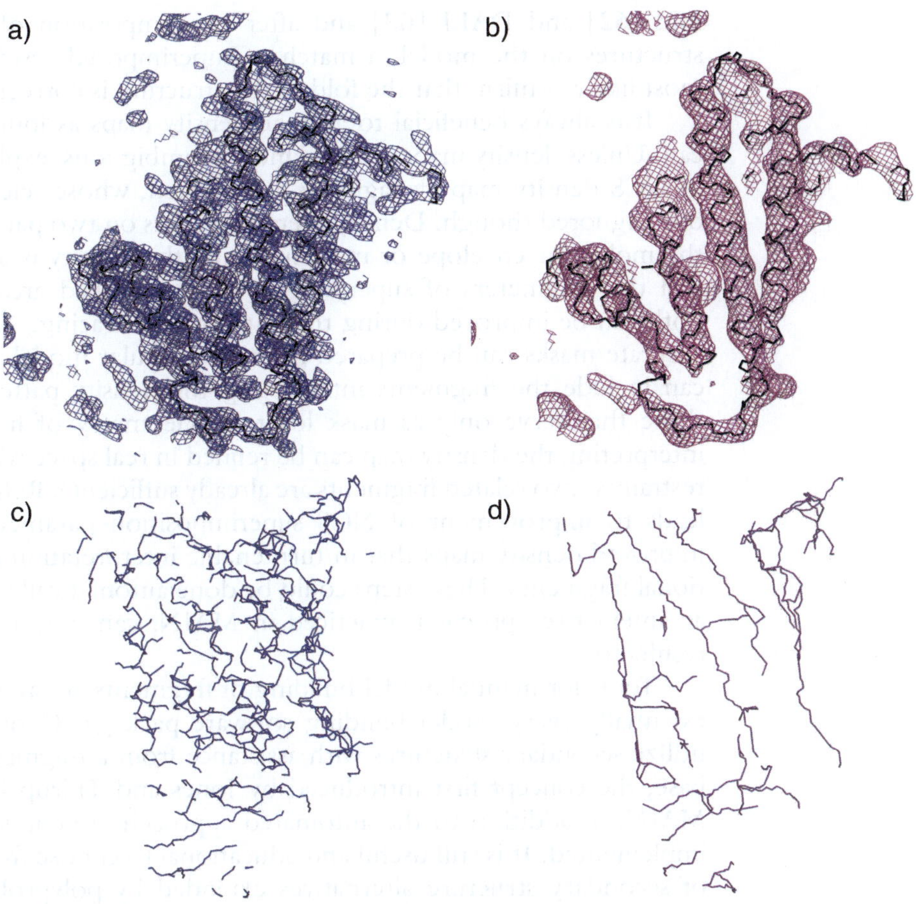

Fig. 9 MAD map presentation of cytochrome C oxidase [128]. At the top, a MAD map and its score map calculated with a 3.6 Å sphere radius are shown as wire frame representation in *blue* and *pink*, respectively. The two presentations below are the skeletons of these two maps

presentation, which can reveal only the features that are present in the map. To facilitate recognition of a chain trace in a noisy map, the score map approach was developed [26]. A score map is in essence calculation of a local similarity to a sphere (convolution between a spherical density and the underlying density map). At different sizes of the sphere (2.0–3.6 Å), different map features get exposed. The size of approximately 3.6 Å exposes alpha-helices, whereas smaller sizes are more useful for recognition of the chain trace. Because score maps filter out the noise, they significantly reduce the number of displayed lines, make maps smoother, and thereby additionally enhance our chain trace recognition capabilities (Fig. 9).

Although there are a number of automated software applications that will try to find only the secondary structure elements—Buccaneer [19], SHELXE [15], PHENIX [18], and ARPwARP [20]—they may fail owing to poor quality of the map. After submission of a partial structure to homology search servers like PDB fold [62] and DALI [63] and after superimposition of related structures on the model, a match of superimposed residues will most likely confirm that the fold of the structure is correctly built.

It is always beneficial to average density maps as long as one can. Unless density maps are extremely unambiguous, exploitation of NCS density map averaging is a dire need, whose relevance is often ignored though. Density averaging relies on two parameters: the molecular envelope or mask in which the density is averaged and the parameters of superposition of the masked areas. They both can be improved during the initial model tracing. The most accurate masks can be prepared from a molecular model. Models can include the fragments interpreting the density patterns and those that serve only as mask locators. Geometry of fragments interpreting the density map can be refined in real space with NCS restraints (two related fragments are already sufficient). Refinement leads to improvement of NCS superimposition parameters and improved density maps that in turn enable interpretation of additional fragments. These steps could be done automatically, though an interactive approach, practiced in MAIN, can quickly deliver results too.

Tools for manual model building of fragments are available in essentially every model building software package. Coot and O utilize secondary structures with assistance from a fragment database, the concept first introduced by Jones and Thirup [64]. In MAIN, in addition to the automated approach, a generic one is implemented. It is still useful and educational to choose from a list of secondary structure alternatives extended by polyproline and collagen elements and I, II, and III turns and their inverses. In practice, one seldom needs to reach outside the elementary secondary structure fragments because there is no need to manually position model parts with a reasonable fit to the density maps. One only needs to get CA atoms in proteins and ribose rings in nucleic acids close enough to the blob of the corresponding density and run energy optimization (atomic and fragment) and fitting tools (side chain and peptide bond rotations). It is astonishing what these tools can achieve.

4.1.4 Finalizing the Structure

When the structure is approaching completion (fold and sequence being established), the last details need to be resolved. The goal is to interpret the largest possible areas of the density map by avoiding interpretations not supported by evidence. This phase begins by addition of known ligands (soaked or cocrystallized compounds, cofactors), covalent modifications like glycosylation and phosphorylation, solvent molecules and double or triple conformations. Once parameters are at hand, ligands can be included in the model. My recommendation is to build them first with zero occupancy and preform a few rounds of automated solvent addition. (No reference is given here because essentially every working environment has such tools at hand.) Building ligands first with zero occupancy has two advantages: solvent molecules do not overlap with ligand atoms, and one can validate the fit of a ligand to density maps through several cycles of model building and refinement with the ligand omitted from calculation of the structure factors before their full incorporation.

Essentials of solvent interpretation are the threshold values of peaks in $F_{obs} - F_{model}$ kind of maps, their consistency with $2F_{obs} - F_{model}$ kind of maps, and a chemically favorable environment with a potential of hydrogen bond donors or acceptors. Normally, sigma-A-weighted maps will do [10]. In the case of doubt, however, I always reach out to the averaged kicked maps [14]. The $F_{obs} - F_{model}$ maps already interpreted by a solvent may still contain high peaks that need to be checked for potential ions. Conservative interpretation requires that after refinement, ions have the same *B*-factors as their surroundings. The problem is that their positions are often only partially occupied. In such cases in MAIN, one can either restrain *B*-factors of ions or ligands to the surroundings by optimizing the occupancy or simply lower occupancy in 0.1 steps and perform parallel *B*-factor refinement. Automated ion and small-ligand recognition is available in such software packages as ARPwARP [65] and PHENIX [66, 67]. Ions should be monitored between the refinement cycles as they may shift around because of inaccurate nonbonding-energy terms. PDB contains a number of ligands in the files not consistent with the elementary criteria of density fitting [68, 69]. A possible reason for such instances is that at the time they were built, the density maps provided evidence for their positioning; however, during later stages of refinement, they were shifted or the density supporting their positioning vanished. There is software that can help resolve the twilight situations [70].

At the solvent interpretation stage, alternate conformations are built. Often, it is forgotten that they may also involve main-chain atoms. The use of automatic tools may be of great assistance in the search for a few alternate conformations, normally overlooked by a human interpreter [71].

Validation helps to confine the structure to the acceptable deviations defined by the validation criteria. MolProbity [36, 72] checks for chiral centers and planar fragments, side-chain rotamers, the Ramachandran plot, packing, and hydrogen-bonding patterns. Coot utilizes external validation tools, whereas in MAIN, they are a part of the distribution. Each validation task generates a list of centers ordered from the worst toward better parts to be explored manually. When in doubt, the interactive work also offers a quick validation approach: the region of interest should be energetically minimized without the density term and with nonbonding interactions confined to the area of interest. When atomic positions remain close to the starting positions, the region is chemically reasonable and consistent with the density maps. Nonetheless, when atoms of groups shift notably, then the region requires additional attention.

Real structures do not entirely conform to the ideals of structure fulfilling all validation criteria. There are also exceptions and ambiguity. Exceptions have clear confirmation in density maps, whereas ambiguous parts are a harder nut. They are indicated by density regions that do not match the superimposed model (It is assumed here that every effort to match density with a proper model has failed and appears to be impossible. This assumption excludes the possibility that the segment in question was modeled poorly due to inexperience or a lack of effort.). I consider the principle that only unambiguous parts consistent with structure validation belong to the final structure a personal decision. If you want to stick to this ideal, then this would be the point to abandon model building. Consequently, such ideal structures will score well in the PDB deposition statistics.

If, however, you decide to move on with interpretation of the ambiguous parts, be aware that real and reciprocal space refinements distort their geometry because the target does not really exist. I am of the opinion that these ambiguous parts should be interpreted too. I want to stress though that this does not justify having most of the model ambiguous because this approach implies that the structure was either not refined or is wrong. However, as long as disorder is confined to the parts whose inclusion would increase the model consistency with the material crystallized or set on the grid, I encourage you to undertake every effort to include the ambiguous parts in the final structure.

To justify attempts to interpret ambiguous regions, the density maps must clearly demark ambiguous parts of the model from their surroundings. It is not important at which contouring level they are recognizable and what kind of map is being examined (residual density or $2F_{obs} - F_{model}$), locally averaged maps (called score maps in MAIN), or FEM maps (in PHENIX). As long as such parts can be interpreted with a recognizable conformation, *B*-factors may be allowed to rise. For the parts in an ambiguous model trusted less (these are parts where conformation cannot be extrapolated or guessed), the occupancy of atoms should be set to zero. When there is no supporting evidence from a density map suggesting a confined area to which a structural part can be assigned, then there is justification to include those parts in the final structure of a single model.

Our recently determined structure of Cwp6 amidase, determined at 1.7 Å resolution [73], contains regions that lack clear density support. I asked Henry van den Bedem for assistance with their tool, which involves the inverse kinematic approach [74]. In this case, no satisfactory interpretation was found; therefore, we struggled further. Finally, the connections were built manually and weighted zero due to the evidence of $F_{obs} - F_{model}$ and score maps

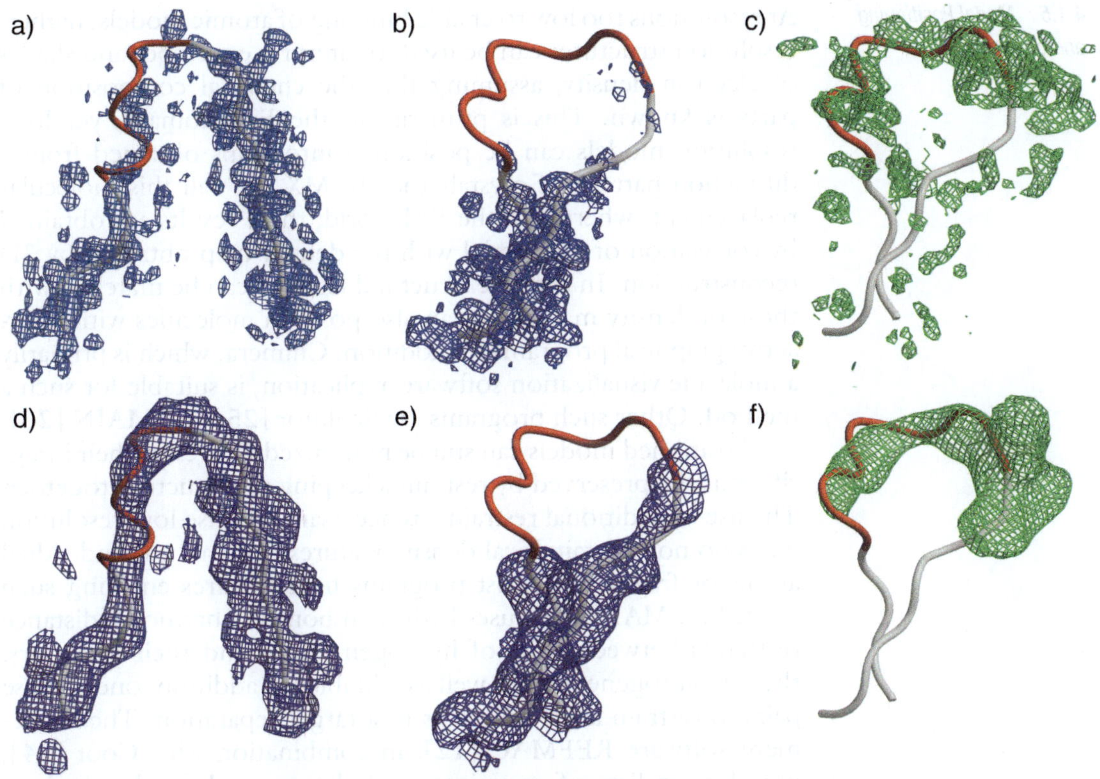

Fig. 10 The disordered loops linking domains in the Cwp6 amidase structure (PDB ID: 5J72). The two panels on the left show a chain trace of the loop between positions 120 and 140. The four figures on the right show the chain trace of the loop between positions 447 and 467. The residues with weighted atoms are shown in *white*, whereas the residues with zero occupancy are shown in red. The map around the first loop at the top is a $2mF_o - DF_c$ map, at the bottom is its score map calculated with the radius of 2.6 Å. The maps around the second loop shown at the top are $2mF_o - DF_c$ and $mF_o - DF_c$ maps, shown in *blue* and *green*, respectively. The maps below are score maps calculated with a 2.6 Å sphere radius

that indicated a path that connected the gaps between domains in the NCS dimers (Fig. 10). These parts were included in the final structure because it was important to elucidate the structure at a monomer level. This would not have been possible if we left the structure at the level of three pairs of separated domains within the NCS dimer.

When disordered density regions lie at a crystallographic rotation axis or there appears to be unusual disorder in a molecule clearly resolved elsewhere, one may make an attempt to lower the space group symmetry. Alternatively, one can consider ensemble models. In such cases, however, we are already stepping out of "the box" of the average structure context into density interpretation by means of multiple models described below.

4.1.5 Model Positioning and/or Placement

At resolutions too low to enable building of atomic models, higher-resolution structures can be used to interpret envelopes and shapes of electron density, assuming that the chemical composition of parts is known. This is primarily in the EM domain, yet low-resolution models can be positioned into maps obtained from a diffraction pattern of crystals too. In MX, we call this molecular replacement, whereas in the EM world, these results are obtained by correlation of the model with the density map obtained by 3D reconstruction. In the case structural features can be matched with those of density maps, one can also position molecules with interactive graphical programs. In addition, Chimera, which is primarily a molecule visualization software application, is suitable for such a method. Other such programs are Sculptor [25] and MAIN [26].

Positioned models can still be minimized; however, their integrity must be preserved by restraints keeping the structure together. The use of additional restraints is necessary because low-resolution maps do not contain local density features that can trap individual atoms or fragments. Most programs have features enabling such modeling. MAIN [26] uses hydrogen bonds as harmonic distance restraints between pairs of hydrogen atoms and their acceptors; they are autogenerated as well as editable. In addition, one can use pairs to restrain any two atoms to a target separation. The refinement software REFMAC [12], in combination with Coot [24], uses longer lists of restraints, including user-defined restraints, which preserve the original structure enhanced by the "jiggle-fit" and "model morphing," which enable fitting to lower-resolution density maps [75]. In contrast, iMDFF [58] relies on molecular dynamics and pulling structural parts around.

More complex problems can be addressed by approaches of integrative modeling described briefly below in a separate section.

4.2 Probabilistic Approaches to Model Building

In a broad context, the probabilistic view also includes multiple models of molecules from crystals with asymmetric units containing several molecules related by NCS, which can also be the result of a reduction in space group symmetry or multiple crystal forms. Furthermore, in these cases, the priority is to determine the average structures first, before differences between subunits are considered. The difference between NCS averaging and the space group reduction is that NCS averaging is usually performed in real space, whereas the space group reduction affects the step in reciprocal space where Bragg spot intensity values are merged. They both are followed by modifications applied to individual molecules. In EM, an equivalent approach is to divide projections of molecules into groups consistent with their separate states.

Clearly, the average structure is the most probable structure. Nevertheless, when a single average model cannot satisfactorily interpret density maps, then the reverse is also true. The average density maps do not represent a single structure. As a consequence, attempts to energetically minimize a model to fit the average density map will distort the model that will still not fit the map and will not result in an overall consistent solution. (This would be the point to abandon structure determination if we aimed to stick with the average structure context.). Nwachukwu et al. [48] tested combinatorial schemes of refinement and concluded that a single model cannot represent the data as well as an ensemble can. There are two assumptions underlying the determination of structures as a single average model: First, the resulting model with the lowest energy (the lowest deviation from the chemical and experimental restraints) is the closest to the real structure. Second, there is a single structure that has the lowest energy. Once we realize that a single structure cannot satisfactorily reproduce experimental data, we need to step out of "the box" of a single average structure. Here, the interactive model building approach hits its limits. Essentially, these are the limits of human perception because the question in actuality is how a human brain can conduct interactive model building of an ensemble in which overlapping conformations cannot be sorted and adjusted to fit the density map on a one-by-one basis. Hence, an ensemble cannot be built manually but must be generated by means of computational tools that can sample through conformation and density map space. There are two general approaches: molecular dynamics and Monte Carlo sampling. With the ensemble approach, MX is approaching the NMR structures [76] that have well-resolved parts held in place by numerous restraints and those that provide a variety of possible solutions spread over areas of space. Clearly, the greater the disorder, the larger is the spread of solutions and the lower is the accuracy of the position of a disordered part of the structure. These models, however, can be inspected together, scrolled through, or examined independently.

Regarding the motion and disorder of crystal structures, there are "in the box" and "out of the box" views. "In the box" refers to the crystal structures determined by means of diffraction measurements of the Bragg spots representing the periodic pattern of the crystal lattice, whereas "out of the box" considers the diffraction pattern of a crystal as a continuum.

4.2.1 Ensemble Models Using the Bragg Spot Data

Providing an ensemble of solutions instead of an average is an alternative almost as old as the introduction of molecular dynamics to crystallography [77]. However, in the current PDB (April 15, 2016), one can find only 71 structures determined by X-ray crystallography with the word *ensemble* in the title. Seventy-one is a relatively low number compared to over 100,000 deposited structures determined by MX, yet there are groups such as George Phillips's devoted to research into protein dynamics [78] because there is a general understanding that average structures do not provide insight into some biologically important events depending on mobility.

A demonstration should be based on an independent data source to confirm whether the ensemble models indeed represent the dynamics of a structure and not its uncertainty and disorder, crystal mosaicity, or a simple lack of data. One study [79] suggests that the TEST portion of data in combination with *R*-free is such an independent measure. My understanding of *R*-free is that its usefulness is vastly overrated and that it is not a parameter to be trusted for its objectivity [47], yet a detailed discussion here would be a digression from the purpose of this chapter. Nevertheless, there are data in the diffraction pattern of crystals on the motions in macromolecular structures that provide information independently of the Bragg spots.

4.2.2 Including the Information Between the Bragg Spots

Normally, crystal structures of biomolecules are determined from Bragg spots of diffraction peaks, yet there is also a continuous diffraction pattern between the peaks and behind them that contains rich information about the two-point correlations of electron density [80]. This is the area of diffuse X-ray scattering, which can shed light on the correlated movements between atoms beyond a single unit cell (these movements do not conform to the average of the crystal lattice). Analysis of motion in a crystal is consistent with the protein motions resembling diffusion in a liquid or vibrations of a soft solid and normal modes [81]; however, it is not consistent with the TLS (translation-liberation/rotation-screw) parameters usually employed to model the crystal disorder in refinement. To make diffuse scattering analyses available for all structures, it is suggested depositing the whole diffraction images and not only structure factors from the Bragg spots.

Information between Bragg spots not only reveals correlation between atoms in the crystal beyond the limitations of the crystal symmetry lattice, but may also be a source of parallel information providing insights into the details of higher resolution than the Bragg intensity values themselves [82]. For me, it was quite a surprise to link the continuous diffraction from a crystal with the phase problem solution, but in the end, a crystal, as large as it may be, is a single particle by itself, and X-rays can penetrate it better than electrons can. During imaging of large particles, fiducial points are required to align separate images. Maybe, at some point in the future, Bragg points will turn into fiducial points of a crystal structure.

4.2.3 Disorder in EM

Cryo-EM approaches are mostly used to study the structures of large macromolecules and molecular machines. These complexes are often characterized by inherent conformational flexibility and compositional heterogeneity, which are linked to their function. Although crystal packing often restricts conformational flexibility in X-ray crystallography, macromolecular complexes are typically examined in a whole ensemble of conformational states in a

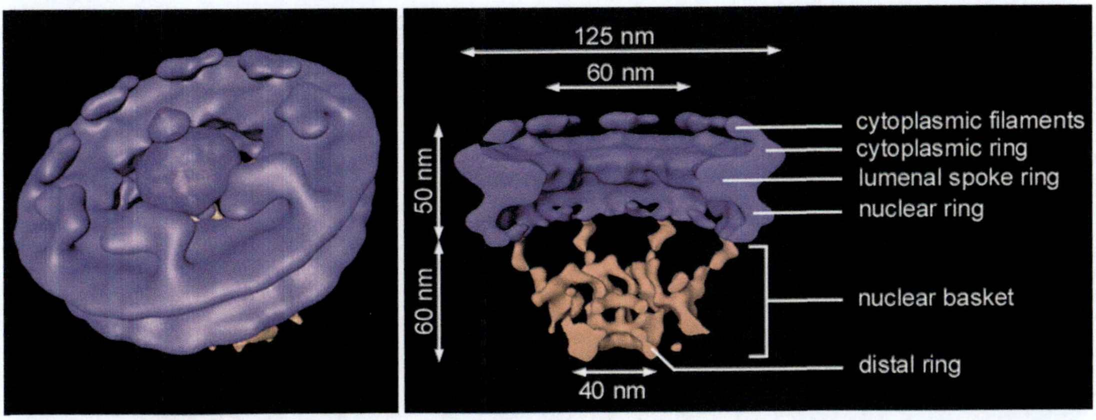

Fig. 11 An isosurface of the *Dictyostelium* nuclear pore complex: dissected and labeled. The figure is courtesy of Jürgen Plitzko

vitrified aqueous solution. Thus, cryo-EM analysis allows for capturing the whole conformational landscape of a macromolecular complex, given the ability to disentangle the different states. Sorting subsets of either conformationally or compositionally heterogeneous particles in cryo-EM is not a trivial task because of the low signal-to-noise ratio, the unknown number of classes, and the often-unbalanced class occupation, but can be achieved by a variety of classification approaches based on multivariate statistical analysis (MSA) or multireference approaches [83].

4.3 Integrative Modeling and Cell Imaging: A New Era

The paper "The Molecular Architecture of the Nuclear Pore Complex" by Albert et al. [84] ushered in a new era of model building by demonstrating that for structure determination, one can combine numerous sources of data (from biochemical to structural) whose integration can provide consistent structures of large macromolecular complexes. Typical restraints for integrative structure determination include subunit localization by protein depletion or fusion experiments, e.g., docking of atomic models from various external sources into the density envelope of the nuclear pore complex (NPC) [85] (Fig. 11). The Integrative Modeling Platform (IMP) [86] now also provides tools to position models in low-resolution maps obtained by EM and SAXS, combined with the use of distance and proximity restraints obtained from other sources such as mass spectrometry. Recent developments in Rosetta [87] indicate development of similar functions too.

A similar breakthrough was made in EM cell imaging. Today, structure determination of even large macromolecular complexes in frozen hydrated cellular environments is possible, thanks to recent technical advances in cryoelectron tomography (CET). The capacity for nearly artifact-free thinning of cells in a vitrified state, using cryofocused ion beam (FIB) micromachining, allows for 3D visualization of cellular structures deeply embedded in the cell

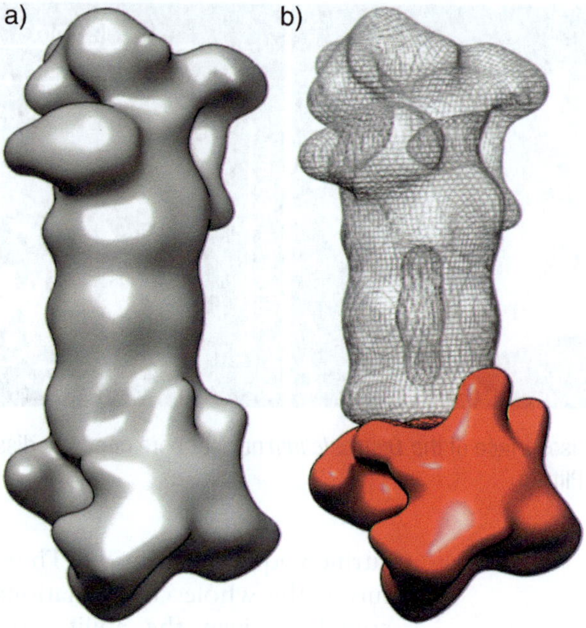

Fig. 12 Subtomogram averages of the 26S proteasome from a cultured hippocampal neuron. On the left is the surface of the double-capped particle. On the right, the single-capped-particle density is shown as a *white* wire frame; it was subtracted from the density of the double-capped particle to obtain the region corresponding to a single-cap region shown in *red*. The maps were obtained by EM tomography. The figure is courtesy of Jürgen Plitzko

body. The quality of tomographic volumes has been substantially improved by the direct detector technology and the recently developed Volta phase plate [88], which considerably increases the low-frequency contrast, particularly crucial for reliable localization of macromolecules in crowded cellular environments. Finally, more elaborate image-processing software designed for CET can utilize the considerably increased information content in tomographic volumes to provide highly detailed insights into the structure and organization of macromolecules in a cellular context. In Fig. 12, the proteasome structure in situ subtomogram shows the density envelopes (EM people call them isosurfaces) of a double-capped 26S proteasome [89]. The capped region at the bottom is shown in red. Tomographic images now allow researchers to visualize large complexes in in situ environments as demonstrated by means of the proteasome 26S structure (Fig. 13) [89, 90].

5 Visualization of Molecules by Molecular Graphics

We cannot directly modify a protein structure, but we can engineer it indirectly by modifying its genetic sequence, express the constructs, and then study their biochemical and biological properties

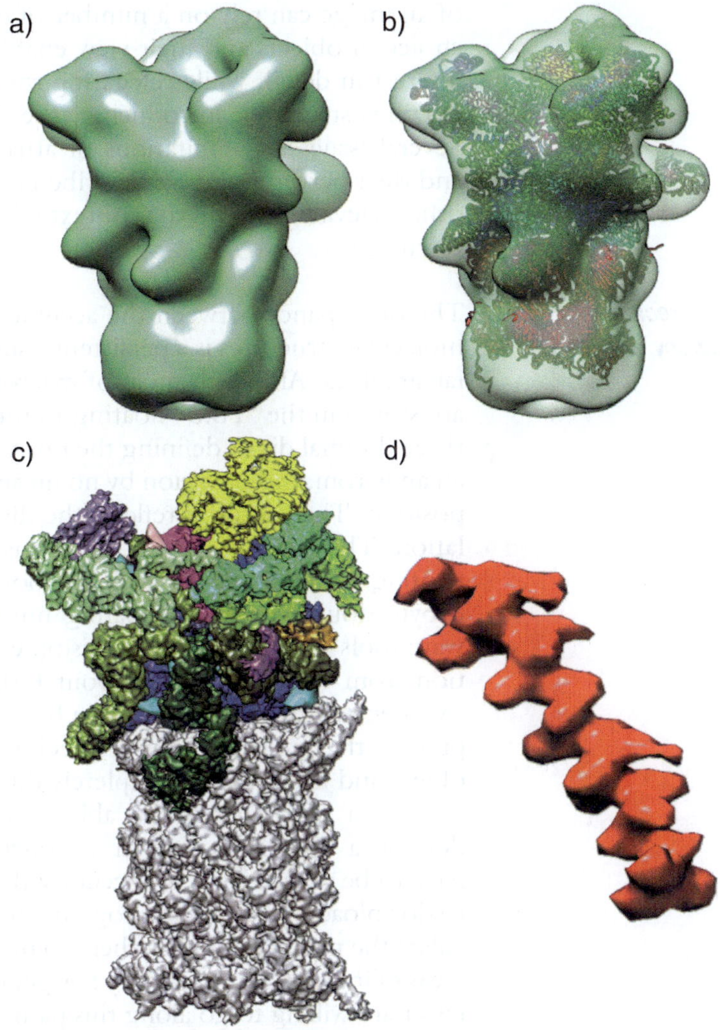

Fig. 13 (**a**) At the top, a ground state average is displayed as a *green* isosurface (single- and double-capped 26S proteasome). (**b**) The top right panel shows a fitted atomic model of human 26S proteasome subunits in the EM density. (**c**) Cryo-EM reconstruction of the human 26S proteasome at 3.9 Å resolution. Colored maps correspond to subunits of the cap, whereas the 20S particle is shown in *white*. (**d**) Magnified features of density corresponding to the helix formed by residues 57–80 of β1 shown in *red*. The figure is courtesy of Jürgen Plitzko

and roles. Molecular graphics give clues to 3D structures of biological molecules. The displayed molecular images have explorational and narrative purposes. Exploration enables us to explore and establish links between sequence and biology, whereas narrative figures present information in the most evident way. The displayed images are translation of biological information into technical composition of geometrical bodies. Their attributes are position, dimensions, shapes, colors, and transparency. Composition of an image can rely on a number of objects of various kinds. The choice of objects and their view enable us to perceive the parts of interest in detail, while relegating the others to the role of a supporting structure introducing the background of molecules. Several issues and limitations of attributes of geometrical bodies and their perception were described in Subheading 3. Below, a few other relevant issues in the context of visualization of structures are discussed.

5.1 Precision, Accuracy, and Shapes

The discrepancy between the accuracy and precision of presented molecular structures is a persistent issue and a challenge in molecular graphics. Although most of macromolecular atomic structures are stored in the "F8.3" floating-number format of PDB files with three decimal digits defining the position of an atom to 1/1000 of an angstrom, this precision by no means reflects the accuracy of the position. The precision reflects the digital form of model manipulation. The current state of the art is the 4-byte single precision floating number, which will be sooner or later replaced by an 8-byte double precision floating number. Most refinement software tools work at double precision even though the target deviation from bond lengths is about 0.02 Å, which suggests that a two-decimal-digit precision may be sufficient. Hence, it is easy to present the molecular object precisely; however, accuracy of the object and its parts is a completely different issue.

For a researcher to be able to explore molecular structure, develop a hypothesis, and draw conclusions, molecular structures have to be perceived at the accuracy they were determined. I prefer to download MX structures together with empirical data, then I calculate the maps and display them to understand the reliability of the areas of interest; however, very few people can do the same and even fewer are willing to go along this path. Besides, this is nuisance and not really enlightening. Hence, the precision of images should give an impression of accuracy of the displayed structure. This requirement imposes restrains on the art of their visual representation.

Technically, we use graphical objects' lines, cylinders, spheres, and other basic geometric bodies composed of polygons (ribbons and surfaces) to project 3D structure of molecules onto electronic media or print it out on paper. An impression of precision of atomic position is decreasing from lines, sticks, spheres, and ribbons to surfaces and envelopes.

Lines connect two points in space. In a computer graphics system run by OpenGL, line thickness is a property of presentation and not of the object it is describing. It is specified in pixels and remains the same regardless of scale/zoom/magnification we choose to inspect the structure. Cylinders and spheres (ball-and-stick representation) are displayed as polygonal objects with defined radii and fines (the number of polygons constituting them). They occlude space around atomic centers, and this property makes them less precise than lines, yet the spheres still indicate atomic positions.

Ribbons and surfaces still rely on atoms, yet they do not directly point to atomic positions. Ribbon presentations facilitate the comprehension of the fold and architecture of molecules [91], as reviewed elsewhere [92]. The first time I have seen ribbons was in FRODO-imported output of John Priestle's Ribbon program. Ribbons are smoothed lines through a fold of a protein structure usually based on spline interpolation. The deviation of ribbons from the underlying details of atomic structure depends on the level of smoothing of the chain trace. Quite often, protein secondary structures are highlighted with ribbons for β-sheets and cylindrical objects for helices, while the rest of the fold is presented as a coil.

A molecular surface is the most intriguing graphical object of molecular images and has numerous potential uses and therefore is also a subject of a large number of developments. A molecular surface describes the accessible space for interactions with other molecules, solvent being the simplest. It gives clues to the shape complementarity and molecular interactions. The details of surface presentations depend on the algorithm used and the smoothing of the surface that excludes inaccessible areas. Accessible surface presentations started with Lee and Richards' [93] drawing contours along dissecting planes. The Connolly accessible surface algorithm [94] described a surface of spheres with distribution of dots, which either belonged to molecular atoms in the convex part of the surface or solvent atoms in the concave parts. Yet another possibility is to calculate the "skin" surface, based on Delaunay tetrahedrization or a Voronoi diagram. Bernstein and Craig, for example, chose to build the surface by its approximation to an electron density map of a single Gaussian [95]. The new release of ChimeraX is replacing the Connolly surface and introducing the Richards surface because of smoother surfaces (Tom Goddard, personal communication). MAIN polygonal surfaces use an implementation of the Richards algorithm, while the dot surface implementation of the original Connolly program remains in use.

As long as we are dealing with a single molecular structure, presentation seems obvious. Nevertheless, as soon as we begin to address disorder and dynamics, the choice of the art of visual representation stops being that obvious. In reality, molecules are jellylike substances whose blur is difficult to comprehend. Adding the time dimension to the image is one of the solutions, but there are limitations in perception of a structure imposed by its disorder on dynamics. The solutions at hand are to superimpose them all on the same image, scroll through them or to utilize transparent objects. The latter approach is demonstrated in Fig. 14. On the left, there is a surface of a single structure of my workhorse structure—ammodytoxin A—presented by means of two surfaces (PDB ID: 3DIH). The inner surface is a standard nontransparent one calculated from standard VdW radii of nonhydrogen atoms. The opacity of the surface gives the impression of a solid hard object. The outer surface

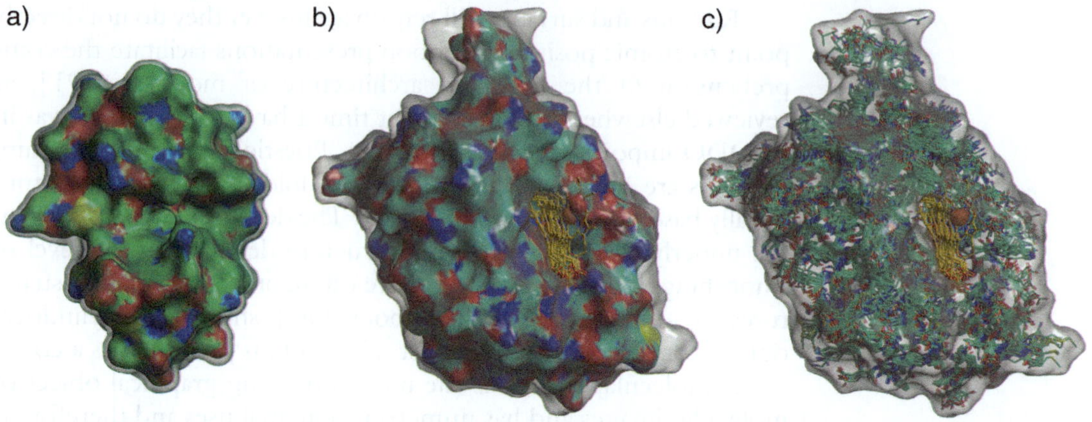

Fig. 14 Uncertainty and flexibility of structures. (**a**) The left panel shows two surfaces. The solid surface was calculated around the ammodytoxin L structure (PDB ID: 3DIH) using the VdW radii. The *green*, *blue*, and *yellow* regions correspond to the surface of carbon, nitrogen, oxygen, and sulfur atoms. The transparent surface was calculated from VdW radii increased by the atomic displacement factor. (**b**) Ensemble models of human heme oxygenase 2 (PDB ID: 2RGZ). The inner surface was calculated from VdW radii of the first model. The *cyan*, *blue*, *red*, and *yellow* regions correspond to the surface of carbon, nitrogen, oxygen, and sulfur atoms. The transparent surface was calculated from VdW radii of all the models combined. The cofactor structure, a heme molecule, is shown as ball-and-stick representation. (**c**) The same as (**b**) only here the inner solid surface is replaced by stick representations of all 16 models

was made transparent to give the impression of a jelly layer covering the molecule. The surface was calculated from the increased atomic radii by means of the isotropic temperature factor (atomic displacement parameter). The other two images show an ensemble model. The middle picture is composed similarly to the one on the left, only this time, the inner hard solid surface is calculated from the first model, whereas the jelly layer was added by calculating the transparent surface from atoms of all 16 deposited models of human heme oxygenase 2 combined [96]. The spread of heme molecules shows that they in part penetrate the jelly as well as the hard surface. The picture on the right shows the same transparent surface as in the middle picture, only here, the spread of 16 models is demonstrated using stick representations.

5.2 Mapping of Molecular Characteristics in Images: The Use of Color

The simplest property for mapping of molecular characteristics is color. Colors can be defined either for a whole graphical object or only for a point in a triangle, which enables smooth transitions in a complex polygonal object like a surface. The color attribute of objects is used for the narrative and mapping purposes.

The elementary use of color is for coding the atom types: blue for nitrogen, red for oxygen, and white for carbon. I often make use of different colors for carbon atoms for narrative purposes. To improve perception, color also serves to code residues, molecules, or their parts.

Spectral colors serve to map a range of values. The uses include rainbow, or a part of it, or a transition between two or three colors where the intermediate is usually white. Rainbows are often applied to color the chain trace from N- to C- or 3′ to 5′ termini or mapping properties such as the temperature factors indicating flexibility and disorder in a molecule, distance from a center or interacting partner, and sequence similarity. The standard way of presenting the magnitude of the electrostatic potential calculated by programs like DelPhi [97] is the blue–white–red color grades.

5.3 Interactive Display, Animation, and Static Figures

Nothing can replace insight into 3D objects by a living picture. The perception of a living picture requires smooth changes of the displayed objects. To achieve this goal, updates (the differences between the two consecutive images) must be small enough to be interpreted as a smooth transition perceived as motion. Different software developers found different solutions. In MAIN, one can use the mouse, keyboard, and dial box to control the changes on the screen. As far as perceptions are concerned, my goal was to optimize the smoothness of image rotation by event autogeneration. Generation of events is the only process that provides smooth motion input. Currently, my favorite motion control device is the mouse. The latter is a device that can be employed not only to repeat a motion event but also to control its step size. The step size and direction can be reconstructed from the difference between the last two sampled positions at the moment when a mouse button was "released". Several users find this disturbing and like to switch it off, yet it allows a researcher to spin, translate, scale, and rebuild molecules via controlled smooth interaction with the program. The last motion event can be repeated until a new event occurs also on a keyboard (here the arrow keys are the intuitive solution for motion events); however, the keyboard can only send preprogramed step size and direction. Unfortunately, I found no solution for repeating the dial box event, where the step size can be controlled by the speed of rotation. Unfortunately, the dial box has another disadvantage. It is an old device programmed to work at the 9600 baud rate, which is too slow for today's computers and consequently triggers jagged motion on the screen. On the other hand, the advantages of the dial box and keyboard are that they trigger an event in only one selected direction, whereas the mouse with three buttons requires to read three-directional inputs from the right mouse button to drive rotation about X, Y, and Z axes, and two-directional input from the middle mouse button to drive the XY translation. Another advantage of the dial box is its eight-channel input, which can simultaneously deliver events from more dials than a user can handle. I can provide no reference for these uses of mouse buttons, but I vaguely remember that I have implemented the application of a single mouse button to rotations in X, Y, and Z directions after a visit to a Silicon Graphics office in Basel in the early 1990s, whereas the spin motion may be my response to the lack of a dial box. Many software packages utilize the scroll button events for a number of commands that require a stepwise response without a requirement for a seemingly continuous response. The usual use is for changing the map contouring level; however, additional applications can be found: for example, for scrolling through a list of structural solutions and colors.

From the narrative point of view, animation has a great advantage over interactively driven changes because it contains a precompiled sequence of images that differ from each other in small precompiled changes in any number of parameters of projections and molecular structure including changes in the image composition. In animations, by defining several viewpoints, one can interpolate a passage through a molecular structure following the narration as for example in CCP4mg [98]. A brief summary of animation software and its challenges can be found in the report on the 2-day workshop on molecular animation in 2010 at UCSF [99]. There, it was noticed that the barriers for the use of excellent software packages are dropping and that now, the challenge is to provide a similar level of accessibility for animation of larger movements.

It will take a while before the Daily Prophet live-picture technology from Harry Potter's world will be possible on printed presentations; however, in electronic media, a living picture either in the form of animation (a precompiled sequence of images) or an interactively driven view is a reality. In an editorial in 2009, Palmer and Matthews [100] announced interactive graphics as a way of communication with the readers of Protein Science. User-generated scripts in hypertext markup language (HTML) provide a link to user-generated Jmol scripts as part of the Supplementary material. Today's version of the interface is called "iMolecules 3D" (http://imolecules3d.wiley.com/imolecules3d/). A number of web servers offer insights into manipulation of objects on the screen starting with PDB. It was a pleasant surprise to read a paper on interactive 3D scientific figures in Portable Document Files [101].

Figures as static images are an important way of broadcasting the structural information. Their advantage is combining a number of components, techniques, and tools in an image with strongest message content. Good informative figures can seldom be prepared with a few clicks on the screen. However, I feel that this part is outside the scope of this chapter; therefore, I decided not to discuss this topic here in any detail.

5.4 Visualization Software Tools

Most model building applications such as O [22], VMD [23], Coot [24], Sculptor [25], and MAIN [26] are used for structure exploration and presentation by a structural biologist. (MAIN is primarily considered model-building software, yet I almost exclusively use it for generating graphical figures and movies for presentations and papers due to its numerous possibilities for construction and addressing the display of structural details and overview.) The software developed with the angle of visualization like PyMol [102], CCP4mg [98], Kinemage [103], DINO (DINO: Visualizing Structural Biology [2002] http://www.dino3d.org), and Chimera [104] acquired a much broader user pool. The list, however, is far from complete. Readers are directed to search the Internet. One of the most comprehensive lists of molecular graphics software packages can be found on the PDB web page (http://www.rcsb.org/pdb/static.do?p=software/software_links/molecular_graphics.html).

There are also specialized tools for viewing and exploring nucleic-acid structures, such as 3DNA [105], and RNA2D3D [106]; Chimera also has some special tools [107] and additional plugins. Specialized software can be found for the studies on binding pockets in substructure sets [108], then there is VASCO [109]: software for calculation of surface properties and visualization of annotated surfaces using PyMol as a graphical display interface. MolSurfer uses a PDB viewer Java program specialized for analysis of contact surfaces [110] and software like UnityMol for visualization challenges exploiting the video gaming graphical approaches [111]. Sculptor [112] seems to be the first molecular graphics program to implement the ambient radial occlusion model. In contrast to the OpenGL fog function, which is usually used to achieve in-depth perception by merging the color of the objects with the background in a distance-dependent manner, the ambient occlusion uses multiple lights from a number of directions that provide a more realistic global illumination model. To demonstrate the effect, Tom Goddard provided Fig. 15, which shows the images of an EM density map of beta-galactosidase [1] presented with the new ChimeraX using the standard illumination model on the left and the ambient light occlusion model with 64 lights on the right.

Apart from individual software applications, there are also attempts for open graphical platform developments such as those by Moreland et al. [113]. More recently, uPy emerged, "a ubiquitous computer graphics Python API with Biological Modeling Applications" [114]. It is an interface to a number of general graphics software applications such as Blender, Maya, Cinema4D, and DejaVu as vehicles for 3D molecular visualization. Based on similar ideas is ePMV (embedded Python Molecular Viewer) [115] using a Python interface to combine professional animation with the structural-biology world.

The web-based interfaces rely on Java. Many of them use Jmol [116] as the driving program. Some of these are specialized, such as the 3V website (a cavity, channel, and cleft volume calculator) [117] and ProVar: visualization of variable binding pockets on protein surfaces by probabilistic analysis of related structure sets [108]. PBEQ-Solver is a web service for online visualization of

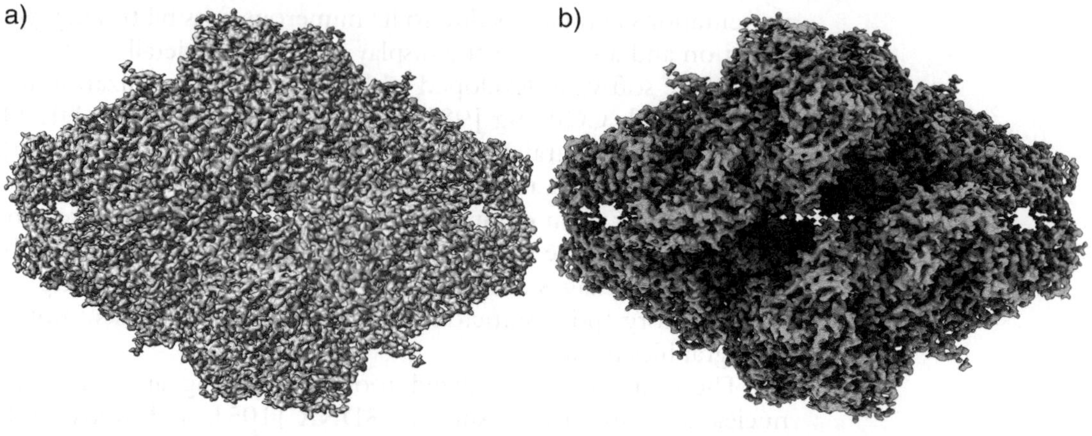

Fig. 15 Ambient occlusion lightning. A 2.2 Å resolution EM map of a beta-galactosidase [1] density map was chosen to demonstrate the difference between (**a**) the standard illumination of a solid object on the left and (**b**) the ambient occlusion effect on the presentation of the same object on the right. The figure is courtesy of Tom Goddard and was prepared in ChimeraX

electrostatic potentials of macromolecules [118]. NGLViewer [119] uses WebGL to access hardware-accelerated 3D graphics in web interfaces. In addition, there are general insights into the macromolecular world as provided by the PDB servers and PROTOPEDIA [120]. The latter combines 3D structural information with text in a narrative manner, which is a relatively young approach aimed at a broad audience.

6 Conclusions and Challenges

The crucial information imbedded in an atomic structure of a macromolecule is the link between the sequence and position of a residue in 3D. Although this information is provided by the average single structure, novel developments are underway that are changing the ways of model building and thus may change our perception of the structure.

6.1 The Growing Role of Computational Tools in Interactive Model Building

The automated approaches from model-building programs to projects like PDB_REDO [121] are slowly but surely closing the gaps in interactive model building, yet interactive density interpretation persists due to the ingenuity of the human brain at solving puzzles and because of the need for oversight in decision-making. The progress in computer resources enabled software developments unthinkable before that increasingly rely on the use of energy optimizations during model building. It is also noteworthy that the boundaries between different steps of structure determination (density calculation, model building, refinement, and structure validation) are disappearing.

Keeping in mind that the human intervention by interactive model building is a push toward the increased congruence of a model with experimental data, it is evident that in the last decade, this push changed its form because radii of convergence of computational tools continued to increase and still do. The automatic model building is delivering better starting models. As a consequence, the need for building from scratch by inserting secondary structure elements is subsiding but has not vanished yet. On the one hand, model building is now cutting into super atomic resolution structures, where these structural features need to be adjusted to shapes of electron density delivered by EM. The shift in model building is toward making choices: which tools to apply and what kind of structures to build. In MAIN, for example, several kinds of minimizer and search procedures are available from rigid body optimizations of large segments, over-restrained fragmented fitting, and real-space all-atom minimization. They are to be tried first before any manual movement of a model takes place. For rebuilding larger structural segments with the combination of external force and energy, a calculation procedure such as molecular dynamics seems the way to go: iMDFF (Fig. 16). For such model building, perception is crucial; thus, optimal use of graphical presentation is mandatory. On the other hand, there are details in the structure that require fiddling around with structural details not yet resolved in an automated manner where insight into the structure can be confined to a smaller area. This is the area well addressed by software like Coot [24].

It is noteworthy that the planning of a structure determination process has also changed. In the prevailing part of MX, the choice of an approach and software tools is based on the lowest threshold of the next step. On the one hand, this appears to be a consequence of the number and complexity of available software tools, which replaced understanding of a structure solution process by providing a stream of easy-to-launch procedures. On the other hand, the complexity of wet-lab and computational approaches used in structural biology does not give scientists and students much time to think and study, but to follow the lowest-threshold path, which may not be optimal overall.

6.2 The Use of Accurate Geometric Restraints

The increasing accuracy of target values obtained from high-resolution structures implicitly suggests that deviations from ideal geometrical characteristics of all structures should correspond to them. However, only these "best" structures are the result of crystallographic data that enable clearly distinguishable interpretation of electron density. The majority of structures are from less ordered crystals that diffracted to lower resolution. Because their diffraction data do not correspond to a single real structure, neither do density maps. Thus, it is wrong to expect that structures determined from such crystals should also have the same geometric

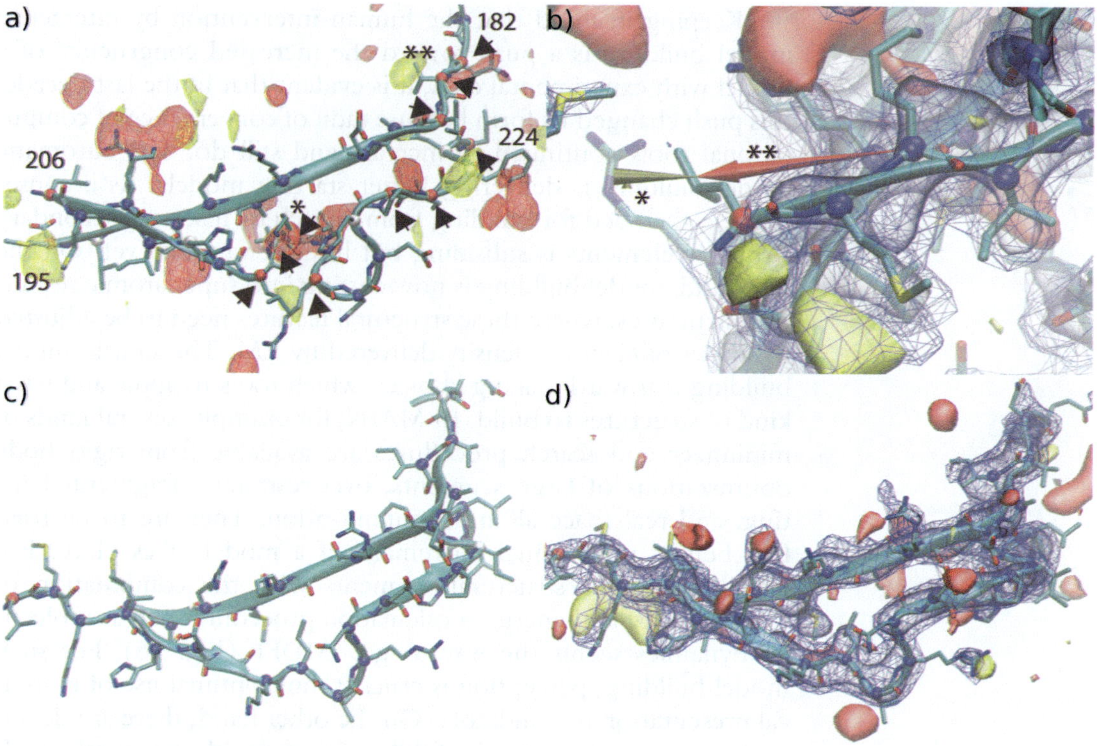

Fig. 16 Annotated screen captures of the iMDFF environment during correction of register errors in the 3.65 Å potassium channel structure (PDB ID: 1p7b). In all panels, blue wireframe $=2mF_o - DF_c$ (1σ), blue surface $=2mF_o - DF_c$ (Bsharp $= -80$ Å^2, 2σ), yellow and red surfaces $= mF_o - DF_c$ ($+3\sigma$ and -3σ, respectively). (**a**) Masking out atoms and maps outside the region of interest and providing information-rich visualization allows for rapid identification of the errors. Clear misthreading is evident in the middle of the strand (*, positions 206–224), while a register error in the adjacent strand (positions 182–195) is suggested by underfitting of the N-terminal density (**), as well as by more subtle clues apparent upon closer inspection. Residues in preferred, allowed, and outlying regions of the Ramachandran plot are indicated in real time by coloring the Cα atoms blue, white, or red, respectively. Ramachandran outliers are indicated by arrows. (**b**) Rearrangements are accomplished by simply pulling on atoms in the context of a running molecular dynamics simulation, using a three-degrees-of-freedom haptic interface. The interface pointer is indicated by a tan cone (*), while the direction and magnitude of the applied force are visualized as a red arrow (**). Pulling on the atoms causes surrounding mobile atoms to respond according to Newton's laws of motion. (**c**) Rearrangement of these strands and their NCS equivalents (not shown) was accomplished in approximately 30 min of interactive modeling. (**d**) Re-refinement in PHENIX reveals a significant reduction in residual density. The figure is courtesy of Tristan Croll

deviations as the best averaged single structures: for example, one Ramachandran outlier per 10,000 residues [37]. A possible solution appears to be the ensemble interpretation. If we want to build models that will equally satisfy experimental data and the strict validation criteria, then a single model will not do. Nevertheless, the ensemble models generated by molecular dynamics or Monte Carlo approaches sample the space; therefore, their geometry does not correspond to validation criteria of a good averaged structure. This observation suggests that the boxes of averaged and ensemble structures are mutually exclusive. At this point, I am unable to provide the ultimate answer, yet the present discussion makes it obvious that this conundrum requires resolution.

Currently, our parameters of geometric restraints are specified by their topology (a covalent-bond network). A deviation from this concept is the parameters separately describing the *cis* and *trans* isoforms of proline residues [27]. Now backbone geometry restraints were added for broad use by the PHENIX user community [29]. This is an obvious deviation from the "one residue–one restraint" target. The next modification of the parameters of concerned atoms is probably hydrogen bonded atoms, and so on. Is it not the time to step out of this box by defining atom types/classes by an approach that combines topology, geometry, and the environment of molecular models and from there to seek the appropriate parameters of geometric restraints and leave the residue and monomer concept to model building, while leaving out of the monomer libraries the parameters of geometric restraints needed for every calculation and validation?

6.3 Structure Accuracy and Precision in Visualization

Visualization of molecular structure is the basis for the human mind to link sequential information with a biological function. We need tools that allow us to explore and present the established links. In most presentations, precision is considered implicitly via selection of the appropriate form representation of graphical objects such as atomic, ribbon, or surface; however, direct display of accuracy or imprecision/uncertainty of structures is seldom used. Usually, atomic displacement parameters indicate the size of the displacement in space. To make them visible for shape and interaction studies, the jelly view based on the double-surface approach is suggested here. The inner surface was calculated from VdW radii and displayed with opaque colors, whereas the outer surface was calculated from the sum of VdW and ADP ($\sqrt{[B/(8\pi^2)]}$) and displayed transparently, giving the impression of a jelly surrounding the molecular structures (Fig. 14). Although the inner surface represents the average structure, the outer surface indicates the area that can be accessed by the structure. Instead of the use of temperature factors, one can also utilize the diffraction precision index [122, 123], which was recently picked up by Kumar et al. [124] who set up a server that will calculate it (http://cluster.physics.iisc.ernet.in/dpi/).

6.4 Averaged, Single, and Ensemble Structure Solutions

Today, the standard view of a molecule is the view of the average single biological molecule. The majority of tools and concepts were developed to support determination and exploration of the average single structure. The problem with the average structure is that it is an average across a vast number of molecules in a diffracting crystal or in EM projections, while the individual molecules (may) appear in a number of different states. Hence, the average single structure is an ideal, whereas every real structure contains parts that are more ambiguously resolved than others. There is a general understanding that the average structure has to be completed first, before one can conclude which parts are ambiguous and make an attempt to interpret them as disordered regions or try alternative solutions such as the ensemble models. Besides, the single structure is static, whereas an ensemble involves dynamics, diversity, and uncertainty. In my opinion, the major obstacles to adoption of the probabilistic models are their presentation, analysis of biological relevance of motion and disorder, and perception of structures. Currently, we present ensembles collectively or individually. Their superimposition does not truly help to see the details, in particular when there is a large spread of models within an ensemble. Analysis of trajectories reveals their normal modes as principle components, yet they need to be visualized. What we seek is a presentation of smooth transitions between principal components of multiple conformation states, which will enable us to get some clues to the individual states and will provide an overview. Such smooth trajectory has to be devoid of the seemingly random choice of small-scale thermal fluctuations. Such tools already exist. To obtain them, we need to reach out into the area of molecular dynamics for such applications as Interactive Essential Dynamics [125], a VMD attachment, or Hybrid Electron Microscopy Normal Mode Analysis (HEMNMA) [126]. Hence, motion phenomena can be examined. It is more difficult to establish links between the dynamics and biological relevance of such studies. Larger structural differences make it easier to establish their biological relevance. Because single-particle analysis is not restrained by crystal packing, it is not surprising that the EM community is striving for flexibility of molecules more often (and is considering making it a part of the standard structure determination [83]) than the MX community is. Currently, ensembles are all generated automatically; it remains to be seen whether in animation of trajectory transition there is still room left for interactive model building.

It is almost needless to say that structural details of protein motions are crucial for many biological processes yet remain hidden for conventional biophysical methods [81]. So far, we tried to collect datasets by optimally resolving Bragg spots. In light of the latest developments, it appears that insight into molecular motion can be obtained by analysis of diffraction of crystals. Therefore, we may reconsider the data collection process and carefully collect and consider not only the Bragg spots but also the space in between. Interpretation of structural motions based on deposition of Bragg spots is a limitation for further research; thus, it is suggested here that the raw diffraction images must be deposited. Their deposition may enable motion analysis at a later stage, after deposition of the original average single structure entry. Building on this, it is worthwhile to make possible depositions of models calculated in various ways, but exploiting the same data collection experiment(s).

6.5 The Need for an Objective Criterion of Structure Correctness

The works of Nwachukwu et al. [48], who tested combinatorial schemes of refinement, and our analysis [47] established the need to end the reign of the *R*-free gap [127] as a measure of structure correctness. Apart from the Fourier series completeness and model bias (absence of structural data, chemical energy interactions, and maximum likelihood function dependence), the *R*-free concept has self-consistency issues with structures containing NCS and twinning. The PDB server does not offer a search criterion that would help to come up with a number; however, the presence of multiple copies of molecules in an asymmetric unit is a common phenomenon in MX. It would not come as a surprise if half of the MX structures belonged to this category. One of the great features of *R*-free is that it is a single number. Nonetheless, because the lowest *R*-free gap in the combination with a chosen TEST set does not ensure the highest accuracy, we need to make an attempt to leave the wishful thinking behind. Thus, we either come up with a novel parameter or use the established rule that a structure is correct when it is consistent with our understanding of biological structures locally and globally and matches electron density. *R*-factor, when calculated from all data within the Free Kick refinement procedure, is not a bad solution after all.

6.6 Concluding Remarks

I have written this review in an attempt to shed light on the model building part and to visualize the process of molecular structure determination. In our endeavor for simplicity, automation, and the black-box approach to molecular structure determination, we tend to forget that the primary goal of our quest is increased knowledge and understanding of molecular structures. To do this, we need to evolve, and if necessary, also revolutionize our approaches, models, and views on the subject. By demonstrating application of "in the box" and "out of the box" thinking, I tried to expose several underlying concepts, limitations, and contradictions of model building and structure presentation that need to be addressed in the future.

Acknowledgments

This work is dedicated to my PhD supervisor Robert Huber at the occasion of his 80s anniversary. Jürgen Plitzko, Tom Goddard, and Tristan Croll are gratefully acknowledged for generation of several figures. Jürgen Plitzko wrote parts of the text concerning the EM structure determination. My coworker Ajda Taler Verčič helped me with formatting the text and sorting out references. Funding was provided by the Structural Biology grant P1-0048 and by the Infrastructural Funds to Centre of Excellence CIPKeBiP, both provided by Slovenian Research Agency.

References

1. Bartesaghi A, Merk A, Banerjee S et al (2015) 2.2 A resolution cryo-EM structure of beta-galactosidase in complex with a cell-permeant inhibitor. Science 348(6239):1147–1151. doi:10.1126/science.aab1576
2. Cheng YF (2015) Single-particle cryo-EM at crystallographic resolution. Cell 161(3):450–457. doi:10.1016/j.cell.2015.03.049
3. Merk A, Bartesaghi A, Banerjee S et al (2016) Breaking cryo-EM resolution barriers to facilitate drug discovery. Cell 165(7):1698–1707
4. Wlodawer A, Minor W, Dauter Z et al (2013) Protein crystallography for aspiring crystallographers or how to avoid pitfalls and traps in macromolecular structure determination. FEBS J 280(22):5705–5736
5. Huber R (2013) How I chose research on proteases or, more correctly, how it chose me. Angew Chem 52(1):68–73. doi:10.1002/anie.201205629
6. Deisenhofer J, Steigemann W (1975) Crystallographic refinement of structure of bovine pancreatic trypsin-inhibitor at 1.5 A resolution. Acta Crystallogr B 31:238–250. doi:10.1107/S0567740875002415
7. Deisenhofer J, Remington SJ, Steigemann W (1985) Experience with various techniques for the refinement of protein structures. Methods Enzymol 115:303–323
8. Terwilliger TC, Grosse-Kunstleve RW, Afonine PV et al (2008) Iterative-build OMIT maps: map improvement by iterative model building and refinement without model bias. Acta Crystallogr D Biol Crystallogr 64(Pt 5):515–524. doi:10.1107/S0907444908004319
9. Wang BC (1985) Resolution of phase ambiguity in macromolecular crystallography. Methods Enzymol 115:90–112
10. Read RJ (1986) Improved Fourier coefficients for maps using phases from partial structures with errors. Acta Crystallogr A 42:140–149. doi:10.1107/S0108767386099622
11. Jiang JS, Brunger AT (1994) Protein hydration observed by X-ray diffraction. Solvation properties of penicillopepsin and neuraminidase crystal structures. J Mol Biol 243(1):100–115. doi:10.1006/jmbi.1994.1633
12. Murshudov GN, Vagin AA, Dodson EJ (1997) Refinement of macromolecular structures by the maximum-likelihood method. Acta Crystallogr D Biol Crystallogr 53:240–255. doi:10.1107/S0907444996012255
13. Rice LM, Shamoo Y, Brunger AT (1998) Phase improvement by multi-start simulated annealing refinement and structure-factor averaging. J Appl Crystallogr 31:798–805. doi:10.1107/S0021889898006645
14. Praznikar J, Afonine PV, Guncar G et al (2009) Averaged kick maps: less noise, more signal... and probably less bias. Acta Crystallogr D Biol Crystallogr 65(Pt 9):921–931
15. Sheldrick GM (2010) Experimental phasing with SHELXC/D/E: combining chain tracing with density modification. Acta Crystallogr D Biol Crystallogr 66(Pt 4):479–485. doi:10.1107/S0907444909038360
16. Afonine PV, Moriarty NW, Mustyakimov M et al (2015) FEM: feature-enhanced map. Acta Crystallogr D Biol Crystallogr 71(Pt 3):646–666
17. Panjikar S, Parthasarathy V, Lamzin VS et al (2009) On the combination of molecular replacement and single-wavelength anomalous diffraction phasing for automated structure determination. Acta Crystallogr D Biol Crystallogr 65:1089–1097. doi:10.1107/S0907444909029643
18. Adams PD, Afonine PV, Bunkoczi G et al (2010) PHENIX: a comprehensive Python-based system for macromolecular structure solution. Acta Crystallogr D Biol Crystallogr 66:213–221. doi:10.1107/S0907444909052925
19. Cowtan K (2006) The Buccaneer software for automated model building. 1. Tracing protein chains. Acta Crystallogr D Biol Crystallogr 62:1002–1011. doi:10.1107/S0907444906022116
20. Lamzin VS, Perrakis A, Wilson KS (2012) ARP/wARP—automated model building and refinement. In: Arnold E, Himmel DM, Rossmann MG (eds) International Tables for Crystallography, vol F: Crystallography of biological macromolecules, 2012th edn. Kluwer Academic Publishers, The Netherlands, pp 525–528
21. DiMaio F, Song Y, Li X et al (2015) Atomic-accuracy models from 4.5-A cryo-electron microscopy data with density-guided iterative local refinement. Nat Methods 12(4):361–365. doi:10.1038/nmeth.3286
22. Jones TA, Zou JY, Cowan SW et al (1991) Improved methods for building protein models in electron density maps and the location of errors in these models. Acta Crystallogr A 47(Pt 2):110–119
23. Humphrey W, Dalke A, Schulten K (1996) VMD: visual molecular dynamics. J Mol Graph 14(1):33–38. 27–38

24. Emsley P, Lohkamp B, Scott WG et al (2010) Features and development of Coot. Acta Crystallogr D Biol Crystallogr 66(Pt 4):486–501. doi:10.1107/S0907444910007493
25. Birmanns S, Rusu M, Wriggers W (2011) Using Sculptor and Situs for simultaneous assembly of atomic components into low-resolution shapes. J Struct Biol 173(3):428–435
26. Turk D (2013) MAIN software for density averaging, model building, structure refinement and validation. Acta Crystallogr D Biol Crystallogr 69(Pt 8):1342–1357
27. Engh RA, Huber R (2001) Structure quality and target parameters. In: Rossmann MG, Arnold E (eds) International Tables for Crystallography, vol F: Crystallography of biological macromolecules. Springer, Netherlands, pp 382–392
28. Parkinson G, Vojtechovsky J, Clowney L et al (1996) New parameters for the refinement of nucleic acid-containing structures. Acta Crystallogr D Biol Crystallogr 52(Pt 1):57–64
29. Moriarty NW, Tronrud DE, Adams PD et al (2016) A new default restraint library for the protein backbone in Phenix: a conformation-dependent geometry goes mainstream. Acta Crystallogr 72(Pt 1):176–179
30. Vagin AA, Steiner RA, Lebedev AA et al (2004) REFMAC5 dictionary: organization of prior chemical knowledge and guidelines for its use. Acta Crystallogr D Biol Crystallogr 60(Pt 12 Pt 1):2184–2195. doi:10.1107/S0907444904023510
31. Schuttelkopf AW, van Aalten DM (2004) PRODRG: a tool for high-throughput crystallography of protein-ligand complexes. Acta Crystallogr D Biol Crystallogr 60(Pt 8):1355–1363. doi:10.1107/S0907444904011679
32. Andrejasic M, Praznikar J, Turk D (2008) PURY: a database of geometric restraints of hetero compounds for refinement in complexes with macromolecular structures. Acta Crystallogr D Biol Crystallogr 64:1093–1109. doi:10.1107/S0907444908027388
33. Lebedev AA, Young P, Isupov MN et al (2012) JLigand: a graphical tool for the CCP4 template-restraint library. Acta Crystallogr D Biol Crystallogr 68:431–440. doi:10.1107/S090744491200251x
34. Jones TA (1985) Diffraction methods for biological macromolecules. Interactive computer graphics: FRODO. Methods Enzymol 115:157–171
35. Ramachandran GN, Ramakrishnan C, Sasisekharan V (1963) Stereochemistry of polypeptide chain configurations. J Mol Biol 7:95–99
36. Chen VB, Arendall WB, Headd JJ et al (2010) MolProbity: all-atom structure validation for macromolecular crystallography. Acta Crystallogr D Biol Crystallogr 66:12–21. doi:10.1107/S0907444909042073
37. Hintze BJ, Lewis SM, Richardson JS et al (2016) MolProbity's ultimate rotamer-library distributions for model validation. Proteins. doi:10.1002/prot.25039
38. Lovell SC, Word JM, Richardson JS et al (2000) The penultimate rotamer library. Proteins 40(3):389–408. doi:10.1002/1097-0134(20000815)40:3<389::Aid-Prot50>3.0.Co;2-2
39. Scouras AD, Daggett V (2011) The Dynameomics rotamer library: amino acid side chain conformations and dynamics from comprehensive molecular dynamics simulations in water. Protein Sci 20(2):341–352. doi:10.1002/pro.565
40. Novotny M, Kleywegt GJ (2005) A survey of left-handed helices in protein structures. J Mol Biol 347(2):231–241. doi:10.1016/j.jmb.2005.01.037
41. Davis IW, Murray LW, Richardson JS et al (2004) MOLPROBITY: structure validation and all-atom contact analysis for nucleic acids and their complexes. Nucleic Acids Res 32(Web Server issue):W615–W619
42. Andreeva A, Howorth D, Chothia C et al (2014) SCOP2 prototype: a new approach to protein structure mining. Nucleic Acids Res 42(D1):D310–D314. doi:10.1093/nar/gkt1242
43. Murzin AG, Brenner SE, Hubbard T et al (1995) Scop—a structural classification of proteins database for the investigation of sequences and structures. J Mol Biol 247(4):536–540. doi:10.1016/S0022-2836(05)80134-2
44. Branden CI, Tooze J (1998) Introduction to protein structure, 2nd edn. Garland Publishing, Inc., New York
45. Brocklehurst K, Kowlessur D, O'Driscoll M et al (1987) Substrate-derived two-protonic-state electrophiles as sensitive kinetic specificity probes for cysteine proteinases. Activation of 2-pyridyl disulphides by hydrogen-bonding. Biochem J 244(1):173–181
46. Musil D, Zucic D, Turk D et al (1991) The refined 2.15 A X-ray crystal structure of human liver cathepsin B: the structural basis for its specificity. EMBO J 10(9):2321–2330
47. Praznikar J, Turk D (2014) Free kick instead of cross-validation in maximum-likelihood refinement of macromolecular crystal structures. Acta Crystallogr D Biol Crystallogr 70(Pt 12):3124–3134
48. Nwachukwu JC, Southern MR, Kiefer JR et al (2013) Improved crystallographic structures using extensive combinatorial refine-

ment. Structure 21(11):1923–1930. doi:10.1016/j.str.2013.07.025

49. Fenn TD, Schnieders MJ, Mustyakimov M et al (2011) Reintroducing electrostatics into macromolecular crystallographic refinement: application to neutron crystallography and DNA hydration. Structure 19(4):523–533
50. Goddard TD, Ferrin TE (2007) Visualization software for molecular assemblies. Curr Opin Struct Biol 17(5):587–595. doi:10.1016/j.sbi.2007.06.008
51. Baugh EH, Lyskov S, Weitzner BD et al (2011) Real-time PyMOL visualization for Rosetta and PyRosetta. PLoS One 6(8):e21931. doi:10.1371/journal.pone.0021931
52. Langer GG, Hazledine S, Wiegels T et al (2013) Visual automated macromolecular model building. Acta Crystallogr D Biol Crystallogr 69(Pt 4):635–641. doi:10.1107/S0907444913000565
53. Subramanian G, Basu S, Liu H et al (2015) Solving protein nanocrystals by cryo-EM diffraction: multiple scattering artifacts. Ultramicroscopy 148:87–93
54. Perutz MF, Rossmann MG, Cullis AF et al (1960) Structure of haemoglobin: a three-dimensional Fourier synthesis at 5.5-A. resolution, obtained by X-ray analysis. Nature 185(4711):416–422
55. Henderson R, Unwin PN (1975) Three-dimensional model of purple membrane obtained by electron microscopy. Nature 257(5521):28–32
56. Henderson R (2015) Overview and future of single particle electron cryomicroscopy. Arch Biochem Biophys 581:19–24
57. Croll TI, Smith BJ, Margetts MB et al (2016) Higher-resolution structure of the human insulin receptor ectodomain: multi-modal inclusion of the insert domain. Structure 24(3):469–476. doi:10.1016/j.str.2015.12.014
58. McGreevy R, Teo I, Singharoy A et al (2016) Advances in the molecular dynamics flexible fitting method for cryo-EM modeling. Methods 100:50–60. doi:10.1016/j.ymeth.2016.01.009
59. Higgins DG, Thompson JD, Gibson TJ (1996) Using CLUSTAL for multiple sequence alignments. Methods Enzymol 266:383–402
60. Jones TA, Kjeldgaard M (1994) Making the first trace with O. In: Bailey S, Hubbard R, Waller D (eds) From first map to final model. SERC Daresbury Laboratory, Warrington, pp 1–13
61. Richardson JS, Richardson DC (1985) Interpretation of electron density maps. Methods Enzymol 115:189–206
62. Berman HM, Westbrook J, Feng Z et al (2000) The Protein Data Bank. Nucleic Acids Res 28(1):235–242
63. Holm L, Rosenstrom P (2010) Dali server: conservation mapping in 3D. Nucleic Acids Res 38(Web Server issue):W545–W549
64. Jones TA, Thirup S (1986) Using known substructures in protein model building and crystallography. EMBO J 5(4):819–822
65. Carolan CG, Lamzin VS (2014) Automated identification of crystallographic ligands using sparse-density representations. Acta Crystallogr D Biol Crystallogr 70:1844–1853. doi:10.1107/S1399004714008578
66. Echols N, Morshed N, Afonine PV et al (2014) Automated identification of elemental ions in macromolecular crystal structures. Acta Crystallogr D Biol Crystallogr 70(Pt 4):1104–1114. doi:10.1107/S1399004714001308
67. Klei HE, Moriarty NW, Echols N et al (2014) Ligand placement based on prior structures: the guided ligand-replacement method. Acta Crystallogr D Biol Crystallogr 70(Pt 1):134–143. doi:10.1107/S1399004713030071
68. Deller MC, Rupp B (2015) Models of protein-ligand crystal structures: trust, but verify. J Comput Aided Mol Des 29(9):817–836
69. Pozharski E, Weichenberger CX, Rupp B (2013) Techniques, tools and best practices for ligand electron-density analysis and results from their application to deposited crystal structures. Acta Crystallogr D Biol Crystallogr 69(Pt 2):150–167
70. Weichenberger CX, Pozharski E, Rupp B (2013) Visualizing ligand molecules in Twilight electron density. Acta Crystallogr Sect F Struct Biol Cryst Commun 69(Pt 2):195–200
71. Keedy DA, Fraser JS, van den Bedem H (2015) Exposing hidden alternative backbone conformations in X-ray crystallography using qFit. PLoS Comput Biol 11(10):e1004507. doi:10.1371/journal.pcbi.1004507
72. Davis IW, Leaver-Fay A, Chen VB et al (2007) MolProbity: all-atom contacts and structure validation for proteins and nucleic acids. Nucleic Acids Res 35:W375–W383. doi:10.1093/nar/gkm216
73. Usenik A, Renko M, Mihelič M, Lindič N, Borišek J, Perdih A, Pretnar G, Müller U, Turk D. The CWB2 cell wall-anchoring module is revealed by the 61 crystal structures of the clostridium difficile cell wall proteins Cwp8 and Cwp6. Structure 2017; 25(3): 514-521. doi: 10.1016/j.str.2016.12.018. Epub 2017 Jan 26.

74. van den Bedem H, Lotan I, Latombe JC et al (2005) Real-space protein-model completion: an inverse-kinematics approach. Acta Crystallogr D Biol Crystallogr 61(Pt 1):2–13. doi:10.1107/S0907444904025697
75. Brown A, Long F, Nicholls RA et al (2015) Tools for macromolecular model building and refinement into electron cryo-microscopy reconstructions. Acta Crystallogr D Biol Crystallogr 71(Pt 1):136–153. doi:10.1107/S1399004714021683
76. van den Bedem H, Fraser JS (2015) Integrative, dynamic structural biology at atomic resolution—it's about time. Nat Methods 12(4):307–318. doi:10.1038/nmeth.3324
77. Gros P, van Gunsteren WF, Hol WG (1990) Inclusion of thermal motion in crystallographic structures by restrained molecular dynamics. Science 249(4973):1149–1152
78. Levin EJ, Kondrashov DA, Wesenberg GE et al (2007) Ensemble refinement of protein crystal structures: validation and application. Structure 15(9):1040–1052
79. Burnley BT, Afonine PV, Adams PD et al (2012) Modelling dynamics in protein crystal structures by ensemble refinement. elife 1:e00311
80. Wall ME, Adams PD, Fraser JS et al (2014) Diffuse X-ray scattering to model protein motions. Structure 22(2):182–184
81. Van Benschoten AH, Liu L, Gonzalez A et al (2016) Measuring and modeling diffuse scattering in protein X-ray crystallography. Proc Natl Acad Sci U S A. doi:10.1073/pnas.1524048113
82. Ayyer K, Yefanov OM, Oberthur D et al (2016) Macromolecular diffractive imaging using imperfect crystals. Nature 530(7589):202–206. doi:10.1038/nature16949
83. Orlova EV, Saibil HR (2010) Methods for three-dimensional reconstruction of heterogeneous assemblies. Methods Enzymol 482:321–341
84. Alber F, Dokudovskaya S, Veenhoff LM et al (2007) The molecular architecture of the nuclear pore complex. Nature 450(7170):695–701. doi:10.1038/nature06405
85. Beck M, Forster F, Ecke M et al (2004) Nuclear pore complex structure and dynamics revealed by cryoelectron tomography. Science 306(5700):1387–1390
86. Russel D, Lasker K, Webb B et al (2012) Putting the pieces together: integrative modeling platform software for structure determination of macromolecular assemblies. PLoS Biol 10(1):e1001244. doi:10.1371/journal.pbio.1001244
87. Baker D (2006) Prediction and design of macromolecular structures and interactions. Philos Trans R Soc Lond 361(1467):459–463
88. Danev R, Buijsse B, Khoshouei M et al (2014) Volta potential phase plate for in-focus phase contrast transmission electron microscopy. Proc Natl Acad Sci U S A 111(44):15635–15640
89. Asano S, Fukuda Y, Beck F et al (2015) Proteasomes. A molecular census of 26S proteasomes in intact neurons. Science 347(6220):439–442
90. Schweitzer A, Aufderheide A, Rudack T et al (2016) Structure of the human 26S proteasome at a resolution of 3.9 A. Proc Natl Acad Sci U S A 113(28):7816–7821
91. Carson M, Bugg CE (1986) Algorithm for ribbon models of proteins. J Mol Graphics 4(2):121. doi:10.1016/0263-7855(86)80010-8
92. Richardson JS (2000) Early ribbon drawings of proteins. Nat Struct Biol 7(8):624–625. doi:10.1038/77912
93. Lee B, Richards FM (1971) Interpretation of protein structures—estimation of static accessibility. Journal of molecular biology 55(3):379. doi:10.1016/0022-2836(71)90324-X
94. Connolly ML (1983) Solvent-accessible surfaces of proteins and nucleic-acids. Science 221(4612):709–713. doi:10.1126/science.6879170
95. Bernstein HJ, Craig PA (2010) Efficient molecular surface rendering by linear-time pseudo-Gaussian approximation to Lee-Richards surfaces (PGALRS). J Appl Crystallogr 43(Pt 2):356–361. doi:10.1107/S0021889809054326
96. Bianchetti CM, Yi L, Ragsdale SW et al (2007) Comparison of apo- and heme-bound crystal structures of a truncated human heme oxygenase-2. J Biol Chem 282(52):37624–37631. doi:10.1074/jbc.M707396200
97. Li L, Li C, Zhang Z et al (2013) On the dielectric "Constant" of proteins: smooth dielectric function for macromolecular modeling and its implementation in DelPhi. J Chem Theory Comput 9(4):2126–2136. doi:10.1021/ct400065j
98. McNicholas S, Potterton E, Wilson KS et al (2011) Presenting your structures: the CCP4mg molecular-graphics software. Acta Crystallogr D Biol Crystallogr 67(Pt 4):386–394
99. Bromberg S, Chiu W, Ferrin TE (2010) Workshop on molecular animation. Structure 18(10):1261–1265. doi:10.1016/j.str.2010.09.001
100. Palmer AG, Matthews BW (2009) Interactive graphics return to protein science. Protein Sci 18(4):677

101. Barnes DG, Vidiassov M, Ruthensteiner B et al (2013) Embedding and publishing interactive, 3-dimensional, scientific figures in Portable Document Format (PDF) files. PLoS One 8(9):e69446
102. DeLano WL (2009) PyMOL molecular viewer: updates and refinements. Abstr Pap Am Chem Soc 238
103. Chen VB, Davis IW, Richardson DC (2009) KING (Kinemage, Next Generation): a versatile interactive molecular and scientific visualization program. Protein Sci 18(11):2403–2409
104. Goddard TD, Huang CC, Ferrin TE (2007) Visualizing density maps with UCSF Chimera. J Struct Biol 157(1):281–287. doi:10.1016/j.jsb.2006.06.010
105. Lu XJ, Olson WK (2003) 3DNA: a software package for the analysis, rebuilding and visualization of three-dimensional nucleic acid structures. Nucleic Acids Res 31(17):5108–5121
106. Martinez HM, Maizel JV Jr, Shapiro BA (2008) RNA2D3D: a program for generating, viewing, and comparing 3-dimensional models of RNA. J Biomol Struct Dyn 25(6):669–683
107. Couch GS, Hendrix DK, Ferrin TE (2006) Nucleic acid visualization with UCSF Chimera. Nucleic Acids Res 34(4):e29
108. Ashford P, Moss DS, Alex A et al (2012) Visualisation of variable binding pockets on protein surfaces by probabilistic analysis of related structure sets. BMC Bioinformatics 13:39. doi:10.1186/1471-2105-13-39
109. Steinkellner G, Rader R, Thallinger GG et al (2009) VASCo: computation and visualization of annotated protein surface contacts. BMC Bioinformatics 10:32. doi:10.1186/1471-2105-10-32
110. Gabdoulline RR, Wade RC, Walther D (2003) MolSurfer: a macromolecular interface navigator. Nucleic Acids Res 31(13):3349–3351
111. Lv ZH, Tek A, Da Silva F et al (2013) Game on, science—how video game technology may help biologists tackle visualization challenges. PLoS One 8(3). doi:10.1371/journal.pone.0057990
112. Wahle M, Wriggers W (2015) Multi-scale visualization of molecular architecture using real-time ambient occlusion in sculptor. PLoS Comput Biol 11(10):e1004516. doi:10.1371/journal.pcbi.1004516
113. Moreland JL, Gramada A, Buzko OV et al (2005) The Molecular Biology Toolkit (MBT): a modular platform for developing molecular visualization applications. BMC Bioinformatics 6:21
114. Autin L, Johnson G, Hake J et al (2012) uPy: a ubiquitous CG Python API with biological-modeling applications. IEEE Comput Graph Appl 32(5):50–61
115. Johnson GT, Autin L, Goodsell DS et al (2011) ePMV embeds molecular modeling into professional animation software environments. Structure 19(3):293–303
116. Herraez A (2006) Biomolecules in the computer: Jmol to the rescue. Biochem Mol Biol Educ 34(4):255–261
117. Voss NR, Gerstein M (2010) 3V: cavity, channel and cleft volume calculator and extractor. Nucleic Acids Res 38(Web Server issue):W555–W562. doi:10.1093/nar/gkq395
118. Jo S, Vargyas M, Vasko-Szedlar J et al (2008) PBEQ-Solver for online visualization of electrostatic potential of biomolecules. Nucleic Acids Res 36 (Web Server issue):W270–W275. doi:10.1093/nar/gkn314
119. Rose AS, Hildebrand PW (2015) NGL Viewer: a web application for molecular visualization. Nucleic Acids Res 43(W1):W576–W579
120. Hodis E, Prilusky J, Martz E et al (2008) Proteopedia—a scientific 'wiki' bridging the rift between three-dimensional structure and function of biomacromolecules. Genome Biol 9(8):R121. doi:10.1186/gb-2008-9-8-r121
121. Joosten RP, Long F, Murshudov GN et al (2014) The PDB_REDO server for macromolecular structure model optimization. IUCrJ 1(Pt 4):213–220
122. Blow DM (2002) Rearrangement of Cruickshank's formulae for the diffraction-component precision index. Acta Crystallogr D Biol Crystallogr 58(Pt 5):792–797
123. Cruickshank DW (1999) Remarks about protein structure precision. Acta Crystallogr D Biol Crystallogr 55(Pt 3):583–601
124. Kumar KSD, Gurusaran M, Satheesh SN et al (2015) Online_DPI: a web server to calculate the diffraction precision index for a protein structure. J Appl Crystallogr 48:939–942
125. Mongan J (2004) Interactive essential dynamics. J Comput Aid Mol Des 18 (6):433–436. doi:10.1007/s10822-004-4121-z
126. Sorzano CO, de la Rosa-Trevin JM, Tama F et al (2014) Hybrid Electron Microscopy Normal Mode Analysis graphical interface and protocol. J Struct Biol 188(2):134–141. doi:10.1016/j.jsb.2014.09.005
127. Brunger AT (1992) Free R value: a novel statistical quantity for assessing the accuracy of crystal structures. Nature 355(6359):472–475
128. Than ME, Hof P, Huber R et al (1997) Thermus thermophilus cytochrome-c552: a new highly thermostable cytochrome-c structure obtained by MAD phasing. J Mol Biol 271(4):629–644

Chapter 22

Structure Refinement at Atomic Resolution

Mariusz Jaskolski

Abstract

X-Ray diffraction data at atomic resolution, i.e., beyond 1.2 Å, provide the most detailed and reliable information we have about the structure of macromolecules, which is especially important for validating new discoveries and resolving subtle issues of molecular mechanisms. Refinement at atomic resolution allows reliable interpretation of static disorder and solvent structure, as well as modeling of anisotropic atomic vibrations and even of H atoms. Stereochemical restraints can be relaxed or removed, providing unbiased information about macromolecular stereochemistry, which in turn can be used to define improved conformation-dependent libraries, and the surplus of data allows estimation of least-squares uncertainties in the derived parameters. At ultrahigh resolution it is possible to study charge density distribution by multipolar refinement of electrons in non-spherical orbitals.

Key words Atomic resolution, Stereochemical restraints, Conformation-dependent stereochemical libraries, H atoms, Multipolar refinement, Charge density, Standard uncertainties

1 Introduction

Strictly speaking, crystallographic resolution refers to the diffraction data and electron density maps (as their Fourier transform) and not to models, which are only interpretations of electron density. It is defined as the minimum *d*-spacing in Bragg's Law ($\lambda = 2d_{min} \cdot \sin\theta_{max}$), corresponding to the maximum glancing angle θ_{max} at which statistically significant reflection intensities are still observed. It can be shown that the d_{min} limit corresponds almost exactly to the minimal separation of two points that can be distinguished in electron density maps generated by Fourier transformation.

On the somewhat arbitrary scale of resolution intervals (Fig. 1), the point at d_{min} = 1.2 Å is defined by Sheldrick [1] as atomic resolution. This choice, also supported by rigorous argument [2], is quite intuitive as it allows resolution of all non-H atoms, including the shortest (1.2 Å) C=O bond. For ultrahigh resolution we use the 0.8 Å mark, corresponding approximately to the limit of Cu *K*α data.

Alexander Wlodawer et al. (eds.), *Protein Crystallography: Methods and Protocols*, Methods in Molecular Biology, vol. 1607, DOI 10.1007/978-1-4939-7000-1_22, © Springer Science+Business Media LLC 2017

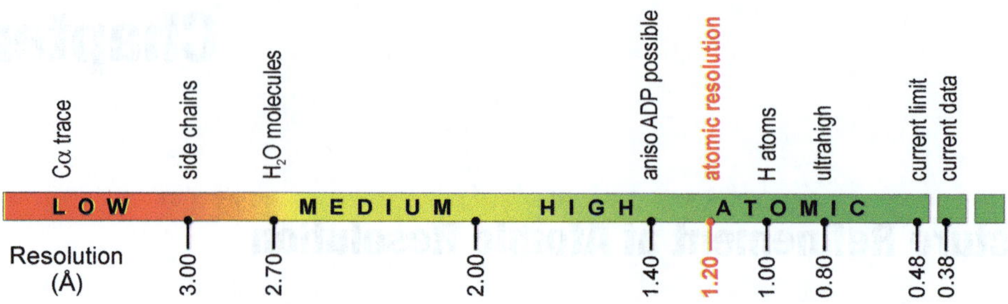

Fig. 1 In an arbitrary division of crystallographic resolution into descriptive ranges, only the criterion of atomic resolution (1.2 Å) has precise definition [1]. The annotations indicate the justified level of interpretation. The highest-resolution structure in the PDB (3nir) is at 0.48 Å for crambin [41]. A 0.38-Å data collection for the same crystal form was announced in the literature [52] but no structure has been reported yet

2 Collecting Atomic-Resolution Data

If a single advice is to be offered, it would be: *always get the highest resolution during your diffraction experiment.* This will ease all subsequent steps, will help reduce model bias, and will authenticate any unusual features discovered in the structure. However, collecting meaningful high-resolution data is not equivalent to "visiting" high-order *hkl* indices without statistically meaningful intensity signal. As a rule of thumb, we used to expect the average signal-to-noise ratio ($<I/\sigma(I)>$) in the highest resolution shell to be at least 2, which is roughly equivalent to having ~50% of the data in that shell with $I > 2\sigma(I)$. The above criteria are rather conservative but will guarantee a high-quality data set. A resolution-oriented approach might, however, push d_{min} to the extreme limit, where adding more observations ceases to add information [3]. Statistically, this would correspond to a correlation coefficient CC_{true} between the experimental and ideal noise-free data of ~0.4. Karplus and Diederichs showed [4] that CC_{true} can be estimated by $CC_{½}$, which measures the correlation between two half-sets and should be acceptable even down to ~0.1. It is also good to have the last resolution shell as complete as possible but incomplete resolution shells should not be rejected! On the contrary, every single reflection is precious and should always be included, particularly at high resolution. If completeness in the last resolution shell is poor, it is possible to estimate effective resolution (as opposed to nominal resolution) by finding that d_{eff} at which a reciprocal-lattice sphere of radius $1/d_{eff}$ would be filled completely. One can also estimate the optical resolution d_{opt} by a Gaussian analysis of the Patterson map based on optical principles, as proposed by Vaguine [5] and implemented in SFCHECK. One should remember that completeness of the lowest-resolution shells is important as well.

The outlier rejection criterion ($I < -3\sigma(I)$) used by data reduction programs should not be manipulated to "improve" the data set. Likewise, no σ-cutoff should be applied to select reflections for the refinement. However, in algorithms that use $|F_o|$ for refinement, the data will be effectively truncated at $0\sigma(I)$, by eliminating negative intensities during the $|F_o| = \sqrt{I}$ conversion. From this point of view, refinement algorithms based on reflection intensities, e.g., in SHELXL [6], are preferred.

The second most important parameter of a good data set is high redundancy, which will always improve data quality unless compromised by radiation damage. In addition to reducing random errors, multiple observations can be used to estimate standard deviations of intensity measurements. R_{merge} as a resolution-limiting criterion is not recommended as it deteriorates at high symmetry and with high redundancy. Better, redundancy-independent parameters (e.g., R_{rim}) were proposed by Diederichs and Karplus [7] and Weiss and Hilgenfeld [8, 9].

If the dynamic range of the detector is insufficient to reliably record very strong and very weak data at the same time, it may be necessary to measure the strongest, low-resolution data in a (first) quick pass. At ultrahigh resolution, three overlapping runs may be necessary, e.g., ∞–2.0 Å, 2.4–1.0 Å and 1.5–0.7 Å, with relative 1:10:100 exposure.

3 Model Refinement at High Resolution Step-by-Step

3.1 Full or Stepping Resolution?

Historically, high order refinements were carried out with careful gradual extension of resolution. This strategy, dictated by limited computer power, is no longer necessary if the starting model is very good. When only an approximate model is available, starting the refinement at ~2 Å and even inclusion of a rigid-body step may be advisable to increase the radius of convergence.

3.2 Atomic Displacement Parameters (ADPs)

The first round of refinement at full resolution is done with isotropic atomic displacement parameters (ADPs, historically called *B*-factors), and is followed by model adjustment in electron density maps and inclusion of the most evident solvent molecules. Switching from isotropic (1 parameter per atom) to anisotropic (6 parameters per atom) model at this stage more than doubles the number of model parameters and brings about a dramatic decrease of the *R* factors (up to 0.05). Individual anisotropic ADPs are used for both the macromolecule and solvent atoms. They should not be mixed with TLS (Translation, Libration, Screw-motion) parameters, which describe concerted anisotropic motions of rigid structural fragments at medium resolution. The subsequent steps of the refinement protocol are listed in Table 1.

Table 1
Stages of macromolecular refinement at atomic resolution

Step	Action
1	Include reflections at full resolution
2	Isotropic ADPs
3	Correction of model errors, evident solvent molecules
4	Bulk-solvent correction
5	Anisotropic ADPs
6	Modeling of disorder
7	Riding H atoms
8	Partial water molecules
9	Refine/adjust occupancies
10	Relax/(remove) restraints
(11)	(H-atoms refined)
12	Include all reflections (work + test)
(13)	(Multipolar refinement)
14	Full-matrix least-squares

The steps listed in parentheses are only possible at ultrahigh resolution

3.3 Prudent Expansion of Model Parameters

If the resolution is 1.2 Å or higher, there is no question of the validity of using individual anisotropic ADPs. However, in the gray zone of 1.3–1.5 Å, the optimal strategy could be less obvious. The Protein Anisotropic Refinement Validation and Analysis (PARVATI) tool [10] and server (http://www.bmsc.washington.edu/parvati/) may be used to guide the optimal choice of strategy in such cases. The question of expanding the ADP model from isotropic to anisotropic is in fact part of a more general optimization problem, namely at what point the expansion of model parameters is no longer statistically justified by the experimental data and thus should be treated as overinterpretation. There is a rigorous "Occam's razor"-type statistical *R*-factor ratio test for such hypotheses introduced by Hamilton [11] but its application to restrained macromolecular refinements is not obvious, although practical solutions have been proposed [12]. Merritt used the Hamilton *R*-factor ratio test to guide the iso/aniso decision [13] and concluded that at 1.5 Å the anisotropic model ceases to be valid, but also warned that proper statistical analysis should not be replaced by this rule of thumb.

3.4 Multiple Conformations

As the increasing resolution permits distinction between closely spaced alternate occupancies, the proportion of fragments that are modeled in dual (or exceptionally triple) conformation increases as well. This only applies to static disorder. Dynamic disorder can be reduced by collecting the diffraction data at low temperature. Fractional occupancies of light atoms (C/N/O) are considered from ca. 0.2, or exceptionally from 0.1 at ultrahigh resolution, i.e., from electron density contribution equivalent to one H atom.

At 0.9 Å resolution or better, stereochemical restraints of well-ordered fragments may be gradually relaxed, or even removed altogether at ultrahigh resolution. However, disordered or multiple-conformation fragments should remain restrained as they are poorly defined by diffraction.

4 Application and Validation of Stereochemical Restraint Libraries

The geometrical (and other) restraints are extra equations that supplement the set of experimental equations, acting as "springs" that tie the model parameters (such as bond lengths) to some predefined targets. The toughness of the spring is dictated by the variance (error estimate) of the target value. The restraints represent, therefore, some prior knowledge, which may be correct or not. It is thus important to be critical of such information and validate it whenever possible. Once wrong information has been fed to the system, it is very difficult to weed out. At lower resolution, the use of stereochemical restraints is absolutely necessary, simply to improve the data/parameter (d/p) ratio. At 1.2 Å, the d/p is ~3 even for anisotropic models and approaches 5 at 1.0 Å, making restraints dispensable from the mathematical point of view. However, while they may be relaxed in well ordered segments, flexible areas (e.g., side chains) still need to be restrained. At ultrahigh resolution, for well ordered structures, the refinement is highly overdetermined and stereochemical restraints may be eliminated altogether, as illustrated, for example, by the structure of Z-DNA at 0.55 Å resolution [14]. Under strict control of stereochemical restraints at lower resolution, model deviations from the target values should not exceed the uncertainties of the target estimates. In the case of protein bond lengths [15, 16], this is on the order of 0.015–0.020 Å [17]. At very high resolution, the results are dominated by the diffraction terms and the root-mean-square deviations (rmsd's) from the target values are likely to reflect errors in the targets themselves. Deviations as high as 0.02–0.03 Å could be still acceptable.

The target values were compiled by analyzing small-molecule databases about 20 years ago, for proteins by Engh and Huber [15, 16] and for nucleic acids by several authors [18–20]. Although they are largely correct, some adjustments might be necessary.

For example, the dictionary entry for the protein C–N peptide bond may need reevaluation [17] and the peptide group planarity is most certainly enforced too strictly, distorting the adjacent φ/ψ backbone torsion angles and deteriorating the overall Ramachandran geometry [21]. The nucleic-acid parameters for the phosphate group and the valence angles at the guanine glycosidic bond also should be reexamined [14]. The situation is now very interesting because not only is the small-molecule CSD database [22] over ten times larger than when originally used for target evaluation, but we now have a subset of ultrahigh resolution structures in the PDB [23] with minimal target bias, from which the targets can be derived independently. Attempts to revise the Engh and Huber libraries have been already published. For example, Malinska et al. showed [24] that the imidazole ring of histidine can be restrained according to its protonation status, deduced from a trial refinement without restraints (at high resolution) or even from its H-bonding pattern (at lower resolution). In addition to covalent geometry, other model parameters, such as the ADPs or non-bonded contacts, are also restrained. Main-chain torsion angles should be left unrestrained to ensure bias-free model validation via Ramachandran plots.

5 Conformation-Dependent Stereochemical Restraints

Macromolecular models refined at ultrahigh resolution are largely independent of the stereochemical targets (even if restraints have been included) and can be used for their validation and improvement. It has been noted in a number of studies that some (especially angular) parameters of such models have surprisingly wide distributions that could be correlated with the conformation and other characteristic features (e.g., H-bonding) of the macromolecules [17]. For instance, the N–Cα–C angle of the polypeptide backbone has a wide spread [21] and is correlated not only with residue type but also with the local φ/ψ backbone conformation [25]. By modeling main-chain bond distances and angles in proteins characterized at 1 Å resolution or better, as functions of the φ/ψ torsion angles, Tronrud and Karplus [26] were able to create a conformation-dependent stereochemical library (CDL) that leads to better models at lower resolution and, when applied at higher resolution, does not distort the models from the diffraction-driven target (r.m.s.d. for bonds ~0.007–0.010 Å) but, indeed, improves the results [27, 28].

6 Unrestrained Refinement vs. Disorder

It is normal to relax the restraints at atomic resolution and restraint-free refinement is mathematically possible at ultrahigh resolution. However, from the point of view of the d/p ratio it is somewhat contradictory that the degree of discrete (static) disorder that can be modeled by fractional-occupancy conformations increases with resolution, thus demanding more parameters. As demonstrated for the case of BPTI, the percent of disordered residues that are seen even in the same crystal structure increases with resolution [29, 21] and reaches 21% at 0.86 Å. In the 0.66 Å crystal structure of human aldose reductase, one-third of all residues were modeled in multiple conformations [30]. This makes the improvement of d/p less spectacular, as many parameters have to be invested in poorly defined fragments, and requires the retention of stereochemical restraints in multiple-conformation areas. The disorder is usually visible in the macromolecule and in the solvent region, and it is often found to form correlated networks, which should be identified and refined with common occupancy.

7 Treatment of H Atoms

The X-ray scattering power of the H atom is very low and, therefore, H atoms are normally omitted in modeling macromolecular crystal structures. Although at high θ angles the scattering cross section diminishes further, paradoxically H atoms can be better visualized using high-resolution data because the disproportion to C/N/O scattering is less drastic. Besides, H atoms in X–H bonds, which are 0.9–1.1 Å long, can be delineated only when the resolution reaches this level. Even if the H atoms are not fully resolved by diffraction, it is still advisable to include their contribution to F_c to improve agreement with F_o and to remove bias in the location of the parent atoms (on which the H atoms are "riding") that otherwise suffer form the "expanded" skeleton syndrome. The positions of most H atoms in proteins and nucleic acids are easily generated from the skeleton of the remaining atoms, and their contribution at atomic resolution will typically decrease the *R* factor by ca. 0.01. The H atoms in $-NH_3^+$ and $-CH_3$ groups, in the ambiguously protonated His residues, –OH groups and (possibly) carboxylic groups cannot be generated blindly and have to be analyzed individually, usually based on logical H-bond circuits. Generation of H atoms with fractional occupancy is not sensible. Generation of H atoms in water molecules cannot be done fully automatically, although there are algorithms that claim to challenge even neutron scattering data. Considering the high proportion of water molecules with fractional occupancy, it is doubtful if en bloc generation of water H atoms would be meaningful. Those special cases where water H atoms are important and are clearly defined in electron density should be dealt with manually.

With the overwhelming overdeterminacy at ultrahigh resolution, full refinement of H-atom parameters (x,y,z,B_{iso}) is possible as in small-molecule crystallography. Such tests have been carried out but the minimal gain (e.g., negligible drop of the R factor) does not justify the massive effort needed to verify the results. It is thus concluded that even at very high resolution, conventional refinement should include riding H atoms and, if necessary, only the key H atoms should be refined individually. Some protocols place or shift H atoms along the X–H bonds to neutron distances [31]. While this procedure yields a geometrically correct model, it is not necessarily compatible with X-ray refinement of spherical atoms. Moreover, normalization of H atoms in very short (and thus of key importance) hydrogen bonds may be simply unjustified [32].

8 Electron Density Maps at Atomic Resolution

Work with electron density maps at better than atomic resolution is very gratifying because they show most of the atoms as well resolved spheres (Fig. 2). The electron density maps can use $2F_o - F_c$ coefficients, or $3F_o - 2F_c$ coefficients as recommended by Lamzin and Wilson [33], but the difference is not very obvious. At very high resolution even F_o maps can be used as series termination effects are negligible. For difference maps, σ_A-derived coefficients are usually used [34]. For methodological correctness, electron density maps should be contoured in absolute $e/\text{Å}^3$ units, but the values from Fourier summation are not absolute because of the missing strong low-order terms and the unknown $F(000)$ term. Owing to the low noise level of accurate maps, the σ unit frequently used for map contouring is usually low and meaningful features correspond to high-level contours.

9 R_{free} Validation

Calculation of R_{free} [35] is the standard way for validating crystallographic models and the process of their generation. Although refinement at atomic resolution is usually not frustrated with profound strategic ambiguities, some decisions are clearly validated by reference to R_{free}. It is enough to set aside 1000–2000 test reflections, rather than applying the 5–10% rule, which could be very wasteful concerning the large data set size at atomic resolution. One should ensure that the test reflections are randomly selected from the entire data set, i.e., including the highest resolution shell as well. When the model has been completed, the test reflections should be included in the working set for a final round of refine-

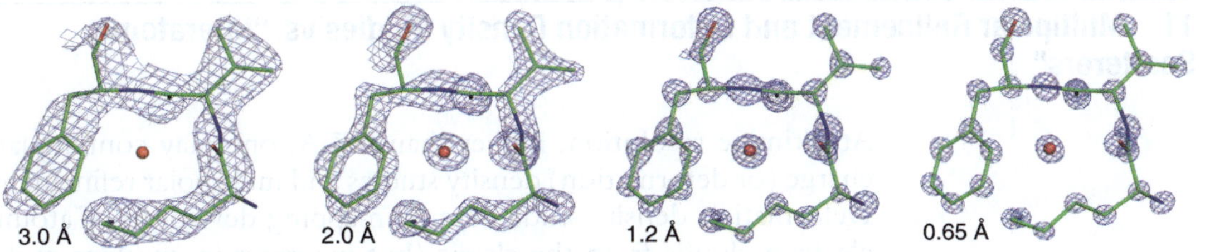

Fig. 2 $2mF_o - DF_c$ electron density map (*blue*) calculated at four levels of resolution (3.0, 2.0, 1.2, 0.65 Å) for the Lys1-Val2-Phe3 fragment (*sticks*) of triclinic lysozyme (PDB code 2vb1) [53], contoured, respectively, at the following level: 1.5σ (0.7 *e*/Å^3), 2.0σ (1.3 *e*/Å^3), 3.0σ (2.2 *e*/Å^3), and 3.6σ (3.0 *e*/Å^3). The absolute contours (in parentheses) were estimated using as *F*(000) the total count of electrons in the model in the unit cell. The following numbers of reflections were used for each map generation: 1909 (3.0 Å), 6519 (2.0 Å), 30325 (1.2 Å), and 185045 (0.65 Å). Note that the electron density of a water molecule (*red sphere*) becomes apparent only at ~2.7 Å resolution. Also note that the presented maps, generated from artificially truncated (1.2, 2.0, and 3.0 Å) but otherwise ultrahigh (0.65 Å) resolution data, are better than typical maps calculated for data extending only to (and thus losing statistical significance at) such a resolution. The σ level for map contouring was estimated from electron density distribution in the entire unit cell, and not around the illustrated fragment. Figure provided by Z. Dauter

ment and for the generation of final electron density maps. This will further improve the final d/p ratio and reduce series truncation errors in the Fourier transform, i.e., will lead to better results, which is the ultimate goal of any high-resolution study.

10 Estimation of Standard Uncertainties

The program particularly well-suited to refining ultrahigh resolution structures is the least-squares oriented SHELXL. Most of the refinement cycles in SHELXL are done using the conjugate-gradient algorithm, which, in the interest of speed, circumvents the inversion of the least-squares matrix. The last refinement cycle (for diagnostic purposes, without application of parameter shifts) should be calculated in the full (or blocked) matrix least-squares mode to estimate the standard uncertainties (s.u.) in the atomic parameters. This is done for all reflections but without restraints and usually for positional parameters only (to obtain s.u.'s of geometrical parameters). If the problem is prohibitively large (over 100 residues), the matrix can be blocked into 50-residue segments (with 5-residue overlap) that will be refined in alternating cycles. Accurately estimated s.u.'s are a treasure trove because they allow meaningful interpretation of model geometry. For instance, it is possible to gauge significant vs. insignificant geometry differences, or even evaluate potential errors in the stereochemical standards. At ultrahigh resolution, the s.u.'s in bond lengths are as low as in small-molecule crystallography. In the 0.55 Å structure of Z-DNA, these values are 0.002–0.004 Å [14], while in the 0.86 Å structure of BPTI they are on the order of 0.005–0.02 Å [29].

11 Multipolar Refinement and Deformation Density Studies vs. "Interatomic Scatterers"

At ultimate resolution, higher than 0.7 Å, one may contemplate charge (or deformation) density studies and multipolar refinement. Deformation density studies aim at mapping deviations of atomic electron clouds from the classic (but incorrect in covalent molecules) spherical independent-atom model (IAM) (Fig. 3). Such studies require data of very high resolution and are rare even in small-molecule crystallography. In multipolar expansion, the atomic electrons are partitioned into core and valence shells, and the latter ones are described by multipolar functions [36]. Depending on the level of multipolar expansion, multipolar atoms require from 3 (for H) to 27 (heavy elements) extra parameters, in addition to the usual three coordinates and six anisotropic ADPs. Usually, the first round of refinement uses reflections from the high-resolution shell only, to distill thermal motion parameters of non-H atoms from the electron distribution functions. Subsequent cycles refine the multipolar parameters of a subset of atoms with excellent order and low thermal motion. Even if ultrahigh resolution is not achieved, deformation density studies are still possible using libraries of transferable multipolar atom models derived from experiment (ELMAM) [37] or by theoretical calculations [38].

Experimental charge-density studies of macromolecules are extremely rare and are limited to aldose reductase, a protein of 316 residues, analyzed at 0.66 Å resolution [39], and to crambin (46 residues), analyzed at 0.54 Å [40] and at 0.48 Å [41]. No charge-density studies for nucleic acids have been published so far, but one study is underway (Fig. 3).

As an alternative to the rigorous multipolar refinement, a simple-minded approximation has been proposed to use "pseudo-atom" scatterers at midpoints of covalent bonds that would take care of the bonding electrons [42]. This simplistic approach is not quite on a par with accurate high-resolution studies.

12 Solvent Structure

As a rule of thumb, one should be allowed to model up to $(3 - |d_{\min}|)$ water molecules per residue [43]. For ultrahigh resolution structures, for which the Matthews fractional volume of solvent [44] is usually low, it is often possible to locate nearly all water molecules, although the count is complicated because many solvent molecules in atomic-resolution structures show a high degree of disorder, populating many sites with partial occupancy. In fact, modeling the outer hydration shell in high-resolution macromolecular structures is usually the most frustrating step, well justifying

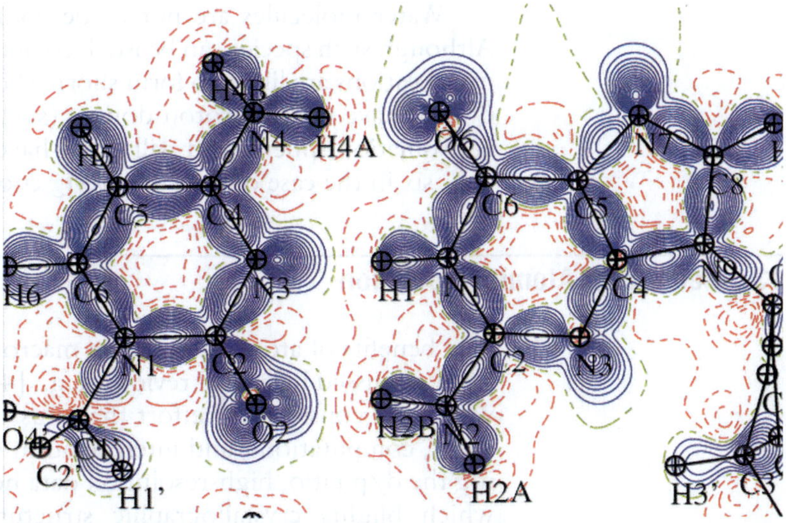

Fig. 3 Static deformation density of a C•G base pair in a Z-DNA structure after multipolar refinement at 0.55 Å resolution. Static deformation density represents charge distribution calculated for atoms at rest as the sum of all multipolar contributions after subtraction of the spherical IAM approximations. The figure is therefore an illustration of the asphericity of real atoms in molecules. Note, for example, the electrons in covalent bonds or in the lone pairs ("rabbit ears") of the oxygen atoms. The contours (*solid blue*—positive, *dashed red*—negative) are drawn with 0.05 *e*/Å^3 increment, starting from 0 *e*/Å^3 (*dashed green* contour). Figure provided by M. Kubicki (unpublished results)

the opinion that "macromolecular refinement against high-resolution data is never finished, only abandoned" [6]. Despite the near-complete atomic interpretation of the solvent region at high resolution, it is common to include in the refinement a bulk-solvent correction, for instance based on Babinet's principle (Fourier transforms of a mask and its complement have the same amplitudes, but opposite phases), which affects only very low-resolution ($d > 15$ Å) data.

The site occupation factors of (even all) water molecules could be refined together with their ADPs but a more prudent approach is to fix them after manual or automatic adjustment. After a round of occupancy (occ) refinement, one would (1) eliminate phantom molecules (occ < 0.2), (2) fix those refined to occ > 0.9 at 1.0, (3) couple the occupancies of alternate sites (O···O distance <2 Å), and (4) let the remaining occupancies refine freely. A water molecule that is retained in the model should have clear $2F_o - F_c$ electron density at the 1σ level, should form at least one reasonable hydrogen bond (2.3–3.2 Å), and should not have prohibitively short contacts, e.g., with C atoms; however, the possibility of forming C–H···O hydrogen bonds (which are usually long, C···O ~ 3 Å) should not be overlooked.

Water molecules are not to be confused with metal cations. Although such species can be isoelectronic (e.g., $H_2O/Na^+/Mg^{2+}$), metal cations are likely to form shorter bonds (e.g., Mg···O ~ 2 Å), do not have typical proton donors (such as amide N–H) in their coordination sphere, and will often have more than four ligands, e.g., six in the case of octahedral Mg coordination.

13 Benefits of Atomic Resolution

The benefits of atomic-resolution macromolecular structures have been discussed in several reviews, e.g., [45–47]. They are certainly worth the considerable effort that must be invested in the experiment, computations, and interpretation of the results. By improving the d/p ratio, high-resolution data help to remove model bias, which blights crystallographic structures solved by molecular replacement. More reflections and better resolving power allow accurate interpretation of multiple conformations, yielding more realistic models and better agreement with the experiment. Unusual stereochemical features are best confirmed at atomic resolution. Refinement with relaxed or eliminated stereochemical restraints is the surest way to the discovery of new phenomena that could be masked by data paucity at low resolution and/or prejudiced ideas about the result. Restraint-free refinement can ultimately produce accurate dictionaries of macromolecular stereochemistry, to be used as restraints at lower resolution. Restraint-free refinement with sufficiently high d/p ratio allows the application of full-matrix least-squares, from which the standard uncertainties of the geometrical parameters can be estimated. The determination of both, the parameters and their error estimates, places the discussion of macromolecular geometry at an entirely new, statistically sound level, and is only possible in crystallography. Although H atoms have only minimal contribution to X-ray scattering and are normally omitted from models of macromolecular structure, they are often of key importance for understanding the functioning of macromolecules, e.g., in enzyme catalysis or fine-tuned intermolecular recognition. Any sensible experimental interpretation of H atoms requires X-ray diffraction data of very high resolution. Indeed, there is evidence suggesting that careful ultrahigh resolution X-ray analysis could be superior in this respect to macromolecular neutron diffraction, which requires prohibitively large crystals (~1 mm^3), deuterated solvent or even perdeuterated protein, and is normally limited to only medium resolution. Even if H atoms are not visualized in electron density maps, their location is easily deduced from the framework of the C/N/O atoms and even in the toughest cases (such as the O–H groups) their placement can be often predicted in atomic-resolution structures not only from H-bond networks but also from the patterns of bond lengths involving the heavier atoms [48]. Also, solvent molecules, which are often disordered and not amenable to accurate modeling at lower resolution, can get sensible interpretation at atomic resolution. Finally, when ultimately high resolution data are available, it is possible to interpret the macromolecular structure at a level of detail that goes far beyond the localization of atoms. Such charge density studies, which involve refinement of multipolar parameters as mock orbitals, are still rare but they are beginning to unveil a fascinating inner world of the macromolecules at the level of electrons in atoms, in interatomic bonds and in intermolecular interactions. Charge density studies also provide a better estimate of the electrostatic properties of biological macromolecules, which are important for understanding their function.

14 Popular Refinement Programs

Refinement at very high resolution is usually carried out in SHELXL, which uses conventional, accurate structure-factor summations [6]. Test calculations with least-squares targets seem to indicate that the results of SHELXL and REFMAC [49] are similar. However, newer versions of programs such as REFMAC or phenix.refine [50] allow refinement against maximum-likelihood targets, not available in the least-squares oriented SHELXL algorithm, and it is yet to be seen if this offers any benefit at high resolution. SHELXL does have, however, its advantages, which include (1) very versatile definition of stereochemical (and other) restraints, (2) flexible definition and refinement of free variables (assigned to selected groups of parameters), (3) strictly enforced refinement on intensities rather than structure factor amplitudes, and (4) the possibility to estimate standard uncertainties in the refined parameters (and in the derived geometrical parameters of the model) through the explicit inversion of the Hessian matrix. Protein multipolar refinement is only possible in a dedicated program, MoPro, developed by Jelsch et al. [51].

References

1. Sheldrick GM (1990) Phase Annealing in SHELX-90: direct methods for larger structures. Acta Crystallogr A 46:467–473
2. Bricogne G, Morris RJ (2003) Sheldrick's 1.2 Å rule and beyond. Acta Crystallogr D Biol Crystallogr 59:615–617
3. Evans P (2012) Resolving some old problems in protein crystallography. Science 336: 986–987
4. Karplus PA, Diederichs K (2012) Linking crystallographic model and data quality. Science 336:1030–1033
5. Vaguine AA, Richelle J, Wodak SJ (1999) SFCHECK: a unified set of procedures for evaluating the quality of macromolecular structure-factor data and their agreement with the atomic model. Acta Crystallogr D Biol Crystallogr 55:191–205
6. Sheldrick GM (2008) A short history of SHELX. Acta Crystallogr A 64:112–122
7. Diederichs K, Karplus PA (1997) Improved R-factors for diffraction data analysis in macromolecular crystallography. Nat Struct Biol 4:269–274

8. Weiss M, Hilgenfeld R (1997) On the use of the merging R factor as a quality indicator for X-ray data. J Appl Cryst 30:203–205
9. Weiss M (2001) Global indicators of X-ray data quality. J Appl Cryst 34:130–135
10. Merritt EA (1999) Expanding the model: anisotropic displacement parameters in protein structure refinement. Acta Crystallogr D Biol Crystallogr 55:1109–1117
11. Hamilton WC (1965) Significance tests on the crystallographic R factor. Acta Crystallogr 18:502–510
12. Bacchi A, Lamzin VS, Wilson KS (1996) A self-validation technique for protein structure refinement: the extended Hamilton test. Acta Crystallogr D Biol Crystallogr 52:641–646
13. Merritt EA (2012) To B or not to B: a question of resolution? Acta Crystallogr D Biol Crystallogr 68:468–477
14. Brzezinski K, Brzuszkiewicz A, Dauter M et al (2011) High regularity of Z-DNA revealed by ultra high-resolution crystal structure at 0.55 Å. Nucleic Acids Res 39:6238–6248
15. Engh RA, Huber R (1991) Accurate bond and angle parameters for X-ray protein structure refinement. Acta Crystallogr A 47:392–400
16. Engh RA, Huber R (2001) Structure quality and target parameters. In: Rossmann MG, Arnold E (eds) International tables for crystallography, vol F. Kluwer Academic, Dordrecht, pp 382–392
17. Jaskolski M, Gilski M, Dauter Z et al (2007) Stereochemical restraints revisited: how accurate are refinement targets and how much should protein structures be allowed to deviate from them? Acta Crystallogr D Biol Crystallogr 63:611–620
18. Parkinson G, Vojtechovsky J, Clowney L et al (1996) New parameters for the refinement of nucleic acid-containing structures. Acta Crystallogr D Biol Crystallogr 52:57–64
19. Clowney L, Jain SC, Srinivasan AR, Westbrook J et al (1996) Geometric parameters in nucleic acids: nitrogenous bases. J Am Chem Soc 118:509–518
20. Gelbin A, Schneider B, Clowney L et al (1996) Geometric parameters in nucleic acids: sugar and phosphate constituents. J Am Chem Soc 118:519–529
21. Addlagatta A, Czapinska H, Krzywda S et al (2001) Ultrahigh-resolution structure of a BPTI mutant. Acta Crystallogr D Biol Crystallogr 57:649–663
22. Allen FH (2002) The Cambridge Structural Database: a quarter of a million crystal structures and rising. Acta Crystallogr B 58:380–388
23. Berman HM, Westbrook J, Feng Z et al (2000) The Protein Data Bank. Nucleic Acids Res 28:235–242
24. Malinska M, Dauter M, Kowiel M et al (2015) Protonation and geometry of histidine rings. Acta Crystallogr D Biol Crystallogr 71: 1444–1454
25. Berkholz DS, Shapovalov MV, Dunbrack RLJ et al (2009) Conformation dependence of backbone geometry in proteins. Structure 17:1316–1325
26. Tronrud DE, Karplus PA (2011) A conformation-dependent stereochemical library improves crystallographic refinement even at atomic resolution. Acta Crystallogr D Biol Crystallogr 67:699–706
27. Moriarty NW, Tronrud DE, Adams PD et al (2014) Conformation-dependent backbone geometry restraints set a new standard for protein crystallographic refinement. FEBS J 281:4061–4071
28. Moriarty NW, Tronrud DE, Adams PD et al (2015) A new default restraint library for the protein backbone in Phenix: a conformation-dependent geometry goes mainstream. Acta Crystallogr D Biol Crystallogr 72:176–179
29. Czapinska H, Otlewski J, Krzywda S et al (2000) High resolution structure of bovine pancreatic trypsin inhibitor with altered binding loop sequence. J Mol Biol 295: 1237–1249
30. Howard ER, Sanishvili R, Cachau RE et al (2004) Ultrahigh resolution drug design I: details of interactions in human aldose reductase-inhibitor complex at 0.66 Å. Proteins 55:792–804
31. Allen FH (1986) A systematic pairwise comparison of geometric parameters obtained by X-ray and neutron diffraction. Acta Crystallogr B 42:515–522
32. Olovsson I, Jaskolski M (1986) Reaction coordinate for the proton transfer in some hydrogen bonds. Pol J Chem 60:759–766
33. Lamzin VS, Wilson KS (1997) Automated refinement for protein crystallography. Methods Enzymol 277:269–305
34. Read RJ (1997) Model phases: probabilities and bias. Methods Enzymol 277:110–128
35. Brünger AT (1992) Free R-value: a novel statistical quantity for assessing the accuracy of the crystal structures. Nature 335:472–475
36. Hansen NK, Coppens P (1978) Testing aspherical atom refinements on small-molecule data sets. Acta Crystallogr A 34:909–921
37. Zarychta B, Pichon-Pesme V, Guillot B et al (2007) On the application of an experimental multipolar pseudo-atom library for accurate refinement of small-molecule and protein crystal structures. Acta Crystallogr A 63:108–125
38. Koritsanszky T, Volkov A, Coppens P (2002) Aspherical-atom scattering factors from molecular wave functions. 1. Transferability and conformation dependence of atomic electron

densities of peptides within the nultipole formalism. Acta Crystallogr A 58:464–472
39. Guillot B, Jelsch C, Podjarny A et al (2008) Charge-density analysis of a protein structure at subatomic resolution: the human aldose reductase case. Acta Crystallogr D Biol Crystallogr 64:567–588
40. Jelsch C, Teeter MM, Lamzin V et al (2000) Accurate protein crystallography at ultra-high resolution: valence electron distribution in crambin. Proc Natl Acad Sci U S A 97:3171–3176
41. Schmidt A, Teeter M, Weckert E et al (2011) Crystal structure of small protein crambin at 0.48 Å resolution. Acta Crystallogr Sect F Struct Biol Cryst Commun 67:424–428
42. Afonine PV, Grosse-Kunstleve RW, Adams PD et al (2007) On macromolecular refinement at subatomic resolution with interatomic scatterers. Acta Crystallogr D Biol Crystallogr 63:1194–1197
43. Wlodawer A, Minor W, Dauter Z et al (2008) Protein crystallography for non-crystallographers, or how to get the best (but not more) from published macromolecular structures. FEBS J 275:1–21
44. Matthews BW (1968) Solvent content of protein crystals. J Mol Biol 33:491–497
45. Dauter Z, Lamzin V, Wilson K (1997) The benefits of atomic resolution. Curr Opin Struct Biol 7:681–688
46. Thaimattam R, Jaskolski M (2004) Synchrotron radiation in atomic-resolution studies of protein structure. J Alloys Compd 362:12–20
47. Jaskolski M (2013) High resolution macromolecular crystallography. In: Read R, Urzhumtsev AG, Lunin VY (eds) Advancing methods for biomolecular crystallography. Springer, New York, pp 259–275
48. Wlodawer A, Li M, Gustchina A et al (2001) Inhibitor complexes of the *Pseudomonas* serine-carboxyl proteinase. Biochemistry 40: 15602–15611
49. Murshudov GN, Skubak P, Lebedev AA et al (2011) Acta Crystallogr D Biol Crystallogr 67:355–367
50. Adams PD, Afonine PV, Bunkóczi B et al (2010) PHENIX: a comprehensive Python-based system for macromolecular structure solution. Acta Crystallogr D Biol Crystallogr 66:213–221
51. Jelsch C, Guillot B, Lagoutte A et al (2005) Advances in protein and small-molecule charge-density refinement methods using MoPro. J Appl Cryst 38:38–54
52. Rosenbaum G, Ginell SL, Chen JC (2015) Energy optimization of a regular macromolecular crystallography beamline for ultra-high-resolution crystallography. J Synchrotron Radiat 22:172–174
53. Wang J, Dauter M, Alkire R et al (2007) Triclinic lysozyme at 0.65 Å resolution. Acta Crystallogr D Biol Crystallogr 63:1254–1268

Chapter 23

Low Resolution Refinement of Atomic Models Against Crystallographic Data

Robert A. Nicholls, Oleg Kovalevskiy, and Garib N. Murshudov

Abstract

This review describes some of the problems encountered during low-resolution refinement and map calculation. Refinement is considered as an application of Bayes' theorem, allowing combination of information from various sources including crystallographic experimental data and prior chemical and structural knowledge. The sources of prior knowledge relevant to macromolecules include basic chemical information such as bonds and angles, structural information from reference models of known homologs, knowledge about secondary structures, hydrogen bonding patterns, and similarity of non-crystallographically related copies of a molecule. Additionally, prior information encapsulating local conformational conservation is exploited, keeping local interatomic distances similar to those in the starting atomic model. The importance of designing an accurate likelihood function—the only link between model parameters and observed data—is emphasized. The review also reemphasizes the importance of phases, and describes how the use of raw observed amplitudes could give a better correlation between the calculated and "true" maps. It is shown that very noisy or absent observations can be replaced by calculated structure factors, weighted according to the accuracy of the atomic model. This approach helps to smoothen the map. However, such replacement should be used sparingly, as the bias toward errors in the model could be too much to avoid. It is in general recommended that, whenever a new map is calculated, map quality should be judged by inspection of the parts of the map where there is no atomic model. It is also noted that it is advisable to work with multiple blurred and sharpened maps, as different parts of a crystal may exhibit different degrees of mobility. Doing so can allow accurate building of atomic models, accounting for overall shape as well as finer structural details. Some of the results described in this review have been implemented in the programs *REFMAC5*, *ProSMART* and *LORESTR*, which are available as part of the *CCP4* software suite.

Key words Refinement, Bayes' theorem, Macromolecules, Low-resolution

1 Introduction

Refinement of atomic structures against crystallographic experimental data is an integral part of crystal structure analysis. Since crystallographic observations are intensities of the corresponding structure factors, and there is no direct way of observing the phases, most crystallographic computations revolve around recovering the lost phases. Hence refinement in general has two purposes: (1) to

Alexander Wlodawer et al. (eds.), *Protein Crystallography: Methods and Protocols*, Methods in Molecular Biology, vol. 1607, DOI 10.1007/978-1-4939-7000-1_23, © Springer Science+Business Media LLC 2017

derive as accurate atomic models as possible, and (2) to improve model phases thus generating the best possible electron density maps. Although the main aim of macromolecular crystallography (MX) is to derive accurate atomic models in order to answer particular biological questions, the importance of improving phases and the resulting electron density maps should not be underestimated. These maps help in automatic [1–3] and manual [4] model building, affecting the quality of final atomic models.

In principle, both electron density maps and atomic models should be considered as statistical models derived from experimental data. Therefore, their quality depends on the quality of these data. There are no computational tools that can replace carefully designed diffraction experiments; computation can only aid experimental design and help increase the amount of information extracted from the data. There have been rapid advances in experimental instrumentation [5], data acquisition [6], and initial data processing, i.e., integration of images and scaling of the resultant intensities [7–11]. However, these can only help if crystals that give sufficient quality of diffraction are available. While there are many techniques to help improve crystal quality [12], in many cases crystals simply do not diffract to high resolution. This could be related to mobility, solubility, purity of proteins, and many other factors resulting in disordered and imperfect crystals. Ultimately, data quality depends on crystal quality.

The purpose of macromolecular crystallography is to answer particular biological questions. In contrast, the purpose of computational tools is to extract as much information as possible from a given dataset. However, information contained in noisy and limited data can be hard to extract. We must develop mathematical and computational tools to help maximize information extraction in such challenging cases.

To achieve accurate models using only low-resolution data it is necessary to use as much available information about macromolecules as possible. The information used must complement the data. Note that X-ray crystallographic data provide information about long-range interactions within and between molecules. At very low-resolution, only information pertaining to shape and crystal packing is present. As the resolution increases, shorter and shorter-range information becomes available. Only at very high resolution—beyond 1.2 Å—do individual atoms become visible; data well beyond 1 Å are needed in order to accurately identify the positions of hydrogen atoms. Therefore, it is almost always necessary to use additional information in order to accurately locate atomic positions. As resolution decreases, longer and longer-range information is needed to complement the data. Basic chemical information, such as "ideal" bond lengths and angles, is usually employed at all resolutions. As resolution decreases, the use of information about torsion angles, secondary structures, domains, and intra-domain interactions might be required. If there are multiple copies of a molecule in the asymmetric unit then non-crystallographic restraints can be used to decrease the effective number of adjustable parameters. One of the problems of using long-range information is the inherent dependence on the structural environments of the molecules. Consequently, special care must be exercised when using such information. Well-known techniques such as robust estimator functions [13] are used in order to improve the application of long-range information derived from known structures.

There are many factors that can reduce the information content of MX data thus reducing the effective resolution. These include crystal growth peculiarities such as twinning and order-disorder. In such cases, although the nominal resolution may be high, not all of the observations are independent. For example, in the case of perfect hemihedral twinning, the number of independent observations is decreased by a factor of two, corresponding to a resolution reduction by a factor of $2^{1/3} \approx 1.26$. Therefore, in the limit, the quality of the electron density map in the presence of perfect hemihedral twinning at 2 Å would correspond to that of a 2.52 Å single crystal case. The refinement of models against data from twinned crystals is now routine [2, 14, 15]. However, statistics after refinement against such data should be interpreted with care [16]. It is important to remember that R-factors and other overall statistics are dependent on the statistical properties of the data, and thus comparison of R-factors from different crystals may give the wrong impression about the comparative quality of the models.

There are many problems that need to be tackled in order to make low-resolution structure analysis routine:

1. The use of chemical and structural information as restraints to increase consistency between available prior knowledge and derived atomic models. The use of chemical information in the form of bond lengths, bond angles, and torsion angles has always been routine. For details on the organization and use of chemical knowledge in refinement, *see* Vagin et al. [17]. Recent years have seen an explosion of approaches toward utilizing structural information [14, 15, 18, 19]. This demonstrates the importance of finding a (and the lack of a unique) solution to the problem of exploiting structural information.
2. Building an accurate likelihood function that reflects noise in the data as well as imperfections in the model. Low-resolution MX data are usually very noisy, and converting them to intensities using the French and Wilson [20] procedure may further reduce whatever tiny amounts of information are available in such data. Using a likelihood function that would bypass this procedure seems to be the best way forward. If the likelihood function is built with care, then noise in the data can be accurately accounted for. However, it seems that integration and

scaling programs do not perform as well as they could when the signal-to-noise ratio is small. One of the directions of research and development must involve the accurate integration and scaling of intensities.

3. Calculation of electron density maps to aid the reduction of errors introduced during manual and automatic model building. Data from low-resolution crystals usually exhibit high isotropic and anisotropic *B*-values. This contributes to observing smeared regions of electron density, with vanishing side chains, secondary structure elements, and even domains. Were this effect removed, the electron density map might reveal more features. Current approaches use only one *B*-value for map sharpening. However, the problem is complicated by the non-negligible influence of contributing factors such as anisotropic diffraction, rigid body oscillation of individual structural units, and correlated motion of whole chains.

Many tools have been developed to aid crystallographic refinement at medium and higher resolutions over the past few decades. One of the current challenges is to develop complementary approaches for dealing with cases where only low-resolution data are available (lower than around 3 Å). Available structural information includes the 3D macromolecular models deposited in the Protein Data Bank [21]. Structural information may be utilized in various forms (such as restraints to secondary structure conformations, homologous reference structures, and homology models) by various modern refinement software packages including *REFMAC5* [14, 22] of *CCP4* [23], *BUSTER-TNT* [24], *phenix.refine* [2], *SHELX* [15] and *CNS* [18]. Many authors have proposed the concept of calculating an electron density map showing more features, e.g., side chains. Notably, Brünger [25, 26] suggests a procedure known in the field of image processing [27] as inverse filtering. However, it is known that such filters can amplify noise, thus masking out real signal. Unfortunately, the electron density always contains noise, which stems from several sources:

1. Noise due to variations in the experimental data.
2. Noise due to errors in the model (e.g., atomic coordinates, model incompleteness, misparameterization, *B*-values, scale factors), and thus in the calculated phases. Such noise correlates with the "true" electron density, and is consequently very hard to address.
3. Noise due to Fourier series termination. When data are collected to the crystal diffraction limit and no map sharpening is used, such noise usually dies out approaching the high-resolution limit. However, when map sharpening is used as an inverse filter then the effects of series termination become pronounced.

There are many different ways of mathematically describing the problem of atomic structure refinement including regularization and energy minimization. In this review we use the language of Bayesian statistics, as it seems to be a good way of combining various sources of information: knowledge about macromolecules and different sources of experimental data including those from crystallography and electron microscopy.

Organization of the review:

In this review, a Bayesian approach to the problem of refinement will be described, including consideration of the likelihood function linking model parameters and experimental data, the use of prior information, and the "best" map calculation procedure. This will be followed by some examples and a discussion of practical usage.

1.1 Notation

In the following sections we shall slightly abuse notation, using the same symbols for random variables as well as for their realizations.

$\langle X \rangle$ —denotes the expectation of a random variable X.

s—a vector in Fourier space (reciprocal space), which has length $|s|$.

$F(s) = (A(s),B(s)) = |F|ei\varphi$—Fourier coefficients of a map with real and imaginary parts A and B, and amplitudes and phases $|F|$ and φ. Fourier coefficients are used interchangeably as complex numbers and two-dimensional vectors.

$I = |F|^2$—intensities of Fourier coefficients.

$F_c(s)$—Fourier coefficients calculated from the current atomic model.

$F_t(s)$—Fourier coefficients of the "true" map.

I_o, I_c, I_t—observed, calculated and "true" intensities, respectively.

Δx—Random error in atomic positions.

ΔB—Random error in atomic temperature factors (B-values).

$D(s) = \cos(2\pi s \Delta x) e^{-\Delta B|s|^2/4}$—the effect of positional and B-value errors on Fourier coefficients; Luzzati's scale parameter [28].

$\Sigma(s) = \langle |F_t(s) - DF_c(s)|^2 \rangle$—the variance of unexplained signal, or variance of the Luzzati distribution for Fourier coefficients.

$\Sigma_0(s) = \langle |F_t(s)|^2 \rangle$—the variance of total signal, or variance of the Wilson distribution for Fourier coefficients [29].

σ_I^2—variance of the noise in the observed intensities.

σ_F^2—variance of the noise in the observed amplitudes.

$$\begin{aligned}\mathrm{cov}(X,Y) &= \langle (X - \langle X \rangle)(Y - \langle Y \rangle) \rangle \\ &= \langle XY \rangle - \langle X \rangle \langle Y \rangle\end{aligned}$$

—covariance between random variables X and Y.

$\mathrm{var}(X) = \mathrm{cov}(X,X)$—variance of the random variable X.

$R(G,\Sigma)=\frac{2}{\Sigma}e^{-\frac{|F|^2+|G|^2}{\pounds}}I_0\left(2\frac{|G||F|}{\Sigma}\right)$—Rice or Srinivasan [30] distribution. Also related to the non-central χ distribution with two degrees of freedom.

$I_0(X)$, $I_1(X)$—modified Bessel functions of the first kind with orders 0 and 1.

$FSC(s)=\frac{\mathrm{cov}(F_1(s),F_2(s))}{\sqrt{\mathrm{var}(F_1(s))\,\mathrm{var}(F_2(s))}}$—Fourier shell correlation between two Fourier coefficients; covariances and variances are calculated in narrow resolution shells.

$P(X;Y)=\frac{P(X,Y)}{P(Y)}$—conditional probability distribution of the random variable X given Y, both of which can be vectors of different sizes.

Relationships and formulas related to the multivariate Normal distribution used in this review can be found in many standard textbooks on multivariate statistics [31, 32].

2 Target Function for Refinement

The problem of atomic structure refinement using crystallographic data can be viewed as a standard statistical problem whereby some knowledge about the internal structure of the system under study is available. Observations are typically incomplete (medium/low-resolution) and noisy (signal-to-noise ratio is small), thus they alone are not sufficient to build and optimize a physically sensible model. There are many models that are consistent with the same set of observations. However, we are looking for a model that is consistent with both our data and our current state of knowledge. One technique for deriving models that are equally well fitted to experimental data and prior information is maximizing the *a posterior* probability (MAP). As the name suggests, this technique maximizes the posterior conditional probability of the model parameters given the current observations. According to Bayes' theorem (*see* for example O'Hagan [33]) the probability distribution of a model given observations may be expressed:

$$P(m;o)=P(m)\frac{P(o;m)}{P(o)}$$

where model parameters are denoted by m and observations by o; $P(m;o)$ is the conditional probability distribution of model parameters given knowledge of the observations (the posterior probability distribution); $P(m)$ is the prior probability distribution of model parameters; $P(o;m)$ is the conditional probability distribution of the observations given the current model parameters (the likelihood function); and $P(o)$ is the probability distribution of the observations (the normalization factor).

It is often convenient to minimize the negative-log of the *a posterior* probability distribution:

$$\begin{aligned} f(m;o) &= -1/2 \log(P(m)) - 1/2 \log(P(o;m)) + 1/2 \log(P(o)) \\ &= f_G(m) + f_D(o;m) + K. \end{aligned}$$

The last term $K = 1/2 \log(P(o))$ is often dropped, as it does not depend on model parameters. Without further justification, we will interchangeably denote model parameters by m in coordinate space and by F_c in Fourier space, and o will correspond to the observed amplitudes ($|F_o|$) or intensities (I_o) of structure factors. All atomic parameters affecting the likelihood function, including positional and thermal parameters, are assumed to be within F_c.

Minimizing the function $f(m;o)$ ensures that as much information is transferred from the observed data to the model as possible (via the likelihood function), and that consistency between the model and our current state of knowledge is maintained (via the prior probability distribution). Obviously, it is impossible (and unnecessary) to use all knowledge about the molecules in designing a prior probability distribution. It is necessary to analyse the information contained within the data, and use complementary prior knowledge. Both crystallographic and cryo-EM techniques produce data pertaining to long-range interactions between atoms in the molecule. As the resolution of the data increases, shorter and shorter-range information becomes available. MX and EM rarely contain information about bond lengths or angles; only data beyond 1.2 Å can contain enough information to allow accurate estimation of bond lengths between well-defined atoms. Consequently, the prior probability distribution must contain at least information about bond lengths and angles. As resolution decreases, longer and longer-range information such as torsion angles and information relating to secondary structures and domains might be needed.

2.1 Likelihood Function

The likelihood function, which links experimental data and model parameters, is an essential ingredient of the MAP and Maximum Likelihood (ML) estimation procedures. The importance of this function cannot be emphasized enough; it is the only link between the model and the data.

In this review, we consider integrated and scaled intensities as observations, and their variances as indicators of their uncertainties. To derive the distribution of observed intensities given model parameters [$P(o;m)$], we first assume that if the "true" structure factors are known then the model structure factors do not say anything new about the observed intensities. In other words, the observed intensities are independent of the "true" intensities, under the condition that the "true" structure factors are known:

$$P(I_o;F_t,F_c)=P(I_o;F_t)$$

Therefore, using the properties of joint probability distributions, we can write:

$$P(I_o,F_t;F_c)=P(I_o;F_t,F_c)\,P(F_t;F_c)=P(I_o;F_t)\,P(F_t;F_c)$$

If we integrate this function over all of the possible (unknown) "true" structure factors then we get the likelihood function:

$$P(I_o;F_c)=\int_{F_t}P(I_o;F_t)\,P(F_t;F_c)\,dA_t dB_t \qquad (1)$$

Thus the likelihood function depends on the distribution of observed intensities given "true" structure factors and the probability of "true" structure factors given the model structure factors. It is generally assumed that the probability of observed intensities is Gaussian with mean value equal to the "true" intensity and variance equal to the estimated variance of the observed intensity:

$$P(I_o;F_t)=\frac{1}{\sqrt{2\pi}\sigma_{I_o}}e^{\frac{-(I_o-I_t)^2}{2\sigma_{I_o}^2}} \qquad (2)$$

Since this distribution depends on σ_{I_o}, estimation of the "observed variances" is very important. Accuracy of observed intensities and variances becomes even more important for weak intensities where the signal-to-noise ratio is small. For very weak intensities, estimation is a challenging problem. This is one of the reasons to use a high-resolution threshold, thus only using the most reliable portion of the data. In future, the problem of estimating experimental uncertainties will need to be addressed carefully if we want to use very noisy data without suffering excessive overfitting into the noise.

An interesting question is: what happens if the observed intensities do not contain any information about the "true" structure factors? In this case, the observed σ_{I_o} should in theory approach to infinity, and therefore the probability distribution $P(I_o;F_c)$ should not depend on the model parameters. However, in practice σ_{I_o} is never infinity. We can consider the problem from a different angle: if there is no information in the data about the crystal we are analysing, then $P(I_o;F_t) = P(I_o)$, i.e., the conditional probability of the intensities given the "true" structure factors does not depend on the "true" structure factors. In this case, Formula 1 shows that the probability distribution of observed intensities does not depend on model structure factors, meaning that there is no information whatsoever in the data about the model. In practice, since σ_{I_o} is always finite, using such data would only result in fitting into the noise, resulting in an overfitted model. Perhaps the integration or scaling programs should analyse data carefully and set the limits where practically useful information ceases to exist. One of the ways of setting this limit would be to use $CC_{1/2}$, as suggested by Karplus and Diederichs [34].

The distribution of "true" structure factors given an atomic model was first derived by Luzzati [28]. Here, we consider only acentric reflections, and assume that structure factors corresponding to different Miller indices are independent:

$$P(F_t;F_c) = \frac{2}{\varepsilon\Sigma} e^{\frac{-|F_t - DF_c|^2}{\varepsilon\Sigma}} \tag{3}$$

where D reflects errors in the atomic positions and B-values, and Σ reflects the variance of the "true" structure factors.

Since phases are not observed, it is usual to work with amplitudes and phases rather than the real and imaginary parts of structure factors. We can integrate over the phases in Formula 3, allowing us to re-express Formula 1 in terms of amplitudes and phases:

$$P(I_o;F_c) = \int_0^\infty P\left(I_o;|F_t|^2\right) P\left(|F_t|;F_c\right) d|F_t|$$

where $P(|F_t|;F_c) = R(D|F_c|,\varepsilon\Sigma)$ follows a Rice distribution (which is a non-central $\chi/\varepsilon\Sigma$ distribution with two degrees of freedom and non-centrality parameter $DF_c/\varepsilon\Sigma$).

Current refinement programs rarely use Formula 1 directly [35]. Rather, intensities are usually converted to the amplitudes of structure factors using the French and Wilson [28] procedure, after which it is assumed that these converted amplitudes with associated uncertainties are the observations. Then Formula 1 is approximated using a Rice distribution with inflated variances $R\left(D|F_c|,\sigma_{I_o} + \varepsilon\Sigma\right)$. While this procedure has worked well for a sufficiently long time [22, 35, 36], it does have certain shortcomings: (a) the conversion from intensities to amplitudes adds a certain bias, which is difficult to overcome; and (b) the distribution of intensities given model parameters is approximated. When the errors in model parameters are large relative to the observed intensities, this approximation works well. However, when they become comparable, the approximation does not work well and it is necessary to search for new approximations. The problem becomes very important when low-resolution crystal structures are analysed; in such cases every bit of information is important. To derive reliable atomic parameters from noisy data it is necessary to use a likelihood function based on Formula 2, or at least a better approximation to it.

2.2 Prior Knowledge

The negative-log prior probability distribution used in crystallographic refinement and fitting into cryo-EM maps has the form:

$$f_{\mathrm{G}}(m) = \sum_{class}\sum_{list} \frac{1}{\sigma^2} w(t_{\mathrm{m}}, t_{\mathrm{i}})(t_{\mathrm{m}} - t_{\mathrm{i}})^2 + \sum_{list} \frac{1}{\sigma^2}(d_{\mathrm{c}} - d_{\mathrm{m}})^2 \quad (4)$$

where *class* is a restraint class (bonds, angles, torsion angles, etc.); *list* is the list of all restraints in the given class; t_{m} denotes the current value calculated from the current model; t_{i} denotes the ideal/target value (either tabulated in a dictionary or calculated from reference homologous structures); and $w(t_{\mathrm{m}},t_{\mathrm{i}})$ is an additional weighting factor that may depend on the current value calculated from the model. Usually, $w(t_{\mathrm{m}},t_{\mathrm{i}}) = 1$ except when robust estimators are used, as well as for non-bonding interactions. In *REFMAC5*, robust estimator weights are used for reference structure and local NCS restraints [14]. For non-bonding interactions, $w(t_{\mathrm{m}},t_{\mathrm{i}}) = 1$ when $t_{\mathrm{m}} < t_{\mathrm{i}}$ and 0 otherwise. σ represents target uncertainty (standard deviation), and is usually listed alongside the "ideal" values in the pre-tabulated dictionary [17].

The latter term in Formula 4 represents the so-called "jelly-body restraints", where d_{m} and d_{c} denote the current and ideal values of interatomic distances in the model. At every cycle of refinement $d_{\mathrm{c}} = d_{\mathrm{m}}$, i.e., the current values of the interatomic distances are always taken as the ideal values. This means that both the value and the first derivative of this term with respect to d_{m} are zero, thus this term changes neither the value nor the gradient of the target function. Since the second derivative of this term is non-zero, it does change the curvature of the parameter space, thus affects how parameters change during refinement. Jelly-body restrained refinement can also be considered as refinement along implicit normal modes [37] at every cycle. This term is similar to the standard regularization terms used in Tikhonov regularization of ill-posed problems [38] or ridge regression [32] in statistics, with the one difference that the regularization term in Formula 4 is applied in distance space instead of parameter space. The purpose of this term is to keep the local conformation of the molecule intact, while allowing groups of atoms (e.g., secondary structures, domains) to move in a concerted fashion. This helps to avoid local minima, increasing the radius of convergence of refinement. Were all pairs of distances restrained with large weights, then the only allowed movement would be a rigid-body movement with six parameters for each domain, i.e., refinement would be reduced to rigid-body fitting. In practice, restraining interatomic distances up to 4.2 Å with $\sigma = 0.01$ or 0.02 Å works sufficiently well for a large class of problems. It should be noted that, despite apparent similarities to Deformable Elastic Network (DEN) refinement [18], it is in fact a markedly different technique. The regularization term in Formula 4 can only be used if the second derivative of the target function or its approximation [39] is used for minimization. If only first derivative methods, such as conjugate gradient, are used as the optimization technique then the minimization iteration must be restarted after a few cycles to account for the updated interatomic distances, making convergence very slow.

Since the models in both crystallography and cryo-EM correspond to macromolecules, the prior knowledge used in both of these techniques is essentially the same [40]. However, one difference is that crystals belong to one of the possible space groups, with corresponding symmetry operators, and therefore symmetry must be accounted for when calculating the interactions between atoms. By contrast, cryo-EM maps are placed in artificial boxes that are large enough to avoid interactions between molecules in neighboring boxes; cryo-EM boxes are not unit cells of a crystal, rather they are used for speed and convenience of calculations only.

2.3 B-Value Restraints

Atomic displacement parameters, or *B*-values, are an integral part of atomic models. They reflect the mobility of atoms. *B*-values have various roles: (1) they reflect atomic mobility, thus refining the *B*-values increases the agreement between the observed data and the refined model; (2) when atoms are incorrectly placed, the atomic *B*-values become large, reflecting errors in the model. It is generally expected for neighboring atoms to have similar *B*-values in regions where modeled atoms are positioned sufficiently accurately. If neighboring atoms have wildly different *B*-values after refinement, it usually means that some of the atoms are either (1) in the wrong place; or (2) otherwise incorrectly parameterized (e.g., the occupancies and/or element types for some of the atoms are wrong). However, when refining atomic *B*-values together with positional parameters it is better to assume that the *B*-values reflect atomic mobility. Therefore, in such cases it is better to restrain the *B*-values of neighboring atoms to be similar to each other. We can also expect that the *B*-values of bonded atoms will be more similar to each other than those of atoms that are spatially close but non-bonded. The prior probability distribution for atomic *B*-values should reflect this intuition.

One of the problems with absolute atomic *B*-values is that if one adds/removes a constant *B*-value to/from all atoms then the average *B*-value will change accordingly. If there are no negative *B*-values as a result of such an operation then it will only affect the scale of observed versus calculated structure factors, as well as causing the resultant density map to be blurred/sharpened. Consequently, one can manipulate the *B*-values of atoms (or structure factors) without changing the information content of the data. This has two consequences:

1. Average *B*-values or Wilson *B*-values are not good indicators of data quality/resolution.
2. Restraints based on *B*-values may not be accurate if they involve anything other than the differences between *B*-values. These are implemented in at least two popular refinement programs: *REFMAC5* [14] and *phenix.refine* [2].

Moreover, if the "real" *B*-values of atoms in a given structure are high, then we can expect the variance of the *B*-values to also be high. Note that shifting all atoms' *B*-values changes the average *B*-value but does not change the variance of the *B*-values. Consequently, the spread of *B*-values is a better indicator of data/model quality than the average *B*-value. This has another implication that if the variance of *B*-values is high then the difference between neighboring *B*-values could also have a higher variation. For example, the difference between *B*-values 10 and 20 Å² should be considered to be much more serious than the difference between 100 and 110 Å².

3 Map Calculation

To calculate the "best" map we need to have the probability distribution of "true" Fourier coefficients given the observations and a model:

$$P(F_t;I_o,F_c)=\frac{P(I_o,F_t;F_c)}{P(I_o;F_c)}$$

If we use this formula then the expected values of the Fourier coefficients that would give the best map (in a certain sense) would have the following form:

$$\langle F_t\rangle=\int_{A_t,B_t} F_t P(F_t;I_o,F_c)\,dA_t dB_t \tag{5}$$

If we assume that the observed intensities and therefore observed amplitudes are exact, i.e., the probability distribution $P(I_o;|F_t|^2)$ is a δ-function $\delta(I_o-|F_t|^2)$ then the expected value (for a particular Fourier coefficient) has the particularly simple form:

$$\langle F_t\rangle=m\,|F_o|\,e^{i\varphi_c} \tag{6}$$

where m is the expected phase error for that particular Fourier coefficient, estimated as:

$$m=<\cos(\Delta\varphi)>=\frac{I_1(X)}{I_0(X)}$$

with $X=2|F_o||F_c|/(\varepsilon\Sigma)$.

However, in general the expected value should be calculated using Formula 5. It is known that Formula 6 produces a map that is biased toward the existing model. To overcome this problem, Main [41] and later Read [42] suggested using $(2mF_o - DF_c)e^{i\varphi}$ type maps. Since this equation uses the observations $|F_o|$ it can be expected that if the data are very noisy then the map will contain observational noise also. Therefore, noise in the map will have two major components: (1) noise due to observational errors, and (2) noise due to errors in the model. Since Formula 5 uses a more accurate form, it should produce a less noisy map. However, it is not clear how to reduce bias in this case in general. Using a $2\langle F_t\rangle - DF_c$ map could serve as the starting point in searching for an unbiased map with less noise.

Considering the behavior of $\langle F_t\rangle$ at different limiting cases can give some insight regarding the behavior of the map:

- *Case 1, when observations are exact*: If the observations are exact then Formula 5 is reduced to Formula 6. Standard refinement procedures use this formula, with a bias reduction technique to reduce model bias, i.e., $2mF_o - DF_c$ coefficients are used for map calculation.
- *Case 2, when observations are absent or very noisy*: The limiting case of noisy intensities can be considered from two different angles: (1) σ_{I_o} approaches infinity, i.e., observations are very noisy. In this case the expected value calculated using Formula 5 approaches DF_c; and (2) Consider the problem as if the probability distribution is $P(I_o;|F_t|^2) = P(I_o)$, i.e., it does not depend on the "true" intensities. In this case, $P(F_t;I_o,F_c) = P(F_t;F_c)$ does not depend on observed intensities, and $\langle F_t\rangle = DF_c$, i.e., a weighted-down version of the calculated structure factors. This approach is used by Murshudov et al. [22] to restore the coefficients for missing observations and free reflections. When the number of missing reflections is small (around 5–10%) then this approach, as a rule, produces smoother maps than would be achieved by excluding those reflections from map calculation. However, when the number of missing reflections becomes substantial, this procedure will produce maps that are biased towards the model. Estimation of D parameters becomes extremely important in such cases. There should be more work directed toward accurate calculation of the D-values, accounting for errors in the model parameters. As a general rule, map quality should be judged by the power to restore missing atoms. Indeed, one should always inspect parts of the map where there is no atomic model built (thus excluded from phase calculation).
- *Case 3, when calculated structure factors are exact*: A less interesting case is when the calculated structure factors are exact, which would never happen in practice. In this case, $D = 1$ and $\Sigma = 0$, and

thus $P(F_t;F_c) = \delta(F_t - F_c)$. Therefore, as expected, $\langle F_t \rangle = F_c$. This simply re-states the obvious fact that if we have a perfect atomic model then no new information can be extracted from the data, i.e., the experiment is not needed.

3.1 Importance of Phases and Amplitudes

It is well known that, for the purpose of map calculation, the phases of structure factors are more important than the amplitudes. To analyse this statement using very simple algebra, we can consider the correlation between the current and "ideal" maps. Since both maps can be calculated using Fourier transformation, we can use the fact that correlations calculated in real and reciprocal space are equivalent:

$$cor(\rho_t, \rho_C) = cor(F_t, F_C) = \frac{\sum |F_t||F_C| \cos(\varphi_t - \varphi_C)}{\sqrt{\sum |F_t|^2 \sum |F_C|^2}} \quad (7)$$

where C denotes the current map, and t the "true" (or "ideal") map.

Formula 7 is the Fourier Shell Correlation (FSC) calculated over all structure factors. One advantage of using the reciprocal space form is that we can perform the calculation in any resolution shell or region of reciprocal space. If the resolution shell is sufficiently narrow then the normalization coefficient (which relates structure factor amplitudes $|F_t|$ and $|F_c|$ to the normalized amplitudes $|E_t|$ and $|E_c|$) will be $\sqrt{\frac{1}{N}\sum |F_t|^2}$ and we can express the FSC in a narrow resolution bin in terms of the normalized amplitudes:

$$FSC = \frac{1}{N} \sum |E_t||E_C| \cos(\varphi_t - \varphi_C)$$

Under the assumption that the reciprocal space points are sufficiently dense (such that the frequency of structure factors in a given region of reciprocal space represents their true frequency) we can replace the average with the expected value. This results in the expected value of the weighted cosine of phase differences:

$$FSC = |E_t||E_C| \cos(\varphi_t - \varphi_C)$$

This assumption seems to work sufficiently well in practice.

It is clear that for "good" maps the FSC would be higher. If we had two maps, and could calculate the FSC between these maps and the "true" maps, then we would prefer the one that had the higher FSC.

Let us consider several limiting cases:

- *Case 1*: *Phases are random, and amplitudes are exact*: In this case, it is clear that the FSC will be 0.
- *Case 2*: *Phases are exact, and amplitudes are random.* In this case we assume that the amplitudes are random but come from a Wilson distribution. In other words, we assume that the amplitudes correspond to a crystal containing atoms, but are not related to the structure at hand.

$$FSC=\langle |E_t||E_C|\rangle=\langle |E_t|\rangle\langle |E_C|\rangle=\frac{\sqrt{\pi}}{2}\frac{\sqrt{\pi}}{2}=\frac{\pi}{4}\approx 0.785$$

Consequently, with random amplitudes but exact phases, the correlation between the current and "true" maps will be ~78.5%. In contrast, when phases are random then the FSC will be 0, irrespective of the accuracy of the amplitudes. We can thus infer that phases are much more important than amplitudes (a fact that has been known for a long time).

If the amplitudes are random, and we know that they are random, then we can replace them with a constant value, e.g., $|E_C| = \langle |E_C|\rangle$ or equivalently: $|F_C|=\sqrt{\langle |F_C|^2\rangle}$. In this case we get:

$$FSC=|E_t|=\frac{\sqrt{\pi}}{2}=0.886$$

Consequently, given that this value is greater than 0.785, if we know that the structure factors have no information about the crystal we are analysing, then it might be better to replace the observations with the expected value of the amplitudes. Note that using anything other than the expected amplitude will cause sudden jumps in the power spectrum of the Fourier series, resulting in ripples in the map. If absent reflections are to be replaced by a constant value then the value must be equal to the expected amplitude (for that resolution). This idea can be exploited to extend resolution beyond observed data [15, 43]. However, extreme care must be exercised when doing so, as accurate estimation of Σ parameters becomes very important; if the Σ values are overestimated, then strong model bias can be expected.

- *Case 3*: *Exact phases and nonrandom amplitudes:* In this case, the FSC will have the form:

$$FSC=|E_t||E_C|=\mathrm{cov}(|E_t|,|E_C|)+|E_t||E_C|=\left(1-\frac{\pi}{4}\right)cor(|E_t|,|E_C|)+\frac{\pi}{4}$$

Here, we used the fact that $\mathrm{var}(|E|) = 1 - \pi/4$. One of the basic questions we can ask is: "how high does the correlation between *E*-values have to be in order to ensure that using the currently available structure factors is better (i.e., gives a better FSC) than simply replacing them with the expected *E*-values?" To answer this question, consider the inequality:

$$\left(1-\frac{\pi}{4}\right)cor\left(|E_t|,|E_C|\right)+\frac{\pi}{4}>\frac{\sqrt{\pi}}{2}$$

It follows that:

$$cor\left(|E_t|,|E_C|\right)>\frac{2\sqrt{\pi}-\pi}{4-\pi}\approx 0.470$$

So if the correlation between the *E*-values of the current and "true" structure factors is larger than 47% then it makes sense to use these data, otherwise it is better to replace the observed amplitudes with $\langle|F_C|\rangle$ for map calculation. It must be stressed that this 47% threshold is a theoretical value, calculated under the assumption that phases are exact, and amplitudes correspond to a crystal. Similar treatment is valid when the phase information is independent of the amplitudes. It is also interesting to note that this result indicates that if the isomorphism between two crystals is 47% then it is possible to use data from one of the crystals to restore missing data from the other.

It is tempting to convert this correlation to a $CC_{1/2}$ on the observed amplitudes. If errors in the amplitudes were additive, and errors were only due to differences in atomic positions, then $cor(|E_t|,|E_C|) = 0.470$ would correspond to $CC_{|F|,1/2} \approx 0.124$. Obviously, in practice these assumptions will not be fulfilled. Therefore, these numbers should be used only as a very rough guide for the limit at which observed data should be used for map calculation. Below this limit, it seems to be better to use the expectation of the amplitudes. The expected values can either be calculated using observed weak data, or interpolated/extrapolated to unobserved regions of reciprocal space using intensity curves [44].

With current map calculation tools, it does seem that using a data threshold (based on the correlation between true and observed amplitudes) should be used. Pragmatically, the value of the optimal threshold is not clear; manually comparing maps, using data at various resolutions, is the best option available at the moment. Even the very approximate numbers based on imperfect assumptions (shown above) indicate that we are still a long way away from exploring/exploiting noisy data fully. Future work should be directed toward better likelihood function estimation, and better map coefficient estimation. All of these improvements would be useless if data integration programs do not produce unbiased data, with good estimates of the observations as well as of the variances of the associated noise.

3.2 Map Sharpening

One of the problems encountered during low-resolution refinement and model building is that the maps are usually blurred, obscuring any finer details of the structure. In general, the overall *B*-values of crystallographic data can be considered arbitrary, depending on data scaling. It is tempting to remove the *B*-values altogether, thus making the power spectrum of the Fourier coefficients flat over all resolution ranges. This can be achieved if one uses normalized structure factors. However, one must remember that removing *B*-values is the same as applying an inverse filter to the map. It is well known that, despite increasing structural details, inverse filters will also increase the noise level. Moreover, amplification of the series termination effect might make the map indistinguishable from pure noise. Nicholls et al. [45, 46] suggest using regularized map sharpening. Although the idea is good, in practice it is tricky to find the optimal parameter values to be used for sharpening and regularization.

More work must be directed toward calculation of the best sharpened maps. Since there is no current technique that can produce the best sharpened map, it is recommended that one works with multiple sharpened maps, switching between them when necessary. Different parts of the map will require different levels of sharpening (related to the signal-to-noise ratio). While the level of noise is more or less constant over the asymmetric unit, the signal level varies depending on local mobility as well as on the presence of the atoms used for phase calculation.

4 Practical Usage

4.1 Automatic Low-Resolution Refinement

We have recently tested various refinement strategies and different *REFMAC5* [14, 22] and *ProSMART* [47] parameters on a test set of more than 100 structures with resolution below 3.0 Å [48] taken from the Protein Data Bank (PDB) [21]. We found that, in cases where high-resolution homologs are available, the best strategy is to first execute a *REFMAC5* refinement using external restraints generated by *ProSMART*, followed by a second round of refinement using jelly-body restraints only. During the first run, external restraints inject additional information into the refinement process. The second jelly-body run allows the target structure to settle and relax into a new conformation. Since bias from the reference homologous models is good only if the models closely represent the actual atomic positions in the target crystal, selection of appropriate high-resolution homologs is important for successful refinement using external restraints. We found that refinement with external restraints is very sensitive to the selection of homologous structures used for restraint generation; for a given target structure, some homologs may perform much better than others. In cases where no homologs are available for a particular protein chain, external restraints representing backbone hydrogen bonds can improve refinement.

In most cases where multiple models of high-resolution homologs are available, using external restraints generated from just one, two, or three homologs with the closest global conformation (lowest global RMSD to the low-resolution model under refinement) produces better results than using all available homologs. Interestingly, sometimes refinement with external restraints generated from homologs with a substantially different conformation (highest global RMSD) from that of the target structure can result in a dramatic drop of R_{free}, substantial structural rearrangement, better geometry and overall improvement of the target model. The fact that the most conformationally different homologs can be the best choice of homologs for external restraint generation implies that, in such cases, the structures of those homologs may better represent the low-resolution crystal contents than the original models in our test set (which had already been deposited in the PDB). This could be the result of suboptimal homolog selection during the initial molecular replacement step. Therefore, in the case of molecular replacement, we recommend trying all available structures (with sufficiently different conformations), subsequently refining each solution, and comparing the results.

The best performing refinement protocols have been implemented in *LORESTR*: an automated pipeline for structure refinement at low resolution [48], distributed as part of the *CCP4* suite [23]. The pipeline facilitates the fully automated selection of optimal external restraints from *ProSMART* for structure refinement by *REFMAC5*. It can automatically run a BLAST search to identify homologs, and download the corresponding models from the PDB. It automatically detects twinning, and finds the optimal scaling method and parameters for solvent modeling. The pipeline runs a number of refinement protocols in order to find the best protocol for each particular case. In our tests, *LORESTR* was able to produce substantially better quality models in the vast majority of cases, improving both R-factors and model geometry for 94% of the test cases. The dramatic improvement in R-factors and stereochemical quality of low-resolution models observed when using the fully automated mode of the pipeline demonstrates its potential utility in low-resolution cases, especially during the initial stages of refinement, or when the refinement process has stalled.

4.2 Automated Re-Refinement of Protein Kinase 1jkt

Here we present an example of the re-refinement of a low-resolution structure taken from the PDB. It should be noted that this kind of scenario, in which we re-refine an already-deposited model, might not be typical of practical structure determination. Here, automated refinement was performed using largely default settings (e.g., no TLS groups), and no attempt was made to achieve "good" final models. Rather, in order to demonstrate practical usage, our example effectively amounts to considering a short snapshot taken during the latter stage of the model building/refinement process.

We consider the re-refinement of the low-resolution 3.5 Å model of Death-Associated Protein Kinase with PDB ID 1jkt (Tereshko et al. [49]), which comprises two chains. The pipeline *LORESTR* was used to automatically optimize the refinement protocol and re-refine the model using *REFMAC5*, aided by external restraints generated by *ProSMART*. It was determined that the data were twinned (two domains with fractions refined to 66% and 34%). The optimal *LORESTR* protocol involved 40 cycles of refinement using external restraints generated from a combination of four of the homologous structures available in the PDB (2x0g_A, 2xuu_A, 4b4l_A and 4tl0_A), followed by a further 20 cycles of refinement using jelly-body restraints (without any external restraints).

Refinement and geometry statistics corresponding to the original and re-refined model are provided in Table 1. Both R_{work} and R_{free} were dramatically reduced, indicating a much better fit of the model to the data after refinement. Furthermore, all geometry statistics were improved, implying the model to be overall more consistent with the prior chemical and structural knowledge.

Figure 1 compares the per-cycle refinement statistics corresponding to three different protocols: standard refinement, refinement using jelly-body restraints, and automated refinement using the optimal *LORESTR* protocol (as described above). Despite the fact that both R and R_{free} were reduced by the default refinement protocol, there was a substantial increase in the difference between the R-factors ($\Delta R = R_{free} - R$) from 4.1% to 9.4%, indicating a high degree of overfitting. In such cases, the use of additional regularizers/restraints is required in order to stabilize refinement and avoid overfitting.

Indeed, the use of jelly-body restraints helps to stabilize the refinement, yielding smaller reductions in both R and R_{free}, but more importantly reducing ΔR from 9.4% to 6.6%. While this is an

Table 1

Refinement and geometry statistics corresponding to the model 1jkt before and after automated refinement using the *LORESTR* pipeline

	R_{work} (%)	R_{free} (%)	Ramachandran outliers (%)	Ramachandran favoured (%)	Clashscore Percentile	MolProbity Percentile
Initial	24.3	28.4	16.2	61.1	3.2	4.6
Final	16.9	20.8	2.6	93.4	73.0	66.1

The optimal *LORESTR* protocol used 40 cycles of *REFMAC5* refinement with main and side chain *ProSMART* external restraints derived from four available homologs (2x0g_A, 2xuu_A, 4b4l_A, and 4tl0_A), followed by 20 cycles of jelly-body restrained refinement. Geometry statistics were calculated using *MolProbity* [55]

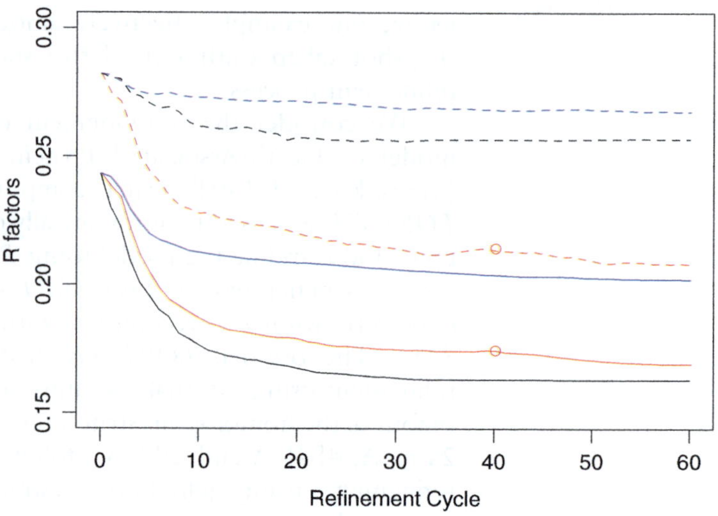

Fig. 1 Refinement statistics corresponding to the re-refinement of 1jkt, which had original R/R_{free} of 24.3/28.4%. Sixty cycles of default refinement (with twinning enabled) resulted in R/R_{free} of 16.3/25.7% (*black lines*), with jelly-body restraints in 20.2/26.8% (*blue lines*), and after automated refinement with the *LORESTR* pipeline in 16.9/20.8% (*red lines*), using the same protocol as in Table 1. R_{work} values are shown as *solid lines*, and R_{free} as *dashed lines*. *Red circles* indicate the point at which refinement using external restraints was stopped and restarted using only jelly-body restraints. The graph was produced using the statistics package *R* [50]

improvement over default refinement without jelly-body restraints, it is still noticeably higher than that of the original model (ΔR = 4.1%). Consequently, we conclude that refinement with jelly-body restraints still suffers from overfitting in this case; ΔR is still too large to inspire confidence in the integrity of the resultant model's consistency with the data. In such cases, additional (or stronger) restraints are required in order to improve the effective data-to-parameter ratio to stabilize the refinement. In this case, refinement could be stabilized by increasing the weight of the jelly-body restraints (i.e., reducing the jelly-body σ below the default value of 0.01) or introducing local NCS restraints.

The refinement protocol from *LORESTR*, which involved a combination of external restraints from four homologs and jelly-body restraints, resulted in substantial reductions in both R and R_{free}, both of which dropped by over 7%. Importantly, the ~4% difference between the R-factors was maintained (ΔR dropped from 4.1% to 3.9%) indicating stable refinement without excessive overfitting. In the first 40 cycles of refinement, the external restraints encouraged the model to adjust local conformation to better agree with prior observations (i.e., the available homologs). This allowed the model to escape local minima in the likelihood function, ultimately resulting in substantial improvement in the

R-factors. Subsequently, the external restraints were released, and jelly-body restraints applied for the latter 20 cycles of refinement. Releasing the external restraints allows the model to relax into the density in its new conformation, rather than repeatedly continuing to pull the model toward the conformation of the homologs, which would reinforce bias toward the reference structures. This release of the external restraints is important so as to allow for any real differences between the structure under refinement and the reference models. However, if we were to simply remove the external restraints, refinement would become unstable and the *R*-factors would diverge. Consequently, jelly-body restraints are introduced in order to stabilize refinement, ensuring that the *R*-factors slowly and stably decrease together.

It is important not to rely solely on refinement statistics when attempting to qualify or quantify model improvement. Various validation tools are available for accessing model reliability given prior knowledge. Figure 2 displays Ramachandran plots corresponding to the model before and after automatic re-refinement using the optimal protocol from *LORESTR*. Overall, we see that the use of external restraints results in greatly improved backbone geometry, indicating a more reasonable model. Note that backbone torsion angles are not explicitly restrained by the external restraints. Rather, the general improvements in their values are a consequence of the stabilization of local structure, which is achieved in interatomic distance space.

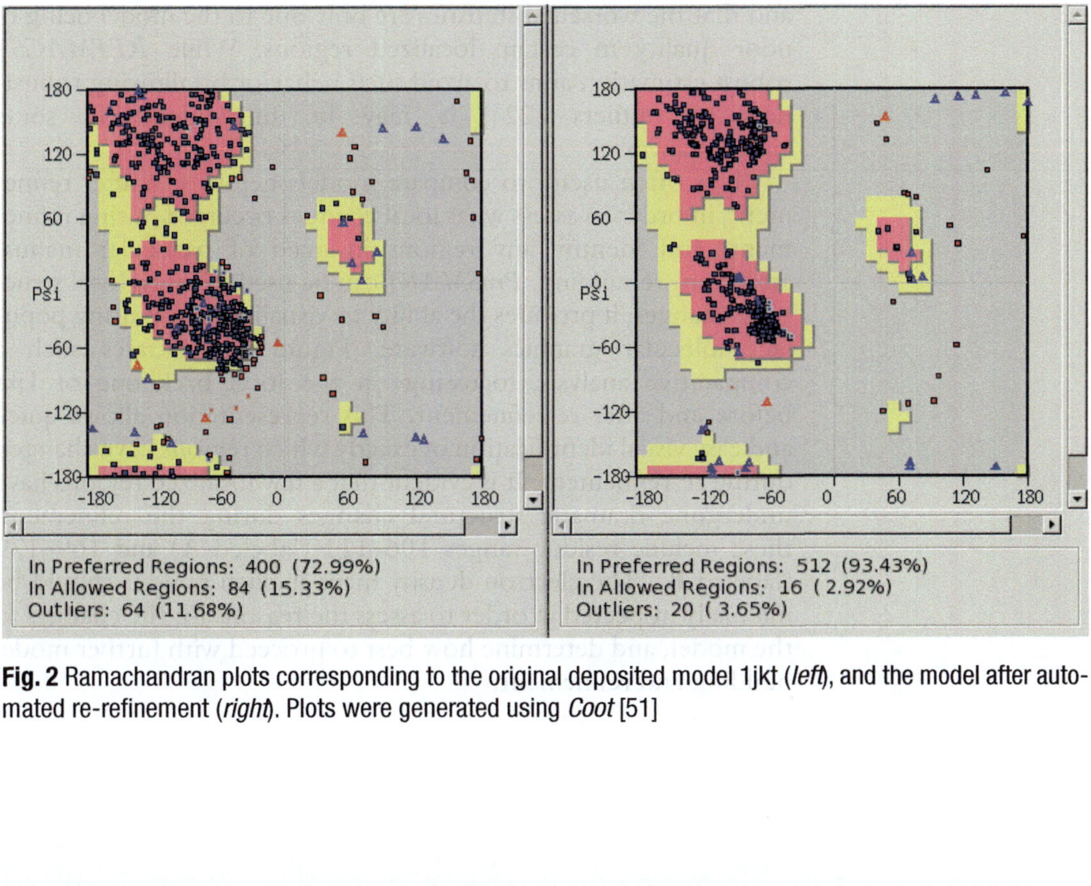

Fig. 2 Ramachandran plots corresponding to the original deposited model 1jkt (*left*), and the model after automated re-refinement (*right*). Plots were generated using *Coot* [51]

Further to looking at global measures of quality/validation—refinement and geometry statistics—it is important to investigate how well localized regions of the model agree with the electron density. Such manual inspection may reveal parts of the model that:

- Can be easily fixed by manual model building and refinement in real space, e.g., side chain flips.
- Are spurious, perhaps incorrectly modeled or disordered, which require careful consideration.
- Have been noticeably improved by the use of external restraints.
- Have been affected by the use of external restraints, changing interpretation of the electron density map, ultimately allowing potential for subsequent improvement by manual real space refinement.
- Have been pulled out of a sensible conformation into an incorrect one by the external restraints, in which case the external restraints should be regenerated excluding these identified regions, and the refinement repeated.

Note that, since external restraints may have a positive effect on some regions of a structure and a negative effect on other regions, overall global statistics may be misleading. For example, in the case of improved refinement statistics, it could be that the external restraints negatively affect some regions, which would need to be fixed. Equally, in the case of worse refinement statistics, it may be that the external restraints positively affect some regions, and that the worsened statistics are only due to the model being of poor quality in certain localized regions. While *REFMAC5*'s robust estimation aims to avoid such behavior by allowing robustness to outliers [22], it may be unavoidable in some circumstances.

It is often useful to compare models before and after refinement, in order to assess what local changes occurred during refinement, and identify any regions in need of particular manual attention/rebuilding. *ProSMART* can be used to assess local structural changes; it provides the ability to visualize results using popular molecular graphics software. Figure 3 illustrates such a comparative analysis, focussing on the local backbone of 1jkt before and after re-refinement. This representation allows quick and easy visual identification of exactly which regions have changed during re-refinement. It is evident that a few localized regions have undergone dramatic structural changes during the refinement; these include residue ranges 106–111 (labeled **A**) and 169–179 (labeled **B**). The electron density maps in such regions should be manually inspected in order to assess the reasons for the changes to the model, and determine how best to proceed with further model building and refinement.

Figure 4 displays the model and electron density maps before and after re-refinement, focusing on the region labeled **A** in Fig. 3. After re-refinement, new features appear in the map. Notably, the density for the side chain of Phe102 (upper left in the image) becomes clearer, reinforced by positive difference density. The refinement was unable to escape local minima and move this side chain into the correct conformation; the side chain would need to be manually flipped into the correct conformation. Also, the model around residues 106–111 (the right half of the image) undergoes substantial conformational change. Indeed, the electron density map in this region is strikingly different before and after re-refinement. This exemplifies how model bias can cause electron density maps to be extremely unreliable, especially at low resolution. In cases such as this, it could be that the region is incorrectly modeled, or that the region is inherently flexible. At this stage, the appropriate strategy might be to attempt to completely rebuild such regions, aided by omit maps created by re-refinement with jelly-body restraints (after removing the region to be rebuilt), although doing so is beyond the scope and purpose of this example.

Figure 5 displays the model and electron density maps before and after re-refinement, focusing on the region labeled **B** in Fig. 3. Again, the interpretation of the map is quite different after

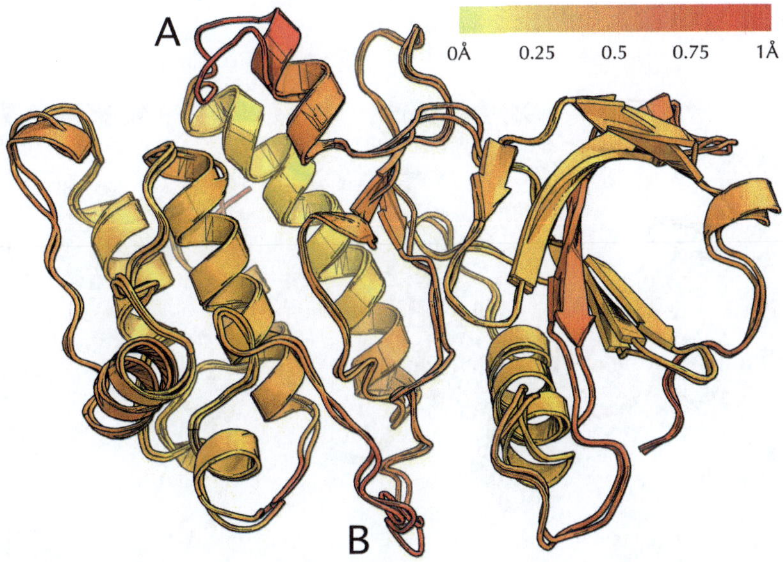

Fig. 3 Superposition of the model 1jkt before and after automated re-refinement. Both models are colored according to local backbone structural conservation, using *ProSMART*'s "flexible" score [51], which is a measure of local backbone RMSD. *Yellow* implies structural similarity, *red* relative dissimilarity. Two regions that have undergone dramatic local structural changes during re-refinement, easily identified as being coloured *red*, are labeled (**A**) and (**B**). The image was generated using *PyMOL* [52]

re-refinement. In well-modeled regions, more clarity and better connectivity is observed in the map. New features appear, such as density for the side chain of His166 (located near the center of the image). Also, difference density is observed near Asp161, indicating that this region may not be optimally modeled—such information would aid manual model rebuilding. The density around residues 169–179 (the upper-right portion of the image) almost completely disappears after re-refinement, indicating this region to be incorrectly modeled. The fact that electron density is visible in this region before re-refinement is further evidence of model bias and map unreliability. In practice, this region might be completely removed at this stage, and rebuilt only if evidence for the loop region appears in the electron density after further refinement iterations.

Both Figs. 4 and 5 demonstrate how inaccurate phases, often due to model bias, can cause density maps at low resolution to be unreliable and extremely misleading. This makes map interpretation difficult and prone to error. There is often a tendency to over-interpret maps during model building, leading to poor quality models and inhibiting progress with refinement. However, it is possible to achieve good quality models in cases where only low-resolution data are available—despite having only data extending to 3.5 Å, we have demonstrated that it is possible to automatically refine 1jkt to produce a model with R/R_{free} of 16.9/20.8%, noting the potential for substantial further improvement (as can be seen in Figs. 4 and 5).

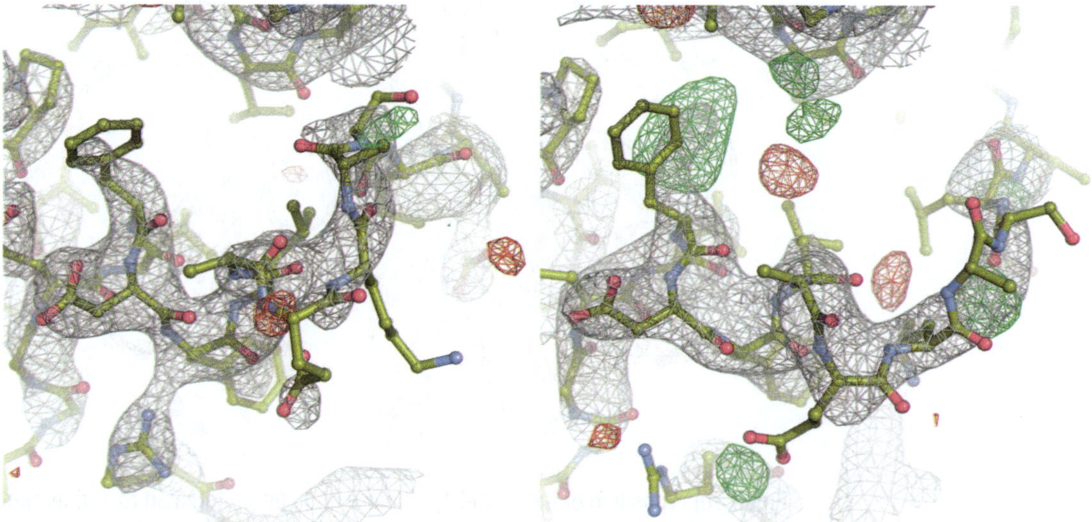

Fig. 4 Illustrations of the model and electron density maps before (*left*) and after (*right*) automated re-refinement using the optimal protocol from *LORESTR*, focusing on residues 100–112 (near the region labeled (**A**) in Fig. 3). The $2F_o - F_c$ maps are shown in *grey*, and the $F_o - F_c$ maps in *green*/*red* (for positive/negative). Models and maps were generated using *Coot* [51] using default contour levels (1.5 for $2F_o$–F_c and 3 σ for F_o–F_c maps), and images rendered using *PyMOL* [52]

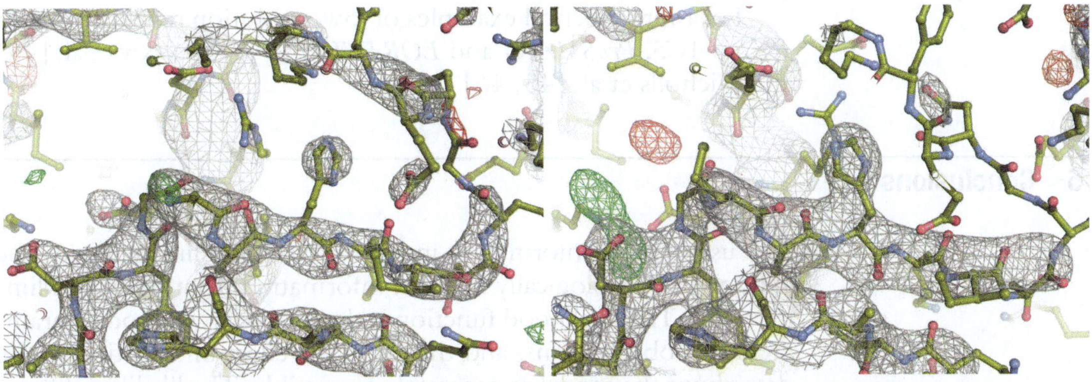

Fig. 5 Illustrations of the model and electron density maps before (*left*) and after (*right*) automated re-refinement using the optimal protocol from *LORESTR*, focusing on residues 160–175 (near the region labeled (**B**) in Fig. 3). The $2F_o - F_c$ maps are shown in *grey*, and the $F_o - F_c$ maps in *green/red* (for positive/negative). Models and maps were generated using *Coot* [51] using default contour levels (1.5 for $2F_o$–F_c and 3 σ for F_o–F_c maps), and images rendered using *PyMOL* [52]

It should be acknowledged that both regions **A** and **B**, which correspond to problematic regions of the model that require manual attention and rebuilding, were easily identified using local comparative structural analysis (as shown in Fig. 3). Indeed, the utility of such analysis to aid the refinement procedure is evident. This example demonstrates that the available techniques for refining at low resolution can aid refinement, allowing new features to be revealed in the map. Improving the model in this way can result in dramatically different interpretations of the electron density, and may in some cases lead to different biological conclusions.

At low resolution, effects such as model bias cause difficulties in qualifying any model improvement during the refinement process. We commonly rely on refinement statistics (e.g., R-factors) to determine model quality, but they are not always conclusive. It is important to complement such measures with independent validation, e.g., from considering model geometry. However, such statistics are also not always conclusive—it is often possible to have worse global scores, but improved local structure in some regions, and *vice versa*. However, the calculated electron density maps may not always be reliable, as can be seen in our example (*see* Figs. 4 and 5). Indeed, suboptimal refinement may lead to incorrect map interpretation. This results in low-resolution structure determination having a high risk of error. Since the use of external restraints will alter global geometry validation statistics, such results should be interpreted accordingly, and the integrity of local structure should always be considered. Indeed, it is important to always manually inspect the electron density to check that the model agrees reasonably well with the data, thus ensuring local suitability of the use of external restraints, despite any apparent improvement or deterioration in overall statistics.

For more practical examples of low-resolution refinement with *REFMAC5*, *ProSMART* and *LORESTR*, *see* Kovalevskiy et al. [49] and Nicholls et al. [45, 46].

5 Conclusions

The use of prior information in low-resolution refinement aids the extraction of biologically relevant information from noisy and limited data. The likelihood function is the link between model parameters and observations, and thus it must reflect information in the data about the model as accurately as possible. The likelihood function currently used works very well when the signal-to-noise ratio in the data is sufficiently high. However, when data are very noisy the current likelihood function performs suboptimally. In future, a more accurate likelihood function will need to be implemented if we want to allow accurate models to be derived from such data.

We have shown, using simple calculation, that if the phases are correct and the amplitudes of structure factors are random, then the correlation between "true" and calculated maps could be as high as 78.5%. However, if the expected values of amplitudes of structure factors are used in map calculation, this correlation could increase up to 88.6%. It is also shown that the correlation between "true" and observed structure factors must be more than 47% in order to see an improvement in electron density maps. Otherwise, it seems that it would be better to use the expected structure factor amplitudes instead of the observed ones for purposes of map calculation. In other words, there is a data quality threshold below which using observed data may not be the best strategy; their expected values have enough information, and using them may result in better maps. These results, while reemphasizing the importance of phases, show that if the data are of exceedingly poor quality then it is better to calculate maps using the expected values of amplitudes, in principle. These types of calculations should be done with extreme care, given that phases are never exact; rather, they are calculated from an imperfect atomic model. If model parameters were sufficiently accurate then such treatment would give a better-looking map. However, the issue of model bias, which is prevalent whenever the model includes errors, should be addressed. In general, whenever new maps are calculated, one always must manually inspect the regions that do not have corresponding atoms built; omit maps or something similar must be used.

At low resolution, bias toward errors in the map is a real problem. Usual statistics such as R-factors are not good indicators of model quality. Agreement between the model and prior knowledge must always be considered. However, as data quality degrades, and thus the effective number of observations decreases, it is much easier to achieve good agreement with the prior knowledge. Note that good agreement with the prior knowledge does not also mean that the derived model corresponds well to the true crystal structure; rather, it only means that the model agrees with our prior knowledge (i.e., that the model is chemically/structurally sensible). At low resolution, all indicators, as well as the predictive power of the model, must be checked. In other words, when maps are displayed/used as evidence, that part of the atomic model (corresponding to the relevant map region) should not be used during the map calculation procedure; omit maps must be used.

The degree of model quality required depends on the questions asked: if questions relate to domain organization, or to mutual orientation of protein molecules in a complex, then probably low-resolution (3–5 Å) data might give sufficient evidence. However, if more specific questions are asked that require a higher degree of model accuracy (e.g., related to specific interatomic distances) then perhaps higher resolution data are needed. In any case, there is always going to be the question as to whether a given model of a crystal is the same as the structure in solution. To answer these concerns, one must perform additional experiments to confirm such hypotheses. One of the techniques that can be used to address such issues involves joint refinement of the model using MX and NMR experimental data such as residual dipolar couplings and perhaps pseudo-contact shifts [53].

Future development will have to be concentrated on designing better likelihood functions. Focussing on Eq. 2 would be a good starting point. It is very likely that if a given crystal diffracts only to low resolution, then it is susceptible to radiation damage. This means that different images correspond to different but related crystal structures. Moreover, data are often collected from multiple crystals; such data are later classified and merged when necessary [54]. If the classification of crystals is necessary then it means that the crystals (and also the contents of the crystals) are different from each other, despite being related. To address this and many other issues, it will be necessary to refine against unmerged data. It might even be necessary to simultaneously perform refinement of atomic models and scaling/merging of intensities. If such a procedure is designed, then it might be possible to address many issues surrounding crystal variability.

Acknowledgments

This work was supported by the Medical Research Council. GNM is funded by the MRC (grant number: MC_US_A025_1012), RAN by CCP4/STFC (grant number: PR140014), and OK by BBSRC (grant number: BB/L007010/1). We would also like to thank the LMB for creating a very active working environment.

References

1. Perrakis A, Morris S, Lamzin VS (1999) Automated protein model building combined with iterative structure refinement. Nat Struct Biol 6:458–463
2. Adams PD, Afonine PV, Bunkòczi G et al (2010) PHENIX: a comprehensive Python-based system for macromolecular structure solution. Acta Crystallogr D Biol Crystallogr 66:213–221
3. Cowtan K (2006) The Buccaneer software for automated model building. 1. Tracing protein chains. Acta Crystallogr D Biol Crystallogr 62:487–493
4. Emsley P, Cowtan KD (2004) Coot: model-building tools for molecular graphics. Acta Crystallogr D Biol Crystallogr 60:2126–2132
5. Helliwell JR, Mitchell EP (2015) Synchrotron radiation macromolecular crystallography: science and spin-offs. IUCrJ 2:283–291
6. Smith JL, Robert F, Fischetti RF et al (2012) Micro-crystallography comes of age. Curr Opin Struct Biol 22:602–612
7. Kabsch W (2010) XDS. Acta Crystallogr D Biol Crystallogr 66:125–132
8. Battye TGG, Kontogiannis L, Johnson O et al (2011) iMOSFLM: a new graphical interface for diffraction-image processing with MOSFLM. Acta Crystallogr D Biol Crystallogr 67:271–281
9. Otwinowski Z, Minor W (1997) Processing of X-ray diffraction data collected in oscillation mode. Methods Enzymol 276:307–326
10. Waterman DG, Winter G, Gildea RJ et al (2016) Diffraction-geometry refinement in the DIALS framework. Acta Crystallogr D Biol Crystallogr 72:558–575
11. Evans PR, Murshudov GN (2013) How good are my data and what is the resolution? Acta Crystallogr D Biol Crystallogr 69:1204–1214
12. Chernov AA (2003) Protein crystals and their growth. J Struct Biol 142:3–21
13. Huber PJ (1981) Robust statistics. Wiley and Sons, NJ
14. Murshudov GN, Skubak P, Lebedev AA et al (2011) REFMAC5 for the refinement of macromolecular crystal structures. Acta Crystallogr D Biol Crystallogr 67:355–467
15. Sheldrick GM (2008) A short history of SHELX. Acta Crystallogr A 64:112–122
16. Murshudov GN (2011) Some properties of crystallographic reliability index—Rfactor: effect of twinning. Appl Comp Math 10:250
17. Vagin AA, Steiner RA, Lebedev AA et al (2004) REFMAC5 dictionary: organization of prior chemical knowledge and guidelines for its use. Acta Crystallogr D Biol Crystallogr 60:2184–2195
18. Schröder GF, Brünger AT, Levitt M (2010) Super-resolution biomolecular crystallography with low-resolution data. Nature 464:1218–1222
19. Smart OS, TO W, Flensburg C et al (2012) Exploiting structure similarity in refinement: automated NCS and target-structure restraints in BUSTER. Acta Crystallogr D Biol Crystallogr 68:368–380
20. French S, Wilson K (1978) On the treatment of negative intensity observations. Acta Crystallogr A 34:517
21. Berman HM, Battistuz T, Bhat TN et al (2002) The Protein Data Bank. Acta Crystallogr D Biol Crystallogr 58:899–907
22. Murshudov GN, Vagin AA, Dodson EJ (1997) Refinement of macromolecular structures by maximum likelihood method. Acta Crystallogr D Biol Crystallogr 53:240–255
23. Winn MD, Ballard CC, Cowtan KD et al (2011) Overview of the CCP4 suite and current developments. Acta Crystallogr D Biol Crystallogr 67:235–242
24. Blanc E, Roversi P, Vonrhein C et al (2004) Refinement of severely incomplete structures with maximum likelihood in BUSTER-TNT. Acta Crystallogr D Biol Crystallogr 60:2210–2221
25. Brünger AT, DeLaBarre B, Davies JM et al (2009) X-ray structure determination at low resolution. Acta Crystallogr D Biol Crystallogr 65:128–133
26. DeLaBarre B, Brünger AT (2006) Considerations for the refinement of low-resolution crystal structures. Acta Crystallogr D Biol Crystallogr 62:923–932
27. Gonzalez RC, Woods RE (2008) Digital image processing. Prentice Hall, NJ
28. Luzzati V (1952) Traitement statistique des erreurs dans la determination des structures cristallines. Acta Crystallogr 5:802–810
29. Wilson AJC (1949) The probability distribution of X-ray intensities. Acta Crystallogr 2:318–321
30. Srinivasan R, Parthasarathy S (1976) Some statistical applications in X-ray crystallography. Pergamon Press, Oxford
31. Eaton ML (2007) Multivariate statistics: a vector space approach. Beachwood, OH
32. Stuart A, Ord K, Arnold S (2009) Kendall's advanced theory of statistics, vol 2a: Classical inference. Oxford University Press, Oxford

33. O'Hagan A (1994) Kendall's advanced theory of statistics, vol 2b: Bayesian inference. A Hodder Arnold Publication, London
34. Karplus PA, Diederichs K (2012) Linking crystallographic model and data quality. Science 336:1030–1033
35. Pannu NS, Read RJ (1996) Improved structure refinement through maximum likelihood. Acta Crystallogr A 52:659–668
36. Bricogne G (1997) Bayesian statistical viewpoint on structure determination: basic concepts and examples. Methods Enzymol 276:361–423
37. Tirion MM (1996) Large amplitude elastic motions in proteins from a single-parameter, atomic analysis. Phys Rev Lett 77:1906–1908
38. Tikhonov AN, Arsenin VY (1977) Solution of ill-posed problems. Winston & Sons, Washington
39. Steiner R, Lebedev A, Murshudov GN (2003) Fisher's information matrix in maximum likelihood molecular refinement. Acta Crystallogr D Biol Crystallogr 59:2114–2124
40. Brown A, Long F, Nicholls RA et al (2015) Tools for macromolecular model building and refinement into electron cryo-microscopy reconstructions. Acta Crystallogr D Biol Crystallogr 71:136–153
41. Main P (1979) A theoretical comparison of the β, γ and $2F_o - F_c$ syntheses. Acta Crystallogr A 35:779–785
42. Read RJ (1986) Improved Fourier coefficients for maps using phases from partial structures with errors. Acta Crystallogr A 42:140–149
43. Altomare A, Cuocci C, Giacovazzo C et al (2008) Minimally resolution biased electron-density maps. Acta Crystallogr A 64:326–336
44. Popov A, Bourenkov G (2003) Choice of data-collection parameters based on statistic modelling. Acta Crystallogr D Biol Crystallogr 59:1145–1153
45. Nicholls RA, Long F, Murshudov GN (2012) Low-resolution refinement tools in REFMAC5. Acta Crystallogr D Biol Crystallogr 68:404–417
46. Nicholls RA, Long F, Murshudov GN (2013) Recent advances in low resolution refinement tools in REFMAC5. In: Read RJ (ed) Advancing methods for biomolecular crystallography. Springer, Netherlands
47. Nicholls RA, Fischer M, McNicholas S et al (2014) Conformation-independent structural comparison of macromolecules with ProSMART. Acta Crystallogr D Biol Crystallogr 70:2487–2499
48. Kovalevskiy O, Nicholls RA, Murshudov GN (2016) Automated refinement of macromolecular structures at low resolution using prior information. Acta Crystallogr D Biol Crystallogr 72:1149–1161
49. Tereshko V, Teplova M, Brunzelle J et al (2001) Crystal structures of the catalytic domain of human protein kinase associated with apoptosis and tumor suppression. Nat Struct Biol 8:899–907
50. R Core Team (2015) R: a language and environment for statistical computing. R Foundation for Statistical Computing, Vienna, Austria. https://www.R-project.org/
51. Emsley P, Lohkamp B, Cowtan KD (2010) Features and development in COOT. Acta Crystallogr D Biol Crystallogr 66:486–501
52. Schrödinger, LLC (2015) The PyMOL molecular graphics system, version 1.8
53. Rinaldelli M, Ravera E, Calderone V et al (2014) Simultaneous use of solution NMR and X-ray data in REFMAC5 for joint refinement/detection of structural differences. Acta Crystallogr D Biol Crystallogr 70:958–967
54. Foadi J, Aller P, Alguel Y et al (2013) Clustering procedures for the optimal selection of data sets from multiple crystals in macromolecular crystallography. Acta Crystallogr D Biol Crystallogr 69:1617–1632
55. Chen VB, Arendall WB, Headd JJ et al (2010) MolProbity: all-atom structure validation for macromolecular crystallography. Acta Crystallogr D Biol Crystallogr 66:12–21

Chapter 24

Stereochemistry and Validation of Macromolecular Structures

Alexander Wlodawer

Abstract

Macromolecular structure is governed by the strict rules of stereochemistry. Several approaches to the validation of the correctness of the interpretation of crystallographic and NMR data that underlie the models deposited in the PDB are utilized in practice. The stereochemical rules applicable to macromolecular structures are discussed in this chapter. Practical, computer-based methods and tools of verification of how well the models adhere to those established structural principles to assure their quality are summarized.

Key words Crystal structure, NMR structure, Ramachandran plot, Bond lengths, Bond angles, Quality check, Geometrical criteria, Protein Data Bank (PDB)

1 Introduction

Biological macromolecules (proteins, as well as other biopolymers, such as nucleic acids and carbohydrates) are composed of atoms belonging to a very limited number of elements (primarily carbon, nitrogen, oxygen, and hydrogen, and to a lesser extent sulfur and phosphorus). A few other elements, for example selenium, are occasionally present in the covalent structure of biologically relevant macromolecules, whereas other elements, primarily metal cations, are frequently coordinated to atoms belonging to the macromolecule, or are located within functional groups that may be either coordinated or covalently attached to their macromolecular targets. The rules of chemical bonding derived from the accurate crystal structures of small organic and organometallic molecules (mainly from over 800,000 crystal structures currently present in the Cambridge Structural Database (CSD) [1]) must apply equally to the macromolecular structures as well. A significant difference between the interpretation of small-molecule and macromolecular structures is due to very different ratios of experimental observations, such as the number of reflection intensities, to the number of model parameters, such as atomic coordinates and displacement

Alexander Wlodawer et al. (eds.), *Protein Crystallography: Methods and Protocols*, Methods in Molecular Biology, vol. 1607, DOI 10.1007/978-1-4939-7000-1_24, © Springer Science+Business Media LLC 2017

parameters (ADPs, formerly known as temperature factors). Except when atomic-resolution crystallographic data (defined as having d_{min} < 1.2 Å; [2]) are available, the vast majority of macromolecular structures (currently ~97.5% of crystal structures and 100% of NMR- and electron microscopy-derived structures in the Protein Data Bank (PDB [3]) could not be properly refined on the basis of the experimental data alone. Thus, as described elsewhere in this volume (Chapter 22 by Jaskolski), stereochemical restraints are almost always applied in some form during the refinement of macromolecules.

Some stereochemical properties of proteins had been elucidated before the first protein structures were experimentally determined. For example, the bond lengths, angles, and the planar nature of the peptide bond had been known since early 1950s. Pauling et al. [4] estimated that deviation of the peptide bond from planarity by 10° should destabilize it by about 1 kcal/mole. In the same paper, they described the stereochemical parameters of the α-helix (not yet named as such), by postulating a twist of the polypeptide with a non-integral number of residues (3.7, later revised to 3.6) per turn. Another classical secondary structure element defined at that time was the β-sheet [5]. Properties of these secondary structures, together with those of other structural elements, were summarized in detail 2 years later [6]. These types of structures were subsequently found in experimentally determined structures of proteins [7–9]. Correct knowledge of the stereochemical properties of nucleic acid polymers (together with other evidence, such as X-ray fiber diffraction) led to the proposed model of the double-helical structure of DNA [10], which changed our understanding of fields even very indirectly related to structure, such as genetics.

2 Restraining Bonds and Angles

Whereas the lengths and angles of the peptide bond were known quite accurately early on [6], knowledge of the corresponding parameters in the side chains of amino acids (and in the nucleic acids and carbohydrates) came gradually, first from only the structures of the individual components of these polymeric compounds, and later from high resolution structures of macromolecules. Bond and angle information was subsequently compiled into stereochemical restraint libraries. Except for the few structures for which diffraction data could be collected to extremely high resolution (0.8 Å or better), all macromolecular refinement procedures utilize such standard stereochemical information [11]. X-ray- and neutron-diffraction structures of individual amino acids were initially used for the construction of restraint libraries for programs such as PROLSQ [12, 13], but the restraints were subsequently improved on the basis of large databases. Almost universally, the currently used refinement programs, such as CNS [14], SHELXL [15], REFMAC5 [16], or PHENIX [17] use the parameters compiled 25 years ago by Engh and Huber [18] and subsequently updated [19] 10 years later. These parameters were obtained by careful analysis of the CSD [1], and were later slightly modified through the analysis of highest-resolution macromolecular structures [20]. It was also pointed out that the values of bond lengths and angles depend, to a certain extent, on the secondary structure and they were modified accordingly to correct for such effects [21]. Other details, such as protonation-dependent variations of the geometry of the imidazole ring of histidine, were analyzed more recently [22]. Some adjustments to the parameters used in the refinement of nucleic acids [23] was proposed on the basis of an ultrahigh-resolution (0.55 Å) crystal structure of Z-DNA [24].

Rms (root-mean-square) deviations from standard stereochemistry indicate how much a refined model departs from the geometrical targets present in the dictionaries. Different parameters can be evaluated by the rmsd criterion, but it is most common to use the values for bond lengths and angles when comparing different models. The allowed departure from the targets depends on the resolution of the diffraction data used in the refinement. Good-quality, medium-to-high resolution structures are expected to have rmsd(bond) values of about 0.02 Å (corresponding to the standard uncertainty of the targets themselves), although numbers half that size are also acceptable [20]. When this number becomes too high (above ~0.03 Å), this may indicate problems with the model. On the other hand, attempts to lower the rmsd further may lead to models that are more idealized, but less accurate. The common values for rmsd(angle) are between 0.5° and 2.0°. These levels of variations in the geometric parameters are in line with the rmsd values averaged for all classes of bonds and angles in polypeptides listed in the original Engh and Huber compilation (0.022 Å and 1.85° for bonds and angles, respectively) [18]. The default target values in different refinement programs are also in the same ranges [20].

3 Ramachandran Plot and Peptide Planarity

One of the most useful tools for validation of protein structures is the Ramachandran plot [25], showing the mapping of pairs of φ/ψ torsion angles of the polypeptide backbone on the backdrop of the "allowed" or expected values. The allowed areas of the Ramachandran plot differ very significantly between glycines and the other amino acids, and to a lesser extent also between different amino acids [26]. The φ/ψ angles have a strong validation power because their values are usually not restrained in the refinement, unless a special torsion-angle-refinement method is used [14]. It was originally suggested that more than 90% of the φ/ψ pairs should be found in the most favored areas of the plots [27], although these areas were later redefined and the more recent estimate is that over 98% of the angles should be found in them [28]. On the other hand, the presence of φ/ψ conformations in the disallowed areas may indicate local problems with the structure. However, it is not unusual to occasionally find very strained torsion angles in some parts of proteins, particularly if the corresponding side chains are involved in multiple contacts and if the distortion has a functional significance. The correctness of the interpretation of such areas will ultimately rely on the appearance of the electron density maps.

The third main-chain conformational parameter of proteins, the peptide torsion angle ω, is expected to be close to 180° or exceptionally to 0° for *cis*-peptides (the latter seen more frequently than originally thought [29]). *Cis*-peptides are most commonly observed in the Xxx-Pro peptides, but are occasionally seen in peptide bonds connecting other types of amino acids as well (estimated at 1 *cis* bond per ~2000 residues [30]). However, due to their expected rarity such bonds are sometimes modeled as *trans* and the incorrect assignment is not detected during structure refinement. A recent global analysis of the structures deposited in the PDB found more than 4000 instances of potential *trans-cis* flips that could be corrected based on stereochemical considerations alone [31]. Thus the assumption that all non-Xxx-Pro peptides should be necessarily in *trans* configuration should be applied with care. The opposite, however, may also be true, since another recent analysis of the PDB indicated a fairly high rate of the presence of non-Xxx-Pro *cis* peptides in structures refined at moderate-to-low resolution [32]. A majority of these nonstandard peptides may represent fitting errors [32].

The peptide planes are usually under very tight stereochemical restraints, although there is growing evidence that deviations of ±20° from strict planarity should not be treated as abnormal if strongly supported by high-resolution electron density [20, 33, 34]). Whereas it was recently proposed that the ω angles could be more tightly restrained even in atomic-resolution structures without deteriorating the model [35], that proposal has been already refuted [36]. Unreasonably tight peptide planarity may lead to artificial distortions of the neighboring φ/ψ angles in the Ramachandran plot. On the other hand, some structures that have been deposited in the PDB quite recently exhibit deviations from peptide planarity exceeding 30° (4oiw, 3ja8, 2j6r, 4zkt, 3j9p, etc.). Models containing such violations should be regarded as highly suspicious, unless created on the basis of atomic-resolution data with absolutely clear electron density maps. For example, structure 4oiw includes several peptides in which the ω torsion angle deviates from planarity by as much as 40°. While some of such peptides are located in the loops with poor electron density and thus are clearly wrong, others are in quite good density, but do not represent a proper fit and may be due to the application of too weak planarity restraints. However, since such problems are highly localized, they may not be critical to the interpretation of the affected structures unless they are found in areas of high significance, such as the active sites of enzymes. On the other hand, the low-resolution structure 5dsv includes almost 200 ω torsion angles deviating from planarity by more than 30° (some by as much as 90°), thus it represents a clear case of improper use of planarity restraints during refinement, or outright wrong modeling.

4 Tools for Validation of Macromolecular Structures

The need to validate the correctness of macromolecular structures was pointed out soon after the progress in the development of computational hardware and software made refinement of large structures possible [37]. Some structure validation tools were built directly into computer programs such as FRODO [38] and O [39] that allowed manual or automated fitting of models to electron density. Programs such as WHAT IF [40], although originally written as general display and modeling tools, contained a number of routines that allowed assessment of the departure of geometrical parameters of macromolecular models from the expected values. A variant of WHAT IF, called WHAT CHECK [41, 42], was specially designed to flag potential deviations of models from the expected geometry. Analysis of the geometry and stereochemical validation of the models is equally applicable to structures determined by either crystallography or NMR, whereas assessment of the agreement of the models with the primary experimental data is, of course, quite different for these two techniques, and is basically absent in the NMR field.

For a number of years the main structure validation tool was the program PROCHECK [27]. Its subroutines analyzed the geometry of protein structures and compared them to other well-refined structures determined at comparable resolution. In addition to analyzing the Ramachandran plots (see above), the programs analyzed the planarity of peptide bonds, bad non-bonded interactions, distortions of the geometry around the Cα atoms, energies of hydrogen bonds, and the departure of the side chain χ torsion angles from expected values. The graphical output of the program allowed its users to quickly identify the most problematic areas. Thus PROCHECK has been extensively used as a tool for guiding the process of structure refinement and rebuilding. An example of a Ramachandran plot for a comparatively large protein structure refined at atomic resolution of 1 Å is shown in Fig. 1a. However, since the database serving this program suite was quite limited as compared to the current situation, PROCHECK is now considered to be obsolete. Nevertheless, since it was used in the past to verify many structures that still serve to explain biologically relevant data, it is still useful to understand its advantages and limitations.

A newer approach to validation of macromolecular structures makes use of various versions of the program suite MolProbity [44–46], used as a web server, in a stand-alone mode, or as part of other program systems, such as Phenix [47]. In addition to analyzing the

a)

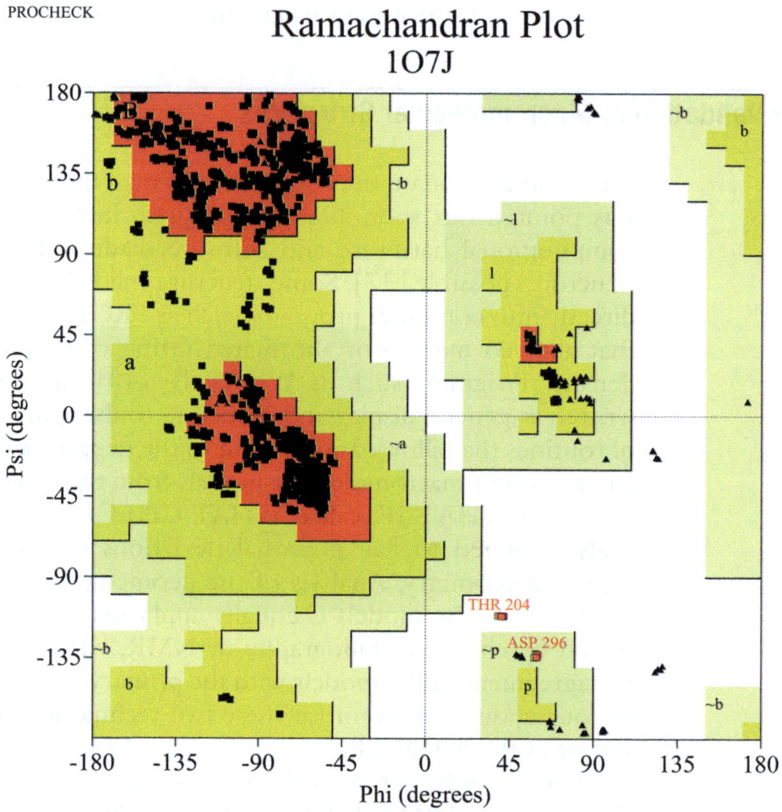

Plot statistics

Residues in most favoured regions [A,B,L]	1013	90.8%
Residues in additional allowed regions [a,b,l,p]	95	8.5%
Residues in generously allowed regions [~a,~b,~l,~p]	4	0.4%
Residues in disallowed regions	4	0.4%
	----	------
Number of non-glycine and non-proline residues	1116	100.0%
Number of end-residues (excl. Gly and Pro)	8	
Number of glycine residues (shown as triangles)	128	
Number of proline residues	48	

Total number of residues	1300	

Based on an analysis of 118 structures of resolution of at least 2.0 Angstroms and R-factor no greater than 20%, a good quality model would be expected to have over 90% in the most favoured regions.

Fig. 1 Ramachandran plots for the tetrameric molecule of *Erwinia chrysanthemi* L-asparaginase [43]. The structure was refined with data extending to 1 Å resolution. (**a**) A plot prepared with PROCHECK [27], showing that almost all φ/ψ torsion angles are found in the favored or additionally allowed regions. However, Thr204 in each subunit is marked as being in a disallowed region, and Asp296 in a generously allowed area. (The unusual conformation of these residues is supported by the original electron density.)

b)

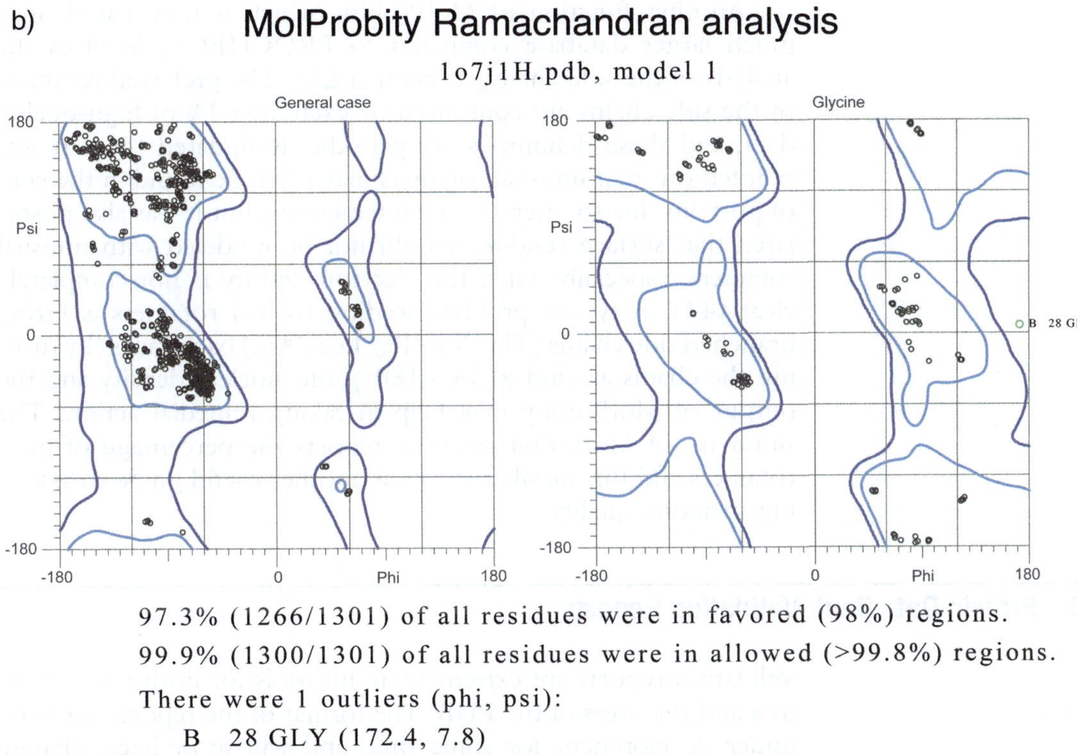

Fig. 1 (continued) (**b**) An analogous plot prepared with MolProbity [44]. The two residues outside of the allowed region in PROCHECK are no longer considered to be outliers, whereas GlyB28 is now marked as such

geometrical parameters discussed above, MolProbity relies very heavily on the analysis of interatomic clashes. For that purpose the program calculates the positions of the hydrogen atoms and adds them to the coordinate files (sometimes replacing the riding H atoms that might already be present there). The side chains of Asn, Gln, and His are subjects of special attention aimed at verification of the most likely orientation of their O/N side chains, or the proper orientation of the imidazole rings. Once these residues have been placed in their most likely orientations and the H atoms have been added, all-atom contacts are analyzed in detail. Close interatomic distances and clashes are shown graphically and in printouts, providing information useful for rebuilding the offending areas, or at least raising a red flag for the users of deposited structures. The program provides plots of the Ramachandran angles, using a much more extensive database than the one utilized by PROCHECK. The *most favored* areas are based on the analysis of quality-filtered data for as many as 100,000 residues, 98% of which are found therein, whereas the *allowed* regions encompass 99.95% of good reference data. This change of definitions of the allowed regions leads to some differences in the interpretation of the Ramachandran plots in comparison with PROCHECK (Fig. 1b). The number of residues in the favored regions of the Ramachandran plot, as well as of the outliers, is listed by the program.

Another function of MolProbity, which is now based on a much larger database compared to PROCHECK, involves the analysis of the side-chain χ torsion angles. The preferred rotamers of the side chains are contoured by excluding 1% of high-quality data, and these definitions are periodically updated [44]. It was pointed out that unusual rotamers may often be found in the core of proteins due to inter-residue interactions, but it was also postulated that surface residues should not be modeled with unusual rotamers, especially when the electron density is not completely clear [44]. A typical problem leading to bad rotamers is fitting branched side chains (Thr/Val/Ile/Leu/Arg) backwards (by turning the chains around $\chi 1$ by ~180°) into unclear density and the output of MolProbity may help in taking remedial action. The summary of MolProbity analysis reports the percentage of poor rotamers and this number serves as another useful guide to assessing structure quality.

5 Protein Data Bank Validation Reports

Validation reports are extremely useful tools for both the depositors and the users of the PDB. The format of the reports has been under development for some time and will most likely change again in the future. The current standard is based on the recommendations of the wwPDB X-ray Validation Task Force [26]. Submission of a validation report is now required by a number of scientific journals as a companion piece to manuscripts that describe crystal structures. The availability of validation reports is expected to help reviewers in assessing whether the structure discussed in a manuscript is reliable and of sufficiently high quality. Of course, since the report had been made available to the depositors first, it should have been scrutinized prior to the final submission of the structure to the PDB. Sadly, as shown below, that seems not always to be the case.

This chapter describes PDB validation reports as applied to crystal structures only—some aspects of the validation process are not applicable to models determined by NMR spectroscopy. The reports are based on the output of several programs and refer to the restraint libraries that are clearly identified in the output. Bond distances and angles in proteins are checked against the latest version of the Engh and Huber library [19], although—quite surprisingly—some refinement programs, such as REFMAC, have been for decades (sic!) using the obsolete early version [18]. However, the PDB validation report takes no account of the conformation-dependent libraries (CDL) [21]. For nucleic acids, the reference library is provided by Parkinson et al. [23] plus two other of the same vintage [48, 49]. The geometrical parameters are checked principally by a recent version of MolProbity [44], whereas the agreement of the model with the diffraction data is monitored by procedures introduced by Kleywegt et al. [50] in their Uppsala Electron-Density Server (EDS; http://eds.bmc.uu.se/eds/).

The first page of the validation report (and the PDB web page for that structure) includes a graphical summary of the "quality indicators" of the structure. While helpful in providing a first-glance impression, this graph provides only fragmentary information about how the structure in question compares to other structures present in the PDB. Five different quality metrics are shown in colors ranging from deep red (poor) to blue (excellent). Black boxes indicate how the structure compares to all structures in the PDB, whereas open boxes provide a more meaningful comparison with structures at comparable resolution.

Examples of such graphs for three recently deposited medium-resolution structures are shown in Fig. 2. Structure 5duy is considered to be quite acceptable, 4xok shows a number of problems, whereas 4ru7 (described in the original publication [51] but now made obsolete and replaced first by 5fc0, and later by 5k1y) exhibits some grave problems that put its validity into question.

Free *R*-factor, in use for practically all structures since its introduction into crystallographic practice in 1992 [52], depicts the relative deviation of the observed and calculated structure factors (in analogy to the conventional *R*-factor) for a subset of (5–10% or ~1000) reflections never used in structure refinement. R_{free} should be higher (by ~4–7 percent points) than *R*. If it is much higher, it indicates serious problems with the model (without pointing out the errors, however), or overinterpretation of the experimental data by too many (unjustified) model parameters. If the $R_{free} - R$ gap is too small, it strongly suggests a compromised test data set (e.g., used, deliberately or not, at some stage in the refinement) and questions the validity of the R_{free} test in such a case. The expected value of R_{free} depends very much on the resolution of the data sets and its interpretation was discussed in detail previously [53]. In exceptionally good cases R_{free} can be as low as 10% for the best-refined structures using ultrahigh-resolution data; typically it is about 20–25% for medium-resolution structures, but should not exceed 30% even for structures refined against low-resolution data. Thus the value of 19.3% for 5duy (refined at 2.12 Å resolution) is in the expected range, it is rather high (32.5%) for 4xok (2.2 Å), and is absolutely random (58.7%) for 4ru7 (2.97 Å; now obsoleted and replaced by 5k1y), suggesting that in the latter case the deposited structure factors might not have corresponded to the coordinates.

R_{free} is a better (even if not ideal) measure of model quality than the standard crystallographic *R* (residual) factor, calculated as $R = 100 \cdot \Sigma||F_o| - |F_c||/\Sigma|F_o|$. The problem with *R* is that it has poor statistical properties and will essentially go down on any, even unreasonable, model expansion. A much better metric would be

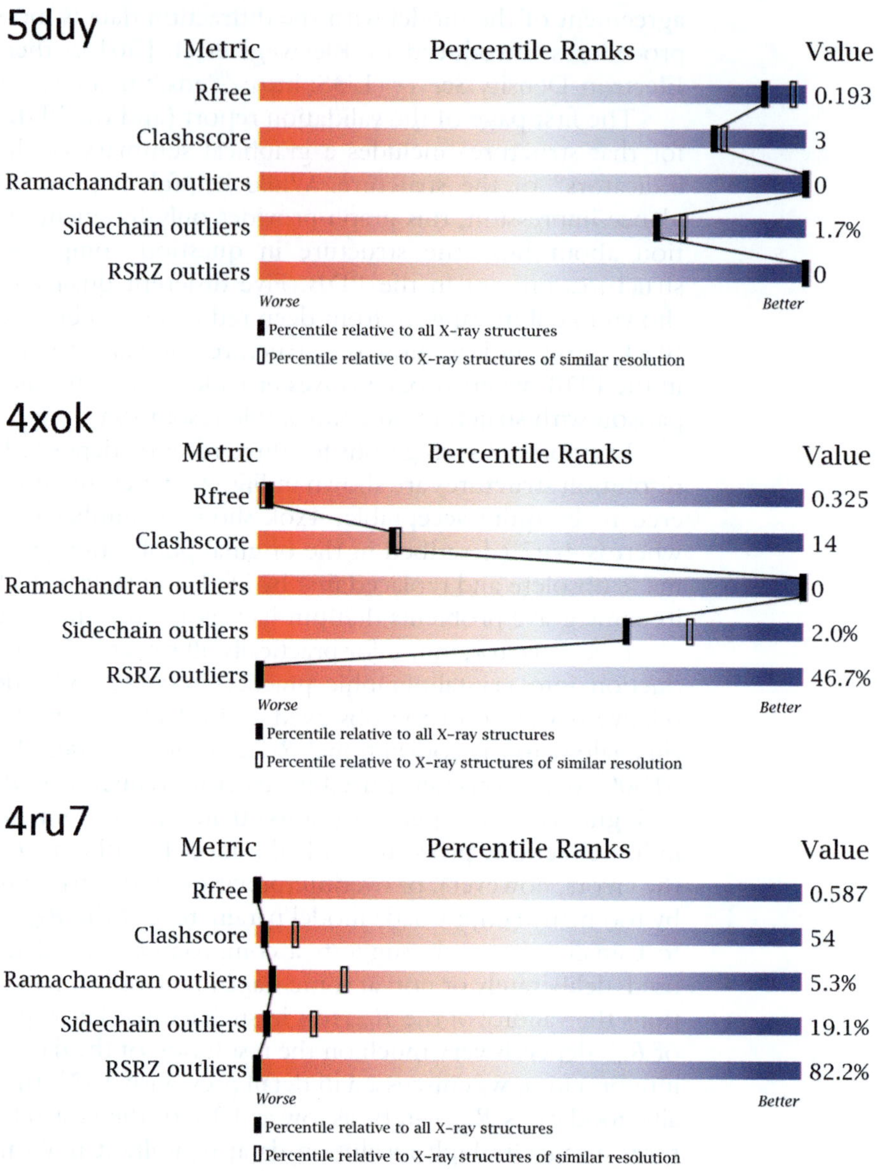

Fig. 2 Summary graphs from the PDB validation reports for three medium-resolution structures, presented here only as examples. The deposit 4ru7 has by now been obsoleted and replaced, but is shown here since it corresponds directly to the original publication [51]

the weighted (note the use of weights, w) $wR2$ factor ($wR2 = 100\cdot[\Sigma(|F_o| - |F_c|)^2/\Sigma|F_o|^2]^{1/2}$) but it is seldom used in macromolecular crystallography.

The second line of the "quality" graph summarizes a parameter called clashscore, introduced in MolProbity. This score is related to the number of interatomic distances that are shorter by more than 0.4 Å than the sum of the van der Waals radii and is expressed as the number of such close contacts per 1000 atoms in the structure. Of course, interatomic clashes do not have any physical meaning (after all, atoms cannot interpenetrate), but since the models cannot provide an error-free description of the structure, some (small) number of such violations is inevitable. The clashscore of 3 for 5duy is almost exactly in the typical range (which seems to be almost independent of resolution). For 4xok this parameter is definitely much worse than the average, and the clashscore of 54 for 4ru7 is certainly unacceptable.

The next bar in the summary picture shows the number of Ramachandran outliers. There are no such violations in 5duy and 4xok, whereas the presence of 5.3% outliers in the case of 4ru7 indicates very poor geometry of the main chain of the model. This number represents the percentage of all residues in the structure that are found in the disallowed areas of the plot (Fig. 1b). It needs to be stressed that whereas the presence of violations (as in 4ru7) is suggestive of problems with the structure, their absence (as in 4xok) does not necessarily prove that the model is of high quality.

Side-chain outliers are defined as the percentage of side chains with a combination of χ torsion angles that are not similar to any combination preferred for that given residue type, calculated in the same way as for the Ramachandran violations. Clearly, some residues will have unusual torsion angles due to packing constraints, but there is no real justification for having outliers in the surface areas where the electron density is relatively weak; thus only proper rotamers should be assumed to be present there. The percentage of side-chain outliers lower or equal to 2% in 5duy and 4xok indicates that the residues were generally modeled with their preferred rotamers and were not distorted during the refinement process. However, more than 19% of outliers found in 4ru7 is a clear indication that the model was allowed to depart very far from a typical conformation, reiterating a real question about the quality of this model.

The final line of the summary plot is related to the number of residues that do not fit well the corresponding electron density. RSR (real space *R* factor), calculated for each residue separately, is a measure of the quality of fit between the coordinates of a residue and the corresponding electron density. The RSR Z score (RSRZ) compares the fit to electron density for each residue with the mean and standard deviation of the fit for all such residues in a similar resolution bin. A residue is considered to be an RSRZ outlier if its RSR is more than 2 standard deviations worse than the average for this residue type, and the plotted score corresponds to the percentage of residues that are considered to be outliers based on this criterion. A well-refined protein structure may show no RSRZ outliers at all, whereas the presence of poorly defined regions in the structure (for example, due to local disorder that was still modeled) will lead to an increase of this parameter. In practice, the number of RSRZ outliers is often correlated with higher-than-average R_{free}. This can be seen in the comparison of the structures in Fig. 2, where 5duy has no RSRZ outliers, whereas almost half of the residues in 4xok do not seem to fit the electron density. Very few residues in 4ru7 appear to fit the electron density, again pointing to serious problems with the structure factor file.

The next summary graph contains two lines for each polypeptide (or polynucleotide) chain present in the coordinate file. The lower bar is colored green, yellow, orange, or red to denote deviation of a particular residue from 0 to 3 (or more) stereochemical quality standards (see below). The top line, if present and colored red, indicates that this segment of the chain exhibits poor fit to the electron density. This plot is followed by a table showing similar outliers for non-polymeric components (such as buffer molecules) of the structure.

As discussed above, a quick glance at the structure quality diagram may be sufficient for the first impression regarding the quality of a PDB deposit, but clean plots are not necessarily sufficient to affirm that the model is excellent. A graph showing large departures from the expected average values, however, should immediately alert the user of such a file (and especially its depositor!) of potentially serious problems.

The next section of the validation report deals with the composition of the entry and is useful for checking the sequence of the macromolecule against relevant databases, what kinds of expression tags are present, and how many atoms are modeled with zero occupancy and/or in alternate conformations.

The third part of the validation report provides the details behind the chain quality plot found on the summary page. It includes residue-by-residue plots colored as defined above for the presence of geometric outliers, with red dots denoting poor fit to electron density (RSRZ > 2). Residues to be present in the sample but not included in the model (most likely because of disorder and/or poor/absent electron density) are marked in gray. Stretches of residues with no apparent problems are marked by a green line.

A table of data and refinement statistics, included in the fourth part of the report, provides selected statistics derived directly from the deposit, or recalculated by the PDB. This table is worth attention, especially if the numbers claimed by the authors are significantly different from the ones recalculated during the validation process. Differences in the resolution limit may be due to the inclusion in the deposited structure factor files of shells that were not actually used in refinement (quite common at low resolution), but very large deviations (such as 1.41 Å computed with EDS vs. 2.98 Å claimed by the depositors of 4ru7) are certainly worth close examination. The values of R_{merge} or R_{sym} (if cited by the depositors) give some indication of the internal consistency of the diffraction data, whereas $\langle I/\sigma(I) \rangle$, computed by the program Xtriage, indicates whether statistically significant observations were present in the outermost data shell. If the value of $\langle I/\sigma(I) \rangle$ is much less than ~2.0, this indicates that the claimed resolution limits may have been overly optimistic.

The values of the refinement R and R_{free} factors are listed as claimed by the depositors, and as recalculated during validation. Some differences, such as R of 13.9% claimed by depositors of 5duy and 15.5% calculated during validation, are not unexpected, due to different assumptions used by different computer programs, e.g., Phenix or REFMAC. The corresponding numbers for 4xok, 30.3% and 30.1%, were most likely calculated by the same software. However, the difference between 21.0% and 57.1% found in the validation report of 4ru7 is a clear indication that the structure factor file does not correspond to the coordinates present in this deposit.

Other parameters listed in that table are helpful in deciding whether the diffraction data could have been twinned or if translational non-crystallographic symmetry is present, and show how the mean B factor for all atoms compares with the Wilson B factor for the diffraction data.

The fifth section of the report is particularly relevant to the subject of this chapter, since it provides a detailed description of the geometric parameters of the modeled coordinates. The four stereochemical criteria are bond lengths, bond angles, chirality (where present), and planarity (where present). All bond lengths and angles with individual Z scores larger than 5 are listed, together with the RMSZ scores for the whole chain. Chiral center volumes (signed volume of the tetrahedron formed by the four substituents of an sp^3 atom) that differ significantly from the expected values are tagged, and potentially planar groups which appear to be nonplanar are similarly marked. Close contacts are evaluated for all-atom models that include hydrogen atoms (either as already present in the deposit or added computationally) and the ones for which the distance is shorter than the sum of their van der Waals radii minus 0.4 Å are highlighted. Additional analysis presents the deviations of the Ramachandran main-chain torsion angles and χ side chain angles from the allowed (or most likely) targets, and summarizes the percentile values with respect to all crystal structures, or structures at similar resolution. A percentile value of 100% means that there are no outliers, whereas a lower number indicates the percentage of PDB structures with more problems (the value of 0% would portend the worst value of that parameter in the entire PDB). The Ramachandran and non-rotameric outliers are explicitly identified, together with the candidate Asp/Gln/His residues that might need flipping. The latter information might be particularly useful if these residues are expected to be important for the function of the protein.

The last section of the validation report deals with the agreement between the coordinates and the electron density maps. The RSRZ values are computed individually for each chain and the number of RSRZ outliers (RSRZ > 2) is given, together with the percentage scores relative to all crystal structures, as well as structures at similar resolution. Thus for 5duy, with no outliers, the percentage is 100%, whereas it is 1% for all entries and 0% for entries at similar resolution for 4xok. The numbers are all 0% for 4ru7.

This section of the report also contains an analysis of any nonstandard residues in proteins, polynucleotides and carbohydrates, as well as for their ligands. An important parameter for the assessment of ligand quality is LLDF. It compares the electron density of the ligand with the electron density of the neighboring atoms of the macromolecule. Values of LLDF that exceed 2.0 indicate potential problems, thus the conformation of such ligands (or even their presence) should be carefully scrutinized. Since the PDB validation report does not provide information which would indicate whether the coordination of metal ions (if present) is plausible, it is worthwhile to obtain such information from a dedicated server CheckMyMetal (http://csgid.org/csgid/metal_sites/) [54].

6 Summary and Conclusions

The stereochemistry of properly determined macromolecular structures cannot deviate very much from standard values of bond lengths and angles, as well as from acceptable torsion angles. Validation is crucial in assuring good quality of the models and such tests should be routinely run during structure refinement. The standard arbiter should be the official PDB validation report. It is recommended that parameters such as the number of residues deviating from favored or allowed regions in the Ramachandran plot should be quoted according to that report. Different programs may report such parameters slightly differently but deferring to validation by the PDB will assure some level of conformity. Of course, depositors should pay real attention to the output of the validation reports; as shown here, this is not always the case. The users of macromolecular coordinates might also benefit from taking a look at such reports before using the data; it is always worthwhile to know how reliable a given structures is, and to adjust the level of confidence accordingly, especially if the user is interested in the atomic details of some specific areas, such as enzyme active sites or intermolecular interfaces. While not perfect, validation reports do contain a lot of useful and crucial information that should never be disregarded. Ultimately, if there is any doubt or controversy, the electron density map should be the final arbiter.

References

1. Allen FH (2002) The Cambridge Structural Database: a quarter of a million crystal structures and rising. Acta Crystallogr B 58:380–388
2. Sheldrick GM (1990) Phase annealing in SHELX-90: direct methods for larger structures. Acta Crystallogr A 46:467–473
3. Berman HM, Westbrook J, Feng Z et al (2000) The Protein Data Bank. Nucleic Acids Res 28:235–242
4. Pauling L, Corey RB, Branson HR (1951) The structure of proteins: two hydrogen-bonded helical configurations of the polypeptide chain. Proc Natl Acad Sci U S A 37:205–211
5. Pauling L, Corey RB (1951) The pleated sheet, a new layer configuration of polypeptide chains. Proc Natl Acad Sci U S A 37:251–256
6. Pauling L, Corey RB (1953) Stable configurations of polypeptide chains. Proc R Soc Lond B 141:21–33
7. Kendrew JC, Bodo G, Dintzis HM et al (1958) A three-dimensional model of the myoglobin molecule obtained by X-ray analysis. Nature 181:662–666
8. Perutz MF, Rossmann MG, Cullis AF et al (1960) Structure of haemoglobin: a three-dimensional Fourier synthesis at 5.5-Å resolution, obtained by X-ray analysis. Nature 185:416–421
9. Blake CC, Fenn RH, North AC et al (1962) Structure of lysozyme. A Fourier map of the electron density at 6 Å resolution obtained by X-ray diffraction. Nature 196:1173–1176
10. Watson JD, Crick FH (1953) Molecular structure of nucleic acids; a structure for deoxyribose nucleic acid. Nature 171:737–738
11. Evans PR (2007) An introduction to stereochemical restraints. Acta Crystallogr D Biol Crystallogr 63:58–61
12. Wlodawer A, Hendrickson WA (1982) A procedure for joint refinement of macromolecular structures with X-ray and neutron diffraction data from single crystals. Acta Crystallogr A 38:239–247
13. Hendrickson WA (1985) Stereochemically restrained refinement of macromolecular structures. Methods Enzymol 115:252–270
14. Brünger AT, Adams PD, Clore GM et al (1998) Crystallography and NMR system: a new software suite for macromolecular structure determination. Acta Crystallogr D Biol Crystallogr 54:905–921
15. Sheldrick GM, Schneider TR (1997) SHELXL: high-resolution refinement. Methods Enzymol 277:319–343
16. Murshudov GN, Skubak P, Lebedev AA et al (2011) REFMAC5 for the refinement of macromolecular crystal structures. Acta Crystallogr D Biol Crystallogr 67:355–367
17. Adams PD, Grosse-Kunstleve RW, Hung LW et al (2002) PHENIX: building new software for automated crystallographic structure determination. Acta Crystallogr D Biol Crystallogr 58:1948–1954
18. Engh R, Huber R (1991) Accurate bond and angle parameters for X-ray protein-structure refinement. Acta Crystallogr A 47:392–400
19. Engh RA, Huber R (2001) International tables for crystallography. Kluwer Academic Publishers, Dordrecht, pp 382–392
20. Jaskolski M, Gilski M, Dauter Z, Wlodawer A (2007) Stereochemical restraints revisited: how accurate are refinement targets and how much should protein structures be allowed to deviate from them? Acta Crystallogr D Biol Crystallogr 63:611–620
21. Tronrud DE, Karplus PA (2011) A conformation-dependent stereochemical library improves crystallographic refinement even at atomic resolution. Acta Crystallogr D Biol Crystallogr 67:699–706
22. Malinska M, Dauter M, Kowiel M et al (2015) Protonation and geometry of histidine rings. Acta Crystallogr D Biol Crystallogr 71:1444–1454
23. Parkinson G, Vojtechovsky J, Clowney L et al (1996) New parameters for the refinement of nucleic acid-containing structures. Acta Crystallogr D Biol Crystallogr 52:57–64
24. Brzezinski K, Brzuszkiewicz A, Dauter M et al (2011) High regularity of Z-DNA revealed by ultra high-resolution crystal structure at 0.55 Å. Nucleic Acids Res 39:6238–6248
25. Ramakrishnan C, Ramachandran GN (1965) Stereochemical criteria for polypeptide and protein chain conformations: II. Allowed conformation for a pair of peptide units. Biophys J 5:909–933
26. Read RJ, Adams PD, Arendall WB III et al (2011) A new generation of crystallographic validation tools for the protein data bank. Structure 19:1395–1412
27. Laskowski RA, MacArthur MW, Moss DS et al (1993) PROCHECK: program to check the stereochemical quality of protein structures. J Appl Crystallogr 26:283–291
28. Kleywegt GJ, Jones TA (1996) Phi/psi-chology: Ramachandran revisited. Structure 4:1395–1400

29. Weiss MS, Hilgenfeld R (1997) On the use of the merging R factor as a quality indicator for X-ray data. J Appl Crystallogr 30:203–205
30. Stewart DE, Sarkar A, Wampler JE (1990) Occurrence and role of cis peptide bonds in protein structures. J Mol Biol 214:253–260
31. Touw WG, Joosten RP, Vriend G (2015) Detection of trans-cis flips and peptide-plane flips in protein structures. Acta Crystallogr D Biol Crystallogr 71:1604–1614
32. Croll TI (2015) The rate of cis-trans conformation errors is increasing in low-resolution crystal structures. Acta Crystallogr D Biol Crystallogr 71:706–709
33. EU 3-D Validation Network (1998) Who checks the checkers? Four validation tools applied to eight atomic resolution structures. J Mol Biol 276:417–436
34. Addlagatta A, Krzywda S, Czapinska H et al (2001) Ultrahigh-resolution structure of a BPTI mutant. Acta Crystallogr D Biol Crystallogr 57:649–663
35. Chellapa GD, Rose GD (2015) On interpretation of protein X-ray structures: planarity of the peptide unit. Proteins 83:1687–1692
36. Brereton AE, Karplus PA (2016) On the reliability of peptide nonplanarity seen in ultrahigh resolution crystal structures. Protein Sci 25:926–932
37. Brändén C-I, Jones TA (1990) Between objectivity and subjectivity. Nature 343:687–689
38. Jones TA (1985) Interactive computer graphics: FRODO. Methods Enzymol 115:157–171
39. Jones TA, Zou JY, Cowan S et al (1991) Improved methods for building protein models in electron density maps and location of errors in these models. Acta Crystallogr A 47:110–119
40. Vriend G (1990) WHAT IF: a molecular modelling and drug design program. J Mol Graph 8:52–56
41. Hooft RW, Vriend G, Sander C et al (1996) Errors in protein structures. Nature 381:272
42. Nabuurs S, Spronk C, Krieger E et al (2004) Computational mechanical chemistry for drug discovery. Marcel Dekker, New York and Basel, pp 387–403
43. Lubkowski J, Dauter M, Aghaiypour K et al (2003) Atomic resolution structure of Erwinia chrysanthemi L-asparaginase. Acta Crystallogr D Biol Crystallogr 59:84–92
44. Chen VB, Arendall WB III, Headd JJ et al (2010) MolProbity: all-atom structure validation for macromolecular crystallography. Acta Crystallogr D Biol Crystallogr 66:12–21
45. Davis IW, Murray LW, Richardson JS et al (2004) MOLPROBITY: structure validation and all-atom contact analysis for nucleic acids and their complexes. Nucleic Acids Res 32:W615–W619
46. Davis IW, Leaver-Fay A, Chen VB et al (2007) MolProbity: all-atom contacts and structure validation for proteins and nucleic acids. Nucleic Acids Res 35:W375–W383
47. Adams PD, Afonine PV, Bunkoczi G et al (2010) PHENIX: a comprehensive python-based system for macromolecular structure solution. Acta Crystallogr D Biol Crystallogr 66:213–221
48. Clowney L, Jain SC, Srinivasan A et al (1996) Geometric parameters in nucleic acids: nitrogenous bases. J Am Chem Soc 118:509–518
49. Gelbin A, Schneider B, Clowney L et al (1996) Geometric parameters in nucleic acids: sugar and phosphate constituents. J Am Chem Soc 118:519–529
50. Kleywegt GJ, Harris MR, Zou JY et al (2004) The Uppsala Electron-Density Server. Acta Crystallogr D Biol Crystallogr 60:2240–2249
51. Schumacher MA, Tonthat NK, Lee J et al (2015) Structures of archaeal DNA segregation machinery reveal bacterial and eukaryotic linkages. Science 349:1120–1124
52. Brünger AT (1992) The free R value: a novel statistical quantity for assessing the accuracy of crystal structures. Nature 355:472–474
53. Wlodawer A, Minor W, Dauter Z et al (2008) Protein crystallography for non-crystallographers or how to get the best (but not more) from the published macromolecular structures. FEBS J 275:1–21
54. Zheng H, Chordia MD, Cooper DR et al (2014) Validation of metal-binding sites in macromolecular structures with the CheckMyMetal web server. Nat Protoc 9:156–170

Chapter 25

Validation of Protein–Ligand Crystal Structure Models: Small Molecule and Peptide Ligands

Edwin Pozharski, Marc C. Deller, and Bernhard Rupp

Abstract

Models of target proteins in complex with small molecule ligands or peptide ligands are of significant interest to the biomedical research community. Structure-guided lead discovery and structure-based drug design make extensive use of such models. The bound ligands comprise only a small fraction of the total X-ray scattering mass, and therefore particular care must be taken to properly validate the atomic model of the ligand as experimental data can often be scarce. The ligand model must be validated against both the primary experimental data and the local environment, specifically: (1) the primary evidence in the form of the electron density, (2) examined for reasonable stereochemistry, and (3) the chemical plausibility of the binding interactions must be inspected. Tools that assist the researcher in the validation process are presented.

Key words Crystal structure, Protein–ligand complexes, Structure model validation, Small molecule ligands, Peptide ligands

1 Introduction

Accurate atomic models of protein–ligand complexes, as deposited in the Protein Data Bank (PDB [1]), are essential for many computational methods including ligand docking and structure-based drug design [2–4]. Rigorous validation of the quality of the atomic model is vital to ensure that these computational methods produce meaningful results [5–8].

Bound ligands, primarily small molecules and short peptides, DNA fragments, or related analogues, have a small mass relative to the protein target, and hence make proportionally minor contribution to the net X-ray scattering intensities. *Global* reciprocal space statistics, or indicators such as the often cited R-values, therefore contain little information specific to the ligand. Additionally, overall real space measures such as r.m.s. deviations (RMSD) from target values, particularly for bond lengths and bond angles, are also often quoted for the structure as a whole, and generally do not specifically reflect the quality of the ligand. Furthermore, ligands

Alexander Wlodawer et al. (eds.), *Protein Crystallography: Methods and Protocols*, Methods in Molecular Biology, vol. 1607, DOI 10.1007/978-1-4939-7000-1_25, © Springer Science+Business Media LLC 2017

bound to a protein are susceptible to a whole host of complicating factors including partial occupancies, enhanced mobility, conformational flexibility, and poor protein stability [9]. All of these factors act to further reduce the discrete scattering contributions of the ligand. It is therefore necessary to conduct a careful inspection of the ligand at the *local* level using *real space* quality indicators. The primary evaluation criteria that should be inspected include:

- The fit of the ligand model to the primary data in the form of the electron density.
- The compliance of the ligand model with prior expectations of reasonable stereochemistry.
- Plausible chemistry within the ligand–protein binding site interactions.

The ability to conduct strict evidence-based validation has revealed a number of instances in which ligands purportedly bound to a target protein are insufficiently supported by the primary evidence (i.e., electron density) [7, 10]. Although there may be other lines of supporting evidence (e.g., biochemical data, database annotations, functional assignments, and protein fold) for the binding of a particular ligand to the protein under study, it is essential to ensure that the placement and conformation (i.e., the pose) of any ligand in a structure model is supported by the electron density. Furthermore, it is essential that the ligand (and the protein environment to which it binds) adhere to well established guidelines of plausible stereochemistry, and suggest meaningful ligand–protein contacts.

Occasionally the quality of the electron density does not allow complete determination of the specific ligand pose, while the presence of some bound chemical entity (frequently disordered) is evident. There is currently no consensus within the structural biology community on how such cases should be described when depositing a structural model in the PDB. Therefore, without the context of the electron density, one should be aware of possible over-interpretation of ligands found in deposited structures.

2 Examination Against the Primary Evidence

The primary evidence supporting the atomic model of a crystal structure is the electron density. The electron density is obtained via Fourier reconstruction from the diffraction data (several tens to hundreds of thousands of X-ray reflection intensities) and the phase angle of each reflection [11, 12].

It is often the case that experimental phases are not used when solving the structure of a protein in complex with a small molecule or peptide ligand. In many cases the preferred method is Molecular

Replacement (MR) in which the structure of an apo-protein is placed in the crystal structure [13, 14]. These MR techniques are traditionally followed by manual adjustments of the model using local real space refinement on a graphics workstation and further rounds of computational reciprocal space refinement against the experimental diffraction data. While the traditional electron density maps obtained by this process (e.g., $2mF_o - DF_c$) are inherently biased, difference electron density maps (e.g., $mF_o - DF_c$) are almost always sufficiently clear to indicate both the presence and location of a bona fide ligand. Therefore, it is essential that difference electron density maps, specifically those calculated with the ligand omitted from the model, are consulted during the model building and ligand validation procedure.

2.1 Displaying Electron Density

Display programs such as *COOT* [8, 15] or *PyMol* [16] allow for the downloading of all the necessary data for electron density reconstruction from data repositories and web sites such as the wwPDB [17], *PDB_REDO* [18, 19], or EDS [20]. The data downloaded from these sites contains the appropriately weighted structure factor amplitudes and phases for each observed X-ray reflection. The reconstructed electron density is displayed together with the atom positions recorded in the associated model file, identified with a unique, 4-symbol PDB ID. The electron density is generally displayed as a three-dimensional contour map of density values, scaled in levels of standard deviations (sigma, σ) from the electron density mean. The standard $2mF_o - DF_c$ map (Fig. 1) is reconstructed from maximum likelihood coefficients obtained during global reciprocal space maximum posterior refinement of the model against the diffraction data [10, 22].

2.2 Real Space Correlation and Real Space R-Value

The fit of the model to the electron density is quantified by the Real Space Correlation Coefficient (RSCC) [23], the Real Space R-Value (RSR) [24], or more complex composite measures such as the Local Ligand Density Fit (LLDF) [19], just to mention a few. These values can be calculated on a local, per-residue basis, and can be obtained from the PDB validation report (http://wwpdb-validation.wwpdb.org/validservice). Additionally, many of these real space values can be displayed on a per-residue basis in *COOT*. Many software packages are available to assist the researcher in the analysis of the real space fit of the ligand to the electron density and several are highlighted in Table 1.

The main output of most real space fit algorithms is a metric that reflects the fit of the model to the electron density. For example, a perfect fit of the model to the electron density would result in an RSCC of 1.0, and values between 0.8 and 1.0 are commonly considered as good. The lower the RSCC, the more uncertainty in the positioning of the model with respect to the electron density. The RSR value is somewhat scale dependent [23], and in an

Table 1
Protein–ligand validation software

Package	Description	URL	Reference
Twilight	Small molecule and peptide ligand validation	http://bit.ly/1shcwu4	[7, 10, 21]
COOT	Molecular graphics package for model building and validation	http://bit.ly/1wJNWBy	[8, 15]
MolProbity	All atom clash scores and other validation statistics	http://bit.ly/1o1PuHW	[25]
VHELIBS	Fit of ligands and binding sites	http://bit.ly/1t5szxl	[26]
ValLigURL server	Compare conformations of ligands in the PDB	http://bit.ly/1v80yoS	[27]
MotiveValidator and *ValidatorDB*	Interactive web-based validation of ligands and residues	http://bit.ly/1tNd7Vs	[28]
PDB_REDO	Updated and optimized X-ray structure models and maps	http://bit.ly/1pVYQQA	[18, 19]
wwPDB	wwPDB Validation Server	http://bit.ly/1xdfoZN and http://bit.ly/1si1ZeL	[19]
PRIVATEER	Checks glycan nomenclature and stereochemistry	http://bit.ly/1Rc6C5e	[29]
LIGPLOT	Ligand–protein interaction diagrams	http://bit.ly/1qwZ7cS	[30]
EDS	Electron density server	http://bit.ly/1Ddnlkg	[20]
EDSTATS	Statistical quality indicators of electron density maps	http://bit.ly/1xv0VI8	[23, 31]
OVERLAPMAP	Average of two maps	http://bit.ly/ZZUSk3	[32]
BUSTER-TNT	Refinement of proteins and ligands	http://bit.ly/1w8NX1Q	[33]
WHAT_IF *WHAT_CHECK*	Protein and ligand verification tools	http://bit.ly/1F0j19Y	[31]

attempt to overcome this issue, the associated measure used by the PDB, the RSRZ value, is a normalized statistic that quantifies how many standard deviations a particular residue deviates from the mean electron density fit. Figure 1 illustrates how the RSCC is generally better for high resolution data and lower B-factors of the ligand (specifically Occupancy Weight Adjusted B-factors; OWAB, as detailed in Fig. 1).

2.3 Difference and Omit Electron Density Maps

Difference electron density maps are essential for the full and accurate validation of any protein–ligand model. A typical difference electron density map, with coefficients of $mF_o - DF_c$, can be simultaneously displayed with a traditional $2mF_o - DF_c$ map using programs like *COOT*. Difference electron density maps reveal *positive*

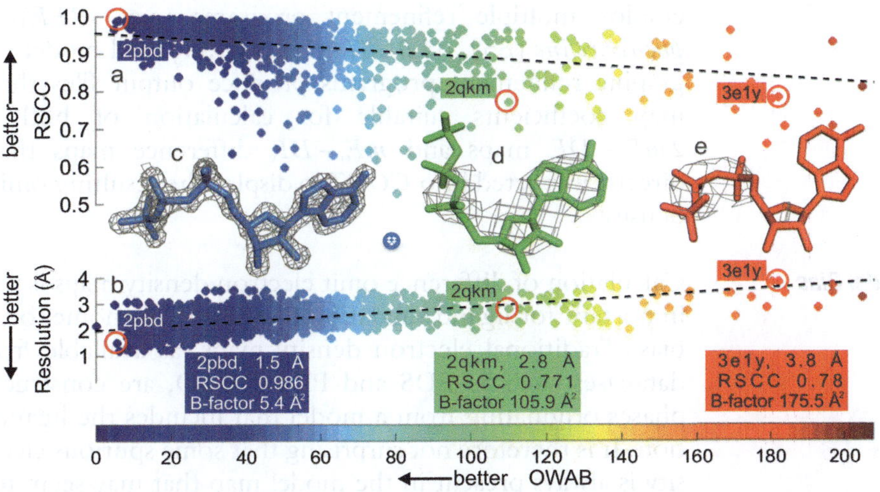

Fig. 1 (**a**) Real Space Correlation Coefficient (RSCC) and typical electron density in relation to (**b**) resolution and occupancy weighted B-factor (OWAB). In general, the higher the resolution of the data and the lower the B-factors of the ligand, the better the fit of the ligand model to the electron density. Data points are shown in each plot for adenosine triphosphate (ATP) bound structures deposited in the PDB (**c**–**e**). RSCC is a metric used to determine the *local* measure of ligand fit to the electron density and OWAB is a *local* measure of the displacement of the atoms of the ligand. RSCC and OWAB values were calculated using *Twilight* [10, 21]. These plots demonstrate how the *local* RSCC and OWAB metrics of the ligand depend closely on *global* metrics such as the resolution of the diffraction data. The ATP is shown as sticks (**c**–**e**) and the maximum likelihood $2mF_o - DF_c$ electron density map is shown as a grey mesh contoured at 2σ. Reproduced with permission from [5]

difference density where parts of the ligand model are missing, and *negative difference density* where the ligand model is present but is not supported by electron density.

Omit difference electron maps are a second type of difference electron density map that can be calculated as a powerful validation of ligand presence. In fact, the most definitive proof positive for ligand density is obtained through inspection of positive omit difference maps. These omit electron density maps are obtained by refinement of the model *sans* the part to be examined, i.e., without including the ligand in question in the refinement. If the data are obtained from a crystal structure with the ligand present, strong positive difference density, in the shape of the ligand, will be present.

Omit difference maps are not directly available from public databases, and therefore need to be generated for each specific protein–ligand complex. While there is nothing particularly difficult about this process, it does require general knowledge of file formats and crystallographic software. The ligand in question is first deleted from the structural model, either by editing the PDB file directly or by using model editing tools available in *COOT* and *PyMol*. The resulting ligand-free model is then subjected to a cycle of standard crystallographic refinement. For thorough examination, and to confirm that the difference electron density for the ligand in question is robustly reproduced, it is recommended to employ multiple refinement programs (e.g., *REFMAC* [34], *phenix.refine* [35], or *BUSTER-TNT* [33]). All modern crystallographic refinement programs produce output files that contain map coefficients suitable for calculation of both standard $2mF_o - DF_c$ maps and $mF_o - DF_c$ difference maps that can be directly imported into *COOT* to display the resulting omit electron density.

2.4 Model Bias

Calculation of difference omit electron density maps is particularly important for ligand validation due to the phenomenon of model bias. Traditional electron density maps, as available from public databases such as EDS and PDB-REDO, are constructed using phases originating from a model that includes the ligand in question. It is therefore not surprising that some spurious electron density is always present in the model map that may seem to confirm the presence of the ligand in the structure. Figure 2 illustrates the phenomenon of model bias by comparing, side-by-side, the electron density maps from the original PDB deposition (biased $2mF_o - DF_c$ map from EDS) with those from corrected sources utilizing difference omits maps (correct $mF_o - DF_c$ map from *PDB_REDO* or *BUSTER-TNT*).

2.5 Incomplete Ligand Models

We define incomplete models as those in which peptides, or any other small molecules, are enthusiastically modeled beyond what is clearly traceable in the electron density. Enthusiastic ligand modeling results in poor quality scores for parts of the ligand, or of the peptide molecule, that are not correctly modeled. As a result, local quality scores, which are typically averaged over the whole ligand, are relatively good, giving the false appearance of a correct model. Furthermore, overall quality measures of electron density, such as ligand RSCC, RSR, RSRZ, and the LLDF, are also artificially lowered by the presence of otherwise correct regions in the model. In addition, the percentile of Ramachandran plot outliers [25, 37] can be very low for short peptide ligands. These examples highlight that the only reliable method for the validation of a protein–ligand complex is inspection of the ligand, on a local basis, and its fit to the electron density. Using these methods it is possible to determine which parts of the ligand model are experimentally defined and consistent with the electron density. It is important to note that limiting the ligand model to the pieces that are visible in the electron density, while crystallographically honest and at first glance in the spirit of parsimony, can lead to incorrect functional assumptions or hypotheses. For example, modeling of only a small portion of a large peptide ligand (e.g., only five modeled residues of a 33 residue peptide) will often result in a model requiring independent supporting evidence to determine the correct register of the peptide [7].

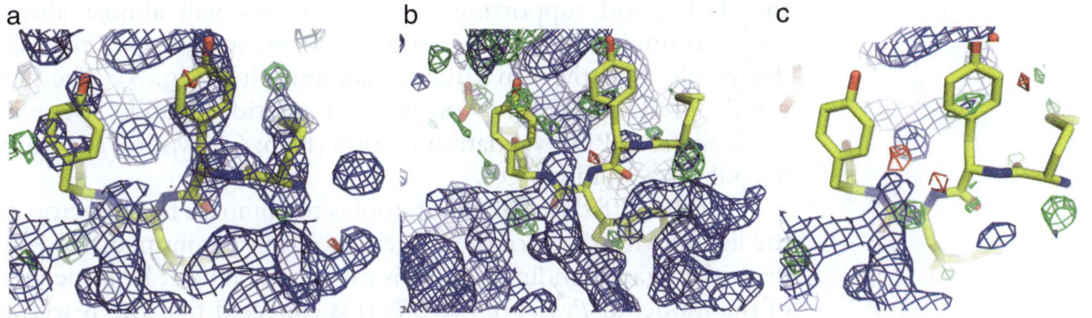

Fig. 2 Evidence of model bias in electron density maps. Shown are (**a**) $2mF_o - DF_c$, (**b**) and (**c**) $mF_o - DF_c$ electron density maps contoured at 1σ (*blue*) of the same region of a concanavalin A/peptide complex at a resolution of 1.93 Å (PDB ID:1jw6) [36]. Electron density maps were calculated by (**a**) EDS, (**b**) *PDB_REDO* and (**c**) *BUSTER-TNT* after refinement of the model with the peptide molecule omitted. The peptide is shown as *yellow sticks* and the protein model has been omitted for clarity. Figures were rendered with *PyMol* [16]. Reproduced with permission from [7]

3 Examination Against Prior Expectations

Basic empirical epistemology requires that an atomic model be examined for agreement with the primary experimental evidence. Also, it is important that the atomic model be weighted by its compliance with independently acquired prior knowledge; a well-refined ligand model built into unambiguous electron density will, in general, exhibit reasonable stereochemistry and plausible ligand–protein interactions.

3.1 Stereochemical Restraints

Presumptions about correct stereochemistry are incorporated in maximum posterior refinement in the form of stereochemical restraints [22, 38]. Stereochemical restraints effectively act as data points, stabilizing the otherwise underdetermined, or only weekly over-determined, macromolecular model refinement. The stereochemistry of the ligand is described in a restraint file. The restraint file contains general well-established information such as the bond lengths, bond angles, planarity restraints, or preferred torsions of the ligand. For small molecules and new chemical entities, particular great care must be taken to ensure that the specific chemistry is also correctly described, including delocalization, tautomerization, and charge state. If the restraints for the ligand are incorrectly described, the crystallographic data, and its contribution to the refinement target function, may not be strong enough to guide the correct placement, pose, and stereochemistry of the ligand; in such cases any resulting deviation from the electron density may not be evident from analysis of standard real space metrics.

In the case of peptide ligands, backbone torsion angles are not restrained, and therefore they provide an excellent geometric real space cross-validation of the model [7]. Peptide ligand models that lack good supporting electron density will almost always refine to implausible high energy conformations; these are easily detectable as outliers in a Ramachandran plot (Fig. 3). The percentile of Ramachandran outliers for peptide ligands is readily available from PDB validation reports (http://wwpdb-validation.wwpdb.org/validservice).

Another effective validation tool is to compare the B-factors of the ligand with those of the protein atoms in the immediate vicinity of the ligand binding site. This can be carried out by inspection of the model in *PyMol* or *COOT*. It is expected that the B-factors of interacting parts of the ligand and the protein will have similar B-factors; significant differences may be indicative of partial occupancy or, in some cases, of incorrectly built ligand molecules.

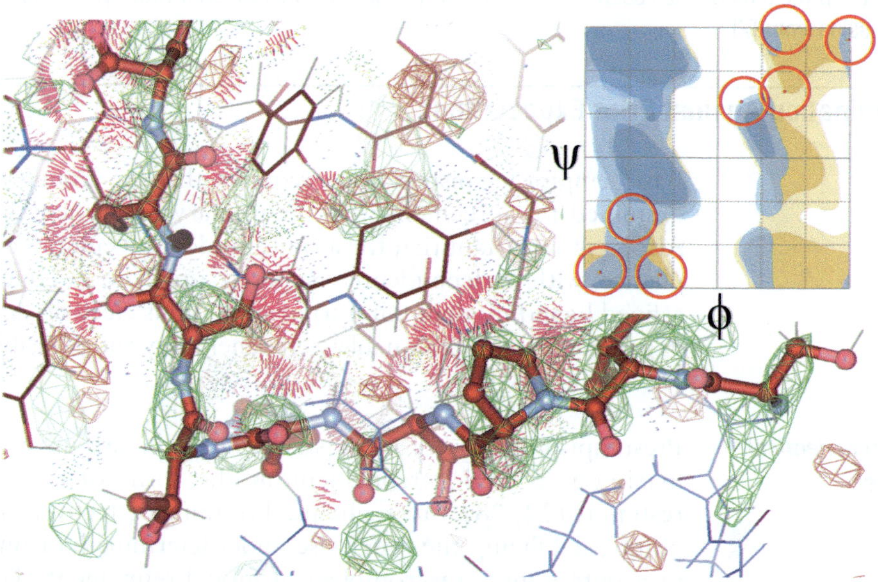

Fig. 3 An improbable peptide ligand. The main panel shows the model of a peptide antigen (ball-and-stick model) bound to a Fab antibody fragment (thin sticks) (PDB entry 2a6i, chain P, [39]). The positive $mF_o - DF_c$ omit difference electron density (calculated with the ligand omitted during maximum likelihood map calculation) is shown as a green grid contoured at 2.5σ above the mean density. In this example, the peptide model should be surrounded by clear positive difference electron density resembling the distinct shape of the peptide. However, only discontinuous fragments of positive difference electron density are visible, which can be explained in part by ordered solvent (the *round green spheres* are typical for water molecules). In addition to the poor fit to the electron density, the antigen model has a multitude of unreasonably close contacts with the Fab (visualized as *red spikes*) and has an implausibly high-energy backbone conformation as denoted by the backbone torsion angles located in unfavourable regions of the Ramachandran plot (*top right insert*). The image was prepared using *COOT* [8] and the $mF_o - DF_c$ map reconstructed from maximum likelihood coefficients computed by *REFMAC* [34]. The backbone torsion angles and interatomic clashes were calculated using *MolProbity* [37]. Updated figure with permission from [40]

3.2 Clash Scores and Binding Site Chemistry

A reasonably placed ligand will have clearly defined interactions with amino acid residues of the target protein. Common problems include steric clashes between a poorly placed ligand and the target protein (Fig. 3), and the absence of sensible contacts between the ligand and the target.

An all-atom close contact analysis (i.e., clash score) can be computed using *MolProbity* and visualized using *COOT* [37]. The program displays a list of bad atom contacts and a visual display of dots and spikes representing the contacts. Figure 3 highlights the crystal structure of an implausible Fab-Antigen complex (PDB entry 2a6i, chain P, [39]). In this example, the combination of poor agreement with the primary evidence (i.e., the electron density), together with violation of any established prior expectations, suggests that the presence of the ligand, as modeled, is unlikely.

Some validation programs, notably *VHELIBS* [26], also evaluate the quality of the interactions in the ligand binding site and rank the ligand on the basis of these quality scores. A summary of validation tools is provided in Table 1.

4 Practice of Ligand Evaluation

The degree of ligand validation that is required depends on the context in which the protein–ligand model will be used. For example, validation of a protein–ligand structure for use in structure-based lead discovery will typically require a higher level of scrutiny, and generally a resolution of around 2 Å or better. Conversely, less scrutiny is required when validating a protein–ligand complex that is being used to analyze global features of the ligand binding site for use in techniques such as structure-guided mutagenesis.

Figure 4 summarizes a useful decision tree for ligand inspection and validation of protein–ligand models. The following steps are a recommended best practice procedure for ligand validation.

4.1 Data and Program Setup

- Does the selected structure model provide enough detail, particularly resolution, to serve your interest?
- Download the full PDB validation report (https://www.ebi.ac.uk/pdbe/).
- Download the coordinates and electron density from EDS or *PDB_REDO* in *COOT*. If the electron density is not available, full validation is not possible.
- Turn on symmetry related atoms.
- Turn on environment distances.
- Select the ligand and zoom-in. If you need an omit map, proceed to the next section, otherwise inspect the $2mF_o - DF_c$ electron density.

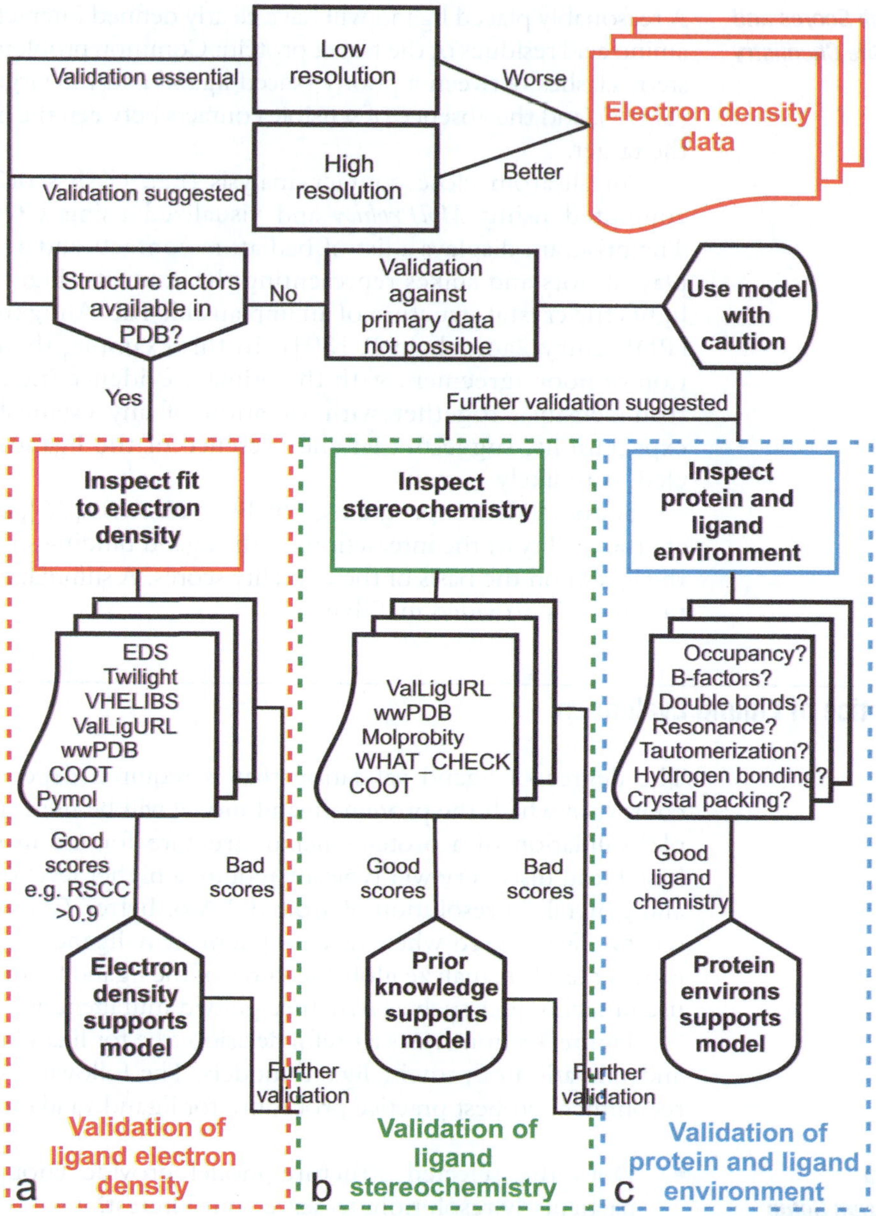

Fig. 4 Decision tree for interpretation of crystallographic data and validation of protein–ligand models. A pathway of recommended examinations starts with the electron density data (*top right*). Important steps include, (**a**) validation of the ligand electron density with a particular focus on ensuring the electron density supports the ligand model, (**b**) validation of the ligand stereochemistry and confirmation that the ligand model is supported by prior expectations and (**c**) validation of the protein and ligand environment and confirmation that the environment supports the ligand model. Reproduced with permission from [5]

4.2 Prepare an Omit Map

- Download the coordinates from the PDB (https://www.ebi.ac.uk/pdbe/).
- Download the reflection data from the PDB and convert it to mtz format using the CCP4i structure factor utilities (http://www.ccp4.ac.uk/examples/tutorial/html/intro-tutorial.html).
- Alternatively, download the mtz file corresponding to the "fully optimized structure" from *PDB_REDO* (http://www.cmbi.ru.nl/*PDB_REDO*). Another tool for converting experimental data from the PDB into the proper input format is phenix.cif_as_mtz from the PHENIX software package (https://www.phenix-online.org/documentation/reference/cif_as_mtz.html).
- Delete the ligand from the model. This can be done by manually editing the text of the PDB file or by using *COOT*. Ligand atom records in the PDB file are identified by "HETATM" and the 3-letter ligand identifier. The PDB file might contain additional LINK/CISPEP records for the ligand in the header section which also need to be removed. Atoms belonging to the protein are denoted by lines beginning with "ATOM", these lines and the "CRYST1" lines need to be retained.
- Run a single cycle of refinement, with the ligand-omitted model, as a starting structure. It is often informative to run several different refinement programs as the results may differ. *REFMAC*, *phenix.refine* and *BUSTER-TNT* are widely used crystallographic refinement programs and simple graphical user interfaces are available for *REFMAC* (http://ccp4wiki.org/~ccp4wiki/wiki/index.php?title=Refinement_with_*REFMAC*5) and *phenix.refine* (https://www.phenix-online.org/documentation/reference/refine_gui.html).
- The output mtz file of the refinement procedure contains the $2mF_o - DF_c$ omit map (default blue) and $mF_o - DF_c$ difference omit map (green, positive difference density, red negative difference density) that can be inspected in *COOT* (https://www2.mrc-lmb.cam.ac.uk/personal/pemsley/coot/).

4.3 Electron Density Inspection

Inspect the $2mF_o - DF_c$ electron density. Does the blue grid surround the model? If not, then it is likely that further validation is required.

- Inspect the omit $mF_o - DF_c$ difference electron density. Import the original PDB model and examine the omit difference density around the ligand molecule. It is common to view the difference electron density at a 3σ level. It may be informative to calculate an omit map after deleting a single protein residue in the binding site—this will provide a good estimate of the magnitude of the difference omit density due to bona fide structural elements.
- Evaluate how much of the ligand molecule can be clearly placed within the electron density map. Partial fits are a fairly

common situation for peptide ligands, lipid molecules and glycosylation sites. In these cases, the most plausible explanation for the partial fit to the electron density is disorder in the ligand, as a result of weak interactions with the protein, or poor occupancy of the ligand binding site.

- Evaluate whether the ligand molecule can be fit into the electron density in a different orientation or conformation.
- Evaluate whether the electron density allows for alternative explanations. Examples include: (1) strings of structured water molecules bound to the ligand binding site, (2) components of the mother liquor or cryo-protectant (e.g., short polyethylene glycol fragments), and (3) terminal fragments or loop regions of a symmetry related protein molecule.
- Examine the "residue density fit" graph in *COOT*, paying particular attention to the ligand and residues within the ligand binding site (https://www2.mrc-lmb.cam.ac.uk/Personal/pemsley/*COOT*/web/docs/*COOT*.html#Validation-Graphs).

4.4 Environmental Examination

- Are symmetry-related molecules involved in ligand binding? If yes, further evaluation using biochemical or biophysical techniques may be required to address if these interactions are real or crystal packing artifacts.
- Are there multiple copies of the same molecule present in the asymmetric unit (ASU, non-crystallographic symmetry)? If yes, then all of the ligand binding sites within the ASU should be examined and validated. Can different conformations or occupancies of the ligand be explained by nonequivalent environments, plasticity, or crystal packing?
- Examine the bond distances within the environment of the ligand binding site. These can be displayed using *COOT*. Are the non-covalent interactions plausible?
- Compute the clash score via the validation menu in *COOT*. Are there any significant clashes between the ligand and the target protein? Can they be explained or perhaps corrected?

5 Tools

There are several tools available for the validation of ligands (Table 1). They include standalone programs, server-based applications on the web, and databases containing collections of ligands scored by various quality criteria. As always, the quality and the suitability of a protein–ligand structure for a given purpose is context-sensitive; not every protein–ligand structure is fit for structure based lead discovery, and not every missing piece of a ligand is reason for panic. As already mentioned before: *trust, but verify* [5].

6 Outlook and New Techniques

Several recent developments in structural biology, such as ultrafast electron diffraction (UED), double electron–electron resonance (DEER), atomic force microscopy, cryo-electron microscopy (cryo-EM), and X-ray free-electron lasers (XFEL) [41, 42] pose an extra set of challenges for ligand validation, as they generally produce structures of lower resolution, typically in the 4–10 Å range. It is essential that special emphasis be placed on the ligand restraints that are applied. New hybrid techniques allow probing the dynamics of ligand binding on new timescales, and in the case of XFEL, analysis of pico- and femto-scale dynamics are possible [43]. Much of the chemistry, and in particular ligand stereochemistry, is yet to be explored on such timescales and new definitions of ligand stereochemistry will be required to ensure that accurate restraints are maintained for structures determined using these techniques.

Structural biology lies at the forefront of biomedical research and drug discovery and it is essential that the structural biology community gets serious about effective ligand validation. It requires effort at all levels to ensure that protein–ligand structures deposited in the PDB, and used by the scientific community, are of the highest quality [44]. It is essential that validation methods continue to evolve and embrace new hybrid techniques. Mandatory validation efforts and task forces, such as those initiated by the PDB, will be essential to this effort [19].

Acknowledgments

BR receives partial support from the Austrian Science Foundation (FWF) under project P28395-B26.

References

1. Berman H, Henrick K, Nakamura H (2003) Announcing the worldwide Protein Data Bank. Nat Struct Biol 10:980
2. Blundell TL, Jhoti H, Abell C (2002) High-throughput crystallography for lead discovery in drug design. Nat Rev Drug Discov 1:45–54
3. Blundell TL, Patel S (2004) High-throughput X-ray crystallography for drug discovery. Curr Opin Pharmacol 4:490–496
4. Congreve M, Murray CW, Blundell TL (2005) Structural biology and drug discovery. Drug Discov Today 10:895–907
5. Deller MC, Rupp B (2015) Models of protein–ligand crystal structures: trust, but verify. J Comput Aided Mol Des 29:817–836
6. Pozharski E, Weichenberger CX, Rupp B (2012) Techniques, tools and best practices for ligand electron-density analysis and results from their application to deposited crystal structures. Acta Crystallogr D Biol Crystallogr 69:150–167
7. Weichenberger C, Pozharski E, Rupp B (2016) Twilight reloaded: the peptide experience. Acta Crystallogr D Biol Crystallogr 72: 211–222
8. Debreczeni JE, Emsley P (2012) Handling ligands with Coot. Acta Crystallogr D Biol Crystallogr 68:425–430
9. Deller MC, Kong L, Rupp B (2016) Protein stability: a crystallographer's perspective. Acta Crystallogr F Struct Biol Commun 72:72–95

10. Pozharski E, Weichenberger CX, Rupp B (2013) Techniques, tools and best practices for ligand electron-density analysis and results from their application to deposited crystal structures. Acta Crystallogr D Biol Crystallogr 69:150–167
11. Rupp B (2009) Biomolecular crystallography: principles, practice, and application to structural biology. Garland Science, New York
12. Rhodes G (2006) Crystallography made crystal clear. Academic Press, London, UK
13. Rossmann M (ed) (1972) The molecular replacement method. Gordon and Breach Science Publishers, New York
14. Evans P, McCoy A (2008) An introduction to molecular replacement. Acta Crystallogr D Biol Crystallogr 64:1–10
15. Emsley P, Lohkamp B, Scott WG et al (2010) Features and development of Coot. Acta Crystallogr D Biol Crystallogr 66:486–501
16. DeLano WL (2008) The PyMOL molecular graphics system. DeLano Scientific, Palo Alto, CA
17. Dutta S, Burkhardt K, Swaminathan GJ et al (2008) Data deposition and annotation at the Worldwide Protein Data Bank. In: Kobe B, Guss M, Huber T (eds) Structural proteomics: high-throughput methods. Humana Press/ Springer, New York
18. Joosten RP, Womack T, Vriend G et al (2009) Re-refinement from deposited X-ray data can deliver improved models for most PDB entries. Acta Crystallogr D Biol Crystallogr 65:176–185
19. Read RJ, Adams PD, Arendall WB 3rd et al (2011) A new generation of crystallographic validation tools for the protein data bank. Structure 19:1395–1412
20. Kleywegt GJ, Harris MR, Zou J-Y et al (2004) The Uppsala Electron-Density Server. Acta Crystallogr D Biol Crystallogr 60:2240–2249
21. Weichenberger CX, Pozharski E, Rupp B (2013) Visualizing ligand molecules in twilight electron density. Acta Crystallogr F Struct Biol Commun 69:195–200
22. Tronrud D (2004) Introduction to macromolecular refinement. Acta Crystallogr D Biol Crystallogr 60:2156–2168
23. Tickle IJ (2012) Statistical quality indicators for electron-density maps. Acta Crystallogr D Biol Crystallogr 68:454–467
24. Brändén CI, Jones TA (1990) Between objectivity and subjectivity. Nature 343:687–689
25. Ramachandran GN, Ramakrishnan C, Sasisekharan V (1963) Stereochemistry of polypeptide chain configurations. J Mol Biol 7:95–99
26. Cereto-Massague A, Ojeda MJ, Joosten RP et al (2013) The good, the bad and the dubious: VHELIBS, a validation helper for ligands and binding sites. J Cheminform 5:36
27. Kleywegt GJ, Harris MR (2007) ValLigURL: a server for ligand-structure comparison and validation. Acta Crystallogr 63:935–938
28. Varekova RS, Jaiswal D, Sehnal D et al (2014) MotiveValidator: interactive web-based validation of ligand and residue structure in biomolecular complexes. Nucleic Acids Res 42:W227–W233
29. Agirre J, Iglesias-Fernandez J, Rovira C et al (2015) Privateer: software for the conformational validation of carbohydrate structures. Nat Struct Mol Biol 22:833–834
30. Laskowski RA, Swindells MB (2011) LigPlot+: multiple ligand-protein interaction diagrams for drug discovery. J Chem Inf Model 51:2778–2786
31. Joosten RP, te Beek TA, Krieger E et al (2011) A series of PDB related databases for everyday needs. Nucleic Acids Res 39:D411–D419
32. Winn MD, Ballard CC, Cowtan KD et al (2011) Overview of the CCP4 suite and current developments. Acta Crystallogr D Biol Crystallogr 67:235–242
33. Smart OS, Womack TO, Flensburg C et al (2011) Better ligand representation in BUSTER protein-complex structure determination. Acta Crystallogr A 67:C134
34. Murshudov GN, Skubak P, Lebedev AA et al (2011) REFMAC5 for the refinement of macromolecular crystal structures. Acta Crystallogr D Biol Crystallogr 67:355–367
35. Afonine PV, Grosse-Kunstleve RW, Echols N et al (2012) Towards automated crystallographic structure refinement with phenix. refine. Acta Crystallogr D Biol Crystallogr 68:352–367
36. Zhang Z, Qian M, Huang Q et al (2001) Crystal structure of the complex of concanavalin A and hexapeptide. J Protein Chem 20(5):423–429
37. Chen VB, Arendall WB III, Headd JJ et al (2010) MolProbity: all-atom structure validation for macromolecular crystallography. Acta Crystallogr D Biol Crystallogr 66:12–21
38. Konnert J (1976) A restrained-parameter structure-factor least-squares refinement procedure for large asymmetric units. Acta Crystallogr A 32:614–617
39. Sethi DK, Agarwal A, Manivel V et al (2006) Differential epitope positioning within the germline antibody paratope enhances promiscuity in the primary immune response. Immunity 24:429–438

40. Rupp B (2016) Only seeing is believing: the power of evidence and reason. Adv Biochem (Postępy Biochemii) 62:250
41. Wakatsuki S (2014) Structural biology applications of synchrotron radiation and X-ray free-electron lasers. In: Jaeschke E, Khan S, Schneider RJ, Hastings BJ (eds) Synchrotron light sources and free-electron lasers: accelerator physics, instrumentation and science applications. Springer International Publishing, Switzerland, pp 1–39
42. Lander GC, Saibil HR, Nogales E (2012) Go hybrid: EM, crystallography, and beyond. Curr Opin Struct Biol 22:627–635
43. Neutze R (2014) Opportunities and challenges for time-resolved studies of protein structural dynamics at X-ray free-electron lasers. Philos Trans R Soc London B Biol Sci 369:20130318
44. Rupp B, Wlodawer A, Minor W et al (2016) Correcting the record of structural publications requires joint effort of the community and journal editors. FEBS J 283(24):4452–4457

Chapter 26

Protein Data Bank (PDB): The Single Global Macromolecular Structure Archive

Stephen K. Burley, Helen M. Berman, Gerard J. Kleywegt, John L. Markley, Haruki Nakamura, and Sameer Velankar

Abstract

The Protein Data Bank (PDB)—the single global repository of experimentally determined 3D structures of biological macromolecules and their complexes—was established in 1971, becoming the first open-access digital resource in the biological sciences. The PDB archive currently houses ~130,000 entries (May 2017). It is managed by the Worldwide Protein Data Bank organization (wwPDB; wwpdb.org), which includes the RCSB Protein Data Bank (RCSB PDB; rcsb.org), the Protein Data Bank Japan (PDBj; pdbj.org), the Protein Data Bank in Europe (PDBe; pdbe.org), and BioMagResBank (BMRB; www.bmrb.wisc.edu). The four wwPDB partners operate a unified global software system that enforces community-agreed data standards and supports data Deposition, Biocuration, and Validation of ~11,000 new PDB entries annually (deposit.wwpdb.org). The RCSB PDB currently acts as the archive keeper, ensuring disaster recovery of PDB data and coordinating weekly updates. wwPDB partners disseminate the same archival data from multiple FTP sites, while operating complementary websites that provide their own views of PDB data with selected value-added information and links to related data resources. At present, the PDB archives experimental data, associated metadata, and 3D-atomic level structural models derived from three well-established methods: crystallography, nuclear magnetic resonance spectroscopy (NMR), and electron microscopy (3DEM). wwPDB partners are working closely with experts in related experimental areas (small-angle scattering, chemical cross-linking/mass spectrometry, Forster energy resonance transfer or FRET, etc.) to establish a federation of data resources that will support sustainable archiving and validation of 3D structural models and experimental data derived from integrative or hybrid methods.

Key words Protein Data Bank, PDB, Worldwide Protein Data Bank, wwPDB, PDBx/mmCIF, Chemical Component Dictionary, Crystallography, NMR spectroscopy, NMR-STAR, NMR Exchange Format, NEF, 3D electron microscopy, Integrative or hybrid methods

1 Evolution of Data Sharing and Data Archiving in Structural Biology

The Protein Data Bank (PDB) was established in 1971 with fewer than ten X-ray crystallographic structures of proteins, becoming the first open access digital data resource in the biological sciences [1]. Soon after X-ray structures of myoglobin [2, 3] and hemoglobin [4, 5] were published, the structural biology community

Alexander Wlodawer et al. (eds.), *Protein Crystallography: Methods and Protocols*, Methods in Molecular Biology, vol. 1607, DOI 10.1007/978-1-4939-7000-1_26, © Springer Science+Business Media LLC 2017

began discussions as to how best to archive protein crystallographic findings and make them broadly available. In 1971, the Cold Spring Harbor Laboratory hosted a symposium on protein crystallography, during which there was extensive discussion of data sharing [6]. Walter C. Hamilton, one of the attendees, offered to provide the first home for what is now the Protein Data Bank (PDB) [7]. Shortly thereafter, the PDB was launched from within the Department of Chemistry at Brookhaven National Laboratory (BNL), building on the Protein Structure Library framework [8]. The importance of scientific data archiving as a global endeavor was understood at the outset, and public announcement of the PDB in 1971 explicitly mentioned collaboration with and the option of data submission *via* the Cambridge Crystallographic Database Centre [1].

When the PDB was launched, data submission was voluntary. In the 1980s, influential members of the structural biology community began to make the case for mandatory data deposition. Various committees were established to define what data should be required and when it should be disseminated. Guidelines were published in 1989 [9], and over time, adopted by virtually all of the scientific journals now requiring PDB deposition of atomic coordinates prior to publication of structural studies. In 2008, further evolution of community mores led to mandatory deposition of crystallographic structure factors and NMR restraints together with atomic coordinates. In 2010, deposition of NMR chemical shifts became mandatory. At the time of writing (May 2016), ~80% of PDB archival entries include experimental data.

2 Growth of the Protein Data Bank Archive

The first 356 structures deposited to the PDB archive were determined by crystallography. In 1988, structures determined using NMR methods began to be deposited, and in 1996 the first structure determined by electron microscopy was deposited. Since 1971, growth of the archive has been decidedly nonlinear (Fig. 1). By 1982, the PDB had reached only ~100 entries. Eleven years later, in 1993, there were 1000 entries. Before the end of the decade (1999), this number had grown to 10,000. Circa fifteen years thereafter, archival contents exceeded 100,000 entries as of May 2014. At the time of writing (May 2016), the PDB archive contains more than 119,000 structures of proteins, nucleic acids, and their complexes with one another and with small molecule ligands. Calendar year depositions in 2015 numbered 10,956 (~900/month). The vast majority of extant PDB archival entries came from X-ray, neutron, and combined X-ray/neutron crystallography (~90%), with the remainder produced by NMR (~9%) and 3DEM (~1%). Among the three experimental methods currently

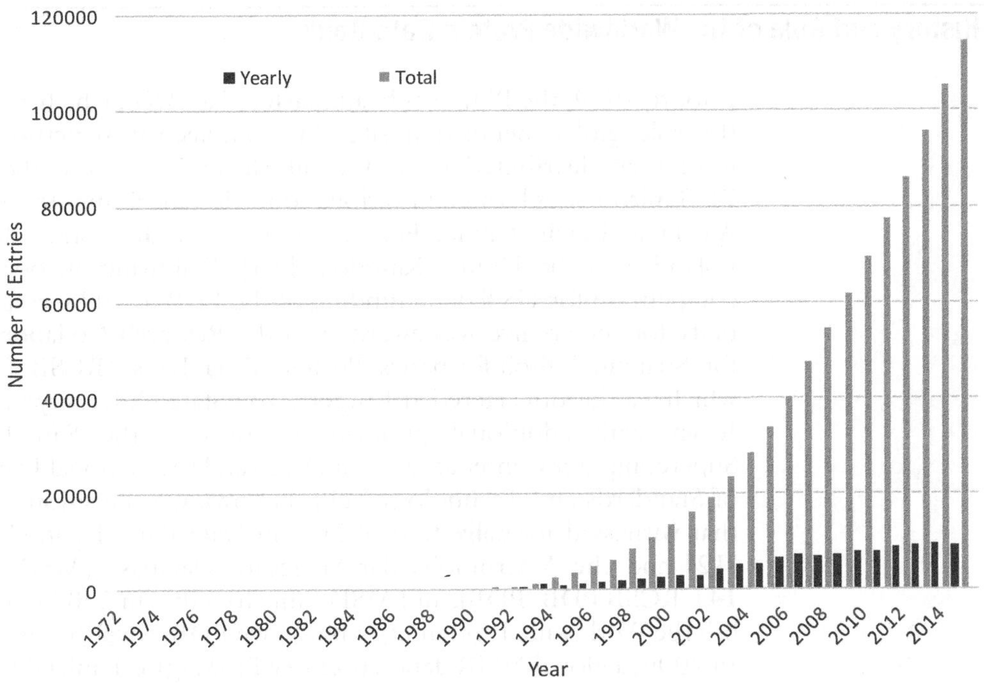

Fig. 1 Growth of the PDB Archive since 1971

Table 1
Proxy measures of complexity for recent PDB archival entries (2012–2015)

Year	Number of new entries with number of polymer chains > 62	Number of new entries with MW > 500,000	Number of new protein–nucleic acid complexes	Number of new compounds added to the Chemical Component Dictionary (CCD)
2012	14	133	~450	1733
2013	32	198	~440	1875
2014	49	164	~690	1767
2015	55	311	~580	1830

represented in the PDB archive, data deposition rates have varied markedly over time. From 2012 to 2015, annual crystallographic depositions have grown slowly year-on-year [9269 in 2012; 10,168 in 2015]. During that same period, 3DEM depositions have increased significantly year on year, rising from 103/year in 2012 to 254/year in 2015. NMR depositions, on the other hand, peaked in 2007 at 1062/year, declining to 510/year in 2015. The PDB archive has also grown considerably in complexity since 1971. Some proxy measures of complexity are provided in Table 1.

3 History and Role of the Worldwide Protein Data Bank

Prior to 1999, the PDB was headquartered at BNL, which acted as the sole global deposition site. Macromolecular structure data were then distributed internationally from BNL by authorized PDB mirror sites located in various countries, including Argentina, Australia, Brazil, China, France, Germany, India, Israel, Japan, Poland, and the United Kingdom [10]. Following an open re-competition for US federal funding of the PDB in 1998, responsibility for the archive was awarded to the Research Collaboratory for Structural Bioinformatics Protein Data Bank (RCSB PDB), which was headquartered at Rutgers, The State University of New Jersey with additional performance sites at the San Diego Supercomputer Center at UC San Diego and the National Institute of Standards and Technology [11]. Following a transition period that witnessed formalization of Protein Data Bank Japan (PDBj) [12] and the Macromolecular Structure Database (MSD) [13, 14], RCSB PDB, PDBj, and MSD came together in 2003 to establish the Worldwide Protein Data Bank (wwPDB; wwpdb.org) [15]. In 2006, a global NMR data repository BioMagResBank (BMRB), founded in 1989 [16], joined the wwPDB organization [17]. BMRB hosts deposition sites in both the US (BMRB; www.bmrb.wisc.edu) and Japan (PDBj-BMRB; bmrbdep.pdbj.org) [18]. (N.B.: MSD was rebranded in 2008 as the Protein Data Bank in Europe or PDBe [14, 19].)

The wwPDB organization is governed by a Memorandum of Understanding (wwpdb.org/about/agreement), which was renewed in 2013. Oversight of wwPDB partner activities is provided by an internationally recognized team of experts in structural biology and bioinformatics comprising the wwPDB Advisory Committee (http://wwpdb.org/about/advisory). As outlined in detail below, wwPDB partners collaborate on "Data In." They are jointly responsible for standardizing, collecting, biocurating, validating, and disseminating macromolecular structure data as a single global archive. At present, RCSB PDB is formally designated as the Archive Keeper, responsible for ensuring disaster recovery of PDB data and coordinating weekly archival updates among partner sites (or regional data centers).

Founding of the wwPDB organization helped to ensure that the PDB has continued to evolve as the single global archive of macromolecular structure data. In contrast, global archiving of nucleic acid sequences is accomplished by three independently operated regional archives comprising the International Nucleotide Sequence Database Collaboration (INSDC), which exchange data nightly.

4 PDB Data Standardization, Deposition, Annotation, and Validation

Following launch of the wwPDB, crystallographic structure depositions to the PDB archive were accepted via two different portals; ADIT, which was operated jointly by RCSB PDB and PDBj [11], and AutoDep, which was developed at BNL [20] and reengineered by MSD/PDBe [21]. NMR depositions were accepted *via* ADIT-NMR at BMRB and PDBj-BMRB, with coordinates and restraint data transferred to RCSB PDB or PDBj, respectively [17]. In addition, PDBe accepted NMR structures via AutoDep, with associated NMR data sent to BMRB for archiving. Early in 2016, the wwPDB partners launched a unified global system for Deposition, Biocuration, and Validation of incoming data supporting crystallography, NMR, and 3DEM (deposit.wwpdb.org). Working to a common set of standards, three wwPDB regional data centers take responsibility for depositions originating from the Americas and Oceania (RCSB PDB), Europe and Africa (PDBe), and the Middle East and Asia (PDBj). The pipeline currently used by the wwPDB to process incoming structures is illustrated schematically in Fig. 2. Approximately 900 depositions are received monthly from every inhabited continent (Fig. 3). RCSB PDB, PDBe, and PDBj refer depositors of NMR data unrelated to 3D structures to BMRB, and, conversely, BMRB refers depositors with atomic coordinate data to the three wwPDB regional data centers. NMR data archived in the PDB are also mirrored in the BMRB archive under a four-digit acquisition code, which in some cases contains additional data on the system supplied by depositors (e.g., NMR relaxation rates, order parameters, and files containing raw time-domain data). Deposited entries are

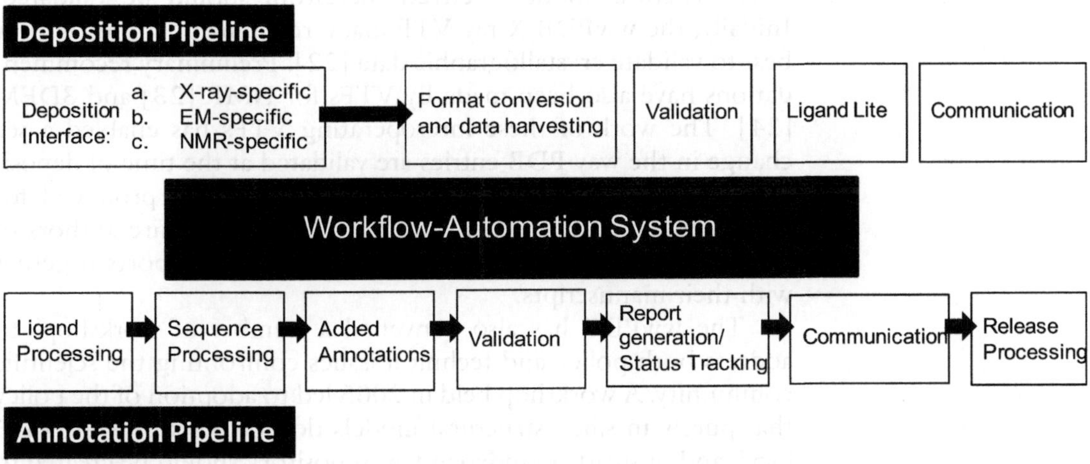

Fig. 2 wwPDB Deposition, Biocuration, and Validation Pipeline. Each box represents a modular component of the data processing workflow

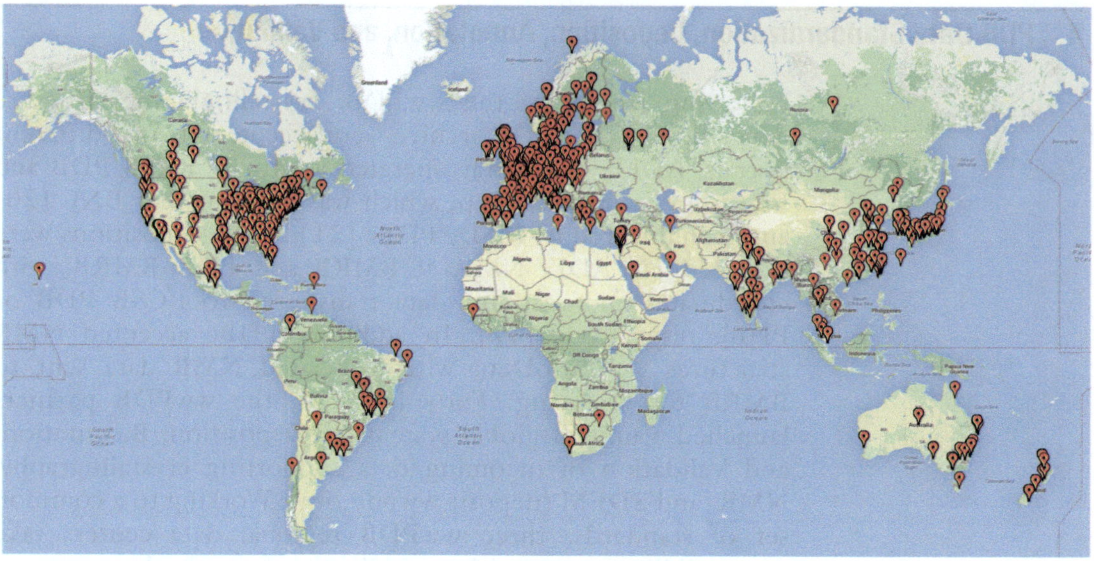

Fig. 3 World map showing global distribution of PDB Depositors (2012–2015)

then validated and annotated by wwPDB biocurators, with wwPDB Validation Reports (wwpdb.org/validation/validation-reports) returned to depositors for review before finalization and data release.

Considerable effort has gone into understanding how best to standardize, biocurate, and validate incoming atomic coordinates and primary experimental data generated by crystallography, NMR, and 3DEM. Over the past decade, the wwPDB has convened a series of expert, method-specific Validation Task Forces (VTFs) to determine which experimental data and metadata from each method should be archived and how these data and the atomic level structural models derived therefrom should be validated. Initially, the wwPDB X-ray VTF made recommendations on how best to validate crystallographic data [22]. Preliminary recommendations have also been made by VTFs for NMR [23] and 3DEM [24]. The work of these interoperating VTFs has enabled a sea change in the way PDB entries are validated at the time of deposition/annotation. A wwPDB Validation Report is produced for every new entry, and more and more journals require authors of structure determination studies to submit these reports together with their manuscripts.

The wwPDB has also convened a number of workshops to address both policy and technical issues confronting the scientific community. A workshop held in 2005 led to adoption of the policy that purely in silico structural models do not belong in the PDB [25], and, instead, an independent repository should be created to archive computed models elsewhere. The Protein Modeling Portal was established in 2007 [26]. In 2012, to address the challenges posed by the presence of a number of non-atomistic structural models of proteins obtained via small-angle scattering (SAS), the wwPDB SAS Task Force was established. This group of community stakeholders met and recommended creation of one or more SAS data repositories that should interoperate with the PDB archive [27]. Subsequently, some 49 PDB entries derived exclusively from SAS methods were transferred into the SAS Biological Data Bank (SASbDB; sasbdb.org) archive [28] and then obsoleted (retired) from the PDB archive. In 2015, the wwPDB partnered with the Cambridge Crystallographic Data Center (CCDC; www.ccdc.cam.ac.uk) [29] and the Drug Design Data Resource (D3R; drugdesigndata.org) to convene a Ligand Validation Workshop, focused on improving the quality and utility of co-crystal structures in the PDB archive. Published recommendations pertaining to representation of small-molecules and validation of co-crystal structures coming from this workshop [30] were endorsed by the wwPDB X-ray VTF in late 2015. Implementation of these recommendations was underway at the time of writing.

5 Data Representation for Biological Macromolecules, Metadata, and Experimental Methods and Results

The PDB archive contains comprehensive descriptions of structural models coming from crystallography, NMR, and 3DEM. Each archival entry is denoted by a 4-character PDB identifier (e.g., 1VTL). In addition to atomic coordinates, details regarding the chemistry of biopolymers and any bound small molecules are archived, as are metadata describing biopolymer sequence, sample composition and preparation, experimental procedures, data-processing methods/software/statistics, structure determination/refinement procedures and statistics, and certain structural features, such as the secondary and quaternary structure. Primary experimental data coming from crystallography (structure-factor amplitudes or intensities) and NMR (restraints and chemical shifts) must be archived in the PDB. Voluntary archiving of diffraction images is currently supported by two resources that operate independently of the PDB, including the Integrated Resource for Reproducibility in Macromolecular Crystallography (IRRMC; www.proteindiffraction.org) and the Structural Biology Data Grid Consortium (SBGrid; sbgrid.org [31]) both of which use digital object identifiers to make the data readily accessible. In addition, some synchrotron radiation facilities now store diffraction images in locally maintained repositories, with data retention and dissemination policies determined by the facility. BMRB [32] has long served as a public repository for NMR experimental data that are not stored in the PDB. Mass density maps used to derive structural models from 3DEM can be archived in EMDB [33]. Voluntary archival deposition of raw 3DEM images is currently supported by EMPIAR [34].

The first data format used by the PDB archive was established in the early 1970s, and was based on the 80-column Hollerith format used for punched cards [35]. Atom records included atom name, residue name, polymer chain identifier, and polymer sequence number. A set of "header records" contained limited metadata. The community readily accepted this format, because it was simple and both human- and machine-readable. However, the format also had limitations that became serious liabilities as structural biologists took the field to new heights. Structural models were limited to 99,999 atoms and relationships among various data items were implicit. These and other weaknesses of the legacy PDB format meant that deep subject matter expertise was required to both create and use software relying on this format. In the 1990s, the International Union of Crystallography (IUCr) charged a committee with creating a more informative and extensible data model for the PDB archive.

In response to the IUCR committee report, the Macromolecular Crystallographic Information File (mmCIF) was proposed [36]. mmCIF is a self-defining format in which every data item has attributes describing its features, including explicit definitions of relationships among data items. Most important, mmCIF has no limitations with respect to the size of the structural model to be archived. In addition, the mmCIF dictionary and mmCIF format data files are fully machine-readable, and no domain knowledge is required to read the files. At inception, the mmCIF dictionary contained over 3000 data items pertaining to crystallography. Over time, data items specific to NMR and 3DEM were added, and the dictionary was subsequently rebranded PDBx/mmCIF [37]. In 2007, it was decided that PDBx would be the PDB Master Format for data collected by the wwPDB. In 2011, major crystallographic structure determination software developers agreed to adopt this data model so that going forward all output from their programs would be available in PDBx/mmCIF.

In collaboration with community stakeholders serving on the PDBx/mmCIF Working Group (wwpdb.org/task/mmcif), the wwPDB continues to extend and enhance archival data representations. As of December 2014, PDBx/mmCIF became the official format for distribution of PDB entries. At the time of writing, the PDBx/mmCIF dictionary contained more than 4400 data items, including ~250 and ~1200 specific to NMR and 3DEM, respectively. PDBML, an XML format based on PDBx/mmCIF [38] and the requisite RDF (Resource Description Framework) conversion have also been developed to facilitate integration of structural biology data with other life sciences data resources [39]. Recently, XML and RDF-formatted BMRB data have been provided as BMRB/XML and BMRB/RDF, respectively [40], by which a federated SPARQL query linking the BMRB is made available to other databases. Finally, other structural biology communities are building on the PDBx/

mmCIF framework to establish their own controlled vocabulary and specialist data items. For example, SASbDB has been working in collaboration with wwPDB partners to develop sasCIF [41], which builds on PDBx/mmCIF. In addition to accelerating development of the SASbDB archive, creation of sasCIF will allow for facile interoperation with the PDB archive using a common exchange protocol based on PDBx/mmCIF.

In 1996, BMRB adopted NMR-STAR (a version of mmCIF) as its archival format [42]. As noted above, this format has been harmonized with PDBx/mmCIF and now serves as the preferred deposition format for NMR structures [43]. Historically, most NMR experimental data have been deposited in "native" format provided by each software package and archived "as is" in the PDB. Format harmonization was addressed in part by the NMR Restraints Grid, which can process restraint files and convert them to the NMR-STAR or CCPN formats [44, 45]. In 2013 and 2014, community stakeholders participating in a pair of NMR format meetings convened by the wwPDB NMR VTF, recommended that an NMR Exchange Format (NEF) be developed for facile data transfer among NMR software packages and faithful conversion to NMR-STAR [46]. BMRB-led efforts are now underway to complete harmonization of NEF with NMR-STAR/PDBx/mmCIF to support NMR data deposition, annotation, and validation using the wwPDB unified global system (deposit.wwpdb.org).

Prior to 2015, reliance on the original PDB format made it necessary for large structure depositions (e.g., ribosomes/ribosomal subunits) archived in the PDB to be "split" into multiple entries, each with its own 4-character PDB identifier and legacy PDB-format file. This stopgap arrangement was entirely suboptimal. Splitting depositions among multiple PDB entries effectively precluded routine visualization of some of the most interesting structural models in the PDB archive, owing to software limitations. With adoption of the PDBx/mmCIF standard, every PDB archival entry is now stored as a single PDBx/mmCIF file, including 277 large structures that had previously been "split." At the time of writing (and for the foreseeable future), archival entries are made available as a public service in "stripped down," best-effort PDB legacy format files wherever possible. In time, visualization, computational chemistry, etc. software providers will need to adjust to the new format and use PDBx/mmCIF files directly.

6 Data Representation for Small Molecules

The PDB Chemical Component Dictionary (CCD) was originally developed [47] to provide a more expressive alternative to the earliest PDB ligand descriptions, which were based purely on atom connectivity records. The CCD embraced data representations for chemical components developed for the PDBx/mmCIF data dictionary [36]. Each new chemical component coming into the archive is identified by a unique three-character alphanumeric code assigned by the wwPDB. The dictionary contains detailed chemical descriptions for standard and modified amino acids/nucleotides, small molecule ligands, and solvent/solute molecules (e.g., chemical properties, such as stereo chemical assignments, chemical descriptors, and systematic chemical names). A set of atomic model coordinates from a selected PDB entry and a computed set of ideal atomic coordinates are provided for each CCD entry. Hydrogen atoms are computationally added to the experimental coordinates and any unobserved heavy atoms, such as leaving groups, are included in the ideal coordinates. Exact matches between the PDB CCD and the Cambridge Structural Database (CSD) operated by CCDC [29] were identified in a collaborative effort, which revealed ~1400 common entries. An External Reference File containing both CCD and CSD descriptors of such matches is available from the PDB Chemical Component Model file (wwpdb.org/data/ccd).

A related PDB chemical reference dictionary is the Biologically Interesting molecule Reference Dictionary (BIRD) [48], which contains information about oligopeptide-like molecules in the PDB archive. BIRD entries include molecular weight and chemical formula, polymer sequence and connectivity, descriptions of structural features and functional classification, natural source, and external references to corresponding UniProt [49] or Norine [50] archived amino acid sequences. BIRD molecules may be represented as a polymer (with sequence information) or as a single compound (with chemical information). Preferred representations are specified in the BIRD file, with a representative PDB identifier. The BIRD resource provides both possible representations; sequence and chemical information are provided in parallel.

7 Distributed Data Dissemination and Value-Added wwPDB Partner Activities

PDB archival data are freely available to the public without limitations on use. Data are released either immediately after they have been fully biocurated/validated or—in most cases—when they are published in a scientific journal. Typically, either the author or the journal informs the wwPDB that the paper describing a given structure is about to be or has been published. At this stage, the primary literature reference for the entry is updated and all data are released together with the wwPDB Validation Report.

PDB data release occurs in two stages. Stage 1: every Saturday at 03:00 UTC the polymer sequences, ligand SMILES strings, and crystallization pH for new entries designated for release are made public (wwpdb.org/download/downloads). Two-stage release is performed as a courtesy to the protein structure modeling and computational chemistry communities to enable two blinded prediction challenges (CAMEO: cameo3d.org [51]; and D3R CELPP: drugdesigndata.org/about/celpp). Stage 2: every Wednesday at 00:00 UTC, all new entries designated for release are made publicly available through four wwPDB FTP sites (wwPDB: ftp.wwpdb.org; RCSB PDB: ftp.rcsb.org; PDBe: ftp.ebi.ac.uk/pub/databases/pdb/; PDBj: ftp.pdbj.org). On average, ~200 structures are released every week, corresponding to ~111,000 structures released/year. Annually, in late December, "snapshots" of the PDB archive are recorded and also made available for FTP download (RCSB PDB: ftp://snapshots.wwpdb.org/; PDBj: ftp://snapshots.pdbj.org/). The wwPDB FTP sites provide core data for many secondary data resources, services, and websites.

When the wwPDB was established in 2003, it was agreed that, to best serve science, wwPDB partner websites would complement one another on "Data Out" and offer many different kinds of services and features (RCSB PDB: rcsb.org; PDBe: pdbe.org; PDBj: pdbj.org; BMRB: bmrb.wisc.edu). Collectively, wwPDB FTP sites and partner websites support in excess of 500 million downloads of data files annually. Simply put, more than one million data files are downloaded by PDB users distributed across all inhabited continents every day of the year. Our records show that FTP downloads of PDB data were made to all but four of the 195 recognized independent states worldwide during the period 2012–2015 (excluding Central African Republic, Cote d'Ivoire, Kosovo, and Swaziland). No PDB FTP download requests were recorded from the disputed territory of Western Sahara during the same period.

8 Future of Structural Biology and the Role of the wwPDB

At present, PDB archival entries come exclusively from measurements using crystallography, NMR, and 3DEM. These mainstay structure determination methods involve the same four basic steps: (1) making measurements from a physical sample of a biological macromolecule(s); (2) utilizing a representation of the measured data that allows encoding of these data for use by a computable scoring function encompassing spatial restraints that directly compares predicted and measured experimental results; (3) construction of structural models of identical composition but differing spatial configurations, followed by identification of one or more models with superior scores from the scoring function; and (4) evaluation of structural models to quantify agreement between prediction and experiment and estimate the uncertainty of each structural model. Notwithstanding the enormous amounts of experimental data measured by structural biologists today, none of the three PDB-supported methods routinely produce sufficient data to serve as the sole source of spatial restraints with which to produce a high quality structural model of a biological macromolecule. Instead, structural biologists combine available experimental data with molecular mechanics force field descriptions of atomic structure for both biopolymers and small molecule ligands. These descriptions represent an essential source of additional spatial restraints corresponding to familiar items such as bond lengths, bond angles, descriptions of chiral centers, aromaticity, etc., which together with experimental data help to ensure that a structural model of a protein or nucleic acid chain makes chemical "sense."

Structural biologists today rely increasingly on complementary experimental measurements to improve research outcomes. For example, it is becoming commonplace to utilize, or "integrate," the results of SAS measurements as an additional source of spatial restraints when computing ensembles of structural models derived primarily from NMR data (reviewed in [52]). Specifically, SAS experimental data serve as a source of spatial restraints reflecting the overall dimensions and shape of the macromolecule, whereas NMR experimental data provide information regarding proximity of different parts of the biopolymer chain with respect to one another. Combined NMR-SAS structure determinations typically yield significant improvements in both accuracy and precision of structural models versus those computed solely with NMR data, particularly for dynamic systems [53, 54].

With the recent advent of direct electron detectors and improvements in sample preparation for electron microscopy under cryogenic conditions, 3DEM is poised to become *the* experimental method of choice for studying larger macromolecular systems, many of which are ill suited to either crystallography or NMR. While the number of 3DEM structural models determined at better than 4 Å resolution and released in the PDB archive is on the rise (3 in 2012 versus 68 in 2015), many 3DEM data sets of biological macromolecules are unlikely to yield atomic level structural models absent integration of complementary experimental data with the mass density map coming from 3DEM. To this end, cryo-electron microscopy studies are increasingly being combined with measurements using one or more of the following methods: crystallography, NMR, chemical cross-linking/mass spectrometry, Forster resonance energy transfer or FRET, and SAS (e.g., [55]). Structural models produced with these integrative (or hybrid) methods have been deposited in the PDB archive, but there is currently no mechanism for PDB archiving of experimental data and associated metadata generated by methods other than crystallography, NMR, and 3DEM. Moreover, there are no universally accepted procedures by which integrative structural models can be validated against experimental data combined from different methods.

In 2014, the wwPDB Integrative/Hybrid Methods Task Force was assembled to assess some of these challenges. Attendees included experts in relevant measurement techniques, integrative modeling, visualization, and experimental data/structural model archiving. The meeting culminated in a unanimous recommendation that the wwPDB work with subject matter experts from complementary experimental methods to ensure that integrative 3D structural models can be deposited to the PDB archive with appropriate bicuration/validation, and that all of the supporting experimental data and associated metadata be made publicly available through a system of federated data resources. An account of this meeting [56] provides guidance as to what experimental data and metadata should be archived, how data should be exchanged among data resources, and how structural models should be validated. Meeting participants quite deliberately decided not to prescribe the makeup of the federation. Instead, an Integrative/Hybrid Methods Working Group (led by Helen M. Berman, Andrej Sali, Torsten Schwede, and Jill Trewella) was established after the meeting to collaborate with the wwPDB partners in establishing the data resource federation. At the time of writing, the SASbDB resource [28] is working closely with wwPDB partners to develop joint data exchange and validation protocols to allow for deposition, annotation, and validation of 3D atomic level structural models determined via crystallography, NMR, or 3DEM combined with SAS data.

9 PDB Archive at 50 Years of Age

The PDB is just 5 years short of its 50th birthday. Based on current deposition rates, archival contents in 2021 will number well in excess of 150,000 entries (i.e., >20,000-fold bigger than in 1971). wwPDB partners are working closely with one another and the global structural biology community to ensure that a federated data resource system is established to enable Deposition, Biocuration, and Validation of 3D integrative structural models of biological macromolecules together with supporting data from diverse experimental methods and associated metadata. By 2021, it is also likely that the wwPDB partnership will have grown to encompass one or more additional regional data centers to help meet the needs of growing structural biology communities in different parts of the world.

Acknowledgments

The RCSB PDB is supported by the National Science Foundation (DBI 1338415), National Institutes of Health, and the Department of Energy; PDBe by the Wellcome Trust, BBSRC, MRC, EU, CCP4, and EMBL-EBI; PDBj by JST-NBDC; and BMRB by the National Institute of General Medical Sciences (GM109046). We thank Christine Zardecki for expert help with manuscript preparation.

References

1. Protein Data Bank (1971) Protein Data Bank. Nature New Biology 233:223
2. Kendrew JC, Bodo G, Dintzis HM et al (1958) A three-dimensional model of the myoglobin molecule obtained by X-ray analysis. Nature 181:662–666
3. Kendrew JC, Dickerson RE, Strandberg BE et al (1960) Structure of myoglobin: a three-dimensional Fourier synthesis at 2 Å resolution. Nature 185:422–427
4. Bolton W, Perutz MF (1970) Three dimensional fourier synthesis of horse deoxyhaemoglobin at 2.8 Ångstrom units resolution. Nature 228:551–552
5. Perutz MF, Rossmann MG, Cullis AF et al (1960) Structure of haemoglobin: a three-dimensional Fourier synthesis at 5.5 Å resolution, obtained by X-ray analysis. Nature 185:416–422
6. Cold Spring Laboratory (1972) Cold Spring Harbor Symposia on quantitative biology, vol 36. Cold Spring Laboratory Press, Cold Spring Harbor, NY
7. Berman H (2008) The Protein Data Bank: a historical perspective. Acta Crystallogr A 64:88–95
8. Meyer EF (1997) The first years of the Protein Data Bank. Protein Sci 6:1591–1597
9. International Union of Crystallography (1989) Policy on publication and the deposition of data from crystallographic studies of biological macromolecules. Acta Crystallogr A 45:658
10. Sussman JL, Lin D, Jiang J et al (1998) Protein Data Bank (PDB): database of three-dimensional structural information of biological macromolecules. Acta Crystallogr D Biol Crystallogr 54:1078–1084
11. Berman HM, Westbrook J, Feng Z et al (2000) The Protein Data Bank. Nucleic Acids Res 28:235–242
12. Standley DM, Kinjo AR, Kinoshita K et al (2008) Protein structure databases with new web services for structural biology and biomedical research. Brief Bioinform 9:276–285
13. Keller PA, Henrick K, McNeil P et al (1998) Deposition of macromolecular structures. Acta Crystallogr D Biol Crystallogr 54:1105–1108
14. Velankar S, van Ginkel G, Alhroub Y et al (2016) PDBe: improved accessibility of macromolecular structure data from PDB and EMDB. Nucleic Acids Res 44:D385–D395
15. Berman HM, Henrick K, Nakamura H (2003) Announcing the worldwide Protein Data Bank. Nat Struct Biol 10:980
16. Ulrich EL, Markley JL, Kyogoku Y (1989) Creation of a nuclear magnetic resonance data repository and literature database. Protein Seq Data Anal 2:23–37
17. Markley JL, Ulrich EL, Berman HM et al (2008) BioMagResBank (BMRB) as a partner in the Worldwide Protein Data Bank (wwPDB): new policies affecting biomolecular NMR depositions. J Biomol NMR 40:153–155
18. Ulrich EL, Akutsu H, Doreleijers JF et al (2008) BioMagResBank. Nucleic Acids Res 36:D402–D408
19. Velankar S, Best C, Beuth B et al (2010) PDBe: Protein Data Bank in Europe. Nucleic Acids Res 38:D308–D317
20. Lin D, Manning NO, Jiang J et al (2000) AutoDep: a web-based system for deposition and validation of macromolecular structural information. Acta Crystallogr D Biol Crystallogr 56:828–841
21. Tagari M, Tate J, Swaminathan GJ et al (2006) E-MSD: improving data deposition and structure quality. Nucleic Acids Res 34:D287–D290
22. Read RJ, Adams PD, Arendall WB et al (2011) A new generation of crystallographic validation tools for the Protein Data Bank. Structure 19:1395–1412
23. Montelione GT, Nilges M, Bax A et al (2013) Recommendations of the wwPDB NMR Validation Task Force. Structure 21:1563–1570
24. Henderson R, Sali A, Baker ML et al (2012) Outcome of the first electron microscopy validation task force meeting. Structure 20:205–214
25. Berman HM, Burley SK, Chiu W et al (2006) Outcome of a workshop on archiving structural models of biological macromolecules. Structure 14:1211–1217
26. Arnold K, Kiefer F, Kopp J et al (2009) The Protein Model Portal. J Struct Funct Genom 10:1–8
27. Trewhella J, Hendrickson WA, Kleywegt GJ et al (2013) Report of the wwPDB Small-Angle Scattering Task Force: data requirements for biomolecular modeling and the PDB. Structure 21:875–881
28. Valentini E, Kikhney AG, Previtali G et al (2015) SASBDB, a repository for biological small-angle scattering data. Nucleic Acids Res 43:D357–D363
29. Groom CR, Bruno IJ, Lightfoot MP et al (2016) The Cambridge Structural Database. Acta Crystallogr B 72:171–179
30. Adams PD, Aertgeerts K, Bauer C et al (2016) Outcome of the First wwPDB/CCDC/D3R Ligand Validation Workshop. Structure 24:502–508

31. Meyer PA, Socias S, Key J et al (2016) Data publication with the structural biology data grid supports live analysis. Nature Commun 7:10882
32. Markley JL, Ulrich EL, Westler WM et al (2003) Macromolecular structure determination by NMR spectroscopy. In: Bourne PE, Weissig H (eds) Structural bioinformatics. John Wiley & Sons, Inc., Hoboken, NJ, pp 89–113
33. Lawson CL, Patwardhan A, Baker ML et al (2016) EMDataBank unified data resource for 3DEM. Nucleic Acids Res 44:D396–D403
34. Iudin A, Korir PK, Salavert-Torres J et al (2016) EMPIAR: a public archive for raw electron microscopy image data. Nat Methods 13:387
35. Bernstein FC, Koetzle TF, Williams GJB et al (1977) Protein Data Bank: a computer-based archival file for macromolecular structures. J Mol Biol 112:535–542
36. Fitzgerald PMD, Westbrook JD, Bourne PE et al (2005) 4.5 Macromolecular dictionary (mmCIF). In: Hall SR, McMahon B (eds) International Tables for Crystallography G. Definition and exchange of crystallographic data. Springer, Dordrecht, The Netherlands, pp 295–443
37. Westbrook JD, Henrick K, Ulrich EL et al (2005) Appendix 3.6.2. The Protein Data Bank Exchange Data Dictionary. In: Hall SR, McMahon B (eds) International Tables for Crystallography G. Definition and exchange of crystallographic data. Springer, Dordrecht, The Netherlands, pp 195–198
38. Westbrook J, Ito N, Nakamura H et al (2005) PDBML: the representation of archival macromolecular structure data in XML. Bioinformatics 21:988–992
39. Kinjo AR, Suzuki H, Yamashita R et al (2012) Protein Data Bank Japan (PDBj): maintaining a structural data archive and resource description framework format. Nucleic Acids Res 40:D453–D460
40. Yokochi M, Kobayashi N, Ulrich EL et al (2016) Publication of nuclear magnetic resonance experimental data with semantic web technology and the application thereof to biomedical research of proteins. J Biomed Semantics 7:16
41. Malfois M, Svergun DI (2000) sasCIF: an extension of core Crystallographic Information File for SAS. J Appl Crystallogr 33:812–816
42. Ulrich EL, Argentar D, Klimowicz A et al (1996) STAR/CIF macromolecular NMR data dictionaries and data file formats. Acta Crystallogr A 52:C577–C577
43. Berman HM, Henrick K, Nakamura H et al (2009) The Worldwide Protein Data Bank. In: Gu J, Bourne PE (eds) Structural bioinformatics, 2nd edn. Wiley, Hoboken, NJ, pp 293–303
44. Doreleijers JF, Vranken WF, Schulte C et al (2012) NRG-CING: integrated validation reports of remediated experimental biomolecular NMR data and coordinates in wwPDB. Nucleic Acids Res 40:D519–D524
45. Doreleijers JF, Vranken WF, Schulte C et al (2009) The NMR restraints grid at BMRB for 5,266 protein and nucleic acid PDB entries. J Biomol NMR 45:389–396
46. Gutmanas A, Adams PD, Bardiaux B et al (2015) NMR Exchange Format: a unified and open standard for representation of NMR restraint data. Nat Struct Mol Biol 22:433–434
47. Westbrook JD, Shao C, Feng Z et al (2015) The chemical component dictionary: complete descriptions of constituent molecules in experimentally determined 3D macromolecules in the Protein Data Bank. Bioinformatics 31:1274–1278
48. Dutta S, Dimitropoulos D, Feng Z et al (2014) Improving the representation of peptide-like inhibitor and antibiotic molecules in the Protein Data Bank. Biopolymers 101:659–668
49. UniProt Consortium (2015) UniProt: a hub for protein information. Nucleic Acids Res 43:D204–D212
50. Caboche S, Pupin M, Leclere V et al (2008) NORINE: a database of nonribosomal peptides. Nucleic Acids Res 36:D326–D331
51. Haas J, Roth S, Arnold K et al (2013) The Protein Model Portal—a comprehensive resource for protein structure and model information. Database 2013:bat031
52. Prischi F, Pastore A (2016) Application of nuclear magnetic resonance and hybrid methods to structure determination of complex systems. Adv Exper Med Biol 896:351–368
53. Cornilescu G, Didychuk AL, Rodgers ML et al (2016) Structural analysis of multi-helical RNAs by NMR-SAXS/WAXS: application to the U4/U6 di-snRNA. J Mol Biol 428:777–789
54. Venditti V, Egner TK, Clore GM (2016) Hybrid approaches to structural characterization of conformational ensembles of complex macromolecular systems combining NMR residual dipolar couplings and solution X-ray scattering. Chem Rev 116:6305–6322
55. Erzberger JP, Stengel F, Pellarin R et al (2014) Molecular architecture of the 40SeIF1eIF3 translation initiation complex. Cell 158:1123–1135
56. Sali A, Berman HM, Schwede T et al (2015) Outcome of the First wwPDB Hybrid/Integrative Methods Task Force Workshop. Structure 23:1156–1167

Chapter 27

Databases, Repositories, and Other Data Resources in Structural Biology

Heping Zheng, Przemyslaw J. Porebski, Marek Grabowski, David R. Cooper, and Wladek Minor

Abstract

Structural biology, like many other areas of modern science, produces an enormous amount of primary, derived, and "meta" data with a high demand on data storage and manipulations. Primary data come from various steps of sample preparation, diffraction experiments, and functional studies. These data are not only used to obtain tangible results, like macromolecular structural models, but also to enrich and guide our analysis and interpretation of various biomedical problems. Herein we define several categories of data resources, (a) Archives, (b) Repositories, (c) Databases, and (d) Advanced Information Systems, that can accommodate primary, derived, or reference data. Data resources may be used either as web portals or internally by structural biology software. To be useful, each resource must be maintained, curated, as well as integrated with other resources. Ideally, the system of interconnected resources should evolve toward comprehensive "hubs", or Advanced Information Systems. Such systems, encompassing the PDB and UniProt, are indispensable not only for structural biology, but for many related fields of science. The categories of data resources described herein are applicable well beyond our usual scientific endeavors.

Key words Database, Repository, Data resource, Structural biology, Archive, Metadata, Information system

1 Introduction

Physics was the driving force of science in the first half of the twentieth century, followed by a shift of emphasis toward the biomedical sciences. We predict that the twenty-first century will be dominated by information technology and that the analysis of data will be even more important than the generation of new data. Massive amounts of data create significant challenges not only for data storage, but also for organization, accessibility, data mining, and analysis. As discussed elsewhere in this book, the Protein Data Bank (PDB) is the crown jewel of structural biology and a well-known example of a data resource used widely in biomedical sciences [1]. Besides the

Alexander Wlodawer et al. (eds.), *Protein Crystallography: Methods and Protocols*, Methods in Molecular Biology, vol. 1607, DOI 10.1007/978-1-4939-7000-1_27, © Springer Science+Business Media LLC 2017

PDB, there are many repositories and databases used in structural biology, chemistry, life sciences, and big pharma, where they are crucial in the drug discovery process.

Modern research produces an enormous amount of primary data. Within structural biology, the primary data come from the various steps of protein production and crystallization, from the X-ray and neutron diffraction experiments, NMR, cryo-EM, or from functional studies aimed at characterizing and/or verifying structure–function relationships. Technological advances and increased computational capabilities permit the collection of terabytes or even petabytes of primary experimental data in a very short time. For example, the Eiger 4M detector at station 30A-3 at the European Synchrotron Radiation Facility (ESRF) [2] can produce 750 diffraction images per second. This would correspond to 64.8 million data frames per day and, assuming 2000 h/year of ESRF operation (for further calculations, we assume that half of the time is used for sample changing, alignment, etc.), 2.7 billion data frames, totaling around 20 petabytes per year when operated at peak efficiency. Assuming conservatively that three data sets containing 1000 images each are sufficient to determine and deposit a macromolecular crystal structure, the output of station 30A-3 would be 270 times higher than the output of all 150 synchrotron stations located at roughly 40 synchrotrons around the world. The disparity between the theoretical data production rate and structure determination rate reflects our limited ability to use information technology to select diffraction-quality crystals prior to diffraction experiments, to find and remove other bottlenecks, and to convert mountains of data into useful information. Structural biology is not unique in this respect. In modern science, we have learned how to generate massive amounts of data, but unfortunately, creating scientific knowledge is a much harder task. Modern scientists quite often forget that "**data is not information, information is not knowledge, knowledge is not understanding, understanding is not wisdom** [3]."

The effective transformation of results into information requires that raw experimental data be associated with metadata—data about data—defining what the underlying numbers represent, what format they take, who collected them, etc. Metadata are crucial for organizing data into "databases." In a narrow technical sense, a database is any organized set of data, a raw material from which information can be "mined." The term "database" is also widely used to describe many data resources, but in our opinion, the term database should be reserved for resources that include some elements of an information system (IS), namely tools for extracting information from data that adhere to the basic paradigm: data-in, information-out. This is in contrast to the functionality of a data repository, which utilizes a data-in, data-out approach. In practice, the capabilities of information extraction tools of "database" resources available in structural biology are very diverse. Arguably, the Holy Grail—a full-fledged "information system"—remains to be implemented yet.

Data resources vary widely with respect to their design, complexity, accessibility, etc. Although data resources of all kinds can be useful, the impact of a resource is partially dependent on its sophistication/complexity and scope. For the purpose of this chapter, we define several specialized categories of data resources, such as (a) Archives, (b) Repositories, (c) Databases, and (d) Advanced Information Systems.[1] Depending on the stage and scale of the project, each of these four categories has its advantages and disadvantages, as summarized in Table 1.

2 Categories of Data Resources

Archives are the least sophisticated type of data resource in terms of complexity and are the most straightforward to implement. In the most simplistic terms, an archive is created when an experimenter collects experimental data on some sort of storage media. Data archives are usually not indexed and possess limited metadata. For that reason, they are usually easy to access and search by their creators, but searching an archive created by somebody else can be a gargantuan task. Due to the low initial set-up cost, archives are usually the most feasible way to store the first batch of primary data or to serve as a reference in computational studies. An electronic laboratory notebook (ELN) is a typical example of a personalized archive for an experimenter, and a set of PDB files is a typical archive used during structural computations.

Archives that are actively used by multiple contributors tend to grow very fast and become difficult to manage, especially when data are more complex than anticipated by the archive creator. This is usually the stage at which the archive has to be converted into a repository.

Repositories are characterized by the use of additional mechanisms (metadata) to index and annotate the primary data. Data in repositories may need to be validated in order to be presented in a more consistent manner and to facilitate searches by different contributors and users. A repository may either remain at small scale with a minimal set of metadata on top of an archive or become very sophisticated, with capabilities similar to a database. Repositories are the most well-known category that implies public accessibility,

[1] Herein we use the term "database" to mean a resource that includes not only a conventional database, but also an interface that facilitates data searches, data retrieval, and data analysis. This alleviates the need to know the internal architecture of the data and the particular query language of the underlying database.

Table 1
Typical characteristics for four specialized categories of data resources

	Archives	Repositories	Databases	Advanced Information Systems (AIS)
Complexity	Low	Medium	High	High
Content	"Raw" (deposited) data with little or no metadata[a]	Probably include some metadata[a]	Extensively curated metadata[a]	Extensively curated metadata[a], integrated with external resources
Searches and Retrieval	Data not necessarily indexed, searching cumbersome	Data is indexed, facilitating searches	Search usually driven by a database (in a technical sense)	Efficient search and retrieval
Data mining	Very difficult	Limited to basic statistics	Built-in data analysis and report generation tools. Precalculated result	Customizable tools for analysis of user data
Data validation	No validation	Limited validation	Full validation	Full validation; Mechanism for moderated user's corrections
Data architecture	No organization; typically just a set of files	Partial organization (e.g., subfolders)	Data is structured and maybe distributed	Data is structured and maybe distributed
Users/Audience	Usually limited to a single lab or institution	Collaborative/Public access	Single lab, Organization or Public	Organization or Public
Cost				
Setup	Low	Medium–High	High	Very High
Storage	Low–Medium	Medium	Medium	Medium
Maintenance	Low	Medium–High	High–Very High	Very High
Annotation	N/A	Medium–High	High	Very High
Curation	N/A	Low–Medium	Medium–Very High	Very High

[a]Metadata is data that describes other data. Meta is a prefix that in most information technology usages means "an underlying definition or description." Metadata summarizes basic information about data, which can make finding and working with particular instances of data easier

which may or may not be true for the other three categories. Repositories are usually less sophisticated than fully fledged databases, but their role and impact cannot be underestimated. For example, the Protein Data Bank is a gigantic repository [4] that handles a wide array of complicated data and serves as a central reference of macromolecular structural models. The UniProt repository plays a similar role for known protein sequences and for that reason is a few orders of magnitude larger in terms of the number of records [5].

The third category of data resource, the **Database**, is characterized by the paradigm of "data in, information out." All data are structured, and there are mechanisms to enforce internal and sometimes external consistency. Well-designed databases use various validation tools to analyze all incoming data and ensure their consistency with external resources. This type of resource has predefined data mining tools available for casual users, as well as sophisticated tools to carry out custom analyses. An example of a predefined report is an automatically generated draft of the methods section of a scientific manuscript providing information about protein production and diffraction experiments [6]. The ability to add a materials/methods section to a manuscript with minimal or no intervention would serve as the ultimate confirmation of the accuracy and usefulness of a database. One feature that distinguishes a database from a repository is the ability to update the data. The inability to change the data in a repository is not only due to technical limitations, but mainly because the repository is not necessarily the owner of the data.

An **Advanced Information Systems (AIS)** will invariably have a database at its core, but will have more connections to other data resources, pulling together information from disparate sources to provide as complete a picture as possible. AIS will have sophisticated tools to allow users to analyze the data, and may include mechanisms to allow others to access the data in an automated fashion. Registered users may have the ability to update information in an AIS, and well-designed systems will have a mechanism to keep track of the changes.

It is also important to keep in mind that boundaries among these four categories are fluid and subjective. A data resource of one archetype may also possess characteristics of other data resource categories. For example, although the PDB exhibits prominent properties of a repository, it also has many properties of a database, such as the ability to perform advanced searches of experimental details and subsequently to combine the results of different queries. The PDB policy of "obsoleting" deposits, which requires the agreement of the deposit's author, is one characteristic that arguably makes it more of a repository than a database. The degree of connectivity with other resources will distinguish AIS. It is also important to note that data resources can evolve into more sophisticated resources, or regress to more primitive forms when maintenance is no longer possible.

The storage and deployment of diffraction images is a typical example used to illustrate the usefulness of different categories of data resources in the life cycle of a project in structural biology. For example, the ftp server of the Center for Structural Genomics of Infectious Diseases (CSGID) [7] (http://www.csgid.org/pages/diffraction_images) was a simple data resource that fell in the "archive" category. This archive was very useful for gathering and preserving all the diffraction images collected by the CSGID (and later Seattle Structural Genomics Center for Infectious Disease) [8] project. The CSGID ftp server was easy to set up and ready for use in a matter of days. Although it was essentially impossible to retrieve data by means other than the target accession code, the CSGID ftp server was successfully used to share diffraction data between research groups and external users. There are other archives of diffraction images in other organizations. Virtually all synchrotron facilities have temporary archives of collected data that are wiped-out after a certain period of time. The Integrated Resource for Reproducibility in Macromolecular Crystallography (IRRMC) (http://ProteinDiffraction.org) [9] in its current form is a mix of such a repository and database, yet the ultimate goal of this project is to evolve into an AIS that has every bit of information structurally organized and fully validated. The transformation of the CSGID ftp server into the IRRMC resource is a good example of resource evolution without change of the data contained in it.

3 Structural Biology Data Resources

3.1 Primary Data Repositories

Despite the fact that many scientists treat the final structural models (both macromolecular and small molecular) as primary data, the models themselves are only interpretations obtained from experimental data and should be treated as derived data. In fact, only diffraction images (in crystallography), recorded spectra (in NMR), and unprocessed images (in electron microscopy) should be considered as primary data. A long-term, large-scale storage of diffraction images has recently become possible due to reduced cost of media and improved data access and storage technologies; however, the cost associated with data management remained steady or even increased due to the complexity of modern crystallography experiments. The anticipated size of a repository has a direct influence on the technical aspects of the design and also on the maintenance of the resource. A resource that has 10 or 100 diffraction experiments will have different issues than a resource that can accommodate 100,000 or one million experiments. Maintaining homogeneity of the data using automated systems becomes more difficult as the scale of a resource grows and the experimental complexity and resulting variability of data types increases.

The first large-scale public archives of macromolecular diffraction images were implemented independently by structural genomics consortia [7, 10]. Some synchrotron facilities have also created large-scale archives, but only a very limited subset of data is publicly available. For example, the Store.Synchrotron [11] implemented at the Australian Synchrotron facility, based on the MyTardis system [12], which claims thousands of archived diffraction experiments, makes only 35 of them publicly available. To ensure standardization of data and metadata, the IUCr established the Diffraction Data Deposition Working Group (DDDWG) [13]. Recently, two repositories that allow deposition of diffraction images of macromolecular structures emerged—IRRMC [9] and SBGrid DB [14]; both explore community needs and assess the technical capabilities and limitations. IRRMC also aims to provide an information system that would allow better data dissemination.

3.2 Reference Repositories

Although structural models are derived data, they serve as a foundation for many further studies and are treated as primary reference data. The reference data resources are usually repositories augmented with database functionality to facilitate data searches. Analysis of a large group of data requires the use of auxiliary tools. The prime example of a data repository used in structural biology is the earlier version of the PDB [15]. This repository is covered in another chapter in this book in detail, but for clarity of presentation, we briefly discuss it also here. The PDB repository [4] has five access sites: the wwPDB site, three data centers (PDBe, PDBj, and RCSB PDB), and an NMR-specific component, the Biological Magnetic Resonance Data Bank (BMRB). While the wwPDB site allows data validation and deposition as well as archive download, the remaining sites have more database capabilities that allow for data dissemination. All three PDB data centers utilize the common mmCIF format [16] to store the same underlying structural data; however, the design, information content, and analysis tools of each site are different. The three data centers create different user experiences and illustrate the different ways for a repository to evolve into an AIS.

Repositories of small molecule structures are indispensable for macromolecular crystallography because they serve as reference resources for all ligands bound to the macromolecules. The most renowned resource of this kind is the Cambridge Structural Database (CSD)—a database of organic and metallo-organic molecules, which comes additionally with a comprehensive package of tools for data mining and analysis [17]. An alternative is the Crystallography Open Database (COD) [18], which is an open-access archive of organic, inorganic, and metallo-organic compounds and minerals. There are also several other, specialized databases available, such as ICSD (Inorganic Crystal Structure Database) [19] and CRYSTMET (metals, alloys, and intermetallics) [20].

In addition to structural information, an enriched set of metadata about small molecules and chemical compounds—including but not limited to identity, chemical properties, and biological activity—can be accessed using the resources forming PubChem [21] and ChEMBL [22]. Similarly, information about proteins can be accessed from the Universal Protein Resource (UniProt) [5], while information about protein location, function and interactions can be accessed from Gene Ontology (GO) [23] or the Kyoto Encyclopedia of Genes and Genomes (KEGG) [24].

3.3 Derived Data Resources

The growing number of macromolecular structures in the PDB provides a solid foundation for and increases the scientific potential of derivative data resources that build upon the data from the PDB. These resources can be divided into three categories: classification/cataloguing, data presentation/analysis/processing, and data aggregation.

Fold classification databases such as CATH [25] and SCOP (Structural Classification of Proteins) [26] are examples of resources that classify the structural data present in the PDB. These databases aim to classify protein folds in terms of evolutionary relationships as well as structure and sequence similarity. The classifications provided by SCOP and CATH have become the de facto standards for describing the fold of a protein that is newly characterized in structural terms. The SCOP database is also a reference database for nonredundant folds and domains used by many structural bioinformatics tools.

SCOP is an interesting example illustrating the necessity of the long-term maintenance of widely adopted databases. The original SCOP classification was last updated in 2009. In 2012, some authors of the original SCOP developed a backwards compatible continuation called SCOPe [27], which is currently up-to-date with the PDB. In 2014, the laboratory which originally developed SCOP renovated the original SCOP classification system and called it SCOP2 [28]. As pointed out by the developers of SCOP2, the simple tree-like hierarchy used in SCOP was replaced by a network of nodes in SCOP2.

There are many other, specialized data resources that try to catalog and classify a different structural aspect beyond fold classifications. For example, several resources for membrane proteins have been developed. One of the first publicly available membrane protein databases was the Protein Data Bank of Transmembrane Proteins (PDBTM; http://pdbtm.enzim.hu) maintained by the Hungarian Institute of Enzymology [29–31]. It continues to provide an up-to-date list of 3D structures of transmembrane proteins based on the detection of transmembrane helices in the structure using the program TMDET [32]. As of April 2016, this resource contained 2771 PDB entries. MPStruc, the database of Membrane Proteins of Known 3D structure [33], established by Stephen White's group, not only identifies unique membrane protein structures (609 as of April 2016) but also provides the MPExplorer tool which can provide hydropathy plots based upon thermodynamic and biological principles, allowing examination of topological properties of membrane proteins (http://blanco.biomol.uci.edu/mpstruc/). The MemProtMD [34] has used the Coarse-Grained Self Assembly Molecular Dynamics simulations to compile the database of structures of over 2000 intrinsic membrane proteins inserted into simulated lipid bilayers. The Membrane Protein Data Bank [35] is still available on the Internet but has not been updated since 2011, also illustrating the problem of database maintenance.

An interesting example of classification of protein features is the KnotProt database [36], which runs a program detecting self-entanglements in protein chains in order to identify proteins whose 3D structures form knots or slipknots. KnotProt currently contains nearly 1400 entries, and it also allows users to submit structural data to check if they correspond to a knotted structure. The KnotProt effectively replaced several previous databases that were still available, but not maintained. The lack of maintenance and curation may have created inconsistencies between resources, leaving users with ambiguous information.

Bioinformatics services are another type of structural biology resources that build upon the structures in the PDB. A prominent example is PDBsum [37]. This service applies different tools and resources in order to present an at-a-glance overview of a protein structure. It includes analysis of structural attributes such as protein surfaces, cavities, and ligands, as well as interaction attributes such as the protein–protein, protein–DNA/RNA, and protein–small molecule interactions. The approach adopted by PDBsum is gradually being incorporated into the PDB access sites such as the RCSB, PDBe and PDBj.

Another well-established resource that supplements the PDB is the Uppsala Electron Density Server (EDS) [38]. In addition to the basic information about the deposit, EDS calculates the electron density maps from the deposited structure factors (if available) and provides the user with a straightforward way to check the real-space model correlation and to download and inspect electron density maps. The EDS is used internally by COOT [39, 40], PyMOL [41] and other programs to download ready-to-view electron density maps for inspection. As with many servers that reach the end of an initial funding period, this server is still available and running in an automatic mode, but it is no longer supported or developed, and there is no mechanism to correct errors, despite its widespread usefulness and appreciation [42].

Another group of structural bioinformatics resources provide data analysis/processing. An example of such data resource is PDB_REDO [43], which automatically re-refines the structures with available structure factors in the PDB using the latest versions of crystallographic tools, and makes the revised structures and re-refinement statistics available for download. PDB_REDO partially implements the concept of the "living PDB" [44]; however, fully automated re-refinement has a long way to go [45, 46]. Somewhat related data resources that provide precalculated results are the repositories of comparative models. The ProteinModelPortal [47] provides a gateway to several repositories of precalculated comparative models, such as ModBase [48] and SWISS-MODEL Repository [49]. Several databases provide precomputed results useful for the analysis of the macromolecular structures deposited in the PDB. For example, PDBFlex [50] provides a database of precalculated structural alignments of different structures of the same protein and analyzes them to explore the flexibility of protein structures. PDBePISA implements a repository of precalculated PISA results for all PDB deposits [51, 52]. These results may be used to analyze protein-protein interfaces and for prediction of energetically favorable assemblies.

The Protein Structure Initiative Structural Biology Knowledge Base (PSI SBKB) [53], which was based on aggregating data from a number of resources created by the PSI programs, has become an important "added value" resource. By combining tools such as KB-Rank and KB-Role with searches through multiple resources, it allows users to identify connections between sequence, structure, annotation, and function. Unfortunately, due to the termination of the PSI, operation of the PSI SBKB is currently scheduled to end in July 2017.

Recently, the NIH Big Data to Knowledge (BD2K) initiative has started building a prototype of the "super-aggregator" DataMed, intended as a one-stop service providing an entry point to all biomedical and health-care related data resources (http://biocaddie.org). Currently, it has indexed a single structural biology resource—the PDB. However, as it also includes data repositories from other domains (e.g., sequence, gene expression, proteomics, clinical trials, and others), it already allows for identifying potential relationships among the PDB structures and datasets from these different domains [54].

4 Data Management in Structural Biology

From the onset of the high-throughput era, different laboratories recognized the need for efficient tools for tracking experimental protocols, parameters, personnel, reagents, remarks, and results. Different tools emerged to serve the particular needs of individual laboratories and their workflows. Apart from generic electronic lab notebooks (ELN) and generic laboratory information management systems (LIMS), several databases specializing in the handling of structural biology pipelines have been developed. The development was mainly motivated and supported by the structural genomics programs to accommodate the needs of laboratories, consortia, and shared user facilities such as synchrotrons, but was later demonstrated to be very productive as tailored LIMS to track experiments in collaborations and laboratories of any size. Examples of databases that are used by structural genomics are SESAME [55] and LabDB [6]. SESAME was initially developed as a database for tracking NMR experiments at the National Magnetic Resonance Facility at Madison [55] and was further enhanced to serve the needs of the Center for Eukaryotic Structural Genomics (CESG) [56]. LabDB, on the other hand, was initially developed as a LIMS for a single laboratory to track crystallization experiments and to serve as a companion database to HKL-3000 [57], which allowed for a tight integration of experimental and computational parts into a unified crystallographic pipeline. LabDB was later adopted as a central LIMS by several structural genomics centers and was enhanced with data harvesting and data analysis capabilities for different experiments. ISPyB is yet another example of a database used to track experiments. It was designed for the ESRF synchrotron to allow users to track their samples and experiments and is now an integral part of the data collection systems at the ESRF and Diamond synchrotrons, and is crucial for further automation.

4.1 Reporting/Data Analysis

The actions and results of structural biology and data mining programs during data processing and structure refinement can also be viewed as data resources for the purpose of keeping track of history, reporting, and data analysis. Simple project management tools exist both in the CCP4 Interface and in PHENIX. Using a text-file based database, both of these project management tools allow tracking jobs, associated files, and runtime parameters. In addition to the text file, other metadata about the history of data harvesting and processing are stored alongside structure factors in MTZ files. Simpler job handling panels that utilize underlying databases are also available in many web servers, such as Robetta [58], MolProbity [59], TLSMD [60], or surface entropy reduction prediction (SERp) server [61].

On the other hand, HKL-3000 runs a full relational database backend (HKLDB) and stores the complete histories of diffraction experiments, data processing, and structure determination. HKLDB seamlessly integrates with LabDB to provide an experimental history of the crystal. The approach of running a full relational database in this scenario also allows for effective project management and collaboration within and between labs and institutions, as well as large-scale data analysis of different approaches and results. The HKLDB/LabDB system has also provided essential data resources to support the statistics and reporting tools used in several structural genomics web portals, such as the CSGID database.

5 Data Resources Used Internally by Structural Biology Software

Many crystallographic software packages use one or several data resources to perform their tasks. These collections of data may include external, unprocessed data sets such as sequence databases, collections of structures, or data prepared for specific application. The data resources used in different applications can be of various sizes, have different sophistication levels, and be characterized by different curation levels. Very often, the fact that the software relies on these resources is not obvious, as the user sees only the interface, and the usage of some kind of a database, either as a data source or data storage, is an internal implementation detail. The quality and type of the data resources used may significantly impact the outcome of the application; therefore, it is important to realize what types of data resources are used internally by specific applications. Here, we would like to highlight several notable, application-specific data resources that form the foundation of various tools used in structural biology.

One reason that an application or data resource may incorporate information from other data resources is to provide additional information that cannot be produced by the program itself or is computationally expensive to recreate. For example, the information content (resolution) of a macromolecular diffraction experiment is typically not sufficient to build a satisfactory model of a macromolecule without prior knowledge about protein geometry; thus, the majority of macromolecular structures are modeled and refined using "restrained refinement". As the name implies, during such refinement atoms are not permitted to move freely but are restricted by various "rules". There are several classes of restraints that are commonly used (covered in other chapters), but for the purpose of this chapter, we will focus on restraints (and their preparation) that are based on different data resources.

The most basic restraints that are used for the refinement of macromolecular structures are restraints that are applied to the monomer residues and small-molecule residues. Programs that perform either reciprocal-space refinement, such as REFMAC [62], PHENIX [63] or SHELXL [64], or real-space refinement, e.g., COOT [40], use an archive of text files to store pregenerated definitions of monomers and some small molecules. Since this is the primary source of correct geometries for the refined residues, any inaccuracies in the restraint library would necessarily propagate to the refined structure and its interpretation [62]. While structural units of macromolecule monomers, such as amino acids and nucleotides, are well defined due to abundant information, small-molecule ligands may have a much wider variation and may be represented by only a limited number of experimental data. Therefore, it was necessary to put a significant effort toward the development of software like LIBCHECK [65], PURY [66], eLBOW [67], and AceDRG [68]. Built on top of prior knowledge of bond lengths and bond and dihedral angles from small molecule crystallography databases such as CSD or COD and ab initio calculations, these data resources perform additional validation and generate custom definitions of the small molecule. Moreover, software such as PURY may also modify or augment the existing data resources to present a more consistent library of restraints for the refinement program. Nevertheless, it must be stressed that any automatically generated ligand restraints should be checked manually for chemical sense because errors, which are still frequent, if uncorrected may lead to lamentable consequences.

With the rapid growth of the PDB, it became possible to develop new applications that efficiently leverage the knowledge gained from existing structures. The applications useful in macromolecular model building are usually supported by underlying data resources that are derived from the PDB. These underlying data resources are usually highly abstracted and curated with only relevant data extracted to serve the algorithm with a significant boost in quality, validity, or speed. For example, as the number of high-resolution structures in the PDB increases, libraries of amino-acid side-chain rotamers evolve over time—leading to popular libraries like the backbone-independent library [69] or the backbone-dependent one [70]. Amino acid side-chain rotamers are commonly stored as a simple combination of the representative χ angles or actual atom coordinates and rotamer occurrence frequencies, and are widely used in model-building and refining software including COOT, PHENIX, or ROSETTA [71]. Sometimes, for specific purposes, especially for exhaustive search as in Fitmunk [72] or Rapper [73], more complicated rotamer libraries containing many curated conformations from representative structures are used. Additionally, some model building programs such as ARP/wARP [74], Buccaneer [75], or RESOLVE [76] use larger fragments of proteins to build the complete model.

Another way a data resource can be used by structural software is to serve as an internal reference of structures or structural features, such as those used for molecular replacement or model building. For example, the BALBES [68] and MORDA [77] molecular replacement pipelines use a prepared database of PDB structures to search for a model suitable for molecular replacement and to carry out structure determination—a tactic that is extremely successful when a simple approach of using a manually selected model fails. Similarly PHASER [78] and AMoRe [79] allow users to create their own small archives of structures that are used to try alternate solutions. Robetta [58] and MODELLER [80] can use prior knowledge in an archive of individual domains or protein fragments to build a complete multi-domain homology model.

Structural biologists rely on various tools not only during structure refinement, but also before and after the experiment, in order to design experiments and analyze structural models, respectively. For example, tools predicting different properties from protein sequence, including protein secondary structure (e.g., JPred4 [81]), intrinsic disorder (e.g., IUPred [82]), and cavities (e.g., SPACEBALL [83]), or protein crystallization propensity (e.g., PDPredictor [84], XtalPred [85]) utilize information derived from protein structures (IUPred, SPACEBALL), protein structures and sequence databases (JPred4), or databases of crystallization trials (PDBPredictor, XtalPred). Tools that predict function from the determined structure, such as ProFUNC [86], combine multiple algorithms and databases.

Utilities that analyze the PDB for various interactions include Bio3D [87], MED-SuMo [88], PDBeMotif [89], and the NEIGHBORHOOD database [90]. Although all were built upon the concept about the interactions between different residues in the protein structure network, these databases have different focuses. While Bio3D and MED-SuMo focus upon the analysis of a single structure, PDBeMotif focuses on the search of a predefined motif, and the NEIGHBORHOOD database focus on the analysis of a group of structures, or even the whole PDB. Running one of these databases in the backend would facilitate various types of front-end applications [90]. For example, CheckMyMetal (CMM) [91] is a web service for the validation of metal-binding environments in macromolecular structures built on top of the NEIGHBORHOOD database.

6 Concluding Remarks

The databases and other data resources used in structural biology are not limited to web portals or servers used to access the databases. The internal use of databases by crystallographic software pointedly illustrates that these resources also underlie various structural biology applications. The use of a database can be completely hidden, because it is utilized only by a particular application. Hence, a scientist may utilize a database without even realizing it. Structural biology data resources are also very deeply interconnected with each other, with reference repositories such as PDB and UniProt being accessed by many others (Fig. 1).

It is difficult to measure the impact of a data resource. The simplest metric is the number of accesses. However, these days this number is skewed by automatic robots that scan these resources monthly or even weekly. For that reason, they all have very similar, high numbers of accesses, despite significant differences in the usefulness of various resources. Another metric is the number of citations; however, researchers quite often do not cite the servers that they used to create hypotheses and citations in supplementary

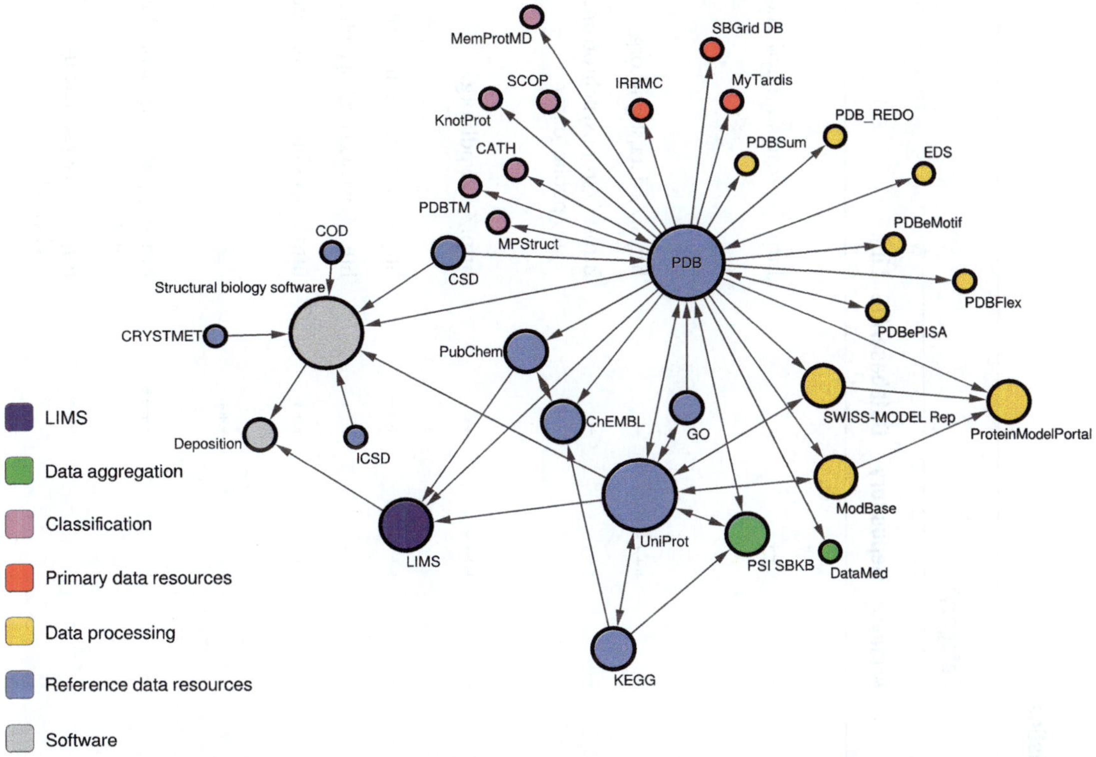

Fig. 1 Visualization of data exchange between data resources mentioned in this chapter. The selected cross-references are visualized as a graph, where each node corresponds to a particular resource and an edge represent data exchange or reference. The radius of each node illustrates the number of resources that use or reference a given data resource. Different roles of the data resources are color coded. The *arrows* show the direction of data flow

materials are not indexed. A good illustration of the disparity between the numbers of citations and access is provided by one of the ribosome structures, 4wqf, which has been downloaded 9375 times, but which has been officially cited just once, according to the Thomson Reuters Data Citation Index [92]. For that reason, some resources create resource-specific metrics to measure their usefulness. For example, the publishers of the articles describing the servers developed in our laboratory, CMM and Fitmunk, report that these articles were accessed/read 2013 and 505 times as of June 2016. The corresponding publications were cited 47 and 1 times, respectively (according to Web of Science in June, 2016). However, internal statistics reveal the servers were used 14,625 and 157 times, respectively.

This chapter presents a survey of the data resources that we find useful for structural biology during sample preparation, experiment planning and execution, structure determination, and structure–function exploration of proteins and nucleic acids, as summarized in Table 2. It also presents the data resources that are sometimes used “behind-the-scenes” but can have significant impact on the

Table 2

The data resources in structural biology described in this chapter and their characteristics

Common/abbreviated name	Full name	Type	Category[a] Archive	Repository	Database/AIS	Link
Primary data repositories						
IRRMC	Integrated Resource for Reproducibility in Macromolecular Crystallography	Protein diffraction images	–	+++	++	http://proteindiffraction.org/
SBGrid DB	The Structural Biology Data Grid	Protein diffraction images	–	+++++	–	https://data.sbgrid.org/
Store.Synchrotron		Protein diffraction images	++	+++	–	https://store.synchrotron.org.au/public_data/
Reference repositories						
PDB[b]	Protein Data Bank	Macromolecular structures and structure factors	–	++++	+	http://wwpdb.org/
CSD	Cambridge Structural Database	Small molecular structures	–	++	+++	http://www.ccdc.cam.ac.uk/
COD	Crystallography Open Database	Small molecular structures	+	+++	+	http://www.crystallography.net/
ICSD	Inorganic Crystal Structure Database	Inorganic molecular structures	–	++	+++	https://icsd.fiz-karlsruhe.de
CRYSTMET		Metals, alloys and intermetallics structures	–	++	+++	
PubChem	PubChem	Chemical substances and their activity	–	+	++++	https://pubchem.ncbi.nlm.nih.gov/
ChEMBL		Chemical substances and their activity	–	+	++++	https://www.ebi.ac.uk/chembl/

UniProt	Universal Protein Resource	Protein sequences	–	+	++++	http://www.uniprot.org/
GO	Gene Ontology	Function of gene products	–	–	+++++	http://geneontology.org/
KEGG	Kyoto Encyclopedia of Genes and Genomes	Integrated resource about biological systems	–	–	+++++	http://www.genome.jp/kegg/
Derived data resources						
Classification						
SCOP	Structural Classification of Proteins	Protein fold classification	+	++	++	http://scop.mrc-lmb.cam.ac.uk/scop/
SCOP2	Structural Classification of Proteins 2	Protein fold classification	–	++	+++	http://scop2.mrc-lmb.cam.ac.uk/
SCOPe	Structural Classification of Proteins—extended	Protein fold classification	–	+	++++	http://scop.berkeley.edu/
CATH	CATH	Protein fold classification	–	+	++++	http://www.cathdb.info/
PDBTM	Protein Data Bank of Transmembrane Proteins	Selection and classification of membrane proteins	–	++++	+	http://pdbtm.enzim.hu
MPStruc	Membrane Proteins of Known 3-D structure	Selection and classification of membrane proteins	–	+++	++	http://blanco.biomol.uci.edu/mpstruc/
MemProtMD	A Database of Membrane Proteins Embedded in Lipid Bilayers	Selection of membrane proteins and MD simulations in lipid bilayers	–	+++	++	http://sbcb.bioch.ox.ac.uk/memprotmd/
KnotProt		Database of "knotted" proteins	–	+	++++	http://knotprot.cent.uw.edu.pl/
Data presentation/analysis/processing						
PDBSum	PDBsum	Macromolecules and their properties	–	++	+++	https://www.ebi.ac.uk/thornton-srv/databases/cgi-bin/pdbsum/GetPage.pl?pdbcode=index.html

(continued)

Table 2
(continued)

Common/abbreviated name	Full name	Type	Category[a]			Link
			Archive	Repository	Database/AIS	
EDS	Uppsala Electron Density Server	Macromolecules and ready to use electron density maps	+	+	+++	http://eds.bmc.uu.se/eds/
RSCB/PDBe/PDBj[b]		Access sites to wwPDB	–	+	++++	http://www.rcsb.org/ http://www.ebi.ac.uk/pdbe/ http://pdbj.org/
PDB_REDO	PDB_REDO	Automatically re-refined macromolecular structures	+	++	++	http://www.cmbi.ru.nl/pdb_redo/
ProteinModelPortal ModBase SWISS-MODEL Repository		Automatically generated comparative models	+	++	++	http://www.proteinmodelportal.org/ http://modbase.compbio.ucsf.edu/modbase-cgi/index.cgi http://swissmodel.expasy.org/repository/
PDBePISA		Analysis of assemblies and interfaces	+++	+	+	http://www.ebi.ac.uk/pdbe/pisa/
PDBFlex		Analysis of protein conformation variability	–	–	+++++	http://pdbflex.org/
PDBeMOTIF		Search of 3D and sequence motifs in proteins	–	–	+++++	http://www.ebi.ac.uk/pdbe-site/pdbemotif/
Data aggregation						
PSI SBKB	Protein Structure Initiative Structural Biology Knowledge Base			+	++++	http://sbkb.org/

DataMed/ bioCADDIE	Biomedical and Healthcare Data Discovery Index Ecosystem		N/A	N/A	N/A	http://biocaddie.org
Data management						
SESAME		LIMS			+++++	http://www.sesame.wisc.edu/
LabDB		LIMS			+++++	https://jurand.med.virginia.edu/labdb
ISPyB	Information System for Protein crYstallography Beamline	Experiment tracking at synchrotrons			+++++	https://wwws.esrf.fr/ispyb

[a]Each resource was assigned a score (from – to +++++) to gauge how well it represents a given category of data resource. As mentioned in the text, the boundaries between particular categories are not well defined, therefore the scores presented here are arbitrary—the presented scores reflect the authors' subjective evaluation. Database and AIS types have been scored together because the boundaries between these two are especially ill-defined

[b]The RCSB/PDBe/PDBj sites add additional database functionalities and data aggregation on top of the wwPDB repository and, therefore, are listed separately

productivity, reproducibility, and validity of the experiments. In the rapidly evolving world of information technology and data science, we expect that parts of this chapter could become outdated very quickly. For example, although we are strong promoters of using DOIs to permanently and unambiguously identify and locate a data resource, in our observation most resources do not currently assign DOIs. When new versions of data resources become available, the old URL may become obsolete and replaced by an updated URL. Therefore, to locate the corresponding data resource mentioned in this chapter, one may want to harness search engines to find the updated location. We would also like to refer the readers to the biosharing.org portal, which hosts references to various data standards, databases, and policies in the life, environmental, and biomedical sciences, and to re3data.org—a global registry of research data repositories, for finding other useful data resources not limited to structural biology.

Acknowledgments

We would like to thank Ewa Niedzialkowska, Kasia Handing, Ivan Shabalin, Esther Sheler, Ethan Steen, Cody LaRowe, and Barat Venkataramany for help and numerous discussions/suggestions. This work was supported by National Institutes of Health Grants HG008424, GM053163, GM117325, GM117080 as well as with federal funds from the National Institute of Allergy and Infectious Diseases, National Institutes of Health, Department of Health and Human Services, under Contract No. HHSN272201200026C.

References

1. Rose PW, Prlic A, Bi C, Bluhm WF, Christie CH, Dutta S et al (2015) The RCSB Protein Data Bank: views of structural biology for basic and applied research and education. Nucleic Acids Res 43(Database issue):D345–D356
2. http://www.esrf.eu/home/UsersAndScience/Experiments/MX/About_our_beamlines/id30a-3--massif-3.html
3. http://todayinsci.com/S/Stoll_Clifford/StollClifford-Quotations.htm
4. Huang YH, Rose PW, Hsu CN (2015) Citing a data repository: a case study of the Protein Data Bank. PLoS One 10(8):e0136631
5. UniProt Consortium (2015) UniProt: a hub for protein information. Nucleic Acids Res 43(Database issue):D204–D212
6. Zimmerman MD, Grabowski M, Domagalski MJ, Maclean EM, Chruszcz M, Minor W (2014) Data management in the modern structural biology and biomedical research environment. Methods Mol Biol 1140:1–25
7. Anderson WF (2009) Structural genomics and drug discovery for infectious diseases. Infect Disord Drug Targets 9(5):507–517
8. Myler PJ, Stacy R, Stewart L, Staker BL, Van Voorhis WC, Varani G et al (2009) The Seattle Structural Genomics Center for Infectious Disease (SSGCID). Infect Disord Drug Targets 9(5):493–506
9. Grabowski M, Langner KM, Cymborowski M, Porebski PJ, Sroka P, Zheng H, Cooper DR, Zimmerman MD, Elsliger MA, Burley SK, Minor W (2016) A public database of macromolecular diffraction experiments. Acta crystallogr D Biol Crystallogr 72:1181–1193
10. Elsliger MA, Deacon AM, Godzik A, Lesley SA, Wooley J, Wuthrich K et al (2010) The JCSG high-throughput structural biology pipeline. Acta Crystallogr F Struct Biol Commun 66(Pt 10):1137–1142
11. Meyer GR, Aragao D, Mudie NJ, Caradoc-Davies TT, McGowan S, Bertling PJ et al (2014)

Operation of the Australian Store.Synchrotron for macromolecular crystallography. Acta Crystallogr D Biol Crystallogr 70(Pt 10):2510–2519

12. Androulakis S, Schmidberger J, Bate MA, DeGori R, Beitz A, Keong C et al (2008) Federated repositories of X-ray diffraction images. Acta Crystallogr D Biol Crystallogr 64(Pt 7):810–814
13. Terwilliger TC (2014) Archiving raw crystallographic data. Acta Crystallogr D Biol Crystallogr 70(Pt 10):2500–2501
14. Meyer PA, Socias S, Key J, Ransey E, Tjon EC, Buschiazzo A et al (2016) Data publication with the structural biology data grid supports live analysis. Nat Commun 7:10882
15. Bernstein FC, Koetzle TF, Williams GJ, Meyer EF Jr, Brice MD, Rodgers JR et al (1977) The Protein Data Bank: a computer-based archival file for macromolecular structures. J Mol Biol 112(3):535–542
16. Westbrook JD, Bourne PE (2000) STAR/mmCIF: an ontology for macromolecular structure. Bioinformatics 16(2):159–168
17. Allen FH (2002) The Cambridge Structural Database: a quarter of a million crystal structures and rising. Acta Crystallogr B 58(Pt 3 Pt 1):380–388
18. Grazulis S, Daskevic A, Merkys A, Chateigner D, Lutterotti L, Quiros M et al (2012) Crystallography Open Database (COD): an open-access collection of crystal structures and platform for world-wide collaboration. Nucleic Acids Res 40(Database issue):D420–D427
19. Belsky A, Hellenbrandt M, Karen VL, Luksch P (2002) New developments in the Inorganic Crystal Structure Database (ICSD): accessibility in support of materials research and design. Acta Crystallogr B 58(Pt 3 Pt 1):364–369
20. White PS, Rodgers JR, Le Page Y (2002) CRYSTMET: a database of the structures and powder patterns of metals and intermetallics. Acta Crystallogr B 58(Pt 3 Pt 1):343–348
21. Kim S, Thiessen PA, Bolton EE, Chen J, Fu G, Gindulyte A et al (2016) PubChem Substance and Compound databases. Nucleic Acids Res 44(D1):D1202–D1213
22. Bento AP, Gaulton A, Hersey A, Bellis LJ, Chambers J, Davies M et al (2014) The ChEMBL bioactivity database: an update. Nucleic Acids Res 42(Database issue):D1083–D1090
23. Ashburner M, Ball CA, Blake JA, Botstein D, Butler H, Cherry JM et al (2000) Gene ontology: tool for the unification of biology. The Gene Ontology Consortium. Nat Genet 25(1):25–29
24. Kanehisa M, Goto S (2000) KEGG: kyoto encyclopedia of genes and genomes. Nucleic Acids Res 28(1):27–30
25. Greene LH, Lewis TE, Addou S, Cuff A, Dallman T, Dibley M et al (2007) The CATH domain structure database: new protocols and classification levels give a more comprehensive resource for exploring evolution. Nucleic Acids Res 35(Database issue):D291–D297
26. Murzin AG, Brenner SE, Hubbard T, Chothia C (1995) SCOP: a structural classification of proteins database for the investigation of sequences and structures. J Mol Biol 247(4):536–540
27. Fox NK, Brenner SE, Chandonia JM (2014) SCOPe: Structural Classification of Proteins—extended, integrating SCOP and ASTRAL data and classification of new structures. Nucleic Acids Res 42(Database issue):D304–D309
28. Andreeva A, Howorth D, Chothia C, Kulesha E, Murzin AG (2014) SCOP2 prototype: a new approach to protein structure mining. Nucleic Acids Res 42(Database issue):D310–D314
29. Kozma D, Simon I, Tusnady GE (2013) PDBTM: Protein Data Bank of transmembrane proteins after 8 years. Nucleic Acids Res 41(Database issue):D524–D529
30. Tusnady GE, Dosztanyi Z, Simon I (2004) Transmembrane proteins in the Protein Data Bank: identification and classification. Bioinformatics 20(17):2964–2972
31. Tusnady GE, Dosztanyi Z, Simon I (2005) PDB_TM: selection and membrane localization of transmembrane proteins in the protein data bank. Nucleic Acids Res 33(Database issue):D275–D278
32. Tusnady GE, Dosztanyi Z, Simon I (2005) TMDET: web server for detecting transmembrane regions of proteins by using their 3D coordinates. Bioinformatics 21(7):1276–1277
33. Moraes I, Evans G, Sanchez-Weatherby J, Newstead S, Stewart PD (2014) Membrane protein structure determination—the next generation. Biochim Biophys Acta 1838(1 Pt A):78–87
34. Stansfeld PJ, Goose JE, Caffrey M, Carpenter EP, Parker JL, Newstead S et al (2015) MemProtMD: automated insertion of membrane protein structures into explicit lipid membranes. Structure 23(7):1350–1361
35. Raman P, Cherezov V, Caffrey M (2006) The Membrane Protein Data Bank. Cell Mol Life Sci 63(1):36–51
36. Sulkowska JI, Rawdon EJ, Millett KC, Onuchic JN, Stasiak A (2012) Conservation of complex knotting and slipknotting patterns in proteins. Proc Natl Acad Sci U S A 109(26):E1715–E1723
37. Laskowski RA (2009) PDBsum new things. Nucleic Acids Res 37(Database issue):D355–D359
38. Kleywegt GJ, Harris MR, Zou JY, Taylor TC, Wahlby A, Jones TA (2004) The Uppsala

electron-density server. Acta Crystallogr D Biol Crystallogr 60(Pt 12 Pt 1):2240–2249

39. Emsley P, Cowtan K (2004) Coot: model-building tools for molecular graphics. Acta Crystallogr D Biol Crystallogr 60:2126–2132
40. Emsley P, Lohkamp B, Scott WG, Cowtan K (2010) Features and development of Coot. Acta Crystallogr D Biol Crystallogr 66(Pt 4):486–501
41. Schrödinger L (2010) The PyMOL Molecular Graphics System, version~1.3r1. Schrodinger, LLC, New York
42. Shabalin IG, Dauter Z, Jaskolski M, Minor W, Wlodawer A (2015) Crystallography and chemistry should always go together: a cautionary tale of protein complexes with cisplatin and carboplatin. Acta Crystallogr D Biol Crystallogr 71:1965–1979
43. Joosten RP, Joosten K, Cohen SX, Vriend G, Perrakis A (2011) Automatic rebuilding and optimization of crystallographic structures in the Protein Data Bank. Bioinformatics 27(24):3392–3398
44. Terwilliger TC, Bricogne G (2014) Continuous mutual improvement of macromolecular structure models in the PDB and of X-ray crystallographic software: the dual role of deposited experimental data. Acta Crystallogr D Biol Crystallogr 70(Pt 10):2533–2543
45. Cooper DR, Porebski PJ, Chruszcz M, Minor W (2011) X-ray crystallography: assessment and validation of protein-small molecule complexes for drug discovery. Expert Opin Drug Discov 6(8):771–782
46. Chruszcz M, Domagalski M, Osinski T, Wlodawer A, Minor W (2010) Unmet challenges of structural genomics. Curr Opin Struct Biol 20(5):587–597
47. Haas J, Roth S, Arnold K, Kiefer F, Schmidt T, Bordoli L et al (2013) The Protein Model Portal—a comprehensive resource for protein structure and model information. Database (Oxford) 2013:bat031
48. Arnold K, Kiefer F, Kopp J, Battey JN, Podvinec M, Westbrook JD et al (2009) The Protein Model Portal. J Struct Funct Genom 10(1):1–8
49. Kopp J, Schwede T (2004) The SWISS-MODEL Repository of annotated three-dimensional protein structure homology models. Nucleic Acids Res 32(Database issue):D230–D234
50. Hrabe T, Li Z, Sedova M, Rotkiewicz P, Jaroszewski L, Godzik A (2016) PDBFlex: exploring flexibility in protein structures. Nucleic Acids Res 44(D1):D423–D428
51. Krissinel E (2010) Crystal contacts as nature's docking solutions. J Comput Chem 31(1):133–143
52. Krissinel E, Henrick K (2007) Inference of macromolecular assemblies from crystalline state. J Mol Biol 372(3):774–797
53. Gabanyi MJ, Adams PD, Arnold K, Bordoli L, Carter LG, Flippen-Andersen J et al (2011) The Structural Biology Knowledgebase: a portal to protein structures, sequences, functions, and methods. J Struct Funct Genom 12(2):45–54
54. Ohno-machado L, Alter G, Fore I, Martone M, Sansone S, Xu H (2015) bioCADDIE white paper—Data Discovery Index figshare. http://dx.doi.org/10.6084/m9.figshare.1362572 Retrieved 01:57, 01 Apr 2015 (GMT)
55. Zolnai Z, Lee PT, Li J, Chapman MR, Newman CS, Phillips GN Jr et al (2003) Project management system for structural and functional proteomics: Sesame. J Struct Funct Genom 4(1):11–23
56. Markley JL, Aceti DJ, Bingman CA, Fox BG, Frederick RO, Makino S et al (2009) The Center for Eukaryotic Structural Genomics. J Struct Funct Genom 10(2):165–179
57. Minor W, Cymborowski M, Otwinowski Z, Chruszcz M (2006) HKL-3000: the integration of data reduction and structure solution—from diffraction images to an initial model in minutes. Acta Crystallogr D Biol Crystallogr 62:859–866
58. Kim DE, Chivian D, Baker D (2004) Protein structure prediction and analysis using the Robetta server. Nucleic Acids Res 32(Web Server issue):W526–W531
59. Chen VB, Arendall WB 3rd, Headd JJ, Keedy DA, Immormino RM, Kapral GJ et al (2010) MolProbity: all-atom structure validation for macromolecular crystallography. Acta Crystallogr D Biol Crystallogr 66(Pt 1):12–21
60. Painter J, Merritt EA (2006) Optimal description of a protein structure in terms of multiple groups undergoing TLS motion. Acta Crystallogr D Biol Crystallogr 62(Pt 4):439–450
61. Goldschmidt L, Cooper DR, Derewenda ZS, Eisenberg D (2007) Toward rational protein crystallization: a Web server for the design of crystallizable protein variants. Protein Sci 16(8):1569–1576
62. Murshudov GN, Skubak P, Lebedev AA, Pannu NS, Steiner RA, Nicholls RA et al (2011) REFMAC5 for the refinement of macromolecular crystal structures. Acta Crystallogr D Biol Crystallogr 67(Pt 4):355–367
63. Adams PD, Afonine PV, Bunkoczi G, Chen VB, Davis IW, Echols N et al (2010) PHENIX: a comprehensive Python-based system for macromolecular structure solution. Acta Crystallogr D Biol Crystallogr 66(Pt 2):213–221

64. Sheldrick GM (2008) A short history of SHELX. Acta Crystallogr A 64:112–122
65. Lebedev AA, Young P, Isupov MN, Moroz OV, Vagin AA, Murshudov GN (2012) JLigand: a graphical tool for the CCP4 template-restraint library. Acta Crystallogr D Biol Crystallogr 68(Pt 4):431–440
66. Andrejasic M, Praaenikar J, Turk D (2008) PURY: a database of geometric restraints of hetero compounds for refinement in complexes with macromolecular structures. Acta Crystallogr D Biol Crystallogr 64(Pt 11):1093–1109
67. Adams PD, Baker D, Brunger AT, Das R, DiMaio F, Read RJ et al (2013) Advances, interactions, and future developments in the CNS, Phenix, and Rosetta structural biology software systems. Annu Rev Biophys 42:265–287
68. Long F, Vagin AA, Young P, Murshudov GN (2008) BALBES: a molecular-replacement pipeline. Acta Crystallogr D Biol Crystallogr 64(Pt 1):125–132
69. Lovell SC, Davis IW, Adrendall WB, de Bakker PIW, Word JM, Prisant MG et al (2003) Structure validation by C alpha geometry: phi, psi and C beta deviation. Proteins 50(3):437–450
70. Shapovalov MV, Dunbrack RL Jr (2011) A smoothed backbone-dependent rotamer library for proteins derived from adaptive kernel density estimates and regressions. Structure 19(6):844–858
71. Rohl CA, Strauss CE, Misura KM, Baker D (2004) Protein structure prediction using Rosetta. Methods Enzymol 383:66–93
72. Porebski PJ, Cymborowski M, Pasenkiewicz-Gierula M, Minor W (2016) Fitmunk: improving protein structures by accurate, automatic modeling of side-chain conformations. Acta Crystallogr D Biol Crystallogr 72(Pt 2):266–280
73. Furnham N, Dore AS, Chirgadze DY, de Bakker PI, Depristo MA, Blundell TL (2006) Knowledge-based real-space explorations for low-resolution structure determination. Structure 14(8):1313–1320
74. Joosten K, Cohen SX, Emsley P, Mooij W, Lamzin VS, Perrakis A (2008) A knowledge-driven approach for crystallographic protein model completion. Acta Crystallogr D Biol Crystallogr 64(Pt 4):416–424
75. Cowtan K (2012) Completion of autobuilt protein models using a database of protein fragments. Acta Crystallogr D Biol Crystallogr 68(Pt 4):328–335
76. Terwilliger TC (2003) Automated main-chain model building by template matching and iterative fragment extension. Acta Crystallogr D Biol Crystallogr 59(Pt 1):38–44
77. MoRDa—Automatic Molecular Replacement Pipeline [Internet]. 2014. Available from: http://www.biomexsolutions.co.uk/morda. Cited 29 Apr 2016
78. McCoy AJ (2007) Solving structures of protein complexes by molecular replacement with Phaser. Acta Crystallogr D Biol Crystallogr 63(Pt 1):32–41
79. Trapani S, Navaza J (2008) AMoRe: classical and modern. Acta Crystallogr D Biol Crystallogr 64(Pt 1):11–16
80. Webb B, Sali A (2014) Protein structure modeling with MODELLER. Methods Mol Biol 1137:1–15
81. Drozdetskiy A, Cole C, Procter J, Barton GJ (2015) JPred4: a protein secondary structure prediction server. Nucleic Acids Res 43(W1):W389–W394
82. Dosztanyi Z, Csizmok V, Tompa P, Simon I (2005) IUPred: web server for the prediction of intrinsically unstructured regions of proteins based on estimated energy content. Bioinformatics 21(16):3433–3434
83. Chwastyk M, Jaskolski M, Cieplak M (2016) The volume of cavities in proteins and virus capsids. Proteins 84(9):1275–1286
84. Babnigg G, Joachimiak A (2010) Predicting protein crystallization propensity from protein sequence. J Struct Funct Genom 11(1):71–80
85. Slabinski L, Jaroszewski L, Rychlewski L, Wilson IA, Lesley SA, Godzik A (2007) XtalPred: a web server for prediction of protein crystallizability. Bioinformatics 23(24):3403–3405
86. Laskowski RA, Watson JD, Thornton JM (2005) ProFunc: a server for predicting protein function from 3D structure. Nucleic Acids Res 33(Web Server issue):W89–W93
87. Skjaerven L, Yao XQ, Scarabelli G, Grant BJ (2014) Integrating protein structural dynamics and evolutionary analysis with Bio3D. BMC Bioinformatics 15(1):399
88. Doppelt-Azeroual O, Moriaud F, Adcock SA, Delfaud F (2009) A review of MED-SuMo applications. Infect Disord Drug Targets 9(3):344–357
89. Golovin A, Henrick K (2008) MSDmotif: exploring protein sites and motifs. BMC Bioinformatics 9:312
90. Zheng H, Chruszcz M, Lasota P, Lebioda L, Minor W (2008) Data mining of metal ion environments present in protein structures. J Inorg Biochem 102(9):1765–1776
91. Zheng H, Chordia MD, Cooper DR, Chruszcz M, Muller P, Sheldrick GM et al (2014) Validation of metal-binding sites in macromolecular structures with the CheckMyMetal web server. Nat Protoc 9(1):156–170
92. Web of Science [Internet]. 2016. Available from: http://wokinfo.com/products_tools/multidisciplinary/dci/. Cited 2 May 2016

INDEX

Alexander Wlodawer et al. (eds.), *Protein Crystallography: Methods and Protocols*, Methods in Molecular Biology, vol. 1607, DOI 10.1007/978-1-4939-7000-1, © Springer Science+Business Media LLC 2017

S

9781493969982